T0186679

PROCEEDINGS
SIXTH INTERNATIONAL CONGRESS
INTERNATIONAL ASSOCIATION OF ENGINEERING GEOLOGY
VOLUME 4

COMPTES-RENDUS
SIXIEME CONGRES INTERNATIONAL
ASSOCIATION INTERNATIONALE DE GEOLOGIE DE L'INGENIEUR
VOLUME 4

Comptes-rendus sixième congrès international association internationale de géologie de l'ingénieur

6–10 AOUT 1990 / AMSTERDAM / PAYS-BAS

Rédacteur
D.G. PRICE
Université de Technique de Delft, Pays-Bas

VOLUME 4
Cinquième thème: Géologie de l'ingénieur souterraine
Sixième thème: Géologie de l'ingénieur des structures hydrauliques sur terre et sur mer
Septième thème: Matériaux de construction

A.A.BALKEMA / ROTTERDAM / BROOKFIELD / 1990

Proceedings
Sixth International Congress
International Association
of Engineering Geology

6–10 AUGUST 1990 / AMSTERDAM / NETHERLANDS

Editor
D.G. PRICE
Delft University of Technology, Delft, Netherlands

VOLUME 4
Theme five: Underground engineering geology
Theme six: Engineering geology of land and marine hydraulic structures
Theme seven: Construction materials

A.A.BALKEMA / ROTTERDAM / BROOKFIELD / 1990

The texts of the various papers in this volume were set individually by typists under the supervision of each of the authors concerned.

Les textes des divers articles dans ce volume ont été dactylographiés sous la supervision de chacun des auteurs concernés.

Complete set of four volumes / Collection complète de quatre volumes: ISBN 90 6191 130 3
Volume 1: ISBN 90 6191 131 1
Volume 2: ISBN 90 6191 132 X
Volume 3: ISBN 90 6191 133 8
Volume 4: ISBN 90 6191 134 6
Volume 5: ISBN 90 6191 135 4
Symposia / Colloques: ISBN 90 6191 136 2

Published by:
© 1990 A.A. Balkema, Postbus 1675, 3000 BR Rotterdam, Netherlands
Distributed in the USA & Canada by: A.A. Balkema Publishers, Old Post Road, Brookfield, VT 05036, USA
Printed in the Netherlands

Publié par:
© 1990 A.A. Balkema, Postbus 1675, 3000 BR Rotterdam, Pays-Bas
Distribué aux USA & Canada par: A.A. Balkema Publishers, Old Post Road, Brookfield, VT 05036, USA
Imprimé aux Pays-Bas

Table of contents
Table des matières

5.4 *Underground storage of energy, liquids and waste*
Stockage souterrain de l'énergie, des liquides et des déchets

7.3 *Testing and classification*
Essais et classification

5 Underground engineering geology
  Géologie de l'ingénieur souterraine

5.1 Tunnels and shafts
    Tunnels et fosses

Keynote lecture: The influence of geological structure on the engineering of underground openings in discontinuous rock masses

Conférence thématique: L'influence de la structure géologique sur l'inginiérie d'ouvertures souterraines dans les masses rocheuses discontinues

Richard Goodman & Yossef Hatzor
Department of Civil Engineering, University of California, Berkeley, Calif., USA

ABSTRACT: Field investigations of underground openings in discontinuous rock masses have revealed an important observation. Block failures that were documented following the excavation could be generally correlated with a small group of joint half-space combinations (**joint pyramids**, or **"JP's"**). Although in a highly discontinuous rock mass with several representative joint sets the number of possibly **removable** JP's is very large, only a limited number was found to create the most common and widespread failures; these JP's consistently determine the most critical blocks, i.e. the **key blocks** of the excavation. Two factors determine the likelihood that a removable JP will create a critical key block for a given free face of the excavation. One is the frequency or mean spacing value of each joint set that creates the boundaries of the given JP. A JP that is made of members of the most frequent sets is more likely to be represented. However this likelihood must be modified by the JP shape. A narrow JP yields blocks that have their apex remote from the free face and therefore is less likely to form critical blocks than a JP that has an apex adjacent to the free face in question. The results of field investigations are given to demonstrate the applicability of the approach.

INTRODUCTION

The role of geological investigations:

For a prudent design of an underground opening the geology of the site must be studied prior to the construction phase. This study typically reveals the stratigraphy and structure of the rock formations which we expect to explore during excavation. The results of the geological investigations are usually summarized in the form of geological maps and cross sections. In a highly discontinuous rock a careful joint analysis is now becoming routine. These procedures are necessary but insufficient to judge the safety of an excavation. The question remains: What is the most critical structure that presents the main risk to the safety of the excavation? And will it appear in a single location or should we expect it to recur?

In this paper we present a method to answer these questions. We use direct field observations in order to define the one or several problematic structures to be encountered during excavation. The key to this approach is the empirical observation that even in a highly discontinuous rock mass, the critical structure for an excavation is formed by a small number of joint half-space combinations. These critical combinations can be predicted before the excavation phase on the basis of the available geological information and therefore can serve as models for design of the excavation.

Please note that we refer to all discontinuities i.e. joints, shears, faults foliation partings, and open contacts under the single name "joints". Although each type of discontinuity is genetically different and represents a different paleostress field, for our purpose using one common name for all discontinuities does not lead to loss of generality and simplifies the description of procedures.

ROCK ENGINEERING IN DISCONTINUOUS ROCK

Modelling rock behaviour - continuous vs. discontinuous approach:

The presence of flexures and complex structures in rocks creates problems not only by virtue of added stratigraphic complexity, but even more importantly due to the discontinuities they generate in the rock mass. Where the rock encountered underground is relatively continuous, its response to the excavation of an opening can be modeled by continuum mechanics. However, when the interaction of discontinuities creates a mass approaching a **discontinuum**, continuum mechanics is not easily applicable. The governing failure mechanism in such environments involves movements of blocks, by opening or sliding along pre-existing joints, sometimes facilitated by fracture of weak or highly stressed rock. Loosening of blocks from the periphery of the opening will be prevented by appropriate and timely emplacement of supports. The type of discontinuity and its gouge material will determine its shear strength, which is mainly frictional. The orientations and spacings of the different joint sets determine the shapes and relative sizes of blocks formed by different excavations.

The validity of empirical design based on rock mass classification methods:

The structure of a given rock mass and the pattern of its discontinuities are commonly known fairly well prior to excavation. Structural data can be obtained utilizing surface mapping or oriented core surveys. If a pilot excavation is driven, excellent data can be gathered by direct observation of the exposed faces. During construction, joint traces and joint surface characteristics can be obtained at the face and exposed periphery. These data are usually compiled in the form of maps and stereographic projections.

In the method of geological engineering that makes maximum use of applied rock classification schemes, the above data are used to produce an estimate of the overall engineering quality of the rock. The methods of excavation and support are then selected according to the classified rock quality. This procedure averages all the data, as if they were thrown together into a giant hopper. The overall quality criterion obtained by conventional classification methods represents an average underground opening and is achieved by averaging rock properties that were encountered in the sampling or pilot excavation process. But this average quality criterion may not represent any specific segment of the opening to be met during actual construction.

The general approach using rock classifications schemes can lead to incorrect conclusions in jointed rock. For example, a bedded and jointed quartzite of high strength, was described by both the Geomechanics system (RMR) and the Q system as "Good Rock". But, the rock did not behave as "good rock" ought, because, tunneling through this formation exposed blocks of various size that were released from the opening periphery resulting in large overbreak and safety hazards.

It is our opinion that in highly discontinuous rock formations, the validity of empirical design methods based upon general rock classifications is questionable. Failure occurs in distinct locations where a removable block is formed by the introduction of a free face to the system. Along other sectors of the opening where the dominant joint sets do not create removable blocks, either because they all belong to one or two sets or because their half-space combinations do not form removable blocks, stability problems may not arise . Rather than an averaging system, what is required is a series of individual analyses of the separate discontinuities, and the blocks they create, along each stretch of the tunnel. Although this might seem tedious, available programs for desk-top computers ease the burden and make such an analysis tractable and feasible.

For rock engineering in a highly discontinuous rock mass it would be more prudent to define a critical joint combination that once encountered creates removable blocks that may endanger the structural integrity of the tunnel. It seems urgent therefore, to establish a convenient and reliable method to find these critical joint combinations. This method must incorporate the orientations of the governing joint sets, their friction angles, and their relationship to the geometry of the opening.

The engineering significance of the "Joint Combination" (JC), "Joint Pyramid" (JP), and "Key Block" concepts:

When excavating through a discontinuous rock mass we are mainly concerned with blocks that may be formed by the half space combinations of the various joint sets that exist in the rock mass. The existence of one or two joint sets in the rock mass does not create an engineering problem since the half space combinations that will be formed by the joint intersections will be infinite and therefore non removable. However, when the number of joint sets in the rock mass is three or more, removable blocks may be formed by the intersection of the tunnel with specific combinations of half spaces of the different joint sets. We will now define the technical terms that will be used in three dimensional treatment of the discontinuous rock mass.

A joint combination (JC) is defined as a point in space where joints of different sets intersect to create a group of half-space combinations. We define a joint pyramid (JP) as the set of points common to a specific choice of half spaces bounded by the plane of each joint of the corresponding joint combination when these planes are shifted to pass through a common origin in the joint combination volume. All the joint planes plotted in stereographic projection pass through the same point, namely the center of the reference sphere. Therefore, the stereographic projection can serve to represent the JP and the JC. The JP is represented as a convex spherical polygon bounded by a set of joint plane great circles. With three joint planes, the JP is a spherical triangle (a mathematical treatment of the JP spherical triangle is discussed by Mauldon and Goodman, 1989), while the JC is represented by the center of the stereographic projection plane.

Block theory (Goodman and Shi, 1985) provides a simple way to determine the removability of a JP utilizing the stereographic projection: A JP yields removable blocks if it plots entirely within the area which represents the space side of the free face. A key block is formed when a given JC creates an unsafe JP defining unsafe removable blocks. A removable block may be safe or unsafe according to the outcome of a stability analysis, with input friction angles for the joint planes and a defined force field, e.e. gravitational forces.

An illustration of the concepts discussed above is shown in Figure 1. The orientations of three joint sets in a Joint Combinations are listed and the stereographic projection of these planes is shown. Consider the right hand, vertical wall of a large underground chamber through this rock mass. The stereographic projection of the wall is given by the dashed line, and the excavated space associated with the wall

is the half plane above this line. The only JP that plots entirely inside the free space is 110, consisting of the intersection of the lower half spaces of joint planes 1 and 2, and the upper half space of joint plane 3. A block formed with JP 110 is shown in sub-figure 1b, and the maximum keyblock of this JP that can be formed in the rock behind the wall is shown in Figure 1c.

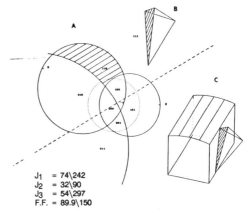

$J_1 = 74\backslash242$
$J_2 = 32\backslash90$
$J_3 = 54\backslash297$
$F.F. = 89.9\backslash150$

Fig. 1: A) A combination of three joint sets, their orientations and the complete stereographic projection of their planes. Upper hemesphere projections are used in all figures. A vertical free face is also shown on the stereographic projection and its orientation is listed. The removable JP from the S.E. side of the free face (JP 110) is shown on the stereographic projection. The JP is removable because it has no intersection with the rock side of the free face (the S.E. side). B) The block that is formed once JP 110 is cut by the free face. C) The maximum key block size this block may have in an opening of given size and shape.

FINDING THE CRITICAL KEY BLOCKS OF THE EXCAVATION.

The Critical Key Block Concept:

Field studies of the performance of underground openings in discontinuous rock masses revealed an important observation. The blocks that failed following the excavation could be correlated with a small number of JC's (and their removable JP's) even where a large number of joint sets was used to fully represent the complex structure of the rock mass. In other words, of the numerous possible removable JP's that had to be considered in the analysis, only a very few were actually involved in failures. This observation provides the motivation to find the particular JC and its removable JP that will create the most expectable key block in the whole population of potential blocks. This key block is defined as the critical key block of the excavation.

We will now explore the different factors that affect: (1) the likelihood of encountering a particular JC of all the possible joint combinations in the rock mass, given a particular geological structure; and (2) the likelihood of actually finding a removable block of a given JP in the excavation, given all the different possible removable JP's. The combined effect of both the JC likelihood of occurrence in the rock mass and the JP likelihood of causing block failure into the given free face will determine the critical key block of the excavation.

Once we determine the expected critical key block of the excavation, we can calculate its maximum size and therefore the optimum support. During actual construction we can mark locations along the opening free faces where the traces of this key block appear, following excavation and before temporary support (particularly shotcrete) is applied. This requires merely the presence of an engineering geologist or trained aide on site, with a trace map of the critical block that is to be expected along the different free faces of the excavation. This map can be produced prior to construction following the procedures discussed by Goodman and Shi (1985) and Goodman (1989). In locations where the critical key block trace is found on the wall, the support should be designed to carry the extra load that may be induced by the release of this block. The block release driving force may be caused by blasting vibrations, earthquake forces, seepage forces, or gravity alone.

Method of analysis:

A step by step procedure for selecting the critical key block of the excavation will be discussed below. The input data for the analysis include both the rock mass discontinuity characteristics and the proposed opening geometry. The rock mass properties that are considered are the governing joint sets in the rock mass, their orientation and their spacing distribution. The data to be input about the excavation include the cross section geometry, and its orientation in space. The analysis out-puts a list of the joint-combinations-outcome

space, deletes the nonhazardous joint combinations, assigns probability values to the hazardous joint combinations, and lists the likelihood of each removable JP to produce actual key blocks in the given opening geometry. A numerical value is assigned to the hazardous-joint-combinations probability and to the removable JP's failure likelihood so that the most probable key block can be found numerically.

The analysis makes the following fundamental assumption. We assume that we can represent the structure of the rock mass by several prominent joint sets that share a common orientation. Thus we model the expected opening behavior on the basis of combinations of ideal joint sets and ignore the combinations that may arise from the intersection of less common orientations. The use of this method therefore can be justified only in regions where the geological structure of the rock mass follows a pattern that can be simplified by the concept of joint sets, using the structural data at hand.

Step 1 - Finding the governing joint sets in the global system:

Careful geological site investigations usually reveal a maze of joints, faults shears and foliation planes. A better insight to the problem may be gained if we use a stereographic projection of the poles of the discontinuities planes, preferably different plots for different structures, i.e. faults and shears, joints and foliation planes each on a different plot.

Fig. 2 shows examples of underground opening trace maps and of stereographic projections of the poles of all the mapped discontinuities. When this technical procedure is complete we expect to find some order in the system. At least in some geological settings this has to be the case because the current structure of the rock mass is the product of a particular tectonic history. In the course of geologic time the rocks have deformed under particular stress fields and therefore they should present corresponding deformation patterns. Stereoplots of normals ("poles") to discontinuities with contours of equal statistical density (Fig. 2) allow us to determine which orientations are the most frequent for a given structure, namely which are the dominant joint sets in each structure. Commonly we will find one to four joint sets for each structure and thus we can represent our global structure with a number of different joint sets, each with a unique orientation. We should examine the results carefully. If we find for example that two different structures share a common orientation, so that the same joint set appears twice , we can safely eliminate one set from the global system since the governing discontinuity property for the kinematic analysis is its orientation, not its genetic relationship.

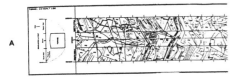

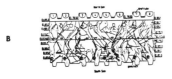

Fig. 2: A) A trace map of the geological structure in a pilot bore construction and a stereographic projection of the poles of the discontinuities that were mapped above. Statistical contouring provides the means to determine the representative joint sets of that structure. Source: Woodward Clyde Consultants. B) A trace map of the geological structure in a powerhouse cavern. After Reik and Soetomo, 1986.

Step 2 - Finding the joint combinations outcome space:

Once we determine the governing joint sets in the system we have to consider their possible intersections because removable blocks are formed by the different possible half-space combinations. If for example we have four joint sets in the system {A;B;C;D} and we wish to consider blocks that will be formed by the intersection of three planes only, than the joint

combinations outcome space is the following group of combinations:

AAA	BAA	CAA	DAA
AAB	BAB	CAB	DAB
AAC	BAC	CAC	DAC
AAD	BAD	CAD	DAD
ABA	BBA	CBA	DBA
ABB	BBB	CBB	DBB
ABC	BBC	CBC	DBC
ABD	BBD	CBD	DBD
ACA	BCA	CCA	DCA
ACB	BCB	CCB	DCB
ACC	BCC	CCC	DCC
ACD	BCD	CCD	DCD
ADA	BDA	CDA	DDA
ADB	BDB	CDB	DDB
ADC	BDC	CDC	DDC
ADD	BDD	CDD	DDD

The list above presents all the possible combinations of three joints that can be formed from a group of four joint sets. However a combinations with repeated joint sets can be excluded here and combinations with three of the same joints, like {AAA}, have no physical meaning. Therefore in a case of three joints JP's we only have to consider those combinations that involve different sets. And since the order of naming the joints is immaterial, these combinations are only the four following:

{A;B;C}, {A;B;D}, {A;C;D} and {B;C;D}

Generally, the number of unordered combinations that involve different joint sets can be found by the relation:

$$[n!]/[k!(n-k)!] \qquad (1)$$

Where k is the number of joint sets that comprise the boundaries of a JP and n is the number of joint sets that comprise the global structure of the rock mass. In the above example n=4 and k=3 therefore the value of (1) is: $4!/[3!(4-3)!] = 4$.

The procedure to find n, the sum of joint sets in the rock mass, was discussed previously. The value of k , the number of joints that comprise the boundary of a JP , must be greater than or equal to 3 and less than or equal to n. As explained above, the k value must start from 3 because one or two joint sets cannot form a finite block in a planar excavation. The value of k can be greater than three but cannot be greater than the value of n unless we wish to consider repetition of a joint set in a JP. Empirically it has been observed that the larger the k value the smaller the recurrence of its JP's. JP's with K = 4 are relatively rare and with k = 5 or more have not been observed as a recurring structure in sites we have studied.

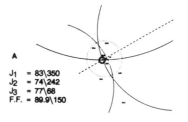

J1 = 83\350
J2 = 74\242
J3 = 77\68
F.F. = 89.9\150

Step 3 - Finding the removable JP's in each joint combination:

Once we define our joint combinations outcome space for a desired k value, we must test each joint combination in order to find the removable JP's it will create with the free faces in question. Using Block Theory (Goodman and Shi, 1985) we can generate an analysis output that indicates the codes of the removable JP's from each free face of the excavation for each separate joint combination. We now want to reduce the problem size by ignoring those joint combinations that may be regarded as "safe", and by further considering only those joint combinations that may be regarded as "hazardous" for the excavation in question. Safe joint combinations will be defined for our purposes as joint combinations that do not create removable JP's against the opening side walls. We use this definition because our sampling of block failures in the field was limited to the side walls of the excavation. Due to technical difficulties the roof was not sampled. Therefore our analysis will attempt to find the critical key blocks of the excavation that will be released from the sidewalls only, on the basis of the given geological mapping. The results of this prediction will then be tested against the actual observations in the field regarding key block failures.

Safe JC's for a particular free face are formed only if one kinematical condition is satisfied: The line of intersection of two joint sets of the JC is included in the plane of the free face in question! The removability of a JP is determined by Shi's Theorem that states that a removable JP has no intersection with the **excavation pyramid** (EP); the EP is the region formed by the block side of each excavation surface, when such surfaces have been translated to pass through a single origin. Fig. 3a shows a safe JC against a free face, a result achieved because an edge (a line of intersection of two joints) plots on the projection of the free face.

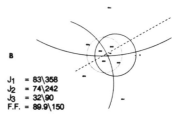

J1 = 83\358
J2 = 74\242
J3 = 32\90
F.F. = 89.9\150

Fig. 3: A) An example of a safe joint combination with respect to a given free face (dashed line). This joint combination is safe because the line of intersection of sets 1 and 3 plots on the free face (solid circle). Therefore, such a joint combination can not create removable blocks with respect to this free face. B) An example of a hazardous joint combination with respect to a given free face (dashed line). No line of intersection between the joint planes has an intersection with the free face, therefore two removable JP's are formed each from a different side of the free face.

This event although rare when only few joint sets are considered, can be quite likely as the number of sets in the rock mass increases, as will be shown in the case history below. If this event occurs for one or more lines of intersection for a given free face then any joint combination that includes the corresponding two joint sets may be regarded as safe from sliding into that free face. Although the joint sets represent ideal orientations that may not be always identical to the individual joints actually encountered in the field, this approximation allows us to neglect the generally safe combinations. All the other joint combinations in the complete JC outcome space that do not have a line of intersection satisfying the above condition, must be regarded as hazardous joint combinations. Fig. 3b shows an example of a hazardous JC against the same free face as in 3a.

Step 4 - Finding the Joint Combination probability P[JC]:

The joint combination was defined above as a point in space where three or more joints intersect at a point. We want to find how often this event will occur along a line of a unit length for the different possible joint combinations. We can expect to find more intersections of the closely spaced sets than of the more widely spaced sets. The chance that a joint of a given set intersects a scan line of a unit length is obtainable from the spacing distribution of each set. There is considerable justification for assuming that the negative exponential probability density provides a good approximation to the distribution of discontinuity spacings measured in the field (Priest and Hudson, 1976; Hudson and Priest, 1979; Wallis and King, 1980). The negative exponential distribution is expressed as:

$$f(x) = \lambda e^{-\lambda x} \tag{2}$$

where f(x) is the frequency of a discontinuity spacing x and λ is the average number of discontinuities per unit length, or the discontinuity frequency.

This distribution is a one parameter (λ) distribution with the mean and S.D. both equal to $1/\lambda$.

Hudson and Priest (1979) have shown that the discontinuity frequency per unit length (λ) depends upon the angle (θ) between the pole of the joint set sampled and the scanline, measured in their common plane. They showed that if the discontinuity frequency is determined along a line at angle (θ) from the original scanline, then the measured value (λ_θ) will be reduced from $\lambda_\theta = \lambda$ when $\theta = 0$ to $\lambda_\theta = 0$ when $\theta = 90$. The relationship between the true frequency and the measured one is given by $\lambda = \lambda_\theta / \cos\theta$. Therefore, the true discontinuity frequency value can be readily determined from scanline survey data oriented in any direction in space.

The discontinuities that intersect a scanline are analogous to arrivals along a timeline. In this analogy, the scanline is the timeline, the discontinuities which intersect the scanline are the arrivals, and the spacing between discontinuities are waiting times. From probability theory we know that if the distribution of the waiting times between arrivals is negative exponential, and if the waiting times between each arrival to the next are independent, then the distribution of the number of arrivals N(I) in a fixed time interval (I) of length t is Poisson (λt) and the number of arrivals in disjoint time intervals are independent. Using our analogy we may say that if the spacing values between discontinuities along a scanline are exponentially distributed, and if disjoint spacings are independent, then the probability to find a specific number of discontinuities N(I) along a fixed length interval I of length (x) along the scanline can be found by Poisson (λx) process:

$$P[N(I) = k] = e^{-\lambda x}(\lambda X)^k/k! \qquad \begin{matrix} k = 0,1\ ..., \\ X \geq 0 \end{matrix}$$

where the only required parameter is the joint set true frequency (λ), a rock mass property that can be obtained from a scanline survey.

Having the above tools we can now ask what is the chance to find more then zero intersections of joints of set i along a very small length interval of a scanline oriented normal to its plane, using the true (λ) value and Poisson equation for non zero intersections:

$$P[N(I)_i > 0] = 1 - P[N(I)_i = 0] = 1 - e^{-\lambda_i x}$$

Similarly we can find the same probability for sets j and k of say joint combination {i;j;k}. Now, since we assume these events to be independent, the probability of all these events occurring together is the product of the three separate events:

$$P(JC) = \{P[N(I_i) > 0]; P[N(I)_j > 0]; P[N(I)_K > 0]\}$$

$$= [1 - e^{-\lambda_i X}] \cdot [1 - e^{-\lambda_j X}] \cdot [1 - e^{-\lambda_K X}] \tag{3}$$

This probability is the JC probability along a line of a given length. We may select any scan line length x for the computation. We must however use the same length x for all the JC probabilities we wish to find. Thus we can compare the relative differences between the probabilities we obtain.

Step 5 - Finding the removable JP's degree of hazard[K]:

Once all the hazardous JC's have been found and the codes have been determined for all the removable JP's from the free face in question, using Block Theory, we need to weigh the degree of hazard offered by each JP. First we define the maximum removable block region behind a free face. This region envelopes all the possible removable blocks that can form behind a free face all of which correspond to a particular removable JP. No block can be larger then the maximum re-movable block region but real blocks can be smaller. We can therefore view the maximum removable block region as a physical failure envelope which corresponds to a particular removable block from a given free face. This envelope includes all the block failures that a particular removable JP will create behind the excavation free face. An important property of the maximum removable block region is the dis-tance of its apex from the corresponding free face, the apex distance. Open JP's, joint pyramids with high angles between the joint planes, will have an apex adjacent to the excavation free face, i.e. they have a small apex distance. These JP's are identified by a small failure envelope and a large spherical triangle area on the stereographic projection (Fig. 4b).

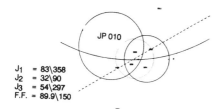

J_1 = 83\358
J_2 = 32\90
J_3 = 54\297
F.F. = 89.9\150

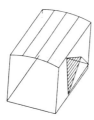

Fig. 4(B): An example of a hazardous removable JP. JP 010 is removable with respect to the S.E. side wall(dashed line). It is hazardous because its apex is adjacent to the free face. This is a reflection of the wide angles between the JP boundary planes.

Closed JP's, i.e. joint pyramids with small angles between the joint planes, will have a remote apex, i.e. with a large apex distance. These JP's are identified by a large failure envelope and a small spherical triangle area on the stereographic projection (Fig. 4a).

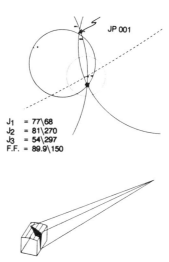

J_1 = 77\68
J_2 = 81\270
J_3 = 54\297
F.F. = 89.9\150

Fig. 4(A): An example of a safe removable JP. JP 001 that is removable from the S.E. side wall (dashed line) and from the roof(reference circle) is safe because its apex is very remote behind the free face. This is a reflection of the small angles between the JP boundary planes.

A hazardous JP is defined as an <u>Open JP</u>. Such a JP, with a short apex distance, has a high potential to produce block failures. A non hazardous JP is defined as a <u>Closed JP</u>, having long apex distance, and a small potential to produce frequent block failures.

A JP can form blocks of different sizes all of which are contained within the maximum removable block envelope that may have a small or large apex distance as was explained above. JP's with a small apex distance will produce more block failures then JP's with a large apex distance for the following reasons:

A) If we assume that all potential block sizes are equally likely to be formed for a given JP, then in the case of a small apex distance a larger proportion of the possible block sizes will exist in the rock mass then in the case of a long apex distance, given a fixed joint persistence value in the rock mass.

B) As the apex distance increases, the removable block experiences more lateral stresses that act as confining pressures on its boundary planes.

C) As the apex distance increases, the area of the side planes also increases and if the joint plane exhibits cohesion, a greater cohesive force resists sliding.

A convenient way to quantify the size of the JP was discovered by Mauldon (in press). Each JP has a particular spherical triangle area that can be measured in the plane of the stereographic projection. The governing factor that controls the degree of hazard of a removable JP is its spherical triangle area; thus we can use this area as a scaling parameter to compare the different JP's. The removable JP degree of hazard [K] can be expressed as the ratio of the JP spherical triangle area to the area of the sphere. Following Mauldon:

Removable JP degree of hazard K

$$K = \frac{\text{area of JP spherical triangle}}{\text{area of a sphere}}$$

$$= \frac{[A + B + C - \pi]R^2}{4 * \pi * R^2} \qquad (4)$$

where A,B,C are the angles between the JP boundary planes for a JP with 3 joints, and R is the radius of the stereographic projection plane.

In the case of openings where there are parallel free faces like a tunnel or an underground chamber, a Block Theory rule applies that if a JP is removable from a free face then its "cousin" is removable from the cousin free face. A cousin to a joint or excavation pyramid is obtained by replacing all half spaces by their opposite half spaces. However, while both the JP and its cousin are removable from two parallel free faces, one JP may be able to slide under gravity while the other, although removable, could generally fail only by being pulled out into the opening space and is therefore safe under gravity. Therefore when finding such a JP degree of hazard only 1 JP area needs to be divided by the area of the sphere. When both JP's are removable by gravity alone from the two parallel free faces, we need to divide 2 JP areas by the area of the sphere to correctly assess the removable JP degree of hazard. When only one free face is considered for the design , we only use 1 JP area, i.e. the area of the corresponding removable JP.

Step 6 - Finding the overall block failure probability P[B]:

Having the JC occurrence probability P[JC] and its removable JP degree of hazard for a given free face or two parallel free faces [K], we can now find the likelihood that a particular JC will actually produce block failures in the opening space. In order to do that we modify the JC probability by its removable JP degree of hazard to obtain the overall block failure probability P[B]. This probability is the product of the JC probability and the JP degree of hazard:

$$P[B] = K * P[JC] \qquad (5)$$

where:

P[JC] = the joint combination probability found in step 4.

K = removable JP degree of hazard found in step 5.

The results of block failure data that were documented in the field will be discussed below.

CASE HISTORY

In this project (name and location are kept with the authors), two exploratory tunnels, were investigated in the field. The tunnels, 12 feet wide and 13 feet thick with a total length of 3,900 feet, were excavated through highly discontinuous but very competent Precambrian crystalline rocks (except where weathered) consisting of Migmatite gneiss and quartz diorite intruded by granite and pegmatite. The global structure of the rock mass contributes open joints, faults and shears, and foliation planes.

Field investigations performed by us following the excavation of the pilot bores for this site revealed that the dominant failure mode was block sliding along pre-existing joint planes. All the block failures that were observed in the field were documented by means of the joint plane orientations, and the corresponding JP, together with description of the free face from which each block was released. An independent analysis of the raw geologic data was also performed following the steps outlined above in order to find the critical key block of the excavation. The validity of the step by step procedure which was outlined above will be demonstrated by means of comparison between the predicted and observed critical key blocks of the excavation. We will now describe the independent step by step analysis procedure.

Finding the critical key block of the excavation:

Step 1: Finding the governing joint sets in the global system

Figure 5 (a,b,c) shows stereoplots of the poles of the discontinuities planes with statistical contours of equal density. Fig. 5a , 5b and 5c show respectively the orientations of the faults and shears, joints and the foliation planes.

FOLIATION ORIENTATIONS   JOINT ORIENTATIONS

FAULT AND SHEAR ORIENTATIONS

Fig. 5: The structural features that were mapped in the excavation of the pilot tunnel rock mass.

The global discontinuity sample population is comprised of 3196 discontinuities of which 319 (10%) are faults and shears, 2746 (86%) are joints and 131 (4%) are foliation planes, the joints being by far the most abundant structure in the system. The foliation planes (Fig. 5c) are very distinct and may be represented safely by two sets using the most frequent orientations namely: dip/dip-direction = 54\297 and 69\246. A less frequent orientation in this structure is 83\175 however, examination of the other two structures quickly reveals that this orientation is common to all structures and therefore must be included in the global joint system. The joint orientations (Fig. 5b) introduce three new sets: 84\356, 32\090 and 81\270 and one that was already found 82\175. The first two frequent orientations in the foliation planes are also

abundant in the joint structure although of lower frequency. Examination of the faults and shears group (Fig. 5a) demonstrates the limitations of trying to fit joints into sets. Of the four orientations picked to represent the faults system, three duplicate others already included, and one new set is introduced at orientation 77/68. The whole system of discontinuities is represented by the attitudes presented in Table 1:

Table 1: The pilot tunnels rock mass structure.

Faults & Shears	Joints	Foliation Planes	Global Sets
82\360	84\356		G₁) 83\358
79\237		69\246	G₂) 74\242
77\68			G₃) 77\068
87\175	82\175	83\175	G₄) 84\175
	32\090		G₅) 32\090
	81\270		G₆) 81\270
		54\297	G₇) 54\297

A plot of the global structure is given in Fig. 6 where the scatter of each set is also included.

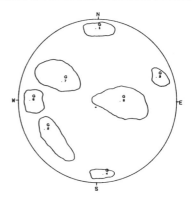

Fig. 6: The pilot tunnel global structure. This structure was used for the correlation between the field failures and the Block Theory analysis predictions.

Step 2: Finding the Joint Combinations outcome space

There are 7 joint sets in the global structure. In this example we will demonstrate the method for k=3, i.e. for potential JP's with three joint sets at their boundaries, the forth being any desired free face of the excavation. For JP's with k=3 we need only consider combinations of three different sets with no repetition of sets in any combination. Using (1) the number of such combinations is 35. The list of these combinations can be found in Table 2.

Step 3: Finding the removable JP's in each joint combination

The results of Block Theory analysis for these 35 possible combinations is presented below (Table 2). Only JP's that were found to be removable are indicated. The joint combinations are numbered according to the global system sets above. Each removable JP has a JP code where 0 means above the plane and 1 means below the plane, so for example for combination 125 a JP code of 001 means that the removable JP is formed by the combination of the half spaces above set 1 and set 2 and below set 5 of the global system. The tunnel axis orientation that was used for Block Theory analysis is horizontal and trending to azimuth 240.

Table 2: Block Theory analysis for removable JP's

Joint Combination		S.E. Wall	N.W. Wall	Roof
JC#	{Gi;Gj;Gk}	JP Code	JP Code	JP Code
1)	123	-	-	111
2)	124	-	-	111
3)	125	001	110	111
4)	126	010	-	-
5)	127	001	-	001
6)	134	-	-	111
7)	135	-	-	111
8)	136	-	-	111
9)	137	-	-	111
10)	145	-	110	110
11)	146	-	-	111
12)	147	110	001	111
13)	156	010	101	-
14)	157	010	101	011
15)	167	001	-	001
16)	234	-	-	110
17)	235	0	110	110
18)	236	100	011	111
19)	237	110	001	-
20)	245	-	-	101
21)	246	-	-	-
22)	247	-	-	-
23)	256	-	011	011
24)	257	110	-	-
25)	267	-	011	101
26)	345	-	001	001
27)	346	-	-	111
28)	347	110	001	111
29)	356	010	-	-
30)	357	010	-	-
31)	367	001	-	001
32)	456	-	-	-
33)	457	110	001	111
34)	467	-	-	101
35)	567	110	001	-

It is important to note that the joint combinations which include lines of intersections of sets 1 and 3 (I_{13}), 2 and 4 (I_{24}), and 4 and 6 ($I_{4,6}$) did not create any removable JP's from the side walls of the excavation. In this excavation, the side walls are parallel trending horizontally to azimuth 240. For this free face orientation the above three lines of intersections are included in the free face. Therefore any JC that includes any of these couples will be safe in the side walls. We define "hazardous JC's" as such joint combinations that create removable blocks from the side walls, because in our field study only the side walls were sampled. Therefore, the hazardous JC's for this excavation are those that do not include the above mentioned lines of intersection. This fact reduces the problem size from an outcome space of 35 possible joint combinations to only 21 hazardous JC's that need to be further analyzed.

Step 4: Finding the joint combination probability

Having completed the Block Theory removability analysis and finding the hazardous JC's, we now proceed to the hazardous JC's probability analysis. Our goal is to find the most probable joint set combinations from the combinations that do create removable JP's. In order to do that we must first find the true intensity values of the global joint sets. These values were recovered from a scanline survey the results of which are given in Table 3 below:

Table 3: True discontinuity frequency values for the pilot tunnels.

G_i	Frequency (1/ft.)	Mean Spacing (ft.)
G_1	1.18	.847
G_2	0.38	2.632
G_3	0.25	4.0
G_4	0.82	1.220
G_5	0.43	2.326
G_6	0.49	2.041
G_7	0.81	1.235

We can now proceed to find the probabilities to encounter each of the joint sets combinations that were found to be hazardous by Block Theory analysis. The formulation which was outlined above in step 4 is used here in order to find the value of the joint combination probability P(JC). The results of this computation can be found in table 4 below.

Step 5 - Finding removable JP degree of hazard [K]:

Having obtained all the removable JP's from the opening parallel side walls by means of Block Theory, we now proceed with the analysis by finding their degree of hazard, namely their ability to create block failures in the excavation. As was discussed above, the JP degree of hazard is a function of the angles between the JP boundary planes. The K value for each removable JP was found following equation 4, the results of which are summarized in Table 4 below.

Step 6 - Finding the overall block failure probability and selecting the critical key block of the excavation:

The overall block failure probability P(B) is found by means of equation 5 discussed in step 6 above. The values of P(B) for all the hazardous JC's are shown Table 4 below.

Table 4: Hazardous joint combination probabilities assuming negative exponential spacing distribution of global joint sets.

JC(#)	{G_i;G_j;G_k}	P(JC)	(K)	P(B)
3)	{1;2;5}	0.08	$2(1.18*10^{-1})$	$1.89*10^{-2}$
4)	{1;2;6}	0.09	$2(3.00*10^{-3})$	$5.40*10^{-4}$
5)	{1;2;7}	0.12	$1(1.40*10^{-3})$	$1.68*10^{-4}$
10)	{1;4;5}	0.14	$1(1.10*10^{-3})$	$1.54*10^{-4}$
12)	{1;4;7}	0.22	$2(1.00*10^{-2})$	$4.40*10^{-3}$
13)	{1;5;6}	0.09	$2(1.46*10^{-1})$	$2.63*10^{-2}$
14)	{1;5;7}	0.14	$2(1.61*10^{-1})$	$4.51*10^{-2}$
15)	{1;6;7}	0.15	$1(8.33*10^{-3})$	$1.25*10^{-3}$
17)	{2;3;5}	0.02	$1(7.00*10^{-5})$	$1.40*10^{-6}$
18)	{2;3;6}	0.03	$2(4.20*10^{-2})$	$2.52*10^{-3}$
19)	{2;3;7}	0.04	$2(1.90*10^{-2})$	$1.52*10^{-3}$
23)	{2;5;6}	0.04	$1(2.40*10^{-2})$	$9.60*10^{-4}$
24)	{2;5;7}	0.06	$2(4.50*10^{-2})$	$5.40*10^{-3}$
25)	{2;6;7}	0.07	$2(6.10*10^{-2})$	$8.54*10^{-3}$
26)	{3;4;5}	0.04	$1(3.47*10^{-2})$	$1.39*10^{-3}$
28)	{3;4;7}	0.07	$2(1.67*10^{-1})$	$2.34*10^{-2}$
29)	{3;5;6}	0.03	$2(1.70*10^{-2})$	$1.02*10^{-3}$
30)	{3;5;7}	0.04	$2(2.50*10^{-2})$	$2.00*10^{-3}$
31)	{3;6;7}	0.05	$1(6.90*10^{-4})$	$3.45*10^{-5}$
33)	{4;5;7}	0.11	$2(1.57*10^{-1})$	$3.45*10^{-2}$
35)	{5;6;7}	0.08	$1(2.50*10^{-4})$	$2.00*10^{-5}$

These values are plotted in Fig. 7.

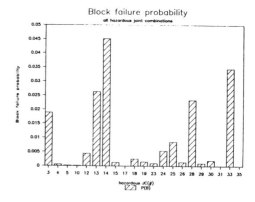

Fig. 7: The Block Failure probability P(B) values of the different hazardous JC's.

From Fig. 7 it can be concluded that JC's # 14, 33, 13, 28 24 and 3 are the most likely to create block failures in the excavation, the other JC's having much smaller P(B) values. However in order to select the critical key block of this excavation we must find a way to consider that joint combination which is most likely to produce the largest number of key block failures among those that share a high P(B) value. If we examine JC # 13 and 14 we find that both JC's include sets 1 and 5 , while JC(13) = {1,5,6} and JC(14) = {1,5,7}. The orientation of Joint set # 6 is 81\270 and that of set # 7 is 54\297. These two different sets have a somewhat similar orientation namely moderately dipping to WNW or say 68\283. If we assume a joint combination that includes sets 1 ,5 and the average trend of sets 6 and 7, then this combination will create the highest number of key block failures.

Fig. 8 shows the trace maps of the key blocks created by JC's 13 and 14 as seen from inside the tunnel looking horizontally towards the south east side-wall. In these two maps only the third set is changed from sets # 6 in JC(13) to set # 7 in JC(14), everything else is kept equal. Thus the effect of the different trend of the third set can be clearly seen. The combined JC with the third set being the average of sets 6 and 7 creates a key block trace map that has the average area shown in Fig. 9. This area, we suggest , should be used for the design.

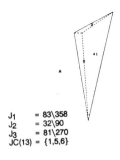

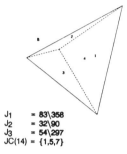

J1 = 83\358
J2 = 32\90
J3 = 81\270
JC(13) = {1,5,6}

J1 = 83\358
J2 = 32\90
J3 = 54\297
JC(14) = {1,5,7}

Fig. 8: A) S.E. wall trace map of the maximum key block size that could be generated by JC(13). B) S.E. wall trace map of the maximum key block size that could be generated by JC(14).

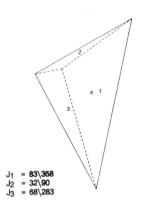

J1 = 83\358
J2 = 32\90
J3 = 68\283

Fig. 9: A S.E. wall trace map of the critical key block of the excavation. This block is formed by global joint sets 1 and 5, the third set is the average trend of sets 6 and 7.

Results of field investigations:

The pilot tunnels have been studied in the field following their construction and before their enlargement to the proposed full size opening.

All the block failures that occurred following the excavation were documented by measuring the orientations of the JP boundary planes and the orientation of the corresponding free faces. Table 5 lists the documented block failures in terms of joint plane orientations, free face from which the failure occurred and the joint half space combination (JP code). The corresponding <u>global</u> joint combination that was used in the independent Block Theory analysis (Table 2) is listed next to blocks where a correlation exists.

Table 5: Observed block failures.

Block#	J1	J2	J3	Face	Code	JC(Global)
1	65\100	80\190	40\340	NW.WALL	001	33:{5;4;7}
2	52\300	84\20	65\128	SE.WALL	001	14:{7;1;5}
3	58\295	80\220	40\120	SE.WALL	011	24:{7;2;5}
4	55\290	55\210	55\85	SE.WALL	0.11	24:{7;2;5}
5		K=4		ROOF		
6	82\85	84\180	40\280	SE.WALL	110	33:{5;4;7}
7	75\285	70\360	24\105	NW.WALL	110	13:{6;1;5}
8	K=2 + 2 blasting induced fractures, released from roof.					
9	60\315	75\205	35\140	SE.WALL	011	24:{7;2;5}
10	32\285	80\190	42\120	SE.WALL	011	33:{7;4;5}
11	44\280	75\005	45\105	SE.WALL	001	14:{7;1;5}
12	80\283	80\10	40\140	NW.WALL	110	13:{6;1;5}
13	74\75	62\200	20\270	NW.WALL	001	-----------
14	50\300	35\180	50\60	ROOF	111	-----------
15	50\96	80\180	65\280	SE.WALL	110	33:{5;4;7}
16	54\90	80\40	13\295	ROOF	101	-----------
17	62\90	88\190	45\250	NW.WALL	001	-----------
18		K=4		NW.WALL		
19	53\280	70\10	40\150	SE.WALL	001	14:{7;1;5}
20	75\280	80\005	20\190	SE.WALL	001	13:{6;1;5}
21	75\290	70\20	20\190	SE.WALL	001	13:{6;1;5}
22	85\280	62\18	30\190	SE.WALL	001	13:{6;1;5}
23	85\100	80\360	15\300	NW.WALL	011	-----------
24	64\70	75\340	20\280	SE.WALL	101	-----------
25	75\110	80\360	10\310	NW.WALL	011	-----------
26	77\120	65\205	38\5	NW.WALL	001	-----------
27	50\290	85\20	42\80	SE.WALL	001	14:{7;1;5}
28	60\260	88\3	25\80	SE.WALL	001	13:{6;1;5}
29	80\270	89\5	17\85	SE.WALL	001	13:{6;1;5}
30	60\305	82\10	25\100	SE.WALL	001	14:{7;1;5}
31	70\70	80\180	25\300	NW.WALL	001	-----------
32	68\255	83\15	40\120	SE.WALL	001	13:{6;1;5}
33	65\265	80\20	24\80	SE.WALL	001	13:{6;1;5}
34	78\270	89\40	26\140	SE.WALL	001	13:{6;1;5}
35	48\335	88\50	5\160	SE.WALL	001	30:{7;3;5}

Comparison between actual field performance and failure patterns predicted in the independent analysis:

The results of the field study reveal several important observations:

* The majority of the blocks that were released into the tunnel space were tetrahedral with a JP of 3 joints only (k = 3) the forth discontinuity plane being a free face of the excavation (See Table 5). Only 2 blocks out of the observed 35 have a JP with k = 4. This observation provides an empirical justification to the preliminary analysis of tetrahedral blocks in order to find the critical key block of the excavation.

* Only one of the observed block failures was created by sliding along blasting induced fractures. The rest of the blocks were released by sliding on pre-existing discontinuities which belong to the global structure of the given rock mass. This observation provides a justification to model the opening behavior on the basis of structural data obtained form geological investigations prior to the construction phase. Although sliding along blasting-induced discontinuity planes is possible it seems to be less likely then sliding over preexisting planes.

* All the correlated blocks that failed belong to the group of hazardous JC's. This observation lends support to the differentiation between hazardous and non-hazardous JC's (step 3 above) at an early stage of the analysis. Recall that non-hazardous JC's are formed when a line of intersection of two planes lies within the plane of the free face in question. Should some deviation from the ideal joint orientations be encountered in actuality, the same line of intersection will be sub-parallel to the free face in question, resulting in slim and narrow blocks which are of little concern due to their limited volume.

* For all the correlated blocks that failed, the particular JP that was released into the tunnel space and the free face from which it was released are exactly those that were predicted by the application of Block Theory to the global joint sets in the independent analysis. This can be verified by comparison between the field data in Table 5 and Block Theory analysis in Table 2. This fact demonstrates the power and applicability of Block Theory in analysis of the response to tunneling, even in a rock mass with such a complex geological structure .

* A considerable number of blocks were formed by joint combinations that could not be directly correlated to any one JC of the theoretical joint combinations outcome space listed in Table 2 above. These blocks are : (13,14,16,17,23,24,25,26,31) . The number of uncorrelated joint combinations could be reduced significantly if we added one more set to the global structure, namely a shallow dipping set to North-West. This would make the

correlation of blocks :13,16,23,24,25,31 possible and leave only 3 uncorrelated blocks. However, the existence of such a distinct joint set orientation could not be predicted on the basis of the available geological mapping.

* The results of the probability prediction (Fig. 7) indicate that joint combinations 13 and 14 are the most likely to create the critical key block of the excavation, if we assume an average trend for the third set (Figs. 8,9). Fig. 10 shows the overall block failure probability predictions against the number of failures that were found in the field for all hazardous JC's.

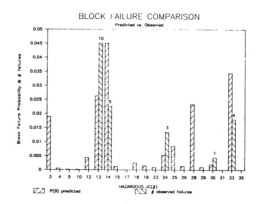

Fig. 10: Block failure probability for all hazardous JC's, compared with the actual failures that were observed in the field. JC's 13 and 14 together are represented by about 65% of the block failures that were observed.

Indeed JC's 13 and 14 form the majority of the key-block failures, 15 out of 23, as was predicted by the independent probability analysis. Fig. 11a shows the critical joint combination which includes joint sets 1 and 5 and the average trend of sets 6 and 7. It can be seen that JP 010 removable from the S.E. wall and its cousin (101), which is removable from the N.W. wall both have a large spherical triangle area as indicated by the large angles between the planes (Check JP 101 for complete view; both JP's have exactly the same area). The frequency of the joint sets in this joint combination is relatively high too, as can be seen in Table 3 above. The combined effect of the relatively high joint sets frequency and the large angles between the joint sets give rise to the high block failure probability of this joint combination. This is supported by field observations.

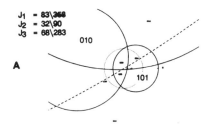

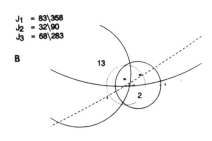

both JP 010 and its cousin JP 101 are removable, JP 101 is much safer than JP 010 and this is reflected in the field results.

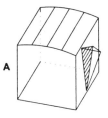

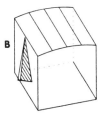

Fig. 11: A) The critical joint combination. The removable JP's it forms with respect to the S.E. and N.W. side walls (dashed line) are JP's 010 and its cousin 101 respectively. Note the wide angles between the removable JP's boundary planes. B) A failure mode analysis for the critical joint combination. The removable JP from the tunnel S.E. wall fails by sliding along $I_{1,3}$, while the removable JP from the N.W. wall fails by sliding along plane 2.

* The summary of the field data (Table 5) indicates that among the 15 key block failures that were correlated with JC's 13 and 14, 13 blocks were released from the S.E. wall (JP 010 of Table 2) and only two cousin JP's were released from the N.W. wall (JP 101 of Table 2). This is not accidental. Fig. 11b shows a failure mode analysis for the critical joint combination of the excavation. It can be seen that while the failure mode of JP 010 is sliding along the line of intersection of planes 1 and 3, the failure mode of its cousin (JP 101) is sliding along plane 2 only. The angle between $I_{1,3}$ and the S.E. wall is only 14 degrees while the angle between plane 2 and the N.W. wall is 75 degrees (Fig. 12). Thus $I_{1,3}$ along which sliding occur in the case of the S.E. wall is about 5 times steeper than plane 2 along which sliding occur from the opposite side wall. Hence we may conclude that although

Fig. 12: A) A three dimensional view of the critical key block of the excavation that is removable from the S.E. side wall of the tunnel (JP 010). The angle between $I_{1,3}$ and the free face is only 14.6 degrees, which makes the failure of this block very likely. 13 blocks that are correlated with this case were observed in the field. B) A three dimensional view of the cousin to the critical key block (JP 101) that is removable from the N.W. side wall of the tunnel. The angle between plane 2 and the free face is 75 degrees, which makes the failure of this block less likely. Only 2 blocks that are correlated with this case were found in the field.

SUMMARY AND CONCLUSIONS

* A prediction of the performance of an underground opening in a discontinuous rock mass must incorporate the orientations of the discontinuities and the opening geometry.
* In underground openings in discontinuous rock masses, the most expected failure mechanism is sliding along discontinuity planes (another possible failure mechanism , rotational failure, is beyond the scope of this work). Therefore an assessment of the suitability of the support measures and method of excavation must incorporate data on the size and shape of

the most probable key blocks to be found in the course of the excavation.

* Several assumptions must be made in order to find the critical key blocks of the excavation:

a) The geological structure must be simplified by the introduction of representative joint sets. The orientations of the joint sets can be found by means of statistical contouring of discontinuities poles on the stereographic projection.

b) The mean spacing value of the different sets must be known. In this work a scanline survey technique is used to find this parameter.

c) In order to create a probabilistic model for joint combinations, the spacing distribution of the joint sets must be determined or assumed. In this work we assume a negative exponential spacing distribution and thus adopt a poisson process to determine the probability of a joint intersection event P(JC).

* Not all joint combinations are of concern. A JC which has one or more lines of intersections that are included in the free face in question is a safe JC that can not create removable blocks behind that free face (Fig. 3a).

* Not all removable JP's from a given free face are of concern. A close removable JP with small angles between its boundary planes (Fig. 4a) is safer than an open removable JP with large angles between its boundary planes (Fig. 4b).

* The overall block failure probability is a function of two factors: 1) the frequency of the boundary joint sets in the rock mass (or their mean spacing value) and, 2) the angles between the boundary joint planes, given that the particular joint combination does create a removable JP from a free face in question.

* The results of field investigations where block failures were observed indicate the applicability of the assumption that a critical key block for a given free face can be found. 15 out of 23 failures that where observed were correlated with the critical key block that was found independently, on the basis of the structural geology data only. This block is formed of joint sets that have relatively high frequency in the rock mass and large angles between its boundary planes.

* The most probable key block for each free face can be found before actual construction takes place on the basis of the geological mapping. The required parameters are the joint set orientations and mean spacing values. The maximum key block size of the critical key block should be used for dimensioning the support.

BIBLIOGRAPHY

Mauldon, M. and R. E. Goodman: Rotational kinematics and equilibrium of blocks in a rock mass, Int. J. of Rock Mech. and Min. Sci, (in press).

Goodman, R. E. and G.-H. Shi: Block Theory and its application to Rock Engineering. Prentice-Hall; Englewood Cliff, NJ (1985).

Reik, G. and S. Soetomo, Influence of geological conditions on design and construction of Cirata powerhouse cavern. in: Saari K. H. O. ed, Large Rock Caverns, Vol.1 , Pergamon Press, 1986.

Priest, S. D. and J. A. Hudson, Discontinuity spacings in rock. Int. J. Rock Mech. Min. Sci. & Geomech. Abstr, 16, 339-362 (1979).

Hudson, J. A. and S. D. Priest, Discontinuities and rock mass geometry. Int. J. Rock Mech. Min. Sci. & Geomech. Abstr, 16, 339-362 (1979).

Wallis, P. F. and M. S. King, Discontinuity spacings in a crystalline rock. Int. J. Rock Mech. Min. Sci. & Geomech. Abstr, 17, 63-66 (1980).

Mauldon, M., Probability aspects of the removability and rotatability of tetrahedral blocks, Int. J. of Rock Mech. and Min. Sci, (in press).

Engineering geology of the Fronhas tunnel, Portugal

Géologie de l'ingénieur du tunnel de Fronhas, Portugal

J.M.Cotelo Neiva & Celso Lima
Electricidade de Portugal, Oporto, Portugal

ABSTRACT: A 8200 m long tunnel with horseshoe cross-section of 4.10 m excavated diameter and 3.50 m concrete lined diameter turns the water of the Alva river into the Mondego river. Mainly Cambrian folded phyllites outcrop in the area. They are cut by dacite porphyry veins, dolerite veins and quartz veins and are covered by unconformed Eocenic sandstone. Several fault sets cut phyllites and a few cut sandstone. The phyllite rock mass is anisotropic and the deformability modulus and the dilatometer modulus increase with depth. Shear strength tests were also carried out in joints. The geological survey helped to drive the tunnel through the phyllites in order to cut appropriately the schistosity, joints, veins and faults which helped very much the tunnel excavation. The measured values during the tunnel excavation and the estimated values for both the Bieniawski classes and primary support are compared.

RESUME: Un tunnel de 8200 mètres de longueur, revêtu de béton, dont la coupe transversale en fer à cheval présente un diamètre caractéristique de 3,50 m et 4,10 m d'excavation, mène l'eau du fleuve Alva vers le fleuve Mondego. Dans la région, il affleure essentiellement des phyllites plissés du Cambrien, qui sont coupés par des veines de porphyre dacitique, de dolérite et de quartz et ils sont couverts par de grès de l'Eocène. Le massif est anisotrope et les valeurs des modules de déformabilité et dilatométrique augmentent en profondeur. On a aussi fait des essais de glissement en diaclases. La reconnaissance géologique de l'aire a aidé à orienter le tunnel à travers le massif de façon à couper convenablement la schistosité, les familles de diaclases, les filons et les failles, ce qui a facilité l'excavation du tunnel. On compare les valeurs mesurées pendant l'excavation et les valeurs estimées pour les classes de Bieniawski et pour le support primaire.

1 INTRODUCTION

A 8200 long tunnel, with horseshoe cross-section, curve walls, flat sill and 3.50 m concrete lined diameter turns the water of the Fronhas reservoir into the Aguieira res ervoir in Mondego river. The level at the entrance of the Fronhas reservoir is of 117 m, while the outlet of the Aguieira reservoir is of 112 m. The dip of the tunnel is about 0.6 m/km. The flow is of 20 m³/s, with maximum inner pressures of 230 kPa upstream and 140 kPa downstream.

This paper presents the engineering geology of the Fronhas tunnel.

2 GEOLOGY

The tunnel area is situated in the surroundings of the Saíl and Cruz do Souto villages of the Arganil and Penacova counties in a plateau of 260-280 m height, in the centre of Portugal.

The Alva river is a tributary of the left bank of the Mondego river. Both rivers incised on a plateau surface, cut the Pliocenic and Eocenic detritic deposits and also the Pre-Ordovician phyllites which they still cut nowadays. The Mondego river incised more quickly than the Alva river. Both show incised harmonious meanders with progressive development.

Dark grey graphitous phyllites, locally slightly corrugated, with thin (0.05-1.00 m) intercalations of grey metagraywackes, form the chlorite zone of a metamorphic complex (Fig. 1b). They may correspond to Algonkian or Lower Cambrian. They are intensively folded. There are nearly isoclinal composite folds with N55-80°W crests plunging 70-75°ESE, but some folds of second and third orders plunge about 57-70°WNW. The axial planes are orientated in average along N68°W, dip 75-80°NNE. The axial schistosity is N60-80°W, 65-85°NNE. They represent the Hercynian folding which is well shown. There is also an early schistosity N20-40°W, 50-77°ENE, which is locally observed and probably represents the Sardinian folding. Generally the joint sets of phyllites have rough surfaces. From the most to the less frequent they are the following: a) N60-80°W, 50-90°NNE, very long, with spacings of 0.08-1.00 m; b) N40-50°W, 40-80°NE, long with spacings of 0.10-1.20 m; c) N20-35°E, 55°WNW-90-60°ESE (mainly dip WNW), slightly long, with spacing of 0.10-1.50 m; d) NO-30°W, 10-35°WSW, slightly long, undulated, with spacing of 0.15-1.70 m; e)N20-35°W, 50-90°ENE, long, with spacing of 0.12-1.40 m; f) N10-35°E, 8-40°WNW, slightly long, undulated, with spacing of 0.20-1.80 m; g) N10°W-N10°E, 50-90°E,long; h) N70-85°E, 50-90°NNW.

Some veins of dacitic porphyry cut the phyllites and three sets were defined: a) N70-80°W, 80-90°NNE, 2.5-3.0 m thick; b) N10°W-N5°E, 65-70°E, 0.20-2.00 m thick; c) N60°E, 70°NW, 0.15-0.20 m thick. There are rare dolerite veins, 0.40-1.20 m thick, along the faults N25-30°W, 75-80°ENE.

In the phyllites there are frequent lenticle milky quartz veinlets. From the most to the less frequent they are the following sets: a) N60-80°W, 65-85°NNE along the schistosity; b) N42-54°W, 55-80°NE; c) N23-38°W, 55-90°ENE; d) N20-35°E, 55° WNW-90-60°ESE; and more rarely e) N10°W-N10°E, 56-90°E. But quartz veins generally of milky quartz and locally of smoky quartz containing some sulfide crystals were also found. Some veins are orientated NW-SE and others NE-SW. The former are the most frequent and dip SW, while the latter dip NW. These veins are 0.20-1.50 m thick.

Separated by an unconformity, Lower Eocenic arkosic sandstones, similar to the sandstone from Buçaco, overlap the phyllites and fossilised a peneplane which cut the Schist-Metagraywacke Complex (Fig. 1a). It is a whitish sandstone containing a cement formed by kaolinite, montmorillonite and colloidal silica. In some sandstone layers there are angular or semirounded fragments of quartz, quartzite, metagraywacke, phyllite and granite.The maximum thickness of the sandstone deposit is of 50-53 m measured in Carapinhal and Estrela de Alva; here there are clay lenticles intercalated in sandstone and exploited for ceramics.

The Upper Pliocenic detritic deposit containing many rounded and semi-rounded blocks of quartzite and quartz in a sandy and clay matrix locally overlaps the Lower Eocenic sandstone. These deposits were not found close to the tunnel.

Some tens of faults cut the Schist--Metagraywacke Complex and a few faults displaced the Eocenic sandstone (Fig. 1a). They form the following sets and the percentage of occurrence is also indicated: a) N60-80°W, 42-90°NNE (28%), rarely dip SSW, 0.05-2.00 m thick (most frequent 0.10-0.30 m thick); b) N40-55°W, 40-80°NE (24%), 0,05-2.00 m thick (most frequent 0.10-0.30 m); c) N20-36°W, 47-90°ENE (23%), rarely dip WSW, 0.05-1.00 m thick (most frequent 0.08-0.30 m thick): d) N70-86°E, 50-90°NNW (8%), rarely dip SSE, 0.05-0.40 m thick (most frequent 0.05-0.10 m thick); e) N8°W-N10°E, 40-90°E (6%), 0,05-0,30 m thick (most frequent 0.06-0.10 m thick); f) N50-66°E, 40-90°NW (4%), rarely dip SE, 0.05-0.70 m thick (most frequent 0.10-0.20 m thick); g) N20-31°E, 55°WNW-90-46°ESE (4%), 0.05-0.40 m thick (most frequent 0.10-0.30 m thick); k) N40-45°E, 35°NW-90-65°SE (3%), 0.07-0.20 m thick.The fault breccias are formed by fragments of phyllite, metagraywacke and quartz with gouge and limonite.

The Pleistocenic river terrace formed by coarse and fine fragments of quartz and quartzite occurs about level 260 m (Fig. 1a). At Carapinhal, Estrela de Alva and Cabecinha the river terrace deposit are 2 to 20 m thick and overlap the Eocenic sandstone.

Talus deposits present a thickness up to 1.50 m, but they were mapped due to their areas. In some places at the bottom of the valleys there are alluvial deposits of thickness up to 7 m.

A geological section of the Fronhas tunnel, which turns the water from Fronhas' reservoir to Aguieira reservoir, is given in Fig. 1b and shows the faults of thickness greater than 0.10 m.

3 MECHANICAL PROPERTIES OF ROCK MASS

Laboratório Nacional de Engenharia Civil (LNEC) carried out the rock mechanical tests for Electricidade de Portugal, EDP. Large flat jacks (LFJ tests) were carried out in slots of chambers of four galleries at heights of 120, 93 and 90 m and excavated in phyllite containing thin intercalations of metagraywackes, in places where the overburden is of 12 to 22 m thickness. Deformations were large during the first cycle, but they were normal during the other cycles. The

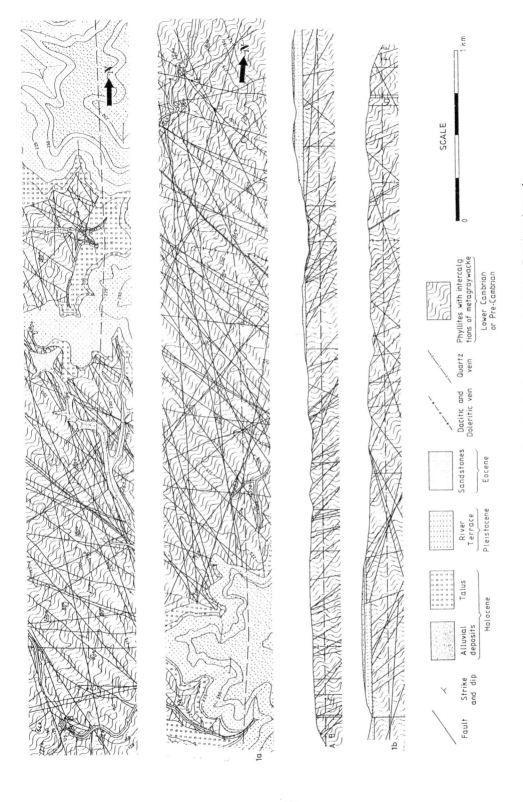

Fig. 1a and Fig. 1b – Geological map of the area and geological section along the Fronhas tunnel

Fault

Strike and dip

Alluvial deposits | Talus
Holocene

River Terrace
Pleistocene

Sandstones
Eocene

Dacitic and Doleritic vein

Quartz vein

Phylites with intercalations of metagraywacke
Lower Cambrian or Pre-Cambrian

SCALE

0   1 km

Table 1. LFJ tests in galleries chambers.
Medium Ex10^6 (kPa)

Galleries	GD120	GD93	GE120	GE90
Perpendicular to schistosity	57	102	63	145
Horizontal	91	90	69	62
Vertical	71	110	40	199

Table 2. Laboratory joint shear tests

Joints	Cohesion (kPa)	Friction angle (°)
Parallel to schistosity A	320-110	34-22
Subvertical B	360-70	37-22
Subhorizontal C	350-90	41-24

A - N60-80°W, 50-90°NNE; B - N20-35°E,
55-90°WNW; C - NO-30°W, 10-35°WSW.

results are given in Table 1.

The deformability moduli vertical and perpendicular to the schistosity are greater at lower heights than at higher heights, because the rocks are less decompressed at lower heights. They present the smallest values where decompression is clearer. The deformability modulus along the horizontal varies much less.

Dilatometer (BHD) tests were also carried out in the chambers where data was obtained from LFJ tests. A positive correlation was found between dilatometric moduli and deformability moduli along the vertical and perpendicular to the schistosity.

Samples were extracted from galleries to study the rock mechanical anisotropy. So uniaxial compressive strength tests were carried out and resistance ranges between 17000 and 107500 kPa, while the modulus of elasticity varies between 16800 and 96400 kPa, showing an ellipsoidal variation for the modulus of elasticity and a quadric variation for resistance with the coefficients of maximum anisotropy of 4.9 and 4.0 respectively. That ellipsoid is negative and of three axes. The acute angle between the minor axis (Z) and the perpendicular to the schistosity (N60-80°W, 65-80°NE) is of about 30°. The axes Z and X lie in a plane of the subvertical joint set N20-35°E.

Four subhorizontal joint shear tests were carried out in three galleries in the field. The cohesion values vary between 330 and 200 kPa and the friction angle ranges between 45° and 24°.

Joint shear tests were also carried out in laboratory to get data from drillcores of galleries. Each sample contains only a joint. Generally the joints have rough surfaces. The results are summarised in table 2. In the subhorizontal joints cohesion is about 270 kPa in the field tests and 190 kPa in the laboratory tests due to the effect of scale. The highest values of friction angle were found in joints without clay film. The friction angle of joints along the schistosity is lower than in the subhorizontal joints.

The values of cohesion and friction angle for the joints parallel to the schistosity are similar to the respective values for the subvertical joints (both are subvertical and approximately perpendicular between them).

4 TUNNEL ALIGNMENT AND ITS EXCAVATION

The geological surveying of the area is shown in Fig. 1a. The tunnel alignment was chosen to have a good overburden (above 50 m and up to 155 m), to be constructed in phyllite containing thin intercalation of metagraywackes where the Bieniawski classes I (very good rock),II (good rock) and III (fair rock) predominated and to cut generally the crest of folds, schistosity, faults, joints and quartz veins (Fig. 1b). This tunnel alignment would make the tunnel excavation easy, although one or another joint set could produce some overbreak.

Some portion of very fissile phyllite, intersection of tectonic structures, fault breccia and gouge and related portion of weathered rocks would imply local primary support. The accumulation of infiltrated water close to those faults and quartz veins, mainly under or close to streams, could produce some difficulties to the excavation. It was estimated that the primary support for mainly the Bieniawski classes IV (poor rock) and V (very poor rock) would not be greater than 25% of the tunnel length.

As phyllite and metagraywacke are very old rocks (Algonkian and Lower Cambrian) which were deformed at least by two foldings earlier than the Triassic period (Sardian movements and Hercynian movements) it was not necessary to consider the residual stress.

The excavation of the tunnel, of characteristic diameter of 4.10 m, was carried out by full-face driving. Two advancing faces were drilled and blasted. About 110 m were excavated per month in each advancing face.

The excavated segments from the upstream advancing face were generally according to the dip of schistosity and of most of the faults. The primary support for the whole of those segments was of 14.2%. The excavated segments from the downstream advancing face were orientated against the dip of schistosity and of most of the faults and the primary support was of 22.0% (Table 3).

Table 3. Excavation of the tunnel

| Advancing face | Length (m) | Primary supported tunnel segments | |
		Length (m)	Percentage (%)
Upstream	4089	578.6	14.2
Downstream	4103	901.6	22.0
Total	8192	1480.2	18.1

In the AB and CD tunnel segments overbreak was found as foreseen in the roof and East wall mainly due to the joint set NO-30°W, 10-35°WSW but also due to the join set N10-35°E, 8-40°WNW which mainly influenced the excavation of the roof and SW wall of the BC segment. The overbreak was not intense due to the use of a good tunnel blasting diagram. So rockbolts and shotcrete were not used.

As the DE and EF tunnel segments were closely orientated according to the fold crest and schistosities than the other segments, it was estimated that more primary support would be necessary for them. In fact it was mainly applied in the DE segment.

Primary support was necessary in the tunnel segments which crossed faults and weathered rocks close to the faults (16.5%). However only 1.1% of the segments needed primary support due to only rock weathering. The total of segments which required primary support correspond to 18.1% of the tunnel length (Table 3).

In some tunnel segments the quartz veinlets were more abundant in phyllite and the water infiltrated through the contact with those veinlets, which locally implied some primary support (0.5%). Most of the water infiltrated through joints and faults. Generally the water was acid, carrying sulfates because it had solubilized pyrite from phyllite. It was necessary to use volcanic cement in the concrete lining and cement injections.

The tunnel segment where Eocenic sandstones, Pleistocenic river terrace and talus slope overlap discordantly phyllite correspond to 3060 m of length. In this segment phyllite is less weathered because the volume of infiltrated water is smaller. So only 9.1% of that segment needed primary support.

The amount of rock classes according to the Bieniawski classification which would be found along the tunnel and primary support to be used during the excavation of some of those classes were estimated (Table 4). The measured values are also given in Table 4. The results are generally close to the estimated values, except for class IV where they are about 50% smaller than the estimated values probably because the average width of the excavated tunnel section was of 4.10 m.

Table 4. RMR and primary support

| Rock mass classes | Bieniawski classification | | Primary support | |
	A (%)	B (%)	A (%)	B (%)
I	27	33	0	0
II	30	32	0	0
III	20	18.6	2	1.7
IV	13	6.2	13	6.2
V	10	10.2	10	10.2

A - estimated, B - measured; I - very good rock; II - good rock; III - fair rock; IV - poor rock; V - very poor rock.

5 CONCRETE LINING

After the excavation, the tunnel was concrete lined in two advancing faces. In the roof and walls concrete lining of 0.30 m thickness was used except if primary support was applied because the deformability modulus of the rock mass was low and required lining involving steel ribs with lagging in a single layer to resist to ground water pressures and to avoid fissures.

The different types of lining are shown in Fig. 2 and 3, while the amounts of concrete used for both advancing faces are given in Table 5.

Grouting of rock mass was carried out two distinct phases (Fig. 2 and 3). In the first phase cement, sand and water were applied at the concrete-rock contacts in the whole tunnel and the absorptions in the downstream advancing face were greater than 35% of those in the upstream advancing face (Table 6). The second

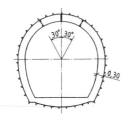

— INJECTION ON A SECTION
-- INJECTION ON THE FOLLOWING SECTION

Fig.2 - Concrete lining and injections schemes.

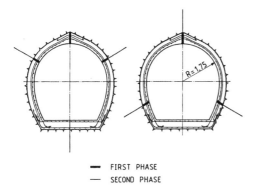

— FIRST PHASE
— SECOND PHASE

Fig.3 - Reinforced concrete lining and injections schemes. A and B alternated cross-sections.

phase comprised grout holes penetrating 1.00 m in the rock mass of only the tunnel segments with reinforced concrete (Fig. 3). Absorptions were greater in the upstream advancing face than in the

Table 5. Concrete lining

Advancing face	A (m^3)	B (m^3)	C (m)
Upstream	12011	21352	0.21
Downstream	12052	27448	0.34
Total	24063	48800	0.27

A - estimated; B - measured;
C - medium overbreak.

downstream advancing face. So in the upstream advancing face the first phase injections at the concrete-rock contacts did not penetrate enough to fill the open

fissures which were less abundant than in the downstream advancing face (table 6). The same cement grout and pressures were used in both faces. Although the downstream advancing face contained more

Table 6. Injections

Advancing face		Tunnel length (m)	Cement (ton)	Sand (m^3)	Cement absorption (kg/m)
Upstream	A	4185	2082	1650	497
	B	578	273	-	472
	C	4165	2355	1650	562
Downstream	A	4007	2701	2073	674
	B	901	211	-	234
	C	4007	2912	2073	727
Total	A	8192	4783	3727	584
	B	1479	484	-	327
	C	6192	5267	3727	643

A - first phase contact; B - second phase consolidation; C - Total.

open spaces they were filled in the first phase which implied that the absorptions were smaller in this face than in the upstream advancing during the second phase.However the total absorption in the downstream advancing face was 29% greater than in the upstream advancing face, which shows that the volume of open spaces was greater in the former than in the latter.

6 CONCLUSIONS

The detailed geological surveying of the area and the mechanical tests of the rock mass indicated the best tunnel alignment and helped its excavation and concrete lining.

Drainage souterrain d'un grand glissement de terrain: Le Thoronet (France)
Underground drainage of a major landslide: Le Thoronet (France)

B. Griveaux
Ministère de l'Environnement, DRAE-PACA, Aix-en-Provence, France

G. Koch Paquier
Ministère de la Culture et de la Communication, CRMH-PACA, Aix-en-Provence, France

G. Colombet
Coyne et Bellier, Bureau d'Ingénieurs Conseils, Paris, France

M. Poosz
Entreprise Migec-Somafer, Marseille, France

RESUME : L'Abbaye du Thoronet (XII° siècle) est au pied du glissement d'une colline dont l'équilibre géomécanique a été modifié par l'exploitation de la bauxite. La communication décrit la réalisation d'un dispositif de drainage dont le coût est de l'ordre de 12 millions de francs et les premiers effets mesurés sur le contexte hydrogéologique et les vitesses de glissement.

ABSTRACT : The Abbey of Thoronet (XIIth century) is at the foot of the landslide of a hill, the geomechanical stability of which was disturbed by bauxite mining. The article describes the construction of a drainage system, the cost of which is around 12 million francs and the initial effects noted in the hydrogeological context and the slide rates.

1. CONTEXTE GEOLOGIQUE ET MINIER : LES RISQUES POUR L'ABBAYE

1.1. Contexte

Située à 180 m d'altitude, la célèbre Abbaye est au pied d'une colline qui culmine à 350 m (fig.1). La structure géologique de la colline est un synclinal très dissymétrique et chevauchant. A la fin de l'Eocène, une puissante série constituée de marno-calcaires du Rhétien surmontés de calcaires du Rhétien supérieur et d'Hettangien a été poussée du Nord vers le Sud. Elle a glissé sur des argiles, cargneules et marnes du Keuper. Elle a ainsi recouvert des plateaux à substratum calcaire du Bathonien tapissés de bauxite. Celle-ci a été piégée sous la série chevauchante (fig.2).

L'exploitation minère de la bauxite a d'abord été souterraine d'Ouest en Est (grisé fig.1). Plus de 80 % de la surface du glissement de bauxite sont concernés par ce type d'extraction qui a concerné 11 millions de tonnes.

En 1966, un grand ciel ouvert entoure le tiers oriental de la colline (fig.1). L'extraction concerne 17 millions de tonnes dont 2 de bauxite. Depuis 1989, toute activité minière a cessé.

1.2. Les risques

Dans une structure prédisposée à l'instabilité, les techniques minières ont sans doute donné l'impulsion à un mouvement de grande ampleur. Le foudroyage systématique des mines souterraines a entraîné l'écoulement des nappes d'eau du Rhétien et de l'Hettangien calcaires au travers de toute la structure et en particulier dans le contact glissant de

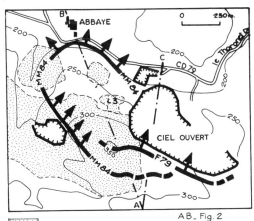

AB _ Fig. 2

░░░ MINES SOUTERRAINES

Fig. 1 - Glissement schématisé

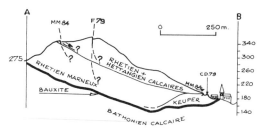

Fig. 2 - Coupe structurale simplifiée

ces calcaires sur le soubassement marneux (faille MM84. fig. 1 et 2) et dans le Keuper.

Dans le ciel ouvert, un effondrement important (F.79, fig. 1 et 2) contribue probablement à pousser la masse glissante vers le Nord-Est et vers le Nord c'est-à-dire vers l'abbaye. Le suivi de sondages inclinométriques et de repères topographiques (fig. 3) montre que cette masse est énorme, au moins 11 millions de m³ dont 7 millions se dirigent vers

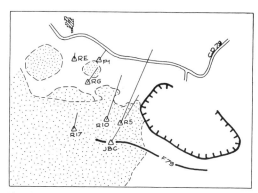

MOUVEMENTS DE JANVIER 1986 A DECEMBRE 1989
△ - REPERE TOPOGRAPHIQUE
0   50cm
└─┴─┴─┴─┴─┘   DEPLACEMENT EN CM.

Fig. 3 - Représentation des mouvements

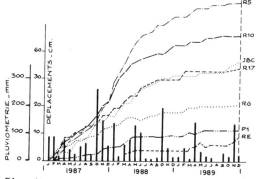

Fig. 4. Comparaison mouvements/pluviométrie

l'abbaye. Dans le secteur proche du monument la vitesse du glissement est en moyenne constante en 1989 et égale à 1,5 à 2 cm/an. Les graphiques des figures 3 et 4 montrent un ralentissement depuis le milieu de 1988 mais le déficit pluviométrique persistant depuis 3 ans explique probablement en grande partie ce phénomène. La masse glissante calcaire pousse devant elle un bourrelet de Keuper lequel déforme la route d'accès à l'abbaye et déplace un bâtiment annexe "la Grange Dimière", entièrement étayée. Outre les épanchements de la nappe du Rhétien-Hettangien dans le plan de décollement des calcaires et dans le Keuper, une nappe existe dans le Bathonien support de l'abbaye et de la structure synclinale de la colline. Le risque existe de voir cette nappe remonter et activer à son tour les contacts glissants.

2.   PROJET DES TRAVAUX DE DRAINAGE

2.1. Principe de la solution retenue

Le Ministère de la Culture et de la Communication a lancé en juin 1986 un appel d'offres pour la conception d'ouvrages susceptibles de s'opposer aux glissements en amont de l'Abbaye. Les membres de la Commission technique et du jury nommés par le Ministère ont retenu la solution drainage proposée par une entreprise associée au bureau d'Ingénieurs Conseils Coyne et Bellier.

Les ouvrages comprennent 2 000 m de forages drainants débouchant dans une galerie de 520 m de longueur placée sous les niveaux de glissement. Les travaux prévus cherchent à drainer gravitairement le Keuper et à augmenter la résistance de la discontinuité Keuper-bauxite, en vue de mettre un terme aux désordres affectant la Grange Dimière. Le système est extensible, si nécessaire.
Cette solution, qui a fait l'objet d'un projet détaillé achevé en avril 1988, comprend essentiellement :

a) un rideau de forages drainants situé à l'aplomb de la portion B-C-D-E de la galerie de drainage (fig.5). Il entoure l'Abbaye à une distance d'environ 80 m ; il est destiné à drainer les horizons perméables du Keuper, en particulier au voisinage de la bauxite.

b) trois forages verticaux pour drainer la nappe perchée des calcaires du Rhétien supérieur, qui alimente au moins en partie celle du Keuper. Ces drains débouchent dans un rameau latéral à la galerie de

2454

drainage (portion CF de la fig.5).

c) le prolongement de la galerie de drainage sur 270 m (portion A-B de la fig.5) pour évacuer les eaux collectées jusqu'au débouché de cette galerie dans un vallon situé au Nord-Est de l'Abbaye.

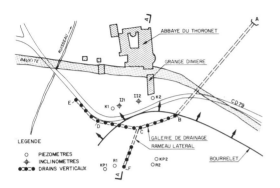

Fig. 5 - Vue en plan des ouvrages

2.2. Nature des forages drainants

a) Drainage du Keuper

Le drainage du Keuper est assuré en premier lieu par 24 drains verticaux de 30 à 45 m de profondeur forés depuis la surface et espacés de 8 mètres (fig. 6).

A la traversée du Keuper, ils sont équipés d'une chaussette géotextile drainante avec graviers roulés et d'un tube central PVC crépiné de 50 mm de diamètre. Le tubage traverse ensuite la bauxite pour déboucher dans la galerie de drainage.

L'action de ces drains verticaux est complétée par des drains inclinés à 20° sur la verticale, forés depuis la galerie. Leur longueur est telle qu'ils traversent le niveau de glissement en toit de la bauxite pour pénétrer de 3 à 4 m dans le Keuper.

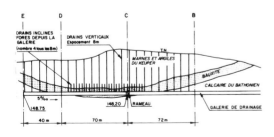

Fig. 6
Profil en long de la galerie de drainage

b) Drainage du Rhétien

Les trois drains verticaux du Rhétien (fig. 7) sont forés depuis la surface en 250 mm de diamètre. Ils ont 50 à 65 m de profondeur.

Ils sont équipés d'un tube métallique de 50 mm de diamètre entouré de géotextile.

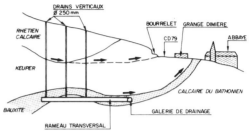

Fig. 7 - Profil en travers du versant - Rameau transversal

Le projet prévoyait que deux drains de ce type seraient réalisés entre deux drains verticaux espacés de 8 m. En cours de travaux leur nombre a été porté à quatre pour augmenter l'efficacité du dispositif. Dans ces conditions, le rideau de forages drainants du Keuper comprend cinq forages tous les 8 mètres linéaires de galerie.

A la traversée du Keuper et du Rhétien, ils sont équipés d'une chaussette drainante avec graviers roulés et d'un tube central métallique crépiné de 50 mm de diamètre. La traversée de la bauxite jusqu'au rameau transversal se fait par un tube de 50 mm de diamètre.

c) Drainage du Bathonien

Le drainage du Bathonien est effectué à partir de la galerie de drainage. Il comprend des captages et des drains dans les zones de circulation d'eau préférentielle (fissures, failles).

3. REALISATION DES TRAVAUX

Les travaux ont été exécutés du 15 mai 1989 au 15 mars 1990 par l'Entreprise MIGEC, la maîtrise d'oeuvre étant assurée par Coyne et Bellier.

3.1. Creusement de la galerie

La galerie a une section en fer à cheval de 8 m² avec une pente montante de 0,5 %. Les 520 m de galerie ont été excavés en méthode traditionnelle à l'explosif. Les tirs ont été effectués sous surveillance sismique pour contrôler les vibrations induites. Le seuil maximum admis sur l'abbaye et les bâtiments voisins était de 1 mm/s pour une fréquence de coupure de 10 Hz.

332 tirs ont été effectués, ce qui représente un avancement moyen de 1,56 m par volée. Aucun tir n'a engendré de vibrations dangereuses. Chaque volée était chargée à 2,2 kg/m^3 et comprenait 52 trous. Les charges unitaires variaient de 0,52 à 0,95 kg par microretard suivant la distance galerie-abbaye.

La galerie a recoupé :

- de 0 à 310 m des calcaires bathoniens fissurés, fracturés et karstifiés à pendage général N140 à 160 inclinés de 10 à 30° vers l'Ouest. Deux failles N115 à 125 inclinées de 10 à 20° vers l'Ouest ont été traversées ;

- de 310 à 430 m, ainsi que dans le rameau transversal sur 66 m, des bauxites tendres très tectonisées avec suintements d'eau ;

- de 430 à 454 m des calcaires bathoniens semblables aux précédents.

3.2. Soutènements

Le soutènement provisoire à l'avancement a été assuré par des cintres sur 55 m et 2 600 boulons au toit, complétés dans les bauxites sur 150 m par un avancement sous enfilage en voûte parapluie suspendue avec béton projeté à prise rapide (cf fig.8).

Le soutènement définitif est en béton projeté avec treillis soudé (épaisseur 0,25 m à 0,05 m) sur 485 m et en béton coffré sur les 35 premiers mètres.

Le béton projeté est systématiquement équipé de forages de décharge.

Le radier est bétonné avec caniveau latéral de section 0,30 x 0,50 m.

3.3. Drains

Les drains subverticaux forés à partir de la surface ont été réalisés au moyen d'un atelier de foration au marteau fond de trou aux diamètres de 250 à 185 mm.

Les drains remontant inclinés à 20° à partir de la galerie ont été forés au supermarteau à air comprimé.

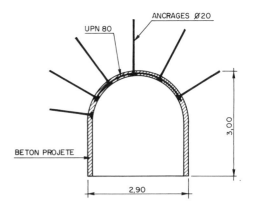

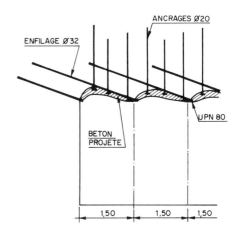

Fig. 8 - Principe de l'enfilage en voûte parapluie suspendue

4. EFFETS CONSTATES SUR L'HYDROGEOLOGIE

4.1. Nappe du Keuper

Les débits drainés dans le Keuper sont relativement modestes. Ceci peut s'expliquer par la faible perméabilité de cette formation géologique à prédominance argileuse et par le faible niveau piézométrique de la nappe consécutivement à une longue période de sécheresse.

Juste après leur percement, les drains ont collecté un débit variant entre 0,01 et 4,79 l/min par trou. Le débit total est passé par un maximum de 16 l/min début décembre 1989.

Un mois plus tard, après rabattement partiel de la nappe, le débit total s'est réduit à 1,3 l/min, dont 40 % provenait de trois drains réalisés au voisinage du point C (mesures de janvier et février 1990).

Deux piézomètres (K1 et K2) et deux inclinomètres (II1 et II2) ont été installés à l'aval du réseau de drainage pour suivre l'efficacité des travaux (fig.5). Des relevés piézométriques sont effectués régulièrement dans chacun de ces forages. Les courbes correspondantes sont reportées sur la figure 9. La nappe se trouve rabattue d'environ 2 m, les niveaux piézométriques se situant actuellement dans la bauxite.

Les mesures réalisées en janvier et février 1990 sur le piézomètre K2 donnent des fluctuations anormales. Le fonctionnement de ce piézomètre est vraisemblablement défectueux. Des travaux pour vérifier son bon fonctionnement sont prévus très prochainement.

4.2. Nappe du Rhétien supérieur

Un des trois forages drainants verticaux traversant les calcaires du Rhétien a évacué un débit de 80 l/min pendant la première journée qui a suivi son percement. Par suite du rabattement de la nappe, le débit drainé a progressivement diminué jusqu'à 1,2 l/min en janvier et février 1990.

Les deux autres forages réalisés de part et d'autre drainent un débit nettement inférieur. Les calcaires du Rhétien présentent une perméabilité de fissures, avec conduits karstiques plus ou moins colmatés d'argile. Ces deux forages ne semblent donc pas avoir recoupé de zones préférentielles de circulation d'eau.

Le piézomètre R1, situé à l'Ouest des travaux de drainage, montre un rabattement sensible de la nappe du Rhétien (fig.9).

Deux piézomètres (KP1 et KP2) ont été installés au voisinage de R1 et R2 (fig.5). Ils permettent de suivre les niveaux piézométriques dans le Keuper, sous la nappe perchée du Rhétien.

Le piézomètre KP1 montre que la nappe du Keuper se situe à cet endroit autour de la cote 160 NGF, c'est-à-dire sous le niveau de cette même nappe au voisinage de l'abbaye (fig.9). Ce piézomètre est situé à proximité d'anciennes exploitations souterraines de bauxite dans lesquelles les niveaux d'eau sont actuellement très bas. La nappe se trouve donc ici rabattue temporairement par les exploitations minières.

A l'inverse, le piézomètre KP2 montre que plus à l'Est, la nappe du Keuper est située vers la cote 180 NGF, c'est-à-dire sensiblement au même niveau que celle du Rhétien susjacent (fig.9). Cette constatation s'est trouvée confirmée par les données du piézomètre KP2 bis foré au même endroit à titre de vérification. Le rabattement de la nappe du Rhétien sur 4 à 5 mètres après les travaux de drainage s'est accompagné d'un rabattement comparable de la nappe du Keuper, confirmant ainsi l'existence de communications entre ces deux nappes superposées.

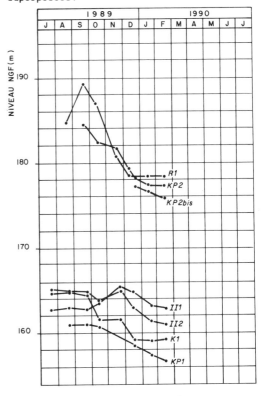

Fig. 9 – Mesures piézométriques

4.3. Drainage du Bathonien

Pendant les travaux, la nappe du Bathonien est toujours restée sous la cote de la galerie d'exhaure. Le calcaire est fissuré, découpé par plusieurs failles probablement liées au chevauchement. Ces zones sont partiellement colmatées d'argile rouge et présentent des signes de circulation d'eau lorsque la nappe phréatique remonte. Le béton projeté placé comme soutènement définitif de la galerie est donc équipé de barbacanes et de drains

5. CONCLUSIONS

La réussite de l'ouvrage de drainage souterrain conçu pour freiner, voire stopper les mouvements de terrain aux abords de l'abbaye du Thoronet est à apprécier sous deux angles : celui de l'exécution de l'ouvrage par rapport au projet et celui des effets attendus.

L'exécution est jugée correcte puisque le creusement de la galerie projetée a été menée in extenso y compris dans un secteur imprévu de bauxite friable et détrempée, probablement pour des raisons tectoniques. Des drains supplémentaires ont pu être forés à partir du toit de la galerie et d'autres pourront l'être ultérieurement si nécessaire.

Quant aux effets attendus, la période de déficit pluviométrique qui sévit depuis 3 ans ne permet pas de les apprécier dans l'immédiat. Certains drains ont recoupé des fissures aquifères et ont montré un phénomène très net de décharge. La plupart des autres drains montrent des écoulements faibles et il conviendra d'observer leur réaction à des évènements pluviométriques significatifs. C'est alors que l'observation fine des mouvements topographiques montrera si le drainage réalisé les réduit sensiblement et si la foration de nouveaux drains peut d'avérer utile.

Rock mechanics investigations on pillar dimensions and roof support at Buckreef gold mine in Tanzania

Mécanique des roches – Etudes sur les dimensions des pilliers et supports du toit à la mine d'or souterraine de Buckreef en Tanzanie

E. Kayogoma
Esamrdc, Section Rock and Soil Mechanics Laboratory, Dodoma, Tanzania

ABSTRACT: Underground investigation at Buckreef Gold Mine was undertaken in view of finding out whether rock bolting technique can be introduced as roof support in the interest of safety and whether the size of the crown pillar can be reduced from 10m thickness to 5m so as to increase the quantity of ore extraction from stopes. Investigations comprised of visual observations, in-situ and laboratory testing of rock samples with the help of schmidt hammer and point load machine. The rock was found to have high compressive strength but fails by splitting at the corners of the pillars. By considering the failure pattern of the rock, rock bolting can only sound well in the second level while the remaining levels can be supported by timbers, while the reduction of pillar dimensions has to be done in few stopes with instrumentation to assess the support behaviour of the pillars.

RESUME: A la mine d'or Souterraine de Buckreef, des études ont été entreprises pour vérifier si on pouvait utiliser la technique dés boulons d'ancrage, pour supporter le toit. On a etudié également la possibilité de réduire la dimension des **piliers** de 10 à 5m et ce, afin d'augmenter l'extraction du minerai. Des examens sur place eurent lieu, ainsi que des essais en laboratoire sur des échantillons de roche avec le marteau Schmidt et la machine produisant une charge ponctuelle. On a trouvé que la roche avait une résistance élevée à la compression mais qu'elle s'effritait aux coins des piliers. Apres avoir examiné le type de cassure de la roche, on peut conclure que l'ancrage par boulons convient au 2eniveau mais que pour les autres niveaux, le soutenement doit être réalisé avec des poutres en bois. Quant à la dimension des piliers elle peut etre réduite dans quelques endroits pour autant que la comportement des **piliers** soit suivi par des instruments de mesure.

1 INTRODUCTION

Investigations in rock and soil mechanics by the way of in-situ, on site and laboratory tests is part of Eastern and Southern African Mineral Resources Development Centre studies. The studies are aimed at generating data needed in scientifically designing mines and also in modifying methods of mining and strata support in mining operation; rock mechanics investigations at Buckreef Gold Mine was undertaken in that line.

1.1 Location and accessibility

Buckreef Gold Mine is located in the Geita district, Mwanza region. Latitudes $32^{\circ}2'E$ and Longitude $3^{\circ}5'S$ defines the location of the mine and the nearest major town is Mwanza which is about 171km from Buckreef. It takes 3 to 4 hours by a four wheel drive vehicle on a non-tarmac road.

1.2 Objective of investigations

It was desired to ascertain whether rock bolting could be introduced in 30m level and 61m level as means of roof support in place of timber support; whether the crown pillars in stopes could be reduced from 10m thickness to 5m.

2 MINE HISTORY

Based on the exploration work carried out during the period 1965 to 1971, an investment decision was taken in 1972 to establish a mine at Buckreef. The investment decision was based on the fact that the equipment available in the nearby abandoned mines would be deployed at Buckreef. In 1973 site preparation and underground development work was taken up and pilot plant tests for gold recovery commenced in the same year. However, the expectation regarding the project implementation could not be met as most of the old equipments in the abandoned mines were seen to be unsuitable for reuse. A help was given by Swedish Government Aid Organisation (SIDA) in 1977 and the mine development and processing plant were completed in 1981 and gold production commenced in 1982.

3 GEOLOGY

Buckreef area lies in the Northern part of the extensive area of granitic rocks which occupy the greater part of Central Tanzania. In general exposure is poor, the granites outcrop as low round hills on lower ground, sporadic deeply weathered outcrops protrude through sandy soils. Through the granitic area there occur irregular outcrop areas of an older series of sedimentary and metamorphosed volcanic rocks, recognized as the Nyanzian greenstone rocks. These have been interpreted as roof pendants to the granitic batholith. Gold deposits of virtually all the goldfields of Norhtern Tanzania are associated with the Nyanzian Greenstone rocks.

The Nyanzian rocks are predominantly metavolcanics but includes other rocks of sedimentary original. A characteristic component of the Nyanzian sequence in Tanzania goldfield is the banded iron-stone which forms prominent strike-oriented hills and is the host to significant gold mineralization.

In contrast to other areas, the banded ironstone does not appear in the Buckreef area. In the Buckreef area the Nyanzian is characterized by basic metavolcanics generally referred to as greenstone due to their colour. The orebody is a series of quartz bodies overlain by laterites. Gold mineralization is associated with disseminated crystalline sulphides.

4 DESCRIPTION OF FIELDWORK AND OBSERVATIONS

Discussions with a number of officers of Buckreef Gold Mine was undertaken at the beginning of the investigation so as to have an appreciation of different operating features of the mine and to appraise the problems concerning rock mechanics.

After getting appraised of the problems and the objectives to the investigations as set out in Chapter 1.2, a study comprising of in-situ, on site and laboratory tests were undertaken. For the in-situ tests schmidt hammer was used to measure compressive strength of intact rock mass. For on site and laboratory tests, point load testing machine was used. The location of tests by schmidt hammer was marked on the longitudinal section of stopes in 30mL, 61mL and 107mL in block 6 and 7 and 70mL and 98mL in block 3 (fig.1). In situ testing were carried out with schmidt hammer kept in horizontal direction at each testing point, four to five readings were taken to get the average rebound number.

On site tests were carried out by using point load testing equipment on rock lumps (irregular specimen). In the laboratory cores were prepared using a laboratory coring machine and rectangular blocks were also prepared.

The values obtained from schmidt hammer called rebound number were converted to in-situ compressive strength with the help of standard chart. While for the results of point load test were multiplied by 24 as per standard practice after size correction to 50mm dia. By visual observations it was found that the rock in 30mL has a tendency to break in irregular small pieces, while in 61mL (Block 6) it was observed the roof to fall in well defined slabs, the overbreak in the roof is also considerable as it extends to as much as 2m.

At 70mL sub level, the quartz reef is granular in texture, and it has a tendency to separate at the corners of the pillar

5 DISCUSSION ON TEST RESULTS

5.1 For 30mL (Block 6):

Wherever the rock was weathered the compressive strength was found to be relatively low (Table 3). The difference in compressive strength by on site point load tests and that by laboratory testings is on account of specimen geometry, for the former the specimens were rock lumps while for the laboratory the specimen were rectangular blocks and cores (Table 2 and 3). The values on rock lumps has also been affected by manual hammering which might have introduced some cracks in the lumps.

5.2 For 61mL (Block 6)

The values of compressive strength from hanging wall and foot wall is low as compared to that from the roof; the basic reason for this lies in the fact that the rock from the sides of the drift have been affected by weathering due to the presence of stagnant water.

5.3 For 70m Sub level (Block 3)

Values of compressive strength are high as is evident from various tests (Table 2) one set of readings were rejected for the on site tests as the value was abnormally low. This was on account of fracture.

5.4 For 98mL (Block 3)

The rock is high in compressive strength (Table 1). It is also the same for 107mL

6 CONCLUDING REMARKS AND RECOMMENDATIONS

Though the rock at Buckreef gold mine is generally good in compressive strength, roof bolting in 30mL may not be tried in the drift for the reason that the rock breaks in small pieces, support of the roof falls with wire net and the present practice of timber support may continue.

Roof bolting in 61mL may be introduced as here the roof falls in slabs on account of well defined stratification. Since overbreak is considerable it would be advisable to install instruments along with roof bolts to monitor their effectiveness.

Since the quartz reef has a granular texture, it has a tendency to crumble with passage of time at the corner of the pillars where we find high concentration of stresses. It is therefore likely that when the crown pillars of less than 10m are experimented will crumble with passage of time.

These pillars are required to provide roof support for the whole life of the mine, if they will crumble some stopes will not be reached by the work force of the mine. If the management will be forced to reduce these pillars on account of economics, then this experiment has to be done with instrumentation to help in ensuring safety.

7 ACKNOWLEDGEMENT

Thanks are due to Director General of ESAMRDC for having sponsored the rock mechanics studies at Buckreef Mine. The cooperation and facilities extended by the management of Buckreef during the studies is greatefully acknowledged. Finally I would like to express my sincere gratitude to Ms. Kacungira of ESAMRDC who typed this paper.

8 REFERENCES

1. Goodman, R. An Introduction to Rock Mechanics, p35-36. John Wiley and Sons.
2. Price, D.G. 1986. An Introduction to Engineering Geology q38 part 1, p28-39 Delft University of Technology
3. Rengers, N. and Verwaal, W. 1988. Collected Manual of testing procedures in rock mechanics laboratory, p28-35 Delft University of Technology
4. Williamson Diamond Ltd., 1971. Report on underground exploration work at Buckreef, Geita District, Tanzania (unpublished report).

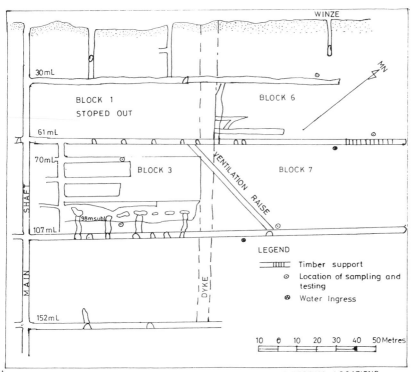

WINZE

30 mL

BLOCK 1
STOPED OUT

BLOCK 6

MN

61 mL

70 mL

BLOCK 3

VENTILATION RAISE

BLOCK 7

SHAFT

98 m sub

MAIN

107 mL

DYKE

152 mL

LEGEND

▭ⅢⅢ Timber support
⊙ Location of sampling and testing
⊗ Water Ingress

10 0 10 20 30 40 50 Metres

Fig 1 LONGITUDINAL SECTION OF STOPES SHOWING SAMPLING AND TESTING LOCATIONS

TABLE 1

IN SITU SCHMIDT HAMMER TESTS FOR COMPRESSIVE STRENGTH

LOCATION				OBSERVATIONS		Compressive strength (MPa)	REMARKS
Level 1 Sublevel (m)	Block No.	PEG No.	Distance from PEG (m)	Rebound No.	Average value		
30	6	1127	41.6	55 63 65 55 67	61	230	Hanging wall side of the reef
61	6	2035	35	55 51 52 49 45	50.4	135	Foot wall side of the reef
				49 50 50 50	49.8	130	Hanging wall side of the reef
70	3	2303	6.2	55 62 56 58 59	58	180	Sub level of crown pillar
98	3	3209	12.3	65 62 63 58	62	235	Draw point pillars
			16.3	68 67 66 66 66	66.6	280	
107	7	3523	11.5	62 60 55 56	58.3	180	For future draw points

Table 2

ON-SITE POINT LOAD TESTS FOR COMPRESSIVE STRENGTH

Serial No.	LOCATION				Point load strength I_s(50mm) MPa	Average value I_s(50mm) MPa	Unconfined compressive strength MPa	REMARKS
	Level 1 sub-level (M.)	Block No.	PEG. No.	Distance from PEG (m)				
1	30	6	1127	41.6	5.5 5.0 6.10 5.0 4.40	5.16	123.8	Greyish Quartz, slightly weathered along the joints
2	61	6	2035	35	6.95 9.00 8.20 6.00	7.54	181.0	Fresh quartz taken from roof
3	70		2303	6.2	0.60 1.00 0.74	0.78	18.7	*
		3			7.6 5.6 6.6	6.60	158.40	Milky vitreous quartz with dark bands
4	98	3	3209	12.3	7.50 6.60 4.86	6.32	151.7	Milky vitreous quartz with greyish bands
5	107	7	3523	5	5.1 6.8 4.6 5.0 6.5	5.60	134.4	Greyish quartz
				11.5	6.8 6.3	6.55	157.2	Metabasalt

* Values rejected as the samples developed fractures during trimming.

NOTE: Type of samples tested: Lumps

TABLE 3

LABORATORY POINT LOAD TESTS FOR COMPRESSIVE STRENGTH

LOCATION				Type of Test	Point Load strength I_s(50mm) MPa.	Average value I_s(50mm) MPa.	Unconfirned compressive strength MPa	REMARKS
Level 1 Sub-level (M)	Block No.	PEG No.	Distance from PEG (m)					
30	6	1127	41.6	Diametral	12.2	12.2	292.8	Fresh greyish quartz with dark bands
					7.37) 5.50)	6.44	154.6	Greyish quartz slightly weathered along weak zones
61	6	2035	35	Rectangular Block	6.2 5.3 4.5 6.0	5.50	132	Quartz, greyish in colour with yellowish strips due to weathering
70	3	2303	6.2	Rectangular Block	4.8*	4.80	115.2	Milky quartz with dark bands randomly fractured
					10.0) 9.2)	9.60	230.4	Milky vitreous quartz with dark bands. No fracture
98	3	3209	12.3	Rectangular block	9.4 7.0 8.4	8.27	198.5	Milky vitreous quartz with greyish bands
107	7	3523	5	Rectangular block	5.4 5.90	5.65	135.6	Greyish quartz
			11.5	Rectangular	11.30	11.30	271.2	Metabasalt

* Values rejected as they do not seem to be representative of the groups.

Material behaviour of coal under high stresses and temperatures
Comportement du charbon sous des hautes températures et des contraintes élevées

H. Meissner & F. Rogmann
University of Kaiserslautern, FR Germany

ABSTRACT: In the field of underground coal conversion in Europe, the linking technology is an important field of interest. The stability of linking channels at great depth , burned in reverse combustion, is the topic of a research projekt at the University of Kaiserslautern. Various experimental studies about the temperaturedependant anisotropic behaviour of coal have been accomplished to evaluate a material law for numerical investigations.

All the coal samples tested are high bituminous and come from the Camphausen-mine near Saarbrücken in the Saarregion: The specimens were bored from coal blocks located in situ at a depth of about 800 m.

The thermomechanical behaviour of coal is derived from triaxial tests. The samples were subjected to both, different stresses and various temperatures from 20^0 C to 200^0C. The diameters of the coal samples differ from 70 mm to 120 mm.

The anisotropic behaviour of the coal is described by the Damage-Tensor Analysis, introduced by Kawamoto (1988). The anisotropy can be described by the three dimensional area density of the cracks in the coal, which is derived from representative coalcubes. It is suggested that the material behaviour of the unfractured coal is isotropic.

The necessary material parameters are derived from triaxial compression tests. To describe the general material behaviour , the model of Mohr-Coulomb is involved.

RESUME : Dans la domaine de la gazéification souterraine du charbon en Europe, le problème de la technique de liaison de gazéification est un centre d'interêt important. La stabilité des galeries de liaison en grand profondeur, brulées en revers, est recherché a l'université de Kaiserslautern. Des differentes études sur le comportement méchanique du charbon, qui depend de la température ont été réalisées pour le devellopment d'un loi de comportement pour des analyses numeriques.

Les échantillons sont du charbon bitumineux et derivent de la mine de camphausen, qui se trouve auprès de Sarrebruck dans la region de la Sarre. Les éprouvettes ont ete prises des blocs de charbon collectionnées dans une profondeur de 800 m.

Le comportement thermomechanique est develloppé sur la base des essais triaxiaux Les echantillons ont été soumis a des differentes contraintes et a des températures des 20^0 C jusqu'a 200^0 C. Les diamètres des éprouvettes variaient de 70 mm jusqu'a 120 mm.

Le comportement anisotropique du charbon de décrivé a l'aide de l' analyse du tenseur d'endommagement, qui a été introduit par Kawamoto (1988). L' Anisotropie peut âre representé a l'aide de la densité des plans de diaclase, qui est relevé des cubes de charbon, représantant la roche discontinue. Un resultat des essais relève un comportement isotropique de la masse rocheuse non fracturée.

La courbe intrinsèque pour la resistance maximale et pour la resistance residuelle dans les essais triaxiaux de pression est definie a l'aide du modèle de Mohr-Coulomb.

1 INTRODUCTION

In the USA as well as in the USSR the Underground Coal Gasification (UCG) is a well investigated exploitation method for energy ressources. In the USSR this method is even economically applied. In western Europe the conditions for underground coal gasification are quite different. The seams are thin and located in relative great depth. Therefore a special technic must be devellopped to succeed the method. Two wells are drilled from the surface to the gasifying seam. One is denoted injection well, the other the production well. A linking channel has to be established to connect the two vertical wells.
The system for the UCG is shown in fig. 1.

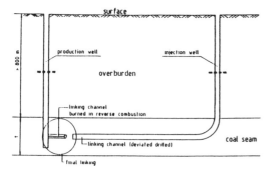

Fig. 1: System of the underground coal gasification

In both the USA and the USSR usually the reverse combustion technique is applied. Tests performed in European countries under different conditions show that this method is not successful for seams subjected to high overburden pressures (Patigny Li, 1986) The problem is the linking of the two wells. Usually nearly the whole length of the linking channel will be performed by continuation of the injection well by the method of deviated drilling. Generally the linking channel will not meet the production well. Therefore only the final distance (some meters) of the linking channel shall be performed by the reverse combustion technic. During gasification an injection gas, which may consist of oxygene and steam, is pumped into the gasifying seam. At well defined places along the linking channel the ignition of the coal is started. In the following an oxidation reaction on the surface of the coal takes place. The reaction gas (H_2 + CO + CO_2 + CH_4) escapes through the linking channel and the production well.
The mechanical stability of a linking channel is the topic of a research projekt at the University of Kaiserslautern. The investigations are performed using the finite element method (FEM) . The used constitutive law describes the mechanical behaviour of coal truly. The parameters for the material model are derived by triaxial tests.

2. SPECIFICATIONS OF THE COAL

The experiments were performed with Camphausen coal samples. The samples were drilled out of big and compact blocks. Figure 2 shows a typical block.

Fig. 2: Typical coal block for sample preparation

The blocks are mainly fabricated in both the seam number 14 and number 16. The depth of sampling was about 750 m until 800 m. Assuming an average unit weight of $23 \, kN/m^3$ for the overburden rock , the vertical primary stress can be estimated to:

$$\sigma_{v,p} = 23 \; kN/m \; \cdot \; 775 \; m \; \approx 18 \quad MPa$$

For the Camphausen coal no special in situ stress measurements are conducted to determine the primary stress state. In our calculations we assume that the lateral stress is equal to the vertikal stress.
The Camphausen coal is high bituminous. It contains 28–35 % volatile matter. Wagner (1986) performed extensive petrographical analysis and showed that this coal is poor in durain ($\leq 5\%$), fusain ($\approx 10\%$) and clarain (5 % – 20 %). The coal is rich in vitrain (30 – 60 %) and in trimacerit (20 – 40 %).

2.1 Sampling

The examined samples were highly fractured. A crack system consists of parallel planes, which include the fissures. The distance between the parallel planes shall be called spacing of the cracks. Three different systems of cracks may be defined. One is parallel to the bedding planes. The spacings of the bedding planes are from 2 to 7 mm. Two cleavages are orientated nearly perpendicular to the bedding planes. The angle between the cleat systems varies from $60°$ to $90°$. The spacing of the major cleat planes is about 5 to 9 mm. The average spacing of the minor cleat planes is about 2 – 5 mm.
The fissures in the coal samples are created by tectonic processes in situ as well as by stress concentrations in the rock during mining activities. In the virgin coal, all the cracks are assumed to be closed. When

the coalblocks are released from the overburden pressure. the cracks arise. Both, the load history of the specimens and the change of the coal structure must be taken into account in the material model. In our tests cores were used, drilled out of orientated blocks. For all samples the angle between the axes and the normal vector of the layer plane are defined. Concerning the evaluation of test results, only cores of the same orientation are immidiatly compared.

The triaxial tests were conducted with constant confining pressures. Water pressures from 1 MPa to 30 MPa were used. The latter value is about twice the primary stress. The diameter of the samples in the high pressure tests was 70 mm or 100 mm In tests using lateral stresses up to 3 MPa (low stress tests) the sample diameter was about 120mm. The ratio height/diameter (h/d) yielded between 1.5 and 2 depending on the friction− conditions at the end plates. Using lubricated end plates, a ratio of 1.5 was chosen.

3. PERFORMANCE OF TESTS

In the high pressure tests, the longitudinal strains were measured by an inductive gauge . The radial strains were measured by a special displacement gauge. In the low pressure tests, all strains were measured by strain gauges.

3.1 Triaxial tests at room temperature

Fig 3. shows a typical stress− strain curve of a sample, subjected to high pressure. The material behaviour can be divided into three significant phases, denoted as hardening, softening and residual part. The hardening part may be approximated by a straight and therefore may be described by an elastic material model. Although the softening part of the stress strain curve may be described by a straight. For the residual state the stress level does not change remarkably with increasing deformations of the sample.

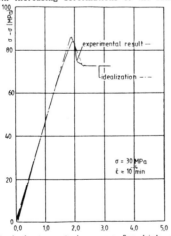

Fig 3: typical stress-strain curve for high confining pressure tests at room temperature

The stress − strain curve in fig. 4 presents a typical triaxial test result at low confining pressures. An idealization of this curve similar to the one of high confining pressures shows a good approximation.

Both tests using low and high confining stresses were performed with cores perpendicular to the bedding planes.

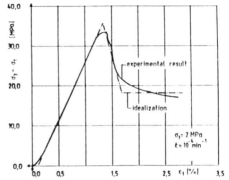

Fig. 4: Typical stress strain curve at low confining pressures at room temperature

3.2. High temperature tests at 200^0 C

In triaxial tests at 200^0 C samples were used with a diameter of 70 mm. In theses tests only the longitudinal strains could be measured .

Fig. 5 shows a typical stress − strain curve of a triaxial test at 200^0 C. The test results are significantly different from those at room temperature. The sample behaviour can be described by a linear elastic − perfect plastic material model

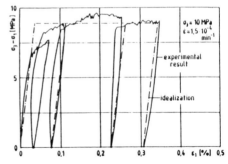

Fig. 5: Typical stress strain curve for coal at 200^0 C and σ_3 =10 MPa.

As can be seen from fig. 5, a softening part of the material may be neglected. Comparisons with both the curves in fig 3 and 4 show a significant higher material stiffness for an initial loading as well as for unloading and repeated loading. In all tests the rate of axial deformations was about $1.5 \ 10^{-4}$/min. The influence of the deformation velocity on the stresses as well as the creep behaviour of coal samples will be presented in a following paper

4. INTERPRETATION OF THE RESULTS

Kawamoto (1986) introduced a sophisticated model to describe the anisotropic mechanical behaviour of a fractured rock mass. The model is practicable for crack systems or bedding planes with small spacings between the crack planes relative to the dimensions of the boundary problem. With this model, an arbitrary number of joint systems can be taken into account.

In the model two cartesian coordinate systems have to be distinguished. In the global system (x,y,z) the calculations are performed. The local system (x_1,y_1,z_1) is the reference coordinate system, in which the orientation of a crack system is defined. The model is dominated by a symmetric Tensor Ω which is calculated in the local system.

By means of the area density of cracks χ_{ei} of a real crack system i, the damage tensor of Ω_i may be defined. For a continuum we obtain $\chi_e = 0$ whereas $\chi_e = 1$ describes a joint extended over the total cross section. We achieve the expression:

$$\Omega_i = \chi_e * (\vec{n}_i \otimes \vec{n}_i) \qquad (1)$$

where $\vec{n}$ is the normal vector on a crack plane in the local coordinate system. For k crack systems, the damage tensor of the rock mass can be calculated as follows:

$$\Omega = \sum_{i=1}^{k} \Omega_i \qquad (2)$$

As can be seen from Eq. (1) the tensor Ω is symmetric. The diagonal form of Ω is called the damage tensor of a discontinous rock mass. The eigenvalues ω_i of this tensor correspond to area densities of three idealized orthotropic crack systems. This ideal local orthotropic crack system is denoted as the main axis system of anisotropy (M.A.S.A).

By an area reduction factor ($1 - \omega_i$) in the (M.A.S.A.), Cauchy's stresses are transferred to stresses on effective areas. The model for three idealized orthotropic crack systems is shown in fig. 6.

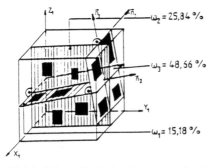

Fig. 6: Model of three idealized orthotropic crack systems with the corresponding damaged cross section areas ω_i.

For only one crack system in a cube, the effective area ($1 - \omega_i$) can be directly calculated by means

of the area density χ_{ei} of the cracks. Then ω_i is equal to χ_{ei}.

The transformation of Ω from the local system into the M.A.S.A. may be achieved by the transformation matrix K_T. Using K_T the damage tensor Ω' in the main axis system is defined as:

$$\Omega' = K_T \Omega K_T^T \qquad (3)$$

The stresses defined in the global coordinate system can be transformed in two steps into the main axis system of anisotropy. Using the transformation matrix T the stress tensor σ in the global coordinate system can be transformed into the local coordinate system, in which the normal vectors $\vec{n}_i$ are defined. T can be derived from the angles between global and local coordinate axes. Figure 7 shows the angles β and δ, which describe the orientation of a local axis in the global system.

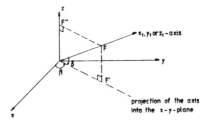

Fig. 7: Orientation of the local system in the global coordinate system.

Introducing the abbreviations l_i, m_i and n_i the matrix T can be written as:

$$T = \begin{bmatrix} l_x & m_x & n_x \\ l_y & m_y & n_y \\ l_z & m_z & n_z \end{bmatrix}$$

with:
$$l_i = \cos \delta_i \cos \beta_i$$
$$m_i = \cos \delta_i \sin \beta_i$$
$$n_i = \sin \delta_i$$

The transformation is described by the formula (4):

$$\sigma_1 = T \sigma T^T \qquad (4)$$

In the local system the corresponding symmetric stress tensor σ_1 can be transformed by means of formula (5) into the main system of anisotropy.

$$\sigma_2 = K_T \sigma_1 K_T^T \qquad (5)$$

This stress tensor can be divided into a part of volumetric stress and a deviatoric stress tensor, respectively.

$$\sigma_2 = \sigma_{2n} + \sigma_{2t} \qquad (6.)$$

In the M.A.S.A. the effective normal and shear stresses can be calculated out of the Cauchy stresses. However, the effective area in a cross section is a

2468

function of the roughness qualities of the cracks and the closure pressure of the cracks. Therefore the ideal area density of cracks ω_i must be modified. The effects of both roughness and closure pressure are taken into account by parameters C_i. For $C_i = 1$, the crack doesn't resist to any stress. If C_i equal 0, the stress will be transmitted through a crack as through the unfractured rock. Applying the parameters C_i to the devellopped concept the Cauchy stresses in the cross section of the idealized crack planes can directly be transformed into effective stresses:

$$\sigma_2^{*'} = \sigma_{2t}' (I - C_t \Omega')^{-1} + \sigma_{2n}' (I - C_n \Omega')^{-1} \quad (7)$$

where C_t is the corresponding factor for the shear stresses and C_n for the normal stresses. Usually we obtain $C_n \neq C_t$. The stresses $\sigma_2^{*'}$ can be retransformed from the main axis system of anisotropy into the local system:

$$\sigma_1^{*'} = K_T^T \sigma_2^{*'} K_T \quad (8a)$$

Using formula (8b)), the retransformation of the stresses $\sigma^{*'}$ from the local system into the global system σ^{*1} can be performed:

$$\sigma^{*} = T^T \sigma_1^{*'} T \quad (8b)$$

The tensor $\sigma^{*'}$ defines the effective stresses regarding the reduction due to the crack systems . This tensor is not necessarily symmetric. For further calculations it has to be symmetrized. The symmetrisation can be performed as follows:

$$\sigma_{1\,ij}^{*s} = \sigma_{1\,ji}^{*s} = \frac{1}{2} (\sigma_i^{*} + \sigma_{ji}) \quad (9)$$

It should be pointed out that the parameters C_i are functions of the roughness in the joints as well as of the stress state. To determine these functions material tests have to be performed.

4.1. Damage tensor of Camphausen coal

The damage tensors of the cleats and the bedding planes were evaluated at several cubes with a side length of 100 mm. On three orthogonal sides the outcrops of the cracks were counted and measured. The general procedure for the determination of the damage tensor is described in details by Kawamoto (1988). On each cube about 1000 outcrops of cracks were counted. Therefore it can be assumed that the cubes are representative for the fractured coal. The area densities $\mathcal{X}_{ei}$ and normal vectors $\vec{n}_i$ of the three crack systems of Camphausen coal read for one cube:

bedding plane	$\mathcal{X}_{e1}$:	18.9 %
face cleat	$\mathcal{X}_{e2}$:	44.5 %
butt cleat	$\mathcal{X}_{e3}$:	26.2 %

bedding plane	$\vec{n}_1$:	(−0.0055, 0.1236 , −0.9923)
face cleat	$\vec{n}_2$:	(−0.1302, 0.8735, 0.4691)
butt cleat	$\vec{n}_3$:	(−0.9912, 0.1024, −0.0834)

Fig. 8. shows the cube with three representative crack planes with the corresponding area densities of the cracks $\mathcal{X}_{ei}$.

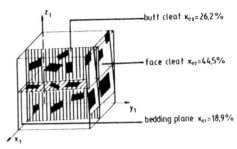

Fig. 8.: Representative crack planes in the cube for which the damage tensor has been determined.

The eigenvalues of the damage tensor, corresponding to the crack densities ω_i of the idealized orthotropic crack system have the following values:

$$\omega_1 = 0.25844$$
$$\omega_2 = 0.4866$$
$$\omega_3 = 0.1518$$

Fig. 5. shows the corresponding main axis system of anisotropy with the idealized crack densities ω_i. The necessary transformation matrix K_T we find to:

$$K_T = \begin{bmatrix} 0.9557 & -0.2179 & 0.1978 \\ 0.2939 & 0.7421 & -0.6024 \\ -0.0155 & 0.6338 & 0.7733 \end{bmatrix}$$

The matrix T depends on the orientation of the cube axis towards the arbitrary coordinate system for the calculations. If both systems are identical, T becomes the unit matrix.

5. MATERIAL PARAMETERS

5.1 Room temperature

In high pressure tests, the confining pressure is higher or equal to 5 MPa. The stress-strain curves of various high pressure triaxial tests at room temperature have been compared with respect to the orientation of the specimens. The secant modulus of these curves for small deformations were evaluated. The differences between the mean values of all tests wth unique as well as different oriented specimen is small in comparison with the scatter of the test results of one orientation. Therefore the elastic behaviour for samples subjected to high stresses can be regarded as isotropic. It can be assumed, that due to the high isotropic pressures all cracks are closed and the rock mass behaves like an unfractured isotropic rock. This means, that there is no reduction of the cross sections. The parameters C_n and C_t become zero. In the hardening part of the stress–strain curve the values for the Youngs modulus E and the Poisson ratio ν of Camphausen coal at room temperature are:

$$E = 4750 \text{ MPa}.$$

$$\nu = 0.418$$

The values for the bulk modulus K and the shear modulus G can be calculated to:

$$K = \frac{E}{1 - 2\nu} = 28\,960 \text{ MPa}$$

$$G = \frac{1}{2(1 + \nu)} = 1675 \text{ MPa}$$

Triaxial tests were also performed with confining pressures from 1 to 5 MPa. In these tests, the former defined secant modulus of the stress–strain curves was lower than in the high pressure tests. In the following this effect shall be explained by the damage Tensor. It is assumed, that the area A_i for the transmission of normal and shear stresses is reduced by the factor $(1 - C_i \omega_i)$. The smaller the effective area is, the higher are the effectve stresses. Therefore only small Cauchy's stresses become relative high effective stresses. In the formula $(1 - C_i \omega_i)$ the idealized area density ω_i is used. In the case of the investigated samples of Camphausen coal the two cleavages are almost perpendicular to the bedding planes. The idealized crack system 3 (ω_3 value) is mainly influenced by the crack system parallel to the bedding planes.

In the case of low stresses, the area density factor $(1 - \omega_i)$ for the idealized crack system 3 is about 0.848. If the measured stresses in the triaxial tests are referred to the effective areas, the mechanical properties of the unfractured coal can be evaluated. The parameter C_n in Eq. (7), which describes the closing behaviour of the crack, can be obtained by the curves in fig. 9. These curves describe the volumetric behaviour of the specimen as a function of the isotropic pressure.

Under isotropic pressures only normal stresses are acting in the crack planes. The curves describe the behaviour of cracks subjected to normal pressures. By the test results presented in fig. 9 the bulk modulus for different stress states can be derived. At confining pressures which are higher than about 10 MPa , the curves are linear. The bulk modulus in this part is identical with the bulk modulus calculated for the hardening part of the stress – strain curves in high pressure triaxial tests. Therefore it can be assumed that at high stresses all cracks are closed and the coal behaves like an unfractured rock.

The shear behaviour of the cracks, which is described by the parameter C_t, is determined in a different way. It is assumed, that the cracks show a frictional behaviour too. Dilatation in the cracks is neglected. The total shear resistance of the cracks depends on the roughness of the cracksurface and the normal force in the cracks. For the bedding planes and the butt cleats the assumption is made that the friction coefficient in the cracks is the same as in the unfractured coal. For the major cleats, the friction coefficient C_T was found to be almost 1. This means, that the major cleats are very smooth and no shear stresses are transmitted in the crack planes. A back-calculation of the triaxial tests with the relaionships introduced above showed a good approximation of the experimental results.

The peak strength of Camphausen coal can be described by the Mohr–Coulomb criterion. For the high pressure tests the cohesion has been evaluated to:

$$c = 8.95 \text{ MPa}.$$

The angle of internal friction is: $\varphi = 28^0$. In the tests with low confining pressures the corresponding values are:

$$\varphi = 56^0$$

$$c = 0.85 \text{ MPa}$$

The residual strength of the high pressure tests can be described with the same angle of internal friction. The cohesion decreases to 5.45 MPa. For the low confining pressure tests the angle φ reduced to 54.4^0 and the cohesion to 0.15 MPa.

5.2 Temperature of 200^0 C

The stress–strain curves at 200^{0}C shown in fig. 4 are quite different from those at room temperatures. The gradient at the reversibel part of the deformations is much higher. The values are about 6 times as high as those at room temperature. An explanation for this phenomenon shall be given in a following paper.

Schloemer e.a. 1984 investigated the dilatation behaviour of coal out of Saar region. They found, that the softening temperature in dilatation tests under confining pressures decreases with increasing

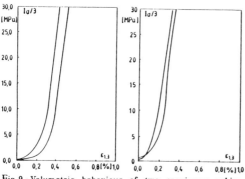

Fig. 9 : Volumetric behaviour of two specimens subjected to isotropic pressure

2470

pressures. The boundary stress conditions in the triaxial tests seem to lower the softening temperature to about 200^0 C. The behaviour of the investigated coal becomes the one of a viscous fluid. Under high stress conditions the gas phase in the coal is completely compressed. The elastic material parameters of coal specimens subjected to high stresses and temperatures respectively shall be presented in a following paper. strain curves.

Under high stress conditions, the behaviour of the coal at 200^0 C is isotropic. In the stress–strain curve, no differences between the peak level and the residual strengths could be noticed. Therefore only one set of material parameters for the Mohr-Coulomb failure criterion is given:

$$\varphi = -5.45^0$$

$$c = 6.3 \ MPa$$

The mechanical behaviour of coal at 200^0 C seems to be similar to a viscous fluid. It is assumed, that the cracks disappear. In the calculations the parameters C_r and, C_1 are reduced to 0.

6 CONCLUSION

Camphausen coal out of the Saar region has been investigated at room temperature and at 200^0 C. Triaxial tests at constant confining pressures have been performed. The mechanical behaviour of the coal changes in this temperature range completely.

At room temperature the investigated coal behaves like a soft rock. In triaxial tests , in which the confining pressures were higher than the primary stress, the coal behaves isotropically. It is assumed that all cracks are closed and the coal behaves like an unfractured rock. The anisotropic behaviour at lower confining pressures can be described by the damage tensor analysis. presented by Kawamoto (1986). This theory assumes, that the stresses are only transmitted in effective areas. These effective areas can be calculated by means of the area density of cracks and their orientation. The backcalculation of the stresses shows, that the material behaviour in the effective areas is the same as in the high pressure tests. With increasing normal stresses the cracks are closing.

At 200^0 C the mechanical behaviour of Camphausen coal is dominated by irreversible deformations. At both high confining stresses and high temperatures the coal behaviour can be described as a viscous fluid. The failure intrinsic curve show a negative angle of internal friction. This can also be explained by an increasing liquefaction with growing confining pressures.

in further investigations the material behaviour at temperatures between room temperature and 200^0 C shall be investigated.

Acknowledgements.

The authors gratefully acknowledge the order of this project by the FR 11.14.02 Technische Mineralogie der Universität des Saarlandes and especially the financial support for this project by the Bundesministerium für Forschung und Technik (BMFT) of the FRG under Grant No. 032 6515 A.

REFERENCES

Patigny.J. & Li, T.-K. 1986. Evolution of the european UCG Project at great depth. Proc. of the 12. UCG Symp. pp. 45 – 52

Wagner, J. 1986 Burning channels at high overburden pressures Proc. 12 UCG Symp. pp. 134 –141

Goodman R.E. 1980 Introduction to rock mechanics John Wiley & Sons

Kawamoto.T. IchikawaY. 1986 Rock mass diskontinuities and damage mechanics. Computational mechanics '86 Tokyo Vol.2. Springer Verlag 1986 pp IX-27

Kawamoto, T. Ichikawa,Y. 1988. Deformation and fracturing behaviour of discontinous rock mass and damage mechanics theory. Int. journ. num. anal. meth. geomech. vol. 1 1988 pp. 1 – 30

Palz K., Schloemer H. 1984. Verhalten von Steinkohlen bei UT-Vergasung im Temperatur – Druckgefälle. Research report T 84- 260 BMFT.

Studies on the corrrelation between the effect of confining pressure and the physical and mechanical properties of weak intercalations

Les recherches sur la corrélation entre l'efficacité de la pression confinante et la particularité physique et mécanique d'intercalation faible

Nie Dexin
Chengdu College of Geology, People's Republic of China

Zhang Xiangong
China University of Geosciences, People's Republic of China

Han Wenfeng
Lanzhou University, People's Republic of China

ABSTRACT: We studied the correlation between confining pressure and the physical properties of fault gouge, and point out that confining pressure is the dominant in controling the void ratio and dry bulk density of fault gouge. There is a fine correlation between confining pressure and the void ratio and dry bulk density. Owing to the controlling effect of the confining pressure on the physical properties of fault gouge, the gouge gets a higher value of shear strength under the higher confining pressure. Here again a fine correlation exists between the confining pressure and the shear strength of the gouge.

RESUME: Les auteurs out étudié la correlation entre la pression confinante et les caractères de la fausse gouge, et montrent que la pression confinante et les facteurs determinants controlent le pore ratio et la densité de secheresse. Car le contrôl la pression confinante sur les caractères physiques de la fausse gouge, lui faisant de trouver une haute valeur de l'intencité tranchante sous la pression confinante d'une valeur plus haute.

Introduction:

Weak intercalations (including fault rock) are the most important discontinuities in rockmass. As their physical and mechanical properties are worse than that of the surrounding rocks, they become one of the main factors which determiner the stability of rockmass, especially that of the dam foundation and abutment rockmass of huge hydraulic power projects. In many engineering rockmasses, there are always lots of weak intercalations. In order to evaluate their physical and mechanical properties, we have to spend much money and labour in exploring and testing. However, the dispersion of the results usually prevents us from determining proper parameters for design from some tests. This results considerable cost for engineering treatment(s) of weak intercalations in the con-struction of many projects. Since 1970s, from the Chinese engineering geology circles, there are many articles concerned with the study of the cause of the mudding and the strength properties of weak intercalations, and there are great advances, especially after the testing and study of crustal stress. Now we know that there are almost always some values of crustal stress owing to the special features of dam sites of huge hydraulic power projects, thus from the confining pressure effect —normal stress generated by ground stress, acting on weak intercalations— to study the physical and mechanical properties of weak intercalations has attracted much attention. From 1980s, the authors have done much research on the physical and mechanical properties of weak intercalations under different confining pressure in some huge projects construction sites in

China. Hence reached some new cognition. That is, confining pressure is one of the controlling factors which determine the physical and mechanical properties of weak intercalations, what is more there are certain correlations between confining pressure and the physical and mechanical properties of weak intercalations, those existing under certain depth always appear with better physical and mechanical properties owing to the higher confing pressure there.

1. Correlation between physical properties and the confining pressure

1.1 Stress state confining weak intercalations. since the measuring of crustal stress in the rockmass of upper crust by N.Hast from 1950s, crustal stress data obtained at home and abroad (Table 1) indicates that:

1. Almost all of the stresses in rockmass in the upper crust are compressive,

2. Most of the values of stresses reach a certain level;

3. In regions of high mountain and deep-narrow valley, there are always high values of crustal stress owing to the existed or existing structure situations there;

4. Weak intercalations, being in rockmass where there is three-dimensional stress, are inevitably influenced by the compressive stress generated from the three-dimensional stress. In this situation, we can regard one weak intercalation as a section. Given the three principle stress components, we can calculate out the normal stress acting on the section from the following formula:

$$\sigma_N = l^2\sigma_1 + m^2\sigma_2 + n^2\sigma_3 \quad (1)$$

in which, $\sigma_1, \sigma_2, \sigma_3$ respectively refers to one of the three principle stress components, while l, m, n to the direction cosine of the normal line of the section, thus

$$l^2 + m^2 + n^2 = 1 \quad (2)$$

from (1) and (2), we have

$$\sigma_N = (1-m^2-n^2)\sigma_1 + m^2\sigma_2 + n^2\sigma_3$$

let $\dfrac{\partial \sigma_N}{\partial m} = 0 \quad \dfrac{\partial \sigma_N}{\partial n} = 0$

then $\sigma_n = \sigma_1$

In the same way, we have

$\sigma_N = \sigma_2 \quad \sigma_N = \sigma_3$

this means that the value of σ_N should be

$$\sigma_3 \leqslant \sigma_N \leqslant \sigma_1$$

As almost all of the ground stresses in the upper crust are compressive, weak intercalations are being pressed by compressive stresses, in other word, being pressed by confining pressures. As to the values of the confining pressures at different parts of one weak intercalation, we can calculate them by elastic mechanics formulae or finite element method of stress inversion according to crustal stress data obtained at the construction site. We can see from calculation results that weak intercalations in some dam sites in China are mostly pressed by comparatively high values of normal stresses (Table 2).

1.2 Correlation between the confining pressure and the physical properties of weak intercalations

Of all the substances in weak intercalations where gouge is the one with the worst physical properties. Physical properties of various gouges in intercalations are similiar to those of soils in many respects. Difference between them lies in that the confining pressure acting on fault gouge in rockmass, in value, by far greater than the self-weight stress, at the same depth, generated by the soil on it. The main compression curve of soil indicates that the correlation between the pore ratio of soil (e) and the pressure (P) is a logarithnic function (Fig.1).

$$e = a - b\lg P \quad (3)$$

Equation (3) tells us that at no matter how much the value of the initial e, there is a linear correlation between e and lgP so long as P is greater than the greatest pressure burdened ever before. This correlation is also right to fault gouge in weak intercalations.

Table 1   Stress state obtained
within China

place	σ_1		σ_2		σ_3	
	value (MPa)	orientation	value (MPa)	orientation	value (MPa)	orientation
Er Tan	26	N34°E∠23°	7.5	S141°E∠19°	2.5	N277°W∠48°
Long-Yang Gorge	5.34	S240°W∠18°	2.55	S115°E∠60°	0.43	N338°W∠22°
Laxiwa	22.7	N338°V∠33°	18.64	N88°W∠27°	13.14	S208°∴∠45°
An kang	σ_{Hi}=4.0	N307° W	σ_{H2}=2.6	N37°E	σ_z=0.52	∠90°
Xi-er river (First stage)	6.36	N305°W∠40°	4.32	N89°E∠3°	2.85	N4°E∠49°
Lubuge	16-18	N60°-80°W ∠20°-30°	10.6-14.0	S10°-20°E ∠40°-60°	3.5-5.2	N30°-50°E ∠20°-30°
Lijiaxia	5.5	N58°E∠43°	4.5	N275°W∠41°	2.8	S168°E∠20°

Table 2   Normal stress values on weak intercalations
(σ_N) in some dam sites in China

No.	Project	weak intercalation		σ_N(MPa)
1	Ankang	f_1——low-angle fault		0.815
2	Longyangxia	fault F_{73}		4.48
3	Longyangxia	T_{66}—— low-angle joint (with gouge)		14.5
4	Longyangxia	F_{215}——low-angle fault		16.0
5		F_{26} vertical fault	at 18m deep under ground	0.34
6			at 36m deep under ground	1.88
7			at 46m deep under ground	3.03
8			at 79m deep under ground	5.65

Comparatively good correlation appears between the field values of void ratios of the gouge at different depths of vertical fault F_{26}—— one of the faults in Lijiaxia dam site —— and the calculated normal stress on the faulted planes corresponding. (Fig.2) We've carried out compression tests on samples made from gouges sampled from weak intercatations with a water-content nearly equals to liquid limit, and

from each we obtained one e—— lgp equation (Table 3). Then, to these equations, we apply the normal stress values in Table 2 into the corresponding equations in Table 3, we obtained corresponding calculated ratios, which are close to the natural void ratios. We can draw from it that one value of pressure pairs with one value of void ratio. As the specific gravities (G) of gouges are mostly about

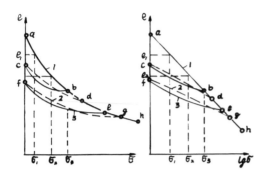

Fig.1 Compression curve of disturbed saturated clayey soil

1. main compression curve

2. recompression curve

3. unburdening curve

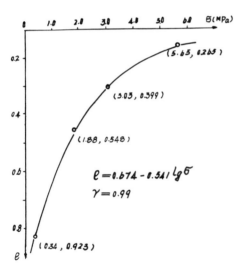

$$\ell = 0.674 - 0.541 \, lg\,\sigma$$
$$\gamma = 0.99$$

Fig.2 Correlation curve between
the normal stress (σ) on
F_{26} and natural pore ratio
(e)

2.7 KN/m³, which can be regarded as
a constant, we can see the correlation between the void ratio (e) and
the dry density (γd) of one fault
gouge from the formula $\gamma d = \dfrac{G}{1+e}$. Now
that there is correlation between e
and its γd of one fault gouge, there
is, of course, correlation between
confining pressure and γd. This correlation is illustrated by the test

data of the intact gouge samples of
fault F26. (Fig3) Void ratio and
dry density of fault gouge are the
most fundamental physical criteria.
Data in Table 2, Fig 2 and Fig 3
indicate that there is a distinct
correlation between confining pressure and the physical properties of
fault gouge, and that confining pressure is the crucial factor which determines the physical properties of
gouge in weak intercalation. From
the data in Table 3, we can also
see that even under similiar confin-

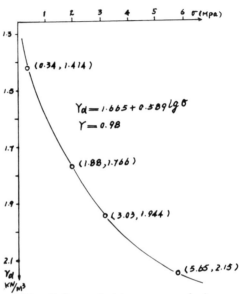

$$\gamma d = 1.665 + 0.589 \, lg\,\sigma$$
$$\gamma = 0.98$$

Fig.3 Correlation curve between
confining pressure (σ)
and dry density (γd) of gouge

ing pressure the values of void
ratios of different fault gouges
differ from each other, which results from that the types of fault
gouges are various and the clayey
minerals in them vary in type and
propotion. Under the same pressure,
gouge containing mainly of montmorillonite has a higher value of void
ratio than that of illite.

2. Correlation between confining
pressure and the strength of
weak intercalations

In one weak intercalation, gouge
is the weakest part, so the strength
of gouge, in general, represents
that of the weak intercalation. The
strength of gouge, however, is determined by the physical properties of
it.

2476

Table 3 Comparison between the natural and calculated
void ratio of gouge in weak intercalations

weak intercalations		the weathering zone of its surrounding rock	Normal stress on the plane of weak intercalations (P) (MPa)	correlation equation	Void ratio (e)	
					natural	calculated
f_1—low-angle fault		slightly weathered zone	0.815	$e=0.414-0.275lgp$	0.41	0.436
F_{73}-fault		slightly weathered zone	4.48	$e=0.668-0.191lgp$	0.541	0.544
T_{66}-low-angle joint (with gouge)		slightly weathered zone	14.5	$e=0.955-0.275lgp$	0.664	0.659
F215-low-angle fault		moderately weathered zone	16.0	$e=0.915-0.265lgp$	0.602	0.596
F26 vertical fault	at 18m deep under ground	slightly weathered zone	0.34	$e=0.662-0.517lgp$	0.923	0.898
	at 36m		1.88		0.548	0.514
	at 46m		3.03		0.399	0.407
	at 79m		5.65		0.265	0.267

2.1 The strength feature of gouge under comparatively high value of confining pressure

Weak intercalations in slightly weathered zones or in fresh rockmass are slightly disturbed while being weathered and in which there are always high values of crustal stress, hence high values of normal stress against weak intercalations, and physical properties of gouges in them are good for strength. Even if gouges are under the level of underground water, the values of the dry bulk density and pore ratio of the gouges will not change much, because the value of confining pressure is by far greater than that of the expansive force of the soaked gouges. The authors have sampled, with great care, some intact samples of gouges in weak intercalations in some dam sites in China. After testing, we see that they have good physical properties and higher shear strength (Table 4).

2.2 The feature of gouge in weak intercalations under low values of confining pressure

Weak intercalations in the severely or moderately weathered zones, compared with those in the slightly weathered zones or in the fresh rockmass, have noticeably lower values of confining pressure owing to the release of crustal stress and the gouges in them also have higher values of void ratio, lower values of dry bulk density, higher values of moisture content in general and noticeably lower values of shear strength. Gouges in weak intercalation exposed in cavities have also very low values of strength owing to the release of crustal stress and the absorbing of water. (Table 5)

2.3 Relation of confining pressure and the shear strength of gouge in weak intercalations

As mentioned above, confining pressure, which is generated from gro-

Table 4   Shear strength of gouges under
          comparatively high confining
          pressure

No.	W (%)	γd (KN/m^3)	e	shear strength		main clayey minerals *	σ_N ** (MPa)
				tg φ	C (MPa)		
f_1	11.85	1.93	0.470	0.44	0.017	I	0.815
F_{73}	16.8	1.78	0.541	0.55	0.065	M	4.48
f_{215}	17.8	1.71	0.602	0.68	0.082	M	16.0
T_{66}	18.8	1.67	0.664	0.61	0.073	M	14.5
F_{26}	5.75	2.15	0.265	0.57	0.092	I	5.65
f_{35}	7.95	2.07	0.300	0.82	0.043	I	6.32

  * M——refers to montmorillonite
    I——refers to illite
 ** σ_N——normal stress on weak intercalation planes

Table 5   Shear strength of the gouge in weak
          intercalations under low confining
          stress

No.	W (%)	γd (K N/m^3)	e	shear strength		main clayey minerals *	σ_N ** (MPa)
				tg φ	C (MPa)		
F_{26}	30.3	1.34	0.83	0.08	0.055	I	0.34
F27	24.8	1.55	0.75	0.16	0.034	I	0.47
f_{215}	49.3	1.26	1.15	0.02	0.043	M	0
F_{73}	27.1	1.52	0.80	0.24	0.037	M	0

 * I —— refers to illite, M——refers to montmorillonite
 ** σ_N —— normal stress on weak intercalation planes

und stress and acts on the weak intercalation plane, is the crurial factor which determines the moisture content and dry bulk density of the gouge in weak intercalations. As an reaction, gouge under certain pressure will change its value of moisture content and dry bulk density to the corresponding state of stress. Hence the higher the value of confining pressure, the higher the value of dry bulk density, which results in the increasing of shear strength. High values of strength appears the weak intercalations in the slightly weathered zone or in fresh rockmass owing to the high value of confining pressure there. As to one high-angle fault which cuts through several weathered zones, its values of strength increase gradually from the surface to deep in the ground owing to the corresponding increasing of the confining pressure. So there is correlation between the shear strength of gouge and the confining pressure. The authors have sampled, with great care, some intact samples of fault F_{26} (at different depth under ground, shown in Table 2) under various confining pressure. Their friction coefficients (laboratory results of direct shear test) is distinctly correlated with their confining pressure in field. (Fig 4) This result is consistent with the correlation between confining pressure and the dry bulk density. Therefore, in the research of the strength of weak intercalations, not only should we sample, with great care, the intact samples of the weak intercalations under the original confining pressure, but also should seriously take it into consideration the determining effect of the confining pressure on the physical and mechanical properties. Current sampling and testing of weak intercalations in underground cavity inevitably results in lowing the value of its shear strength, for the confining pressure effect is neglected and the samples to be tested are disturbed── released and saturated (or moisture absorbed).

Conclusion

As ground stress in the rockmass in the upper crust is dominately compressive, weak intercalations are inevitably influenced by the normal stress──confining pressure──generated by ground stress. Confining pressure is the crucial factor which determines the physical and mechanical properties of weak intercalations. There is certain correlation between the confining pressure and that of the dry bulk density and the friction coefficient of weak intercalations.

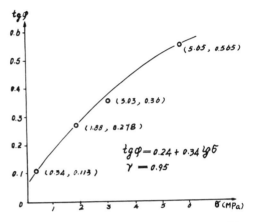

Fig. 4 Correlation between confining pressure (σ) and tgφ

REFERENS

Zhang Xiagong et al. 1986, Engineering geological classifications of fault rocks. Proc. of the 5th Inter. Cong of IAEG.

N.Hast, The gate of stress in the upper part of the earth crust, Tectonophysics, Vol. 8, No.3, P169-175.

6th International IAEG Congres / 6ème Congrès International de AIGI, © 1990 Balkema, Rotterdam. ISBN 90 6191 130 3

Tunnel rock movements and loads in squeezing sheared shale
Mouvements des rochers de tunnels et pressions de schistes argileux cisaillés

N. Phien-wej
Asian Institute of Technology, Bangkok, Thailand

E. J. Cording
University of Illinois at Urbana-Champaign, Ill., USA

ABSTRACT: Squeezing and ravelling of sheared shale caused severe difficulty to TBM tunnelling. Squeezing was caused by both face advance effect and time-dependent behavior of the shale, but their contributing effects depended on methods of excavation and support. The former was greater in sections with slow face-advance and early support erection, and less in fast face-advance TBM excavated section. Long-term rock load on concrete segments was up to 7 % overburden pressure and rate of rock closure decreased linearly with time in a log-log scale.

RESUME: La compression et l'effritement de schiste argileux cisailler ont causé de sérieurs difficultés au tunels TBM. La compression a été créé d'une part par un effet d'avancement de front et d'un comportement de la schiste argileux au fonction du temp. Cependant ces effets sont liés aux méthods d'excavation et de soutainement. Le premier effet était plus important dans les sections d'avance de front bute lent et aux supports installés tôt dans le projet. Cet effet était moins important dans les sections d'avance frontale rapide creusés par TBM. Les masses rocheuses pesant à long tenue sur les segments en béton atteignaient une pression jusqu'à 7% de la surcharge maximale et le taux d'affaissement des roches allait diminué de manière linéaire avec le temps suivant une échelle logasithmique.

1 INTRODUCTION

The 13-km-long Stillwater Tunnel, with 3.5-to 4-m excavated diameters, is a part of the transmountain Strawberry Aqueduct of the Central Utah Project of the U.S. Bureau of Reclamation (USBR). It is located on the southwestern side of the Uinta Mountain of the Rocky Mountain System, which is an anticlinal fold that has been subjected to major faulting. The tunnel passes through a high ridge, upto 780 m. high. Except for a short distance (approx. 600 m) from the inlet portal the tunnel penetrated entirely in flat lying sedimentary rocks of the Red Pine Shale formation (Fig. 1). The rocks consisted primarily of a dark gray to black, medium to hard, thinly bedded to laminated, well indurated illitic shale with numerous thin (few centimeters to meters) interbeds of sandstone and siltstone. The intact shale had unconfined compressive strength between 50 to 88 MPa.

Besides the South Flank fault which is the major geologic features intersecting the tunnel alignment, a numerous high

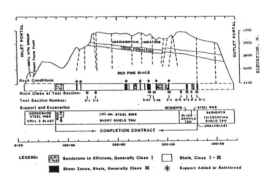

Fig. 1 Geology, construction and instrumentation of Stillwater Tunnel

angle fault lineations were exposed in the younger rocks on the ground surface.

On the basis of the pre-construction investigation on geological condition and engineering properties of the rocks, the USBR believed that the tunnel could be successfully excavated using mechanized tunnelling which would save time and construction cost. Therefore, in the initial contract, the contractor chose to excavate the tunnel with a TBM incorporated with precast concrete segments installation. The selected Robbins TBM, which had a telescoping, long, full-barreled shield, was not so compatible with the heavy ground of the tunnel. The TBM stalled at least 14 times throughout its 2100-m advance in jointed and sheared shale even in the low cover area. It eventually ran to a standstill in a wide fault zone at a depth of 660 m. In this section it stalled several times with increasing difficulties to become free as the ground conditions progressively worse toward a fault seam where the TBM was ultimately binded. In addition, a number of concrete segmented ring, (4-piece, unbolted, 125 mm-thick) were heavily damaged by bending cracks caused by the combined effect of high rock load and difficulty in obtaining effective backfilling with pea gravel and by compression cracks caused by excessive thrust in attempt to free the TBM. It was then that the USBR decided to stop the TBM excavation and launched a ground re-investigation by excavating an exploratory reach using drill-and-blast method with steel rib support in the fault zone ahead of the stalled TBM. A large amount of instrumentation were employed in the section to evaluate the behavior of the sheared shale. Subsequently, un-anticipated heavy ground squeezing was blamed as the cause of the failure of the TBM and the USBR decided to terminate the contract for the convenience of the government.

To complete the tunnel, the USBR solicited proposals that were evaluated on the basis of both technical merit and cost (Sinha and Schoeman, 1984). The successful bidder chose to excavated the remaining length of the tunnel using 2 shrinkable shield TBM's designed principally to advance in squeezing ground. The TBM's shields could shrink diametrically up to 75 mm and 300 mm, respectively to accommodate ground movement in order to prevent high squeezing pressure from building up on the shields. One of the TBM's was an innovative "blade-gripper" TBM that had a flexible full-barrelled shield, made of 12 separate steel blades, which also acted as a gripping unit to provide forward thrust to shove the cutterhead. The TBM was designed to accommodate the installation of concrete segments inherited from the initial contract. The other TBM consisted of short shield and allowed installation of light expanded steel ribs or rock bolts as an initial support. Both TBM's successfully excavated through numerous faulted shale areas encountered in the remaining length of tunnel without any drastic problems. The short-shield TBM was particularly successful due to its capability of fast face advancing and early support installation close to the tunnel face after the passage of the shield.

Due in part to numerous unanswered questions regarding the nature of ground deformations and loads which resulted in the stalling of the TBM in the First Contract the USBR specified an instrumentation program as a part of the contract for completion of the tunnel. Between the two contracts, considerable amounts of instrumentation data were obtained which helped shed light on the response of the sheared shale at depth excavated by the TBM's. The summary and interpretation of the instrumentation results obtained in both contract are presented in this paper. The detailed outline of the performance of the TBM's and supports has been already presented elsewhere (Phien-wej and Cording, 1989).

2 GROUND CONDITIONS AND RESPONSES

A classification shown below was developed as a means for grouping the shale masses and aiding the determination of their potential ground responses.

Class I   Siltstone to sandstone, silty to sandy shale, widely jointed (1-3 m spacing).

Class II   Shale to siltstone, moderately jointed (0.3 to 1 m spacing), some sheared, slickensided surfaces.

Class III Shale, closely jointed (0.05 to 0.3 m spacing), some thin shears.

Class IV   Shale, closely jointed and sheared shear zones have some width (several centimeters) and contain crushed or soft materials. Wide shear zones with large amount of clay gouge are classed as IVb

In general, the shale encountered at tunnel level was moderately jointed (Class II) to closely jointed, often with tight shears spaced less than 0.3 m. (Class III) while the sandstone and siltstone interbeds tended to be widely jointed massive (Class I-II). The shale in closely jointed or sheared areas (Class III-IV) was very fissured and fractures.

The shale exhibited a variety of tunnel ground responses, namely, mild stress slabbing, loosening and ravelling upon excavation, light to moderate squeezing in shear zones, minor swelling in wet shear zones, and minor slaking of wet fissured and fractured shale. The siltstone and sandstone were generally massive and experienced only minor stress slabbing. Squeezing and ravelling of the shale were the two significant ground responses that caused some difficulties.

Ravelling was very significant in the closely jointed and sheared shale due to the combined effect of high degree of fissuring and fracturing (from both TBM cutting and stress slabbing) and the orientation of the very closely spaced near-vertical joints that cut across throughout most of tunnel alignment. Ravelling led to formation of large-sized overbreak or trapped small rock pieces in the tunnel roof, making it difficult to advance the blade-gripper TBM and to backfill the segments. Approximately 26 percent of the tunnel length experienced light to moderate squeezing. Moderate squeezing only occurred in shear or fault zones containing considerable amounts of clay gouge infills (Class IV). The time-dependent behavior of this ground is largely controlled by the clay gouge. In an early reach of the tunnel excavated with the short shield TBM, moderate ground squeezing resulted in severe distortion and failure of the light steel ribs (100 mm x 100 mm).

3 ROCK MOVEMENTS

3.1 Rock movements around advancing face

Instrumentation data indicated that the displacements of shale masses were induced by both the initial stress changes from tunnel face advance and the time-dependent behavior of the shale. In the exploratory drill-and-blast section where the face advance was slow, the early increase in rock displacement was primarily related to the face advance. The horizontal rock closure history (Fig. 2) recorded in the 7.5 m area ahead of the fault seam where the first TBM was ultimately stalled showed that the effect of tunnel face advance on displacement of the sheared shale (Class IVb) at each measurement location ceased to exist after the tunnel face had advanced approximately 10.8 to 12.6 m (3 to 3.5 tunnel diameters) beyond the location, which took approximately 10 to 15 days. Any interruption in face

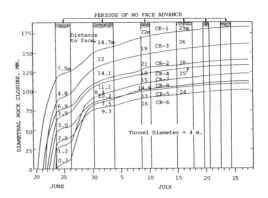

Fig. 2 History of horizontal rock closure in exploratory section

advance during this period resulted in an abrupt decrease in the rate of rock closure. Upon resuming face advance the rate of closure increased abruptly to the level comparable to that occurring prior to the interruption of face advance.

The zone of influence of face advance observed in the sheared shale extended wider than that predicted by the elastic analyses (1 to 1.5 tunnel diameters) because the shale in a zone adjacent to the tunnel wall experienced yielding. FEM analyses, e.g. Ranken et al (1978) and Panet and Guenot (1982) etc., have shown that the extent of zone of face advance influence varies with the level of ground overstressing. The multiposition borehole extensometer data at the test section suggested that a plastic zone developed to an approximate depth of 1 to 1.8 tunnel radii from the wall (diameter of plastic zone from 2 to 2.8 tunnel diameters). Therefore, the extent of the zone of face advance influence is slightly less than the diameter of the plastic zone.

Fig. 3 summarizes the time-related horizontal rock closure expressed in the relation-ship between elapsed time and the normalized rock closure to the total closure at 35 days in the exploratory section. The diametral rock closure recorded at 35 days ranged from 100 to 180 mm (2.7 to 4.6 percent of tunnel diameter). The figure shows that the patterns of rock closure during the first 10 days were erratic because of the still existence of the effect of face advance. However, after the effect of face advance had ceased (approximately at 10 days), the normalized relationship of rock closure and time at all locations followed the same pattern. This suggests that the long term tunnel closure in the area can be predicted by any creep laws that specify

2483

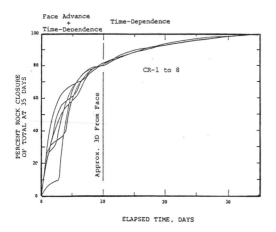

Fig. 3 Percent rock closure of total at 35 days in relation with time

Table 1 Summary of measured diametral rock closure at test sections in completion contract

Test Section	Ground Class	Support	Diametral Closure, $\Delta D/D$, %			
			1 Day	1 Week	1 Month	3 Months
ITS1	II	SR	0.07-0.18	0.16-0.23	0.20-0.32	0.23-0.38
OTS2	II	CS	-	0.04	-	0.07
ITS2	II-III	SR	0.07-0.22	0.09-0.41	0.16-0.53	0.23-0.81
OTS4	III	CS	-	0.14	-	0.14
OTS1	III	SR	0.21	0.49	0.77	0.95
ITS3b	III	SR	0.39-0.47	0.83-1.08	1.44-1.62	2.0-2.2
OTS3	IV	CS	0.49	0.67	0.71	0.74
ITS4	IVb	SR	0.81	1.8-2.5	2.9-4.0	NA

RS = Steel ribs,   CS = Concrete segments

One percent diametral closure ($\Delta D/D$) is equivalent to 30 mm

creep rate as a function of a stress level. The difference in the magnitude of rock closure between these locations was due to the difference in stress levels that the ground at each location experienced. A larger portion of the increase in the rock displacement during the first 10 days was caused by initial stress relief from face advance, not the creep behavior of the shale. A creep closure analysis by Phien-wej (1987) suggested that the portion of the face advance-induced rock displacement was up to 75 percent of the total displacement recorded at 10 days. Therefore, the observed moderate squeezing of ground in this dry sheared shale was largely due to the process of slow face advance and fast support installation close to the tunnel face.

3.2 Time-dependent displacement

The amounts of rock closure measured at various test section are summarized in Table 1. All classes of the shale (Class II to IV) exhibited time-dependent displacements. Even a massive shale (Class II) showed the rock closure to continue to increase with time, but at a very small rate.

Because of the limited access to the face imposed by the shields of the TBM's, the initial readings on the rock closure measured in the Completion Contract were taken 2 to 3 tunnel diameters behind the tunnel face, 6 to 12 hours after excavation when most of the face advance-induced rock displacements would have mostly developed. The measured diametral rock closure in the heaviest squeezing area of the tunnel (Class IVb, ITS4), which was supported with the expanded steel ribs, was 4 percent of the tunnel diameter at one month. This amount was 15 times that measured in the massive shale area (Class II, ITS1).

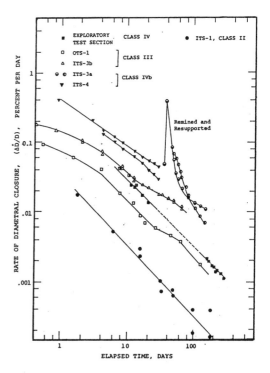

Fig. 4 Relationship between rate of diametral rock closure and time in log-log plot

2484

The rate of diametral rock closure at all test sections are summarized in log-closure rate vs. log-time plots as in Fig. 4. Rates of diametral rock closure in the areas supported with expanded steel ribs were as high as 0.09 percent of diameter/day at 110 days in the heaviest ground of the tunnel (Class IVb, ITS4) and were as low as 0.00036 percent/day in a massive shale (Class II, ITS1). The initial closure rates at one day at ITS4 where the tunnel became unstable shortly after excavation was greater than 0.4 percent/day. The tunnel closure rates continued to decrease with time and the relationship was generally linear on a log-log scale, except when there occurred support distortion or rock loosening. The continued decrease in the rate of closure with time as shown in the figure reflects creep of the ground under stable condition.

In Class IVb ground at ITS3a, following the failure of the support and remining, the rate of rock closure increased significantly to a level comparable to those occurring early during initial excavation in a similar ground condition at ITS4. The closure rates at ITS3a increased from 0.045 to 0.38 percent/day due to the remining. The remining involved a removal of support, which resulted in a large increase in stress level that the ground was subjected to. After finishing resupporting, these increased closure rates, when plotted versus elapsed time after excavation on a log-log scale, showed a decreasing trend that with time seemed to approach the slopes of the curves of ITS4.

At all test sections except ITS3b, the closure rates decreased by 90 and 82 percent per log-cycle of time for Class II and Class IVb grounds, respectively. The smallest decrease in the closure rate per log-cycle, observed at ITS3b (Class III ground) was 75 percent which could be attributed to the effect of pronounced swelling and slaking of the shale in the area. The area was constantly wet and the invert was submerged in 0.3 m of water.

4. ROCK LOADS

Supports installed at the rear end of the TBM's, when most of the initial stress change from face advance had mostly occurred, were subjected to a smaller load than those installed close to the tunnel face in the drill-and-blast sections. In similar squeezing ground condition (Class IV), the 3.9 -m-diameter tunnel excavated by the drill-and-blast method in the exploratory section of the First Contract required heavier steel ribs (W150mmx150mm at 0.45 m o.c. - support capacity of 0.70 MPa equiva-lent to 5 percent of overburden pressure) than the 3-m-diameter TBM-excavated sections (ITS4) in the Completion Contract where light expanded steel ribs (M100mmx100mm at 1.2 m o.c. - support capacity around 0.30-0.40 MPa equivalent to 2 to 3 percent of overburden pressure) were capable of stabilizing the ground with only minor reinforcement. The instrumentation data indicated that rock pressures on both supports reached yielding limits soon after installation (1-3 days). Both supports were capable of supporting the ground without failure.

In the light squeezing area at OTS4 (Class III), the maximum support pressure on the segments was as high as 1 MPa (7 percent of overburden pressure) at 3 months due to high ring stiffness, however, it was still smaller than 30 percent of the ultimate capacity of the segments at the corresponding load eccentricity. The delay in erection of the segments by 4.5 days and the use of a compressible grout backfill (sand-cement-flyash with styrene foam beads) in the moderate squeezing area at OTS3 (Class IV) resulted in a smaller rock pressure on the segments (below 0.8 MPa at 3 months). The segmented rings, which were designed to be capable of carrying a uniform rock pressure up to 2 MPa (under 25 mm eccentricity), were stable despite the formation of numerous bending cracks early after erection.

Bending cracks occurred in a large number of segmented rings installed in the Completion Contract. However, these cracks did not hamper the performance of the rings because the contractor shortly carried out neat cement grouting. The cracks were induced by the deflection of the segmented rings into soft zone of overbreak or trapped fallen rock pieces which remained above the rings in the crown after erection due to difficulty to obtain early completion of ring backfilling with sand-cement-flyash grout.

In such a condition the segmented rings expanded diametrically in the vertical direction while contracted horizontally as rock load exerted onto the rings in the side wall area (Fig. 5). The soft zones created high load eccentricity in the arch segments. Bending cracks usually formed shortly after segments erection and the instrumentation data indicated very low rock pressure around 0.1-0.3 MPa but with high load eccentricity. Fig. 6 shows

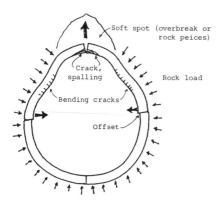

Fig. 5 Schematic depiction of segment distortion due to soft zone above crown

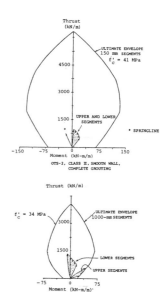

Fig. 6 Moment thrust paths in concrete segments, up to 2 to 4 months

moment-thrust history for well and poorly backfilled rings. The load eccentricity in the arch area of the poorly backfilled rings where bending cracks developed was greater than 100 mm while it was smaller 45 mm in the invert segments (OTS3). The well backfilled ring at OTS2 experienced eccentricity between 30-45 mm. The designed eccentricity assuming uniform loading was only at 25 mm.

The instrumentation data indicated that light circular steel ribs expanded in good contact around the perimeter with the TBM-excavated rock surface and having adequate longitudinal bracing showed excellent performance. The ribs experienced minimal bending stresses and were capable of continuing to support the squeezing ground long after yielding without experiencing severe distortion and failure. Under high rock load bracing of the steel ribs to prevent out-of-plane buckling became very crucial for the continued performance of the ribs. The ribs continued to deform after being yielded without severe distortions, preventing high rock pressure from building up.

The instrumentation data showed that the support pressures on both expanded steel ribs and concrete segmented rings, continued to increase with time but at decreasing rates in a similar fashion to that of the rock closure.

CONCLUSIONS

The observed ground squeezing of the sheared and faulted shale were caused by the initial stress changes from face advance and the time-dependent property of the shale mass. The role of each depended on the construction procedure. When the support was installed right behind the slowly advancing drill-and-blast face, a larger portion of the observed ground squeezing in the early period was due to the effect of face advance not the time-dependent behavior of the shale mass. For the TBM excavated section where the supports could be installed only after the passage of the shield, time-dependent behavior of the ground in the form of creep and progressive weakening is the prime mechanism causing the observed ground squeezing.

The sheared Red Pine shale showed significant ravelling and squeezing responses. However, squeezing only occurred at light to moderate intensities even the tunnel was at a great depth, probably due to the favorable effect of the small size of the tunnel.

Contact conditions between support and rock surface had profound effect on the actual performance of the support. TBM provided smooth rock surface which allowed steel ribs to be expanded tightly against, thus the perform near their optimum capacity. This allowed the use of light steel ribs (100 mm, 1.2 m. o.c.) to safely support most of the tunnel till the placement of final concrete liner.

Backfilling of concrete segments in fractured and sheared shale was very difficult to complete early due to overbreak and loosened rock pieces fallen onto the segment crown, even though carefully backfilling operation was employed. The lightly reinforced segments were unavoidingly experienced bending cracks even though the rock loads were much lower than the designed segment capacity due to high load eccentricity resulted from ineffective backfilling. However, the cracks did not jeopardize the long term carrying capacity of the segments because a better backfilling was subsequently obtained.

The maximum rock pressure on segment ring in the light squeezing ground was less than 7 percent of the overburden at 3 months. The rates of rock closure generally decreased linearly with time in a double logarithmic scale.

ACKNOWLEDGMENTS

The authors would like to express their thanks to Mr. Glen Traylor of Traylor Brothers-Fruin Colnon Joint Venture for providing the opportunity and financial support for them to conduct the study. Thanks are also extended to all the Traylor Brothers personnel on the project site for their kind assistance.

REFERENCES

Cording, E.J., Mahar, J.W. and Fernandez, G. 1982. Evaluation of ground conditions and their effect on excavation and support, Stillwater Tunnel Completion Project, Report for Traylor Brothers-Fruin-Colon Joint Venture.

Marushack, J.M. and Tilp, P.J. 1980. Stillwater Tunnel-A progress report, U.S. National Committee on Tunnel Technology, Newletter No. 31, September.

Panet, M. and Guenot, A. 1982. Analysis of convergence behind the face of a tunnel, Tunnelling'82 187-207.

Phien-wej, N. 1987. Ground response and support performance in a sheared shale, Stillwater Tunnel, Utah. Ph.D. Thesis, University of Illinois at Urbana-Champaign.

Phien-wej, N. and Cording, E.J. 1989. TBM Tunnelling and supports in ravelling and squeezing rock, Int. Symp. on Excavations in Soils and Rocks, Bangkok, Thailand.

Ranken, R.E. Ghaboussi, J. and Hendron, A.J. Jr. 1978. Analysis of ground-liner interaction for tunnel, Report No. UMTA-IL-06-0043-78-3, U.S. Department of Transportation.

Sinha, R.S. and Schoeman, K.D. 1984. Case history-Stillwater Tunnel, Central Utah Project, Utah, U.S.A., Int. Sym. on Case Histories Rolla, Missouri, U.S.A.

U.S. Bureau of Reclamation 1981. Construction and foundation materials test data and Stillwater Tunnel instrumentation data for Stillwater Tunnel Completion, Bonneville Unit, Utah, Central Utah Project, Specifications 40-C2035.

6th International IAEG Congres / 6ème Congrès International de AIGI, © 1990 Balkema, Rotterdam. ISBN 90 6191 130 3

Geological approach for the design of tunnels at large depth

Approche géologique pour la mise au point de projets de tunnels à grande profondeur

W. Riemer
Ehner, Luxemburg

A. Thomas
Santiago de Chile, Chile

ABSTRACT: For the design of tunnels below up to 1500 m of cover a project specific system of geotechnical categories was derived from field surveys. Rock mass strength indices corresponding to these categories were evaluated in parametric studies of ground reaction reflecting the possible range of variation in the geological input. Routing and designing the tunnels under consideration of geotechnical categories and local ground reaction assures suitable adjustment to geological conditions.

RESUME: Lors de la mise au point d'un projet de tunnels sous un recouvrement de quelque 1500 m un système spécifique de catégories géotechniques fut établi sur la base de reconnaissances de terrain. Les indices de résistance de la masse rocheuse correspondant à ces catégories furent évalués au moyen d'études paramétriques des réactions du sol, tenant compte des variations présumables des données géologiques. La méthode appliquée a démontré que le choix de l'alignement et l'étude du projet de tunnels basés sur des catégories géotechniques et sur la réaction du sous-sol permet de tenir compte au mieux des conditions géologiques existantes.

1. INTRODUCTION

Owing to morphological conditions and geological hazards, hydroprojects in alpidic mountain ranges require preferential underground routing of waterways. The tunnels frequently are to be drilled at depths where high ambient stresses are likely to cause serious construction problems and, therefore, the prediction of geological and geotechnical conditions gains fundamental importance for layout and design of the project. However, access in the project area and high cover above the tunnel routes impede direct investigation by core drilling and even seriously interfere with geophysical explorations. Thus, in many cases the project has to rely essentially on geological field surveys. The selected example is intended to demonstrate how classical geological methods can be geared to produce relevant information allowing to:

1. give a qualitative rating for suitably defined engineering geologic rock mass units,

2. quantify strength parameters for such units with a minimum of complementary investigations.

This permits to assess rock mass perform-

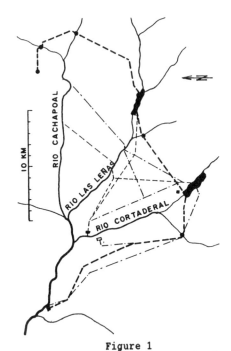

Figure 1
Cortaderal project. Investigated alternatives and selected tunnel routing.

PRESSURE TUNNEL LAS LEÑAS, CORTADERAL

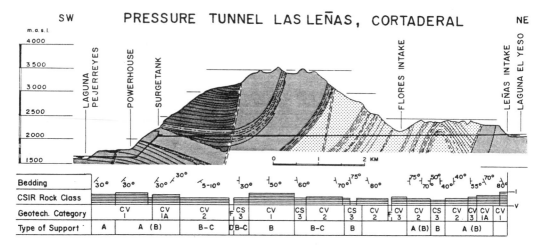

Bedding	30°	30°	30° 30°	5-10°	30°	50°	60°	70° 75° 80°		75° 50° 40° 70° 40°	55° 70° 80°					
CSIR Rock Class																
Geotech. Category	CV I	CV IA	CV 2	CS 3	CV I	CS 3	CV 2	CS 3	CV 2	CV 3	CV 2	CS 3	CV 2	CV 3	CV IA	CV I
Type of Support	A	A (B)	B-C	D B-C	B		B-C		B		A (B)	B	A (B)			

Figure 2

Geological section, rock classification and support allocation for headrace tunnel. Legend see table 1 and clause 5.

ance during tunnel construction for various stress levels and to optimize tunnel routing, construction methods and project cost accordingly.

2. THE CORTADERAL PROJECT

The Cortaderal project in Chile is particularly suited to illustrate the geological approach in the design of deep lying tunnels.

The project area is located 85 km south of Santiago in the Cordillera near the El Teniente mine. The objective of the project is to develop the hydropower potential in the upper Cachapoal catchment approximately between elevations 2500 and 1000 m. Mountain peaks exceed 5000 m and the area is notorious for snow avalanches, rock slides and glacier outburst floods.

A comprehensive study of alternatives (fig. 1) adopted a two stage development with 7 intakes, one underground and one surface power house, 42 km tunnels for the main water ways plus tunnels for temporary and permanent access. Obviously, the underground works are a decisive component for the technical and economical feasibility of such project.

3. GENERAL GEOLOGICAL SETTING

The principal objectives of the geological investigations were to establish a detailed litho-stratigraphic column and to derive an accurate structural model of the project area. Complex tectonics imposed a process of mutually linked progressive refinement for these two tasks.

Table 1 gives a condensed version of the stratigraphic column. The formation names are adopted according to the regional geological framework but subdivisions had to be introduced to reflect variations in engineering geological characteristics in detail.

The sequence starts with marine sediments of tithonian age, followed by two continental-volcanic formations dating from Cretaceous to Tertiary, which can be lumped together under the general name of porphyrite formation of the Andes. (The geologically quite interesting quaternary deposits are neglected in this context because one of the basic aims of the geological investigations was to avoid tunneling in these unconsolidated deposits.)

The marine sediments and the older of the volcanic formations (Coya-Machali) display intricate folding and faulting, as shown schematically in the section on figure 2. The younger Farellones formation has undergone only slight deformations in the project area.

The complicated tectonic structure, combined with frequent abrupt facies changes in the volcanic-continental formations made the clarification of the geological conditions a challenging task. Field work and evaluation took about one year to bring the geological concept to the required confidence level. The evaluation of aerial photographs and satellite images provided important support.

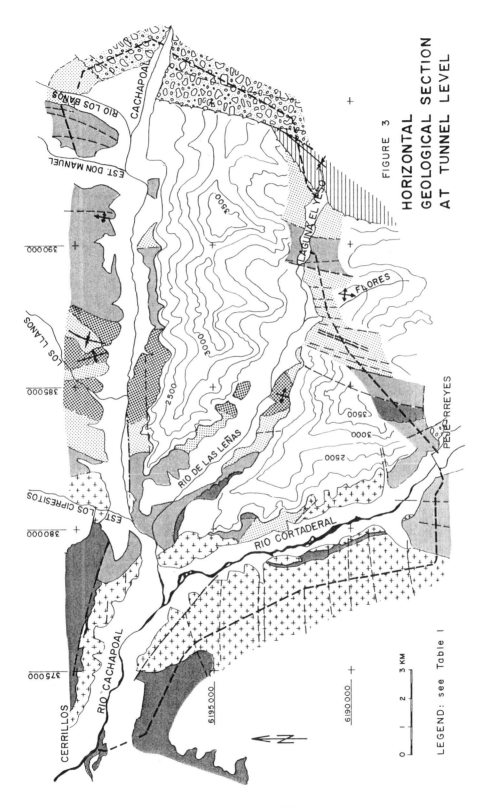

FIGURE 3

HORIZONTAL
GEOLOGICAL SECTION
AT TUNNEL LEVEL

LEGEND: see Table I

0 1 2 3 KM

CERRILLOS
RIO CACHAPOAL
EST. LOS CIPRESITOS
RIO CORTADERAL
RIO DE LAS LEÑAS
LOS LLANOS
RIO LOS BAÑOS
EST. DON MANUEL
CACHAPOAL
LAGUNA EL
FLORES
PEÑERREYES

375000
380000
385000
390000

6195000
6190000

2500
3000
3500
2500
3000
3500

TABLE I   LITHO–STRATIGRAPHY, ROCK MASS CLASSIFICATION AND MEDIAN STRENGTH PARAMETERS

AGE	LITHOSTRAT. UNIT	SUBUNIT LITHOLOGY	SYMBOL	THICKNESS [m]	CSIR RATING	GEOTECHN. CATEGORY	EL.MODUL [MPa]	POISSON COEFFIC.	Q cyl. [MPa]	STRENGTH COEFFTS. s	STRENGTH COEFFTS. m	COVER [m]
NEOGENE	CORTADERAL PLUTON	**GRANODIORITE**			II	CC-1	12000	0.2	190	0.1	2.5	800
		Coarse, massive intrusive rock										
EOCENE TO MIOCENE	FARELLONES FORMATION	**UPPER ANDESITES**		300								
		Andesite, subordinate pyroclastics										
		INTERMEDIATE TRACHYANDESITES		500								
		Laharic aglomerates, moderate lithification			II	CV-1B	8000	0.25	40	0.004	1.5	1000
		Tuffs and laves, highly lithified		300								
		Tuff, lava moderately lithified		450	III	CV-2	5000	0.25	175	0.05	0.3	500
		LOWER ANDESITES										
		Lava and recrystallized tuff		350								
		UNCONFORMITY										
MAESTRICHTIAN TO PALEOCENE	COYA-MACHALI FORMATION	**MAIN ANDESITES**			I-II	CV-1	12000	0.2	200	0.1	3.0	1350
		Porphyric massive lava		250								
		Aphanitic, layered lava		150								
		Aglomerates, flow breccies, massive		500								
		Red Sandstone and shale		50								
		Coarse porphyric lava, massive		300	II	CV-1A	10000	0.2	160	0.05	1.7	1000
		INTERMEDIATE TUFFS										
		Dacitic, andesitic tuffs with intercalated red beds		400	III	CV-2	5000	0.3	100	0.0001	0.34	200
		Dacitic welded tuffs		300	II-III	CV-3	5000	0.3	150	0.004	0.5	
		Red sandstone, volcaniclastics		100	III-IV	CV-2A	3500	0.3	75	0.0001	0.15	
		LOWER SEDIMENTS										
		Red sandstones and tuffs		200	III-IV	CS-3	3000	0.3	75	0.0001	0.2	300
		Red Cachapoal conglomerate		600	II	CS-1	10000	0.25	120	0.004	1.5	1300
		UNCONFORMITY										
TITHONIAN	LENAS-ESPINOZA FORMATION	Limestone		400	II	CS-2	5000	0.25	90	0.004	1.0	400
		Siliceous sandstone, shale		350								
		Calcareous sandstone		300								
		Coloured sandstone, shale		500								

4. ENGINEERING GEOLOGICAL DATA BASE

Field Survey:
In the course of the field work, in addition to a comprehensive lithological description of the rock mass, data on the rock mass fabric were collected. For the engineering geologic application a redundance of field data on rock mass fabric is needed in order to permit a reliable definition of homogeneous zones and to correlate the geometric characteristics of the fabric with lithology and with regional structural units. These are necessary conditions for the projection of surface data to the level of the underground structures which is of decisive importance for the viability of the adopted methodology for the project.

At selected locations sample geotechnical rock mass classifications were made. An adequate number of locations had to be sampled in order to establish the median geotechnical quality as well as the normal range of variation in rock mass quality. The CSIR classification system was preferred because it evaluates primarily intrinsic rock mass properties. (Other factors which control the rock mass performance, as for instance the ambient stresses, can be superimposed as independent variable). Compressive strength of the rock was determined in the field with a Schmidt hammer, evaluated statistically and calibrated by compression tests in the laboratory.

Exploratory Works:
Site conditions definitely ruled out global investigation by deep core drilling. However, several refraction seismic profiles were shot which contributed general information on rock mass properties. Finally, at the site of the Pejerreyes underground powerhouse an adit was excavated, mainly with the purpose to demonstrate the conception of rock mass performance which had been deducted from the geological survey data.

Laboratory Tests:
In addition to compression tests with determination of elasticity parameters the laboratory tests comprised petrographic analysis in thin sections. The objectives of the petrographic analysis were a precise definition of lithology, the search for components which might influence rock performance (as for instance minerals resulting from hydrothermal alteration) and an evaluation of the abrasiveness of the rock with regard to the application of a TBM.

5. GEOTECHNICAL EVALUATION

Index Properties of Rock Mass:
The above described field surveys, marginally complemented by subsurface explorations and laboratory tests, allow a comprehensive rock mass classification for which the concept of the CSIR system was used. Taking the Rock Mass Rating as qualitative scale permits to distinguish favorable or difficult tunnel routes in a horizontal geological section near tunnel level and to trace promising tunnel routes in this section.

The next step in project design then requires a quantified comparison of the tentatively selected alternatives in terms of technical viability, cost and time of construction. In this respect the ground reaction to the tunnel excavation constitutes the crucial factor when rock mass shear strength is exceeded and a plastic zone forms around the tunnel (in other circumstances e. g. groundwater conditions may be controlling). The CSIR system suggests values for the shear strength but these are rather of an indicative nature. Resorting to the "empiric strength criterion" offers substantially more specific data. Combining the empiric strength criterion with the rock mass rating and the litho-stratigraphic column a project-specific system of "geotechnical categories" is obtained which evaluates shear strength as a function of the geological and geotechnical input. For the Cortaderal project 5 main categories were adopted, designated:

CS for sedimentary,
CV for volcanic,
CC for intrusive rocks and
F   for highly tectonized zones.

Subdivision of the categories reflects subordinate variations in lithology, fracturing, etc. Table 1 gives an overview of the system and states typical index properties for median quality of the rock mass types.

Estimates of the elasticity modulus are based on the rock mass rating, seismic velocity and rock strength. Following experience from other projects an anisotropic ambient stress field with major horizontal stress was assumed for shallow depth and isotropic stresses for large depth.

Design Evaluation:
The "geotechnical categories" define the parameters which are needed to evaluate the tunnel design, especially as regards support requirements, but the methodology of

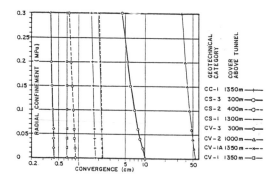

Figure 4
Ground reaction curves for selected geotechnical categories and stress levels

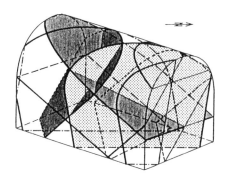

Figure 5
Block diagram of discontinuity system at Pejerreyes underground power house

evaluation must take the essentially geological nature of the input into consideration. It is, therefore, appropriate to conduct parametric studies which demonstrate the impact of variations and of uncertainty in the input and to adopt independent approaches for mutual verification.

A suitable method to evaluate the performance of the rocks as described by the geotechnical categories follows the concept of convergence-confinement for a pseudo-continuum. Closed form analytic models on the convergence-confinement principle with the empiric strength criterion constitute a convenient method for rapid parametric analysis, probing, for instance, the effects of variations in stress levels and in shear strength indices. Figure 4 gives samples of ground reaction curves for selected geotechnical categories. Favorable and potentially critical conditions are readily recognized in such curves. For critical segments of the tunnel route a more elaborate analysis by finite element modeling is in order. Again, the geotechnical categories and the empiric strength criterion offer the required input.

For extreme conditions (weak rock, high stress levels) the empiric strength criterion tends to render exceedingly conservative results. For such cases a cross check with directly estimated strength parameters or with alternative classification systems has to be made to assist in geotechnical judgment. For this purpose the NGI procedure was found to be particularly useful.

Finally, because the classification systems only indirectly take this factor into consideration, the effect of the geometry of the discontinuity system has to be studied, either by merely qualitative assessment of strength anisotropy aided by fabric diagrams or as e. g. for the Peje-

rreyes power house by combined graphic and computational methods on the principle of rigid wedges (see fig. 5 for discontinuity orientation). Meanwhile, integrated software is available which allows for rapid and convenient analysis of discontinuity controlled instability (Eppler & Graf 1989).

According to the results of the geotechnical evaluation 4 support types were adopted and allocated to the respective stretches of the tunnel (see fig. 2). The support types represent:

A   stable rock,
B   bolts,
C   bolts and shotcrete,
D   steel ribs, bolts and shotcrete.

In practice, during construction intermediate types will be applied following the concept of NATM.

The allocation of support and respective excavation types to the tunnel routes provides the concluding input for final determination of the tunnel routing.

6. CONCLUSIONS

Tunnels at large depth can be properly designed without extensive subsurface exploration provided the geological field surveys furnish a clear image of the lithostratigraphy, the tectonic structure and the rock mass fabric. Spot checks of fabric and sample rock mass classification must be done in sufficient number to describe local and regional variation in rock mass properties.

For the Cortaderal project preference was given to the CSIR system. Its advantages are the neutrality to site conditions and its wide application which offers ample possibility for cross reference. For quantified evaluation of rock mass performance the RMR in combination with the empiric strength criterion was found to be a good base to build a project specific set of "geotechnical categories" directly related to rock mass strength. There is a certain risk of overemphasis for compressive strength in this approach because this parameter is used in the rating system as well as in the strength criterion. Therefore, a more general basic rating system could prove advantageous (e. g. the "Rock Condition Types" suggested by John, 1981).

Table 1 shows different indices of rock mass strength for one CSIR rock mass class, depending on the geotechnical category to which the class refers. Thus, the geotechnical categories present a constructive amendment to the basic rock mass rating because they offer more specific input for the design of underground works.

The strength characteristics have to be allowed a range of variation which has to be probed by parametric studies. Straightforward analytic models were found appropriate for this purpose.

A horizontal engineering geologic section always proves an important tool in planning of the tunnel layout. Applying the results of the parametric analysis of ground reaction to the horizontal section suitable tunnel alignments can be readily selected.

Despite its obvious merits the pseudocontinuum approach does not fully cover all geological implications on ground response to tunneling. The effect of the discontinuities, their properties and orientation on the design and stability of an underground opening has also to be evaluated. Such evaluation should in any case involve geology-orientated graphic methods.

The described geological approach for the design of tunnels gave satisfactory results also in other Andean countries (Bolivia, Ecuador). In more vegetated regions (Philippines, Taiwan) notably more time had to be dedicated to the field work. Under adverse field conditions and in the case of special design problems (e. g. pressure tunnels at shallow depth) the professional judgment of the engineering geologist will be called for to recommend the necessary complementary subsurface investigations.

REFERENCES

A.F.T.E.S. Groupe de Travail No.7 1983. L'emploi de la méthode convergence/confinement. Tunnels et Ouvrages Souterrains, 59, Septembre-Octobre 1983: 218-236.

Eppler, Th. & N. Graf 1989. Geodatplus. PC-Programm zur Auswertung geologischer Daten. Ruhruniversität Bochum, Institut für Geologie, November 1989.

Hoek, E. and E. T. Brown 1980a. Empirical strength criterion for rock masses. J. Geotechn. Eng. Div., ASCE, Vol. 106, No. GT9, September 1980: 1013-1035.

Hoek, E. and E. T. Brown 1980b. Underground excavations in rock. Institution of Mining and Metallurgy, London, 1980. 527 pages.

John, K. W. 1981. A compromise approach to tunnel design. 22nd US Symp. Rock Mech. 1981: 333-341.

Kirkaldie, L. 1988. Rock classification system for engineering purposes. ASTM STP 984, Ann Arbor 1988: 167 pages.

Klohn, C. 1960. Geología de la cordillera de los Andes de Chile Central. Instituto de Investigaciones Geológicas, Chile, 1960

Acknowledgement: The study of the Cortaderal project was conducted by the Joint Venture Electrowatt-ARA-Minmetal on behalf of Codelco and Chilectra.

6th International IAEG Congres / 6ème Congrès International de AIGI, © 1990 Balkema, Rotterdam. ISBN 90 6191 130 3

The Grange Subway: An old military tunnel in chalk
La Grange Subway: un vieux tunnel militaire dans la craie

M.S. Rosenbaum
Imperial College of Science, Technology and Medicine, London, UK

ABSTRACT: In the First World War, mining carried out by the belligerents of the Western Front created an extensive network of tunnels and underground chambers. Many of these tunnels still exist and a few remain accessible. Part of one, the Grange Subway at Vimy, has recently been renovated thus providing not only an instructive insight to the horrors of military mining but also evidence for the long-term stability of small tunnels within chalk rock. The generally stable state of the 75 year old tunnels at Vimy is discussed in the light of the prevailing character of the chalk and its environmental setting.

RESUME: Lors de la Première Guerre mondiale, des creusés par les belligérants du front occidental forment un extensif réseau des tunnels et des salles souterraines. De nombreux galeries souterraines existent encore et quelques-uns sont toujours accessibles. Une partie de l'un d'eux, la Grange Subway à Vimy, a été restauré et il a été ainsi possible d'avoir non seulement un aperçu instructif des horreurs des souterrains militaires mais aussi l'evidence de la stabilité à longue échéance des petits tunnels creusés dans la craie. L'état généralement stable des tunnels vieux de 75 ans à Vimy est discutée à la lumière du caractère dominant de la craie et de l'environment.

1 INTRODUCTION

Collapse of old mine workings has been a major concern across the chalk outcrop of northern France in recent years since there has been extensive urban expansion of towns such as Arras and Lille, together with associated new road construction. There has been a long history of tunnelling in this region, records of workings extending back to pre-Roman times (Bivert 1988). There was a surge of activity from medieval times with the rapid development of towns in what is now la Région Nord-Pas-de-Calais such as Arras, Béthune and Cambrai. Further west in northern France the study of the risks associated with such abandoned underground workings has been described by Evrard (1987) based on work in the limestone country of Normandy.

The vast number of accessible, or partially accessible, artificial cavities in the Pas-de-Calais have been described by Bivert (1988). Here the distribution of old mine workings can be seen to be concentrated around the 17th and 18th century fringes of towns such as Arras. Primarily the purpose of the mines was to provide a better source for building stone than could be obtained from quarries. Because of the poor transportation systems of the day this necessitated their close proximity to the building sites. Other uses of chalk mines included the supply of agricultural lime and of flints.

Engineering geological aspects of the chalk mines located in the south of the Netherlands are described in detail by Price and Verhoef (1988). Other difficulties associated with underground chalk workings have also recently been

discussed during the 1987 Conference on Underground Movements and a useful summary has been presented by Edmunds (1988). His work is continuing with the compilation of a database containing all known natural and artificial cavities occurring within the British chalk outcrops under funding from the Department of the Environment.

Mining for quarrying purposes ended with the 19th century, but soon after underground excavation of the chalk took on a new perspective with the development of both protective and offensive workings as military mining developed along the Western Front during the 1914–18 war. The engineering geological aspects of military mining have not been covered hitherto but a summary of the geological influence along the Western Front has recently been presented (Rosenbaum 1989). A recently renovated example is the Grange Subway, located 8km north of Arras beneath Vimy Ridge, and this now provides access to a typical military mine in the chalk terrain of northern France.

Difficulties of excavation and risks of countermining led to ever greater delays and the role of civilian consultants proved ineffective owing to problems of limited time and levels of communication. This led within two years of the outbreak of the Great War to the appointment by the British Army of a full-time geological adviser, the first time that a geologist was employed as such within any modern army. Geological advice was sought for predicting likely ground conditions for trenches and mines and was initially provided by Lt (later to become Professor) W.B.R. King for the British, Major (also later to become

Professor) Edgeworth David for the Australians, and a number of geologists (names unknown) for the Germans. A detailed account of some of this work, illustrated with coloured maps and cross–sections, was subsequently presented in a special volume published by the Institution of Royal Engineers (anon. 1922a).

Knowledge of the underground geological structure in the Vimy Ridge area was not easy to come by at this time but was not altogether lacking. Much of the information was housed in university libraries or behind enemy lines. A general understanding of the regional geology could therefore be compiled but this required a considerable effort from the geologists together with advisers who were able to access and interpret the diverse less accessible records and maps. It was still found at the local level that the geology was insufficiently well defined for practical purposes. Therefore field mapping of available exposures, particularly those revealed by new craters (!), supplemented by information from shallow borings and existing tunnels was undertaken. Once the tunnelling commenced, there was an ongoing requirement to check on the excavation progress and in particular to warn of likely hazards ahead, notably faulted ground and adverse water conditions. Limited use was made of advance boring to supplement geological measurements within the tunnels, but the close proximity to the ground surface, not to mention the enemy counter–mining activities, made this a hazardous task.

2 GEOLOGY

The Ridge at Vimy is an extensive NW–SE trending feature overlooking the Douai Plain to the NE. It is effectively a scarp slope developed in the Senonian and the underlying less flinty Turonian chalk. The Senonian and Turonian are also the principal Cretaceous Stages of northern France which have been worked by mining methods mainly for building stone and agricultural lime, especially from the 17th until the end of the 19th centuries (Leplat 1973).

The ground drops away to the SW at a much gentler angle as it follows the regional dip of the chalk. The presence of the Ridge is largely due to a major disturbance known as the Marqueffles Fault which runs in a north westerly direction. This disturbance has brought the underlying Hercynian basement close to the surface on the north–eastern side and it is this which is responsible for the extensive coalfield centred on Lens to the north of Vimy.

The excavations for the Grange Subway lie for the most part in the Senonian chalk. This is a weak rock characterised by numerous flint nodules mostly forming bands sub–parallel to bedding. The near surface zones are considerably weakened by periglacial frost shattering and solifluction, the shattering giving rise to the development of many irregular fractures in the chalk penetrating to depths of 15 to 20 m below ground level, and these are especially frequent in the shallowest 5 m. Additional shattering of the chalk accompanies minor normal faults which are thought to be associated with the Marqueffles Fault, and during the War locally intense shattering was induced as underground mines were detonated.

Sandy Lower Tertiary Landenian sediments once covered the area but none have been positively identified in this part of Vimy, the nearest outcrop being 1/2 km to the north. However, a mantle of silt about one metre thick blankets much of the area and this can sometimes be found as deep as 15 m below ground level where it has been washed into solution cavities or open fractures within the chalk. The silt is predominantly loessic in origin, but may also contain a significant component of remnant Landenian sediments.

Groundwater is always a major concern with tunnelling projects and was considered to be one of the most important parameters requiring description for the First World War excavations. From an historical study of the well records it was established that water levels in the chalk rose and fell seasonally by as much as 10 m between the summer and winter months. The levels thus defined the lower limits to which the tunnels could be effectively driven.

The top few metres of the ground was, and still is, highly disturbed by earthworks associated with protective military excavations. As might be expected, considerable war disturbance is also evident, particularly arising from the effects of mortar and artillery shells as well as the detonation of underground mines.

3 TUNNEL CONSTRUCTION

Lengths of small diameter fighting tunnels had already been excavated under Vimy Ridge when the British took over from the French in April 1916. These had initially led from the trench system and were effectively a downward projection of those trenches. Later in that year shallow inclines were constructed to counter German mining activities and then a chain of defensive listening posts became established. These excavations were later connected up by laterals which together formed part of an integrated underground complex.

Subways provided underground access routes leading from the relatively safe rear assembly areas to the trench and tunnel systems at the Front. The construction of these various tunnels was conducted under conditions both of haste and of hardship, but followed the established tunnelling principles of the day. The technology employed was adopted from the coal mining industry since much of the manpower had been drawn from there. The methodology was written up in detail after the War and published in another special volume by the Institution of Royal Engineers (anon. 1922b).

At Vimy the Grange Subway was constructed in just four months by 172 Tunnelling Company of the Royal Engineers, commencing towards the end of 1916, in preparation for the assault on Vimy Ridge by the Canadian Corps in the Spring of 1917. This subway was 800 metres long and not only provided covered approaches to the front line trenches but also incorporated a command post, a first aid dressing station, a water supply point, an electricity power station, magazines and accommodation. It was equipped with tramlines (wooden so as to minimise the noise) for assisting removal of spoil and the supply of munitions. The interconnection of these various excavations together formed part of an underground complex extending for over 6 km. In addition there were also connections, partly for ventilation and partly for access, along laterals to some of the eleven other similar subway systems along Vimy Ridge, most of which are now blocked by debris deliberately placed to prohibit access.

During the tunnelling at Vimy the presence of gas, especially methane, carbon dioxide and carbon monoxide, passing into the workings was regarded as being a major threat. One major source was considered to be the fracture system leading from the Carboniferous rocks beyond the Marqueffles Fault. The presence of carbon dioxide was already a well known phenomenon within the old chalk mine workings of the Pas–de–Calais. Its widespread occurrence could be expected arising from natural solution as acidified groundwater containing humic acids percolated down through the surface soil and attacked the calcium carbonate of the chalk. Workings of this period employed short drives or multiple shafts to cope with such foul air, but within the deeper workings an extensive system of air lines was installed in order to cope with the gases within the short time available.

The release and absorption of carbon monoxide through the pores of the chalk and its associated fractures was a particularly difficult phenomenon to predict since it depended greatly on the prevailing atmospheric pressure, the moisture content of the exposed chalk rock, and the degree of air circulation within the tunnel workings.

4 TUNNEL STABILITY

The chalk has been largely self supporting so the tunnels have remained reasonably stable during throughout their 75 year history. However there are some zones where stability has not been maintained and these are associated with the following specific features. Stress relief of the chalk associated with erosion of the overburden has induced a pronounced set of fractures sub–parallel to ground surface which tend to become progressively more closely spaced as ground level is approached. The sub–horizontal joints have often taken advantage of clay or marl seams within the chalk. Perpendicular to these fractures other cracks have developed due to stress relief of the chalk slabs generating a blocky fabric. This locally has led to poor stability, especially of the roof and haunches at the intersection of galleries and the sites of underground chambers where spans are larger.

In the shallowest 15 m the fracture pattern has been opened out by ice heave associated with periglacial weathering during the Quaternary. This has effectively loosened the chalk rock mass and to some degree has opened up the aperture of the fractures thereby reducing the strength and increasing the permeability. Additional fracturing has been generated in close proximity to detonated mines where severe slabbing perpendicular to the propagating shock waves has developed. Spalling of the chalk is particularly evident here, notably at the corners of tunnels which is where stresses tend to be most concentrated and horizontal stress relief is at a maximum. Such instability was of little concern to the military miners since once a mine had been detonated the military objective had been achieved (or thwarted!) and further

access was unlikely to have been required.

The tunnel dimensions for the mines leading beneath the front line trenches were largely dictated by the access required by the miners while at the same time minimising the quantity of material that had to be removed in order to maintain an acceptably rapid rate of advance.

Uncontrolled collapse of the excavations had to be avoided during the operational life of the tunnels and the chalk fracture spacing was a major control for determining the local design of each tunnel. The stress relief pattern tended to favour falls of rock slabs which could come directly from the roof, thus the tunnel width was in general limited to 0.8 m. This was so as to provide sufficient bridging maintained by friction along the joint surfaces. As a matter of standard procedure based on then prevailing coal mining practice. The tunnels were supported by timbers for every one metre of advance, the supports consisting of two pillars driven into small recesses in the wall supporting a roof beam, the whole being driven home by wooden wedges. The tunnels were 1.3 m high and so working conditions were very cramped.

Most of the mine timbers were salvaged after a successful attack due to the chronic shortage of wood in the war zone. Those mine timbers which were left behind have now rotted completely but the tunnel profile has nevertheless survived largely intact through the succeeding 75 years. However, the standard tunnel size was too small for the assaulting infantry to pass through rapidly so the main access subways were built to a larger profile of one metre width and two metres height. This has been too great for long term stability of the roof and 10 years after the War concrete beams had to be installed in order to maintain the roof against further collapse.

The other major hazard encountered in tunnelling was due to the inflow of water. Indeed, military mining was ineffective below the water table because of the interference to excavations and the risk of detection that would be created if a pumping system were to be installed. Another important consequence of tunnelling beneath the water table was that the chalk there would be fully saturated. In such a state chalk can lose as much as a half of its strength (Bonvallet 1979) thus leading to rapid collapse if no change was made to the tunnel dimensions or the support systems.

Groundwater flow was also concentrated in the shallower depths along the fractures dilated by ice heave associated with a periglacial environment. Loess was subsequently washed into some of these fractures as the ice thawed and this has maintained a wide aperture infilled with silt which is comparatively permeable. Rainwater percolating down from the ground surface thus becomes concentrated as it seeps towards the water table leading to further local softening of the chalk. The water inflow problem would therefore be a particular problem if a highly fractured zone were intersected by the tunnel. The remedy was to excavate all tunnels at a slope of at least 1 in 100 and to provide sumps in side passages capable of storing any excess water until it could be either pumped out or carried away. The need to maintain as dry conditions as possible underfoot as well as in the roof and walls of the tunnel need not be elaborated in view of the well known conditions of the trenches at that time!

5 PRESERVATION

The Canadian Government has established a Memorial Park on Vimy Ridge to commemorate those who had fallen during the War. Most notable is the construction of a prominent twin–spired War Memorial but the Authorities have also planted an individual tree for every Canadian who fell. In addition a small part of the defensive earthworks have been preserved, including the main tunnel of the Grange Subway. Here the effort was directed towards installing concrete walls and ceilings where the chalk was particularly closely fractured or spalling seemed likely from the roof or walls. The initial work was completed in 1926 and so the tunnel was available for inspection by visitors until the outbreak of the Second World War. The Grange Subway was then quickly returned to military use to provide protection for members of the British Expeditionary Force; later it was similarly employed by the German Army.

The subway tunnel was reopened to the public after the War but the passing of time and the humid atmosphere was leading to deterioration of the concrete supports, especially where the reinforcement cover was too thin to ensure continuous protection. The need for remedial measures led to the necessary work being undertaken by local contractors in 1986. This mainly consisted of installing wire netting in the roof, supported by steel bars, to prevent minor rock falls arising from spalling rock.

Further renovation has been undertaken by the British Army as a Military Aid to the Civil Community task, appropriately by the Royal Engineers since it was their Tunnelling Companies who originally excavated the larger part of the Grange Subway system. The basis of the remedial works required to protect the underground excavations has been to ensure good drainage of groundwater and maintaining the strength of the rock roof and walls. This has been done by preventing further deterioration or loosening of small rock blocks so ensuring that they remain locked together.

Elsewhere within the tunnel system, away from areas accessible to the public, the stability can be related to the geological influences outlined above. Where the depth of cover is less than 10 m a number of collapses have occurred within the last 10 years. Such collapses have tended to be relatively sudden and commonly give rise to a crown hole, a steep–sided subsidence of between 1 and 2 m depth and 2 to 4 m diameter. This represents total collapse of the tunnel roof with the void migrating directly up to the ground surface. Here there has been insufficient cover above the tunnel to enable bulking of the rubble from the roof falls forming sufficient support to prevent further upward collapse.

Coal mining experience in tunnels of similar size to those at Vimy (Littlejohn 1979) suggests that collapses which develop at depths in excess of 20 m induce bulking which is unlikely to permit voids to migrate right through to the ground surface. The behaviour of the weaker chalk rock would not be expected to be exactly the same, but Evrard (1987) notes that cover in excess of 25 m above the rather higher tunnels of the chalk mines in Normandy has been sufficient to prevent breakthrough of collapse

voids right to the ground surface. Instead shallow dish–shaped depressions form there with diameters some 3 to 5 times their depth above the points of roof collapse.

6 VIMY RIDGE TODAY

Anyone who would like to visit the old tunnels and defences of the Western Front at Vimy may do so by visiting the Canadian Vimy Memorial Park which is open daily from 1 April to 15 November. Guides (English and French speaking) are available to escort visitors around the 250 m of the Grange Subway which has been preserved, starting from the Information Kiosk near the car park at the southern end of the Park.

In Arras, 8 kilometres to the south of Vimy, the underground chalk workings beneath the Town Hall are also open to the public. These old chalk mines date largely from the 16th and 17th centuries and were used to extract building stone, yet minor excavation continued until the start of the First World War. Then the last major phase of activity commenced as the mines were put to use for providing protection against enemy artillery fire.

REFERENCES

anon. (1922a). The Work of the Royal Engineers in the European War, 1914–19: Geological Work on the Western Front. 71pp. Institution of Royal Engineers, Chatham.

anon. (1922b). The Work of the Royal Engineers in the European War, 1914–19: Military Mining. 148pp. Institution of Royal Engineers, Chatham.

Bivert, B. (1988). Les Souterrains du Nord–Pas–de–Calais. 358pp. Conseil Général Département du Nord.

Bonvallet, J. (1979). Une classification géotechnique des craies du nord utilisée pour l'étude de stabilité des carrières souterrains. Revue Française de Géotechnique, No.8, p.5–14.

Edmunds, C.N. (1988). Induced subsurface movements associated with the presence of natural and artificial underground openings in areas underlain by Cretaceous Chalk. In Bell, F.G., Culshaw, M.G., Cripps, J.C. and Lovell, M.A. (eds.). Engineering Geology of Underground Movements, Geological Society Engineering Geology Special Publication No.5, p.205–214.

Evrard, H. (1987). Risques liés aux carrières souterrains abandonées de Normandie. Bulletin de Liaison des Laboratoires des Ponts et Chaussées, No. 150/151, p.96–108.

Leplat, J. (1973). Les cavités souterrains de la craie dans le Nord de la France. En "La Craie", Bulletin des Laboratoires des Ponts et Chaussées, Special V, p.123–148.

Littlejohn, G.S. (1979). Surface stability in areas underlain by old coal workings. Ground Engineering 12, 222–230.

Price, D.G. and Verhoef, P.N.W. (1988). The stability of abandoned mine workings in the Maastrichtian Limestones of Limburg, The Netherlands. In Bell, F.G., Culshaw, M.G., Cripps, J.C. and Lovell, M.A. (eds.). Engineering Geology of Underground Movements, Geological Society Engineering Geology Special Publication No.5, p.193–204.

Robinson, E. (1988). Geologists at War. Geologists' Association Circular No.867, p.30–31.

Rosenbaum, M.S. (1989). Geological influence on tunnelling under the Western Front at Vimy Ridge. Proceedings of the Geologists' Association 100, 135–140.

Engineering geological aspects of tunnels constructed for a new high-speed railway line in the Federal Republic of Germany

Aspects géotechniques de tunnels construits pour une nouvelle ligne à grande vitesse de la 'Deutsche Bundesbahn'

V. Schenk
Lahmeyer International GmbH, Consulting Engineers, Frankfurt/M, FR Germany

ABSTRACT: For the 320 km long New Hannover-Würzburg High-Speed Railway Line to be commissioned in 1991 some 118 km twin-track tunnels were constructed which rank among the largest traffic projects in Europe. Examples of two tunnels, their difficult tunnel sections below thin cover a method of fault classification/prediction and limitations of the application of NATM will be reported on.

RESUME: Pour la ligne à grande vitesse Hannover-Würzburg d'une longueur de 320 km, qui sera prise en marche en 1991, 118 km de tunnel ont été construits. Ce rapport contient la description des 2 tunnels de sections difficiles sous couverture mince. En plus il traite une classification de failles, ainsi que sa prédiction et les limites de NATM.

1 INTRODUCTION

In Germany there are presently two new twin track railway lines under construction to be commissioned 1991 to satisfy the demand for high-speed public and freight transportation. For the northern 320 km long Hannover-Würzburg line (Fig. 1) which is subject of this paper twin track tunnels totalling around 118 km were constructed.

Besides 263 km of free track, embankment dams and bridges, 147 km (36%) of the two routes are through tunnels, ranking among the largest tunnel projects worldwide.

The track designed for 250 km/h requires a minimum curve radius of 7.0 km and a gradient of max. 1.25%. Due to the mountainous terrain this can only be realised by constructing a high proportion of structures and numerous tunnels frequently under shallow cover.

In the past neither in urban tunnel construction nor in the German Mountains experiences with tunnels of very large cross sections (110-180 m²) under low cover and in a rock mass of high degree of separation were made.

Thus especially in the entrance sections the sub-horizontally bedded sandstone layers and sandstone/siltstone interbeddings of the middle Buntsandstein (lower Triassic) are affected by deep reaching

figure 1
New high-speed railway lines

loosening and weathering (incl. permafrost effects) resulting in a rockmass up to soillike conditions. The bearing behavior of such a rockmass may considerably be reduced by high deformability, stress release and low shear strength, affecting the tunnel's stability under a cover of

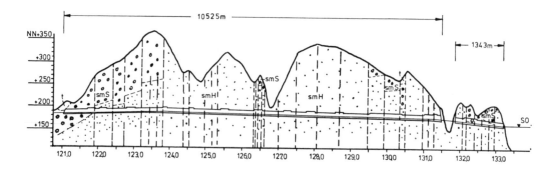

figure 2
Mündener- and Mühlenkopftunnel
geological longitudinal section, t = tertiary, smS = Solling sequence,
smH = Hardegsen sequence (middle Buntsandstein)

more than two tunnel diameters.

2 ENGINEERING GEOLOGICAL EXPERIENCES DURING TUNNEL CONSTRUCTION

Against this background of 12 tunnels (totalling 29.3 km) experiences selected case histories from difficult sections of Mündener- (10,5 km length) and Mühlenkopf-tunnel (1.3 km length), will be dealt with. Generally drill and blast excavation by calotte and bench heading was applied using initial support of steel arches, pattern bolting and shotcrete (NATM method). Usual cross section see Fig. 8.

2.1 Mündener Tunnel

The tunnel, the second longest in Germany was predominantly driven through thinly to medium banked sandstone with thinly banked clay/siltstone interbeds (Fig. 2).

Originally 2 separate tunnels were designed. Environmental restrictions at the centre of the tunnel, however, required an open cut (25 m in depth) used for attacks and cut & cover section as an alternative. During open-cut excavation at both slopes continuous creeping of scree (max. 15 m thick) up to 0.5 m within a year was measured (Fig. 3).

Drillings showed a thin layer of clay forming the interface between periglacial creeping talus and rock line, which was not recognized during site investigations, triggered the slides. Prestressed anchors (40 MPa) were subsequently necessary to stabilize the slope.

Calotte heading revealed a rock mass weakened by numerous master joints and faults cutting the alignment perpendicu-

larly (Fig. 3).

First at chainage 75 m shear fissures and within 10 hours double the primary deformations were encountered.

In order to avoid the collapse of the tunnel backfill into the central access ramp and closing of concrete floor was provided. Afterwards 2 rows of timber beams were placed as roof support (Fig. 4). The development of settlements which nearly reached 0.6 m and the counteraction of support measures are shown Fig. 5.

An analysis of the damage revealed ground failure due to intense and close separation of the rockmass by means of master joints and mylonitic faults. The trenching and heavy rainfalls resulted in critical loads at floor of calotte such as up to a cover of 30 m a good portion of the deformations was recognized at ground surface.

Before reexcavation systematical grouting was provided.

Due to saxonic tectonics the rock mass proved to be separated by wide and close graben and horst systems often dissected by internal swarms of mylonitic faults. Especially in the southern tunnel lots in every 300 m length such significant fault systems were met crossing the alignment diagonally. These faults always initiated higher deformations up to 15 cm resulting in fissures and damages of shotcrete sealing (20-30 cm thick). Even under a cover of more than 100 m it was obvious that after passing the faults the calotte section mainly reached almost convergence (some 2 diametres behind) and later were reactivated up to distances in the order of 60 m.

SW

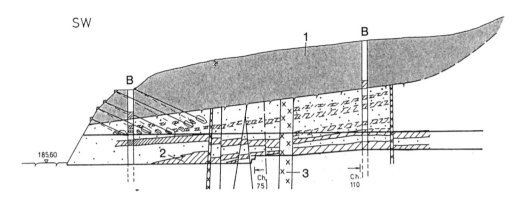

figure 3
Section of damage
1 = creeping scree, 2 = claystone/clay, 3 = faults, B = borings

2.2 Approach for a prognosis of defects of rock mass

To improve the prognosis of geotechnically relevant defects of rock mass during construction, of the Mündener Tunnel the following approach was made:

- Evaluation of the current engineering geological tunnel documentation with regard to a classification of faults and faulted zones (G-index).
- Indication of lineations by air photo interpretation.
- Determination of joint porosity of rock mass by means of resistivity geo-electrics.

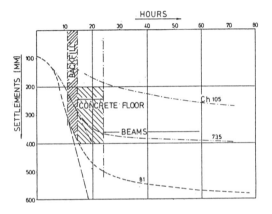

figure 5
Settlements of roof and
effect of counter measures

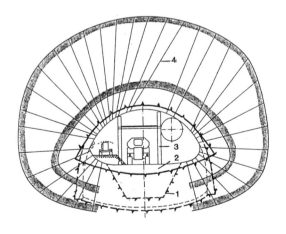

figure 4
Section of damage, counter measures
1 = access trench,
2 = closure of concrete floor,
3 = timber beams, 4 = grout holes

According to the big size of the cross section it was obvious that geometry and geotechnical behaviour of faults/faulted zones are imperative for tunnel stability in a rock mass of large scale separation i.e. time dependant reaction of such defects and interaction of initial support.

With this respect the following empirical equation was applied:

$$G = (S \times L) + (T \times W)$$

G = Tunnelling degree of difficulty (G-index) of rock mass defects according to geological criteria

S = Fault category (Fig. 6).

L = Extent of fault along tunnel track

2505

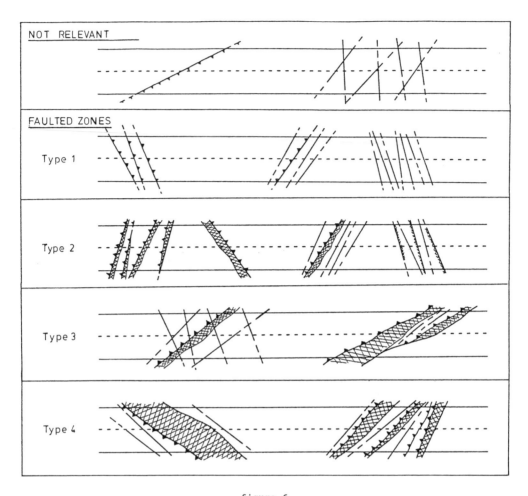

figure 6
Fault categories, type 1: swarms of faults, master joints, closed;
type 2: swarms of faults/master joints, single fault,
all filled with fines, all combinations with type 1;
type 3: faulted zone, separated by master joints; sheared zone, mylonitic,
more than 0.5 m thick, closely jointed, throw, all combinations with type 1 and 2;
type 4: mylonite zone more than 0.5 m thick, bundle of mylonite zones
more than 0.1 m thick each, all combinationswith types 1, 2, and 3

T = Percentage of clay/siltstone volume in the area of rock mass defects

W = Water factor 1 = rock mass, dry to moist
              2 = continuous inflow of water

The classification of rock mass defects proved to be a practicable help in addition to the results of convergency measurements in the decision to apply the appropriate support category in the calotte (e.g. thickness of shotcrete, rock bolting) and for design of bench heading (central ramp or not).

As experienced, rock mass defects of an index higher than 100 proved to be difficult from the constructional point of view, whereby fissures resp. larger deformations at the means of support could frequently be stated.

To gain information in advance an investigation was also performed, to estimate the geotechnical relevance of faults with regard to their location as well as

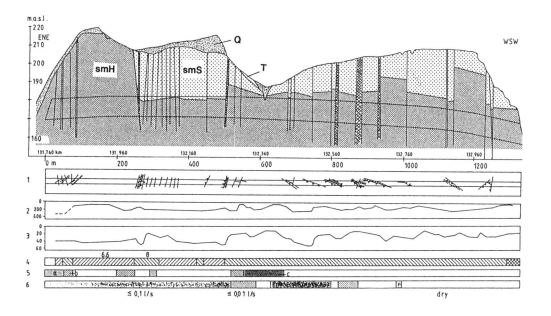

figure 7
Mühlenkopftunnel, geotechnical longitudinal section,
Q = scree, t = tertiary, smS = Solling sequence, smH = Hardegsen sequence,
1 = faults, 2 = settlements of roof, mm; 3 = percentage of silt/claystone
4 = support categories, 5a = rehabilitation of portal section,
5b = consolidation grouting, 5c = cut & cover section, 6 = inflow of water

by means of geoelectrical measurements for the tunnel sections still to be driven.

It can be summarized that some 90% of the faults predicted by this method could be confirmed and verified during excavation with sufficient accuracy.

Finalizing, it has to be noted, that the actual position of faults in underground structures diverts from the forecasted chainage in dependence of the dipping angle of discontinuity and the covering height. On the basis of the measured Land S-values and the estimated T- and W-values G-indexes were determined for the forecasted faults, which according to their size could largely be confirmed by the later effected drive.

2.3 Mühlenkopf Tunnel

According to its alignment parallel and near by the deeply incised Fulda valley and a maximum cover of 30 m only, construction problems had to be taken into account from the commencement (Fig. 7).

During excavation it was confirmed that weathering frequently associated with caolinitic clays (tertiary weathering products) led to high desintegration of the rock mass until well below the tunnel gradient. Moreover, the rock mass (thinly banked siltstone/sandstone alterations and medium banked sandstone sequences) is separated by widely spaced tension cracks (so called "Hangzerreissung", Stiny 1929) frequently filled with silty material and running under acute angles or subparallel to the tunnel.

Tunnel gradient and a rock mass partially weakened by caolinitic weathering of sandstone and siltstone decomposed up to stiff clay required a short cut and cover portion in the centre of the track. In addition a pocket of tertiary clays and sand fills the core of the tectonical graben.

At the northern portal roof settlements up to 30 cm were measured when drive reached chainage 50 m.

Accordingly, placement of timber beams was the only measure to avoid the collapse of the portal section. The thinly banked siltstone/sandstone interstratification showed bedding internal joint bodies in the order of 1 dm, of which silt layers

2507

were often deteriorated to clay. Nearly the whole settlements were measured on the ground surface at a cover of 20 m.

The lack of residual stresses also favoured loosening of rock mass such as full load of cover acted on initial support.

The analysis of the damage revealed ground failure of the calotte floor such as plastification of siltstone up to stiff consistency was observed below the steel arch foundation. Reexcavation was started after extent grouting in advance and subsequent strengthening of rock mass by longer rockbolts and 35 cm shotcrete inclusive floor concrete.

When approaching the Fulda valley subparallel to alignment open tension cracks up to some cm wide occured in a rockmass rich in siltstone layers (30-50%) resulting in roof settlements until 20 cm (chainage 250-350 m). Consolidation grouting and close pattern bolting was necessary to avoid progressive loosening of rock mass.

In between chainage 350-500 m a sheet of claystone was met in the floor where even after placement of lining small long term settlements lasted about 1 year to be believed as an effect of consolidation.

During the excavation of cut and cover section a slope sliding was initiated in periglacial slope wash when the clay layer of tertiary age was undercut.

According to tender design a bored pile wall of 120 m length was constructed to stabilize the slope.

In the cut and cover section deep seated slope movements until valley floor were feared by the contractor. For slope control some exploration boreholes were sunk partly below valley floor and installed with inclinometers. Fig. 8 presents a measuring cross section consisting of extensometers and inclinometers which was placed at the NW portal proved to be sensitive to slope movements (supported by prestressed anchors).

Immediately after construction of the tunnel pipe the open cut was closed by backfill to rebuild the original slope conditions.

Up until now after several years of monitoring including geodetical precise measurements no significant movements along and across the tunnel alignment were recorded not exceeding the measuring accuracy.

3 CONCLUSION

3.1 Geological features

Tectonic pattern, evidence of gravitational tension cracks and deep seated penetration of weathering result in a rock mass of a high degree of separation, with poor and faulted rocks occurring frequently, especially under low cover and at slope sections.

Such a rock mass of low shear strength, stiffness and lack of primary stresses is very sensitive to deformations particularly in view of the large cross section and thin cover as several damages also in other tunnels of the track revealed in the mean time.

Sandy to clayey slope wash up to 15 m thick often loosened the top layers of the rock when creeping during permaforst conditions.

3.2 Engineering geological and constructional criteria

The experiences obtained during construction of tunnels especially with thin cover can be summarized as follows:
- Limitation of the process of loosening and desintegration such as creation of discrete planes of separation has to be aimed at or activation of primary joints has to be reduced. Otherwise full load of cover progressively affects the support.

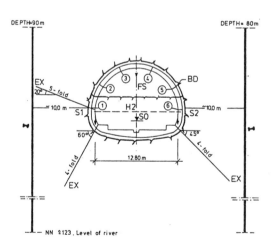

figure. 8
Geotechnical measuring cross section
I = inclinometer, Ex = extensometer,
1-6 = bolts for convergency measurements,
FS = measuring of roof settlements,
S1/S2 = measuring of calotte settlements,
BD = pressure gauge

More rapidly acting rock bolts should be installed in future.

- Progressive deformations may require a rapid and stiff closure of initial support or even lining. Eventually grouting and strengthening of shotcrete skin.

Long term deformations (time dependant behavior) particularly experienced in thinly bedded sandstone/siltstone sequences and in siltstone beds originated in release fissuring and consolidation phase afterwards may require precautionary placement of concrete lining. These processes also increase distribution of seepage paths reducing the rock mass strength considerably.

Particularly during bench heading the stand up time behavior is affected to be proved sensitive to ground failure. Thus the construction of bench trenches should be decided cautiously.

Grouting in advance proved to be very useful to improve the interbedding sequences and portions of high degree of separation.

Interruptions of drive affected the stress conditions and load distribution of initial support such as in sequences sensitive to deformations, its bearing conditions were reduced.

Consequently in sections of low cover and given geology the method of controlled deformation as an essential part of the shotcrete method (NATM) cannot be admitted i.e. revealed its limitation. A progressive loosening of rock mass therefore has to be prevented by means of rapid placement of stiff support.

On the other hand a flexible application of support dependant of rock mass behavior, which is an imperative criterion of the shotcrete method plays a decisive role to govern the middle Buntsandstein.

A rigid layout of suppport classes is a contradiction to this demand.

An essential base of prognosis is a detailed geological mapping a specific rock mass classification and its engineering geological judgement throughout the drive.

For future improvement of geological prognosis and its geotechnical relevance further development of geophysics field measurements such as recent promising results of infrared thermography, resistivity geoelectrics, ground penetrating radar and others showed might be an aid for optimization. Moreover, probabilistic methods are regarded to be a field of future application in underground works.

Anyway, in a booming construction time of large traffic tunnels and sizes of its cross section it is to be noted, however that damages can be minimized but not prevented absolutely.

4 ACKNOWLEDGEMENTS

The author would like to thank Project Manager, F. Schrewe and Mr. H. Wesemüller, both Bundesbahndirektion Hannover for their kind permission to publish this paper.

Thanks are also due to my collegues, K. Kaiser and H. Schranz for valuable discussions.

REFERENCES

Kaus, A. (1987). Die Entwicklung des GDH-Verfahrens, Geoelektrische Durchlässigkeit- und Hohlraumbestimmung. Ber. 6. NAT. Tagung für Ingenieurgeologie Aachen, Essen 1988.

Lux, K.H. & Schrewe, F.W. (1986). Tunnelbaumaßnahmen im Rahmen der DB Neubaustrecke Hannover-Würzburg zwischen Werra und Fulda. Eisenbahntechnische Rundschau 35, Heft 3.

Schenk, V., Winkler, E., Schranz. H.& Paul D. (1983). Geotechnische Begutachtung für 3 Tunnel mit freier Strecke der NBS Hannover-Würzburg (PA4) im Raum Hann.-Münden, Ber. 4. Nat. Tag. Ing. Geol., Goslar, 1983, Essen 1984.

Schenk, V., Effler, M., Kaiser, K., Kaus, A. & Wilhelm. R. 1989. Ingenieurgeologische Erkenntnisse beim Bau flachliegender Tunnel der NBS Hannover-Würzburg im Bereich der Projektgruppe Nord und Mitte. Ber. 7. Nat. Tagung für Ingenieurgeologie, Bensheim, Essen, 1989

Schrewe, F. (1987). NÖT/NATM - Erfahrungen aus dem Bereich der Neubaustrecke der DB Tunnel 3.

Stegemann, K.D. (1988). Tunnelling for the New German Railway System. World Tunnelling, Bexhill-on-Sea.

Universität Hannover (1987). Standsicherheitsuntersuchungen von Bauzuständen im Mündener Tunnel unter Berücksichtigung eines zeitabhängigen Materialverhaltens des Spritzbetons. Interimreport, Inst. für Unterirdisches Bauen, Universität Hannover, unpublished report, Hannover 1987.

Preliminary Q-system assessment of the tunnelling conditions of the Pala-Maneri hydel scheme (Uttar Pradesh/India)

L'évaluation préliminaire du système-Q des conditions de la construction de tunnels d'arrangement 'Hydel' de Pala-Maneri (Uttar Pradesh/Inde)

S.K.Shome & S.C.Srivastava
Geological Survey of India, Dehra Dun, India

ABSTRACT: The major and critical component of the Pala Maneri Hydel Scheme is the 11.3 Km long horse-shoe shaped equivalent section of 6m diameter headrace tunnel.

The rocks encompassing the scheme have been grouped as (i) the Garhwal Group comprising quartzite and basic intrusives, and (ii) the Central Crystallines containing a variety of gneisses and schists. The two groups of rocks (both Proterozoic age) are separated by the Main Central Thrust (MCT). In order to evaluate the tunnelling conditions, the Q-system has been used. In view of the rugged relief and unapproachable terrain a deliberately biased sampling of the best and the worst rocks have been carried out by studying 13 blocks in detail. On the basis of the studies, it has been concluded that in 93% reach no support shall be required, in 6% reach spot bolting with shotcreting shall be required and only in 1% reach heavy cast concrete lining with tensioned bolts shall be necessary.

RÉSUMÉ: Le componant majeur et delicat d'arrangement 'Hydel' de Pala-Maneri est un tunnel 'headrace' ayant un tunnel de la section équivalent d'une forme du fera cheval, 6m en diamètre et 11.3Km en longuer.

Les roches d'entourer d'arrangement sont gropués comme (1) Le groupe de Garhwal constituant des quartzites et de l'intrusion de la base et (2) les cristallins centrale ayant les diversité des gneisses et des schists. Les deux groupes des roches (de l'age Proterozoique) sont separées par la Poussée central (Main Central Thrust, MCT), Pour faire l'évaluation des conditions du tunnel, le systeme de Q est utilisé. Rappalont le terrain inaccessible et rude, un échantillonnage décentré délibéré des roches mouvais et de bien ont été éffectué étudiant les 13 blocs en détail. Il est estimé baser sur les études faite qu'aucune support besoin pour 93% d'atteindre, il sera nécessarire d'avoir un boulon tache arec le 'Shotcreting' pour 6% d'atteindre et la béton armé avec le boulon tensioné sera nécessaire pour 1% d'atteindre seulement.

INTRODUCTION

River Bhagirathi appearing from Gomukh-the ice cave at the foot of Gangotri glacier at an elevation of 3894m meets river Alaknanda at Devprayag at an elevation of 415m with a net drop of 3479m in a distance of 221Km. This enormous potential for hydropower still remains to be harnessed, except Maneri Bhali Scheme, Stage-I commissioned in 1984 for 90 MW. A number of schemes proposed to tap the power potential are either under advanced stage of investigation or in construction stage. Pala Maneri Hydel Scheme, the fourth in the series of systematic development, is a run off river scheme just upstream of Maneri Bhali Scheme Stage-I. The Scheme envisages construction of a 74m high concrete gravity dam near Pala village, a 11.3Km long and 6m equivalent diameter horse-shoe shaped headrace tunnel on right bank and a surface power house at Aungi village. The power house with an installed capacity of 416MW will utilise a gross head of 364m. The scheme is in the detailed investigation stage.

GEOLOGY OF THE AREA

Two main groups of rocks (both of Protero-zoic age) - the Central Crystalline Group and the Garhwal Group are exposed in the area. The rocks of two groups are separated by the Main Central Thrust (MCT) - Figure-I.

The rocks of the Central Crystallines occupying the area upstream of the confluence of Kumalti and Dugadda gads with river Bhagirathi, show considerable migmatisation giving rise to various types of gneisses and schists. Quartz-biotite, chlorite and sericite schists with bands of porphyritic gneiss have been found to occur near the MCT zone. The Garhwal Group comprising quartzite and metabasics (epidiorite) occupies the area west of Kumalti gad. Quartzite in general, is massive and moderately to closely jointed. However, crushing and shearing in the rock has been noticed near the MCT zone. Metabasics (epidiorite) have intrusive relationship with the host rock and is a metamorphic derivative of dolerite. The medium to coarse grained rock is, in general, massive and hard. However, foliated nature of rock has also been noticed at a few locations. Porphyroblasts of felspars have been seen upstream of power house site.

The MCT is a major regional tectonic discontinuity intersecting, besides the Bhagirathi valley, the Bhillangana, Mandakini and Alaknanda valleys. The extent of crushing and brecciation of rocks along the thrust is, however, variable in different areas (Pant and Kumar, 1972). In Bhagirathi valley, the thrust has been interpreted to be along Kumalti and Dugadda gads. In this section the MCT is nowhere clearly exposed and is generally concealed by terrace gravels/landslide deposits and the zone appears to be healed up by basic intrusion. In order to explore the width of this poor zone and its tunnelling conditions, an exploratory drift from right bank of Kumalti gad was driven at an elevation of 1452m (the tunnel grade being El±1599m). The drift has intersected basic rock, breccia-ted quartz felspathic rock, sheared quartz-sericite and chlorite schists and xenoliths of quartzite. It has so far proved 100m width of the zone. In another 15m of its proposed extension, it is anticipated that the quartzite will be pierced through and the total width of the crushed/sheared zone will be known (Shome & Narula, 1969; Dhanota, 1971).

GEOTECHNICAL APPRAISAL OF HEADRACE TUNNEL

The proposed 11.3Km long headrace tunnel alignment on the right bank of river Bhagirathi crosses three major perennial drainages Viz., Papar gad, (cover 51m), Nahar Gad (cover 520m) and Kumalti gad (cover 75m). The rocks to be encoun-tered in the tunnel upto Kumalti gad comprise gneisses/migmatites with schistose interbeds and metabasics. The rest of tunnel will be driven through quartzite and metabasics. The zone at tunnel grade is expected to contain massive to foliated metabasic interpreted to be a post thrust intrusive body, brecciated quartz-felspathic rock, sheared quartz-sericite and chlorite schists impregnated with intense silicification, and xenoliths of quartzite and gneisses in metabasics.

The length wise distribution of rock units expected to be met along the proposed tunnel alignment and maximum and minimum cover above the tunnel grade in each rocktype is tabulated in Table-I.

The strike of foliation in gneisses is variable due to folding. In Pala dam site-Papar gad section, the foliation

TABLE - I

Rocktype to be Intersected at Tunnel Grade.	Length (% of total)	Maximum cover (Elevation)	Minimum cover (Elevation)
Gneisses with Schistose interbeds	7175 m (63%)	750 m (2360 m)	51 m (1680 m)
Quartzite	780 m (7%)	365 m (1960 m)	102 m (1700 m)
Metabasics including MCT zone	3345 m (30%)	425 m (2020 m)	170 m (1760 m)

TABLE - II

S.No.	Discontinuity set.	Strike	Range of and quadrant of dip.	Sets identified in Plate - II.
1.	I	N45°-70°W	55°-90°/SW&NE	$Jg_1, Jg_2, Jb_1, Jb_2, Jq_2$.
2.	II	N15°E	75°-90°/SE	Jg_5, Jq_6 ,
3.	III	E	25°-35°/N	Jq_5, Jb_4
4.	IV	N40°W	35°-90°/NE&SW	Jq_1, Jb_3, Jb_5
5.	V	N	35°/E&W	Jq_3, Jq_4
6.	VI	N50°E	65°/NW	Jg_4
7.	VII	N60°W	20°/NE	Jg_3

strike will subtend an angle of 17° to 78° with the tunnel axis. The tunnel axis between Paper gad and Kumalti gad will be parallel or 27° skewed to the strike of foliation. The schistose interbeds and foliation shear seams/zones will be daylighted for longer length of the tunnel where foliation plane is subparallel to the tunnel axis. Such zones particularly when saturated are likely to present acute problems of tunnelling. Cavity and high overbreak formation are most likely in these zones. The rocks of the Central Crystallines are traversed by various sets of joints, in addition to foliation and shear zones. As many as five major sets of joints have been deduced with the help of stereonet (Fig.1 left inset A). The strike of bedding in quartzite undulates from N 35°W-S35°E to N40°E-S40°W to E-W with rolling dips of 10° to 40° towards north-east, north-west and north. The tunnel axis in quartzite will, therefore, be at an angle of 13° to 88° to the strike of bedding. The discontinuities in the rock have

been plotted on stereonet and six major joint sets deduced (Fig.I left inset B). The foliation in metabasics trends in N75°E-S75°W to N40°W-S40°E with 20° to 35° dips towards NNW and north-east. The tunnel in metabasics will intersect the foliation at 23° to 88°. Five major sets of joints have been established with the help of stereonet plottings (Fig.1 left inset C) (Srivastava, 1985,86).

These 16 major sets present in the three different rocktypes, if combined, give 7 major sets. These sets are tabulated in Table-II.

It would be rare that all these sets are present at a particular location in a rock unit. Even so, the discontinuity sets have been displayed in plan and section of the tunnel in gneisses, quartzite and metabasics for a better appreciation (Fig.1 right inset ABC). One or more of these joint sets will positively be present at each location of the tunnel in their respective rock units. Fig.I right inset ABC picks out the vulnerable joints which are likely to form wedges and be responsible for over-breaks. For instance, presence of joint set Jg-1, Jg-2 and Jg-3 in gneisses will be prime cause for over-breaks in the crown region. Set Jg-3 being low angle joint may work as plane of mobilisation of huge rockmass during tunnelling operations from inlet end. If the joint plane contains saturated infilling reducing the frictional force of the plane, the mobility of the rockmass may be instantaneous. In quartzite, joint set Jq-1 and Jq-6 are likely to subtend wedges at the crown. The overbreaks due to these will be continuous in the longitudinal direction. Sets Jq-3 and Jq-4 are low dipping and skewed at 37° to the tunnel alignment. Therefore, these are likely to cut wedges of large width and length. In metabasics joint sets Jb-1 and Jb-4 may be responsible for considerable overbreaks at the crown, while plane of set Jb-3 being low angle joint will have sliding tendency, in case of tunnel being driven from inlet end.

ESTIMATION OF ROCK STRESSES

It is evident from Table-I that the headrace tunnel on the proposed alignment will have minimum and maximum rock cover of 51m and 750m in gneisses, 102m and 365m in quartzite and 170m and 425m in metabasics. Assuming average densities of gneisses, quartzite and metabasics to be 2.7gm/cm^3 , 2.6gm/cm^3 and 2.9gm/cm^3 respectively, the maximum rock cover of 750m, 365m and 425m shall develop vertical stresses (σ_1) of 203Kg/cm^2 in gneisses, 96Kg/cm^2 in quartzite and 123Kg/cm^2 in metabasics. The compressive strength (σ_c) of gneisses ranges from 800 to 2500Kg/cm^2 (average **1650** Kg/cm^2). The vertical stresses exerted over the cavity in each rock type are much less than the respective values of compressive strength and therefore, but for loosening pressure, chances for the development of squeezing pressure along the vertical component is remote. The stresses along horizontal direction is expected to be around 20% of the vertical stresses; therefore, squeezing pressure on the side walls is also not anticipated. The competence factor 'FC' ($\frac{\sigma_c}{\sigma_1}$) in each rock type is around 10 or more. This is the situation wherein ground is capable of standing unsupported (Alan Marshall, et al. 1972). However, schistose rocks, shear zones, clay seams and

the MCT zone in particular, under saturated conditions, are most likely to exhibit acute tunnelling problems.

TUNNELLABILITY ON THE BASIS OF 'Q' VALUES

Quantification of rockmass quality for tunnelling of the proposed headrace tunnel has been attempted on the 'Q' System of Barton et al. (1974). Joint set frequency, spacings and other characteristics have wide range of variations from place to place in each rocktype, ratings of the geotechnical parameters will thus vary. Keeping this in view, a number of blocks each of 1m^2 have been taken in each rock unit, care being taken to deliberately sample the best and worst variants and characteristic details with regard to the parameters recorded. Out of 13 blocks mapped, 6 blocks have been selected in gneisses, 3 blocks in quartzite and 4 blocks in metabasics of which 2 each are shown on Fig.II. The ratings of geotechnical parameters recorded in each block, the respective rockmass quality index 'Q', expected support pressure at the roof of the tunnel and general tunnelling conditions have been depicted in these figures.

The above studies in case of schistose rocks could not be carried out on

TABLE - III

Rock Type	Schistose Interbeds within Gneisses	Crushed/Sheared/Breccia-ted rock in MCT		Zone
RQD	30	30	To	10
Jn	2	2	To	20
Jr	2	1.5	To	1
Ja	4	4	To	6
Jw	0.66	0.66		
SRF	2.5	2.5	To	5
Q	2	1	To	0.01
P(roof)	0.4Kg/cm^2	0.6Kg/cm^2	To	9Kg/cm^2
Tunnelling condition	Very poor	Very poor To Extremely poor		

blocks. A few surface outcrops of schists in gneisses and the crushed/ sheared/brecciated schistose rocks met with in the exploratory drifts at Kumalti gad have been studied and suitable ratings of the geotechnical parameters assigned for evaluating 'Q' values. These are given in Table-III.

Gneisses with Q-Values of 6 to 43 (excluding schistose interbeds) offer fair to very good tunnelling conditions. Schistose interbeds corresponding to Q-values of around 2 are likely to present poor to very poor tunnelling media in case of enhanced ingress of water. The maximum width of schistose rocks expected to be intercepted by the tunnel along the proposed alignment is of the order of 700m i.e., about 10% of the central crystallines in the area encompassed by the scheme. Tunnel- ling conditions in quartzite with Q-values ranging from 8 to 39 is expected to be fair to very good. The Q-values of 6 to 41 in 90% of metabasics suggest that tunnelling medium will be fair to very good while 10% of the rock

which comprises foliated metabasics (6%) and MCT zone (4%) would represent very poor to extremely poor tunnelling conditions having standup times of even less than 1 hour. Such type of rocks have Q-values of 1 to 0.01. In order to identify support system for the tunnel, Barton's graph depicting relation between equivalent dimension 'De' and rock mass quality index 'Q' has been used, taking 'De' as 4 for the proposed tunnel. It is clear that for Q-values of 6 to 43, no support is required for the proposed tunnel. However, for Q-values between 2 and 6, the nearest support category corresponds to Box 21 of Barton under poor rock conditions with spot bolting to bolting with shotcreting (20-30mm) and for Q-values between 0.01 and 1 the rock can be classified as extremely poor to very poor and the support category corresponds to Box-31 and 34 for which cast concrete lining (20-60cm) with tensioned bolts have been recommended. The results of all the foregoing studies have been synthesised and presented in Table-IV.

TABLE - IV

| S.No | Parameters | GNEISSES | | QUARTZITE | METABASICS | |
		Massive	Schistose		Massive	Schistose
1.	Q	6 to 43	2	8 to 39	6 to 41	1 to 0.01
2.	Jn	3 to 6	2	3 to 12	6 to 12	2 to 20
3.	P-roof	0.16 to 0.5Kg/cm^2	0.4Kg/cm^2	0.1 to 0.66Kg/cm^2	0.23 to 0.88Kg/cm^2	0.6 to 9Kg/cm^2
4.	Category	I (Fair to very good)	II (Poor)Box-21	I (Fair to very good)	I (Fair to very good)	III (Very poor to Extremely poor) Box 31, 34
5.	Length of tunnel	6458 (57.15%)	717m (6.34%)	780m (6.90%)	3210m (28.39%)	135 (1.22%)
6.	Anticipa- ted support require- ments.	Unsupported	Spot bolt- ing to bolting with shotcre- ting (20 to 30mm)	Unsuppor- ted	Unsupported	Cast concrete lining (20-60cm) +tensioned bolts.

CONCLUSIONS

Thus on the basis of a purely geological evaluation nearly 93% of the tunnel length shall be self supporting where no support system shall be required. In nearly 6% of the tunnel length, only spot bolting and shotcreting is required whereas in 1% length i.e., close to the MCT zone, cast concrete lining with tensioned bolts shall be required.

In addition to the support requirements in the jointed gneisses, quartzite and metabasics overbreaks due to wedge failures (as a result of joint intersections) may occur. The authors feel that in such situations concurrent shotcreting (though not required as a structural support) may perhaps be beneficial and obviate holdups due to rock falls. They have had similar experience during the Maneri stage II tunnel wherein the overbreaks have been recorded only in the hard and brittle rocks like quartzites and such phenomenon has not been observed in the hard and massive metabasics.

However, in this analysis there is no input regarding the likely hydrogeological conditions. It is an universally accepted fact that water is the single largest contributory factor for tunnel stoppages and hold ups (Like flowing conditions, heavy ingress of water leading to flooding etc.). At this stage it can be said that below the 2 major drainages as discussed earlier, the cover is low and all being perennial drainages, certain quantum of water ingress at these locations cannot be ruled out. During actual tunnel excavations at least from a distance of 40m on either side of these drainage crossings advance probe hole be drilled to assess the conditions. It is also suggested that during actual tunnelling a careful documentation bylogging of the tunnel be concurrently carried out for evaluating the departures, if any, from the conditions forecast in Table-IV which could help in refining the techniques adopted for evaluating the different parameters.

ACKNOWLEDGEMENT

The authors are grateful to the Director General, Geological Survey of India, for his kind permission to publish this paper. They are also thankful to the project authorities for their assistance during the data collection.

REFERENCES

Barton, N., R. Lien and J. Lunde, 1974, 'Engineering classification of rock masses for the design of tunnel supports', Rock Mechanics, vol.6, No.4, pp. 189-236.

Barton, N., R. Lien and J. Lunde, 1976, "Estimation of support requirements for underground excavations". Proc. of 16th Symp. on design Methods in Rock Mechanic Minnesota, ASCE, N.Y.1977, pp.163-177. Discussion pp. 234-241.

Alan Marshall, Huir Wood, 1972, "Tunnels for Roads and Motorways", Quarterly journal of Engineering Geology Vol.5, No.1 & 2.

Shome, S.K. & Narula, P.L. 1969, "Reconnaissance geological report on the Loharinag Pala Hydel Project and Pala Maneri Hydel Project, Uttarkashi dist. Uttar Pradesh. Unpublished report of Geological Survey of India.

Dhanota, A.S., 1971, "Progress report No-1. On preliminary geological investigation of the Pala Maneri Hydel Schemes, Uttarkashi dist. U.P." unpublished report of Geological Survey of India.

Pant, C. & Kumar, V. 1972 "Progress report No-2 on the geotechnical studies for the Pala Maneri Hydel Scheme Bhagirathi river, Uttarkashi distt. U.P.", Unpublished report of Geological Survey of India.

Srivastava, S.C., 1985, "Progress report No.14, on the geotechnical studies for the Pala Maneri Hydel Scheme, Bhagirathi river, Uttarkashi distt., U.P. Unpublished report of Geological Survey of India.

Srivastava, S.C., 1986, "Progress report No.15, on the geotechnical studies for the Pala Maneri Hydel Scheme, Bhagirathi river, Uttarkashi distt., U.P." Unpublished report of Geological Survey of India.

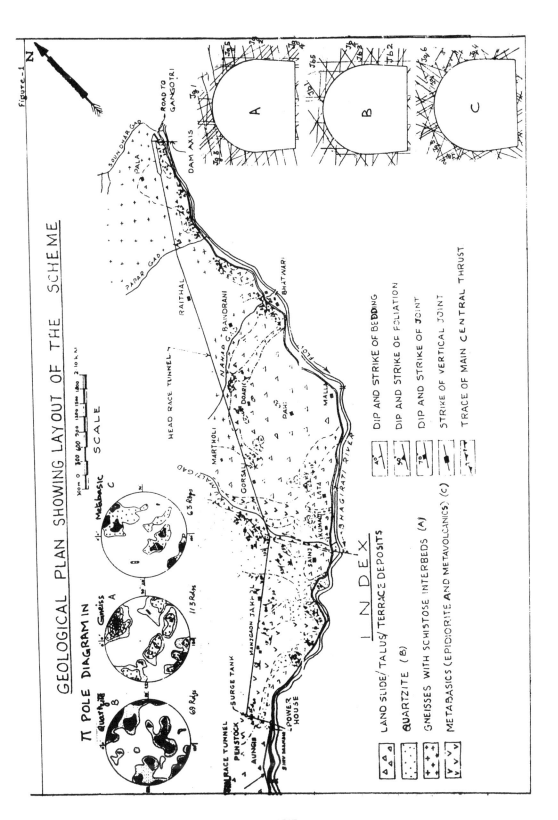

GEOLOGICAL PLAN SHOWING LAYOUT OF THE SCHEME

Figure-1

INDEX

△ LAND SLIDE/TALUS/TERRACE DEPOSITS

QUARTZITE (B)

+ GNEISSES WITH SCHISTOSE INTERBEDS (A)

V METABASICS (EPIDIORITE AND METAVOLCANICS) (C)

DIP AND STRIKE OF BEDDING

DIP AND STRIKE OF FOLIATION

DIP AND STRIKE OF JOINT

STRIKE OF VERTICAL JOINT

TRACE OF MAIN CENTRAL THRUST

π POLE DIAGRAM IN

Quartzite B — 69 R₃ψs

Gneiss A — 113 R₃ψs

Metabasic C — 63 R₃ψs

SCALE
300m 0 300 600 900 1200 1500 1800 2 10 KM

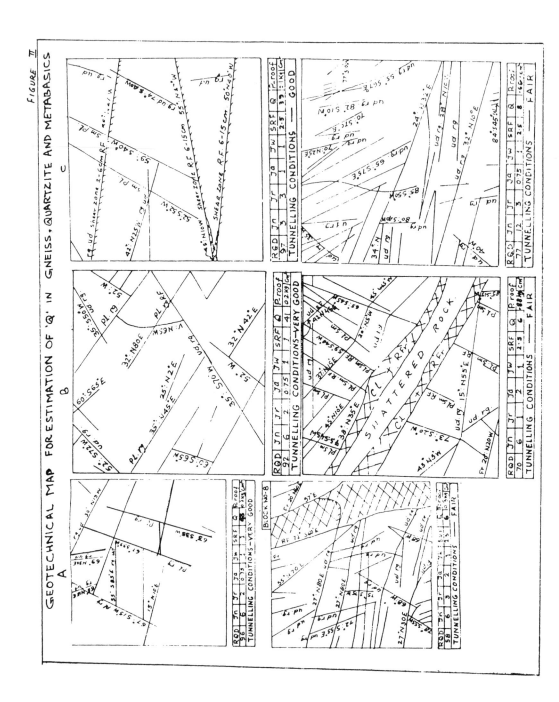

GEOTECHNICAL MAP FOR ESTIMATION OF 'Q' IN GNEISS, QUARTZITE AND METABASICS

FIGURE II

Srinagar thrust in the Maneri state-II tunnel – The anticipated tunnelling problems and the actual behaviour (Uttar Pradesh/India)

La poussée Srinagar à Maneri étape-II tunnel – Les problèmes anticipés de la construction de tunnel et le comportement réel

S.K.Shome, J.S.Rawat & R.S.Negi
Geological Survey of India, Dehra Dun, India

ABSTRACT: The Maneri Bhali stage-II hydel project presently under construction involves tunnelling of 16Km long, 6m diameter headrace tunnel. The tunnelling is being done by drill + blast, and nearly 50% of the length has been completed. The tunnel pierces through rocks of Garhwal group (Older Proterozoic) and the Tehri formation/Simla slates (Jaunsar-Simla Group–younger Proterozoic). The contact between these two groups of rocks is marked by a regional tectonic dislocation, designated as the Srinagar thrust.

The paper presents the apprehensions of tunnelling in this zone (Srinagar thrust) through flowing rock conditions with heavy inrush of water over an anticipated 250m (approx.) wide stretch. These assessments were based on surface mapping in a largely landslide debris and scree covered area and meagre sub-surface explorations by way of drilling.

Now that the tunnel has crossed this zone the actual tunnelling conditions have been documented. An attempt has been made to present the reasons for the variations in the actual tunnelling behaviour from that earlier predicted and to enumerate the lessons learnt from this exercise.

RESUME: Le project 'hydel' de Maneri Bhali étape-II, actuellement sous construction enveloppe la construction de tunnel de 16Km long et de 6m en diamètre. La construction de tunnel est faite par du forage et de l'exlosion, complétant 50% de longuer environ. Le tunnel a pénétré les roches de Groupe Garhwal (Proterozoique-ancien) et les formations de Tehri/ ardoise de Simla (Le Groupe de Jaunsar-Simla, Proterozoic-Jeune). Le contact entre les deux groupes des roches sout indiqué par de disloquer techtonique régionale, désigné comme la poussée Srinagar.

La publication donne une apprehension de la construction de tunnel de cette zone (poussé Srinagar) pardes conditions découlement des roches arec l'affluence de l'eau sur létandre anticiper de 250m (environ). Les évolution out été base sur le plan de surface d'aire au débris déboulement et du talus et très peu l'exploration sous-surface par le forage. Maintenant, le tunnel a travarcé celte zone donc les conditions présent de la construction de tunnel sont documentés. Nons avonsessayés de présenter les raisons pour des variations au comportement de la construction de tunnel réel par centre l'anticipé anciennement et d'énumerér les laçons prie de cet exercise.

INTRODUCTION

Srinagar Thrust is a regional tectonic lineament in the Garhwal Himalaya, U.P., India, having a strike continuity of over 100 of kilometre, a zone of disturbance/shattering on the surface nearly 200-300 m in width and a ruling dip of the order of 40° to 60° in the south-westerly quadrant. It juxtaposes the rocks of the Garhwal Group (Older Proterozoic) against the rocks of the Tehri formation/Simla slates (Jaunsar-Simla Group, younger proterozoic). It is believed to be the strike extension of the North Al-mora thrust in the east i.e. towards the Kumaon hills and the Tons thrust (believed to be the folded counterpart of the Krol thrust, Auden 1934) in the Tons valley to the west.

The Maneri hydel project had been conceived of in the early part of sixties when the total hydraulic head available between Maneri and Bhali was proposed to be developed in one single phase. However, in March 1981 it was split up in 2 phases, the Maneri stage I between Maneri-Tiloth (opposite Uttarkashi) and the Maneri Stage II betweenJoshiyara-Dharasu (Fig.1). From the initial stages the various workers have highlighted the apprehensions and problems of crossing this regional tectonic lineament based on purely geologic and other related evidences to problems of actual tunnelling behaviour

At this stage it may not be out of place to record certain basic factual information which have imposed serious and often crippling limitations in the extrapolations. The rugged relief

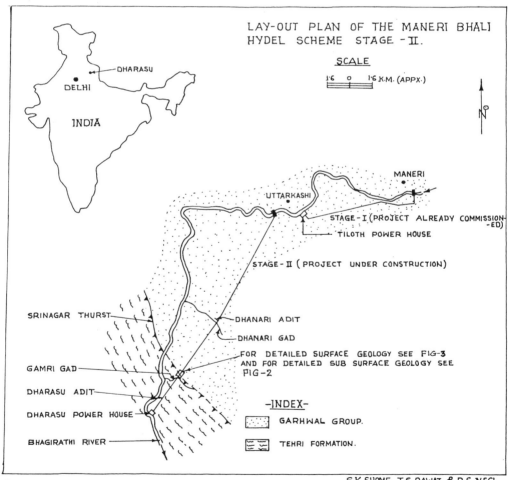

LAY-OUT PLAN OF THE MANERI BHALI HYDEL SCHEME STAGE - II.

SCALE

1·6   0   1·6 K.M. (APPX.)

DHARASU

DELHI

INDIA

MANERI

UTTARKASHI

STAGE-I (PROJECT ALREADY COMMISSION-ED)

TILOTH POWER HOUSE

STAGE-II (PROJECT UNDER CONSTRUCTION)

SRINAGAR THURST

DHANARI ADIT

DHANARI GAD

FOR DETAILED SURFACE GEOLOGY SEE FIG-3 AND FOR DETAILED SUB SURFACE GEOLOGY SEE FIG-2

GAMRI GAD

DHARASU ADIT

DHARASU POWER HOUSE

BHAGIRATHI RIVER

-INDEX-

GARHWAL GROUP.

TEHRI FORMATION.

FIG. 1

S.K.SHOME, J.S.RAWAT & R.S.NEGI.

2520

of the proposed tunnel area:1) the Bhagirathi(R) flows at a level of EL ± 850m in the vicinity of the tunnel layout, the hills rise to over EL 2000 m, (2) the projected tunnel route on the ground even today when the tunnel is halfway to completion is still largely inaccessible, (3) the area is moderately to heavily forested, and (4) extensive deposits of scree, landslide debris mask the area of exposure of the Srinagar thrust along the tunnel route on the ground. To further confound the issue the performance of drilling to explore the zone of the Srinagar thrust during the investigation stage has been very much below average with core recovery being seldom above 30-40%. This made any reliable interpretation very difficult and only helped to broadly delineate the zone.

APPREHENSIONS

In the early sixties when the project was conceived there was a dearth of experience in actual tunnelling conditions through comparable setup in Himalaya with the result that the recommendations then made were not only generalised but had a greater bias on the conservative side (Srivastava 61-62, Pant 63-66, Pant and Narula 66-67 and Shome and Narula 67-68).

However, by the late sixties and early seventies the Yamuna and Giri tunnels having been initiated lead to a revolutionary change in the geological appraisal and engineering expertise in the field of tunnelling in the country for tackling squeezing ground and negotiating active faults/thrusts. Dhanota (69-73), Dhanota and Battacharya (72-73), Jaitle (73-77), on the basis of ongoing explorations had summarised different aspects of the anticipated tunnelling problems. These were considered to range from moderate to heavy inrush of water, flowing conditions in a zone of nearly 100m width at the tunnel grade.

Shome & Kumar (78), after reviewing the behaviour of a part of the rock types of Garhwal group met in the Maneri Stage-I, extrapolated the tunnelling conditions likely to be met during Maneri Stage II purely on the basis of the geological appraisal. They report the width of the Srinagar thrust zone likely to be met in the tunnel to be of the order of 250m. On the basis of the anticipated crossing of this feature below a perennial drainage

(the Gamri Gad) they apprehended flowing rock conditions coupled with heavy inrush of water and suggested further detailing by way of drilling in the Gamri Gad area. They had put this zone (the Srinagar thrust zone) in the category C3 i.e. intensely jointed and sheared with an anticipated standup time of less than 4 hrs. They further anticipated a rock load of over 8m and had recommended heading and benching method of excavations with immediate shotcrete on chain link fabric followed by steel sets at 0.2 m c/c with initial concrete.

Subsequently, Rajgopalan and Sondhi (78-79) and Rajgopalan and Agarwal (79-80) Kumar, Rajgopalan and Agarwal (80-81) reported the progress of explorations in the Gamri Gad. In all, four drill holes aggregating to a cumulative depth of 160m were put down, but no conclusive interpretations were possible about the Srinagar thrust zone. Vinay Kumar, Rajagopalan, Sushil Kumar and Agarwal (1981-82) report that the full recommended explorations have not been completed.

Rawat & Srivastava (83-87) attempted to decipher and delineate the nature and width of thrust zone in the Gamri Gad area and also reiterated the need for completing the pending explorations. They inferred that the Srinagar thrust zone shall be met between RD 1550m (the running RD of the tunnel) and 1900m.

However, a critical reappraisal of the map by Rawat (Fig.3) shows that assuming a ruling dip of the order of 40°-60°/SW the maximum width of the thrust zone at the tunnel grade would be about 375m taking the maximum width of surface (outcrop) separation of the two rock units. Moreover, the Srinagar thrust zone should have been encountered between RD 1520m and 1900m approx. The total downdip projection being of the order of 300m even a minor change of dip would make a significant variation in the RD's forecasted.

ACTUAL BEHAVIOUR

The tunnelling work for negotiating the Srinagar thrust in the Maneri Bhali Stage II project was started in middle of 1982 from the Dharasu and (Fig.1). Till the end of the May'89 a total of 8.930m has been completed out of the total length of 16 km. The tunnelling work is being carried out of the total length of 16 km. The tunnelling work is being carried out through the five headings (1 at Joshiyara, 2 at Dhanari Gad adit and 2 at Dharasu adit). Due to labour strike and holdups to tackle a collapse the work of the Dhanari Gad and Dharasu adit were suspended for nearly 2 years and 1 year respectively. The annual rate of progress for the project (excluding the holdups) works out to 1.68 km/yr. from the 5 headings which is roughly 30m/month/face. This rate of progress even in the Indian context of tunnelling by drill + blast with old equipment is very slow. Now with the length of haul distances gradually increasing coupled with the need for more and more efforts on ventilation the rate of progress is bound to fall, unless more input by way of new machinery and equipment is provided.

From RD 1500m onwards in view of the apprehensions of serious tunnelling problems advance in the tunnel excavations were carried out cautiously. The tunnel went on advancing smoothly without any problem raising serious doubts about the existence of the thrust itself in the minds of engineers. Rock conditions started deteriorating from RD 1700 onwards (Fig.2) when the greywackes and phyllites of the Tehri formation were gradually becoming more and more closely jointed, shattered and even sheared till a nearly 34m wide zone of sheared rock was met between RD 1766 and 1800m. The zone was a chaotic assemblage of crushed phyllite, sheared metabasics and closely jointed to sheared milky white to grey ortho-quartzites. This marked the trace of the Srinagar thrust wherein the phyllites/greywackes of the Tehri formation were juxtaposed against the quartzites of the Garhwal Group. There was minor dripping of water. The rock was of poor quality but had sufficient standup time for the full circular ribs spaced at 0.5m apart to be installed. The most astonishing feature was that after RD 1800m (Fig.2) the tunnel entered into closely jointed metabasics which gradually with each advance became more and more massive. The rock improved to such an extent that the supports were dispensed with and stood unsupported between RD 1860 and 1905m (Fig.2) for nearly 40-60 days.

A closely jointed to moderately sheared zone of about 20m width was again met between RD 1890 and 1910m. In this zone the metabasics had thin bands of sheared chlorite-talc schist. At RD 1908 to 1911m the other end of the Srinagar thrust was met, wherein the metabasics are in a faulted contact with the massive white quartzites of the Garhwal Group. The rocks in this reach between RD 1890 and 1910m, though closely jointed to moderately sheared, behaved quite well as a tunnelling media. It was also observed that the zone of shearing associated with this thrust zone was more pronounced in the metabasics (nearly 18m wide and was only nominally affecting the quartzites, only 2m wide). The rock met in the tunnel between RD 1700 and 1920m have been logged on 1:200 scale; but the log has been replotted on 1:1000 scale (Fig.2). The NGI or the 'Q' classification of the various sectors have been carried out which clearly shows that in the reach between RD 1700 and 1770m support by way of systematic bolting at 1 to 1.15m apart is necessary. However, in the absence of the facility for installing systematic bolting circular steel ribs at 0.5m c/c have been installed. In the rest of the reach the 'Q' values being very good practically no support is required; but as a measure of abundant caution ribs have been installed 40 to 60 days after excavation, as the permanent lining shall take at least 150-180 days more for emplacement.

DEPARTURE FROM THE FORECAST

Thus it is seen that the apprehension of a nearly 250m wide zone of sheared and shattered rock zone with trapped water was completely belied. In fact, what was actually met was nearly 34m wide poor rock zone between RD 1766 and 1800m, followed by a 90m wide zone of moderately jointed to massive metabasics, followed again by a closely jointed to moderately sheared rock zone between RD 1890 and 1910m. Thus the entire Srinagar thrust zone was only 144m wide at the tunnel grade met between RD 1766 and 1910m, as against the forecasted between RD's 1550 and 1990m. The dip of the Srinagar thrust as recorded in the two separate shears were of the order of 48° to 60° in SW quadrant. The tunnelling conditions can broadly be classified as fair to good in this reach with no ingress of water.

EXPERIENCES GAINED AND DISCUSSIONS ON POSSIBLE REASONS FOR DEPARTURE.

Analysing the data of the tunnel the following broad conclusions can be drawn.

1. The Srinagar thrust zone is only about 144m wide at the tunnel grade.

2. The thrust has a rather steep dip of the order of 48° to 60° towards SW.

3. It has been healed up in the Central portion by a later volcanic intrusive of about 90m width. In view of limited exposures along the tunnel route this fact could not have been visualised during the investigations.

4. There appears to have occurred some amount of reactivation of this megashear in so far as at both ends zones of 34m and 20m width are closely jointed to heavily sheared in which even part of the central metabasics are also involved.

5. The fact that it has been healed up by the later volcanic intrusive (which has since been metamorphosed) has made it to have a more or less similar overall permeability to that of the unsheared rocks. This has not allowed the water to be entrapped on one side of the megashear (as is normally wont to happen e.g. Krol thrust in the Yamuna stage-II tunnel).

6. As far as the forecasted RD's are concerned one end has matched quite well i.e. RD 1900 against the actual RD 1910m.

7. In so far as the forecast of the tunnelling conditions are concerned, the departure has been very drastic; but fortunately the problem having been overestimated there was smooth sailing during the actual crossing of the thrust zone in the tunnel.

8. In hindsight as against the attempt to explore the Gamri Gad area by a series of drill holes if an exploratory drift of nearly 400m length was driven on the northeastern side of the Gad it could have very precisely delineated the tunnelling conditions. However, this is more easily said than done because at the preliminary stage to get to convince the engineers-in-charge of planning to drive a 400m long exploratory drift is next to impossible (authors' opinion).

9. Possibly with the increasing knowhow of geophysical exploration more positive inferences could have been drawn by carrying out geophysical surveys around the Gamri Gad.

ACKNOWLEDGEMENTS

The authors are grateful to the Director General, Geological Survey of India for his kind permission to submit the paper for publication. They are thankful to the project authorities for their assistance during the data collection.

REFERENCES

Auden, J.B. (1934) - The Geology of the Krol Belt, Rec.Geol.Surv.Ind., V.67, pp. 357-454.

Dhanota, A.S. (1968-1973) - Unpublished GSI progress report on Maneri Bhali Stage-II.

Dhanota, A.S. and Bhattacharya, S.K. (1972-73) - Unpublished GSI progress report on Maneri Bhali Stage-II.

Jaitle, C.N. (1973-77) - Unpublished GSI progress reports on Maneri Bhali Stage-II.

Kumar, Vinay, Rajagopalan, G. and Agarwal, K.K. (1980-81) - Unpublished GSI progress report on Maneri Bhali Stage-II.

Kumar, Vinay, Rajagopalan, G. Kumar, Sushil and Agarwal, K.K. (1981-82) - Unpublished GSI progress report on Maneri Bhali Stage-II.

Krishnaswamy, V.S., Jalota, S.P. and Shome, S.K. (1970) - Recent crustal movements in Northwest Himalaya and Gangetic foredeep, and related pattern of seismicity. Proc.4th Symp.Earthquake Engineering, Roorkee University, Roorkee, pp. 419-439.

Pant, G. (1963-66) - Unpublished GSI progress reports on Maneri Bhali Stage-II.

Pant, G. and Narula, P.L. (1966-67) – Unpublished GSI progress reports on Maneri Bhali Stage-II.

Rajagopalan, G. and Sondhi, S.N. (1978-79) – Unpublished GSI progress reports on Maneri Bhali Stage-II.

Rajagopalan, G. and Agarwal, K.K. (1979-80) – Unpublished GSI progress reports on Maneri Bhali Stage-II.

Srivastava, K.N. (1961-62) – Unpublished GSI progress reports on Maneri Bhali Stage-II.

Shome, S.K. and Narula, P.L. (1967-68) – Unpublished GSI progress reports on Maneri Bhali Stage-II.

Shome, S.K. and Kumar, Vinay (1978-79) – Unpublished GSI progress reports on Maneri Bhali Stage-II.

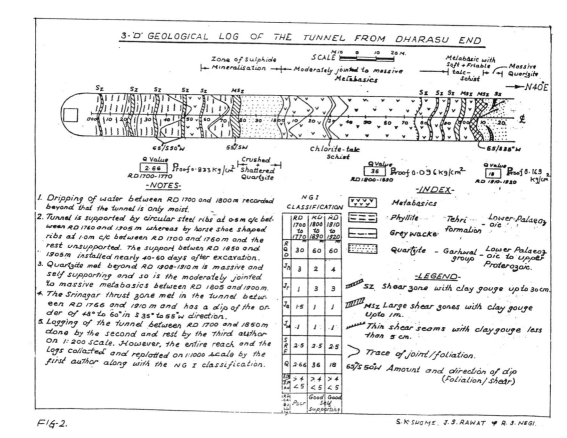

FIG. 2.

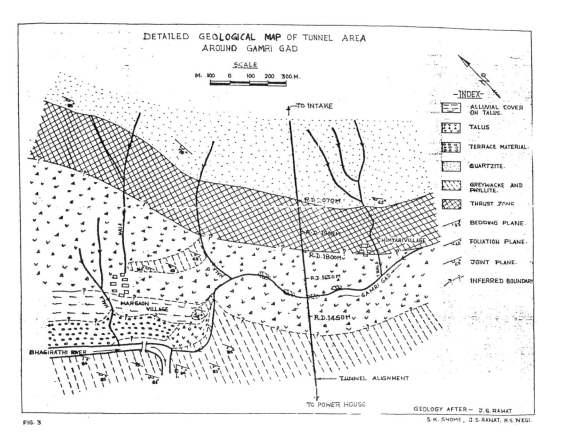

DETAILED GEOLOGICAL MAP OF TUNNEL AREA
AROUND GAMRI GAD

FIG. 3

GEOLOGY AFTER— J.G. RAWAT

S.K. SHOME, J.S. RAWAT, R.S. NEGI.

Difficulties in deep foundation of underground sewer in granodiorite massif of Brno

La réalisation d'une galerie de canalisation dans le corps de terrain granodiorite de Brno

J.Šmíd
Geotest, Brno, Czechoslovakia

ABSTRAKT: In this paper an example of the drivage of a sewer channel is described situated within most complex geologic conditions. In spite of a comparatively detailed and perfect preliminary geotechnical exploration a breakdown with very serious, fatal consequences took place. In this paper some accepted and realized recovery measures are described.

RÉSUMÉ: Dans cet article un exemple de la percée d'une galerie de canalisation est introduite située en comditions géologiques très complexes. Malgré la prospection géologique initiale très détaillée et qualifiée un incident avec des conséquences catastrophiques avait lieu. Cet article décrit des mesures d'assainissement acceptées et réalisées.

The problems joined with the inconvenient state of the sewer network together with the necessity of its extensive reconstruction continue to come into the viewpoint of czechoslovak experts more frequently in accordance with the interest of the environmental protection.

One of the examples where a very serious breakdown occurred during the reconstruction of a sewer network was the drivage of a collector channel for a sewer built within a granodiorite rock of Brno. The channel was driven for to transfer the surface waters of the little river Ponávka from the Svratka-watershed into the channel of the Svitava-river. The river Ponávka was one of three streams crossing the town of Brno was vaulted in the past and today it flows through Brno under the surface of the ground. Some sewer collectors lead into its underground flow. Heavy showers used to flood the whole sewer system the water ascending above the ground and overflowing parts of the urban park. The realization of a drainage channel separating the surface waters of the Ponávka-river would prevent dilution and cooling down of the sewage, in the same time the effect of flood waters of the Ponávka-river, influencing the capacity of the incouncenvenient urban collector network would be excluded - Fig.1 demonstrates the overall situation.

During 1982 and 1983 an engineerring geological study was realized. The situation of the relevant area is seen of Fig.2. For the channel of 3010 m lenght an extensive complex of test boring and laboratory and geophysical work was realized. It could be prooved that the designed channel line crossed a very complex geologic and tectonic structure where sound, weathered and entirely loose rocks alternated very frequently. On the basis of an exploration made in variable depths under the ground surface the surface of the granodiorite, granite and aplite bedrock of the proterozoic massiv of Brno was found. The deep bound neogene sediments (clays and sands) reached sporadically up to the channel level. The continuous groundwater table layd 32 to 38 m above the channel level. The exploration work revealed numerous expresive tectonic lines and faul-

FIG. 1   SITUATION
- - - - - UNDERGROUND BED OF THE RIVER PONÁVKA
✗   CITY SEWAGE PLANT
•••••••••• DESIGNED TUNNEL

ing was adapted and the drivage continued without any problems.

In august 1986 when the last 240 m of the total lenght had to bei driven a breakdown occurred suddenly. At about 3 m from the heading a part of the channel roof laying 55 m under the terrain collapsed. A great water-and sand afflux begun to flow into the driven work with total capacity of approx. 70 l/sec. The situation at the moment of the breakdown showes Fig.4. The consequences were fatal. By the influence of liquefying and washing of sandy sediments into the channel an extensive underground cavern resulted, the roof of which was not able to bear the weight of the overlying strata and collapsed. On the ground surface a deep crater of 20 m diameter appeared. The roadway, the pavement and the closed terrain crashed down and adjacent buildings, in the first place the building of the building of the policlinic were endangered. A special established committee accepted following objectives:

- to review and determine the safety of the policlinic and of other buildings

- to secure the saving of the underground
- to secure the saving of the surface
- to find some suitable solution for to finish the whole work.

To this purpose the area was closed for the public and the traffic was excluded. Within all exposed buildings measuring marks were situated and regular geodetic control was made. The crater was filled with soil and the engineering networks were transfered. Around the crater wells were deeped for grouting which gradually saved the underground caverns. In the same way a part of the crashed channel was saved and ultimately closed with a concrete wall. In the same time a detailed engineering-geological exploration was initiated for to complete the informations partly about the original line of the nondriven section of the channel, partly about the new suggested line by-passing the crashed part of the channel.

ted zones laying vertically to the channel direction. Fig.3 showes the longitudinal geologic section. The attention was called to the pressure of the rock and to the possibility of water discharges into the driven work.

The drivage of the channel using classic methods was opened in 1984. After initial difficulties caused by single outbursts and discharges of groundwater and sands into the driven work the technology of driv-

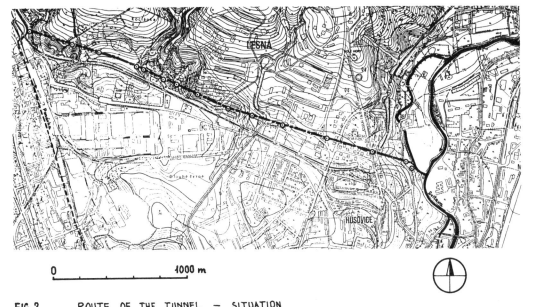

FIG. 2   ROUTE OF THE TUNNEL — SITUATION

-------- UNDERGROUND BED OF THE RIVER PONÁVKA
—·—·— ROUT OF THE TUNNEL
⊙   INVESTIGATION BOREHOLE

The complex exploration work consisted of drilling and some geophysical work joined with special control of the walls of the boreholes by aid of a periscope, drilling of fan holes together with measurements and indications of the hydrostatic pressure of the saturated zones so as detailed control and evaluation of the movement of the groundwater table within all boreholes. The detailed exploration confirmed fully the conclusions and presumptions of the initial exploration of 1983. On the other hand some new and relevant informations were acquired.

Within the interested area a flat terrain with moderat e slope of 3-4 southward was found. The elevation above sea level varied between 258 and 264 m. The mass of the solid rock showed signs of heavy tectonic processes. Above all there were expresive radial cracks pointing fron N to S, up to NNE to SSW and WSW to ENE. Along these disturbances some faults occurred in the past and an inclination of several blocks of rock

eventually. The surface of the northern solid bedrock did not up-raise practically. Within the neogene the surface of the terrain levelled at the influence of the sedimentation of sand. The local neogene sediments within the area of the channel line have a relative variable height depending on tectonic predisposed depressions of the granodiorite, differring as to their depth and width. This geologic structure finally did not influence the morphology of the terrain owing to neogene sedimentation and final deposition of loess drifts. The groundwater here was found within two water-bearing horizons- the first horizon at the interface of quarternary loesses and neogene clays, the second horizon within the layers of neogene sands. After the breakdown an important dewatering of an extensive territory and consequently the washing out of neogene sands into the driven work occurred i.e. it came to an undergroud suffosion. In this connection the overlying clays were destructed and teared up. Finally they sinked, as the pe-

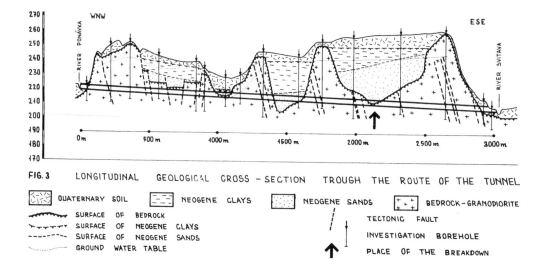

FIG.3   LONGITUDINAL   GEOLOGICAL   CROSS – SECTION   TROUGH   THE   ROUTE   OF THE   TUNNEL

QUATERNARY SOIL	NEOGENE CLAYS	NEOGENE SANDS BEDROCK – GRANODIORITE

SURFACE   OF   BEDROCK
SURFACE   OF   NEOGENE   CLAYS   TECTONIC   FAULT
SURFACE   OF   NEOGENE   SANDS   INVESTIGATION   BOREHOLE
GROUND   WATER   TABLE   PLACE   OF   THE   BREAKDOWN

riscope control of the walls proved. Consequently the clays stopped to fulfil their function of
the dividing layers. Between both
existing saturated zones local in-
terconnections appeared. Comparing
the results of the measurements
of the groundwater table and its
fluctuation it could be stated,
that the water level within the
whole area layed 8 up to 12 m
deeper in relation to the situation
found in 1983.

It was evident that both the initial and newly suggested line
(the by-pass) passed through a very
complex geologic structure. The
surface of the solid bedrock was
found to lay 242,32 m up to 207,68
m above the sea level and within
some boreholes the solid bedrock
was not found at all. In consideration of the seam floor level of
about 209 m above the sea level,
of the seam height of 3,5 m and
of the recommendations given in
the expertises, the surface of the
solid bedrock should not sink under 220 m above the sea level.
Nevertheles this condition seemed
not to be fulfilled by any of
both discussed variables of the
channel line. As to the myocene
sands they were classified as in-
groutable according to the laboratory tests. The partial compaction of these sands consists in
grouting, disturbing the environ-

mental soil - the soil cracking.
The perfect compacting would be
attained only by freezing.
The goals of the emergency
committee were fulfilled entirely.
After the saving following facts
could be stated:

a/ After the evaluation of the
gedetic measurements the stability of the policlinic and
of all controlled buildings
was demonstrated. No vertical
or horizontal movements were
recorded.
Even after several months no
cracks or other deformations
could be found.
b/ The crashed section of the
channel was closed with a
concrete dam, the caved space
was compacted by grouting.
In the same time the caverns
build during the breakdown
were grouted too. Compacting
the channel stopped the
discharge of the groundwater.
By longtermed measurements
of the water table the stabilization of the hydrogeological conditions was proved.
c/ The crater on the surface was
filled with soil, gravel and
concrete. The roadway and the
pavement were renewed and
according to the geodetic
control no deformations were
stated. The traffic was reo-

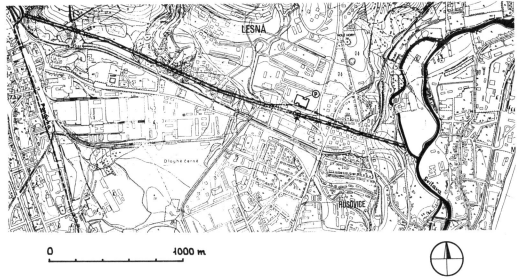

FIG. 4   SITUATION IN THE TIME OF THE BREAKDOWN AND THE REPLACEMENT
OF THE ROUTE

═══════════ DRIVEN TUNNEL
┅┅┅┅┅┅ NONDRIVEN SECTION OF THE TUNNEL
⊏ HEADING
↑ PLACE OF THE BREAKDOWN
⊙ SURFACE DEFORMATIONS
—·—·—·— REPLACEMENT ROUTE OF THE TUNNEL — BY-PASS
Ⓟ POLICLINIC

pened fully in the spring 1987.

The complex exploration showed considerable complicated geological conditions within the original channel line so as within the new suggested by-pass line of the crashed section. The presumptions

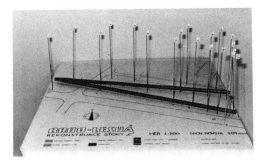

FIG. 5   SPACE MODEL OF THE GEOLOGICAL STRUCTURE BOREHOLES AND BOTH ROUTES OF THE TUNNEL

outlined by the geophysical prospecting from october 1986 i.e. the impossibility to by-pass the trenches within the solid bedrock passing vertically to the channel and filled with myocene sands so as the non-existence of a continuous surface of the solid bedrock at a suitable height, were confirmed in full extent.

The proposal to vault the crashed section of the channel and to finish the work pursuing the original line under the protection of freezing the rocky area was not accepted. The enterprise charged with the drivage choosed the alternative of a by-pass. Works continue up to now under great difficulties. The most critical point was not overcome up to this time.

On the basis of this example the evidence of the repeated necessity to realize the very particular engineering work in very complicated engineering-geological conditions was documented.

Every particular and complex geo-technical prospecting should tend towards this goal, of course in tight collaboration with the designed and the executive enterprise. Only this way the success would be ensured and the damage anticipated.

Effect of cement grout on fractured Vindhyachal sandstone

L'effet de jointure du ciment sur le grès fracturé du Vindhyachal

R. K. Srivastava, A. V. Jalota & Ahmed A. A. Amir
M.N.R. Engineering College, Allahabad, India

K. S. Rao
Indian Institute of Technology, Delhi, India

ABSTRACT : Rock grouting is commonly carried out to improve the strength properties and to decrease the permeability of the rockmass. There are many unknowns associated with the art of grouting. Mainly because it is difficult to evaluate the performance of the grouting after it has been carried out. With this in view a study has been carried out under controlled conditions). In laboratory prepared specimen (Nx size, L/D = 2), planar joints have been created oriented at 0°, 30°, 60° and 90° from major principal stress direction. Cement-sand grout (w/c = 0.7) of 3 mm thickness has been used as a filler. Ungrouted and grout jointed rock have been tested under triaxial stress condition using confining pressure up to 100 kg/cm^2 to study and compare shear behaviour of ungrouted and grouted rocks. Further applicability of Hoek-Brown and Rao et. al. strength criteria proposed for anisotropic rocks has been assessed for grout jointed rock.

RÉSUMÉ : Jointoiement des pierres est en commun exécuté pour fortifier les propriéts et de diminuer le permeabilité de la masse rocheuse. Il y a beaucoup d'inconnus associées avec l'habileté de jointoyer. Principalement parceque c'est très difficile d'éraluer l'accomplissement du jointoiement après qu'il est executé. Avec cette consideration on a fait une étude dans une condition controllee, acec du specimen préparé dans la laboratoire (grandeur Nx, L/D = 2). Des joints plats étaient crée et orientee à 0°, 30°, 60° et 90° dans la sens de tension majeure et principale. Caitance de ciment-sable (w/c = 0.7) d'épaisseur 3 mm était utiusé comme un remplisseur les roches jointes de coulis et les roches qui ne sont pas jointes de coulis étaient essayés dans une condition de tension à trois axes, en utilisant une pression jus'qu'à 100 Kg/cm^2 pour etudier et comparer l'allure des roches coules et des roches qui ne sont pas coulées. L'applicabilité du critere de resistence de Hoek-Brown et Rao proposé pour les roches (anisotropique) étaient eralués pour les roches jointes de coulis.

1 INTRODUCTION

The presence of joints, shear planes, cracks, fissures, faults and bedding planes greatly influence the strength and deformation behaviour of rock and rock masses. These geological features one one hand render the site, for construction of massive and important structures (e.g. dams, tunnels, underground power houses), unsuitable and on the other hand, evaluation and prediction of strength and deformation behaviour with limited scope because of the several unknowns and uncertainties involved.

But with ever increasing demand of civic facilities, the number of favourable sites have considerably reduced and many structures are required to be built in unfavourable geological conditions. This requires special measures to be undertaken to improve the properties of rock and rock mass. There are several geotechnical

processes in practice for this purpose. Rock grouting is one of the oldest and very commonly used process to accomplish improvement in the strength properties (consolidation grouting) and to decrease permeability of rock (curtain grouting). The whole activity of construction of massive and complicated structures in relatively unfavourable conditions (made suitable and safe by use of ground improvement techniques) makes the project capital intensive and requires techno-economically viable designs. Even small variations in assessment, prediction and designs would result in very large economic impact. Any adverse cost variations, which are in the range of millions of rupees, are very important for a developing country like India. Thus there is a need to develop the understanding and evaluation of the performance of ground improvement techniques, the present endeavour being for the rock grouting.

The information available on grouting performance is rather scarce. It has been practiced more as an art than engineering and has been carried out depending upon the personnel and particular project. In fact, to actually evaluate the performance of a grout, large scale field test would be required before and after grouting work has been carried out. But this is very complicated, expensive and time consuming with relaibility of the data obtained still in doubt (with the available techniques). Thus with the intention to develop the feel of the problem, the present laboratory study has been carried out.

2 EXPERIMENTAL PROGRAMME

The sandstone rock specimen used are from Vindhyachal region of Uttar Pradesh, India and belong to Bhander series of upper Vindhyans. The rocks is isotropic, light yellowish in colour due to presence of more silica and feldspar. Physical properties (e.g. density, sp. gravity, water absorption and porosity) and strength indices (e.g. Brazilian and point load strength, UCS of intact rock) have been determined first to characterise the rock. Subsequently a series of tests have been carried out under triaxial stress conditions on various types of specimen as shown schematically in figure 1.

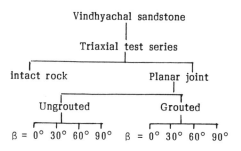

grout used cement : fine sand = 1 : 1
w/c = 0.7 (by weight)
Confining pressures used = 50, 75 and 100 Kg/cm²

Fig. 1 Schematic diagram-specimen tested

3 STRENGTH CRITERION FOR ANISOTROPIC ROCKS

Anisotropic rocks are more complicated, because another variable, i.e. orientation angle 'ß', comes into picture. A number of failure criteria have been proposed. Some of the widely used criteria for anisotropic rocks are by Walsh and Brace (1964), Jaeger (1960) (single plane of weakness theory-assuming that the rock behaves linearly with the applied load), Jaeger (1960) (variable cohesive strength theory) and Mclamore and Gray (1967) criterion (predicting nonlinear behaviour of anisotropic rocks).

Mclamore and Gray (1967) assume that the material fails in shear and has a variable cohesive strength τ_o, but constant values of internal friction tan. ϕ But Walsh and Brace (1964) assume that the failure is tensile in nature and that the body is composed of long, non randomly oriented cracks that are superposed on an isotropic array of randomly distributed smaller cracks or Griffith's cracks. They further assume that fracture may occur through the growth of either the long or small cracks depending upon the orientation of the long crack system to the applied stress, σ_1.

The above strength criteria[1] can not be used for all types of rock because of their limitations. The practical utility is very less as a large number of tests are to be performed at different confining pressures and orientations, to evaluate the anisotropic rock strength.

Hoek and Brown (1980) developed empirical failure criterion for anisotropic rocks using the non-linear failure envelope predicted by Griffith's theory for plane compression and through process of trial and error. It can be written as

$$\sigma_1 = \sigma_3 + (m_a \ \sigma_{ca} \ \sigma_3 + s_a \ \sigma_{ca}^2)^{\frac{1}{2}}$$

where σ_1 and σ_3 are major and minor principal stresses respectively.

σ_{ca} = uniaxial compressive strength of rock with weak plane.

m_a and s_a are dimensionless constants which characterise the degree of interlocking between particles in rock mass containing a weak plane. For intact rock s = 1 and for completely broken rock mass s = 0. The range of variation of m is very wide and is believed to be a function of rock type and rock quality.

Yudhbir et. al. (1983) have also proposed an empirical criterion by modifying Bieniawaski's (1974) criterion to take into consideration the anisotropic rock behaviour and can be expressed as :

$$\sigma_1 / \sigma_{ca} = A_a + B_a \ (\sigma_3 / \sigma_{ca})^{\alpha_a}$$

where A_a = constant depending on rock mass quality

B_a = rock material constant.

α_a = slope of the plot between ($\sigma_1/\sigma_{ca} - A_a$) vs. σ_3/σ_{ca} on log-log scale.

σ_{ca} = uniaxial compressive strength of rock with a weak plane.

Rao et. al. (1985) suggested a criterion for anisotropic rocks as

$$(\sigma_1 - \sigma_3)/ \sigma_3 = B_j (\sigma_{cj} / \sigma_3)^\alpha$$

where σ_{cj} is the uniaxial compressive strength of rock with a weak plane or joint oriented at ß greater than zero degree and B_j = material constant for joint orientation.

α = slope of the plot between ($\sigma_1 - \sigma_3)/ \sigma_3$ and σ_{cj}/σ_3 on log-log scale.

Rao (1984) has critically reviewed and evaluated the applicability of the various strength criteria proposed by using data generated from carrying out tests on four Indian rocks and analysis of published data for more than 100 rocks. He has concluded that a better prediction of strength is possible by use of Rao (1984), Rao et. al. (1985) proposed strength criteria. In the present study two strength criteria viz. Hoek and Brown (1980) and Rao et. al. (1985), have been used to assess their applicability for the case of grout jointed rocks.

4 RESULTS AND DISCUSSIONS

4.1 Physical properties

The mean value of the physical properties of Vindhyachal sadstone are presented in table 1.

Table 1 Physical properties of sandstone

Property		Value
Water absorption		2.95 %
Sp. gravity		2.65
Density	(a) dry	2.45g/cm³
	(b) saturated	2.52g/cm³
	(c) bulk	2.47g/cm³
Porosity	(a) apparent	7.24 %
	(b) Total	7.36 %

4.2 Strength indices

The mean value of UCS, Brazilian and axial and diametral point load strength tests are presented in table 2. The classification of this sandstone on Deere and Miller's (1966) chart is 'CM'. The ratio of σ_c/σ_{tb} $\simeq$ 14, $\sigma_c/\sigma_{(tp)d}$ $\simeq$ 25 and $\sigma_c/\sigma_{(tp)a}$ $\simeq$ 13, which are in the range of values published in the literature (e.g. Broch and Franklin 1972).

Table 2 Strength indices-Vindhyachal sandstone

Strength Index		Value (Kg/cm²)
σ_c		689.6
σ_{tb}	air dry	49.7
	saturated	44.5
$\sigma_{(tp)d}$	air dry	27.9
	saturated	16.9
$\sigma_{(tp)a}$	air dry	52.5
	saturated	32.3

4.3 Intact rock behaviour

The table 3 presents the values of peak stress (σ_1), elasticity modulus E_t (obtained at 50% peak stress value) and Poisson ratio for various confining pressures. From the analysis of triaxial test data, the value of c = 105.47 kg/cm² and ϕ = 36.3° is obtained.

Table 3. Variation of σ_1, E_t and ν with confining pressure – Intact rock.

σ_3 Kg/cm²	σ_1 Kg/cm²	E_t x10⁵Kg/cm²	ν
0	698.60	1.493	0.077
50	960.61	1.976	0.099
75	1135.26	4.979	0.079
100	1222.59	6.573	0.065

4.4 Variation of principal stresses

Figures 2 and 3 show variation of principal stress for different joint orientations for ungrouted and grouted specimen respctively. It is observed that the variation of σ_1 with σ_3 is non-linear in all the cases. The curve corresponding to ß = 0° orientation shows highest strength ot higher confining pressure ranges in both the cases. The peak strength shown is higher for grouted specimen as compared to ungrouted specimen at all joint orientations. The curve for ß = 30° is lowest amongst all the curves, indicating minimum strength for this joint orientation. The strength obtained for ß = 0° are very close, the former being slightly higher in both the cases.

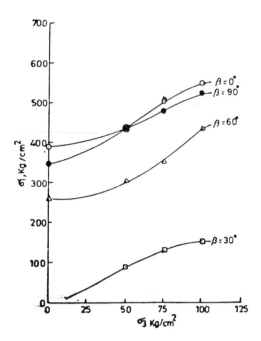

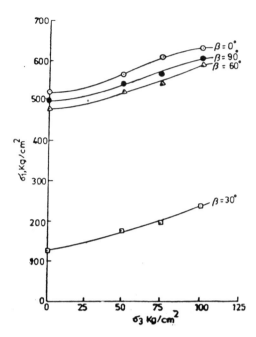

Fig. 2 Variation of principal stresses at different joint orientations-ungrouted specimen.

Fig. 3 Variation of principal stresses at different joint orientations-grouted specimen.

4.5 Strength parameters

The variation of cohesion and friction angle with joint orientation is shown in figures 4 and 5 for both ungrouted as well as grouted specimens. It is observed that the cohesion value variation with joint orientation is quite substantial, being minimum for β = 30° and highest for β = 90°. The cohesion values for grouted specimen is higher as compared to those for ungrouted specimen. In case of friction angle, the variation with joint orientation is small, but in this case also, it has smallest value for β = 30° and highest value for β = 60° joint orientation. The friction angle values for grouted specimen is smaller as compared to ungrouted specimen. This indicates that the higher strength shown by grouted specimen as compared to ungrouted specimen, is because of cohesion only.

4.6 Variation of peak stress with joint orientation

Figures 6 and 7 show variation of peak stress with joint orientation for different σ_3 values for ungrouted and grouted specimens. It is observed that with increasing confining pressure the peak stress values go on increasing at each joint orientation for both the cases.

In case of ungrouted specimen the variation in the values with joint orientation is very marked. It is lowest for β = 30°. The peak stress values are highest for β = 0° and slightly less for β = 90°. The difference between β = 60° and β = 0° or 30° is quite substantial.

In case of grouted specimen, the trend of variation of peak stress values with joint orientation is similar but the values in each case is higher as compared to ungrouted specimens. In this case also, the peak stress values go on increasing with increase in confining pressure. The lowest peak stress value for grouted specimens also occures at β = 30° joint orientation for all the confining pressures. For grouted specimen, the difference in the values of peak stress for β = 0°, 60° and 90° is not much. However when these stress values are compared with that of β = 30°, the difference is quite substantial.

4.7 Applicability of strength criteria for jointed rocks

In the present study, two strength criteria proposed by Hoek and Brown (1980) and Rao et.al. (1985) have been considered for assessment of their applicability in case of

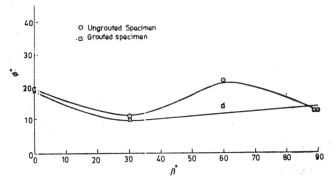

Fig. 4   Variation of friction angle ϕ with joint orientation β .

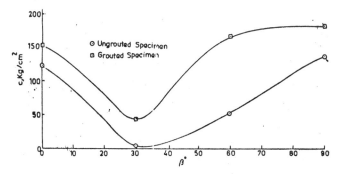

Fig. 5   Variation of cohesion c with joint orientation β .

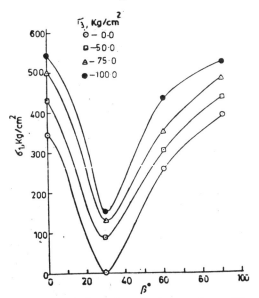

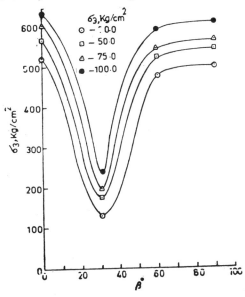

Fig. 6   Variation of peak stress values with joint orientation β at different confining pressures-ungrouted specimen.

Fig. 7   Variation of peak stress values with joint orientation β at different confining pressures-grouted specimen.

grout jointed rocks. Calculations have been carried out to obtain the values of empirical constants in the proposed strength criteria for different joint orientations. Then an average general value is proposed for these constants applicable for all joint orientations. From the calculations, it is proposed that in the present case, a value of m_a = 0.62 and s_a = 0.83 may be used for Hoek-Brown yield criterion and B_j = 1.16 and α = 0.90 may be used for Rao et.al. yield criterion. Using the value of these constants, variation of σ_1 with β at different σ_3 is obtained and compared with actually observed values (figures 8, 9 and 10). It is observed that experimentally obtained values and predicted values are very close for these two strength criteria. However it is felt that there is a need to carry out large number of such studies to generate more data and to propose general values of these constants for different rock and grout combinations.

5 CONCLUSIONS

The endeavour in the present study has been to develop an understanding of strength behaviour and possibility of strength prediction in case of grouted rocks. The study has been carried out on a sandstone with artificially created smooth joints at different orientations. A cement-sand grout of 3mm thickness (w/c = 0.7) has been used as a filler. Both ungrouted and grout jointed rock behaviour has been studied. Following observations have been made from the present study.

1. The physical and strength properties indicate that the sandstone used in the present study is a medium strength rock that may be classified as CM from Deere and Miller (1966) chart. The ratio of σ_c/σ_{tb} = 14, $\sigma_c/\sigma_{(tp)d}$ = 25 and $\sigma_c/\sigma{(tp)a}$ = 13 are obtained which are in the reported range of values. In case of intact rock, variation of σ_1 with σ_3 is non-linear. The value of E_t also increases with increasing confining pressure as expected.

2. In case of jointed rocks, both ungrouted and grouted, the variation of σ_1 with σ_3 is non-linear. The highest strength is observed for β = 0° orientation and lowest for β = 30° orientation.

3. It has been observed that the cohesion values varies very markedly with joint orientation β . It is observed to be substantially higher in case of grouted rock as compared to ungrouted rock. It is minimum for β = 30°. The variation in ϕ is comparatively small.

It has been concluded that the higher strength obtained for grouted specimen is

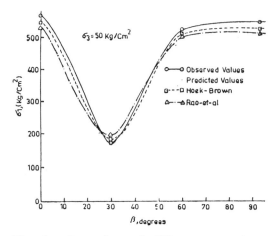

Fig. 8   Comparison of different strength criteria, σ_3 = 50 Kg/cm².

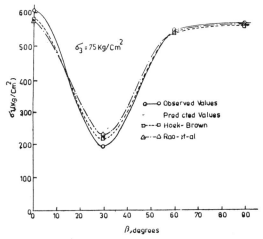

Fig. 9   Comparision of different strength criteria, σ_3 = 75 Kg/cm².

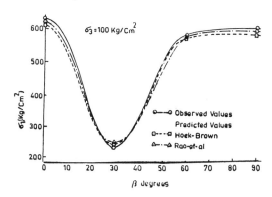

Fig. 10   Comparison of different strength criteria, σ_3 = 100 Kg/cm².

2538

because of increased contribution of cohesion only.

4. In case of ungrouted specimen, the peak stress is lowest for $\beta = 30°$ and high for $\beta = 0°$ and $90°$.

In case of grouted specimen, the trend is similar but all the peak stress values are correspondingly higher. Further the difference between $\beta = 0°$, $60°$ and $90°$ also is reduced markedly as compared to ungrouted rocks.

5. The empirical strength criteria proposed by Hoek and Brown (1980) and Rao et.al. (1985) can prove to be very useful in strength prediction for grout jointed rocks. There is good agreement between observed and predicted values. But there is a need to generate more data for different grout and rock combinations to propose a range of general values for the empirical constants in these criteria.

REFERENCES

Bieniawaski, Z. T. 1974. Estimating the strength of rock materials. J. S. Afr. Inst. Min. Metall. Vol 74, No. 8 : 312-320.

Broch, E. and Franklin, J. A. 1972. The point load strength test. Int. J. Rock Mech. Min. Sci. Vol. 9 : 669-697.

Deere, D. U. and Miller, R. P. 1966. Engineering classification and index properties for intact rock. Tech. report, A. F. W. lab. New Mexico.

Hoek, E. and Brown, E. T. 1980. Underground excavations in rock. Institution of Mining and Metallurgy, London.

Jaeger, J. C. 1960. Shear failure of anisotropic rocks. Geol. Mag., Vol. 97 : 65-72.

Mclamore, R. and Gray, K. E. 1967. The mechanical behaviour of anisotropic sedimentary rocks. Trans. Am. Soc. Mech. Engr. Series B, Vol. 89 : 62-76.

Rao, K. S. 1984. Strength and deformation behaviour of sandstones. Ph. D. Thesis, submitted to I.I.T. Delhi, India.

Rao, K. S. Rao, G. V. and Ramamurthy, T. 1985. Rock mass strength from classification. Paper no. 3, Proc. Workshop on Engineering classification of rocks, CBIP, New Delhi, India : 27-50.

Walsh, J. B. and Brace, W. F. 1964. A fracture criterion for brittle anisotropic rocks. Geophysics Research J., Vol. 69 : 34-49.

Yudhbir, Lemanza, W. and Prinzl, F. 1983. An empirical failure criterion for rock masses. 5th Int. Cong. Rock Mech., Vol. 1 : B1 - 8, Melbourne.

Engineering geological problems encountered during construction of Incegez tunnel

Les problèmes de géologie de l'ingénieur rencontrés au cours de la construction du tunnel d'Incegez

T. Topal & V. Doyuran
Middle East Technical University, Ankara, Turkey

E. Aşçıoğlu
Sial Ltd. Co., Ankara, Turkey

ABSTRACT: İncegez tunnel is designed to convey the water of the Eğrekkaya reservoir to the reservoir of the Kurtboğazı dam. The tunnel is approximately 6 kms long and throughout its alignment agglomerates with occasional basalts (Miocene) and uncon-solidated to semi-consolidated clastic sediments (Pliocene) dominate the lithology. The agglomerates constitute almost 94% of the bedrock and are locally intensely sheared.

Problems encountered during tunnel construction include large groundwater inflows and overbreaks associated with the shear zones. Locally, initial groundwater dis-charges upto 200 lt/sec are noted. Dewatering efforts include deep well pumping and drainage through pilot holes drilled ahead of the face.

RESUME: Le tunnel d'İncegez a été conçu pour transmettre l'eau du réservoir d'Eğrekkaya au réservoir de Kurtboğazı. Le tunnel a une longueur de 6 km et sur zon trajet affleurent les agglomérats d'âge Miocene intercalés de niveaux basaltiques et les sediments clastiques consolidés et semi-consolidés du Pliocene. Les agglomérats forment les 94% du substratum et montrent par endroit importantes zones de cisaillement.

Pendant la construction du tunnel on a eu principalement des problèmes d'eau souterraine avec des débits élevés et aussi des effondrements dûs aux zones de cisaillement. On a mesuré par endroit des débits d'eau souterraine de l'ordre de 200 l/s. Pour l'evacuation d'eau on a eu recours aux pompes de puits profonds et aussi au drainage d'eau obtenue par des forages pilotes horizontaux effectués au miroir.

1 INTRODUCTION

In order to meet the increasing water de-mand of Ankara, a series of water supply projects were planned by the State Hydrau-lic Works (D.S.İ.). The so-called "Ankara Municipal Water Supply Project" is de-signed in 1969 to meet the municipal water requirement of Ankara until the year 2010, at which time the population of the city is expected to be around 3 650 000. The project includes numerous dams, aqueduct tunnels, and pipeline systems (Figure 1). Within the context of the project eight dams were planned of which five are pres-ently in service. These include Çubuk I (1936)*, Çubuk II (1964)*, Bayındır (1965)*, Kurtboğazı (1967)*, and Çamlıdere (1985)* dams. The construction of Eğrekka-ya dam is started in 1986 and its comple-tion is planned for 1990. The Akyar and
* Date of completion

Işıklı dams are planned for 1992 and 2010, respectively. The Kınık tunnel, which is designed to convey the water of the Çamlıdere dam to the İvedik Water Treat-ment Plant via pipelines, is 15.7 km long. It is the longest tunnel of the project and completed in 1988. The water of the Eğrekkaya dam will be transmitted to the reservoir of the Kurtboğazı dam through a pipeline system and the İncegez tunnel. The same tunnel and the pipeline system will also convey the water of the Akyar dam.

The İncegez tunnel is 5888.90 m long. It has a horse-shoe cross-section having di-mensions 4.60 m (height) and 4.10 m (width). However, the tunnel will be cir-cular after final concrete lining. It has a hydraulic slope of 0.0004 so that the water will flow by gravity.

In this paper, engineering geological investigations performed along the İncegez

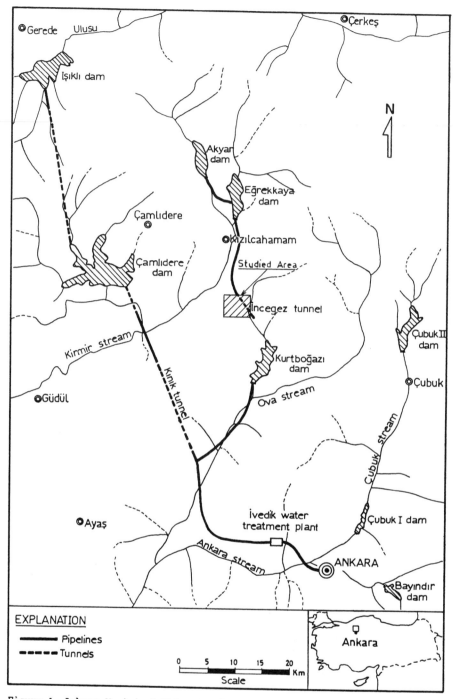

Figure 1. Ankara Municipal Water Supply Project (After D.S.İ., 1984).

tunnel, the problems encountered during its construction, and their solutions will be discussed.

2 SITE GEOLOGY

Along the tunnel alignment volcanic rocks consisting of agglomerates, basalts, and andesites (Miocene); unconsolidated to semi-consolidated clastic sediments (Pliocene); and surficial deposits (Holocene) constitute the main lithological units (Figures 2 and 3).

2.1 Volcanic rocks

Volcanic rocks form extensive outcrops throughout the study area. The stratigraphic relation among the agglomerates, basalts, and andesites are highly complex (Topal, 1987).

The agglomerates constitute the dominant lithology along the İncegez tunnel alignment. They are composed of andesitic and basaltic fragments and blocks embedded within a tuffaceous matrix. Individual fragments are mostly angular to subrounded. The size of the fragments ranges between 2 cm and more than a meter. Occasionally almost flat-lying tuff layers may be observed within the agglomerate.

The basalt is dark gray, porphyritic, highly jointed, and shows flow-layering. The phenocrysts of plagioclase, pyroxene, and rarely olivine are embedded within a groundmass consisting of plagioclase microlites.

Field observations, core drilling, and construction stage investigations revealed that basalts occur as diverse blocks within the agglomerates.

Andesite is gray, highly jointed, and shows flow-layering. The rock has a porphyritic texture. The phenocrysts of plagioclase, hornblende, and rarely pyroxene are observed within a groundmass of plagioclase microlites.

The andesites probably correspond to the latest stage of volcanic activity. They generally occupy the highlands and overlie the agglomerates. For this reason they are not going to be encountered during tunnelling.

The age of the volcanic rocks is Miocene (Erişen and Ünlü, 1980). They are unconformably overlain by the unconsolidated to semi-consolidated clastic sediments (Topal, 1987).

2.2 Unconsolidated to semi-consolidated clastic sediments

The clastic sediments consist of clay, silt, sand and gravel derived mostly from the volcanic rocks and dip 18-20° northeast. Among the clastic sediments clays and silts dominate the lithology. The sands and gravels generally occur in the form of lenses (Topal, 1987). The age of the unit is Pliocene (Erişen and Ünlü, 1980).

Talus and alluvium are observed in the form of local surficial cover (Figure 2). Talus is composed of semirounded to angular fragments and blocks of andesite. It is well observed at the inlet portal area of the tunnel. The alluvium is mostly observed along the stream courses at the outlet portal area.

3 ENGINEERING GEOLOGY

3.1 Engineering geological properties of volcanic rocks

The agglomerate is moderately strong to weak. Its uniaxial compressive strength and the modulus of elasticity average around 26 MPa and 1.77×10^4 MPa, respectively. The strength of the rock, however, varies considerably based on the degree of saturation and the grade of weathering.

The field observations as well as core drilling confirmed that weathering is generally restricted to the uppermost levels of the natural rock exposures. The rock is slightly weathered at the tunnel elevation, except at the portal areas where moderately weathered rocks are observed. Locally, highly weathered zones are also encountered along the shear zones, which are rather frequently observed.

The initial porosity and permeability of the agglomerates are rather low. Although no well-developed systematic joint sets are observed, large amounts of groundwater inflows are anticipated where shear planes and/or zones are intersected.

The field evidences suggest that the rock has self supporting and good arching capabilities. However, locally heavy supports are needed where the rock is saturated and sheared.

Basalt is highly jointed and very strong (174 MPa) where fresh. The modulus of elasticity of the rock averages around 5.25×10^4 MPa. It is moderately weathered at the surface. The joint line surveys performed following the procedure described by Hoek and Brown (1980) revealed

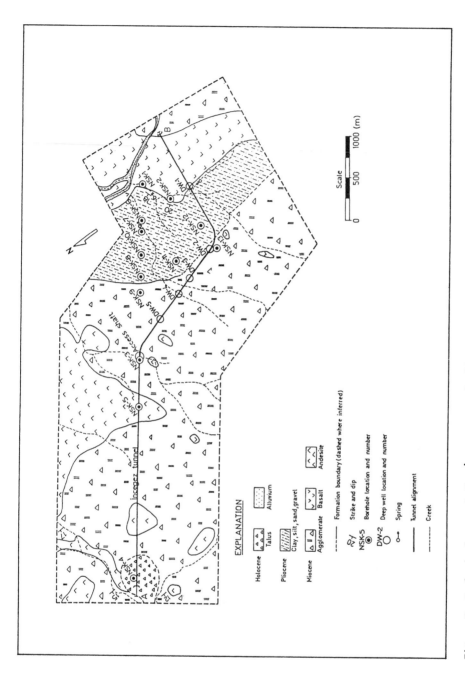

Figure 2. Geological map of the İncegez tunnel alignment.

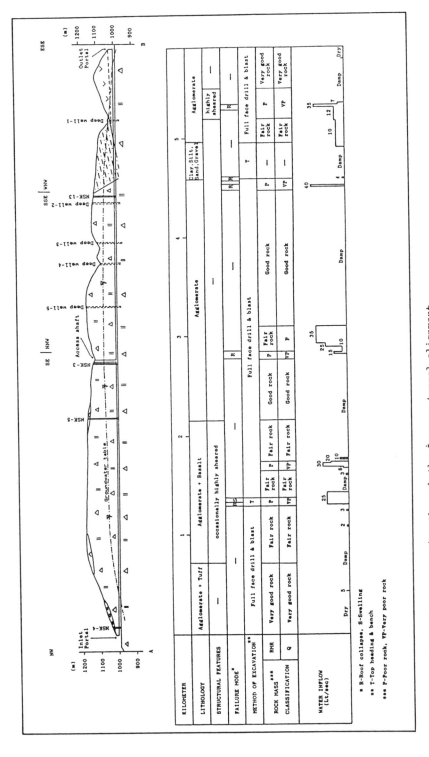

Figure 3. Geological cross-section through the İncegez tunnel alignment.

two dominant joint sets having attitudes N55°W/85°NW and N12°W/83°NE. However, there are also randomly oriented fractures, shear planes, and zones. The joints give rise to a blocky structure. The block sizes are small to medium. The rock possesses secondary porosity and permeability due to the joints it contains.

3.2 Engineering geological properties of unconsolidated and semi-consolidated clastic sediments

The clay and silt are soft and light brown. The X-ray diffraction of the clay samples taken from the borehole NSK-8 yielded quartz, feldspar, smectite, chlorite, illite, and vermiculite (Sial, 1986). Due to low clay/silt ratio, and low smectite content, no swelling ground problems are anticipated during tunnelling.

Information obtained from the boreholes revealed that sand and gravel are mostly encountered above the red-line.

3.3 Subsurface investigations

The subsurface investigations comprise electrical resistivity survey using Schlumberger arrangement, and core drilling. The geophysical survey was performed to determine the subsurface distribution of various lithological units and to locate faults and/or shear zones, if present.

No obvious structural features could be detected by the geophysical surveys. However, distinct resistivity contrasts between the volcanic rocks and the sedimentary sequence helped to define the subsurface boundary of the two.

Drilling activities were started along the initially proposed tunnel alignment, which runs straight through boreholes NSK-4 and NSK-1. In boreholes NSK-1 and NSK-2 very shallow groundwater table is encountered within the basalts. In borehole NSK-6, however, groundwater under confined pressure is noted within the basalts, below the clastic sediments. Due to poor core recovery it was difficult to assess the characteristics of the clastic sediments. Thus, in order to avoid uncertainities associated with the clastic sediments, the tunnel alignment is shifted towards south to cross them at shortest possible distance. The topography at the outlet portal of the new alignment was also highly favorable for portal excavation.

Water pressure tests performed in the boreholes have shown that the agglomerates are practically impervious (<1 lugeon); basalts and granular units of the clastic sediments are slightly pervious to highly pervious (1-25 lugeon); and the silts and clays are impervious to slightly pervious (<1-4 lugeon).

3.4 Engineering classification of rock masses

The rock masses are classified according to Barton et al. (1974) and Bieniawski (1979) using the information obtained from surface geological observations and borehole logs. The rock mass quality (Q) and the rock mass rating (RMR) values are determined separately for rocks having significantly different engineering properties.

Rock mass classes determined by Q and RMR systems have yielded consistent results. Rock classes ranging from very good to very poor are found at different sections of the tunnel (Figure 3).

4 PROBLEMS ENCOUNTERED DURING CONSTRUCTION

Construction works have been started in 1986. However, due to unexpected adverse ground conditions completion of the tunnel is delayed considerably. Major construction problems involved heavy groundwater inflows and roof collapse.

The groundwater was first noted at distances of 430 m and 100 m from the inlet and outlet portals, respectively. However, the discharge rates were not high enough to delay the construction works. At a distance of 587 m from the outlet portal a roof collapse has occurred. Groundwater under high hydrostatic pressure rushed in from the tunnel face and swept the collapse material for a distance of 6 m. The initial rate of discharge was measured as 65 lt/sec. In about two weeks the discharge rate is reduced and stabilized around 35 lt/sec. The collapse zone is temporarily stabilized through shotcrete application. Five inclined holes (12° from the horizontal) were drilled around the face to drain the water behind it. A pilot gallery is driven to explore the ground conditions behind the face as well as to by-pass the collapse zone. The gallery has penetrated a shear zone of 6 m wide at the agglomerate-basalt boundary.

Another roof collapse has occurred at the boundary of the agglomerate and the sedimentary sequence. A peaked roof of about 3 m high was formed at the crown. There was, however, no significant groundwater inflow. The zone is stabilized and reinforced through shotcrete, wiremesh, rock bolts, and closely spaced steel ribs.

Major groundwater inflow was encountered at a distance of 1384 m from the outlet portal. Here, the initial rate of discharge was measured as 200 lt/sec, which has also produced a large roof collapse. The collapse material is washed into the tunnel for a distance of 40 m from the face. The discharge rate is stabilized around 40 lt/sec in about one month. Major drainage efforts involved deep well pumping and nearly horizontal drainage holes. Two by-pass galleries were driven around the collapse zone and approximately 400 tons of cement was injected into the collapse area.

At the inlet section of the tunnel similar roof collapse and groundwater inflow problems were encountered within the agglomerates at distances of 1297 m and 2815 m from the portal. Additionally, swelling ground conditions were also encountered within the hydrothermally altered tuffaceous matrix. In the remaining parts of the tunnel deep well pumping proved rather effective in draining the groundwater.

Method of excavation within the agglomerates involved full face drilling and blasting. For the critical sections of the tunnel top heading and bench method is employed. In crossing the sedimentary sequence this technique has proved very effective in eliminating construction and support problems.

5 CONCLUSIONS

Along the İncegez tunnel alignment agglomerates with occasional basalt alternations and unconsolidated to semi-consolidated clastic sediments dominate the lithology. The agglomerates constitute almost 94% of the bedrock, and are locally intensely sheared and/or fractured. Based on Q and RMR systems the rocks are classified as very good to very poor.

Unexpected groundwater inflows and occasional roof collapses associated with the shear zones considerably delayed the construction works. Locally, initial groundwater inflows of upto 200 lt/sec are noted.

Dewatering of groundwater is accomplished through deep well pumping and drainage through pilot holes drilled ahead of the face.

Full face drill and blast method is adopted within the agglomerates; whereas, top heading and bench method is preferred for the excavation of highly sheared zones and the clastic sediments.

During the construction of the tunnel, rock bolting, where required, wiremesh, shotcrete (5 cm-15 cm), and steel ribs (at 1 m-2 m intervals) are used. Depending on the ground conditions, spacing of the steel ribs is reduced down to 0.50 m.

REFERENCES

Barton, N., Lien, R., and Lunde, J., 1974, Engineering classification of rock masses for the design of tunnel support. Rock Mechanics, vol.6, no.4, pp.189-236.

Bieniawski, Z.T., 1979, The geomechanics classification in rock engineering applications. Proc. 4th. Int. Congr. Rock Mech., I.S.R.M., vol.2, pp.41-48.

D.S.İ., 1984, Ankara su temini projesi. D.S.İ. V. Bölge Müdürlüğü (Brochure-in Turkish).

Erişen, B., and Ünlü, R., 1980, Ankara-Çubuk-Kızılcahamam-Kazan alanının jeolojisi ve jeotermal enerji olanakları. M.T.A., 68 p (unpublished).

Hoek, E., and Brown, E.T., 1980, Underground excavations in rock. The Institution of Mining and Metallurgy. London, 527 p.

Sial, 1986, İncegez tüneli jeoteknik raporu, 56 p (unpublished).

Topal, T., 1987, Engineering geological investigations along the İncegez tunnel alignment. M.E.T.U., M.S. Thesis, 107 p (unpublished).

Geotechnical measurement campaign for designing underground works through modelling under unfavorable geomechanical conditions

Campagne de mesures géotechniques pour dessiner des travaux souterrains par modélisation sous conditions géomécaniques défavorables

A. Tzitziras & D. Vardakastanis
Institute of Geology and Mineral Exploration (IGME), Athens, Greece

A. Katsifos
Aegean Metallurgical Industries S.A. (METBA), Greece

ABSTRACT: A campaign of geomechanical measurements has been carried out in a case of unfavorable geomechanical conditions. It consists of convergence, deformation, inclinometer and load measurements. Classical rock mass classification and laboratory testing form an essential part of this approach. The results are to be used for modelling trial exploitation methods. Basic characteristics of the model to be used are derived from these measurements which prove to be an essential part of modelling preparation procedures. Displacement measurements seem to be more exact and useful than load measurements. Indications are derived also for the nature and orientation of in situ stresses.

RESUME: Une campagne de mesures géomécaniques est realisée dans le cas des conditions géomécaniques défavorables. Elle se constitue de mesures de convergence, de déformation, par inclinomètre et de mesures de charge. Classification classique de la masse rocheuse et essais de laboratoire ont été réalisées. Les résultats vont être utilisées pour modeliser l'exploitation expérimentale. Caractéristiques essentiels du model à utiliser sont mis en évidence par ces mesures, qui se confirme comme étant une partie essentielle de la modelisation. Les mesures de déplacement sont plus exacts est plus utiles que les mesures de charge. Quelques indications sont aussi derivées concernant la nature et l'orientation des contraintes naturelles.

1 INTRODUCTION

A number of geotechnical measurements with installation of necessary instruments have been carried out in an attempt to design underground works at the Molai mine (Greece) through modelling. This campaign is included in the project "Optimization of the exploitation of vein type polymetallic sulfide deposits through mathematical modelling and rock mechanics - An application at the Molai mine" which is an R & D project partly financed from E.E.C. program of Raw Materials and Recycling.

The above mentioned measurements are decided because the rock mass presents low mechanical characteristics due to the geological and tectonic conditions in connection with underground water existence.

2 AREA MORPHOLOGY

The Molai mine is situated at the southeastern part of Pelloponeese in Greece, east of Kourkoulas mountain at +165 altitude. Shallow torrents cross and drain superfi-

cially the mine area.

3 GEOLOGY OF THE AREA

Geological formations that form the wide area belong to the geotectonical zone of Tripolis and are the following:

1. Marbles gray, compact with coarse crystals.
2. Volcanic rocks namely basaltic andesites, dacites, ryolites and tuffs. They are intensively fractured from tectonic forces developed in the area and at the same time altered from hydrothermal solutions that created the orebody.
3. Dolomites and dolomitic limestones, compact and gray.

The only geological formations that are found in the vicinity of the mine and in which all underground openings have been excavated are the volcanic rocks.

4 GEOLOGY OF THE DEPOSIT

The mineralization has been created in the

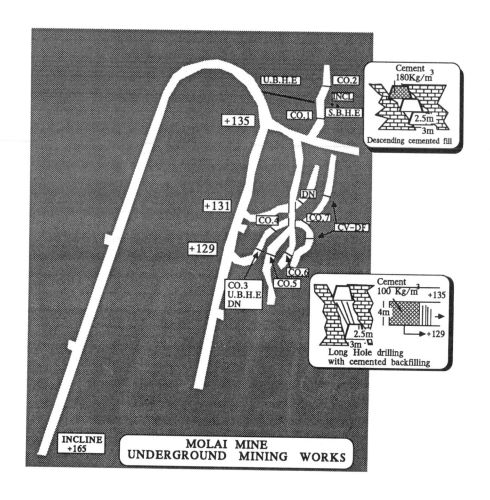

S.B.H.E =Bore hole extensometer (from surface)
U.B.H.E =Bore hole extensometer(Underground installation)
CO.n =Convergency station No n
DN =Dynamometer
CV-DF =Curvometer-Deformeter
INCL. =Inclinometer

FIGURE 1

volcanic tuffs and its nature is subsea volcanosedimentary deposit. Its direction is mainly N-S with general inclination 60° East.

5 HYDROGEOLOGY

The geological formations at the area are classified as permeable (Marbles, Dolom-

ites, Dolomitic Limestones) and partly permeable to locally impermeable (Volcanic Tuffs).

Boreholes drilled during mineral exploration revealed initial artesian or sub-artesian water. Places with developed water storage capacities are faults and mineralized zones.

Geological conditions in combination with the permeability and area tectonics create a steady, low capacity, water level in the mine area. First indications of water in the access inclined gallery start at a distance of 45 m from the entrance. Water quantity pumped out of the mine is between 5.5 and 25.5 m³/h through the year.

6 TECTONICS

In the area there have been developed intensive tectonic phenomena during Alpic orogenese and are distinguished in folds, overthrusts and normal faults with direction NW and NE .

This important tectonic action together with hydrothermal solutions that have been manifested caused intensive fracturing and alteration, resulting in low mechanical characteristics for the rock mass.

7 UNDERGROUND MINING WORKS

The geometrical characteristics of the orebody, the ore nature and the contained metal values as well as the mechanical and physical characteristics of the mineralization and the hoist rocks led to the experimental application of two mining methods with cemented fill:

1. Descending method with cemented fill.
2. Long hole drilling with cemented backfilling.

Access to the deposit is done by a 400 m long inclined gallery (-12%) which has been excavated at the footwall of the orebody (Fig. 1). Subsequently cross-cuts have been excavated at three altitudes (+135, +131, +129) from the incline and reach the deposit.

Exploitation by the first method is to be done by galleries in the direction of the orebody, northern of the cross-cut +135 in three slices (+135,+131,+129) at a length of 50 m .

The second method is to be applied at the southern part of the mine. It will be done by two galleries excavated in the direction of the orebody at a length of 20 m between altitudes +129 and +135. After the extraction of the pillar through gallery +129 the void will be filled with cemented fill from gallery +135.

8 GEOMECHANICAL MEASUREMENTS

Due to the unfavorable geomechanical conditions and the uncertainty of the deposit's economic attractiveness, it has been decided to study the final exploitation through numerical modelling in order to examine all possible solutions and to

choose the best as far as economic return and safety are concerned.

The major problem that one faces when applying modelling methods, from the moment that a suitable algorithm has been chosen, is the correct simulation of rock mass behavior.

Particularly, in cases like this , where the hydrogeological and tectonic context is complex, great care is requested in choosing suitable parameter values lest the results would be of an academic interest not suitable for the particular practical application.

The choice of the suitable type of measurements and the sites to be executed, always in the direction of data collection for the modelling, constitutes a fine point which is not dealt always with the care it merits because of financial and/or designing limitations.

Choosing site criteria differs from the ones usually applied during geotechnical measurements that are executed for purposes of control and/or rock characterization. In this case the target is not to survey the most dangerous sites but the most sensible parameters at the sites most influenced by the progress of the works. Also the type of these measurements must allow the extension of rock characterization to the whole mass to be exploited and at the same time to permit a check of the modeling results.

The complex geological and tectonic environment places the first problem, namely of the expected difference between rock properties measured at the laboratory on specimen and rock mass characteristics in situ. Mechanical properties of the rock required for modelling usually are the elasticity modulus, Poisson ratio, failure criteria and discontinuity properties. These properties are known from laboratory tests on samples. Their application at the rock mass requires a certain classification of the rock mass so that transition is possible. As such classification method the one proposed by Bieniawski has been applied.

Modelling results can be checked and the quantity and nature of the required modifications established by displacement and load measurements. These measurements must be of such a type as to provide elements for the global response of the rock mass to load transfer and at the same time of such an extent as to reflect local variability of behavior. For these purposes measurement of gallery walls' convergence and deformation at the interior of the rock have been selected . At the same time locations were selected so as to be affected by development and exploitation works.

Another restriction exists because of the need to represent in the model the geometry of mining works near the sites of the measurements' execution (distribution of mine openings as simple as possible).

8.1 Rock mass classification

At a number of sites the rock mass characteristics have been measured and the rock has been classified according to the Bieniawski criteria. Deriving from measurements and observations the rock mass is classified as being of "poor" to "very poor" quality. Exceptionally, at few locations the classification has characterised rock mass as of "fair" quality.

8.2 Laboratory tests

A number of laboratory tests have been done on suitable specimen from the formations of the area of the mine, namely:
1. Uniaxial tests (with elasticity modulus measurement).
2. Triaxial tests.
3. Shear tests on rock mass discontinuities.
These tests showed an extented variability of the mechanical properties. This variability is a major restraining factor in efforts for proper modelling. It can be raised by proper assumptions so as to classify the rock in a limited number of groups that may be represented by the model. This grouping can only be done through the use of the classification results which include also discontinuities up to a certain scale.

8.3 Convergence measurements

Convergence measurements have been executed in order to survey the galleries' walls during the development stage.
Six locations were chosen so as the measurements' plane would be as much as possible perpendicular to the part of the orebody to be exploited and as near as possible to its center. The exact choice of the locations was done by taking into consideration the sites where the greatest values of displacements are expected. Only in the footwall of the orebody it has been possible to place instruments. Future modelling is to be applied on these sections.
From these stations CO3, CO4, CO5 and CO6 are to record also convergence that is due to the experimental mining of the south sector.
The effect of the application of the filling method at the northern part of the orebody will be surveyed by the use of

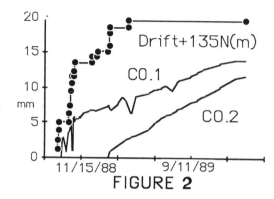

FIGURE 2

electrical convergence measurement devices that are to be implemented in the filling material.
From the measurements results that a certain stabilization has occurred for most of the stations after a rather long period (Fig. 2), while stations CO5 and CO6 do not present stabilization due to deterioration of the wooden support. In particular for station CO6 (Fig. 3) a stabilization trend

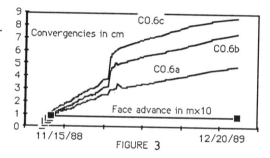

FIGURE 3

is observed for the length a and not for b and c. The step recorded is due to the passing of gallery +129 under +135 in the vicinity of the station.
Because of this long period of stabilization a number of stations have been installed on steel sets of the support (Fig. 4). Results show a short stabilization period as well as vertical loading prevailing in most of the cases.

8.4 Borehole extensiometer measurements

Rock deformation measurements have been done in five stations on sections perpendicularly to the exploitation panels.
At the south sector three stations were installed in one section (Fig. 1) with three borehole extensiometer (BHE) of three

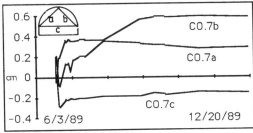

CO.7b

CO.7a

CO.7c

6/3/89   12/20/89

FIGURE 4

points each (Fig. 5). Results were obtained during the development stage. In order to get the best results during the exploita-ˊ on stage, one of the three BHE has been c.. ected towards the center of the panel to be exploited.

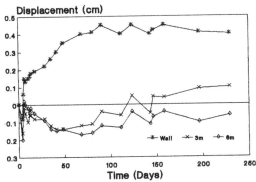

FIGURE 5. 3-point BHE at 3,6 and 9 m.

At the north sector two BHE of two points each have been installed in the hanging wall and the footwall of the orebody.

It is concluded from the measurements that the zone affected by the gallery is limited at about 1.5 times the gallery width.

8.5 Inclinometer measurements

In order to measure horizontal displace-ments at the northern sector an inclinome-ter station was installed (Fig. 1) which covers the area from surface to a depth of 40 m at 10 m distance from the panel in the hanging wall.

From measurements executed during the excavation of gallery +135 no significant influence was detected.

8.6 Load measurement on supports

Load measurement on supports has been done by two methods.

1. Dynamometers were installed in two positions (Fig. 1) and measurements were taken during the advance of the faces. From the results it is derived that the effect of the front, as far as load is concerned, is limited between 10 and 15 m (Fig. 6).

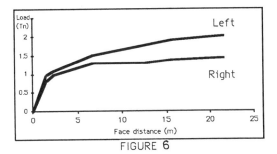

FIGURE 6

2. Curvometer and deformeter stations were installed in two locations at the south sector of the mine. Results verified the above mentioned distance of 15 m. One of these stations is to be used for load transfer recording during exploitation of the south panel.

9 RESULTS

Research and measurements carried out at the Molai mine gave the following results:
1. Convergence measurements on the gal-lery's walls revealed variations in meas-ured values for long periods (16 months). This is true either the front has ceased advancing or not. On the contrary, for the steel sets, stabilization occur quite shortly (1 month). This means that all procedure of load transfer is completed much more rapidly than convergence.
2. From load measurements a distance of face influence has been detected.
3. From deformation measurements it has been established that acting pressure is the result of decompression over a limited zone of the surrounding rock.
4. With an overall synthesis of results from convergence and deformation measure-ments the existence of lateral stress from east to west has been established.
Points 1, 2 and 3 prove that the pressure is of the loosening or creeping type.

10 CONCLUSIONS

Geomechanical works up to now have given

the possibility to form a number of concluding remarks regarding rock mass modelling:

1. The model to be used to simulate rock
mass behavior must be of the strain softening type.

2. Trial modellings should be carried out
with and without time depended behavior in
order to check wether the adding effect of
creeping is of importance.

3. Initial conditions should take into
consideration the existence of lateral
pressure.

Finally some general remarks are self
imposed:

1. Displacement measurements provide more
and broadly evaluated data than load
measurements in case of complex unfavorable
geomechanical conditions.

2. Rock mass classification is a guiding
tool for the final choice of rock parameters and forming rock type groups for modelling purposes.

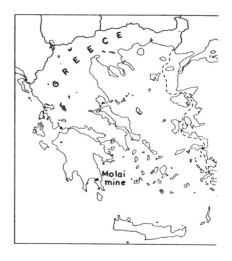

REFERENCES

Agelopoulos, K. & Constantinidis, D. Zinc-
Silver-Lead Molai deposit.Internal
report (in Greek).

Andronopoulos, B. & Tzitziras, A. (March
1988). Geotechnical study of the engineering behavior of formations of Molai
mineralization area. E.E.C. report.

Andronopoulos, B. & Tzitziras, A. (August
1988). Geotechnical study of the engineering behavior of formations of Molai
mineralization area. E.E.C. report.

Andronopoulos, B. & Tzitziras, A. & Vardakastanis D. (June 1989) Geotechnical
study of the engineering behavior of formations of Molai mineralization area.
E.E.C. report.

Andronopoulos, B. & Tzitziras, A. & Vardakastanis D. (November 1989) Geotechnical
study of the engineering behavior of formations of Molai mineralization area.
E.E.C. report.

Georgoulis, J. (1988). Hydrogeological
study on the mineralization area of Molai
- Lakonia. E.E.C. Report.

Doutsos, T. & Koukouvelas, J. (1985).
Structural analysis of Molai deposit.
Mineral Wealth 47, 7-16 (in Greek).

Katsifos, A. (1988). Technical study of
experimental exploitation at Molai mine.
METBA Internal Report.

Behavior of rock mass during underground excavation in the Victoria Power Tunnel

Le comportement des masses de roche pendant l'excavation souterraine du tunnel 'Victoria'

D.L.C.Welikala
Central Engineering Consultancy Bureau, Colombo, Sri Lanka

ABSTRACT: The Victoria Project in the Central Highlands of Sri Lanka comprises a double curvature arch dam and a power house of 3*70 MW capacity, linked by a lined pressure tunnel (5.6km long with an internal diameter of 6.2m) forming the largest power project in Sri Lanka.
  The tunnel was excavated through gently dipping gneisses, granulites, quartzites and marbles, belonging to the Precambrian Highland Series. During the construction, cavity formation and rock falls took place in certain locations and one major fault which was not detected by field investigations, resulted in realignment of the original tunnel route.

RESUME: le Projet Victoria, dans les Central Highland du Sri Lanka, comporte un barrage en arche a double courbure, et une centrale de 3*70 de puissance, reunis par un tunnel en charge betonne (5.6 km de long, 6.2 m de diametre interne), constituant le projet de centrale hydroelectrique le plus grand du Sri Lanka. Le tunnel a ete creuse a travers des gneiss peu inclines, des granulites, des quartztites et des marbres, appartenant aux series precambriennes de Highlands. Au cours de la construction, des formatin de cavites et des chutes de pierre se produisirent a certain endroits, et une faille majeure, non reconnue par les observation de terrain, a necessite le realignement du trace original du tunnel.

1 INTRODUCTION

The Victoria Power Project incorporates a double curvature arch dam and power house of 3*70 MW capacity linked by a 5.6 km long, 6.2m diameter concrete lined pressure tunnel.
  The Victoria tunnel intercepts high grade metamorphic rocks of precambrian age, which have been folded and fractured in Pre Jurassic time. The rock mass is composed of gently dipping gneisses, granulites and crystalline lime stone.
  Data obtained from the subsurface investigation revealed that one major and several minor faults could be predicted crossing the tunnel line. During excavation, a second major fault was detected and consequently the original tunnel route was realigned.
  Excavation of the tunnel commenced in September 1980, and in November 1982 the tunnel was completed through four headings. Poor progress in excavation was recorded in some locations due to bad ground conditions

and breakdown of mechanical plant. Eighty four percent of the total tunnel length was driven in good quality rock with a maximum progress of 70m/week.

2 TUNNEL HEADINGS

The tunnel was completed in four headings, namely
  1) Intake heading
  2) North heading
  3) South heading
  4) Downstream heading (Outfall drive)
  These headings are shown in fig 1. Tunneling through Intake and South headings did not encounter any major problems. The predicted fault intercepted the tunnel at a 100m distance upstream from the adit junction in the north heading. It was 3 - 4m wide, associated with series of minor faults and produced no water.
  Unfavorably oriented discontinuities and an unforseen major fault were detected in the Downstream heading. This paper will

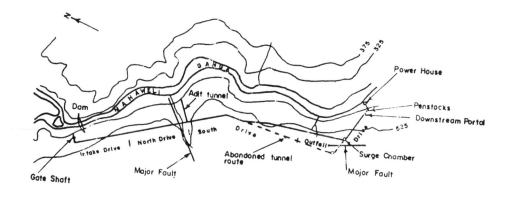

Fig.1 Plan showing Victoria Power Tunnel

highlight the Downstream heading as the most interesting geotechnical features of the tunnel were found in it.

The tunneling condition through in lime stone was excellent as compared with the other rock types.

3 DOWNSTREAM HEADING

3.1 Geotechnical Features

The first 10m distance of the tunnel excavation from downstream heading was in a highly fractured quartz gneiss. Tunnel supports were installed in the affected area comprising of 5 Nos. steel arch ribs at 1m spacing with a 50mm layer of shotcrete and 3m long patterned rock bolting. There were three major types of discontinuities present in this heading, namely:

1) Joints (three major sets, subvertical)
2) Foliation (gently dipping)
3) faults (minor/major)

As these discontinuities were unfavorably oriented, a stereo plot analysis was carried out and potential wedge failures were confirmed. Based on the analysis, the affected areas were supported by 25mm diameter 4m long tensioned rock bolts with 50mm layer of shotcrete and mesh.

Five point convergence array was installed to monitor the behavior of the tunnel profile at 150m intervals. Observed behavior for the period of four months was very satisfactory with tunnel closures recorded as low as 0.02 per cent of the tunnel diameter.

3.2 Tunnel Collapse

About 600m of the Downstream heading was completed on a steady 1 in 20 uphill slope through the quartz gneiss and limestone satisfactorily.

A probe hole driven ahead of the face at 637m distance detected water under high pressure of 6-7 bars. Cement grout was pumped in, but after advancing further few meters more grout had to be injected. A total of 233 tons of grout was used to form a shell round the tunnel. Tunnelling was resumed but 20 minutes after firing the second round, the roof of the tunnel became very unstable with water rushing through the roof. Nearly one hour later about 300m^3 of rock fell from the crown with a heavy inflow of water 800 l/s. The flow gradually reduced to 400 l/s on the following day and to 100 l/s after two days. The water inflow which was charged with sediments, breccia, rounded pebbles and greenish sandy clay. Nearly 0.5m^3 of cement blocks were also found in a muck pile indicating the presence of grout filled cavities. The following support measures were recommended immediately in order to reestablish stability.

1) Four meter additional rock bolts
2) Hundred millimeter thick layers of shotcrete with wire mesh

The muck pile was gradually removed as supports progressed. Seven ribs were installed in the process and back filled with concrete. During this period a further 500-600 m^3 of rock fell from the crown including previously bolted area. A chimney was then exposed on the left hand side of the crown and was estimated to be some 30-40 m in height (fig 2). Three exploratory bore holes drilled from the tunnel wall confirmed the existence of a 6m wide fault zone consisting of sheared and broken rock

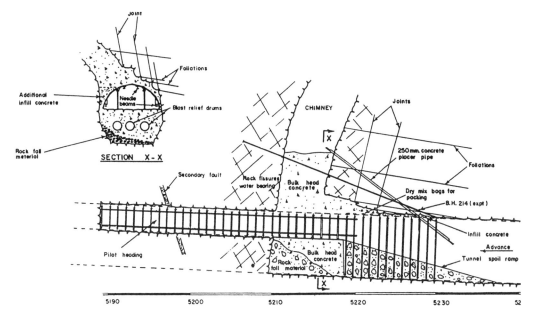

Fig.2 Longitudinal section of Downstream heading showing tunnel collapes area

trending 45 degrees to the east of the tunnel axis. Bore hole BT 213 and 215 initially produced water inflow of 24 l/s and 5 l/s respectively, and diminished with time. No water was encountered in the third bore hole BT. 214 (fig.3).

3.3 TREATMENT OF THE FAULT

The following steps were taken to continue the heading through the major fault.

1) Construction of a stopend shutter at the 7th rib and pumping concrete over the shutter to form a concrete plug. Installation of void formers to facilitate the subsequent excavation through the plug.

2) Raising of the shutter to the crown and pumping 500 m³ of concrete into the cavity and chimney through 300m long placer pipe in a bore hole drilled into the chimney from the tunnel (fig.3).

3) Advancing the tunnel full face for 5m through the concrete plug in 1m rounds and installing another five arch ribs.

4) Drilling of four drainage holes to prevent water pressure building up behind the fault zone.

5) Continuation of tunnelling through the concrete plug and into the fault zone with a 3m*3m crown heading.

6) Timber boarding against the fault face to support loose material as the excavation proceeds.

The crown heading was then advanced with steel supports for four meters by excavation of loose material using hand held tools. It was continued for a further 28m with steel supports using pneumatic hammers. Very poor progress was recorded in this heading due to the fault and the work schedule of the project was affected considerably.

4 RELOCATION OF SURGE CHAMBER

Due to continued delays in tunnelling through the fault, two bore holes were drilled to study a proposal to relocate the surge chamber downstream of the fault. The result of this study led to the conclusion that it will be feasible to construct a shaft at the proposed location, and the construction methods proposed at the original location were adequate to deal with the range of foreseen conditions at the proposed location. The decision was then taken to abandon the 60m length of tunnel excavated through the faulted ground.

6 GROUND CONDITIONS FOR TUNNEL REALIGNMENT

The relocation of the surge chamber required that an alternative tunnel alignment be considered: A tunnel aligned between the new surge chamber position and

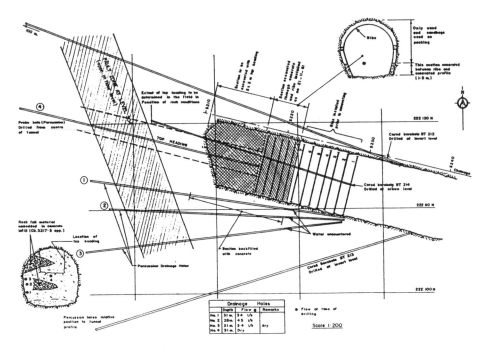

Fig.3 Plan showing area of the tunnel fault

the south drive could potentially result in substantial saving in tunnel length and avoid further delays at the Downstream heading. However, there was concern that the proposed realignment would run in or close to the fault. The author recommended a drill bore hole BT 217 to investigate this possibility and demonstrated that the fault diverged from the first part of the proposed realignment. A detailed reappraisal of the possible location of the fault using all the available evidence led to the conclusion that there are two possible zones in which it might occur. These are shown in fig.4 with the most probable location.

Other considerations affecting a realignment of the tunnel are;
1) The requirement for sufficient ground cover
2) The presence of the lime stone band
3) The influence of other possible faults
4) The need to allow for construction of a second power tunnel and a surge chamber. These features are also shown in fig.3.

Ground cover is low at two stream valleys and this imposed an easterly limit on any realignment.

The main significance of the limestone band arises from the possibility of large water inflows being encountered particularly when a fault occurs. In fact, there appears to be no surface indications

of widespread karsticity such as the travertine deposits encountered above the Downstream heading. Two bore holes in limestone in the South drive showed no evidence of widespread karstification. However, the two possible zones of the fault could both intersect in the limestone horizon and it was considered desirable to avoid this area. Experiences gained in the Downstream heading suggested that if it is unfaulted, limestone offered the best tunnelling conditions.

Taking these constraints into account, the optimum realignment was selected, and this is shown in dotted line in the Fig 4.

The proposed realignment would however involve an estimated tunnel length of 450m in limestone compared with 180m on the original alignment, and within this area other faults may still be encountered.

7. TUNNEL REALIGNMENT

The following was taken into account in selecting the new tunnel alignment as shown in fig.4 to avoid any further delays in Downstream heading
1) The new surge chamber was in a known ground through the excavation tunnel.
2) Avoidance of the major fault crossing the new tunnel line as interpreted by the exploratory bore hole data.

2558

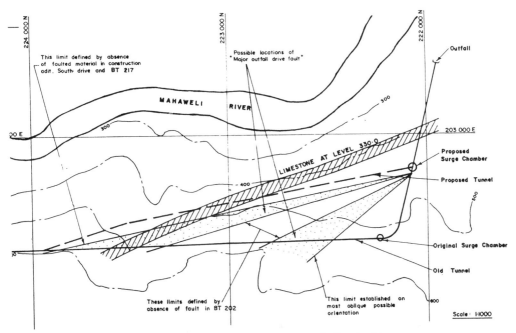

Fig.4 Plan showing various predictions of locations of the fault

3) The new alignment would shorten the tunnel length by 202m.

4) The absence of surface indications of widespread karsticity on the slopes above the portal area, and two local bore holes in to the limestone showing no evidence of such solution cavities at the tunnel level.

5) Subsurface investigations revealed that the original alignment will encounter the fault a second time and may pass through faulted limestone area.

Excavation was carried out satisfactorily through the realignment and no major problems were encountered.

CONCLUSION

An unforeseen major fault resulted in rescheduling of the project programme, relocation of the surge shaft and Tunnel realignment. Far more comprehensive investigations, time permitting, may have discovered the fault.

Subsurface and surface investigations are an important part of project planing and such investigations must be carried out in sufficient detail to ensure cost minimisation and completion on schedule.

ACKNOWLEDGMENT

The author was a member of the geological team of Sir Alexander Gibb & Partners, Consulting Engineers for the Victoria Hydro Electric Project. He wishes to thank the geological team of the Victoria Project for providing technical advice and support.

Some of major distinctions between the two kinds of rock movements caused by open pit mining and underground mining

Des distinctions principales entre deux types de mouvement rocheux à cause de l'exploitation minière ouverte et souterraine

Xu Jiamo

Institute of Geology, Academia Sinica, Beijing, People's Republic of China

ABSTRACT: In the study of the regularity of rock movement caused by mining, many of papers in the past were concerned with undergraund mining and few of them with open pit mining. Summarizing the more than 30-year date of monitoring of mining in Fushun coal mine, liaoning Province, China, using experiences of some other mines, and contrasting one with the other, in the paper the author has discussed and expounded some major distinctions between the two kinds of rock movements caused by open pit mining and underground mining from the following 7 principal aspects.

RESUME: Recapitulant les donnèes de surveillance plus de 30 ans au Mine de Charbonnage Fushun, ainsi des expèriences venant d'autre mines, le papié ici discute et conclure avec la mèthode xomtraste, 7 distinctions principales entre deux types de mouvement rocheux à cause de l'exploitation minière ouverte et souterrain, au niveau de la géologie ingénieur.

1 CURVE OF SUBSIDENCE

Wherher open pit mining or underground mining, generally, the range of the rock movement caused by them is much bigger than the area of excavation. According to elastic theory, all the two kinds of mining can cause the subsidence and rise of ground. But in underground mining, the long-term rise of ground is seldom measured. This is because the behaviour of deformation of rockmass as a whole doesn't coincide with the costitutive relation of linear elastisity due to dense jointing in it . And so the ground movement caused by underground mining is usually subsidence, which is vividly called "subsidence basin ". The fandamental curve of vertical displacement given in Fig.1 is based on the model tests (Xu Jiamo 1983) and the measuring data in several open pit mines. Fig . 1 Shows that, open pit mining can cause the rise at the bottom of a open pit and lower parts of the two slopes; but in underground mining, the ground over the excavation area subsides. In the plane model (Fig .1) , there are two subsidence points of maximum , whose positions coincide with the two highest points in the slopes . A point at which only horizontal displacement occurs should exsit between the subsidence zone and the rise zone , and the position of the point goes down with an increase of depth of the open pit . Naturally, from the point into a inside of the slope, a curve in which no horizontal displacement occurs must also exist, and its position should also go down with an increase of depth of the open pit .

In fact, there are also many small benches of a sublevel in a slope of an open pit . Therefore, a real

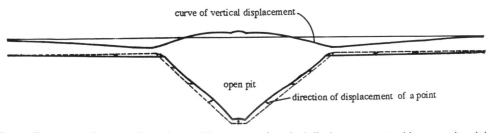

curve of vertical displacement

open pit

direction of displacement of a point

Fig. 1. The sketch of slope deformation and the curves of vertical displacement caused by open pit mining

curve of the subsidence and the rise caused by open pit mining is more complex in detail. In comparison, a curve of ground subsidence in underground mining is simpler and smoother, and so it is mose easy to be expressed in a mathematical form .

No matter which mining method , if the dip angle of stratification plane on the profile is not equal to zero, the symmetenrization of the subsidence curve will vary correspondingly . The distance of ground movement in the direction of the dip of strata due to excavation effect is longer than all the other directions . And when stratification plane is steep and dips into the hill , under conditions of open pit mining the appearance above mentioned is more obvious than that under conditions of underground mining . The reason may be that , under conditions of underground mining , the rock movement toward the extracted space is more difficult than that in the open pit mining due to much bigger limitation.

2 RANGE OF GROUND MOVEMENT CAUSED BY MINING

The range of ground movement caused by underground mining is also affected by many factors . But a definite range can be determined theoritically by means of the concept of limit angle . The limit angles in Fushun mine zone are over 50 °. According to it , the influence radius of underground mining is smaller than the depth of mining. For the range of ground movement caused by open pit mining, if a line is drawn from the slope foot in pit to the margin of the ground morement, the angle between the line and a horizontal plane is much smaller than the limit angle . Such the angle in W200 section of the mine is only about 10°. It shows the range of influence of open pit mining on ground movement is much bigger than that of underground mining . This can also be understood

in this way : The angle of slope in open pit is generally smaller than limit angle for underground mining , and the ground movement caused by open pit mining generally occurs in a distance outside the open pit . Fig. 2 shows the relationship of the layout of measuring points in open pit mining with the depth of mining.

According to the chinese norm for measuring in mine , k=1; and the Soviet Union, k=1.5. In light of experience in measuring , especially in a area of the rock slope of topping failure , K can be chosed as 2. The range of layout of control points, is chosen between 30-50m for China, 100-150m for Soviet Union. In fact , the range has a relation with the depth H of mining, which is denoted as ηH in this paper, and $\eta < 1$. From the paramaters of measuring norm for mine under conditions of open pit mining shown in Fig. 2 , a conclusion that the range of ground movement caused by open pit mining is bigger than that by underground mining in a same depth and with a critical or subcritical area of extraction can be drawn . The reason for the situation is relevant not only to the non-bound laterial movement of rockmass toward the open pit in open pit mining , but also to the more enough development of ground movement characterized by creep (because open pit mining often continues for a comparative long time). And so as long as a long-term monitoring is kept , a ground movement caused by open pit mining can be detected out in the distance awy from the pit .

The conclusion above mentioned is correct at least for the ground movemend in a zone of the rock slope of typical toppling failure.

3 REGULARITIES OF THE DISPLACEMENT AND DAMAGE TO ROCK MASS WITH DEPTH OF MINING

For the rockmass in which there is no fault, and

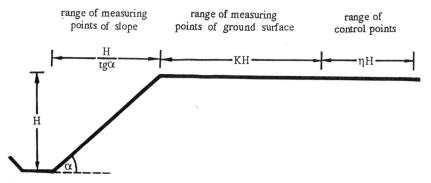

Fig. 2   Sketch of relationship of the layout of measuing points inside and outside the open pit with the depth of mining

therefore the deformation is coordinative, one of the regularities of its displacement due to mining is that, in a same displacement vector trajectory , the nearer a point is away from the excavation space , the bigger the module of the displacement vector of the poit is, conversely, the smaller. In underground mining, if a measuring shaft is from ground surface to the empty along a same displacement vector trajectory, the magnitutes of displacements in various depth incerase gradually. And under conditions of open pit mining, the magnitutes of displacements from the surface to the inside decrease gradually. Since 1987, the measured displacements of many ground surface points and ones in the open pit have shown the regularity above mentioned.

It is necessary to point out that in underground mining, even though in spacial distribution the displacements are in coincidence with the regularity mentioned alove, the shape of the space where the rock movement ocrrus is in a characteristic that the upper is wider and the lower is narrower, like a normal fan, as viewed from a profile .

As to the regularity of damage to rock mass , a lot of experience in underground nining show that the damage starts at the boundary of the extraction space (espesially, the roof), and then develops toward the upper rock mass. If caving does not occur , the shape of the fractured rockmass is characterized by that the upper is smaller and the lower is bigger, like a reversed fan . The reversed fan takes up only a small part in the normal fan mentioned above . In this meaning, it can be said, for underground mining, the damage starts in the deep rockmass and become slight up to graund surface . But if developing up to ground surface , the damage and deformation of the rockmass is more or less serious than that near the surface , because the stress state at the surface is more close to that of plane stress, in which the confiniment of rockmass is smaller than that in the deep. This is also a kind of surface effect

in the deformation and damage. But under conditons of open pit mining, according to a lot of experience, the damage to rockmass, generally takes place at first from the surface of rock mass , than develops into the inside of rockmass. This is in coincidence with the distribution zones of tensile strain in slope rockmass, whether the direct tensiled ones or indirect ones derived under the action of compression and squeezing.

4 THE EFFECT OF FAULTING ON ROCK MOVEMENT AND ROCKMASS DEFORMATION CAUSED BY MINING

The existence of faults in mining zones, will break down the continuity of rockmass, and also often disturb the normal regularity of rock movement, which makes the displacement field caused by mining complex. It can be called the fault effect on rock movement caused by mining. This kind of fault effect can appear in many aspects. For faults with same conditions, the effects are different from each other due to different relative positions to a extraction area. The factors, such as fault size dip angle , and direchon of fault dip, the strength of fault plane and the compliance of rock mass, can affect the effect .

Under conditions of open pit mining, the pit is usually over the level of underground mining and the faults outside the pit are always in a tensile zone. According to the analysis of stress field in slope , the tensile stress in the rock mass increases with the reduction of depth. The relative (open) displacements between the two walls of a fault decrease gradually from ground to inside, and at the time result in a more and more relative subsidence of the upper wall. The maximum open width of fault and of the maximum displacement occur in ground

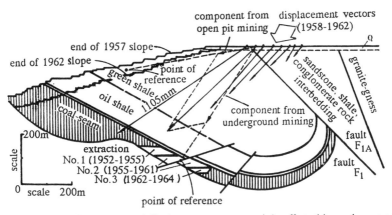

Fig. 3 . Directions of the measured displacement vectors mainly affected by underground mining

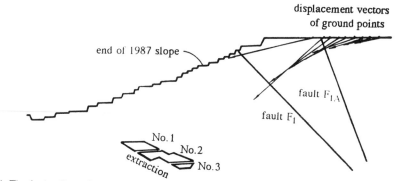

Fig. 4. The fault offect of open pit mining on a change of the displacemtnt directions of the ground points.

surface and the width and displacement decrease as the depth increases, which is different from that under conditions of underground mining. For the latter, a part of fault can be opened at first in a deep positon near the extraction area, and ground surface steeps do not occur in a short time. when caused by underground mining, the bend of fault plane dipping reversely into extraction area is in a convex shape toward the extraction area; and when caused by open pit mining, in a concave shape toward the pit .

The two kinds of mining methods can cause different fault effect separately, and so the mining methods can also be distinguished in the light of the different fault effects withe special features. As mentioned above , the fault effect of ground movement caused by open pit mining is characterized by that the directions of displacement vectors of the measuring points in the two walls of a fault are different from those in underground mining, which, of course, is an important aspect of the distinction between the fault effects caused by the two kinds of mining methods and is verified by the measured result of the ground movement in Fushun west open pit mine which is mined by underground mining in lower levels . Fig, 3 shows the measured displace ment vectors of groand points in underground mining and open pit mining at the same time, At that time, the slope was very gentle , the ground movement was mainly controlled by underground mining, and so the effect of fault on the displace-ment vectors is not obvious. After 30 years, the slope became steep due to open pit mining, which plays an important role in the ground movement (at that time, the underground mining had been stopod for 20 years) . The displacement vectors measured from 1979 to 1987 are shown in Fig. 4. It can be found from Fig. 4 that the open pit mining has an obvious fault effect on the directions of displace-ment vectors of the ground points. Besides, themeasured vertical displacements of the points in the upper wall of fault F_{1A} are bigger than those of

the lower wall in the two years after 1986 , which also indirectly demonstrates that in the condition of open mining, the effect of the faults on the change of the displacement direction of the points is more obvious than that in underground mining.

5 DESTRUCTIBILITY

The extraction amount in underground mining is generally smaller than that in open pit mining. But the destructibility of the underground mining is bigger than that of the open pit mining. This kind of destructibility is mainly from the strong gravity potential energy of a part of rockmass above the level of mining, partly from the elastic potential energy of rockmass around the excavating zone. After a extracrion space of a certain size is formed in rockmass, this kind of gravity polential energy will turn into a strong work making the rockmass deform and fail. The destructibility of underground mining is bigger than that of open pit mining. This is because the volume of rockmass displaced by this kind of mining is bigger than that by open pit min-ing . According to the theory on energy and energy transformation, it is not difficult to calculate these two amounts of work , into which the gravity poten-tial energe is transformed in two kinds of mining methods under comparable geometrical conditions, and a conclusion that the works will be very differ-ent can be drawn. For the open pit mining, besides landslide, generally the deformation of slope rock-mass is mainly lateral movement, and the work transformed gravity potentiel energe into is smaller.

Another obvious characteristic of undergraund mining is that it can lead to failuer of a rock mass in a short time (e. g., several months) of mining.

The destructibility of undergraund mining for the failure of ground and buildings is generally stronger than that of open pit mining, for which there is another reason. This is because the effect of underground mining on stress state of a ground point, shape of the subsidence trough and the position of Its center can change correspondingly due to the continous change of position of stope, and so can turn a point in a tensile zone (or compression zone) into it in a comperssion zone (or tensile zone) . Strain in a point also can change in nature, and so makes buldings in the corresponding zone undergo damage easily. Similarly because of a hight rate of strain, rapid damage to buildings is more easily caused. But the ground outside an open pit is always in a tensile state. Exception of the special condition of the slope rockmass deformation caused by toppling , generally the rate of strain is lower, damage to buildings is also slower occasioned.

6 THE RETARDATION TIME OF THE ROCK MOVEMENT

In excavation of a large-size, obvious viscosity of rock mass is noticeble. It is relavent to the factors: the mechanical effect (especially, the shear effect) of weak planes in a rockmass, the mechanical properties of rock and the load condition (such as mining depth, mining thickness, methods of roof control and the velocity of working face advancing). From the mining zone, exactly from the working face, the movement of deformation occurs gradually from the close to the distant. Under otherwise equal conditions , starting time and amount of displacement of a point in rock-mass is relevant to the position of the point relative to the working face. The deformation of rockmass caused by excavation, generally speaking, is a process of progressive deformation. Therefore, a measured displacement of a point at the monent does not depend on the state of mining at the monent, but depend on a excavation history. As a matter of experience of the mine, the destructibility time of the deformation appearence of ground point in different positions can be roughly estimated.

The rockmass deformation caused by open pit mining is also different from that caused by underground mining in continued time. The continued time of rockmass deformation caused by a same kind of mining method also depends on many factors. But generally, the continued time for open pit mining is shorter than that for underground mining.The reason is that, for a excavation element of a some volume in a same depth of mining, the volume for rockmass deformation from underground mining is much bigger than that from open pit mining . Such a larger volume of the deformation of rockmass is due to the transformation of gravity potential energe of rockmass above the extraction area into plastic work, and the elestic potential

energy is the secondary. The plastic work is mainly wasted on the overcomming of the internal friction (mainly, friction between discontinuous srufaces in rockmass) resistance when the rockmass deforms. In underground mining, for the rockmass with stratified structure, the bending-shearing occurs in first shase. So the plastic work is wasted in larger amounts and the continued time is also long. In the rockmass deformation from open pit mining, generally, the plastic work is wasted in smallert amounts with the exception of the slope rockmass of toppling failure . Both ground point 9 and point 3 in the open pit are in the section W200. The two points are 270m apart. The difference of elevation is 32m. A rock movement of a peak displacement rate propagates from point 3 to 9 in about a month. Under conditions of underground mining, the time need from the end of excavation to that of movement (or deformation) is generally about 1-2 years or 2-3 years. But in open pit mining, the time for deformation is never continued so ling.

7 DIRECTIONS OF DISPLACEMTNT VECTORS OF MEASURING POINTS

If there is no fault near measuring points that does not make rockmass deformation coordinational, the displacement vectors of measuring points caused whether by open pit mining or by underground mining , are characterized by directing sluggishly towards mining position or working face. It can be said that, therefore, under conditions of open pit mining, the displacement vectors of measuring points are generally characterized by that their horizontal component is bigger than their vertical one , i . e . , lateral movement towards the open pit is main. Reversly, in a very big range (relevant to mining depth) near mining position, the vertical component of displacement is bigger than the horizontal one, i . e ., subsidence is main . This kind of displacement regularity of rockmass in mining engineering, can be verified not only by the measured result in open pit mining or underground mining, but also by the model tests of underground mining and of a soft material model of slope rockmass (Xu Jiamo, 1983).

THE MEANING OF STUDY

From the point of view whether as an engineering geological environment factor or an engineering dynamic grological function , a deep study of the characteristics and differences in the rock movement caused by oper pit mining and underground mining, is of importance in theory and application for the

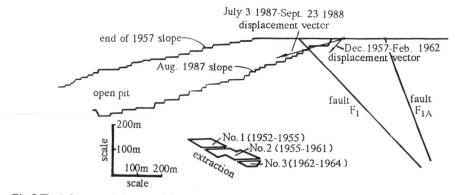

Fig.5. The influence of the two minings of different space-positions on the displacement direction of the point at a same position

development of engineering geology.

For example, after the conception mentioned above in the aspect 7 is understood, when we need to distinguish and determine the major cause for a rock movement in a mining zone, a method (Xu Jiamo, 1988) can be proposed to determine the cause of ground movement of deformation by using the vectors of displacement of measuring points. Briefly speaking, in two mining spaces (or pits), whether in a same kind of mining method or different ones, for the determination of degree of influence of each space on displacement of a measuring poit. Firstly, two rays from the measuring point to two reference point (e . g., the geometrical centre of a space. But sometimes, we did not choose the centre) in the two space need to be drawn in the section through the poit and the two reference points in the two mining spaces, in general, the displacement voctor of the point also is in the section. And then we have two angles included between the displacement voctor of the point and the two rays. So according to the parallelogram method of resolution of vector, we will obtain the two component of the displacement vector of the point separtely in the two rays. The two components respectively and quantitatively represent each influence of the two mining spaces on the displacement of this point. In the past, it is very difficult quantitatively to deter-

mine the major effect factor of the rock movement for the zone influenced by underground mining and open pit mining at the same time. But this method can be applied approximately to quantify each one of the two effects and to determine the major effect factor (see Fig. 3). According to the measured results in Fushun west open pit mine, the displacement voctors are shown in Fig. 5. It is easy from the figure to find out which mining acting on the displacements of measuring points in different terms.

REFERENCES

Xu Jiamo. 1983. An experimental study of deformation of rock mass slope with the use of a soft photoelastic material model. Selection of theses of Master degree, 1981. Edited by the Institute of Geology, Academia Sinica, PP. 191-195. Sciences and technology Publishing house, Beijing, China. (in Chinese)

Xu Jiamo. 1988. A method and examples for the determination of cause of ground movement according to the characteristics of displacement vectors. The selection of the articles in the third Chinese National Engineering Geological Congress, PP. 1205-1211, Chengdu Univ. Publishing house. (in Chinese)

Treatment of tension developed in the roof of mine enlarged openings
Le traitement de tension developpée dans le toit des ouvertures agrandies de mines

M.A.Yassein
Al-Azhar University, Cairo, Egypt

ABSTRACT: To study the tensile stress developed in the roof of mine enlarged openings, photoelastic model was prepared and examined. The model was loaded in two directions to simulate the real case in the underground, then the opening was excavated. The stress field around all the cross-sections generated during the enlargement stages was studied. The driving starts with a central pioneer heading, and enlarged through three stages to reach a horseshoe cross-section. All the cross-sections generated during the enlargement stages are free of tension.

  The results of the present study indicate that, it is possible to arrange the sequences of the enlargement of underground openings to avoid the tension developed in the roof during the enlargement stages.

RESUME: Afin d'etudier les contraintes de tension developpees dans le toit des ouvertures de mine, lors de l'elargissement, un modele photoelastique a ete prepare et examine. Celui-ci est charge dans deux directions pour simuler le vria cas sous-terrain, lorsque l'ouverture est creuse. Le cheminement des contraintes autour de la coupe en travers, qui se produit pendant les successions d'agrandissement, est etudie. Le debut du creusement a ete effectue par un faconnement en tete, a partir du point central. Il s'agrandit par trois etaps, atteignant une forone de fer a cheval. Toutes les coupes en travers se produisant pendant les successions d'agrandissement ne continnent pas de tension dans le toit.

  Les resultats de l'etude indiquent qu'il est tout a fait possible d'arranger les successions d'agrandissement de l'ouverture du sous-terrain, afin d'eviter la tension developpe dans le toit pendant les etaps de l'enlargissement.

1 INTRODUCTION

It is generally known that, because of the discontinuities rocks in natural state are relatively weak in tension. Reduction of tension developed in a rock mass during the driving of underground structures is important for the stability of roofs of underground openings. Mining engineers always try to reduce the tension developed in the roof of an underground openings by generating a geometrical cross-section which gives minimum tensile stresses in the roof during the enlargement stages.

A horseshoe-shaped section has been adopted by Tatsuo and Yoechi(1985) for the first time in Japan for covern stabilty and retionalization of construction. The behaviour of the surrounding rocks during excavation with particular emphasis on details of deformation behaviour is given. In the construction of Bochom Ring Road, Steineheuser and Maidl(1985) discribed a top heading method of excavation which is used to control the encountered geological difficulties. Andrasky, Ramer and Berger (1983) discribed different methods of excavation used in the construction of a tramway in Zurich. The evaluation of the project was made on the basis of vibration and nois caused by the different excavation methods and on the basis of their respective construction time and cost. In the design and construction of the Timpagrande powerhouse, Brosetto and Giuseppetti (1983) showed that the design was carried out in three stages using different geomechanical parameters. The final design was based on a back analysis from a trail enlargement. In the design and construction of the Furka base tunnel, Amberg (1983) presented that the assesment of mechanical properties of various rock series led to the choice of a hourseshoe profile with straight sides. In U.S.S.R. the methods of driving horizontal excavations are treated by Pokrovsky (1980).

In the present study photoelastic model is prepared to study the tensile stress developed in the roof of mine openings when enlarged. The driving starts with a central pioneer heading, and enlarged through three stages to reach a horseshoe cross-sectional shape. All the shapes generated during the enlargment stages are free of tension in the roof.

2 EXPERIMENTAL WORK

The experimental procedures of the present study are carried out as follow.

2.1 Model material

The plastic used in the present study was obtained by mixing a clear epoxy resin commercially known as Araldite D with Hardener HY951 and Dibutyl Phathlate. The suitable composition for this study was 100 parts Araldite D, 12 parts Hardner HY951 and 28 parts Dibutyl Phathalate by weight. It is found that this composition takes its semi cure state after 14 hours, and is like a hard rubber mass. In this state working operations such as cutting and drilling do not cause any working stresses. The mechanical and photoelastic properties of this material were determined by Yassein (1980). The model dimensions were 10.5 cm. hieght, 10.5 cm. width and 4 cm. thickness.

2.2 Loads Applied to the model

Denlhans (1958) stated that, A theoritical consideration of the virgin stress which can occur in rock which is free from tectonic forces suggests that the horizontal stress may be of the order of one quarter of the vertical stress, depending upon the poisson's ratio of material and the degree of horizontal restraint imposed by the surrounding rock.

Under these applied stress conditions, tensile fracture of the roof and the floor of excavation such as haulages can be anticipated. Consequently, in planning this model study, it was considered necessary to apply stresses of 0.25 ratio of lateral to vertical applied stress.

2.3 Excavation method

After the model has been loaded in two directions, the following excavation procedure was employed using an electrical drilling machine:

1. A square cross-sectional shape at the centre of the opening was excavated through the thickness of the model.

2. The full width of the roof was enlarged in two steps progressing from top arch and the two side arches.

3. The full width of the floor was enlarged in two steps progressing from invert arch in the bottom and the two straight sides.

The scheme of driving stages and the isochromatics of each stage are shown in figure 1.

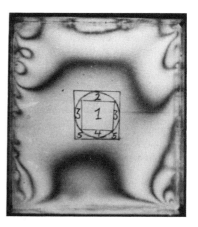

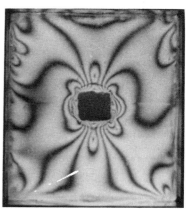

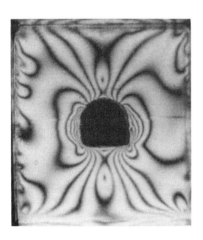

Figure 1
Isochromatics of both primary state
and stages of enlargement

3. RESULTS AND DISCUSSION

The results of this study are presented in the form of stress distribution curves. The stress concentration factor is expressed as the ratio between the tangential stress σ'_ϕ and the vertical applied stress σ_V. From the isochromatics representing the primary state of stress field applied to the model it is found that the vertical applied stress is equivelant to 3.3 fringes.

Figure 2 shows the stress distribution around the pilot opening which has a square cross-sectional shape. This figure indicates a tensile stress concentration factor of about (-0.5) on the roof and floor of the opening.

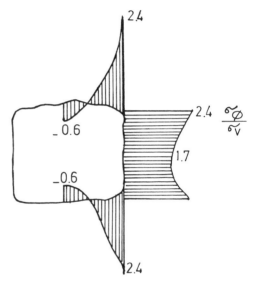

Figure 2
Stress distribution around
pilot opening

Furthermore, it indicates a compressive stress concentration factor of about +2.42 on the corners of the opening. The decrease in the stress concentration factor on the corners due to the curvature which happened during the drilling operation. The choice of large square pilot opening was based on some considerations which are required in practice such as

-dewatering of the rock.

-rock stress relief for the subsequent careful blasting work with rock falling into the free opening of the pilot opening.

-easier ventilation during main excavation.

-accurate survey of the continually changing geological conditions.

The tension developed in the roof of the pilot opening can be avoided by choosing another cross-sectional shape during the enlargement stages. Figure 3 shows the stress distribution around the first stage of enlargement which has a segmental arch cross-sectional shape. The figure indicates that the roof of the opening becomes free of tension. It indicates also that the compressive stress concentration factor remains the same as that which occurs around the pilot opening.

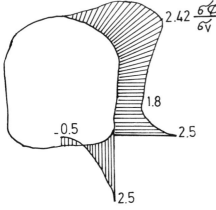

Figure 3
Stress distribution around first stage
of enlargement

Figure 4 shows the stress distribution around the second stage of enlargement which has a semicircular arch cross-sectional shape.

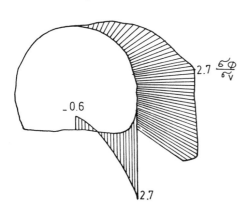

Figure 4
Stress distribution around
second stage of enlargement

This figure indicates an increase in the stress concentration factor on the side walls by about 11.5% more than the stress concentration on the side walls of both the pilot opening and the first stage of enlargement, where the roof is still free of tension.

The stress distribution around the final stage of enlargement which has a hourseshoe with straight sides cross-sectional shape is shown in figure 5.

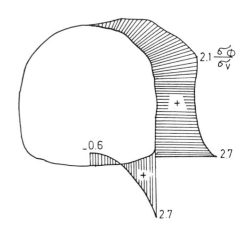

Figure 5
Stress distribution around
final stage of enlargement

This figure indicates that the roof of the opening is free of tension and the maximum stress concentration occurs at the junction of the wall and the floor. Fracture of rock at this point would not seriously affect the stability of the ordinary mine opening because the fractured rock would not be a hazard and would not require support as would rock; which broke from the roof or walls. This means that it is possible to continue the enlargement of the final shape in down ward direction without increasing the critical stress concentration.

4 CONCLUSION

The results of the present study indicate that, it is possible to arrange the sequences of enlargement stages of mine openings to avoid the tension developed in the roof. A horseshoe with straight sides cross-sectional shape is generated from a central pioneer square cross-sectional shape through three stages of enlargement each of them is free of tension in the roof.

REFERENCES

Amberg, R. (1983). Design and construction of the Furka base tunnel: Rock Mechanics and Rock Engineering Vol. 6 No. 4.

Andraskay, E., Ramer, E. & Berger, E. (1983).Choice of rock excavation method for tramway tunnel in Zurich: Rock Mechanics and Rock Engineering, Vol. 16 No. 2.

Borsetto, M., Giuseppetti, G. Martinetti, S., Ribacchi, R.& Silvestri, T. (1983). Design and construction of the Timpagranda power-house: Rock Mechanics and Rock Engineering, Vol.16 No.2.

Denkhaus, H.G. (1958). The application of the mathematical theory of elasticity to problems of stress in hard rock at great depth: Ass.Min. Mngrs.S. Africa, Vol.1958/59.

Pokrovsky, N.M. (1980). Driving horizontal workings and tunnels. 257 pp. Mir Publishers, Moscow.

Steineheuser, G. & Maidl, B. (1985). Bochumring road: planning and choice of construction method: Rock Mechanics and Rock Engineering Nol.18 No.2.

Tastuo, M.& Youichi, M.(1985).Deformation behaviour of a large underground cavern: Rock Mechanics and Rock Engineering Vol.18 No.4.

Yassein,M.A.(1980). Application of photoelasticity to the design of multiple and multi-cross mine openings. Ph.D. Desertation, Hungarian Academy of Sciences.

Rockbursting characteristics and structural effects of rock mass

Caractéristiques de cassures de roches et effets de structure de masses rocheuses

Yian Tan
China International Engineering Consulting Corporation, Beijing, People's Republic of China

Guangzhong Sun
Institute of Geology, Academia Sinica, Beijing, People's Republic of China

Guochang Liu
Xian College of Geology, Xian, People's Republic of China

ABSTRACT: This paper is concerned with the rockbursting characteristics and the structural effects of the rock masses in the diversion tunnel of Tianshengqiao Hydropower Station. The controling effects of rock mass structure on the occurrence and intensity of the rockburst is demonstrated.

Rockbursts do not take place in the rock mass with crushed structure and multicrevice, but in Class 1 and Class 2 rock mass, which are intact and / or brittle.

In either of the two latter cases, even for the same class of rock mass, the rockburst intensity varies because of the aeolotropism of the rock mass in deformation, strength and failure. When angle β between major set of joint and the maximum principal stress σ_1 is within $30°-0°$, the smaller the β is, the higher the rockburst intensity will be; When $\beta = 30°-45°$, no rockburst happens in the rock mass but there may be shear failure, and when $\beta > 45°$, weak rockburst is usual.

The above structural effects on rockbursting have been proved by rock mechanic tests carried out by the authors.

RESUME: Ont ete exposees dans cet article des caracterlstiques de roche—eclat et des ejjets de structure de roche concernant des roches environantes autour des tunnels de la Centrale hydraullque a Tian Shengqiao.

Des roche—eclats ne se produient pas dans des roches cassees ou jendues, Ils arrivent plus souvent a des roches en grand bloc, compactes et de nature jragile de classe 1 et 2. Dans un de ces derniers cas, a cause de l'anlsotropie sur la dejormation 'I' intensite et d'autre jacteur, des intensites de roche—eclats se presentent dijjerentes, meme pour les roches de classe indentiques. Lorsque I' angle β entre le joint principal et la contrainte maxi est de 30 a 0 degree, plus β est petit, plus intense le roche—eclat. Et l' intensite de roche—eclat se change de jaible a importante. Lorsque β est de 30–45 degree, ne se produit pas le roche—eclat, mais Il arrive blen possible parjols quelque endommagement de cisaillage. Lorsque $\beta > 45$ degree, on vois produire souvent des roche—eclats moins importants.

Les constatations susmentionnees ont ete demontrees par des experimentations sur la mecanique de roche—eclat.

1. ROCKBURSTING CHARACTERISTICS IN THE TUNNEL OF TIANSHENGQIAO HYDROPOWER STATION

Tianshengqiao Hydropower Station with an installed capatity of 1320 MW is located on the River Napan, China. There are three 9776 meter long diversion tunnels with a diameter of 10.8 m between the dam and the power house. The rock type from the dam site to 2000 m in the direction to the power house is Triassic limestone and dolomitite, and then changes into interstratified sandstone and shale all the way to the power house. The average thickness of the overburden of the tunnels is 400 m, the maximum is 800 m. By the end of june, 1988, 26 segments in the 1390 m long No.2 adit and the completed 637 meters of No.1 diversion tunnel subjected to rockbursts of weak to moderate intensity. The total length of the segments is 330 m. The greatest depths of the rockbursts are 1.5–2 m and the largest volume of bursted rock fragments is 65 m^3. Construction equipments were damaged, workers' lives were threatened and reduced the tunnelling speed.

The maximum principal stress (σ_1) measured in–situ was 21.0–26.0 Mpa, the strike is NW$310°-340°$ and the dip is $45°-65°$.

Rockbursting characteristics in the tunnels are as follows:

• Rockbursts all occur in the limestone segments of the tunnel where rock is brittle, no rockburst occurs in sandstone interstratified with shale.

• During tunnelling not every part of the limestone segments subjected to rockburst, no rockburst happened in Class 3 rock mass and in those which are worse than Class 3, according to geomechanical classification of rock masses introduced by Z. T. Bieniawski; Class 4 and Class 5 rock masses subjected to rock fall or to plastic failure due to the fractures therein, filled with mud and ground water. So the elastic energy can not easly concentrate in those rock masses.

• No rockburst has occured in Class 1 rock mass owing to its high compressive strength, and the lack of initial stress or of the maximum tangential stress in

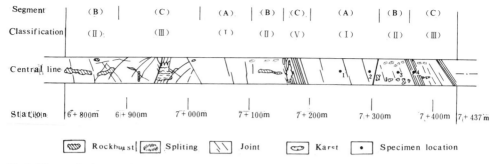

Segment	(B)	(C)	(A)	(B)	(C)	(A)	(B)	(C)
Classification	(Ⅱ)	(Ⅲ)	(Ⅰ)	(Ⅱ)	(Ⅴ)	(Ⅰ)	(Ⅱ)	(Ⅲ)

Central line

| Station | 6+800m | 6+900m | 7+000m | 7+100m | 7+200m | 7+300m | 7+400m | 7+437m |

| Rockburst | Spliting | Joint | Karst | • Specimen location |

Fig.1 The overlook map of geology of No.1 Diversion Tunnel (from 6800 m to 7437 m)
A. Complete segment B. Rockburst segment C. Multicrevice or crushed segment

the tunnel periphery. The rockburst mainly occurs in the sectors of Class 2 rock mass, namely, in those sectors which are neither intact nor crushed, and in the parts of tunnel where the rock masses are dry and joints are moderately developed.

From the above observation, it can be concluded that the tunnels of this hydropower station appear to have as a role 'intact sector—rockbursting sector—rupture (or rock fall) sector' arranged orderly and repeatedly, as shown in Fig.1.

• Rockbursts were mainly encountered in Class 2 rock mass. The intensities are very different since rockburst is not only related to the earth's crustal stress, rock properties and the presence of ground water, but also to the structure of rock masses, according to the geomechanical classification mentioned above and the angle β between the major set of joint and the initial maximum principal stress σ_1 in the tunnels. A smaller β corresponds to a more intensive rockburst in the sectors of Class 2 rock mass of the tunnel. When β increases upto $20°-30°$, the degree of rockburst reduces to lower level; When β is in $30°-45°$, no rockburst occurs, but shear failure may take place sametimes; When $\beta > 45°$, weak rockburst usually occurs.

• Rockburst occurs usually at the upper left and the lower right parts of the tunnel section, and the angle between the connecting line of the two spliting or cracking centres and horizontal line is usually $50° \pm 10°$. The centric line is just perpendiclar to the direction of maximum initial principal stress σ_1, as shown in Fig.2.

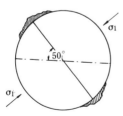

Fig.2 Rockburst section at station 820 m of No.2 adit.

• The bursting rock fragments shoting into th tunnel have a definite initial velocity and a scattering angle.

In addition, the scattering fragments on the bottom of the tunnel form a gradation, namely, the fragments in larger sizes are distributed   mainly near to its centre

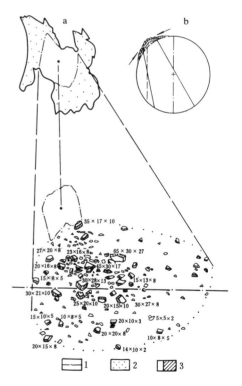

Fig.3 Scattering fragments of rockburst and the section of N0.1 tunnel at station 7310 m
1. Central line of the tunnel 2. First burst face 3. Second and Third burst face

2574

and the smaller the fragments are distributed, the farther from the centre. In other words, rockburst is dynamic in its charaters, see Fig.3.

It can be seen that rockbursting activity is different from rock fall, which is only a free—fall movement caused by gravity.

2. STRUCYURE EFFECTS OF ROCK MASSES ON ROCKBURSTING

The rockbursting characters mentioned above show that rock mass structure has controling effect on the occurrence and intensity of rockburst. In other words, under the initial of the same intensity not every part of the tunnel with same rock type will sufer from rockbursting. In certain segments rockburst may occur, in some other segments rockburst may not occur. Rockburst in some segments may be intensive, but it can be much less intensive in some other segments. The reason is that different sectors of the tunnel have rock masses with different structures and there exists aeolotropism in the same rock mass. In this connection, the structural effects of the rock mass have been proved by rockbursting tests carried out by the authors.

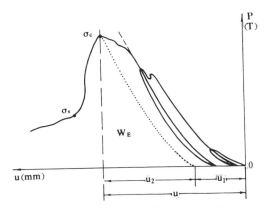

Fig.4 The Complete load—displacement Curve

2.1 Rock mechanical test of rockbursting

A number of limestone samples, mechanical classified as Class 1,2 and 3 rock masses, were collected from different segments of No.1 tunnel. The limestone specimens were prepared in form of cylinder with a diameter of 50 mm and a height of 100 mm.

The specimens were uniaxially compressed till its faliure and flew away from a hydraulic serve controlled rock testing machine. From the equation $V_o = s / (2h / g)^{1/2}$ we calculate the initial velocities of flown fragments and from the equation:

$$W_{eff} = \sum_{i=1}^{n} \frac{1}{2} m_i v_{oi}^2$$

where, n is number of rock fragments, we obtained

the total effective bursting energy (W_{eff}) during rockbursting, and at same time the complete load—displacement curve and the other relevant mechanical indexes, as shown in Fig.4. So it is possible to approach the operating rules in the rockburst, such as the relationship between the effective bursting energy and the rock mass structure, etc..

2.2 Relationships between W_{eff} and the Classes of rock masses and strain rate

Tests were made with the strain rates of $10^{-3} - 10^{-7} /$ sec. upon 11 specimens of Class 1 rock mass. The complete load—displacement curves, for five typical samples for example, are shown in Fig.5. The relationships between the effective bursting energy and the strain rate, and brittleness index of rock Ku $(= u / u_1)$, and the stress drop $\sigma_{dr}(= \sigma_c - \sigma_s)$ on the post—failure curve, etc., are shown in Fig.6 and Table 1.

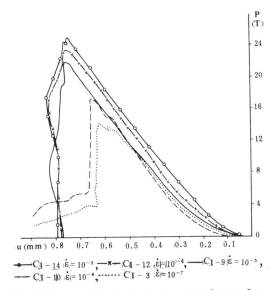

Fig.5 The complete load—displacement Curves for Class 1 rock mass under different strain rate

These results show that:
1. Under the strain rate of $10^{-3} - 10^{-6} /$ sec 10 specimens of Class 1 rock mass subjected to rockbursting but no rockburst has occured in the other one of the 11 specimens under $\dot{\varepsilon} = 10^{-7} /$ sec.
2. The brittleness indexes (Ku) of the specimens are all larger than 4, and it increases quickly with an increase in strain rate. In other words, the faster the strain rate is, the greater the brittleness index will be. And the more brittle rocks subject more intensive rockbursts.

Table 1 Relationships among W_{eff}, $\dot{\varepsilon}$, K_u and σ_{dr}, etc.

No.spec.	$\dot{\varepsilon}$	σ_c MPa	σ_{dr} Mpa	E 10^4 MPa	E_o 10^4 MPa	K_u	W_{eff} 10^4 erg
C_{1-14}	10^{-3}	133.6	133.6	4.5	1.8	9.4	1475
C_{1-12}	10^{-4}	119.2	119.2	4.4	1.6	6.1	1205
C_{1-9}	10^{-5}	113.0	93.7	4.3	1.5	5.6	433
C_{1-10}	10^{-6}	88.5	60.1	4.2	1.4	5.4	175
C_{1-3}	10^{-7}	73.8	44.8	3.8	1.2	4.5	0

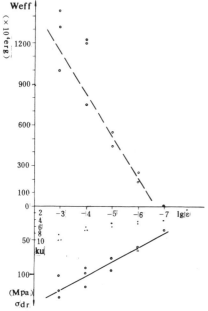

Fig.6 Relationships among W_{eff}, $\lg\dot{\varepsilon}$, K_u and σ_{dr}

3. The peak strengths of the rocks are mostly greater than 100 MPa and increase with an increase of strain rate. The elastic modules (E) are mostly more than 4.0×10^4 Mpa and also increase with an increase of strain rate. It can be seen that the elastic energy in the speciments increase with increasing strain rate; When both $\sigma_c > 88.5$ Mpa and $E > 4.2\times10$ Mpa, rockburst usually occurs.

4. The stress drop (σ_{dr}) after the peak streagth increases with an increase of strain rate, and the capacity of releasing energy will increase with increasing strain rate. When $\sigma_{dr} > 60$ Mpa, rockburst usually occurs. In short, the bursting property indexes of the rocks, such as K_u, σ_c, E and σ_{dr}, etc., increase with an increase of strain rate, resulting in a rapid increase in the total effective bursting energy, so that the intensity of the rockburst gets greater.

By testing on 12 speciments of Class 2 rock mass, the results similar to Class 1 rock mass were obained. Very different form Class 1 and Class 2 upon specimens of Class 3 rock mass the testing results,

however, were obtained.

Upon 14 specimens of Class 3 rock mass, which are of multicrevice, tests were conducted with the strain rate of 10^{-3}–10^{-7}/sec. The results show that the indexes K_u, σ_c, σ_{dr} and W_{eff}, etc. increase slightly with an increase in strain rate, but no rockburst occurred in most of the specimens of the Class 3 rock mass, except the two specimens in which rockburst of low intensity occurred under the strain rate $\dot{\varepsilon} = 10^{-3}$/ sec. The reason for this is that the elasitic energy in Class 3 rock mass is very poorly accumulated, there has not been enough energy to release, therefore no rockburst could occur.

From the above analyses of the results of the tests, we can see that the different classes of rock masses have different features of deformantion, strength and failure, and have different capacities of accumulating and releasing energy under the same testing condition, therefore they have different structural effects on the rockbursting of rock masses. This is in line with the obsevation on the rockbursts happened in the tunnels of Tianshengqiao Hydropower Station.

2.3 Relationship between effective bursting energy W_{eff} and the angle β

Taking 40 specimens of the same Class 2 rock mass (d = 50 mm, L = 100 mm), shetched before being used for tests to determine the major set of the joint and to measure the angle β between the major sets of joints in the specimens and the axial stress applied to the specimens, we conducted rock mechanical tests on the rcokbursting with the same strain rate 10^{-5}/ sec. but different structures at $\beta = 0$–$90°$.

The results are shown in Table 2, and in Fig.7 and 8. From the Figures and table containing the results of the tests, a few conclusions are drawn further:

1. When $\beta = 30°$–$45°$, both the peak strength σ_c and the brittleness index $K_u(=u / u_1)$ are of their minimum values, and the permanent deformantion u_1 reaches its maximum value, and the elastic deformantion u_2 is of its minimum value. In this case the storage of the elastic energy (We) is the smallest in the specimens, and the stress drop after the peak strength is the least; Thus the capacity of accumulating or releasing energy is far from resulting in rockburst. At this time, the effective bursing energy

(W_{eff}) equals to zero, and the shear failure generally occur only along the pre-existing joints in the specimen.

2. When $0° < \beta < 30°$, a $10°$ decrement of the β will usually lead to an increment of 15–30% in the peak strength σ_c. The greatest increment can up to 50%, and when $\beta = 0°-10°$, σ_c reaches its maximum. The brittleness index (Ku), elastic energy storage (We) and stress drop (σ_{dr}) increase with decreasing β. So a decrease in β results in an increase in the effective bursting energy with a great rate of 400%. Such that

the intensity of the rockburst also increases rapidly, and when $\beta = 0°-10°$, the rockburst becomes the most intensive of all cases.

3. When $45° < \beta < 90°$, all the values of σ_c, Ku, We and W_{eff}, etc. increase with increasing β, but rockburst are usually not intensive.

The results described above are also in line with the observation on the rockbursting taken place in the diversion tunnel of Tianshengqiao Hydropower Station during excavation.

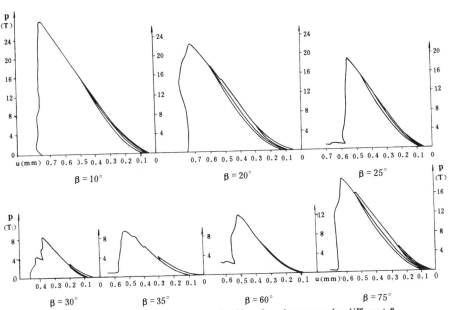

Fig.7 The complete load–displacement curves for Class 2 rock mass under different β

Table 2. Relationship between effective burst energy and the angle β

Indexes	0–10°	$0° < \beta < 30°$ 10–20°	20–30°	$30° \leqslant \beta \leqslant 45$	$45° < 90°$ 45–60°	60–90°
$W_{eff}(10^4 erg)$	500	138	33	0	24	50
Ku	10	6.9	6.1	3.5	3.8	4
$We(10^6 erg)$	655	395	318	175	265	315
σ_c (Mpa)	126.7	97.9	83.7	55.6	78.8	82.9
σ_s (Mpa)	0	0	70	51	40	77
σ_{dr} (Mpa)	124.3	90.4	68.3	42.3	70.3	67.8
Num. of spec.	5	8	5	11	5	6

2577

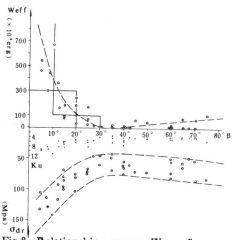

Fig.8. Relationships among W_{eff}, β, σ_{dr} and Ku of Class 2 rock mass

3. SUMMARY

Rockburst is one of the geological catastrophes that takes place in the surrounding rock masses of tunnels, shafts and underground power houses, and that releases accumulated energy violently during excavation. It is generally considered that the rock bursting is related to the stress in the earth's crust, rock properties and presence of ground water and structures of the rock masses. Even if the first three factors are favourable for rockbursting, not every part of the tunnel will suffer from failure, because of the controling effect of the rock mass structure, including the differences in mechanical properties among the classes of the rock masses and the aeolotropism of the rock masses, on rockburst occurrence and its intensity. Therefore, the detailed investigation of the rock mass structure, the correct classification of rock masses and the understanding of the rules in the structural changes of the rock masses are of essential importance.

By taking account of the structural effects of the rock masses on rockburst, underground power house and tunnel sites could be successfully fulfilled to occur no rockbursting, and by adjusting the axes of the underground projects the rockburst intensity can be reduced, and the rockburst in tunnels under excavation can be predicted. So it is obvious that the structural effect theory of rock masses on rockbursting is of practical value.

REFERENCES

Yian Tan, (1987). A fuzzy mathematical comprehensive method for evaluating stability of caverns of hydralic projects, Journal of Chinese Water Resources and Hydropower Engineering, No.2.

Yian Tan, (1988). Application of system engineering in synthetic analysis of stability of rock masses surrounding caverns, Journal of Xi'an College of Geology, Vol.10, No.1.

Yian Tan, Guochang Liu, Shili Chuei, (1988). System engineering classification of rock mass stability and the fuzzy mathematical compreheasive evaluation, Proc. 3rd Cong., CAEG, Chengdu, China

Yian Tan, (1989). The study of rockburst mechanism, Journal of Chinese Hydrogeology and Engineering Geology, No.1.

Yian Tan, (1989). Analysis of fractured face of rockburst with scanning electron microscopy and its progressive failure process, Journal of Chinese Electron Microscopy Society, Vol.8, No.2.

Yian Tan, (1989) The application of fuzzy comprehensive evaluation to rockburst forecast for the underground caverns, Proc. Second Cong., on Rock Mech. and Eng.,CSRME, Guangzhuo, China.

Studies on principal factors effecting shear strength of weak intercalations and the correlation between them

Les recherches concernant le facteur principal qui influence l'intensité anticoupe de l'intercalation faible et la corrélation entre eux

Zhang Xiangong
China University of Geosciences, People's Republic of China

Nie Dexin
Chengdu College of Geology, People's Republic of China

Han Wenfeng
Lanzhou University, People's Republic of China

ABSTRACT: In this article we discuss the influence of factors of fault gouge upon its shear strength. We set forth an index ——— consistence state index Ws/Wp, which indicates the effects of these factors on the fault gouge. A fine correlation appears between Ws/Wp and the friction coefficient, hence we get an equation reflecting the correlation between each factor of the gouge and the friction coefficient of it, which is a formula of fine applicability and exactitude.

RESUME: Dans cet article on a discuté l'influence des facteurs de la fausse gouge sur l'intensité tranchante. Les auteurs presentent ici l'indice Ws/Wp de l'état consistent, qui peut indiquer les effets de ces facteurs sur la fausse gouge. On y trouve une correlation très fine entre Ws/Wp et le coefficient frictionnel. Ainsi on produit une équation qui peut se refléter la correlation entre chaque facteur de la gouge et le coefficient frictionnel. Et celle-là est justement la formule parfaite en applicabilité et exactitude.

1 Introduction

Weak intercalations is principle structure plane which controls the stability of rock mass. Studing principle factors influencing its strength is foundational to determine its strength criteria. Predecessors did much in this aspect, but many studied it more from single aspect than from synthetical influence, let alone finding out the correlation between these factors and the strength, nor hence to gain appropriate interequation. If we able to obtain correlations of this kind, without doubt it will have important consequence to mitigate the workamount of explorations and texts during early explorations, and to obtain better believable strength criteria of weak intercalations. This article will explore the influence of single factors—— water content, (dry) bulk density, clayey mineral component, consistence state-upon its shear strength. On this basis, we'll explore the synthetical influence of these factors upon its strength, and give on equation indicaling the relationship between each factor and shear strength.

2 Principle factors influencing the shear strength of weak intercalations

Intercalated mud in weak intercalation is its lowest strength portion. Studying and determining its strength can better decide the strength of weak intercalations. Generally speaking, the principle factors influencing its strength are its bulk density, water content, clayey mineral component, consistence state and capacity of sticky clay particles, etc. Their changement can avert the strength of intercalated mud. Since capacity of sticky clay particles in intercalated mud is almost between 15 to 40 per cent in this article, we'll only discuss former four factors' influence upon shear strength.

2.1. Influence of density of intercalated mud on its shear strength

Density of intercalated mud, especially dry density not only reflects the results of all kinds of geological actions (including environmental change of ground stress), which weak intercalations experiences in the long geological environment, but also reflects its present existing environment. Because major weak intercalations are the product of geological structure process, they experience higher process of structure stress in their history, and at present undergo that process with differenct amount in different parts. The intercalated mud which has experienced higher process (action) of structure stress, has higher bulk density, lower water content. By contrary, it has lower bulk density, and higher water content. The higher its bulk density, the bigger interparticle connective areas, the bigger connected force becomes obviously-that force which is not only the force produced by water-sticked connect, what is more, is interparticle chemical affinity and static electrical attraction. In the state of saturated water, when (dry) density lowers, water content of intercalated mud rises up, interparticle connective areas reduces, bond-water is full of pore and its strength obivously lowers. By various patterns of tests about intercalated mud, from the material of 64 groups of direct shearing tests in the state of various dry densities, we are able to find out that the friction coefficient of the same kind of intercalated mud becomes bigger while bulk density becomes higher, and they have abvious correlation. If the patterns of intercalated mud are not considered, we shall analyse the correlation of all the material in Fig 1, obtaining their interequation between bulk density and friction coefficient. It's that

$$f=0.9 \log \gamma d + 0.139 \quad (1)$$
$$r=0.234$$

in the equation, f-friction coefficient of intercalated mud;

γd-dry density (KN/m³)

r-correlation coefficient

Obviously, in the state of the patterns of intercalated mud not being considered, there is not correlation between its dry density and its friction coefficient.

2.2. The influence of water content, clayey mineral component and consistence state upon shear strength

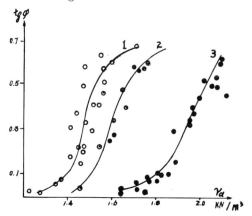

Fig.1 Correlation curve between friction cofficient dan (dry) bulk density of intercalated mud

1. Capacity of montmorillonniste in white mud, 95%
2. Capacity of montmorillonniste in grey mud 65%
3. Capacity of montmorillonniste in read mud 25%

It's known to all of us that shear strength of intercalated mud increases with rising water content. Fig.2 is a free-dotted figure of correlation between friction coefficient and water content of various kinds of intercalated muds. From it, we can find out while water content lowers, friction coefficient extends of intercalated mud to increase, but they have not better correlation. This is because the plastic limit and liquid limit differ very much in different intercalated mud. Thus when we research the influence of its water content upon shear strength, we should consider consistence state in certain water content. From three groups of besting meterial, we are able to account for such a problem that when water content is higher than or equal to liquid limit, intercalated mud exists in liquid state, and shear strength is lowest (see Table 1); when it is between plastic limit and liquid limit, its state character is plastic state and its shear strength is still lower (see Table 2); when it is

lower than liquild limit, its shear
strength obviously develops (see
Table 3).

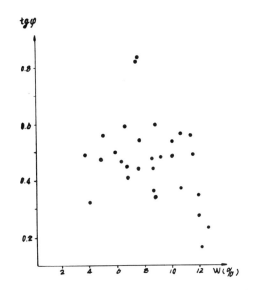

Fig.2 Relation figure between variant intercalated mud water
content (w) and friction coefficients (tg Φ)

Above material shows although intercalated mud water content is a prin
ciple factor to strength, if we separate it from consistence state
it's very difficult to reueal influence ability of water content to
shearing.

Because of various clayey mineral
component, it's plastic limit is
not the same as another. Intercalated mud's plactic limit living mainly on montmorillonniste is usually
higher than liquid limit living
mainly on illite. Therefore, whe
the former is still in solid state,
the latter has gone into liquid state, so that their strengths differ
very much (see Table 4). To a certain extent, the patterns and capacity of clayey mineral one controllable factors to strength, especially to that of montmorilloniste.
When capacity of montmorilloniste
becomes higher, plastic limit becomes higher. Therefore, with the
same water content, intercalated
mud with higher capacity of montmorillonite has higher shear strength.

3 Synthetic influence of all factors
upon shear strength and correlation analysis

Table 1. shear strength of intercalated mud when water
content equal to WL

character of in-tercalated mud	WL	W_p	shearing water content (%)	shear strength	
				tg Φ	C (Kpa)
red mud	33.3	16.1	33.3	0.0192	2
red mud	29.5	16.2	29.5	0.034	1
fracture mud	60.3	23.0	60.3	0.009	14
fault mud	59.1	24.5	59.1	0.052	3

We analysed in quantities synthetic influence each of there factors-
density, water content, the patterns
of clayey mineral, consistence state,
etc-upon shear strength of intercalated mud in the state of saturated
water. And on this basis, we suggest to use consistence state index

Ws/Wp to research synthetical influence of all these factors upon
shear strength: Here Ws stands for
saturated water content in the state
of non-expansion, its value respre-
senting value of water content, and
reflecting the value of intercalated mud pore rate or dry density.

Table 2. shear strength whenwater content is
higher than WL but lower tha WP

character of in-tercalated mud	WL	W_p	shearing water content (%)	shear strength	
				tg φ	C (KPa)
red mud	30.1	16.5	17.5	0.28	25
grey mud	50.8	28.5	31	0.297	41
white mud	72.3	32.7	34	0.195	21

Table 3. shear strength when water content is
lower than Wp

character of in-tercalated mud	WL	W_p	shearing water content (%)	shear strength	
				tg φ	C (Kpa)
grey mud	53	30.3	28.7	0.24	89
grey mud	53	28.9	25.9	0.283	88.6
grey mud	60.2	29.1	20.8	0.546	150
white mud	70.5	36.5	24	0.665	140

Table 4. shear strength of interclated mud with def-
ferent clayey mineral and the same water content

character of intercalated mud	clayey mineral component	WL	W_p	shearing water content (%)	shear strength	
					tg φ	C (Kpa)
white mud	M-90%, K.ch-minin	80.6	34.3	33.2	0.324	650
read mud	M-24%, I-64% others-K.Ch	33.3	18.1	33.3	0.019	20
grey mud	M-65% I-15% ch-18%	53.1	28.8	31.0	0.29	430

M-montmorillonite,   I-illce

ch-chlorite,   K-Kaolinite

Because water content becomes higher, pore rate become bigger, and density grows smaller. By contrary, when water content becomes smaller, pore rate gets smaller, and bulk density becomes bigger. Wp stands for plastic limit of intercalated mud. It may reflect the patterns of intercalated mud clayey mineral. Ws/Wp may indicate consistence state. When Ws/Wp is bigger tnan 1, intercalated mud is in plastic state or liquid state. when Ws/Wp is smaller than 1, it is in solid state. Therefore, Ws/Wp can connect a few principle factors influencing shear strength. We have finished 64 groups of shearing tests in different intercalated mud, and there is a fine correlation between friction coefficients and Ws/Wp. (see Tig.3). There is an equations:

$$tg \phi = 0.27-1.5034 lg\ Ws/Wp \quad (2)$$

$$r=0.91$$

In the equation, $tg\phi$ -friction coefficient

Wp-plastic limit

Ws-saturated water content

r - correlation coefficient

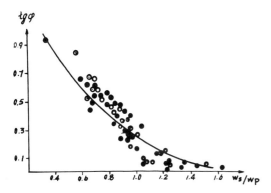

Fig 3. correlation curve between friction coefficient and Ws/Wp

- red mud (M-24%,I-64%, others-K,ch)
- grey mud (M-65%, I-15%, ch-18%)
- white mud (M-90%, minim.K,ch)

Through examining to weak intercalations on some engineering spots in China,formula (2) has finer apprvability. During early engineering explorations, if only water content, special gravity, dry density plasticvlimit and Liquid limit of intercalated mud, are able to be abtained on the point, we can use formula (2) to abtain friction coefficient of intercalated mud easily. With the increasement of test data in future, formula(2) will graduaually develop and become a principle experimental formula to judge shear strength of weak intercalation.

4 Conclusion

Principle factors influencing shear strength of weak intercalation are their density (or dry density), water content ,clayey mineral compenent, consistence state. And the patterns and capacity of clayey mineal are the most important factors controlling strength. In this article, consistance state index Ws/Wp was got by us. It may indicate synthetical influence of all above factors upon shear strength of weak intercalation, and there is a fine correlation between it and friction coefficient of intercalated mud. Formula (2) is a mathmatic formula of relationship of this sort.

REFERENCES

Zhang Xiangong, 1979. Engineering Geology Vol.1, Beijing: Geol. Pub. Hou.
Gu Dezheng, 1979. Fundamentals of rock engineering geomechanics. Beijing: Scientific publishing House.
Zhang Xiangong et al., 1986. Engineering geological classification of fault rocks. Proceedings 5th international IAEG Congress.
Nie Dexin et al., 1989, Controlling effect of the montmorillonite content to the physical and mechanical properties of fault gouge, New advances in engineering geology. Chengdu, China.

5.2 Large permanent underground openings
Grands vides souterrains

Construction of large underground caverns for the storage of crude oil in ductile rocks of Miocene volcanism

Construction de larges cavernes souterraines pour le stockage de pétrole cru en roche volcanique miocène ductile

Kazuo Hoshino
Shimizu Corporation, Tokyo, Japan

Toshiaki Makita
Japan Undergound Oil Storage Co., Ltd, Tokyo, Japan

ABSTRACT : The huge caverns in Kushikino base are being constructed in autobrecciated parts of lava flows along the margin of submarine volcanoes. Although this pyroclastics are composed of various kinds of volcanic fragments, patient studies succeeded in finding out the most suitable places for construction. The following excavation insured the results of previous survey and is revealing interesting inside-structure of the Miocene submarine eruption.

Resumé : Les cavernes gigantesques en la base de Kushikino se construisent dans les parties autobréchiformes des coulees de lave en marge des volcans sous-marins. Malgré que les roches pyroclastiques se composent des fragments volcaniques compliqués, les études patientes laissent voir les premiere parties à la construction. L'excavation suivante s'est assurée des résultat des ètudes avancées et a montrée des structures internes intéressantes des eruptions sous-marines miocènes.

1 INTRODUCTION

Since the recent oil crisis, the Japanese Government has been forced to establish the means to maintain the storage of enough amount of crude oil within the country. The Ministry of International Trade and Industry (MITI) decided a policy to build 10 national storage bases with total amount of 30 million kls as shown in Fig.1. Among them are included 3 sites of underground cavern system, Kushikino, Kikuma and Kuji.

The underground system is a new technique in Japan. Some people were concerned about possibility of construction of such large scale caverns in the Japanese rock-mass because that Japan was subjected to severe orogenic disturbance in geological ages. MITI made preliminary experimentation in Cretaceous granite in Kikuma to ensure geological stability and engineering feasibility in Japan from 1979. This experimentation proved that the rock-mass like Kikuma granite has good quality to built so large caverans comparable with European storage sites. The working group in MITI selected around 170 sites as suitable places for the construction. In 1981, the above mentioned three sites became the tragets of the feasibility study by MITI.

The construction of three sites started in 1986 and will be completed in a few years ahead. Both Kikuma and Kuji sites are located in Mesozoic granites. However, the Kusikino site was selected in the Tertiary volcanic rocks.

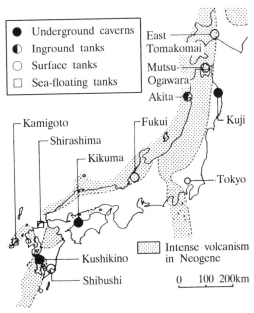

Fig.1 Location of national oil storage bases.

2 FEASIBILITY STUDY IN KUSHIKINO

The most sites for the underground oil storage in the world were chosed in the homogeneous and hardest rock-mass such as linestones or granites.

The Kushikino site, however, is an exceptional case that inhomogeneous and ductile pyroclasitic rocks are used for such construction.

The following is the main items of the site survey of the three underground storage areas including Kushikino during several years from 1981 to 86.

1) surface geological survey with aerophoto-interpretation of fracture system
2) seismic refraction survey
3) borings (20-25 holes each site)
4) well logging (electrical, density, velocity, caliper, temperature, etc.)
5) hydrological survey (Lugeon test, ground water table, pore pressure-JFT)
6) laboratory measurement (density, porosity, atmospheric and triaxial strength, elastic constant, wave velocity, etc.)
7) in-situ measurment (state of stress, shearing stress, defromation, creep)
8) analytical studies (structual stability of cavern, hydraulic behavior around cavern, water balance and water level)

3 GEOLOGY

As indicated in Fig.1, the district where the Kushikino site is included was the places of violent and contineous volcanic sctivity through Neogene and Quaternary. As the basement rocks beneath this Neogene, we have Cretaceous and late Paleozoic rocks.

The main part of the volcanic rocks are called the Hokusatsu volcanics and divided into three stages as shown in Tab.1.

The early volcanism was dated in Miocene, which is consisted of the andesitic lava flows and pyroclastic rocks, resulted from submarine eruption at the bottom of the sea or lake. The second volcanic activity in Pliocene and the third in Pleistocene are characterised by andesite and ryolite. Both are called respectively middle and late Hokusatsu volcanics by local name.

In the Miocene volcanic rocks, which is named as the early Hokusatsu andesites, there are the lava flow (LA) in upper, while, the pyroclastics or volcanic breccia in lower parts. The most parts of the volcanic breccia are possibly auto-brecciated parts of the lava flows (LB). It is interesting that some parts of the lowest layers of the volcanic breccia contain the round pebbles that are derived from the sedimentary rocks of the basement. Although that the geological situation of this pebbles-containing volcanic breccia facies is not clear yet, we separate this facies from the other breccia facies and conventionally call it as conglomerate (LBg). The caverns of the Kushikino site are constructed in the pyroclastics, either LB or LBg, because of disqualification of LA as explained next.

4 MECHANICAL PROPERTIES AND HYDROGEOLOGY

The detailed study of the well logging and laboratory measurement made clear the mechanical properties of these volcanic rocks.

Tab.1 Stratigraphy.

Age and Formation		Geology	Typical Section
Quaternary		pumice flow, welded tuff	
Hokusatsu Volcanics	Pleistocene	andesite, ryolite	
	Pliocene	andesite, ryolite	
	Miocene	andesitic lava flows (LA) pyroclastics autobrecciated parts (LB) volcanic conglomerate (LBg)	
Mesozoic and Paleozoic basement rocks		sandstone, shale chert, slate	

Fig.2 shows uniaxial strength of both lava (LA) and pyroclastics (LB and LBg) in relation to porosity. As that it is clearly shown in the figure, lava is tight and strong as ranging 240 to 20 MPa in strength, on the other hand, pyroclastics is porous and comparatively weak as ranging between 80 and 10 MPa in strength. The contrast of mechanical properties between two facies of the Hokusatsu andesites is also remarlable in stress strain relation of Fig.3. The hard lava behaves brittle deformation, while the soft pyroclastics is quite ductile on stress strain curves. As matter of fact, the andesitic lava is easy to be split in layer by layer along the planes of flow, and has frequent joints. There-fore, the brittle andesitic lava (LA) is not suitable for construction of the large caverns. On the other hand, in the ductile pyroclastics there is no remarkable faults and joints. The RQD is 90-100% for the most core samples. Therefore, the rock-mass of the pyrocalstics is extremely low in permiability ranging 10^{-8} to 10^{-10} m/s.

The groundwater table is at an elevation of approximately 50-120m from the sea level. As the caverns lay at the depth of 20 to 40m below the sea level, the under-ground water near the caverns has pressure of approximately 100 or 150m of water head.

In Tab.2, some mechanical properties of the rock-mass (LB and LBg) are shown.

Classification of the rock-mass was made mainly on basis of hardness and the interval of fractures.

5 CONSTRUCTION OF THE STORAGE CAVERNS

The construction of the Kushikino under-ground oil storage base has started in 1986 after careful and patient feasibility studies

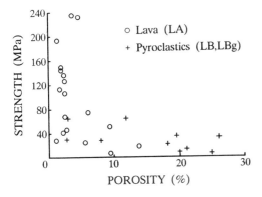

Fig.2 Uniaxial strength versus porosity.

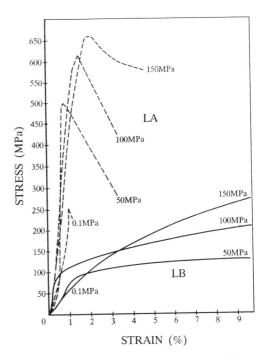

Fig.3 Representative stress-strain curves of lava (dotted-line) and pyroclastics (solid-line).

Tab.2 Mechanical properties of the rock-mass.

Classification of rock-mass		Hv	H	M	L
Unit weight	KN/ m³	2.5	2.5	2.5	2.5
Poisson ratio		0.20	0.25	0.30	0.35
Modulus of deformation	MPa	6X10³	4X10³	2-6X10³	1-6X10³
Shear strength	MPa	2.1	1.8	1.5	1.2
Tensile strength	MPa	0.42	0.36	0.30	0.24
Ratio of horizontal to vertical stresses		1.0	1.0	1.0	1.0

of 5 years. The final layout of the base is shown in Fig.4. The storage site is consisted of 10 caverns, which occupy an area of about 500m × 600m. Each cavern has length of 555m, width of 18m, and height of 22m. The caverns trend northwest to southeast, which is almost parallel to the direction of the maximum compressional stress. The height of the earth surface over the caverns is roughly 200 to 100m and the groundwater table is approximately 120 to 50m above from the sea level. In order to maintain enough hydrostatic pressure around the caverns, each cavern is laid at depth of 20 to 40m below the sea level.

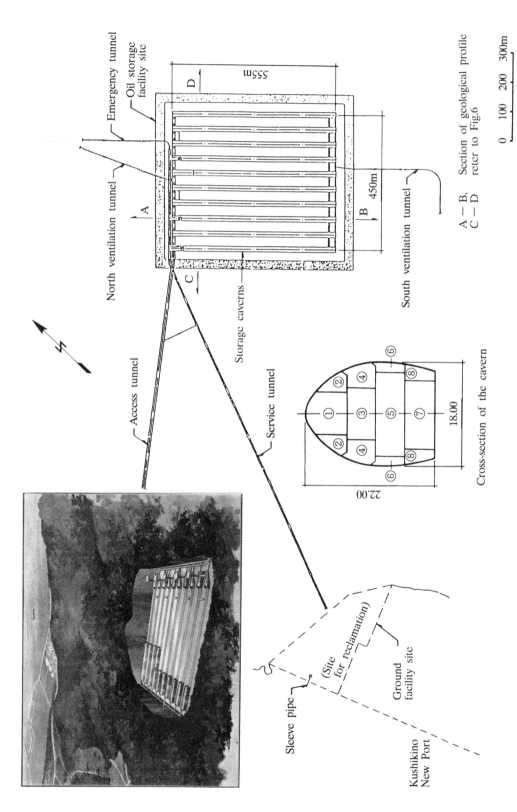

North ventilation tunnel

Emergency tunnel

Oil storage facility site

555m

450m

D

A

B

C

South ventilation tunnel

A — B,   Section of geological profile
C — D   reter to Fig.6

0   100 200 300m

Access tunnel

Storage caverns

Service tunnel

Sleeve pipe

(Site for reclamation)

Ground facility site

Kushikino New Port

22.00

18.00

① ② ③ ④ ⑤ ⑥ ⑦ ⑧

Cross-section of the cavern

Fig.4   Layout of the Kushikino oil storage base.

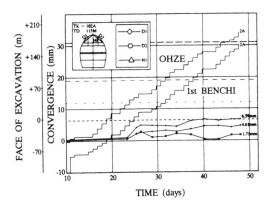

Fig.5 Convergence of the pyroclastic rocks at a top drift.

The 10 caverns have total capacity of 1.75 million kls of crude oil. As the storage site is distant about 1 or 1.5km from the coast, we have a service tunnel to connect with the ground facility site and the tanker.

Because that the cross-section of the caverns is so large, excavation was done through 4 stages, top drift and arch, 1st, 2nd, and 3rd benches. The construction began from the entrance of the access tunnel in the end of 1986. In spring of the next year, the operation advanced to the caverns site. It is expected that most excavation in the caverns site will be finished by early 1991.

Through the excavation of the caverns and tunnels, it has been confirmed that the main features of geology are almost same as expected by the previous geological survey.

The volcanic breccia (LB, LBg) have shown excellent characteristics in mechanical stability after the excavation, except the narrow zone near some faults. Fig.5 shows an example of the convergence of the autobrecciated pyroclastics at the central part of the cavarn site. The dislocation is quite small even when the lower excavation was advanced close to the measurement point in the arch. The water table has been quite stable during the excavation.

6 STRUCTURE OF THE MIOCENE VOLCANOES

The network-like excavation of the large caverans and tunnels in the rock-mass of the old volcanoes revealed the secret image of submarine eruption.

Fig.6 shows geological cross sections in two directions, parallel and perpendicular to the caverns. The volcanic conglomerates (LBg) lay in the lower, inclining slightly towards west and north. The southeastren part of the caverns area are ocupied by these volcanic conglomerates, (Fig.4).

The tuff breccia or autobrecciated pyroclastics (LB) are found to overlay the conglomerate, on the geological map. In most boundaries, however, there ia no plane to separate the both facies and the matrix is quite same.

The most parts of the construction area are included in this pyroclastics. The lava covers the pyroclastics in whole area. In some places, thin layers of lava flows are found in the auto-brecciated pyroclastics.

The lava is main facies of Hokusatsu andesites. It distributes in a wide areas in Kushikino, whereas, the pyroclastics including conglomerates are found only local places.

The model of the Miocene volcanism in the Kushikino underground oil storage area is presented as Fig.7. In early Miocene the submarine eruption of the andesitic magma came up through the basement of Mesozoic sedimentary rocks. In marginal area of eruption, the hot, flowing lava sheets touched to cold sea water and suddenly cooled. Then, the massive lava sheets were broken into fragments. It seems that there were the hill made of the conglomerate at the southwestern places in the site. The lava-flow coming from the northeast stopped at the hill and then flowed over the hill. It is possible that around the hill auto-brecciation of the lava develop in wider places. As layers of the pyroclastics buries the sea, the later eruptives emerged on land. The most upper parts of the LA facies are perhaps the results of the flows on land.

It is quite interesting that such Miocene volcanic rocks of good quality in mechanical properties as in Kushikino is few in other areas even that the same rocks develop in many areas in southern Kyushu.

Already shown in Fig.1, fissure-type eruption prevailed in almost every places along the Japan sea and the East China Sea. It produced tremendous amounts of the volcanic sediments, which is very familiar with the Japanese geologists by name of green tuff. It is true that the most green tuff is inhomogeneous and less strong in mechanical properties. The green tuff is quite famous geological unit among the Japanese geologists because that it is much altered to yield many metal mines or fractured to have reservoirs for oil and gas accumulations. In this points, the discovery of the Kushikino site should be attributed to surprising effort of the every person with patient and challenging mind, who has been joined in this national project.

ACKOWLEDGEMENTS

The authors are much oblized to the Engineering Advancement Association of Japan, the Agency of Natural Resources and Technology, Ministry of International Trade and Industry, and National Petroleum Corporation for kind arrangement for preparing our manuscript. We thank to Dr. Ryoichi Kouda of Geological Survey of Japan and Mr. Kuniichiro Miyashita of Shimizu Corporation for kind asistance in many ways.

REFERENCES

Geological Survey of Japan 1977. Geology and mineral resources of Japan.
Geological Survey of Japan 1980. 1,500,000 geological map "Kagoshima".
Hoshino K. 1980 Site investigation of underground oil storage** Engineering Geology 21-4, p.36-43.
Hoshino K. 1985 Studies on the porosity of reservoirs in deep seated volcanic rocks**, Jour. Jap. Asso. Petrol. Tech. 50, P.269.
Hoshino K. and Makita T. 1988 Underground storage of crude oil in submarine volcanic rocks, Proc. Kagoshima Intern. confer. volcanoes. p.770-773.
Makita T., Miyanaga Y. and Shozakai T. 1990 Construction of the Kushikino underground oil storage base** Doboku-Seko or Civil Eng. Jour. 2, p.13-24.
Sato E. 1990 On the Magome Breccia at Kushikino City, Kagoshima, Japan, Commemoration volume for Prof. Yukitoshi Urashima's retirement*.(in press)
** written in Japanese
* written in Japanese with English abstract

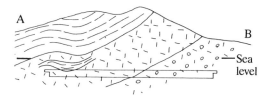

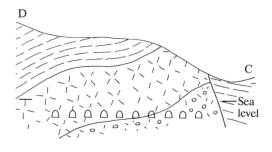

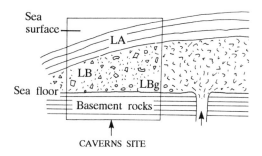

Fig.7 The inside structure of submarine volcanism at Kushikino area (imaginative sketch). LA : lava flow, LB : autobrecciated lava, LBg : volcanic conglomerate.

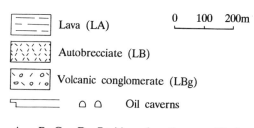

Fig.6 Geological sections of the cavern site, A-B is parallel and C-D is vertical to the elongation of the caverns.

Concepts of dimensional analysis applied to underground engineering geology

Conceptions d'analyse dimensionnelle appliquées en géologie de l'ingénieur souterraine

B. Indraratna & S. Naguleswary
Asian Institute of Technology, Bangkok, Thailand

ABSTRACT: Physical modelling is a useful approach to solve the problems in Geomechanics. It provides direct experiences of both expected and unexpected deformations and failure modes which could not be contemplated in the field (i.e. prototype scale). In contrast, concepts of mathematical modelling would incorporate many variable parameters as well as various similitude quantities, thereby providing a better and more fundamental basis for both experimental design and numerical analysis. The application of this methodology in modelling the stress-strain behaviour, for example in underground excavations is also discussed in order to facilitate comprehension of the proposed model in practice. The results are justified by the utilization of these quantities in the past by several investigators. A computational method for dimensional analysis is also discussed briefly.

RESUME: La modélisation physique est un moyen d'approche utile pour résoudre les problèmes de mecanique des roches et sols. Elle procure une experience directe des deformations et ruptures, aussi bien attendues que inattendues, alors que celles-ci ne peuvent être observées sur le terrain (échelle du prototype). A l'oppose, les concepts à la base de la modélisation mathémetique permettent d'incorporer de nombreux paramètres variables ainsi que des quentités similaires, procurant de cette façon une base fondamentale pour l'établissesment de modèles expérimentaux et l'analyse numérique. La discussion de l'application de cette méthode de modélisation des comportements en contrainte, par exemple pour les excavations souterraniennes, est aussi destinée à faciliter la compréhension et l'application du modèle proposé. L'utilisation antérieure de ces données par de nombreux chercheurs corrobore les résultats obtenus par la méthode proposée. Une méthode de calcul pour d'analysis dimensional est discuté, en bref.

1. INTRODUCTION

Mathematical modelling combined with the concepts of physical modelling can incorporate many variable parameters as well as various similitude quantities. The fundamental principles which determine the proper design, construction operation and interpretation of test results of those models comprise the theory of similitude. The theory of similitude includes a consideration of the conditions under which the behaviour of two separate entities or systems will be similar and the techniques of predicting the behaviour of one system from the other.

In this research, mathematical modelling combined with the concepts of physical simulation is introduced. This mathematical model which is based on dimensional analysis is formulated in order to determine similitude quantities which can be computed and arranged in specific formats to simulate accurately the field conditions and subsequent behaviour. The phases of the proposed model are illustrated in Figure 1. As shown in this figure, any geotechnical problem can be defined by the principal criteria and uniquity conditions (e.g. boundary conditions and governing equations). These can be combined to form a geomechanics model with appropriate simi-

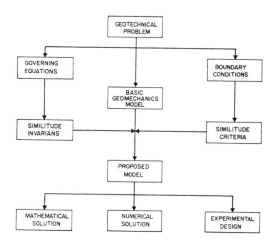

Fig. 1 Phases of the model

litude invariants and criteria. The proposed model can incorporate many variable parameters, hence would provide a fundamental input base for analytical and numerical computations as well as for experimental design.

2. DIMENSIONAL ANALYSIS

2.1 Selection of variables

The dimensions of any variable U_j can be written as a function of the basic terms, mass, length and time as follows:

$$[U_j] = [M]^{a_{1j}} \cdot [L]^{a_{2j}} \cdot [T]^{a_{3j}}$$

where M, L and T represent mass, length and time respectively.

The dimensions of a set of variables U_1, U_2, U_3, ...U_n may be described by the exponents of the dimensional matrix (Table 1). In Table 1, the first subscript or a row number indicates the dimension; the second subscript or column number indicates the variable. In this way an exponent a_{ij} specifies the dimension and the variable associated with it. For example, a_{ij} is the exponent of L of the variable U_3. Some values of a_{ij} may be zero, depending upon the units of the variable.

Table 1 Dimensional matrix

Dimension	U_1	U_2	U_3	...	U_n
Mass (M)	a_{11}	a_{12}	a_{13}	...	a_{1n}
Length (L)	a_{21}	a_{22}	a_{23}	...	a_{2n}
Time (T)	a_{31}	a_{32}	a_{33}	...	a_{3n}

In the analysis, Table 1 may be written simply in the form of a matrix. Therefore a dimensional matrix can be written as follows:

$$[a_{1j}] = \begin{bmatrix} a_{11} & a_{12} & a_{13} & .... & a_{1n} \\ a_{21} & a_{22} & a_{23} & .... & a_{2n} \\ a_{31} & a_{32} & a_{33} & .... & a_{3n} \end{bmatrix}$$

If the variables U_1 to U_n are known, then the functional form of the governing equation can be given by:

$$F(U_1, U_2, ...U_n) = 0$$

Even if the function F is unknown, it can be considered as dimensionally homogeneous, provided that all parameters which influence the system are included in the analysis.

3. APPLICATIONS IN GEOMECHANICS

3.1 Material similitude parameters

In case of modelling stress-strain behaviour, similitude of material properties is of paramount importance. The following analysis (Table 2) demonstrates the method of computing the relevant dimensionless terms to establish stress-strain similarity between two materials.

Table 2 Dimensional matrix for material similitude

Dimension	υ	ϕ	σ_c	σ_t	E
Mass (M)	0	0	1	1	1
Length (L)	0	0	-1	-1	-1
Time (T)	0	0	-2	-2	-2

The symbols are defined by:

υ, ϕ – poisson's ratio and friction angle

σ_c, σ_t – compressive strength and tensile strength respectively

E – Young's modulus

For the given five parameters, the rank of the matrix shown in Table 2 is unity. Hence, there should be at least four independent dimensionless terms to represent complete similitude. The following independent dimensionless terms can be regarded as appropriate:

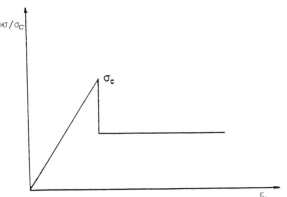

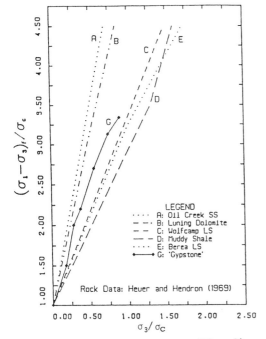

σ/σ_c

σ_c

$\varepsilon.$

**Fig. 2 Elastic brittle plastic model
(after Indraratna, 1987)**

$(\sigma_1-\sigma_3)_f/\sigma_c$

LEGEND
.... A: Oil Creek SS
— · — B: Luning Dolomite
— — C: Wolfcamp LS
— — D: Muddy Shale
.... E: Berea LS
•—• G: 'Gypstone'

Rock Data: Heuer and Hendron (1969)

σ_3/σ_c

**Fig. 3 Comparison of gypstone with sedimen-
tary rocks in triaxial compression
(after Indraratna, 1987)**

poisson's ratio : υ
friction angle : ϕ
critical strain : $\varepsilon_c = \sigma_c/E$
uniaxial strength ratio : σ_t/σ_c

For the model and the prototype these
dimensionless terms should be identical,

hence,

$(\upsilon)_m = (\upsilon)_p$, $(\phi)_m = (\phi)_p$, $(\sigma_c/E)_m = (\sigma_c/E)_p$ and $(\sigma_t/\sigma_c)_m = (\sigma_t/\sigma_c)_p$

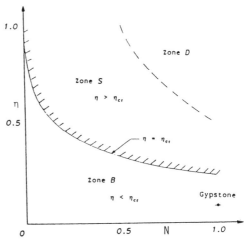

**Fig. 4 Stability curve for failure
mechanisms (after Vardoulakis)**

Poisson's ratio and friction angle are
selected automatically because of their
dimensionless form. Critical strain has
been considered as an important parameter
in the development of the elastic, brittle
plastic model for rock as shown in Figure
2 (Indraratna & Kaiser, 1990). The elastic,
brittle plastic model is a simplification
of strain weakening behaviour, which is
generally observed in most rocks where the
post-failure behaviour is strain-dependent.

The application of these material similitude parameters can be elucidated by the
development of an artificially simulated
rock (gypstone) which satisfies the material similitude quantities (Indraratna,
1987). Gypstone is characterized by the
properties: $\phi = 32^\circ$, $\upsilon = 0.25$, $\varepsilon_c = 0.24\%$,
$\sigma_t/\sigma_c = 7.4\%$ which closely resemble an
array of sedimentary rocks ($25^\circ < \phi < 40^\circ$,
$0.25 < \upsilon < 0.35$, $0.2\% < \sigma_c/E < 0.4\%$, $5\% <$
$\sigma_t/\sigma_c < 10\%$). Figure 3 illustrates the
triaxial behaviour (normalized form) of
gypstone in comparison with several sedimentary rocks. In such a dimensionless
represenation, it is evident that the
above artificial material resembles the
stress-strain behaviour of some prototype
rocks, particularly sandstone.

The relevance of the uniaxial strength
ratio (σ_t/σ_c) has already been recognized
in stability analysis (bifurcation theory).
The theory of bifurcation and the associated failure mechanisms for a given
material is illustrated in Figure 4
(Vardoulakis, 1984). The vertical axis represents the ratio of the tensile strength
to the compressive strength ($\eta = \sigma_t/\sigma_c$) of
an uniaxial strength specimen, and the

2595

horizontal axis represents the hardening parameter (N) which is determined from the stress-strain relationship. The stability curve separates distinct regions where shear failure predominates in zone B, surface instability predominates in zone S and only surface instability occurs in zone D. Rocks which exhibit a Mohr-Coulomb behaviour generally have a magnitude of (σ_t / σ_c) between 0.05 and 0.1, and a hardening parameter (N) close to unity, hence indicating predominant shear failure characteristics. Gypstone (N = 1 and η = 0.074) falls on zone B and as expected has revealed predominant shear fracture during failure (Indraratna, 1987).

3.2 Tunnel stability

a) **Two dimensional analysis** - Plane strain condition (longitudinal strain ε_z = 0) can be assumed in tunnelling, because the tunnel lengths are much greater than their diameters. For this condition the tunnel stability depends on the geometry of the opening, the depth of the tunnel and the properties of the rock. Furthermore the support pressure is an important parameter which controls stability. The above relevant parameters are tabulated in Table 3, and a diagrammatic illustration is given in Figure 5.

Table 3 Dimensional matrix for two dimensional tunnel stability analysis

Dimension	σ_s	σ_c	γ	D	h
Mass (M)	1	1	1	0	0
Length (L)	-1	-1	-2	1	1
Time (T)	-2	-2	-2	0	0

The symbols are defined by:
σ_s - support pressure
σ_c - unconfined compressive strength
γ - unit weight of the rock
D - diameter
h - depth of the tunnel
Due to the fact that the rank of the above matrix is two, the number of minimum independent dimensionless terms required is three, which are given by:

σ_s/σ_c - support pressure to strength ratio
$\gamma(h+0.5D)/\sigma_c$ - overburden to strength ratio
h/D - cover to diameter ratio

For the model and the prototype these non-dimensional terms should be identical.

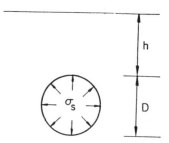

Fig. 5 Two dimensional tunnel section

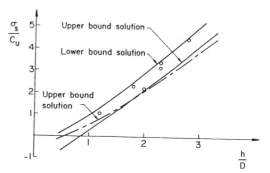

Fig. 6 Predicted and observed tunnel support pressure at collapse of model tunnels (modified after Mair, 1979)

According to the principles of dimensional analysis, (σ_s/σ_c) can be expressed as a function of the other two independent dimensionless products in the following way:

$$(\sigma_s/\sigma_c) = f(\gamma h+D/2/\sigma_c, h/D)$$

For undrained soft ground tunnelling these parameters can be used by replacing the uniaxial compressive strength by the undrained shear strength (C_u).
Mair (1979) have conducted some model tests and has found the variation of (σ_s/C_u) with (h/D) as illustrated in Figure 6. In this model test Mair (1979) has not considered the influence of the overburden pressure $\gamma(h+D/2)/C_u$. However, Davis et al, (1980) have determined the analytical solution which is illustrated in Figure 7, in which it can be seen that, the term ($\sigma_s-\sigma_a$)/C_u depends on the independent overburden terms, (h/D) and ($\gamma D/C_u$), where σ_a is the surcharge pressure.

b)   **Three(3) dimensional analysis**
Although the plane strain condition can be generally assumed for tunnels, the conditions at the face differ significantly and

2596

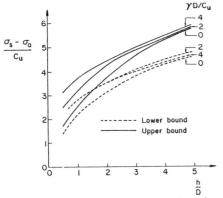

Fig. 7 Analytical solution for the plane strain problem (modified after Davis et al, 1980)

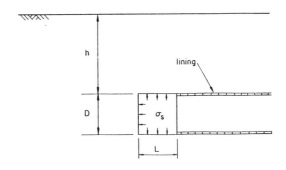

Fig. 8 Tunnel section

the three dimensional effects should be considered. In addition to the parameters which are listed for plane strain condition, the unlined length of the tunnel has significant effect on the tunnel face stability. The parameters which are relevant for tunnelling in rock are tabulated in Table 4 and described in Figure 8.

Table 4 Dimensional matrix for tunnel stability

Dimension	σ_S	σ_C	γ	D	h	L
Mass (M)	1	1	1	0	0	0
Length (L)	-1	-1	-2	1	1	1
Time (T)	-2	-2	-2	0	0	0

The symbols are defined by:

σ_S – support pressure
σ_C – unconfined compressive strength
γ – unit weight of the rock
D – diameter
h – depth of the tunnel
L – unlined length of the tunnel

The rank of the above matrix is two. Therefore the number of minimum independent (dimensionless) terms required is four for a comprehensive mathematical model, which are given by:

σ_S/σ_C – support pressure to strength ratio
σ_V/σ_C – overburden to strength ratio
h/D – cover to diameter ratio
L/D – unlined length to diameter ratio

The following dimensionless quantities should be made equal for the model and the prototype:

$(\sigma_S/\sigma_C)_m = (\sigma_S/\sigma_C)_p$, $(\sigma_V/\sigma_C)_m = (\sigma_V/\sigma_C)_p$, $(h/D)_m = (h/D)_p$ and $(L/D)_m = (L/D)_p$

Due to the fact that the above dimensionless terms constitute a complete set, they can be combined to give the following equation:

$$\sigma_S/\sigma_C = f_1(\sigma_V/\sigma_C, h/D, L/D)$$

The overburden pressure (σ_V) at the tunnel axis can be written as:

$$\sigma_p = \gamma(h+D/2)$$

By normalizing the overburden stress with the uniaxial compressive strength, the following expression can be obtained:

$$\sigma_V/\sigma_C = 1/2(\gamma D/\sigma_C).(1+2h/D)$$

Hence,

$$\gamma D/\sigma_C = (\sigma_V/\sigma_C).f_2(h/D)$$

Combining the above equations:

$$(\sigma_V-\sigma_S)/\sigma_C = f(h/D, L/D)$$

The factor $(\sigma_V-\sigma_S)/\sigma_C$ can be regarded as an indicator of tunnel wall stability.

As explained in the two-dimensional case, these parameters can also be used in soft ground tunnelling (undrained) by replacing the uniaxial compressive strength (σ_C) by the undrained shear strength (C_u). Some of these parameters have been used by several investigators in the past. Broms & Bennermark (1967) have discussed an empirical design factor for soft ground tunnel stability given by $(\sigma_V-\sigma_S)/C_u < 6$, which is the same factor proposed as explained earlier. Kimura et al (1981), Mair (1979) have also used the same dimensionless factor as shown in Figures 9 and 10, where $N = (\sigma_V-\sigma_S)/C_u$.

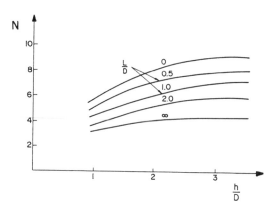

Fig. 9 Influence of heading geometry and depth on tunnel stability ratio at failure (modified after Mair, 1979)

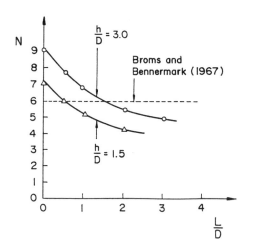

Fig. 10 Influence of unlined length on tunnel stability ratio at failure (modified after Kimura et al., 1981)

4 CONCLUSIONS

This paper has presented a methodology based on dimensional analysis to derive normalized (similitude) quantities which are prudent in physical modelling as well as to develop mathematical models representative of the actual behaviour of earth structures. The use of dimensional analysis significantly reduces the number of variable terms required for proper experimental design.

In tunnelling, normalized terms have been used often in both design and data analysis, but mainly in an intuitive manner rather than deriving them specifically from a theoretical basis. As a result in many situations only a few dimensionless similitude terms have been considered. It is pertinent that all variables such as the support pressures, overburden, tunnel geometry, strength and frictional properties of the ground etc. must be systematically modelled for accurate predictions. Similitude parameters must accurately represent not only the actual deformation response, but also the failure modes under the given stress conditions. However, modelling difficulties associated with complex geologic environments and dynamic loading conditions impose a limitation on the proposed analysis.

The methodology proposed in this paper can be extended to large scale mass movements such as rock avalanches and landslides, where a large number of variables have to be considered. For this purpose, a computational methodology has been developed by the authors (see Appendix). However, to make acceptable predictions some variable (dimensionless) terms may not be relevant. Fundamental or essential similitude parameters can be selectively incorporated into the analysis, thereby optimizing the computational efficiency. The remaining terms, in general, are of secondary importance, and therefore, in most cases, do not significantly influence the predictions (Naguleswary, 1989).

ACKNOWLEDGEMENTS

The authors wish to express their gratitude to the Norwegian Government for sponsoring this research study during the period 1988-1989. Sincere thanks are also due to N. Kuganenthira for assisting in the development of the computer program DAGE, capable of dimensional analysis of a very large array of variables. The authors also wish to extend their gratitude to Mrs. Uraivan Singchinsuk for her efforts during the preparation of this mauscript.

REFERENCES

Indraratna, B. N. and Kaiser, P.K. 1990. Design of grouted rock bolts based on convergence control method. Int. J. of Rock Mech. & Min. Sci. & Geomech. Abstr., (in press).

Indraratna, B. 1987. Application of fully grouted bolts in yielding rock. Ph.D. thesis, Deptartment of Civil Engineering, University of Alberta, 286p.

Vardoulakis, I. 1984. Rock bursting as a surface instability phenomenon. International Journal of Rock Mechanics and Min. Science & Geomechanics Abstr., Vol. 21, No. 3, pp. 137-144.

Mair, R.J. 1979. Centrifugal modelling of tunnel construction in soft clay. Ph.D. thesis, University of Cambridge.

Davis, E.H., Gunn, M.J., Mair, R.J. and Seneviratne, H.N. 1980. The stability of shallow tunnels and underground openings in cohesive material. Geotechnique 30, No. 4, pp. 397-416.

Broms, B. B. and Bennermark, H. 1967. Stability of clay at vertical openings. ASCE J. Soil Mech. & Foundations Division, Vol. 93, SM1, pp. 71-94.

Kimura, T. and Mair, R.J. 1981. Centrifugal testing of model tunnels in soft clay. Proc. 10th Int. Conf. on Soil Mech. and Foundation Engineering, Stockholm, Vol.1, pp. 319-322.

Naguleswary, S. 1989. Mathematical modelling applicable to geomechanics. M. Eng. thesis, Asian Institute of Technology, Bangkok, Thailand, 86p.

Hoek, E. and Brown, E.T. 1980. Underground excavation in rock. Inst. Mining and Metallurgy, London, 527p.

APPENDIX: COMPUTER SIMULATION

A micro-computer program (DAGE) for complex dimensional analysis has been developed in this study. DAGE is based on microsoft Fortran 77, and its methodology is presented briefly by the flow chart shown in Figure 11. The determination of the rank of the dimensional matrix is the primary objective, in order to form the required number of similitude terms. In the initial step, the (3x3) determinants are evaluated. For non-zero determinants, the corresponding sub-matrix is removed from the global dimensional matrix and the 'subroutine solve3' is called to determine the normalized (similitude) terms. If all the (3x3) matrices are singular, the determinants of the (2x2) matrices are computed. For non-zero determinants, the associated similitude terms are computed by the 'subroutine solve2'. If all the (2x2) determinants vanish, then the rank of the matrix is unity, and the similitude numbers are determined by the 'subroutine solve1'. Due to the limited length of the paper, the user manual and the Fortran

listing of DAGE are not presented, however, these are available upon request from the authors.

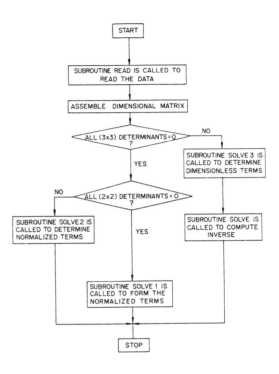

Fig. 11   Computation of dimensionless terms

A survey of some engineering properties of coal bearing strata from South Africa in relation to the stability of roof rocks in coal mines

Une étude de quelques propriétés géotechniques de la position des couches de charbon de l'Afrique du Sud, en rapport avec la stabilité des plafonds rocheux dans les mines de charbon

C.A.Jermy & F.G.Bell
University of Natal, Durban, South Africa

ABSTRACT: Core material obtained from coal bearing rocks of the Vryheid Formation in north Natal has shown the existence of numerous distinct sedimentary facies. These facies have widely differing geotechnical properties which obviously influence the design and development of coal mines. It would appear that most stable strata belong to the fine grained arenaceous facies, the coarse grained arenaceous facies being slightly less stable. By contrast poor roof and floor conditions exist in argillaceous rocks. The latter deteriorate rapidly on exposure with consequent reduction of strength and potential instability.

The uniaxial compressive and tensile strength of the rocks concerned was determined along with their free swelling coefficient. The latter, together with the uniaxial compressive strength, was used to assess the geodurability of the material tested. The geodurability classification indicated that some argillaceous material, in particular, was of very poor quality.

The durability and tensile strength provide an indication of the potential behaviour of the roof rocks when mining takes place, the direct tensile strength being a good index property in this respect. The uniaxial compressive strength is required for pillar design and an understanding of the failure of floors in coal mines.

RESUME: Des échantilons venant des rochers contenant du charbon de la formation du Vryheid an nord du Natal a demonstré l'existence de nombreux faciès sedimentaire distinctes. Ces faciès ont des propriétés géotechniques très differentes qui influencent le dessin et développement des mines à charbon. Il semblerait que les stratum les plus stables appartiennent à le faciès arenaceouse à fin grain, le grain rude étant moins stable. En contraste les conditions du plafond et du sol sont pauvres dans les rochers argillaceouse. La detérioration du dernier est rapide en découvertement avec une consequente réduction de la force et de l' instabilite potentiel.

La compressive uniaxiale et la force tensile des rochers concernés one été déterminées avec leur coéfficient d'enflure libre. Celui-ci avec la force compressive uniaxiale été utilisé pour déterminer la géodurabilité du matériel examiné. La classification a indiqué que quelque matériel argillaceouse, en particulier, était de très pauvre qualité.

La durabilité et force tensile fournissent une indication du comportement potentiel du rocher due plafond quand on mine, la force tensile directe étant une bonne indication dans ce respect. La force compressive uniaxiale est nécessaire pour la construction des piliers et pour la compréhension de l'échec des sols dans les mines du charbon.

INTRODUCTION.

Although major advances have been made in methods of predicting unstable roof conditions in Australian and American collieries, little is known about the geological factors affecting roof and floor stability in South African coal mines. This is mainly because of the dearth of precise data about local rock properties related to instability in mines.

The results of an investigation into the geological and geotechnical factors affecting underground stability in South African coal mines are reported. By relating the geotechnical properties of the various sedimentary facies present in coal bearing strata to their observed behaviour during mining it may be possible to predict in advance of mining the type of roof and floor conditions which may be expected. Thus, if hazards can be identified by geotechnical testing of drill core before specific areas are mined, then either the hazards can be avoided or methods of mining

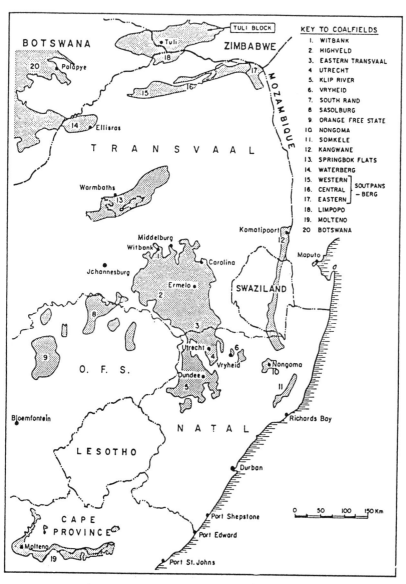

Fig.1. Location of coalfields in South Africa.

can be devised to cope with them.

Core material was obtained from the Waterberg, Witbank, Highveld, Eastern Transvaal, Klip River, Utrecht and Vryheid coalfields (Fig. 1). The coal seams of these areas occur mainly within the Vryheid Formation which forms part of the Ecca Group of the Karoo Sequence. The Vryheid Formation (formerly the Middle Ecca) is a sedimentary unit consisting of a succession of alternating sandstones and subordinate shales. The coal seams are well defined, usually with sharp lower and upper contacts between them and the clastic sediments. The coals vary in thickness with two economically exploitable coal seams occurring in the Klip River coalfield, five in the Vryheid area and three to four in the Highveld, Witbank and Eastern Transvaal coalfields.

COAL BEARING STRATA IN SOUTH AFRICA.

The major coal bearing strata in South Africa are associated with the Karoo Basin primarily occurring on the southern and eastern flanks of the Kaapvaal Craton. The strata range in age from Permian to mid-Triassic and occupy a wide range of structural and sedimentary settings. Unlike the Coal Measures of the northern hemisphere which accumulated in low-lying swamps in hot humid climates, the Permian swamps of the southern hemisphere existed under cold to cool temperate climatic conditions associated with the waning of a massive ice age. The coal bearing strata were deposited in relatively stable continental depressions. Two types of environments have been recognized in which the coal bearing strata were formed, namely, the paraglacial environments and the epicontinental environments. The former developed, in late Dwyka-Lower Ecca times, around the north west and northern side of the Karoo basin, as a result of the deposition of glaciolacustrine and fluvioglacial sediments on the moraine covered surface of the Kaapvaal Craton. The Witbank Coal Basin provides an example.

The epicontinental environments formed across the north-eastern stable shelf area of the Karoo Basin. They continue southwards to the edge of the shelf area where the sediments overlie and interdigitate with deeper water argillaceous sediments of the southern basin.

This retrogressive wedge of deltaic sediments (Vryheid Formation i.e. Mid-Ecca)

is characterized by several episodes of transgression and regression, the latter frequently closing with a coal seam. The cyclic sedimentation of the Vryheid Formation probably represents deposits which accumulated in alluvial plains, upper and lower deltas and associated shallow lagoon and coastal swamps.

The climate was warmer during the Mid-and Upper Ecca and this warming continued into the Beaufort times so that a savannah and mudflats flora then existed. Accumulation of coal bearing strata appears to have taken place in mountain bound and intercratonic basins where slow continuous subsidence was occurring, as exemplified by the Waterberg Basin. The Permian closed with a dry period when no coal bearing strata were deposited. Then a reversal to a wetter climate occurred in the mid-Triassic and coal and associated deposits formed in the Molteno Basin. It has been suggested that deposition of the Molteno Formation occurred in a warm temperature fluvio-lacustrine environment.

SEDIMENTARY FACIES AND GEOTECHNICAL TESTING.

Twenty four sedimentary facies have been identified within a fixed stratigraphic interval, defined to include 15m of hanging wall and 5m of footwall associated with any economically exploitable coal seam. The twenty four facies types are described in Table 1. Some additional varieties do occur but these are considered as modifications of the more important facies types. The facies may be broadly divided into agrillaceous and arenoceous lithotypes with some intermediate members.

The first phase of the geotechnical testing programme involved detailed sedimentological and geotechnical logging of the core. Specimens for determination of uniaxial compressive strength, direct tensile strength, Brazilian disc strength, point load index and swelling coefficient were prepared for testing according to the recommendations of the International Society for Rock Mechanics (ISRM, 1981).

To ensure reproducability and accurate comparison of the geotechnical properties of the various sedimentary facies, it is most important to impose two strict experimental controls on the preparation of samples before testing. Firstly, since the exploratory drill core diameter varies from different exploration sites, it is necessary to control the length to diameter

Table 1. Description of sedimentary facies and summary of their underground behaviour.

Facies	Description	Behaviour of strata underground
1	Massive carbonaceous mudrock	Very poor roof and floor strata due to low tensile strength and poor durability - deteriorates rapidly upon exposure. Rock falls common and floor heave occurs when depth of mining exceeds 150m.
2	Lenticular bedded mudrock	
3	Alternating layers of mudrock and sandstone	
4	Flaser bedded sandstone	Reasonable roof strata which deteriorates upon exposure, giving rise to spalling from the roof.
5	Ripple cross laminated sandstone	Reasonable roof strata, although localised roof falls do occur due to parting along silt drapes. Good durability.
6	Ripple cross laminated sandstone with grit bands	
7	Massive fine grained feldspathic sandstone	Very competent floor and roof strata due to low porosity and high tensile strength.
8	Cross laminated fine grained feldspathic sandstone	
9	Massive medium grained feldspathic sandstone	Good roof and floor strata with fairly high tensile strengths. Sometimes creates problems due to poor goafing ability in longwall areas.
10	Cross laminated medium grained feldspathic sandstone	
11	Massive coarse grained feldspathic sandstone	Good roof and floor strata. May disintegrate under prolonged saturation giving rise to stability problems.
12	Cross laminated coarse grained feldspathic sandstone	
13	Bioturbated siltstone or sandstone	Deteriorates rapidly upon exposure and saturation to give roof and floor instability.
14	Sandstone with carbonaceous draped and slump structures	Potentially very poor roof conditions. Very unpredictable.
15	Carbonaceous silty sandstone	No information available.
16	Calcrete	Surficial material - not applicable.
17	Weathered mudrock (Grootegeluk Formation)	No underground workings developed in these facies types.
18	Unweathered mudrock (Grootegeluk Formation)	
19	Massive grey mudrock (diamictite)	
20	Gritty diamictite	
21	Pebbly diamictite	
22	Coal, mixed dull and bright	Generally mined, but when left in roof or floor are more stable than facies 1 to 3.
23	Mixed coal and mudrock	
24	Carbonaceous mudrock (associated with coal seams)	

ratio of test specimens. The ISRM recommends that with the exception of the Brazilian disc tensile test, the length to diameter ratio of test specimens be between 2.5:1 and 3:1 (Brown, 1981). In this work the length to diameter ratio of all test specimens except the Brazilian disc was 2.5:1. The length to diameter ratio of Brazilian disc specimens was 0.5:1. Secondly, it is essential that the moisture content of test specimens is strictly controlled. Figures 2a and 2b show the effect of variations in moisture content on the uniaxial compressive strength of facies 11 sandstone and on the direct tensile strength of facies 11 sandstone respectively. In both cases the strength is significantly reduced by increasing moisture content. This reduction has been explained by Colback and Wild (1966) as due to a reduction in free surface energy and an increase in pore water pressure within the rock specimen. Similar reductions in strength were noted in all of the tests performed on the various facies when the moisture content of specimens was not standardised.

Although testing saturated specimens would provide more realistic strength estimates for rocks under in situ mining conditions this cannot always be accomplished since some rock types breakdown when immersed in water. For example, this led to disintegration of argillaceous specimens (facies 1 - 4). Attempts to saturate the argillaceous specimens by placing them in a desiccator at 98% relative humidity also proved unsuccessful. Consequently most specimens were tested according to the moisture control recommendations of the ISRM (i,e, after they were cut, the specimens are stored for 5 to 6 days at 20°C and 50% humidity before they are

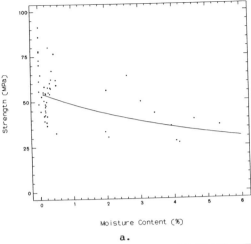

Moisture Content (%)

a.

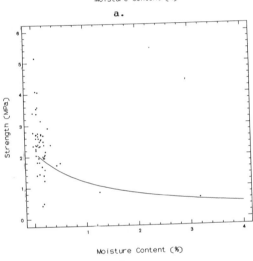

Moisture Content (%)

b.

Fig.2. Example of influence of moisture content on (a) uniaxial compressive strength (b) direct tensile strength, of facies 11.

tested; Brown, 1981). Under these conditions, the moisture content of specimens was approximately 0.3%.

RESULTS.

The geotechnical properties of facies 1 to 24 are summarized in Table 2. Fig. 3a represents the results of 551 tests carried out on the different facies types. It is apparent that there is a wide scatter of strengths for all the facies present and that the argillaceous facies possess as high, and in come cases higher, strengths than the arenaceous facies. The medium and

coarse grained sandstones have some of the weakest uniaxial compressive strengths probably because of the low cement that binds the grains together.

The results of 1608 Brazilian disc tests are illustrated in Fig 3b, the strengths obtained showing a surprising consistency. The median values of samples from the argillaceous and finer grained arenaceous facies are very similar, ranging from 7 to 8.5 MPa. Again the coarser grained sandstones are, on average weaker, having medians ranging between 3.5 and 5.5 MPa. Although the bioturbated siltstone has a strength range very similar to that of the argillaceous facies it tends to decompose rapidly.

Some 241 direct tensile tests were carried out, the results of which are shown in Fig 3c. The argillaceous facies have very similar direct tensile strengths, with median values around 0.2 MPa. The argillaceous facies also exhibit a very small range of strengths indicating that they have a consistently low tensile strength. The arenaceous facies (7 - 12) exhibit a wide range of strengths, the lowest ones being attributable to the presence of any argillaceous or carbonaceous partings, whilst the highest values represent the strength of the mass of the feldspathic sandstone, without any partings.

Figures 3d and 3e represent the data from 337 diametral and 194 axial determinations of point load strength. The argillaceous facies exhibit low diametral point load strengths, the specimens failing along planes of lamination. The arenaceous facies show a very wide range of strengths, low strengths being due to the presence of argillaceous partings, higher strengths failure occurring in the more massive feldspathic sandstone. The axial point load strengths are all higher than the diametral strengths. This is to be expected, considering the anisotropic nature of the sediments. The anisotropy index of ten facies is presented in Table 3. The high anisotropy indices are indicative of the presence of parting planes, notably lamination, within the rock and characterise the argillaceous facies. The more arenaceous facies have anisotropy ratios close to unity.

Investigation of underground behaviour of the various facies has shown that the argillaceous and intermediate groups generally give rise to some form of sub-surface instability while the arenaceous

Table 2. Geotechnical properties of the various facies.

Rock properties (MPa)

Facies	Uniaxial compressive strength				Direct tensile strength				Brazilian disc tensile strength				Diametral point load strength				Axial point load strength			
	Min	Max	Mean	Std dev	Min	Max	Mean	Std dev	Min	Max	Mean	Std dev	Min	Max	Mean	Std dev	Min	Max	Mean	Std dev
1	13.17	74.24	26.72	18.18	0.02	0.07	0.04	0.01	0.21	10.20	3.34	2.12	0.03	1.59	0.33	0.23	0.49	4.58	1.57	0.81
2	24.16	35.47	30.56	4.57	-	-	-	-	1.00	10.66	4.59	2.83	0.30	1.29	0.75	0.50	-	-	-	-
3	27.70	27.70	27.70	0.00	-	-	-	-	3.23	8.27	5.75	1.81	0.03	0.54	0.27	0.25	1.20	2.26	1.87	0.58
4	30.25	68.25	40.09	12.88	0.02	0.21	0.08	0.08	2.24	13.86	6.57	2.61	0.03	5.09	1.16	1.25	1.77	6.68	3.66	1.59
5	20.45	98.50	52.98	18.17	0.04	2.89	0.88	0.84	4.09	15.25	7.59	2.45	1.80	4.79	3.04	0.96	2.84	6.32	4.84	1.26
6	20.87	68.56	37.97	14.59	0.11	0.91	0.59	0.38	4.11	7.42	5.79	1.65	0.03	4.19	1.31	1.67	2.18	4.58	3.53	0.83
7	34.46	101.27	56.16	17.27	0.07	1.45	0.57	0.59	5.18	10.11	7.08	1.43	0.03	0.03	0.03	0.00	2.19	2.19	2.19	0.00
8	27.38	92.13	57.80	14.13	0.18	4.95	1.83	1.41	3.41	14.54	8.11	2.28	0.30	2.09	1.29	0.79	2.09	5.52	4.07	1.21
9	26.79	77.07	47.85	11.57	0.28	5.25	2.76	1.30	1.57	12.20	4.91	1.63	0.30	5.39	3.35	1.18	2.65	5.91	3.78	0.92
10	25.49	85.97	50.44	13.88	0.18	3.47	2.01	0.90	2.58	11.89	5.82	2.04	0.66	3.74	2.42	0.84	2.07	3.91	3.27	0.55
11	17.68	101.90	44.48	13.57	0.42	5.15	2.33	0.86	1.99	13.94	4.59	1.88	0.57	4.49	3.03	0.91	1.79	6.43	3.75	1.15
12	19.12	69.12	47.67	15.67	1.91	2.14	2.02	0.16	2.42	10.94	5.15	2.30	2.99	3.29	3.14	0.21	-	-	-	-
13	54.74	71.84	62.34	8.70	0.18	2.28	1.34	1.06	5.61	10.10	7.82	1.40	3.59	3.59	3.59	1.00	-	-	-	-
14	29.62	64.71	43.07	12.95	0.04	1.65	0.32	0.48	3.79	6.25	4.88	0.91	1.65	5.24	3.10	1.21	2.71	5.10	3.73	0.82
15	49.49	50.11	49.80	0.43	-	-	-	-	3.17	6.18	4.17	1.40	-	-	-	-	-	-	-	-
16	6.30	39.22	19.87	14.82	1.96	1.96	1.96	0.00	1.36	4.49	2.58	1.05	0.24	1.17	0.65	0.36	0.39	2.20	1.05	0.99
17	26.93	54.80	40.38	13.98	0.05	0.09	0.70	0.02	0.79	13.87	5.51	3.71	0.06	3.77	0.33	0.45	0.13	1.81	0.48	0.35
18	11.07	94.90	45.27	24.63	0.02	0.40	0.17	0.11	3.76	11.47	6.03	2.16	0.48	2.87	1.65	0.68	2.67	3.95	3.61	0.48
19	12.12	90.13	42.55	20.70	0.04	0.84	0.23	0.26	1.78	15.01	6.42	2.81	0.01	5.98	0.66	1.19	0.62	4.41	2.07	0.94
20	14.60	75.50	40.01	14.51	0.05	1.19	0.45	0.40	1.79	14.48	5.76	2.69	0.03	4.25	0.72	0.81	0.31	4.95	2.74	1.20
21	22.76	60.05	41.22	11.06	0.07	0.70	0.26	0.22	1.30	10.20	5.11	1.93	0.06	2.69	1.18	0.96	1.26	4.31	2.97	1.11
22	3.85	33.83	15.53	6.03	0.02	0.21	0.06	0.04	0.39	7.67	1.83	0.87	0.03	2.69	0.48	0.26	0.17	2.96	1.05	0.57
23	3.01	33.20	15.34	7.32	0.02	0.14	0.04	0.03	0.73	6.39	2.79	0.98	0.03	1.20	0.44	0.21	0.30	2.70	1.28	0.48
24	16.39	33.55	25.20	6.83	0.07	0.07	0.07	0.00	1.38	9.62	4.56	1.96	0.12	2.69	0.61	0.40	0.66	4.85	2.17	1.14

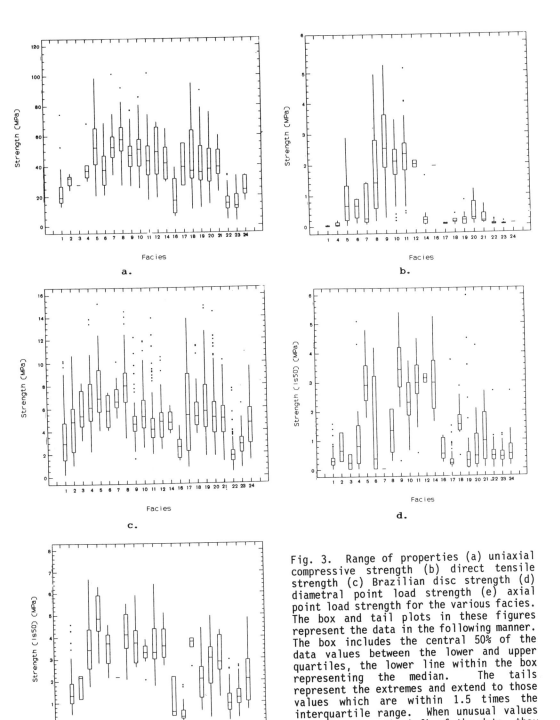

Fig. 3. Range of properties (a) uniaxial compressive strength (b) direct tensile strength (c) Brazilian disc strength (d) diametral point load strength (e) axial point load strength for the various facies. The box and tail plots in these figures represent the data in the following manner. The box includes the central 50% of the data values between the lower and upper quartiles, the lower line within the box representing the median. The tails represent the extremes and extend to those values which are within 1.5 times the interquartile range. When unusual values occur far from the bulk of the data, they are plotted as separate points.

Table 3. Some examples of anisotropy indices obtained from the point load test.

Facies	Strength (MPa) Diametral	Axial	Anisotropy index
1	0.11	2.3	20.9
2	0.31	2.6	8.4
3	0.52	2.7	5.2
4	0.84	3.5	4.2
5	2.14	5.0	2.3
7	5.38	5.6	1.0
8	2.30	4.3	1.8
9	2.23	3.5	1.6
11	1.71	2.3	1.3
14	1.2	3.9	3.2

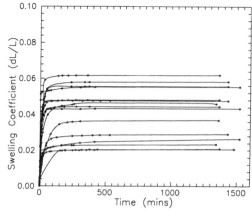

Fig.4. An example of free swell for facies 1.

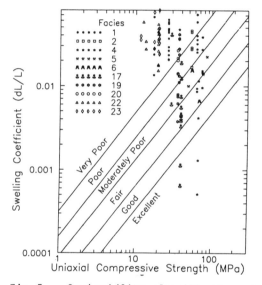

Fig 5. Geodurability classification of material from argillaceous facies.

facies provide good roof and floor conditions. Table 1 provides a summary of underground behaviour of the different facies. It must be borne in mind, however, that the distinction between good and poor hanging and footwall strata is difficult to define, since a roof that is considered good in bord and pillar mining may have poor goafing properties and be considered poor in longwall mining. Table 1 refers specifically to bord and pillar mining. A comparison of Tables 1 and 2 shows a close relationship between the underground properties of a particular facies and its direct tensile, point load and to a lesser extent Brazilian disc tensile strengths. In particular, the direct tensile strength of a particular facies provides a good basis for predicting its underground behaviour. Those facies with low tensile strengths usually promote some form of subsurface instability, while those with higher tensile strengths provide stable roof and floor strata.

No relationship is immediately apparent between the uniaxial compressive strength of a particular rock type and its underground behaviour.

The swelling coefficients of rocks belonging to various facies was determined and in the great majority of cases most of the swelling took place within the first hour, although tests generally lasted for 25 hours (Fig. 4). In the vast majority of cases the amount of swelling which occurred was less than 6%. The swelling coefficient was used with the unconfined compressive strength to determine the geodurability of

the specimen material. As can be seen from when exposed by mining. It is the combination of relatively poor durability with low tensile strength that gives rise to most dangerous and unstable underground conditions.

CONCLUSIONS.

The mode of roof failure in coal mines is essentially tensile. Underground investigations at several collieries in South Africa have shown that those facies with lower direct tensile strengths generally give rise to unstable roof conditions. The low direct tensile strengths of the argillaceous facies is of prime importance when considering the behaviour of those rocks underground. Although the arenaceous facies have a higher average direct tensile strength, it can be dramatically reduced by the presence of argillaceous or carbonaceous partings within the rock, thus affecting underground stability. Such argillaceous partings are generally erratic and discontinuous within the rock mass and they may therefore result in localized falls in a roof which would otherwise be considered stable. The Brazilian disc strength does not provide such a good indication of the weakness of the rock in tension. Indeed, the average Brazilian disc strength of the argillaceous facies are greater than those of some of the arenaceous facies.

The uniaxial compressive strength of a facies is of importance when considering heave associated with punching of pillars into the floor, calculating the size of mining pillars, and in the design of underground development in general.

Low durability facies are more prone to floor heave than the more durable types. Facies that exhibit poor durability and low tensile strengths are considered very unstable under mining conditions.

ACKNOWLEDGEMENTS

The authors would like to thank BP South Africa (Pty) Ltd, and Gold Fields of South Africa for their financial assistance during the course of this study, and for making borehole core available for study.

REFERENCES.

BROWN, E.T.(Ed). 1981. Rock Characterization Testing and Monitoring. Pergamon Press. Oxford. p.211.

COLBACK, P.S.B. & WILD, B.L. 1965. Influence of moisture content ont eh compressive strength of rock. Symp Canadian Dept Min. Tech. Survey, Ottawa, 65-83.

OLIVIER, H.J. (1978). A new engineering geological rock durability classification. Eng. Geol. 14, 255-279.

6th International IAEG Congres / 6ème Congrès International de AIGI, © 1990 Balkema, Rotterdam. ISBN 90 6191 130 3

A tourist development scheme using underground spaces and earth covered structures

Un schéma de développement touristique utilisant les espaces souterrains et structures couvertes

P.M. Maurenbrecher
University of Technology, Delft, Stichting Nova Terra, Netherlands

E.J. von Meyenfeldt
Trimbos & von Meyenfeldt Architekten, Stichting Nova Terra, Netherlands

M.H. Keyzer
Gemeente Schiedam (Formerly: Trimbos & von Meyenfeldt Architekten), Netherlands

ABSTRACT: In the southern part of the Province of Limburg in the Southeast of the Netherlands, extensive exploitation of the Cretaceous calcarenite plateau has taken place since pre-historical times. Historically most excavations were underground whereas present excavation is by open pit quarrying. South Limburg is very popular with tourism and is also subject to continual urban and industrial encroachment. A solution is sought through combining the skills of architects, civil engineers and engineering geologists to more effectively use surface and underground excavations in favour of further sacrificing the landscape. An example is given of a tourist resort proposal as part of a master plan for further developing the tourist, infrastructural and industrial potential of Valkenburg through the use of excavated spaces.

RESUME: Dans le sud de la province Limbourg, dans le sud-est des Pays Bas une exploitation tres intensive de choix vive a eu lieu depuis la prehistoire. Dans l'histoire les plus des excavations etaient sous la surface, bien que maintenant tous les excavations sont surface. La Sud de Limbourg est tres connue par son tourisme et a cause de la progressation industrielle et les grandissements des villes. Une solution pour les problemes des anciennes excavations est recherchee par l'architect, l'ingenieur civil et l'ingenieur geologue grace a leurs experience. En utilisant l'environnement et l'excavation sous la surface d'une facon plus efficace on peut sauver le paysage. Un example present de l'idee c'est d'utiliser les excavations dans la region de Valkenburg pour le tourisme.

1 INTRODUCTION

In the south eastern extremity of the Netherlands extensive exploitation has taken place of the Cretaceous calcarenite sediments since pre-historical times. Earliest evidence of exploitation, the fashioning tools from flintstone in Neandertal times, was found in the Belvederre quarry (Louwe Kooijmans 1989). Since the last ice-age underground mining of flintstones for tools occurred during Neolithic times at Rijkholt (Bosch, 1979). Flintstone is still mined for use as building stone near Eben Emael (Engelen, 1989).

Since Roman times the calcarenites have been exploited for building stone and for use as lime in agriculture. The calcarenite exploitation followed specific horizons because of the uniformity and strength. Room and pillar mining techniques were used. The room ceilings often follow a calcitic layer, of much higher strength then the surrounding rock. The layer is attributed to a reduction in deposition rate which occurred in cycles throughout the Maastrictian (up to 12 horizons are found). Only from the Sibbe mine, near Valkenburg, are calcarenite blocks still extracted for building purposes. The stone is used for making ornaments and for restoration work on existing buildings and structures.

Today three open pit quarries extensively excavate the calcarenite for cement and lime production. In the St. Petersberg quarry the calcarenites have been excavated to well below groundwater levels and into the underlying flintstone Gulpen formation. Numerous square cave portals have appeared on the quarry face as excavations intersected the old mine workings. The excavated material is used for the production of cement in the adjoining cement works of the ENCI Groep, the largest cement works in the Netherlands.

The two other quarries are situated to

the east of the River Maas, 't-Rooth and Curfs both being exploited for lime by the Ankersmith milling company. Both quarries are nearing the end of their concession area and hence further expansion is subject to planning permission. The 't-Rooth quarry is excavated in the dry whereas the Curfs quarry is subject to inflow, most from surface drainage and to a lesser extent groundwater flow. The overburden at 't-Rooth consists predominantly of the Valkenburg gravel terrace of the Maas River (Felder et al 1989). At Curfs the overbur- den is at present levels about 60 m thick and consists of Oligocene glaucanitic sands and about 10 to 15m thick gravels of the St. Geertruid Maas terrace (Felder et al 1989). The gravels are exploited commer- cially whereas the glaucanitic sands are run to spoil as well as, in both quarries, a four to five meter thick loess top layer.

2 POST-QUARRYING

Disused quarries either remain open or are infilled by a combination of overburden run to spoil and waste material. In both instances the quarries are then landscaped to become a recreational facility. This is being done on sections of the St. Peters- berg quarry and the 't-Rooth quarry. The recreational facility usually is a combina- tion of sporting facilities, park and nature reserve. In the latter instance fauna and flora seem to thrive which would not be otherwise found in the more natural environment of south Limburg. Such micro- ecological units are then subject to protection ordinances lobbied for by vociferous groups of conservationists often the same people who opposed scarring of the landscape by the digging of the quarries in the first place.

Older parts of the 't-Rooth quarry have been landscaped into a nature reserve of which one area is used as a rifle range. Excavation spoil is concealed behind relic quarry faces purposely left standing to act as retention walls. The quarry is only worked during work days so that during weekends fossil collectors can pry at the quarry faces.

The Curfs quarry is less accessible than 't-Rooth. Its entrance is well disguised as it can only be entered through a 75 m long tunnel from a country road. The entrance forms part of the extensive underground building stone mines extending along the Geul River escarpment at Valkenburg. There are many entrances along the escarpment usually shuttered or barred by padlocked gates. Some of the shuttered

entrances are concrete block walls with special openings for a species of bat whic' occupy the spaces by day. The Curfs quarr' tunnel opens into a steep sided gorge whic' then widens into an elongated open pit quarry. After a distance of 200 m the quarry turns right into what appears like large deep amphitheatre. The dramatic visual impact of this quarry has inspired concept for a visitors rest camp and natur' park. The landscape would be preserved by placing the rest camp chalets underground as earth covered structures on the quarry' west slope.

3 A SCHEME FOR VALKENBURG

3.1 Motivation and inspiration

An early initiative of a newly established National Research Centre for Innovative Uses for Underground Spaces in the Nether- lands (Nova Terra) in partnership with the Municipality of Valkenburg, Grabowski en Poort Consulting Engineers and Trimbos and von Meyenfeldt in 1986 was to set up a town planning scheme to explore possibilies for extending tourism, the infrastructure and industry. Further development above ground was not possible owing to a stalemate situation. Popular entertainment facili- ties are a major earner but could not be developed further as these facilities were at the expense of Valkenburg's traditional reputation of quiet charm. Hence more attention is given to developing existing and new underground spaces.

The location of Valkenburg is given in figure 1. A plan for developing Valkenburg and its surrounding is shown schematically in figure 2. Part of this scheme shows the development of new underground spaces. Examples of the types of uses that can be made of such spaces in the existing room and pillar mines at Valkenburg are shown in figure 3. The University of Technology Delft is currently carrying out research into the stability of these mines to give guidelines for construction and structural design codes based on use, size, rock for- mation and age of the underground opening. Classification systems have been developed from which mine stability maps have been drawn. An example of this is given in van Steveninck, (1987).

As part of the general plan for Valken- burg's future development a proposal, called "Holiday in Caves", was made by Keyzer (1989) as part of a final year study project. The proposal develops an alterna- tive recreational facility to extend the range of possibilities Valkenburg can offer

facilities of Valkenburg, such as bars, discoteques, amusement arcades, fun parks, casinos and cafes.

The proposal is intended for those holiday seekers who prefer the more quiet nature park setting of the Geul valley. An example of increasing the range is the Thermae 2000 health-spa facilities using natural warm mineral waters pumped from the Carboniferous Limestone formations at 350 m depth (Sobczak, 1989).

3.2 Holiday in Caves

The scheme consists of two parts: its general setting with respect to the overall plan for Valkenburg, and the residential, sporting and meeting spaces of the recreation park. In its general setting the recreation park forms the terminus of a walkway/ nature trail extending from Valkenburg to Geulhem along the steep side of the Geul valley. The trail will make use of existing caves to act in part as covered walkways. At several focal points there would be places of interest such as an underground exhibition centre, a museum, a cottage industry in addition to existing hotels and restaurants and a wild life sanctuary (special cave to view a rare species of bat).

3.3 Walkway

The trail would make use of the existing underground building stone mine passages nearest the valley escarpment, and may require extra openings as well as

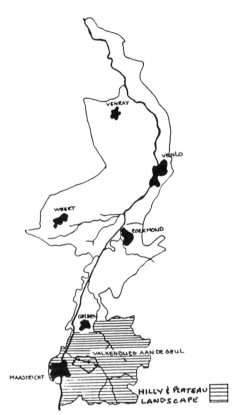

Fig. 1 Location of Valkenburg a/d Geul in the Province of Limburg, The Netherlands

to the tourist. The present balance favours the urban entertainment oriented

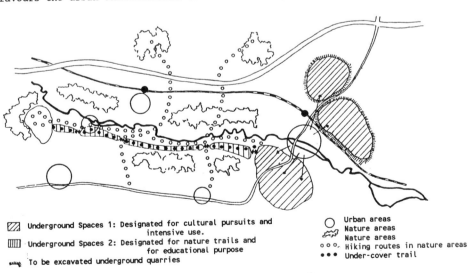

Underground Spaces 1: Designated for cultural pursuits and intensive use.
Underground Spaces 2: Designated for nature trails and for educational purpose
To be excavated underground quarries

○ Urban areas
⌇ Nature areas
Nature areas
o o o Hiking routes in nature areas
••• Under-cover trail

Fig. 2 Schematic map of Valkenburg showing master plan

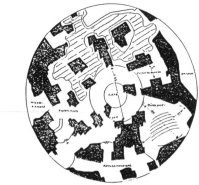

ORGANIC-SHAPED COLUMN PATTERN The spaces between the rather jagged and random spaced columns can be used for recreation such as sporting facilities, cinema, saunas, swimming, amusement arcade, cafes, nightclubs.

CHECKER-BOARD COLUMN PATTERN Spaces between these regular column can be used for recreational services of educational in nature such as museums, exhibition centre, night-time wildlife zoo, night-time wildlife zoo, mushroom farming, cottage industry.

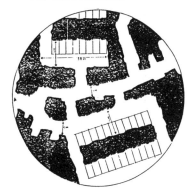

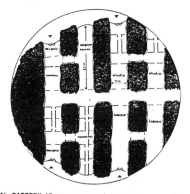

T-JUNCTION COLUMN PATTERN This pattern has relatively large rectangular spaces and hence can be used for several functions such as for recreation, parking and small industry.

FUNCTIONAL PATTERN When a new underground quarry is excavated regular column pattern should be planned allowing multi-functional uses especially for industry, laboratories, storage, offices and shops.

Fig. 3 Plans showing possible uses of underground spaces with respect to typical room-pillar patterns which exist in the building stone mines of Valkenburg.

connecting passages. This part of the scheme would be the most difficult to realize. The caves have different owners. The calcarenite rock is not of high quality (Engelen, 1989) and collapse should not be ruled out as this happened in 1988 at the Heidegroeve in Valkenburg (Price 1989). Furthermore some of the mines have been turned into wildlife sanctuaries for a rare species of bat, though for these sanctuaries special viewing facilities could be constructed. Much of the trail would have to be either open or have constructed coverings such as earth roofs with plant growth so as to blend with the existing flora.

3.4. Entrance

The park itself will be off limits for motorised vehicles. Parking is proposed underground in existing passages. This would require increasing the size of any existing mine galleries and excavating new spaces. A general view of the entrance and parking area is shown in figure 4. Historically only better quality rock has been excavated so that un-mined areas should be approached with caution as they may contain poorer quality rock and therefore be less stable. Even in good quality rock it is doubtful if the larger spaces would be self supporting and hence reinforcement would have to be introduced as little advantage can be gained from the calcite horizons when the spans become too large. The advantages to be gained is that the excavated material has a ready market at either the Ankersmit (limestone) Milling Company (the present owners) or the ENCI (cement manufacturers). The second advantage is that the natural beauty of the Geul valley will not be compromised by a car park for the rest camp. A perspective of the main entrance is given in figure 3.

At the quarry portal a reception area will be contructed of which part will be recessed into the calcarenite face. This area also would contain a shopping precinct for provisions.

3.5 Recreation Park

The park consists of several units extending along the the quarry length as shown in plan in figure 5. The park is integrated by connecting the units by landscaping the existing excavation with a lake along the length of the quarry floor and through a system of walkways. The latter also double up as hiking trails which would connect with trails in existing woodland parks surrounding the present quarry.

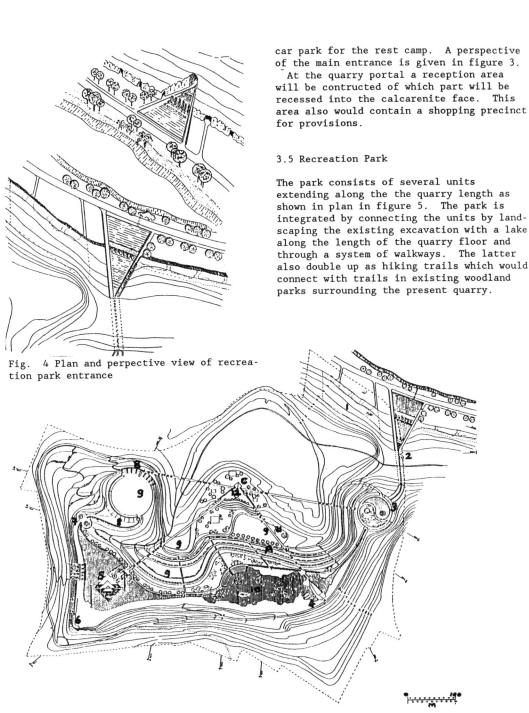

Fig. 4 Plan and perpective view of recreation park entrance

Legend: 1. Parking area (overflow outside as well as underground). 2. Entrance from Geul valley. 3. Reception and shopping precinct 4. Boat house and restaurant. 5. Bath house/ swimming pool. 6. rock face for rock climbing type outcrop for geological excursions. 7. Stables. 8. studio and meeting/conference halls. 9. sport fields. 10. boating and fishing. 11. childcare centre. 12. central theatre square and coffee shop.
Accommodation: A. Chalets versions for 2, 4 and 6 people. B. 10 bed dormitories total 14 number and C. camping and ablution blocks

Fig. 5 General plan of the Curfs Quarry recreation area.

The principal accommodation consists of earth covered chalets built into the earth fill. The earth fill consists of quarry spoil, recompacted to sufficient strength to ensure both the support of the chalets and to avoid any slope instability. Hence the existing spoil will have to excavated and replaced in stages. The recompacted spoil would require internal drains. Most of the spoil will be originally from the loess and green sands which overlie the calcarenites. Figure 6. shows the various phases proposed to construct the "cut and cover" chalets as well as the proposed drainage layers. Soil reinforcement, "Terre Armee", is suggested as possible in-expensive substitute for constructing the rear retaining wall.

The chalets have been specially designed to ensure adequate lighting and ventilation so the the occupants would not be aware they are in a buried structure. Various views of the layout of a four person accommodation chalet is given in figure 7.

4 FUTURE DEVELOPMENTS

The proposals above are primarily intended to stimulate ideas for improving the economy of Valkenburg by using and improving existing natural assets. The Thermae 2000 baths, for example, make use of geological conditions. The Curfs recreation park proposal not only makes use of the the geology but also puts to good use both natural and existing man-made landforms, without trying to create a new but false landscape. In fact it endeavours to preserve historical features by putting them to new uses.

Since the preliminary initiatives further studies are anticipated for the general development of Valkenburg. It will still be a number of years before the Ankersmit concession expires and that such a scheme can be developed. What is essential is to make municipal planners aware of the potential uses of underground space. It is intended to increase this awareness by setting up a competition inviting archi-tects and engineers to submit proposals within the terms of reference of the development plan. Hence efforts are being made to set up rules and regulations for the competition and hopefully such a competition will help increase awareness not only amongst the architects, engineers and geologists but also stir public interest.

The main aim is not essentially to provide recreational facilities but to ensure employment for the local population.

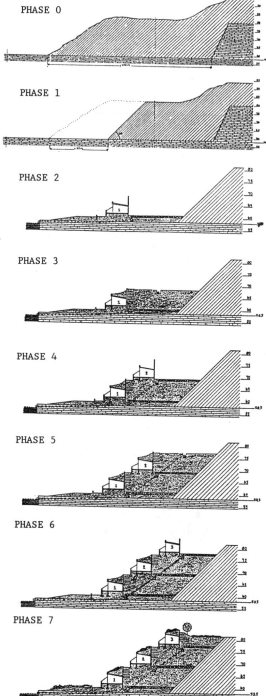

Fig. 6 Construction phases of chalets embankment.

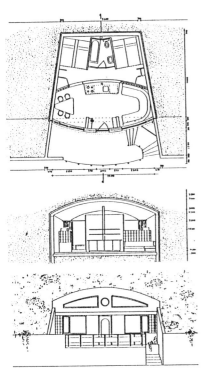

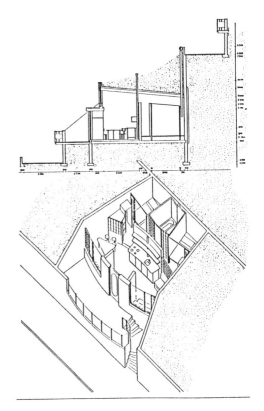

Fig. 7 Underground chalet: various views

Much of the employment is due to tourism.
By providing a comprehensive scheme as
suggested above employment opportunities
are increased especially if visitors are
attracted throughout the year. This
certainly will be possible if, as the above
example shows, combined conference and
residential facilities are provided.

REFERENCES

Bosch, P.W. 1979 A Neolithic flint mine
    Scientific American:240 6 98-104 New
    York
Engelen, F.H.G. 1989 De kalkstenen uit het
    Boven-Krijt en Onder Tertiair als
    Delfstof (The Upper Cretaceous & Lower
    Tertiary limestone as material source).
    'Delfstoffen in Limburg', Grondboor en
    Hamer, p29-44 N.G.V., Beek, Limburg
Engelen, F.H.G. 1989. Vuursteenwinning en
    toepassing (Flintsone mining and
    applications), 'Delfstoffen in Limburg',
    Grondboor en Hamer, p79-82 N.G.V., Beek,
    Limburg
Felder, W.M., P.W. Bosch & J.H. Bisschops
    1989. Geologische kaart van Zuid Limburg
    en Omgeving: Afzettingen van de Maas

1:50 000 (Geological Map of South
    Limburg, Maas deposits) : Haarlem, Rijks
    Geologische Dienst (Netherlands
    Geological Survey)
Keyzer, M. 1989. Holiday in Caves, diepbouw
    in Valkenburgse Kalksteengroeven
    (underground construction in a Valkenburg
    limestone quarry) Faculteit der Bouwkunde
    Afstudeerprojekt, 1:32p & 2:80+19p,
    Technische Universiteit, Eindhoven
Louwe Kooijmans, L.P. 1989. Hoe modern was
    de Neandertaler? Diescollege, Rijks-
    universiteit Leiden
Price, D.G. 1989. The collapse of the
    Heidegroeve: a case history of subsidence
    over abandoned mine workings in Creta-
    ceous calcarenites, pre-prints Int.
    Chalk symposium 4-7 Sep.1989 Brighton :
    p221-227 London, Thomas Telford Ltd.
Sobczak, M. 1989. Termae 2000, Valkenburg
    a/d Geul 'Delfstoffen in Limburg',
    Grondboor en Hamer, p277-280 N.G.V.,
    Beek, Limburg
Steveninck, R. van, 1987. Past mining in
    the Maastrichtian Limestone, Memoirs of
    the Centre for Engng. Geology in The
    Netherlands, 48:137p Technische
    Universiteit, Delft

An engineering geological appraisal of the progress of construction of the Lakhwar underground power house

Une évaluation géotechnique du progrès de la construction de la station à puissance souterraine de Lakhwar

S.K.Shome, G.K.Kaistha & A.K.Jain
Geological Survey of India, Dehra Dun, India

ABSTRACT: The construction of Lakhwar underground power house on the river Yamuna in a basic pluton is under advanced stage of construction. The 130m(L) x 20m (W) x 46m (H) cavern is located on the right bank in a moderately to highly jointed basic body. For designing the support system the rock loads were computed by using the available techniques. The values of rock loads computed by RMR classification are within reasonable limit with reference to the ones established by the wedge method. The correlation between the RMR and Q system suggested by Bieniawski is found to be more realistic than the one suggested by Rutledge and Preston. The P-roof for good and the poor rock was computed based on the Q values (Barton et al.). Keeping in view the location of the cavern in the close vicinity of 180m head of dam reservoir and lack of facilities as well as expertise in bolting, a conservative approach for the support design was adopted.

The excavation of the cavern has been completed upto the roof arch by conventional drilling and blasting method and has progressed in predetermined steps making use of the existing exploratory and approach adits. No major problem of rock falls was faced during the excavation and the rock behaved exceptionally well allowing ample time for the support erection. About 22,000m^3 of rock has already been excavated in a period of about two years. Necessary instrumentation has been planned and is being installed to monitor the rock behaviour during the after the construction.

RESUME: La construction de la station à la puissance sous-terrain de Lakhwar sur la rivière Yamuna dans les roche 'pluton' de la base est dans une étape avancé de la construction. La caverne de 130m (1) x 20m (W) x 46m (H) est située sur la rive à droite dans la substance de base avec les jointure variant de moyen à grand. Pour désigner une système de support, le poid des roches à été éstimé par les techniques disponible. Les valeurs de poid des roches calculé par la classification de RMR sout dans les limites raisonable par rapport de Celles établie par le méthode de Wedge. Le correlation entre le système RMR et Q proposé par Beiniawski est trouvé le plus réel par rapport de celle propesé par Rutledge et Preston. Le 'P-roof' pour les roche bien et les pouvres à été estimé basé sur les valuers de Q (Barton et al). Reppelont la position de la Caverne près du reservoir du barrage. Rappalont la location de la caverne près du reservoir de barrage de l'hauher 180m et sans la facilité ainsi que l'expertise au boulon, une approache conservateur a été adoptée por le dessin du support.

L'excavation de la caverne a été complétée jusqu'au toit de la arche par des méthodes conventional du foret et du souffle at elle est progrèss, dans une manière décidée en avance utilisant l'approche de l'exploration existant. Il n'y avait pas de problème de la chute des roches pendant l'excavation et les roches ont été exceptionnellement bien, permettant une dure comfartable pour l'erection du support. 22,000m³ environs des roches sont déjà fouillés pendant une période de deux ans. Le mécanisme necessaire est envisige and il est bien instale pour moniteur le fonctionnement des roches pendant et après la construction.

Introduction

The Lakhwar dam, a 204 m. high and 454 m. long concrete gravity dam having an underground power house (on the right flank) with an installed capacity of 3 x 100 MW across the river Yamuna is under active stage of construction.

Geology

In the project area and vicinity the rock types exposed belong to Mandhali, Chandpur and Nagthat formations represented by slates, phyllites, quartzites and minor limestones. They have been intruded by a nearly 280 m. (across the river) wide basic body, ranging in composition from dolerite to hornblende rhyolite. The dam and power house complex has been endeavoured to be located mostly within this intrusive (Fig. 1).

Layout

The power house cavity is 130 m. long, 20 m. wide and 46 m. high. The general floor level of the power house is at 628.5 m. and the intake has been kept at an elevation of 740 m. (Fig. 2). Besides the usual collection chamber, an additional collection chamber of 65x10x34 m. and a 155 m. long, 8.5 m. diameter tail race tunnel (open channel) have also been provided. From the intake the waters shall be led to the turbines through three, 4.8 m. diameter penstocks. The underground power house being practically below the reservoir, two drainage galleries at different levels have been provided.

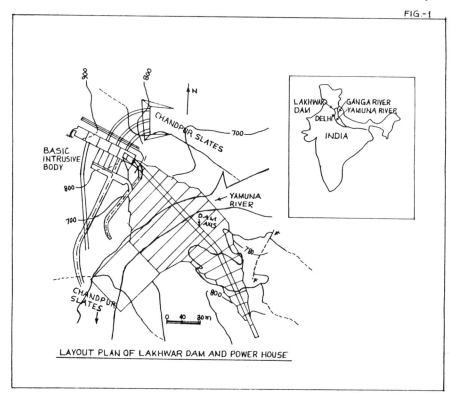

FIG.-1

LAYOUT PLAN OF LAKHWAR DAM AND POWER HOUSE

In the initial stages of power house construction the excavation of two approach adits was taken up and completed. The adit to the roof arch of control room (A_1) which gave the best possible picture of the cavern rock was used for the detailed studies. It had a 'D' shaped section (6.2 m. base and 6.25 m. height) and remained unsupported for more than eighteen months, till its enlargement. This adit had been thoroughly sampled for the joint data and 246 joint readings were plotted on the standard Schmidt's net. This analysis identified 9 sets of discontinuities (4 more prominent, first order & 5 less prominent, second order). Their intersection over the cavity roof gave wedges of height varying between 1.6 m. & 17.4 m. (Fig. 2). The former was formed by the intersection of the first order joint sets and the latter one by the intersection of the first and second order joint sets (Fig. 2).

Rock load estimation

The powerhouse and dam complex has been explored by 69 drill holes aggregating a total length of about 5100 m. However, the overall performance of drilling was very poor which made usage of the modern RMC systems difficult (Shome & Kaistha 1988). Later on Shome, Kaistha & Sharma (1989 in press) devised a method of computing RQD values realistically.

Therefore, in the initial stages the RSR concept of Wickham et al. (1972) and the RMR of Bieniawski (1974) were used after qualitatively dividing the rocks likely to be met in the cavity into good and poor and assigning a percentage of their areal extent. Later on the 'Q' values as per Barton et al. (1974) were computed.

The different ratings are tabulated below :

These values have been computed by the first two authors during the preparation of their paper Shome & Kaistha (1988). From the RMR classification as per Unal (1983) as referred in Manual in Rock Mechanics (CBIP 1988) the rock load computed for the good and poor rock works out to be (100-71)/100 x 20 = 5.8 m., (100-32)/100 x 20 = 13.6 m. which is in fairly good agreement with the values of 3.2 m. and 17.4 m. as worked out by intersections of critical joint sets.

Similarly the correlation between the RMR and Q system suggested by Bieniawski (1976) of (RMR = 9 $\log_{nn}$ Q + 44) was found to be a better fit than the one suggested by Ruttedge & Preston (1978).

The calculated permanent support pressure P-roof calculated from the Q values varied from 0.15 kg/cm^2 for good rock and 0.33 kg/cm^2 for poor rock. Generally for this order of support pressure the support system comprising rock bolts and shotcrete would have sufficed. The power house cavern is however, located in the close vicinity and below 180 m. reservoir head of the dam. In the absence of any guarantee about the long term efficacy of bolting under such a sustained hydraulic head and also in the absence of the facilities as well as expertise of systematic bolting with the project authorities the authors as a measure of abundant caution suggested conventional steel set supports for the roof arch for a design load of 1 kg/cm^2.

Excavation sequence

Two exploratory drifts for lengths of 180 m. and 133 m. respectively along the alignments of the proposed adits A_1 (to the roof arch) and A_2 (to the erection bay) were excavated. Subsequently they were enlarged to the full size adits. The adits had

S.No.	Type of rock	RSR rating	RMR rating	Q rating
1.	Good (90 %)	82	71	25.5
2.	Poor (10 %)	51	32	8

Figure - 2

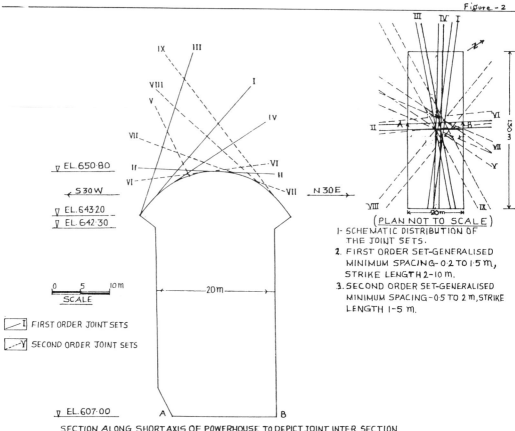

SECTION ALONG SHORT AXIS OF POWERHOUSE TO DEPICT JOINT INTER SECTION

(PLAN NOT TO SCALE)
1 - SCHEMATIC DISTRIBUTION OF THE JOINT SETS.
2. FIRST ORDER SET-GENERALISED MINIMUM SPACING- 0.2 TO 1.5 m, STRIKE LENGTH 2-10 m.
3. SECOND ORDER SET-GENERALISED MINIMUM SPACING- 0.5 TO 2 m, STRIKE LENGTH 1-5 m.

FIRST ORDER JOINT SETS

SECOND ORDER JOINT SETS

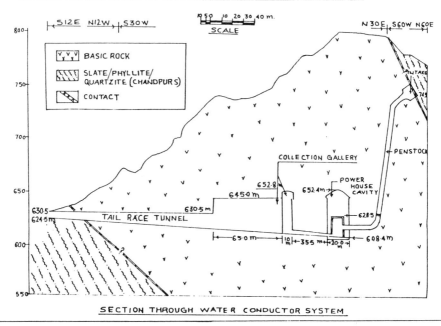

SECTION THROUGH WATER CONDUCTOR SYSTEM

FIG-3

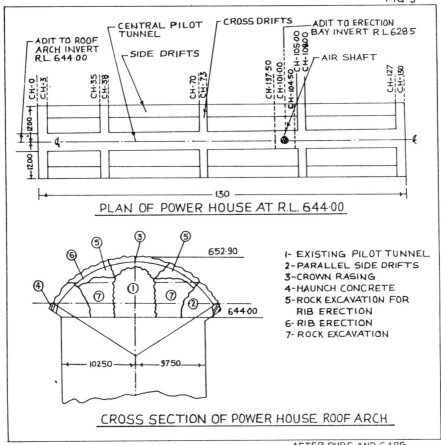

PLAN OF POWER HOUSE AT R.L. 644·00

CROSS SECTION OF POWER HOUSE ROOF ARCH

1- EXISTING PILOT TUNNEL
2- PARALLEL SIDE DRIFTS
3- CROWN RASING
4- HAUNCH CONCRETE
5- ROCK EXCAVATION FOR
   RIB ERECTION
6- RIB ERECTION
7- ROCK EXCAVATION

AFTER DUBE AND GARG

'D' shaped sections (6.2 m x 6.25 m. and 7.8 m x 8.9 m. base x height) and lengths of 185.5 m (A$_1$) and 213 m (A$_2$). The adit A$_1$ was then extended in the power house cavity up to the length of 130 m. The adit A$_2$ was extended further by 13 m. to reach the centre line of power house cavity. The two adits were connected by a vertical shaft of length 8 m. which was used for mucking and ventilation.

The drifts on both sides of adit A$_1$ were excavated at RD's 0, 35, 70, 105 & 130 m. Then two longitudinal drifts of 3 x 3 m. size were excavated at the end of these points and they were used for the excavation of the haunches and the haunches were laid. Further the crown of the main central drift was raised to the crown level of the power house. The rock columns between the haunch, drifts and the central drift were strengthened with rock bolts and chain link and then shotcreted to avoid any destressing of the rock. Next the columns on both sides were removed gradually by face excavation to achieve a full section of 3 m (Fig.3). Immediate shotcreting was done on the rock surface and four numbers of ribs were installed at a spacing of 0.75 m. centre to centre, the props were errected between the face and outer periphery of the ribs. The empty space between and above the ribs was filled with concrete blocks and blocking concrete. The excavation of the roof arch section has been completed. After completing the grou-ting, drilling of drainage holes etc. the bench excavation shall be taken up. These and other engineering details have been discussed by Dube and Garg (1988).

2623

The excavation in the power house cavity has been done by conventional drilling and blasting method, and the roof arch involving excavation of 22,000 m^3 of rock has been completed in a period of about two years. During excavation, the cavity rock has behaved exceptionally well. Barring a few local falls no major hazards were recorded. The overexcavation beyond the payline can be attributed to the improper drilling and blasting which has been of the order of 2.31 m. maximum and in only less than 5% reach. A visual estimation of destressing in the rock columns, in the cavity revealed that in a period of ± 30 to ± 60 days the maximum destressing was observed to be more pronounced along the short axis as compared to the long axis.

Monitoring/Surveillance

In order to monitor the behaviour of the roof arch instrumentation was recommended which comprises load cells, strain meters, bore hole extensometers (single & multipoint) and tape convergence meters. Except the tape convergence meters the remaining instruments have already been installed in approach adits and partly in the main cavity. In adits A_1 they have been installed at RD's 135 m, 184 m & 135 m, and adit A_2 at RD 213 m. In the main powerhouse cavity the installation is in progress and till date it has been done at RD's 0, 36, 69.32, 87.37, 104.63, 108.32, 125.62 and 129.86 m.

The regular monitoring is in progress. However, so far the data are not sufficient to carry out any detailed synthesis. From the study of the available data till to date nothing unexpected has come to light. In situ stress field measurements have also been recommended and are yet to be carried out.

Acknowledgements.

The authors are grateful to the Director General, Geological Survey of India for his kind permission to submit the paper for publication. They are thankful to the project authorities for their assistance during the data collection.

REFERENCES

Barton, N., Lien, R. & Lunde, J. (1974) – Engineering classification of rock masses for the design of tunnel support. Rock Mechanics (Springer Verlag) Vol. 6 No. 4, pp. 189-236.

Bieniawski, Z.T. (1974) – Geomechanics classification of rock masses and its application in tunnelling. Proc. 3rd Int. Cong. Rock Mech., ISRM, Denver, USA Vol.II A, pp. 27-32.

Bieniawski, Z.T. (1976) – Rock mass classification in rock engineering. Proc. Symp. Exploration for Rock Engg. Zohannesburg, A A Balkewa. Vol.I, pp. 93-106.

Dube, M.D. & Garg, M.M. (1988) – Lakhwar underground powerhouse construction aspects of roof arch. Proc. All India seminar on underground structures in rock. Lucknow, 1989, pp. 76-90, Institution of Engineers India, UP State Centre.

Rutledge, J.C. & Preston, R.L. (1978) – New Zealand experience with engineering classification of rock for the prediction of tunnel supports. Int. tunnel Symp. Tokyo.

Shome, S.K. & Kaistha, G.K. (1988) – Some questions & doubts about the Rock Mass Classification (RMC) System. Proc. Ind. Geotechnical Conf. (IGC – 88), Allahabad, Dec. 1988, Vol.I, pp. 107-109.

Shome, S.K., Kaistha, G.K. & Sharma, K. (1989) – A method for computing RQD in jointed rock not amenable to good core recovery. Symposium on Rock Mechanics organised by the Indian Geotechnical Society at Roorkee Nov. 89.

Unal (1983) – as referred in Manual on Rock Mechanics, CBIP, New Delhi 1988, pp. 152.

Wickham, G.E., Tiedman, H.R. & Skinner, E.H. (1972) – Ground support prediction Model (RSR concept). Proc. Ist Rapid Excavation Tunnelling Conf., AIME, New York, pp. 43-64.

5.3 Engineering and environmental problems caused by subsidence brought about by underground extraction of minerals, oil and gas
Problèmes techniques et de l'environnement causés par la subsidence à la suite de l'extraction souterraine de minéraux, de pétrole et de gaz

6th International IAEG Congres / 6ème Congrès International de AIGI, © 1990 Balkema, Rotterdam. ISBN 90 6191 130 3

Stability assessment of the Hoorensberg room and pillar mine, Southern Limburg, using separate shape factors for intact and cracked pillars

L'évaluation de la stabilité de la 'room and pillar' mine Hoorensberg au Limburg-du-sud par l'usage des formules, reliant la résistance à la compression simple, à la forme de l'échantillon, séparées pour piliers intacts ou crevassés

R. F. Bekendam & W. G. Dirks
Engineering Geology Section, Faculty of Mining & Petroleum Engineering, Delft University of Technology, Netherlands

ABSTRACT: The Hoorensberg mine is one of the abandoned room and pillar mines, excavated in Upper Cretaceous calcarenites, which is affected by intensive pillar fracturing. Subsidence following a large scale collapse would severely damage constructions at the surface above. A support plan was developed, which was based on stability analysis using the tributary area method as outlined by Goodman (1980). Shape factors were determined experimentally for intact and cracked pillars respectively. The results of the stability analysis reflect the actual state of the mine reasonably well, but there is a need for some basic improvements of the method.

RESUME: La mine Hoorensberg est une des "room and pillar" mines abandonnées, ayant été excavées dans les calcarenites du Cretacé supérieur, qui montre un fracturement intense dans presque toute son étendue. L'affaisement résultant d'un écroulement á grande échelle endommagerait sérieusement les constructions situées sur la surface au-dessus de le mine. Un project de soutien fut développé, basé sur l'analyse de la stabilité suivant les principes de la methode de "tributary area", telle qu'elle fut esquissées par Goodman. A titre d'expérience, des formules, reliant la résistance á la compression simple á la forme de l'échantillon, furent déterminés respectivement pour les piliers intacts et crevassés. Les résultats de l'analyse de la stabilité reflètes assez bien l'état actuel de la mine, mais la méthode demande une mise au point sur certaines points fondamenteaux.

1 INTRODUCTION

In Southern Limburg, The Netherlands, Maastrichtian rocks occur near the surface. They are composed of a series of calcarenites and calcisiltites containing bands of flint nodules. Some layers are free of flint and have been mined for building stone using room and pillar methods.

The average porosity of these rocks is 50 percent and the uniaxial strength generally does not exceed 5 MPa, allowing mining entirely by sawing out blocks.

The mining is believed to have begun in Roman times, but was carried out mainly in the nineteenth century and the first half of this century. Now mining is only undertaken intermittently on a small scale for restoration purposes. In the vicinity of Maastricht and Valkenburg many kilometers of open tunnel are to be found (figure 1).

Several mines are used for tourism, storage and mushroom growing, and one of them even as a nuclear hiding place.

Most workings are reasonably safe for the greater part, but many areas are collapsed or show a critical degree of instability due to intensive fracturing of mine pillars. It has become evident that most of these stability problems have resulted from long term deterioration often in combination with unskilled mining or local geological conditions.

For an example, the Heidegroeve near Valkenburg (figure 1) appeared to be safe for a period of forty years after the last mining activities (Price 1989). A few years ago a deterioration of the mine was observed and in June 1988 more than half of the area collapsed within one minute, as has been recorded by a seismograph. More mines have shown a slow decrease in pillar stability followed by a sudden collapse. This sensitivity to long term deterioration has been confirmed by a series of creep tests on calcarenite.

The collapse of the Heidegroeve also showed that instability did not only put users of the mine into great danger, but also people and buildings at the surface: the collapse underground was immediately followed by subsidence of an area of 100 by 70 meter with a downward displacement of up to 0.8 meter. Subsidence has also been observed directly after a major collapse in the nineteentwenties in the Jesuitengrot near Maastricht.

Both the sensitivity of calcarenite to creep and the possibility of subsidence due to an underground collapse made it necessary to investigate the long term stability of the Hoorensberg mine.

2 THE HOORENSBERG MINE

The Hoorensberg mine is situated near Valkenburg (figure 1) and adjoints an extensive underground network with many entrances, the Gemeentegroeve. The area of the mine measures about 150 by 120 meter (figure 2). Towards the west it passes into the Gemeentegrot. Also in the north a connection between the two mines exist. The entrance way is situated in the south-eastern part of the studied area.

Figure 2 also shows the amount of overburden, ranging from 24 meter in the northeast to 50 meter in the southern part. The overlying strata consist almost entirely of calcarenite, with only 0.5 to 1.5 meter of loess on top.

Dates written on the walls suggest that mining started in the beginning of the nineteenth century. The workings were abandoned probably before the second world war.

Stratigraphically, the calcarenite of the Hoorensberg mine forms part of the Formation of Maastricht, which is subdivided into six lithological units: the units of Valkenburg, Grondsveld, Schiepersberg, Emael, Nekum and Meerssen. The rocks of the entrance way are situated within the flint bearing unit of Schiepersberg and lower part of the unit of Emael (figure 3). The main part of the mine is cut out in the unit of Nekum, which is homogeneous and without flints. Roof and floor are formed by hardgrounds. In the western and southwestern part also the upper 1.5 meter of the unit of Emael has been mined.

The mine pillars are more or less rectangular but irregularly shaped. The average height is 2 meter and the horizontal dimensions vary from 0.5 to 80 meter.

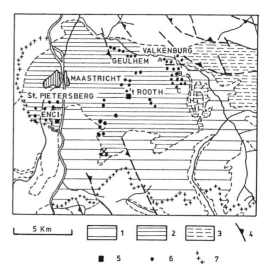

Fig. 1. Geology and locations of mines in the Upper-Cretaceous limestones of Southern Limburg, The Netherlands. 1.Formation of Gulpen 2.Formation of Maastricht 3.Formation of Kunrade 4.Fault 5.Major open quarry 6.Mine entrance 7.Frontier A.Heidegroeve Mine B.Hoorensberg Mine C.Sibbe Mine.

In the western part of the mine pillars reach a height of 3.5 meter.

Only a few joints occur, running generally subvertical from east to west, but other orientations are to be found as well (figure 4).

Figure 4 also shows the fractures in the roof and in the pillars due to stress concentrations as a result of the mining. It is obvious that, because almost all pillars are affected by fracturing in various degrees, the overall stability of the mine is not sufficient. Failure of a certain pillar can bring about failure of surrounding pillars. Within minutes or even seconds the collapse can spread over the whole mine by this domino effect, as has been recorded in the Heidegroeve. If all pillars between the entrance ways in the south and the northern boundary of the mine collapse, an unsupported span of about 80 meters wide is left. This will result in caving of the overlying calcarenite strata, which might well reach the surface 30 to 40 meters above, assuming an angle of break equal to 45 degrees.

Constructions at the surface above the mine will certainly be damaged severely by a large scale collapse underground. To prevent such damage a stability analysis has been carried out resulting in an support plan of the mine.

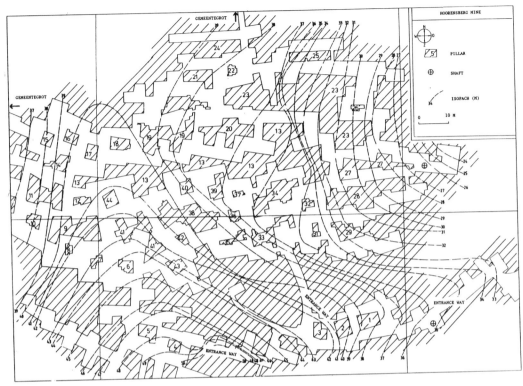

Fig. 2. Map of the studied area of the Hoorensberg Mine.

3 METHOD OF STABILITY ASSESSMENT

Pillar damage, as recorded during the mine survey (figure 4), was classified following the system devised by Van Steveninck (Price & Van Steveninck 1988).In this system each individual pillar is classified relative to the fracture pattern it shows. Five classes are distinquished:

Class 1 : no fractures
Class 2 : fractures only on top or bottom corners
Class 3 : fractures on top and bottom corners
Class 4 : fractures running from top the bottom
Class 5 : failed

For pillars of highly irregular shape a subdivision has been made (figure 4). Projections which are small relative to the pillar of which they form part are not regarded as individual units because of their minor support role.
In table 1 pillar classification numbers are presented. Although no failed pillars occur, 65 percent of the pillars show a classification number of 3 or more. Obser-

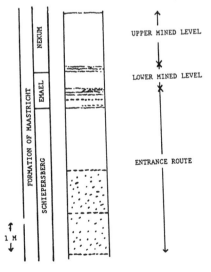

Fig. 3. Statigraphy and mined levels of the Hoorensberg Mine.

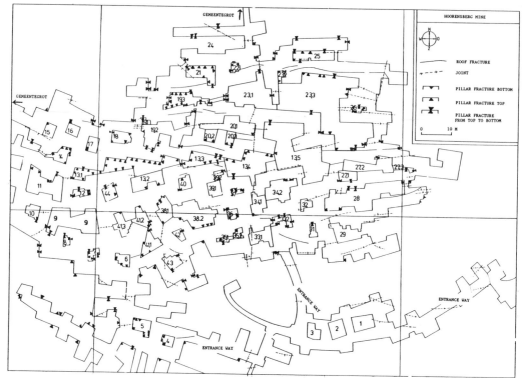

Fig. 4. Cracks and joints in pillars and roof of the Hoorensberg Mine.
Pillars are subdivided into more regular areas.

vations made during uniaxial compressive experiments on calcarenite block samples (Vreugdenhil 1988) revealed that the strength of class 3 pillars are all well below their peak strength. From this evidence it appears that the Hoorensberg mine is threatened by a large scale collapse.

To be able to calculate the locations and amount of additional support a stability analysis was performed using the tributary area method (Goodman et al. 1980). This method considers each pillar to be loaded by the column of material above the pillar and its surrounding room. The average stress S on each pillar is calculated using the formula:

$$S = P * At / Ap \quad (1)$$

where P is the initial vertical stress at the level of the roof of the opening, At is the tributary area and Ap is the cross-sectional area of the pillar. The strength Sp of each pillar can be expressed by the formula:

$$Sp = UCS * Nshape * Nsize \quad (2)$$

where UCS is the unconfined compressive strength of the rock material. Nshape and Nsize relate the strength of the cylindrical specimen of dimensions twice as long as its diameter used for the determination of the unconfined compressive strength to the strength of a specimen with a rectangular base of a certain width/height ratio and size.
The ratio Sp/S then gives a safety factor for each pillar.

Nsize was assumed to be equal to unity because no size effect was recorded for uniaxial compression tests on cubical samples of calcarenite ranging from 5 to 15 cm.

The calcarenite is almost unjointed relative to the dimensions of the pillars, so intact rock strength can be considered.

However, the assessment of Nshape poses some problems. Most empirically determined shape factors relate only to coal. Hustrulid (1976) reviewed size and shape factors and concluded that the following equation described experimental data for several rock types very well:

$$Spr/Sc = 0.778 + 0.222 * W/H \quad (3)$$

Pillar number	Pillar Class	W/H	Nshape	Pillar Strength MPa	Pillar Stress MPa	Safety Factor
1	1	1.67	1.29	2.78	1.97	1.41
2	1	2.00	1.38	2.96	1.79	1.65
3	1	1.25	1.19	2.56	2.51	1.02
4	2	1.31	0.60	1.60	3.62	0.44
5	3	1.38	0.62	1.67	2.87	0.58
6	3	1.58	0.69	1.86	4.22	0.44
7	4	0.80	0.41	1.11	5.62	0.20
8	4	0.85	0.43	1.16	4.73	0.24
9	2	1.48	0.66	1.77	1.62	1.09
10	2	1.40	0.63	1.69	2.86	0.59
11	2	1.67	0.73	1.95	1.50	1.30
12	4	0.53	0.31	0.85	3.42	0.25
13.1	4	1.55	0.68	1.83	1.36	1.35
13.2	2	2.09	0.88	2.36	1.87	1.26
13.3	4	1.36	0.61	1.65	1.64	1.01
13.4	3	2.18	0.91	2.44	1.18	2.07
13.5	3	2.18	0.91	2.44	1.06	2.31
14	4	1.42	0.64	1.71	2.57	0.66
15	2	1.49	0.66	1.78	3.16	0.56
16	2	1.55	0.68	1.83	3.28	0.56
17	4	1.36	0.61	1.65	2.93	0.56
18	4	1.57	0.69	1.85	2.24	0.83
19.1	4	0.55	0.32	0.87	2.91	0.30
19.2	4	1.09	0.52	1.39	1.48	0.94
19.3	4	1.09	0.52	1.39	2.44	0.57
20.1	1	1.73	1.31	2.81	1.55	1.82
20.2	2	1.73	0.75	2.01	1.66	1.21
20.3	4	1.27	0.58	1.56	2.06	0.76
21	4	1.60	0.70	1.88	1.93	0.98
22	4	1.16	0.54	1.46	4.05	0.36
23.1	4	2.31	0.96	2.57	1.28	2.01
23.2	4	1.07	0.51	1.37	3.12	0.44
23.3	4	2.67	1.09	2.92	0.99	2.95
24	2	1.78	0.76	2.06	1.27	1.62
25	4	1.33	0.60	1.62	1.71	0.95
26	4	0.53	0.31	0.85	2.87	0.30
27.1	4	1.51	0.67	1.80	1.31	1.37
27.2	2	1.33	0.60	1.62	1.15	1.41
27.3	4	0.44	0.28	0.76	2.32	0.33
28	3	1.60	0.70	1.88	1.05	1.79
29	1	1.33	1.21	2.60	1.62	1.60
31	4	0.75	0.39	1.06	4.06	0.26
32	2	1.09	0.52	1.39	3.29	0.42
33.1	1	1.20	1.18	2.53	1.80	1.40
33.2	4	0.96	0.47	1.26	3.06	0.41
34.1	2	1.80	0.77	2.08	1.63	1.27
34.2	4	2.00	0.84	2.27	1.35	1.68
35.1	2	0.92	0.46	1.22	3.85	0.32
35.2	4	0.67	0.37	0.98	2.70	0.36
36	4	1.60	0.70	1.88	3.22	0.58
37	4	0.55	0.32	0.87	1.28	0.68
38.1	4	2.55	1.04	2.80	1.86	1.51
38.2	2	1.73	0.75	2.01	1.37	1.47
39.1	1	1.87	1.34	2.89	2.03	1.42
39.2	4	0.67	0.37	0.98	2.38	0.41
40	4	1.90	0.81	2.17	2.87	0.76
41.1	3	1.57	0.69	1.85	2.50	0.74
41.2	4	1.04	0.50	1.34	2.12	0.63
41.3	2	1.13	0.53	1.43	2.78	0.51
42	4	0.69	0.37	1.00	3.76	0.27
43	4	1.33	0.60	1.62	2.61	0.62
44	4	1.87	0.80	2.14	3.00	0.71

Table 1. Pillar classes, shape factors and safety factors for the Hoorensberg Mine pillars.

which relates the strength of a prism Spr of a certain width/heigth ratio W/H to the cube compressive strength of the material Sc. Goodman used a revised version of this formula in which prism strength is related to the uniaxial compressive strength UCS of a cylindrical core:

$$\text{Nshape} = \text{Spr/UCS} = 0.875 + 0.250 * \text{W/H} \quad (4)$$

In recent investigations in calcarenite mines (Price & Verhoef 1988) this formula was used, but it could not be verified accurately enough by field data from calcarenite mines. It was decided to check experimentally if equation (4) really applies to calcarenite.

Furthermore, pillars which have safety factors of less than one are considered by Goodman not to carry any load. In reality such pillars still have a residual strength and carry some of the overburden load. Hence it is proposed not to make this differentiation between pillars with a safety factor of less and those of more than one. Concerning the strength the actual state of the pillar is of more importance. Cracking will obviously reduce pillar strength. For this reason different shape formulas will be applied to cracked and uncracked pillars. In the next section also a relation between width/height ratio and residual strength is presented for pillars which show cracks but no failure.

4 DETERMINATION OF SHAPE FACTORS

At one location in the mine of Sibbe (figure 1) a quantity of calcarenite was sampled, which was sawed into blocks of about 20 by 50 by 40 cm of homogeneous rock material. These blocks were immediately packed in plastic foil to keep the material at the natural moisture content of approximately 8 percent. From each block 5 to 6 samples with a square base were sawn. This resulted in 48 prism shaped samples of width/heigth ratios varying from 0.25 to 4, which is within the range showed by the actual pillars in the Hoorensberg mine. The smallest dimension, either width or height, came to 5.0 cm for all samples. The upper and lower surface were ground flat. From each block 5 to 6 cylindrical cores were produced following the ASTM standard D4543-85 to be able to relate data measured on samples from different blocks. The UCS tests were in accordance with the ASTM standard D2938-86. The samples were tested in two closed loop uniaxial compression machines which are capable of producing loads up to 50 kN and 600 kN respectively. The experiments were performed with constant strain rates ranging from $5.6 * 10^{-6}$ s-1 to $1.3 * 10^{-5}$ s-1 to obtain a time to failure of about 10 min for all samples. Sample-platen interfaces were not treated by any means.

Amount of Samples	Prism Size mm	W/H	Prism Strength/UCS	Residual Prism Strength/UCS
5	50*50*200	0.25	0.86 + 0.03	0.24 + 0.10
5	50*50*150	0.33	0.91 + 0.01	0.22 + 0.09
5	50*50*100	0.5	0.98 + 0.07	0.28 + 0.12
16	50*50*50 - 150*150*150	1	1.16 + 0.10	0.48 + 0.07
6	100*100*50	2	1.40 + 0.15	0.90 + 0.14
6	150*150*50	3	1.58 + 0.05	1.21 + 0.05
4	200*200*50	4	1.86 + 0.07	1.53 + 0.04

Table 2. Summary of experimental results.

A summary of the results is presented in table 2. Figure 5 shows that the experimental data fit the Hustrulid-Goodman relation very well. The validity of this equation for calcarenite has been proved now.

The development of the first crack was observed to occur close to the peak strength. Then in a relative short time the crack is running through the whole heigth of the sample while more cracks are growing simultaneously. Eventually a final residual strength is reached, which was found to be determined by the following equation (table 2, figure 6):

$$N_{shape,res} = 0.124 + 0.360 \, W/H \quad (5)$$

It was not possible to establish different shape factors for each individual crack pattern ranging from class 2 to 4. Hence relation (5) will be applied to all pillars which show a degree of cracking but no failure, while relation 1 will be used for unfractured pillars only.

5 STABILITY ANALYSIS

A major shortcoming of the method of stability assessment as presented so far is that results of short term experiments performed under conditions of constant strain rate and variable stress are applied to a situation of constant stress acting over a long time span. Experiments on various rock types have revealed that due to creep processes the long term strength is generally only 80 percent of the short term uniaxial strength (Singh & Banaras 1977). From preliminary results from creep tests it appears that this also holds for calcarenite.

For this reason a factor of 0.8 is incorporated into equation (2) for class 1 pillars and pillar strength is determined by the equation:

$$S_p = UCS * 0.8 *(0.875 + 0.250 * W/H) \quad (6)$$

Once pillars are beyond their peak strength due to fracturing no creep factor

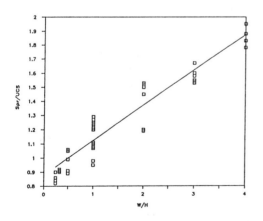

Fig. 5. Experimental results showing the relationship between prism uniaxial compressive strength and width/height ratio for calcarenite.

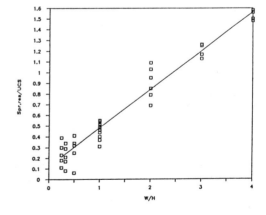

Fig. 6. Relationship between width/heigth ratio and residual prism compressive strength.

is applied anymore. For pillars of class 2, 3 an 4 equation (5) for residual strength can be used without creep factor. The final equation for class 2, 3 an 4 pillars is:

$$Sp = UCS * (0.124 + 0.360 * W/H) \quad (7)$$

Another problem is that it is not clear which horizontal dimension of an irregular pillar should be chosen as the width for the calculations. However, a subdivision of pillars into more regular parts has been made and relatively small projections are not considered. The smallest horizontal pillar dimension is then used as the width.

A series of tests on the calcarenite of the Hoorensberg mine gave UCS values of 2.69 ± 0.20 MPa. The value of 2.69 MPa is used in the calculations.

The average density of the overburden, consisting almost entirely of calcarenite, was estimated to be 1.75 * 10e3 kg/m3.

In table 1 the results of the stability calculations are presented. 37 out of the 62 pillars have a safety factor of less than one. Figure 7 gives the relation between safety factors and pillar classes. All unfractured pillars have safety factors greater than one. Pillars of class 2,3 an 4 all show safety factors within more or less the same range below and above the line of safety factor = 1, but for class 4 pillars the percentage of safety factors less than one is considerably greater than for class 2 and 3 pillars.

The stability of the whole area can be calculated by dividing the bearing capacity of all pillars together by the total overburden load. This gives an overall safety factor of 1.32.

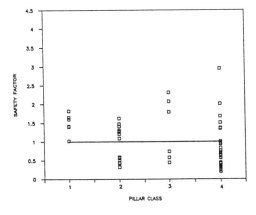

Fig. 7. Safety factor versus pillar class

6 ADDITIONAL SUPPORT

The support plan is characterised by the following elements:

1. Reinforcement of the underground network should be done by placing extra pillars, consisting of brickwork. The stress will be divided over the new and the original calcarenite pillars.
2. The elasticity modulus of the building material should be within the range of values measured on calcarenite. Material with a much lower elasticity modulus would not give enough support, material with a considerably higher value would cause intolerable point loads in roof and floor. A kind of aerated concrete with a elasticity modulus of 2100 MPa and a UCS of 3.75 MPa was chosen. Also for this material a UCS creep correction of 0.8 was applied.
3. The aim of the extra support is to attain a safer overall situation of the mine. A large scale collapse involving a majority of the pillars in the Hoorensberg mine should be prevented. It has already been stated above that such an event is likely to trigger subsidence of the overlying area. Regarding this subsidence potential the additional support should be built in the area between the entrance ways in the south and the northern boundery of the mine. Pillars 1 to 5 situated outside this zone are not included in the calculation for additional support. No attention will be paid to small scale roof collapses.
4. The overall safety factor of the mine should be at least 2.00.
5. The additional support should be built against existing pillars to create extra pillar surface and more favourable pillar shapes at the same time.

The same methods were applied as used in section 5. By trial and error the support plan presented in figure 8 was obtained. An area of 712 m2 gives an overall safety factor of 2.10.

6 DISCUSSION

1. It is generally known that friction between platens and specimen ends and different elastic properties of rock and steel tend to prevent expansion of the specimen at its ends. Due to this lateral constraint shear stresses arise at the specimen-platen contact, resulting in a non-uniform stress distribution within the specimen affecting its compressive

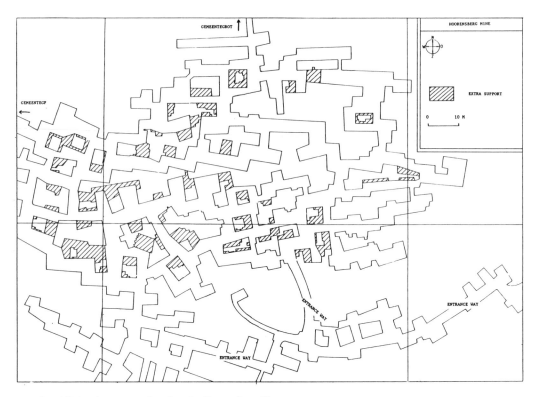

Fig. 8. Additional support plan for the Hoorensberg Mine.

strength. This stress distribution varies as a function of specimen geometry, e.g. the width/heigth ratio. This means that the experimentally found shape factors are entirely determined by end effects.

The similarity of experimentally determined strength-shape relations from various authors mentioned by Hustrulid (1976) is probably a result of identical specimen end-platen configurations: rigid steel platens directly in contact with the specimen ends.

However, this experimental specimen end-platen configuration does not necessary apply to real pillars. In the mine pillars are bounded by a roof and a floor of the same rock material instead of steel platens. Furthermore, in the calcarenite mine pillars, roof and floor generally form part of one and the same rock mass, while in the experimental set-up a smooth steel-specimen interface exist.

Because the calcarenite roof and floor represent "platens" of much lower stiffness than steel constraining shear stresses at the pillar ends tend to be lower than those resulting at the experimental specimen-platen interface. But on the

other hand the absence of a discontinuity plane as present in the experimental configuration will result in an increase of the constraining shear stresses at the pillar ends. It is not known which of those two opposing effects will dominate.

Uniaxial compression tests on specimens representing actual pillar geometry could be performed. The thickness of "roof" and "floor" should be at least three times as large as the heigth of the "pillar". In this way specimen end-platen effects are not experienced in the roof and the floor near the pillar. However, using this method some practical problems will be encountered, e.g. the difficulty of producing a large series of specimen and the geometrical constraints of the testing machine.

Hence finite-difference methods are applied now to this problem. First results look promising.

2. Nsize was assumed to be equal to one. However, many authors do report significant strength reductions with increasing specimen size. Hoek and Brown (1980) give a compilation of the influence of diameter upon strength of cylindrical cores for

various rock types. These data all fit more or less the same relationship, according to which a strength reduction of 18 percent should occur if specimen diameter increases from 5 to 15 cm. The relationship for coal cubes, resulting from a compilation by Singh (1981) indicates even a strength reduction of about 45 percent. Experiments on cubical calcarenite specimens did not show any size effect within this range.

However, to be able to apply a size factor equal to one with more certainty experiments with calcarenite specimens larger than 15 cm should be performed.

3. Pillar strength reduction due to joints is not considered in the stability analysis, because joints are not abundant and generally subvertical. Experiments on aerated concrete, performed by Vreugdenhil (1988), showed that vertical joints do not bring about a significant reduction in strength, unless a joint is situated closer to the pillar edge than one tenth of its width. This results in a strength reduction of about 10 percent.

Experiments on jointed calcarenite specimens should be carried out and finite-difference methods should be applied to obtain more reliable estimations of the influence of joints on pillar strength. It is possible that the calculations presented above give a slight overestimation of pillar strength.

4. If the strength of a pillar is reduced due to fracturing, a redistribution of pillar load occurs resulting in an increased load on adjacent pillars. Fairhurst (1973) clearly illustrated this pressure arch formation. Such effects have not been taken into account in calculating pillar stresses. Without the presence of pressure arches a majority of the pillars should already have been failed, according to their safety factors of less than one. Probably the unmined areas to the north, east and south of the Hoorensberg mine have taken over part of the loads on the pillars in the enclosed area.

Research is needed to get a better idea of the redistribution of pillar loads.

5. Most pillars of the Hoorensberg mine are irregular in shape, which made it difficult to apply the shape factors derived above. Studying this mine working gave a good impression of all kinds of problems related to stability assessment of calcarenite mines, but the methods outlined above should be tested on more regularly excavated mines as well.

Compression tests on non-square based calcarenite specimens and finite-difference experiments are planned for the near future.

REFERENCES

Fairhurst, C. (1973). Laboratory testing of rock and its relevance to mine design, Chapter 13.3. SME Mining Engineering Handbook, 13-16 to 13-51.

Goodman, R.E., Korbay, S.& Buchignani, A. (1980). Evaluation of collapse potential over abandoned room and pillar mines. Bulletin Association of Engineering Geologists 17, 27-37.

Hoek, E. & Brown, E.T. (1980). Underground excavations in rock. Institution of Mining and Metallurgy. London.

Hustrulid, W.A. (1976). A review of coal pillar strength formulas. Rock Mechanics 8, 115-145.

Price, D.G. (1989). The collapse of the Heidegroeve: a case history of subsidence over abandoned mine workings in Cretaceous calcarenites. International Chalk Symposium, Brighton, september 1989.

Price, D.G.& Verhoef, P.N.W. (1988). The stability of abandoned mine workings in the Maastrichtian limestones of Limburg, The Netherlands. Engineering Geology of Underground Movements, Geological Society Engineering Geology Special Publication no. 5, 193-204.

Singh, M.M. (1981). Strength of rock. Physical properties of rock and minerals. MCGraw-Hill. New York, 83-121.

Singh, D.P.& Banaras (1977).Long-term strength of rock. Colliery Guardian 25 (11), 861-866.

Vreugdenhil, R. (1988). The influence of natural discontinuities and material strength on the safety of calcarenite mine pillars. Memoirs of the Centre for Engineering Geology in the Netherlands No. 61.

6th International IAEG Congres / 6ème Congrès International de AIGI, © 1990 Balkema, Rotterdam. ISBN 90 6191 130 3

Engineering geological investigations for the Don Valley intercepting sewer (Sheffield/England)

Investigations géologiques de génie civil pour l'égout d'interception de la vallée du Don (Sheffield/Angleterre)

A. P. Deaves
Sheffield City Council, UK

J. C. Cripps
University of Sheffield, UK

ABSTRACT: As a part of a major upgrading of sewage disposal and treatment within the drainage area of the City of Sheffield a system of low level intercepting sewers is under construction. The works are located in the valleys of the rivers Don and Sheaf with excavations to depths of between 20 and 30 m below ground level. The geological and hydrological setting of the Lower Don Valley together with the effects of past mining activities in the area give rise to a highly variable and, in parts, adverse and potentially hazardous tunnelling prospect.

The paper briefly reviews the geological, hydrogeological and geotechnical appraisal of ground conditions for sub-surface excavations. The generation and presentation of engineering geological information are illustrated with reference to particular features identified during the investigation in the Lower Don Valley.

RESUME: - Un système d'égouts d'interception de niveau inférier est actuellement en construction dans le cadre d'une amélioration majeure apportée à l'assainissement et au traitement des eaux usées dans l'aire de drainage de la ville de Sheffield. Les travaux se situent dans les vallées des rivières Don et Sheaf et exigent des excavations jusqu'à des profondeurs allant de 20 à 30 m au-dessous du niveau du terrain. Le site géologique et hydrologique de la basse vallée du Don ainsi que les effets d'une activité minière antérieure donnent lieu à une perspective de forages hautement variables et qui s'annoncent par endroits difficiles, voire potentiellement dangereuses.

Cette communication donne un bref aperçu de l'évaluation géologique, hydro- géologique, et géotechnique des conditions du terrain en matière de forage et d'excavations. La collecte et la présentation des informations géologiques de génie civil sont illustrees par rapport aux traits particuliers identifiés au cours de l'investigation de la basse vallée du Don.

1.0 GROUND INVESTIGATIONS FOR TUNNELS

Ground investigations for tunnels are concerned with providing data to assist with the choice of appropriate methods of excavation and support of the ground and, in the case of excavation below the water table, controlling the ingress of ground water into the workings. As Dumbleton and West (1976) point out, it is necessary to characterise the materials through which the tunnel is to pass both in terms of their lithology and their performance as a tunnelling medium. The positions of particular rock and soil units with respect to the excavation are usually determined by standard geological mapping and correlation techniques. An assessment of the performance of the rockmass needs to be based on quantitative determinations of strength, intactness and other properties of the material.

The methods and procedures of site investigation are well established in the literature (see Clayton et al. 1982) and national codes of practice as defined, for example by British Standard BS 5930:1981 (Anon 1981). Site investigation for tunnels can be a more difficult prospect than for, as Dumbleton and West (1976) indicate, many tunnels are constructed in mountainous areas of complex geology, beneath stretches of water, where the problems include the possible presence of a buried valley, and urban areas. In each of these cases it can be very difficult to site boreholes in optimum positions. In addition, a narrow corridor of ground is investigated although, in order to provide an adequate appraisal of the ground conditions, the regional geology might need to be taken into consideration.

In most cases the most efficient way of conducting a site investigation is to undertake the work in stages, namely the desk study, main investigation and constructional stages. There are several advantages attached to this strategy, for example at the early stage of a project, the actual route and design philosophy for the works may have to be considered in light of the ground conditions. Hence data gathered during the desk study may allow certain of the alteratives to be eliminated, and also ensure that later stages of the investiation are well designed and focussed in approach.

1.1 Desk Study and Reconnaissance

During this stage all available published and unpublished relevant data including all topographic, geological, soils, land classification and engineering geological maps, reports and memoirs, mining data, aerial photographs and previous investigations are reviewed. Topographic maps are useful for indicating past land uses and all editions of these and geological maps should be examined. On occasions older geological interpretations based on features later obscured by urban development, have proved more reliable than recent editions.

In the UK British Coal and the Abandoned Mines Records Office are the primary sources of data regarding past mining activities. Similar organisations exist in other countries. Mining information also exists in private collections, public records offices and local museums. Unfortunately, in many cases, information relating to past mining activities is found to be incomplete, inaccurate, ambiguous and lacking in adequate points of reference.

All of the available data are considered in the light of the complete geological history of the area as ascertained from relevant geological memoirs and learned papers. It is normal to include a site reconnaissance within the remit of a desk study. The purpose of this is to examine any outcrops of soils or rocks at or near the site, record the drainage conditions and look for evidence of past land uses and land instability. Depending on the individual requirements engineering geological or geomorphological mapping geophysical surveys, trial pitting, borehole drilling or in situ testing may be undertaken to supplement or clarify details of the desk study.

1.2 Main Investigation Stage

The main part of the site investigation is designed to confirm or determine the basic geological interpretation and to provide a quantitative assessment of the ground conditions so that optimum methods for ground pre-treatment where required, the excavation operation, ground support and control of ground water can be chosen. The various rock and soil units present in the area are thus positioned with respect to the tunnel works. The depth of rockhead below any superficial cover and the ground water levels are also determined. It should be the intention of the investigation to derive a realistic geological interpretation consistent with the observations made in outcrops, trial pits and boreholes and the geological structure of the area and to present these data as a continuous ground profile for the works. Boreholes should be sited with this in mind, close to the works and also sufficiently deep to provide appropriate samples for examination and testing. In order to correlate strata between holes in steeply dipping beds, heavily faulted ground or strata lacking distinctive marker horizons, boreholes may need to be located at close centres and drilled to depths considerably greater than the two tunnel diameters below the invert level figure often set for the depth limit of investigations.

In order to derive a quantitative appraisal of the ground as a tunnelling medium it is necessary to determine various engineering properties of the material. Rocks and soils need to be fully described in terms of engineering geology (Anon (1970, 1977a, b). Particular attention is paid to weathering grade, strength, the character of discontinuities and ground water conditions in these descriptions. It is also usual for an assessment of core recovery and rockmass intactness to be made and for various structural data, such as the angle of dip to be recorded.

In addition to the visually determined strength values mentioned in engineering descriptions of soils and rock, various laboratory and in situ tests may also be carried out, depending on the lithologies present and the possible tunnelling method. Where tunnelling is to take place below the water table, estimates of the water ingress would normally be based on the results of in situ permeability tests and observations of the ground water levels in the area.

1.3 Investigations During Construction

Site investigation is carried out during construction to confirm that the conditions encountered accord with those adopted for the design. Departures from the expected tunnelling conditions may require modification to the design and method of construction being used. Investigation at this stage entails making frequent and detailed engineering geological descriptions of the ground present in excavations, noting water ingress and observing the performance of the material as a tunnelling medium.

In order to safeguard the excavations against the hazards associated with tunnelling, it is usual to drill probe holes into the face to provide a bulkhead of proved ground and to forewarn of any troublesome ground. Anon (1975) do not envisage that probing should curtail pre-construction site investigation since usually it would give insufficient warning of impending changes in the ground conditions. It should be regarded as a final safeguard and a means of pinpointing particular features. In areas of complex geology or where a lack of adequate surface access has precluded a full determination of the conditions it may be necessary to adopt a 'design and build' strategy. In this case the results of probing are very important. Under these circumstances a small diameter pilot tunnel may be driven, prior to enlargement to the required dimensions.

2.0 THE DON VALLEY INTERCEPTING SEWER SCHEME

As indicated in Fig. 1, phase 1 of the Don Valley Scheme comprises a large diameter, low level, intercepting sewer extending 5.7 km from shaft AO at the Blackburn Meadows Sewage Treatment Plant to shaft A14 near to Sheffield City Centre together with eight subsidiary branch sewers. The main tunnel is constructed in bolted segments, approximately 20 m below ground level. Between shafts AO and A6 the tunnel was driven by a Dosco SB 600 boom mounted roadheader tunnelling machine

mounted within a 6 m external diameter Lawrence shield. At shaft A6, gasketts were fitted to the segments, the tunnel diameter was reduced to 4.65 m and the tunnel was driven from within shields by a Dosco SB 400 to shaft A11 and thence by drill and blast techniques to shaft A14. The scheme also included 2.6 km of 1.37 to 2.44 m diameter branch tunnels twenty eight deep 3.00 to 10.00 m diameter shafts and twelve flow transfer chambers.

In order to guard against, and minimise the dangers from, inrushes of gases and water, and to indicate the presence of uncharted cavities, forward probing was carried out along all tunnel sections. For this 50 mm diameter holes were advanced by rotary drilling techniques for distances of between 50 and 80 m to provide a minimum bulkhead of two tunnel diameters.

2.1 The Geological Conditions

The strata of the Lower Don Valley belong to the Lower and Middle Coal Measures of Carboniferous Age (see Eadon et al. 1957). These are overlain by alluvium of Recent and Late Pleistocene age comprising deposits of clayey, silty or sandy loam in the upper part with gravelly material at the base. The sequence through which the main tunnel was driven extends from the Parkgate Rock at shaft AO upwards to above the Barnsley Coal at shaft A4, and then down through essentially the same sequence to the Thorncliffe Coal at shaft A14 (Fig. 1). These rocks comprised cyclothemic sequences of interbedded mudstone, siltstone, sandstone, and rocks of intermediate grain size, together with coal seams and seatearths. Individual cycles varied in thickness between 10 and 15 m although one cycle over 50 m thick was recorded. Commonly, individual members of cycles were absent and both repetition of some individual members and lateral variations in the thickness and composition of others were noted. Occasional transgressive beds of sandstone were encountered forming a 'wash out' or channel feature. The coal seams varied in thickness between a thin film and beds up to 2.5 m (see Table 1) with most falling within the range of 0.60 to 1.50 m. Most coal seams were bituminous in type with carbon contents of between 82 and 87 per cent, although occasional seams of cannel coal were encountered. Throughout much of the sequence iron carbonate occurred in the form of sideritic clay ironstone nodules, lenses and bands in mudstone or seatearth beds and occasionally in sandstone or siltstone units. The geologi-

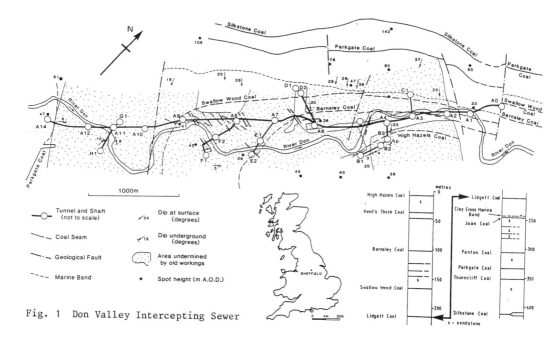

Fig. 1   Don Valley Intercepting Sewer

cal structure in the Lower Don Valley is dominated by the Don Monocline, a strongly asymmetric synclinal fold, which trends NE-SW (see Fig. 1). On the north west flank of this fold the angle of dip generally ranges between 20 and 35 degrees but to the south then it decreases to between 5 and 10 degrees while the strike swings to NW-SE.

A major fault trending NNW-SSE was encountered near shaft AO with six individual planes of movement in a zone of disturbed ground approximately 20 m wide (see Fig. 1). The individual faults were normal in type with hades varying between 5 and 15 degrees and throws varying between less than 1 m to over 6 m. Overall it is possible that the total displacement exceeded 90 m. Another major fault trending NNW-SSE was encountered near shaft A10 with one plane of movement, a throw of about 40 m and a hade of 10 degrees. Elsewhere a number of lesser generally normal faults with throws ranging from less than 0.5 m to over 5 m and hades of between 10 and 30 degrees were recorded. Occasional low angle dip-step faults were also encountered.

2.2   Past Mining Activities

The published geological information, mining abandonment plans and exploratory boreholes indicate the presence of extensive abandoned old mine workings at various levels throughout much of the Lower Don Valley. Coal is the principal material extracted although ironstone and fireclay workings are also known.

The main coal seams extracted are the Silkstone, Parkgate and Barnsley coals details of which are given in Table 1. At deep levels coal has been mined by pillar and stall and short wall mining techniques and workings including essential access and ventilation shafts, connecting roadways and cross measure drifts. Close to outcrop workings include largely uncharted bell pits, adits and localised pillar and stall developments. In general, these workings do not extend significantly below the water table.

TABLE 1   Principal coal seams within the Don Valley

Coal Seam	Seam Thick-ness	Approx. depth (m) below ground level at shaft location					
	(m)	A2	A4	A6	A9	A11	A14
Barnsley	2.00			59	30		
Parkgate	1.70-2.20	233	281	252	165	64	
Silkstone	1.80-2.50	310	369	350	264	160	90

2.3 The Hydrogeological Conditions

Observation of groundwater levels in the area indicates that two main water levels are present. The upper level corresponds with a water table in the alluvium while the lower level stands within the underlying bed rock formations, beneath the weathered, often clayey zone at rockhead. Locally the standing level in bedrock and drift formations was the same, indicating a probable connection. Occasional, depressed water levels standing well below the base of the drift deposits were observed indicating downward drainage, possibly through old abandoned mine drainage systems.

The granular made ground and unconsolidated alluvial open textured sand and gravel deposits of high porosity and permeability acted as aquifers with a high ground water storage capacity. Recharge of these beds occurs generally by infiltration from the surface water courses. The clayey and silty horizons within the drift deposits displayed relatively low permeability and acted as aquicludes and aquitards, occasionally producing confined or semi-confined conditions within the drift formations.

In the rockmass intergranular permeability was only observed in the coarser grained rock types and was insignificant compared with the fissure flow along joint and bedding plane discontinuities. The mass permeability and porosity were dependent upon the character of discontinuities and these were closely related to the lithology, bedding characteristics, weathering/alteration state and structural setting. Within the sequence of rocks present sandstone, siltstone, coal and fractured ground, which displayed high mass permeability and water storage capacity, acted as aquifers. The mudrocks, in which open discontinuities were usually blocked by degradation products, and zones of weathered/altered ground formed aquicludes or aquitards. The recharge of aquifers occurred directly at outcrop, on the high ground bounding the valley, by infiltration from the drift or by leakage through aquicludes and aquitards.

The overall geological structure has a significant bearing on the hydrology of the area. As already explained the rock sequence is folded into a synclinal structure and is traversed by numerous faults. Some confined aquifers outcrop on the relatively high lying areas to the north west and south east of the valley. Direct recharge at this level has given rise to sub-artesian peizometric water

levels in some formations and has contributed to the variable pressure head present throughout much of the sequence. Depending whether zones or fault breccia or fault gouge had been produced faults acted both as conduits for, and barriers against, the flow of groundwater in the rockmass (see Cripps et al. 1988).

The hydrology is further complicated by the presence of extensive mineral workings and mines drainage. Most of abandoned mine workings are in a waterlogged condition and the uncollapsed cavities and associated fractured ground constitute a large water storage capacity. Pumping operations, which form part of a mines drainage system operating in the area abstract groundwater from the level of the Parkgate and Silkstone coals. This abstraction of groundwater has produced localised unsaturated conditions within the general sequence and given rise to overlying perched water tables.

3.0 GEOLOGICAL INTERPRETATION

The predicted geological conditions were presented on a series of vertical long sections drawn for the tunnel centre line and shaft axes. These showed the geological conditions anticipated to exist between the ground surface and at least two tunnel diameters below the proposed excavations. In order to derive these sections it was necessary to project information from off-line holes on to the section. Since the drift deposits were essentially flat lying, these measures were projected horizontally from boreholes but for the solid geological formations, it was necessary to adjust bed levels in accordance with the geological structure. Fig. 2 illustrates the manner in which the structural interpretation was made, and the structural geological information presented, for part of the drive between shafts A11 and A12.

The correlation of the strata was carried out by carefully examining the records for each borehole to determine whether recognisable features of the strata were found in nearby holes. It was found useful to allocate identification letters to particular marker horizons (in this case P, Q and R) used for correlation. The interpreted geological structure was presented on a structural plan (Fig. 2) which displays these data in terms of strike lines giving the height above Ordnance Datum (sea level) of the bases of marker horizons and the implied dip of the beds. Primary data presented included the heights above datum of

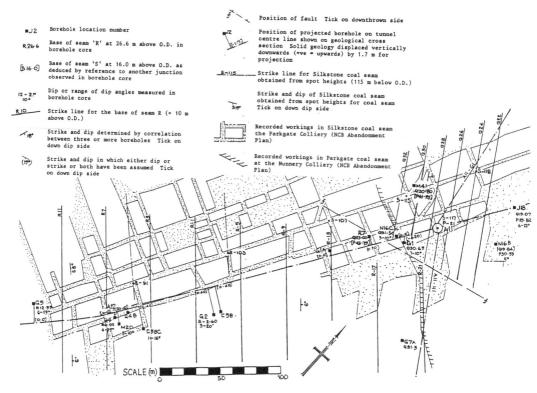

Fig. 2 Geological structure plan for drive between shafts A11 and A12

marker horizons, the angle of dip of the beds and any relevant mining information. Some of the heights of the marker beds were deduced by reference to other junctions and known bed separations observed in adjacent holes. The plan also shows the projected position of boreholes on the vertical geological section together with the required vertical displacement of the solid geology.

Where it was possible to correlate the beds in three or more holes, the strike direction and angle of dip were as derived from that correlation. However, should a particular interpretation not conform reasonably well with data from geological maps, mining records and measured dips, it was not accepted. In this case a strike direction and/or dip angle which corresponded reasonably well with other evidence was adopted.

Faults were postulated only if positively identified in rock cores or where demanded by a lack of correlation between adjacent boreholes. Owing to imprecision in correlation, faults with throws of less than 2 m could not be predicted. Although some apparent faults may be due to bed thickness variation or minor folding of

the strata, this method of prediction minimizes the number of faults to give a reasonable explanation of the evidence. Some faults are thought to intersect boreholes. However, although non-intact zones, losses of core, surface polishing, slickensiding and mineralisation are all features seen in fault affected cores, such features were not, in themselves, taken as being indicative of faulting.

Fig. 3 shows the geological data presented for shaft A12. This diagram also shows the manner in which the geological section observed in the nearby borehole has been projected onto the shaft axis. As in the case of the boreholes given on the long sections for the tunnel drives, the pictorial lithological log was accompanied by mechanical indices logs and strength values were also presented.

4.0 CONCLUSIONS

As with all civil engineering operations, tunnel construction should be achieved in a controlled and economic fashion such that the structure performs its intended function over its design life. The successful construction of tunnels is dependent on a detailed knowledge and understanding of the ground conditions during the planning, design and construction phases. Unexpected events, such as collapse or flooding of underground excavations constitutes a considerable hazard which may result in delay and extra cost. Similarly, delays to construction may also result from ground conditions varying from those assumed during the design of the works, particularly where an inappropriate method of construction has been adopted. Minor deviations from the expected conditions may cause delay, especially if an onerous feature runs with the tunnel.

Of course constructional difficulties may arise even if tunnelling medium is fully and accurately characterised should the ground behave in an unexpected manner or the conditions not be fully taken into account in the design. However a thorough site investigation should reduce the

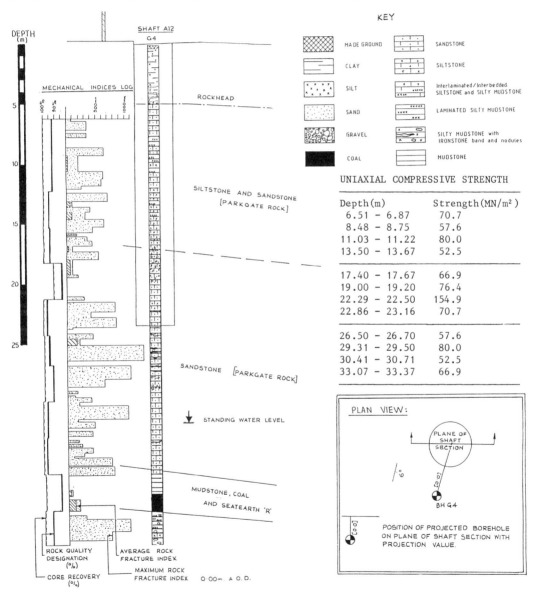

Fig. 3   Geological section for shaft A12

2643

possibility of such problems. Generally speaking, the most efficient means of assembling the required ground characterisation data is to carry out a phased programme of ground exploration.

The case study considered underlines the importance of the desk study in site investigation work. Although little geotechnical data were available at this stage, published geological maps and memoirs indicated the basic structural framework and lithologies present in the area. Where available, abandoned mines records provided a further, more detailed, appreciation of these features. Data from mineral boreholes and shaft sinkings served as valuable aids to establishing the basic stratigraphy. Information collated during these tunnelling works indicate geological interpretations that are at variance with published geological maps.

During the main investigation, boreholes were sited to provide the data needed to derive a continuous ground profile for the works. Attention was also paid to the geotechnical properties and likely engineering behaviour of the ground. Accordingly, detailed engineering geological and geomechanical logging of borehole cores was carried out. Particular attention was paid to discontinuities each of which was described individually in terms of its type, orientation, curvature, persistence, surface roughness, openness and nature of infill. Standpipe piezometers installed in both drift and bedrock formations provided data about ground water levels and in situ permeabilities were also determined. Laboratory tests on samples included uniaxial compressive and tensile strength, abrasivity and specific energy and Duncan free swelling determinations and mineralogical studies of rocks, the grading, consistency limits and natural water content of soils and the chemical aggressiveness of ground waters.

The data derived during the desk study and main investigation were used to provide a statement of the performance of the ground as a tunnelling and shaft sinking medium for all the excavations. The possible incidence of ground instability, breakback, difficulty with excavation and trimming, sliding, spoil and water handling, product degradation water ingress and the presence of gas were all assessed. It is believed that such an approach provides clear guidance as to the ground conditions without favouring any particular construction method, and limits delays and design variations. Also it provides for controlled and economic tunnelling within a highly variable and in part hazardous tunnelling environment.

ACKNOWLEDGEMENT: The authors are grateful to Mr. A.J. Wood, Director and Mr. R. Jolly, Chief Engineer, Department of Design and Building Services, Sheffield City Council for permission to publish this paper.

REFERENCES

Anon, 1970. The logging of rock cores for engineering purposes. Geological Society Engineering Group Working Party Report. Quart. J. Eng. Geol., 3, 1-24.

Anon, 1975. Probing ahead for tunnels : a review of present methods and recommendations for research. The Building Research Establishment/Transport and Road Research Laboratory Working Party. Sup. Report 171 UC, Trans. and Road Res. Lab., Crowthorne, Berks.

Anon, 1977a. The logging of rock cores for engineering purposes. Geological Society Engineering Group Working Party Report. Quart. J. Eng. Geol., 10, 45-52.

Anon, 1977b. The description of rock masses for engineering purposes. Engineering Group Working Party Report. Quart. J. Eng. Geol., 10, 355-388.

Anon, 1981. Code of Practice for Site Investigations. BS5930:1981, British Standards Institution, London.

Clayton, C.R.I., Simons, N.E. and Matthews, M.C., 1982. Site Investigations - A handbook for engineers. Granada Publishing, London.

Cripps, J.C., Deaves, A.P., Bell, F.G. and Culshaw, M.G., 1988. Geological controls on the flow of groundwater into underground excavations. 3rd Int. Mine Water Cong., Melbourne, 77-86.

Dumbleton, M.J. and West, G., 1976. A guide to site investigation procedures for tunnels. Lab. Report 740. Trans. and Road Res. Lab., Crowthorne, Berks.

Eadon,R.A.,Stevenson, I.P. and Edwards, W., 1957. Geology of the Country around Sheffield. Memoirs of the Geological Survey, H.M.S.O., London.

Influence of surface load on the equation of mine subsidence
Direct and inverse problems

Influence de la charge du terrain sur l'équation des déplacements de la cuvette d'affaissement: Problèmes directs et inversés

V.I. Dimova
Higher Institute of Mining and Geology, Sofia, Bulgaria

ABSTRACT: The problems for determination of the surface load (buildings, equipment, etc.) on the equation of the subsidence trough displacements (direct problem) and these for protection of the surface items (buildings, equipment, etc.) by proper performing of the underground activities (inverse problem) are reduced to correct and incorrect (after Hadamar) problems for Laplace's equation respectively. The solutions are obtained analytically and numerically.

RESUME: Les problémes de détermination de l'influence de la charge du terrain (batiments, constructions, etc.) sur l'équation des déplacements de la cuvette d'affaissement (probléme direct) et ceux de protection des batiments, constructions, etc. par un tracage commode des travaux miniers (probleme inverse), sont réduits vers des problémes directs ou inverses (selon Hadamar) pour l'équation de Laplace. Les solutions sont obtenies par une maniére analytique et numerique.

The underground geotechnologic or construction activities are the reason for the disturbances of the initial stressed and deformation state of the rock masses and the ensuing redistribution in the fields of stresses and displacements. As one of the directly observed effects of this phenomenon is the subsidence of the earth's surface (mine subsidence, subsidence trough, depression), which creates unfavourable conditions for the functioning of the surface buildings and equipment.

As a base in the description of the above mentioned phenomenon we will make use of a continious medium, the particles of which have one degree of freedom (the only displacements possible are these perpendicular to the earth's surface) and the tensor of stresses is asymmetric, as seen, from the mechanic model and the infinitesimal volume presented in Fig. 1 and Fig. 2 (Switka 1968), (Muravskii 1977), ("approximated theory of elasticity", elastostatics of H.A. Rakchmatulin).

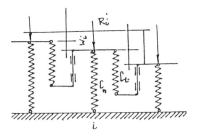

Fig. 1

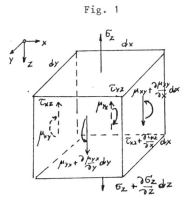

Fig. 2

$$\sigma_x = \sigma_y = \tau_{zx} = \tau_{zy} = \tau_{yz} = \tau_{xy} = 0$$

$$\sigma_z = E_0 \varepsilon_z, \ \tau_{xz} = \mu_0 \gamma_{xz}, \ \tau_{yz} = \mu_0 \gamma_{yz}$$

$$\gamma_{yz} = \frac{\partial W}{\partial y}, \ \gamma_{xz} = \frac{\partial W}{\partial x}, \ \varepsilon_z = \frac{\partial W}{\partial x} \qquad (1)$$

$$\varepsilon_x = \varepsilon_y = \gamma_{xy} = \gamma_{zx} = \gamma_{zy} = \gamma_{yx} = 0$$

$$\frac{\partial \tilde{\tau}_{xz}}{\partial x} + \frac{\partial \tilde{\tau}_{yz}}{\partial y} + \frac{\partial \sigma_z}{\partial z} = 0$$

$$\tau_{xz} - \frac{\partial \mu_{xy}}{\partial x} = 0 \qquad (2)$$

$$\tau_{yz} = \frac{\partial \mu_{xy}}{\partial y}$$

where E_0 and μ_0 are the characteristics of the medium, μ_{xy} - microtorques (stress dipoles) and the last two equations are identically satisfied due to the presence of guides in the mechanic model (Fig. 1)

On the basis of (1) and (2) the basic equation for the unique displacements is obtained

$$\frac{\partial^2 W}{\partial x^2} + \frac{\partial^2 W}{\partial y^2} + k^2 \frac{\partial^2 W}{\partial z^2} = 0, \ k^2 = \frac{E_0}{\mu_0} \quad (3)$$

to which the boundary conditions should be added, corresponding to the specific problem (the makes sense only when vertical load is applied).

The relationship between the characteristics of the medium E_0 and μ_0 and the classical elastic constants E and ν is obtained by the comparison of the results of Boussinesq's problem, derived by the methods of linear theory of elasticity and on the basis of the model herein being considered.

Thus, for $G = G_0$ it is obtained (Muravskii 1977)

$$E_0 = E/2(1+\nu)(1-\nu^2) = G/(1-\nu^2), \ k = 1/(1-\nu)$$

Now in the light of the presented theory, the solutions of some definite problems will be described.

Problem 1. Consider a plane problem for determination of the equation of mine subsidence, caused by mining out of horizontal layer of mineral resources in the boundaries $x \in [-a, a]$ when the earth's surface is uniformly loaded ($\sigma_0 = $ const) in the boundaries $x \in [b, c]$.(Fig. 3)

The sought mine subsidence equation is obtained as a result of the solution of the following boundary problem (Fig. 3)

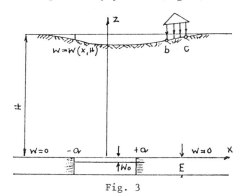

Fig. 3

$$\frac{\partial^2 W}{\partial x^2} + k^2 \frac{\partial^2 W}{\partial z^2} = 0 \qquad (4)$$

$$W(x,0) = \begin{cases} -W_0 = m\eta, & x \in [-a, a] \\ 0, & |x| \in (a, \infty) \end{cases} \quad (5)$$

$$\sigma_z = E_0 \frac{\partial W}{\partial z} \bigg|_{z=H} = \begin{cases} -\sigma_0, & b \leq x \leq c \\ 0, & x < b, x > c \end{cases} \quad (6)$$

where η = 0.2-1.6 is the fill factor and we will determine the horizontal displacement by S.G. Aversin's empirical relationship

$$u = -B(z)\frac{\partial W}{\partial x}, \ v = -B(z)\frac{\partial W}{\partial y} \quad (7)$$

where B = B(z) is the characteristic of the rock masses determined from the geodesic surveys (Salustowicz 1955).

In the solution of the problem (4) – (6) we apply Fourier's integral transformation

$$\widetilde{W}(\alpha, Z) := \frac{1}{\sqrt{2\pi}} \int_{-\infty}^{+\infty} W(x,z) e^{i\alpha x} dx \quad (8)$$

$$W(x,Z) := \frac{1}{\sqrt{2\pi}} \int_{-\infty}^{+\infty} \widetilde{W}(\alpha, z) e^{-i\alpha x} d\alpha \quad (9)$$

that leads to the desired aim (Dimova, Dimov 1989),(Dimova 1990)

$$W(X,H) = -\frac{W_0}{\pi} \left\{ \operatorname{arctg}\left[\operatorname{sh}\pi k\left(\frac{a+x}{2H}\right) + \operatorname{arctg}\left[\operatorname{sh}\pi k\left(\frac{a-x}{2H}\right)\right] \right] - \frac{\sigma_0}{E_0}\left\{ \frac{4H}{\pi^2}\sum_{l=0}^{l=\infty} \frac{(-1)^l}{(2l+1)} \sin\overline{\Pi}\left(\frac{2l+1}{2}\right) \cdot \exp\left[-\frac{(2l+1)(b-x)}{2H}\cdot\Pi\right] - \frac{H}{4} \right\} - \quad (10)$$
$$\frac{\sigma_0}{E_0}\left\{ \frac{4H}{\pi^2}\sum_{l=0}^{\infty} \frac{(-1)^l}{(2l+1)} \sin\overline{\Pi}\left(\frac{2l+1}{2}\right) \cdot \exp\left[-\frac{(2l+1)(C-x)\pi}{2H}\right] - \frac{H}{4}\right\}$$

The verification of the formula (10) (for $\sigma_0 = 0$) shows that (Jedrzejczyk 1974) k = 1.8, i.e. ν = 0.4.
The example results are shown in Fig. 4

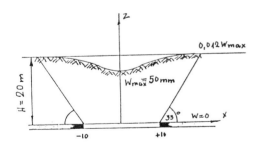

Fig. 4

In the same way the basic three dimensional problems which occur in the mining and construction practice were solved (Dimova 1990).

Problem 2. Let us solve Problem 1. in case of mining out a inclined layer (Fig. 5)

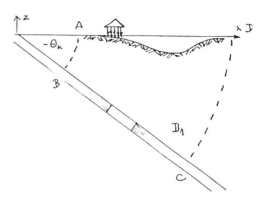

Fig. 5

In this case due to the limited influence of the heading (tunnel) as well as of the surface load the problem is reduced to the following boundary problem for Laplace's equation in a region with a form of a part of circular ring (for properly defined r_1 and r_2)

$$r_1 \leqslant r \leqslant r_2 \ , \ r = \sqrt{x^2 + \overline{z}^2}$$

$$-\Theta_k \leqslant \Theta \leqslant 0 \ , \ \Theta = \operatorname{ctg}\frac{\overline{z}}{x}$$

$$\frac{\partial^2 W}{\partial x^2} + \frac{\partial^2 W}{\partial \overline{z}^2} = 0 \ , \ \overline{z} = \frac{z}{k} \ , \ r_1^2 \leqslant x^2 + \overline{z}^2 \leqslant r_2^2 \quad (11)$$

$$W(x,\overline{z})\Big|_{\Theta = -\Theta_k} = \varphi(r) \quad (12)$$

$$W(x,\overline{z})\Big|_{r=r_1} = W(x,\overline{z})\Big|_{r=r_2} = 0 \quad (13)$$

$$\frac{\partial W}{\partial \overline{z}}\Big|_{\Theta = 0} = \psi(r) \quad (14)$$

By means of the transformation formulas

$$T: \begin{vmatrix} x = e^{\rho}\cos\Theta \ , \ \rho = \ln r \\ \overline{z} = e^{\rho}\sin\Theta \end{vmatrix} \quad (15)$$

the region D_1 is turned into D_1^* (Fig. 6) determined by the inequalities

$$\mathbb{D}_1^*: \begin{vmatrix} \ln r_1 \le \varphi \le \ln r_2 \\ \\ -\theta_k \le \theta \le 0 \end{vmatrix} \qquad (16)$$

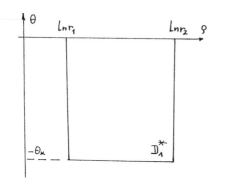

Fig. 6

and the problem (11) – (14) becomes

$$\frac{\partial^2 w}{\partial \varphi^2} + \frac{\partial^2 w}{\partial \theta^2} = 0 \qquad (17)$$

$$w(\varphi, -\theta_k) = \psi(\varphi), \ln r_1 \le \varphi \le \ln r_2 \qquad (18)$$

$$w(\ln r_1, \theta) = w(\ln r_2, \theta), -\theta_k \le \theta \le 0 \qquad (19)$$

$$\frac{\partial w}{\partial \theta} = e^{\varphi} \psi'(\varphi) \qquad (20)$$

which was solved analytically and numerically (boundary differences)(Dimova, Dimov 1989).

The example results are shown in Fig. 7

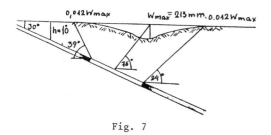

Fig. 7

Problem 3. (Inverse promlem). For a given mine subsidence and surface load equation to be determined the way of mining out a layer (Fig. 8),i.e. how to carry out the underground activities so as to protect the surface items (buildings, equipment, reservoirs, etc.).

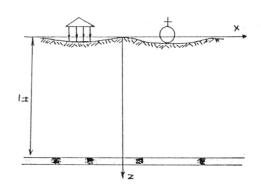

Fig. 8

This problem can be formulated as a incorrect (after Hadamar) problem for Laplace's equation (Cauchy's problem)

$$\frac{\partial^2 w}{\partial x^2} + \frac{\partial^2 w}{\partial \bar{z}^2} = 0, -\infty < x < +\infty, 0 \le \bar{z} \le \bar{H} \qquad (21)$$

$$w(x,0) = f(x), \frac{\partial w(x,0)}{\partial \bar{z}} \frac{\bar{g}(x)}{E_0} = g(x) \qquad (22)$$

which solved by B. Ivanov's method (Ivanov 1965) leads to following approximate solution

$$W_\alpha(x\bar{z}) = \frac{1}{2\pi} \int_{-\infty}^{+\infty} R_\alpha(x-t,\bar{z}) f(t) dt - \frac{1}{2\pi} \int_{-\infty}^{+\infty} T_\alpha(x-t,\bar{z}) g(t) \frac{dt}{t} \qquad (23)$$

where α is Tichonov's regularization parameter

$$R_\alpha = \frac{1}{2\alpha\sqrt{\pi}} \exp\left[\frac{\bar{z}^2 - x^2}{4\alpha^2}\right] \cos \frac{x\bar{z}}{2\alpha^2} \qquad (24)$$

$$T_\alpha = \frac{1}{2\alpha\sqrt{\pi}} \cdot \exp\left[-\frac{x^2}{4\alpha}\right] \int_{0}^{\bar{z}/2\alpha} \exp(t^2) \cos \frac{tx}{s} dt \qquad (25)$$

2648

In the numerical solution of the problem (21) – (22) we apply the method of L.S. Frank (Frank 1968) who constructed the tri-layered difference scheme with a high degree of accuracy for Laplace's equation (21) and proved its stability. The function for the first two layers should be known as required by the scheme. The zero layer is defined by the first of the conditions (22) and for the first layer calculation a scheme is constructed, which at fixed ν (see below) has the same error of approximation as the scheme of equation (21).

The obtained results by L.S. Frank are the following
- difference scheme for equation (21)

$$\frac{W_h(\bar{z}+\tau,x)-W_h(\bar{z}-\tau,x)}{2}=\sum_{|k|\leq\nu}\alpha^k W_h(\bar{z},x+kh) \tag{26}$$

$$\alpha_k=\frac{(-1)^k}{(\nu-k)!\,(\nu+k)!}\cdot\frac{1}{r^2+k^2}\left.\prod_{p=0}^{\nu}(r^2+p^2)\right. \tag{27}$$

- difference scheme for first layer calculation

$$W_h(\tau,x)=\sum_{|k|\leq\nu}\alpha_k f(x+kh)+h\sum_{|k|\leq\nu}\beta_k g(x+kh) \tag{28}$$

$$f(x)=W(0,x)\ ,\ g(x)=\frac{\partial W(0,x)}{\partial \bar{z}} \tag{29}$$

$$\alpha_k=\frac{(-1)^k}{(\nu-k)!\,(\nu+k)!}\cdot\frac{1}{r^2+k^2}\left.\prod_{p=0}^{\nu}(r^2+p^2)\right. \tag{30}$$

$$\beta_k(r)=\int_0^r \alpha_k(\xi)d\xi=\frac{(-1)^k}{(\nu-k)!(\nu+k)!}\int_0^r\left(\prod_{0\leq p\leq\nu,\,p\neq k}(\xi^2+p^2)\right)d\xi \tag{31}$$

where h and $\quad$ = r.h (r = const) are the steps along x and z axes, respectively at fixed $\quad$.

Let us note that the problem for protection of surface items in another situation was reduced to other incorrect problems of mathematical physics (problem for heatconductivity equation in "back time" and Cauchy's problem concerning Laplace's equation)(Dimova 1990).

REFERENCES

Dimova, V. & I.Dimov (1989). Involving Rakchmatulin's elastic base model in geomechanics. Annual of the Higher Istitute of Mining and Geology, v.XXXIV, 347-350, Sofia.(in Bulgarian)

Dimov, I. & V.Dimova (1989). About the application of coordinate transformation when solving problems of potential theory, arising in geomechanics. Annual of the Higher Institute of Mining and Geology, v.XXXIV, 339-345, Sofia.(in Bulgarian)

Dimova, V. (1990). Some direct and inverse problems in Rakchmatulin's elastostatics and their applications in geomechanics. Ph.D. Dissertation, Sofia.(in Bulgarian)

Frank, L.S. (1968). O scheme vajsokovo porajdka totchnosti pri tchislenom reschenie nekorektnoj zadatchi Cauchy. Trudaj VTz, University of Moskow, 181-196.(in Russian)

Ivanov, V.B. (1965). Cauchy's problem for Laplace's equation. Differetial equations 1, 87-96.(in Russian)

Jedrzejczyk, A. (1974). Warunki stosowania rozwiazan aproksymovanej przestrzeni sprezystej. Arch.Gorn. t.XIX, z.1, 67-79. (in Polish)

Muravskii, B.A. (1977). O modeli uprugovo osnovania,Stroitelnaja mechanica i rastchot sourugenii 6, 14-17.(in Russian)

Salustowicz, A. (1955). Mechanica gorotworn. 458pp. WG-H, Stalinogrod.(in Polish)

Switka, R. (1968). Aproksymovana polprzestrzen sprezysta. Politechika Poznanska, Rozprawy.31.(in Polish)

Engineering geological mapping of sedimentary rocks in Hungarian settlements having cellars

Elaboration des cartes géotechniques des roches sédimenteuses dans les habitats hongrois avec celliers

D. Cyóry, E. Kóvári-Gulyás, R. Petz, Gy. Scheuer & L. Szentirmai
FTV Consulting Engineering, Budapest, Hungary

ABSTRACT: In Hungary in earlier times a lot of wine-cellars were made on the edges of hilly settlements. Nowadays they are found everywhere under the inhabited areas because of the advanced urbanization. The known or unknown cellar- and cave-systems mean a real danger in the everyday life of towns. The solution of problems needs a good preparation in form of suitable engineering geological mapping. We show our experiences about it in some settlements.

RESUME: Il y avait beaucoup de celliers en bord des habitats monteux en Hongrie anciennement. De nos jours on peut les trouve partout sous les régions habitées en raison de l' urbanisation avancé. Les cellier- et cave-systèms connus et inconnus signifient un réel danger dans la vie quotidienne des villes. A la solution des problèmes nous trouvons nécessaire la bonne préparation en forme d'élaboration des ingénieur-géologiques carte-séries convenables. Nous présentons nos expériences de ce thème dans quelques habitats.

INTRODUCTION

The engineering geological mapping has started in the framework of the cellar-investigations at the most problematical areas of Hungary, like Eger, Pécs, later on Szentendre, Szekszárd and in small villages Nagymaros, Noszvaj, Novaj Ostoros /Fig.1./. The map-series are made on sheets covering the administrative territory of the settlements, in the scale 1:4 000 and summerizing 1:10 000.

According to the hilly landscape geological and hydrogeological conditions can be complicated and require a careful, detailed exploration. Generally near the surface performed by tectonic movements, there are Neogene or younger sediments. To this natural situation adds the anthropogenic effects positive or negative way in aspect of cellar stability. Caves are classified from that point of view by means of new borings.

Fig. 1. Sketch of Hungary with settlements endangered by cellars.

OBSERVATION OF CELLARS IN SOME SETTLEMENTS

Cellar damage in the town Szekszárd has only affected that ones situated on dissected slopes.

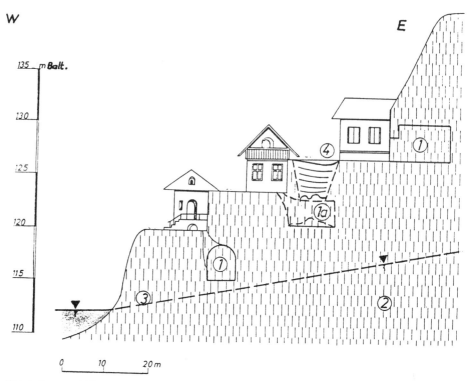

Fig. 2. **Sketch** profile at Szekszárd
1: cellar, 1a: collapsed cellar; 2: loess sequence; 3: groundwater;
4: road subsidence

The excavation of cellars proved to be more favouruable in the hill-sides because they are composed of considerable thick loess and the groundwater level is at greater depths /Fig.2./. The flat areas covered by alluvial sediments,were unfavourable for cellar construction either because of high groundwater level or because of the coexistence of loose sands and gravels.

Under the Pleistocene loess Upper pannon aleurit was deposited with sandy strata. These two formation determines the hydrogeological conditions which have important role in the slope stability. During any kind of construction work the cutting needs attention to avoid landslides /Petz 1988./ On the other hand due to water seepage from public utilities and poor drainage the loess locally could be moistened, and it can loose it's stability endangering establishment overlied.

The geomorfologically charecteristic loess deep-ways due to the effect of water rush down, cause block-collapses of the cellar entrances. Similar problems have arosen at Nagymaros where cellars can be found also in the loess. In that village the great highway traffic and its dynamical effect because of starting the construction of the Danube barrage had a harmful effect on the consistence of cellars.

The morphological, geological and hydrogeological feature of Szentendre was extramely good for creating cellars. As far as it is known a veritable labirynth of caves is hidden under the city. They were in use for wine storage but nowadays they are mostly abandoned and their situations are often forgotten. In that way caves in good or bad conditions exist above each other at several kilometers length /Fig.3./.

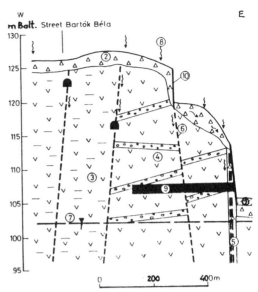

Fig. 4. Characteristic engineering geological profile
1: upfill; 2: slope detritus; 3:

Miocene clayey tuffite; 4: Miocene tuffaceous sand, sandstone; 5: fault, broken, zone; 6: fractures; 7: average water level; 8: water-infiltration; 9: cellar, 10: supporting wall

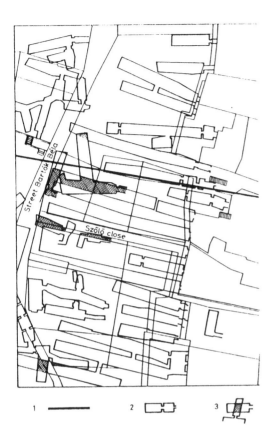

Fig.3. Cellar-system on Szamár-hill of Szentendre
1: direction of engineering geological profile; 2: cellar; 3: cellars above each other

The considerable part of stability problems is caused by the deterioration of the rock material. In Szentendre that is Miocene sediment: clayey tuffite, tuffaceous sand, sandstone. Usually these are covered with detritus on slopes, which are dissected by supporting walls. The water-infiltration from the surface has influence on the intensity of weathering. That process can lead to decrease the rock solidity to 25-30 % of original. The collapse of cellar-ceilings can reach the above-lied caves or foundations of buildings, sometimes the surface itself. The more or less stratified Miocene layers have been separated blocks by fractures and faults, with broken zones along them. / Fig.4./.

In an old town like Szentendre primary attention was given to save historical houses and churches, altogether to save certain cellars utilizing their advantages from the tourtistical aspects. In such cases fortifying them can be more important than their elimination by filling-in. The task of engineering geological mapping is to make easier the choice of suitable geotecnical methods mentioned above.

Special situation is characterizing one part of the caves in the villages around Eger. At Ostoros and mainly at Noszvaj there are cellars used in the past or even recently as a flat. The rock of cellars of Ostoros is the quite fractured Miocene tuff which is therefore sensitive to the rain or other surface-water infiltration. The case is similar at Noszvaj with cellar-flats still in use on the SE-side of the village. The geological sketch of it is presented on the Fig.5.

2653

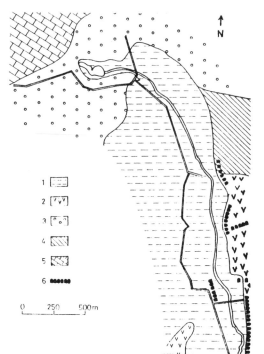

Fig.5. Geological sketch of Noszvaj
1: Holocene-Pleistocene silty clay;
2: Miocene tuff; 3: Miocene gravel,
conglomerate; 4: Oligocene clay;
5: Eocene limestone; 6: cellars

As this figure shows one part of
the cellars is excavated in Oligocene clay. In such case only the
inappropriate human intervention
can cause problems where the cellar
mesures exceed the load-bearing
capacity of the rock. It results
the partial collapse of ceiling.
The most of cellars can be found
in the Miocene tuff, is in good
condition.

The North part of Novaj the Miocene
layers are tuffaceous sand and clay
with several artifical caves in
it. Ostoros village is characterized by cellars created in fractured tuff from the same age,
where the lack of drainage is the
main endangering factor.

METHODS OF ENGINEERING GEOLOGICAL
MAPPING CONCERNING CELLARS

As it is presented above, the engineering geological mapping of even
one settlement means a very complex
task,which exclaimes the collaboration of all geosciences. The map-
series con ist of geomorphological,
covered and uncovered geological,
hydrogeological, different engineering geological variants based on
borehole location, hydrogeological
observation map and a map of state
of engineering development. This
last variant presents the built-up
areas with row or detached houses,
green areas, monuments, buildings
with historical or townscape value,
areas provided with utility services altogether with location of
cellars, building-damages, and water inrushes /Petz 1989./

The detailed engineering geological description of cellar condition
summerizes the most characteristic
features according to shape, form,
dimension, situation, and the following geological and sedimentological aspects: age and formation of
rock, stratification type, solidity, degree of weathering, number
and orientation of fractures, location of water infiltrations etc.

All of these in situ observations
supplying with laboratorical analysis make possible to decide the
types of technical interventions.

CONCLUSIONS

In the course of engineering geological mapping different map-series
are prepared to render an assistance for projecting of settlement
development, reconstruction,elimi-
nation of cellar problems. Methods
of such type mapping of sedimentary
rocks are presented above with some
examples of Hungarian towns and
villages.

REFERENCES

Petz, R. - Scheuer, Gy. /1988./
    Slope stablility problem at Szek-
    szárd, in the Barátság street.
    Engineering geological review 37,
    193-205.
Petz, R. - Scheuer, Gy. - Mrs.
    Szentirmai, L. - Győry, D. - Kő-
    vári-Gulyás, E. /1989./ Tasks and
    problems of the engineering geo-
    logical mapping of settlements
    with cellars. Acta Geologica Hun-
    garica 32/1-2, 257-268.

Engineering geological mapping of settlements underbolstered with cellars cut into rocks

Cartographie géotechnique des sites avec des caves creusées dans la roche

B. Kleb
Technical University, Budapest, Hungary

ABSTRACT: From the middle of the sixties a dangerous situation originating from damaging and rupturing of cellars cut into rocks has arisen on more and more settlements of Hungary as a very specific problem. This study demonstrates the engineering geological investigation of caves and their surroundings formed in vulcanic tuffs, loess and rough lime environment.

RESUME: C'est le danger de l'affaisement et de l'écroulement des caves qui se manifest depuis des années mi-soixante aux plusieurs sites de l'Hongrie. Cette étude s'occupe de l'analyse ingénieur-géologique des caves exploitées dans le tuf volcanique dans le loess et dans le calcaire grossier.

The rapid development of urbanization commenced in Hungary in the sixties. The law on construction control was issued in 1964 prescribing the completion of a general arrangement plan for the future area utilization of settlements.

The engineering geological mapping in 1966 commenced with surveying areas having priority in the development plan in the region of Balaton and Budapest. From 1968 on, however, the examinations were focussed on towns endangered by cellar collapse - Eger, Pécs, Szekszárd, Szentendre, Paks - and on smaller settlements as Nagymaros, Ostoros, Noszvaj, Novaj (Fig. 1).

The detailed engineering geological mapping of the settlements in large towns is under process by the professional supervision and financial support of the Central Geological Authority, with uniform principles on a scale of 1:10000 in downtowns and smaller settlements on scale 1:5000 and recently on 1:4000.

1 EXPLORATION OF CELLAR NETWORK

The exploration and examination of cellar networks is a complex problem partly with engineering geological tasks. This is partly due to the great number of the hollows, partly to their establishing time, condition, proprietary rights and leasing forms.

1.1 Instrumental detection of unexplored caves

At the preliminary surveying it became obvious that under the cities, there are numerous cellars or caves that are out of use, water loaded or closed with rockburst and their opening and spatial is unknown.

According to international experience, there is a real chance of detecting a cave by geophysical methods, only, when the depth and cross-sectional dimension of the cave are of the same order of magnitude. Since this applies to most of the cases, the idea of making geophysical cave research has arosen. The measurement was carried out by the Eötvös Loránd Geophysical Institute.

Surface measurements:
The geophysical measurements (seismic-small-refractional and geoelectric method) on the surface were encumbered by the disturbing effect of traffic, the public works and that the area was built up.

"Transillumination" between the bores. Since the surface measurements carried out in different place with different methods were not successful, we tried instrumental measurements in bores drilled nearby known cellars (seismic, geoelectric test).

According to the above mentioned there is a possibility to detect the cave. For this method a 6 x 6 meters drilling network would be needed knowing the section of the

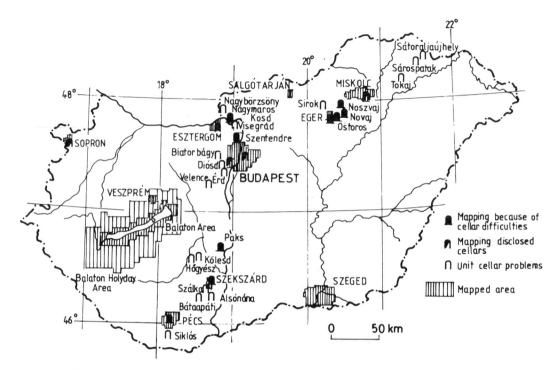

Fig 1 The engineering geological mapping and cellar difficulties in Hungary

cave (average 2,5 x 2,5 m) and the measurable anomaly. Since its costs are so high and cannot provided in a built-up area, the application of this method was not adopted.

1.2 Detection of caves with "destructive" method

The insufficient result of instrumental test made necessary the application of other expensive methods.

Cave exploration with drilling
With relatively small expenses, we achieved quick and reliable result with horizontal bores started from known cellar caves (50--100 m).

Cave exploration with mining method
In cities struggling with cellar problems, there are many collapsed perilous caves, that should be accessed or explored with precautious measures.

Next to the strongly ruptured sections there are flooded cellars in great extent. That is why we consider dewatering as one of the most important method carried.

1.3 Surveying of cellar networks

The necessity of the exact surveying of the cellar network nowadays is inevitable, and it has got multiple purpose:
- by the longitudinal-and cross-section it determines the position of the under-ground caves to the related surface constructions,
- provides bases for planning: at stemming and supporting for the volumetric calculations, and also for the exploration of further unknown caves.
- provides information for the future town--planning program, and the actual situation is recorded.

The mapping of Eger that is mostly under cut by cellar network was made on the scale of 1:200. Later, on other settlements scales of 1:250 and 1:100. The exploration work done in the framework of engineering geological mapping is remarkable, even in international level (Table I).

Table I. Information about the extent of cellar networks

Settlement	Length of cellar network km
Eger	135
Pécs	55
Szekszárd	15
Szentendre	6
Budapest	
Budafok (1)	14
Kőbánya (2)	18
Várhegy (3)	10
Paks	10
Nagymaros	8
Ostoros	15
Noszvaj	5
Novaj	3

(1) Originally underground stone-pit now-days vine cellar, mushroomcellar, storage place
(2) Originallya underground stone-pit new-days cellar network of a brewery
(3) Partly natural hostpring cave partly medieval cellar.

2 THE STABILITY PROBLEMS OF CELLAR NETWORKS

The ambivalent of the cellars being in small depth under the examined cities and villages, should be considered as stable, since those caves for decades or even for centuries are stable. The deterioration that happens from time to time and place to place may be due to the recent upset of the balance. Since no new caves were cutted and the pillars were not diminished, the causes of the earlier upset of the balance can be the following:
- by the installation of water supply network the water consumption rapidly increased, but the sewage disposal was not solved: the groundwater level significantly raised. Due to the deterioration of the supply network, pipe breakage is more frequent.
- by the development of the city more and more building were built above the caves, due to this the static load increased
- by the increasing number of quick and heavy trucks, the static and dynamic load increased on the roof and pillars of the cellars
- by the changes of the way of life and the economic system, the cellars got out of use, so the maintenance work was neglected, the construction material worn out, the decay of rock ambient became gradually more advanced.

The cause of the deterioration, obviously, may be the full or partial coincidence of the above factors.

2.1 The shape and location of the cellar

The considerable change of the environmental effects should be significant in the deterioration of the condition, because the depth under the surface and location of the collaps is unfavourable.

On the basis of a great number of examinations carried out in different settlements the caves can be classified into three depth categories: 5 m shallow-or crust cellar, 5-10 m medium depth cellars,   10 m deep cellars.

On the hand, the depth of the location can be unfavorable, because the rock ambient of the cellars is sensitive to the static and dynamic surface loads.

The unfavourable characteristic is worsened by the fact, that - first of all in Eger - the cellars established in different ages are situated above each other sometimes even in more levels (Fig. 2).

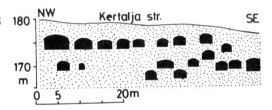

Fig 2 Tuff-cellars on multiple level situated within the range of effect in Eger

From the point of view of stability the interrelation of the location of the cellars has got primary importance. Namely, the cellars and cellar arms located usually on a way, that the stresses and shape transforming forces induced in the ambient of the caves are superposed, i.e. the caves are within the range of the affect - that is why from the point of view of rock mechanics we call it cellar network.

The thickness of the stripe pillars in many cases does not exceed 10 to 20 cm and due to the weathering, drenching and the increased loads they are not stable any more, so the rupture and brust of the pillars is rather frequent.

From the point of view of safety of the cellars and underground caves, their cross--sectional size is also of importance.

According to the surveying work that extended to more thousands of caves, the width and internal height of the caves is rather variable:
- Classification according to width: 1 m narrow, 1-3 m average, 3-6 m large, 6 m hall like
- According to heigh: 2 m low, 2-3 m average, 3 m high.

The wide span in the losed rock ambient resulted in broken og inadequate strength in many place.

2.2 The rock ambient of the cellars

The deteriorating and movement phenomena developed in the rockambient of the underground caves greatly depend on the strength characteristic and structure of the rock ambient, and similary on the stratification and fissurance.

At the examined settlements the cellars and caves were cutted to lowstrength, easy-cut rocks: riodacittuff, andezittuff, raw-limestone, loess, coherent sand (Fig.3).

Fig 3. The geological ambient of the examined cellars and caves

1-infilling, 2-clay, 3-clayey regolith, 4-clayey sandy gravel, 5-loess, 6-limnetic-stone, 7-tuffy-clay, 8-coherent sand, 9-clay, 10-calciferous clay, 11-rough limestone, 12-rhyodacittuff, 13-andesittuf, 14-buda-marl, 2-7 Pleistocene, 8 Pliocene, 9-13 Miocene, 14 Eocene

It is generally valid that if a cave formed in small depth is arched and thus considered as a stable - the cellars are like that - and its superincumbent rock is stable, its thickness above the highest point of the cave is not less than half of the span (1/2) then the maximum rupture will not break through the superincumbent rock, i.e. the rupture will not affect the surface.

It theexamined areas we experienced that the cellars cutted to low-strenth rocks, in their original condition are stable. The structure - the loose - pass beside the fissurance and stratification - effects the deterioration volumetrically, break down of the roof takes place in tuff and loess in these places. That is why the mocrotectonical tests are important.

2.3 Variation of the environmental effects

On the basis of the soil mechanical tests it should be concluded that in a previously stable cellar ambient, a great extent rupture, sometimes breaking through up to the surface occurs, when, the state of the rocks is fundamentally changed - e,g. drenched - or the superincumbent rock cannot stand the increased static or dynamic surface load, any more.

Knowing the rock ambient of the cellars - vulcanic tuffs, rough-limestone, loess, sand - it is obvious that the primary factor of the deterioration is the water. Due to the drenching the strength of the tuff is reduced to 1/3rd or 1/5th of ist original value (Fig 4).

The cellar waters cause not only stability

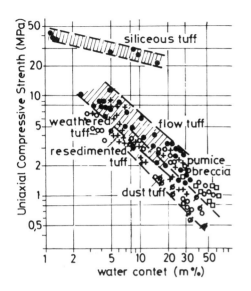

Fig 4   The decrease of strength of rhyo-
        dacittuff as effect of water (Eger)

problems, but from the point of view of
public health the degree of their pollution
is also not acceptable.

The stability of the numerous loess cellars with local characteristic is undermined by the periodic thorough drenching
that is due to the faults of the water
supply network (Fig 5).

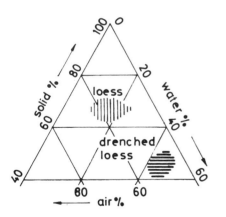

Fig 5   The variation of phase composition
        of loess due to drenching
        (Szekszárd, Nagymaros)

The cellars and cellar networks under our
historical cities and orther smaller settlements with their deteriorated state, endanger the safety of the life and economy,
the very valuable ancient historical and
architectural monuments and the development
of the settlement. To bring forth safety in
these settlements a complex and efficient
preparation and continuous accomplishment
of security works is needed.

3.1   Supporting works with the liquidation
      of the caves

Most of the cellar networks that are
under our historical cities are not utili-
zied any more, their ambient is drenched
and some places is ruptured - the liquidation of thouse is necessary. The method
applied for liquidation is different from
place-to-place (stow; tuff, sand, flue-dust,
hydraulic sand stowage).

3.2   Supporting with maintaining the caves

The cellar support does not mean that all
the caves must be liquidated. Under the
settlement there are numerous caves, that
before deterioration were utilized, and
their further partial utilization can be
benefical. By the supporting works one must
make efforts to get the reinforced cellars
and underground stopage places usable.
First of all, regional dewatering, is to be
ensured and after that we should select the
supporting method which showws a wide
variety according to the experience.

Previously, the concrete bricklaying -
that is common in the mining - was applied,
but this is very expensive. In the great
extent cellar networks the monolith-conc-
rete method is applied providing attractive,
smooth surface.

The spurt concrete method is also common
in certain places. It is greatly mechani-
zable, there is no need for shuttering, its
strength and loading capacity is better
than that of the momolith-concrete. Its
disavantage is that necessary wall thickness should be achieved by sucessive layers,
and there are great fall-back losses.

At great extent cellar surface the rock
screw reinforcement is applied. It is simple
quickly applicable, the quiantity of utilized material is small, and it is important
that it does not reduce the cross-section
of the cellar.

3.3 Regional cellar support

The reinforcement of cellar arms ruptured or deteriorated from the point of view of life-transportand establishment safety in necessary. The supporting works of the territories undercut by interconnected cellar networks demand high expensis, that is why at the beginning of the work an economical evaluation had to be made.

In this type of area at the new constructions, the foundation of building and roads, the safety of public works arise the necessary of pre-planning, and the application of extraordinary constructional methods, that are more expensive and the constructional period is longer.

It is resulting from the above that cellars beneath the cities should be taken into consideration not only from the point of view of exploration and supporting works, but they should be handled as part of the city reconstruction plan. Because of these, the moral and technical support of the caves should be done together with the road and public works reconstruction.

After the expensive supporting works the next important question is the adequate utilization of the caves - that can be traditional as vegetable, fruit, package, storage place or according to latest ideas entertaining place, museum.

REFERENCES

Habberjam,G.M.1969. The location of spherical cavities using a tripotential resistivity technique. Geophysics, V.34.N.5. p.780-784.

Kleb, B.1979. Engineering geological mapping of a city undercut by cellar network. Bull.of IAEG, N.19. p 128-134.

Kleb, B.1978. Engineering geological map series of Eger. M=1:10000, Budapest, Kartográfia. p.36.

Kleb, B.1978. The past in the present life of Eger. A settlement historical and building geological survey on the caves under the city. Budapest, KÖZDOK, p.399.

Snyder, D.D. & Merke, R.M. 1979. Analytic models for the interpretation of electrical surveys using buried current electrodes. Geophysics, V.38.N.3.p 513-529.

The main problems on environmental geology of some coastal cities in China

Principaux problèmes de la géologie de l'environnement de certaines villes côtières en Chine

Ling Zemin & Hu Ruilin
The Institute of Hydrogeology & Engineering Geology, CAGS, Zhengding, Hebei, People's Republic of China

ABSTRACT, Summerarizing the works in environmental geology of the coastal cities of China, the authors put forewords ten models of geological environmental change in these cities, and point out the main problems on environmental geology. This paper not only deep presents their occurrence and development in these cities, but also analyse their reasons for occurrence and harms.

RESUME, Après avoir fait le bilan de l'étude réalisée durant ces dernières années dans le domaine de la géologie de l'environnement de certaines villes côtières, les auteurs ont résumé dix modeles de modification de l'environnement géologique dans ces villes et montré les principaux problèmes existants. Non seulement cet article présente totalement leur production et développement, mais encore analyse profondément leur cause et nuisibilité.

1 INTRODUCTION

Having a long history and excellent natural conditions, coastal cities in China have been advanced in economics and technology all along. The convenient traffic and advanced industry make them become the important window opened to abroad. Chiefly these cities include Dalian, Qinhuangdao, Tianjin, Yantai, Qingdao, Lianyungang, Nantong, Shanghai, Ningbo, Wenzhou, Fuzhou, Guangzhou, Zhanjiang, Shenzheng, Xiamen etc.(Fig.4). Their economic developments are of the utmost importance in the national economy.

With the rapid development of economic construction in these cities recently, many problems concerned with environmental geology are also laid bare. To some extent these problems hold up their development. It has become a urgent task to expose and solve all latent problems in these cities.

2 MODEL OF GEOLOGICAL ENVIRONMENTAL CHANGE

The change of geological environment in coastal cities is the origin bringing about environmental problems. Different changing forms, ranges and origins will give the environmental change a appropriate feature , which will bring varying degrees of actions or influence to cities' construction and development , too. In these coastal cities , we summarized ten kinds of models of geological environmental change shown in table 1.

To identify different models help us not only have a good grasp of law of these environmental changes

but also chose reasonable countermeasures or measures. For the Strengthening Model and the Man-made Model, for example, we should eradicate their conditions bringing about the environmental changes, but for the Natural Developing Model and some Cataclysm Models, it was reasonable to avoid them or strengthen our construction structures.

3 MAIN PROBLEMS ON ENVIRONMENTAL GEOLOGY IN SOME COASTAL CITIES

The occurrence of the problem on environmental geology in the cities is involved in two respects of sources, one from the change of cities' environment which was induced by the natural development or man's destruction, the other from the contradictory between the demand for water or land and the reasonable supply from their environments. The first one chiefly include water pollution, land subsidence, regional crustal stability as well as harbor outwashing and clogging, and the second one apears as the shortage of water resource and the stability of foundations on special rocks and soils. Here are respective discussions about the characters of these problems.

3.1 The shortage of water resource

Water resource is one of the most important conditions to keep the people's life as well as the development of industry and agriculture of the city. With the scale of city and construction increasingly

Table 1. Model of geological environmental change and its characters

Basis for classification	Name of model	Characters	Example
form of change	Dynamic equibibrium model	The utilization of environment and the compensation of men or natures are basically in a state of imbalance. Geological environment is suited to city's development without more environmental problems.	The precipitations within Zhanjiang area can fill up exploited yields so that groundwater system is stable.
	Cataclysm model	It was suddenly that geological environmental change took place, which often bring about extreme disasters.	Famous Tangshan earthquake and its influence on Tianjin is a typical example.
	Strengthening model	It was from weakly to strongly that geological environment changed, which was caused by natural or man's actions. Its harm is often great, which we often neglect.	Water pollution, land subsidence, seawater intrusion etc.. It was the most frequent model.
	Weakening model	Geological environmental change is a process from strength to weakness. It makes the engineering geological characters more and more favorable.	The conversion from valley flat to terrace.
range of change	Partial change model	The change took place only in some sites of city.	slope collapse, landslide. landslide
	Linear change model	Geological environment changes in line.	harbor outwashing and clogging, pollution of river.
	Regional change model	The changes took place within large areas of city, sometimes out of urban areas. The harm is extreme.	land subsidence, earthquake.
origin of change	Resource exploited model	The change was caused by exploiting resources or land	land subsidence, seawater intrusion .
	Natural developing model	The change was the result of natural development.	earthquake
	Man-made model	The change was harmful results from destroying natural equilibrium by men's engineering activity	water pollution

enlarging, water shortage also become more and more severe (Table 2). Beside such social factors as over-growth of population and over-concentration of industries, unfavorable geological environments are also important factors causing the situation.

Most of large or middle cities along the coast of China, except Tianjin, Shanghai and Nantong, are around hills. Being thin and less, the artesian aquifers fall short of natural water resources. Although aquifers in Tianjin, Shanghai, Nantong are thick, they still have no more than needs because of less natural supplies. Multiple marine transgressions in history had mixed local groundwater with seawater to some extent so that the saline water is frequently involved in groundwater. Tianjin Plain, for example, the area of groundwater is 11305 km^2, but which the saline water covers a area of 8980 km^2, making up 79% of total areas. Therefore, in coastal cities groundwater resources, especially fresh water, cannot meet the need of development of these cities at all. Surfacewater resources should be fully exploited.

Since these coastal cities lead over temperate, subtropical and tropical zones, they differ so greatly in climate that their surfacewater resources have not been well-distributed. Located in the temperate zone, the northern cities, such as Tianjin, Dalian, Qinghuangdao, Yantai and Qingdao, are much short of surfacewater because of low precipitation and less rivers(Table 2). While making full use of local groundwater, the municipal government of Tianjin built the diversion works from Nuan River to

Table 2. The situation of water resources in coastal cities of China

City	Surfacewater (x10⁶m³/y)	Groundwater (x10⁶m³/y)	Exploiting Resources (x10⁶m³/y)
Dalian	8.88	1.38	2.4
Qinhuangdao	3.02	0.51	1.01
Tianjin	11	7.6	33.7
Yantai	6.53	0.6	0.58
Qingdao	0.48	0.28	0.66
Nantong	9600	0.39	4.63
Shanghai	10192.8	4.28	82.4
Ningbo	33.18	0.08	0.44
Guangzhou	434.7	1.75	15.14
Xiamen	82.6		0.62
Zhanjiang	0.633	9.5	2.09

Tianjin which mitigate the serious contradiction between supply and need for water to some extent during the period of normal precipitation. However, according to normal guarantee to development of this city, annual shortage of water is 430000000 m³. If effective measures can't be taken, the situation will become more critical.

In southern cities, although precipitation is large and rivers are flourishing, surfacewater resources are still in a state of imbalance. Those nearby large rivers, such as Shanghai, Xiamen and Shantou, have strong and stable streams so that the shortage of water isn't present. Nevertheless, Lianyungang, Zhuhai, Shenzhen and Ningbo have no large-scale surfacewater but weak and unstable streams, so water resources fail to meet all needs of these cities. It is very possible to guarantee water supplies to building reservoirs or diversion works to rivers nearby.

3.2 Water pollution

Cities' groundwater and surfacewater are not only the water resources but also the important factors protecting their ecologic environment serious attention wasn't paid to the water protection, almost all the water in coastal cities are polluted to certain extent, sometimes which even become a environmental disaster. In the light of pollution origin, two kinds of water

pollution exist in these cities,

1. Waste pollution, the main pollutants are consumer wasters, industrial wastes and liquid wastes which give surfacewater and shallow groundwater a direct pollution. The pollution of the Huangpu River in Shanghai belongs to this origin. Within a year the river has kept black and smelly for more more than 150 days. The river's length below the standards of water quality reaches 65 km, making up more than 57% of total length of the river (113.4 km). In Nantong's shallow groundwater polluted, the degree of mineralization is over 1.0 g/l and the contents of Fe^{2+}, organic matter and other harmful compositions go far beyond the allowable limit, in which the oxygen consumption is 1.4—27 times over the national standard so that they can't be taken as drinking water. Besides, Nantong's Nei River, Shantou's Han River, Wenzhou's Tang River, Zhuhai's Qiansan River as well as Dalian's Mala River, which are all the sources of water for these cities, are all polluted to some extent by wastes.

2. Seawater pollution, As the result of the overextraction of groundwater, there is a marked drop in the groundwater tables, and the water funnel is increasingly becoming large, which not only induced the land subsidence but also removed the defense against the seawater . Therefore seawater intruded into the inland through the underground permeable layer and surface rivers, with the result that the fresh water of inland were mixed with seawater and the source's condition was worsened. In Qinhuangdao the intruded areas had amounted to 3—5Km² and in the irrigation water the contents of Cl^- are over 1 g/l so that the crops were killed. In Dalian, the areas intruded by seawater was 4.2 km² before 1969, but had arrived up to 208.6 km² by 1986. The seawater intrusion also takes place in Tianjin, Nantong, Shanghai, Fuzhou, Guangzhou, Zhanjiang, Shantou, Xiamen. In Ningbo, there take place a typical event for the seawater intrusion into river. The level of

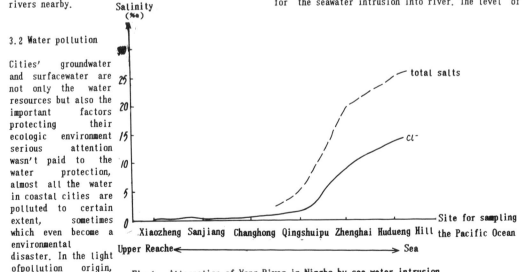

Fig.1 Alternation of Yong River in Ningbo by sea water intrusion

tide has risen 0.3-0.35 meters and can effect 18 km away from the Pacific Ocean so that the river water below Qingsuipu is always salty throughout the year (Fig.1).

3.3 Land subsidence

Soft clay with high compressibility is widespread over the coastal cities. It is common that overextraction of groundwater leads to the land subsidence, which appears more serious in Shanghai, Tianjin, Ningbo and Zhanjing (Table 3). In 1930, the

Table 3. The general situation of land subsidence

city	areas of sub-sidence (km²)	maximum rate of subsidence (cm/y)	maximum settle-ment (m)	main period	cause
Shanghai	>200	10.1	2.37	1930-1965	over-
Tianjin	>135	2.62	2.69	1966-1985	extr-
Ningbo	65		0.29	1974-1986	act-
Zhanjiang	<140	4.09	0.11	1956-1984	ion

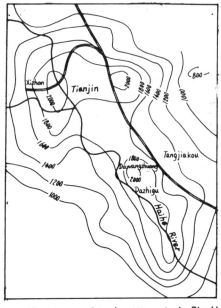

Fig.2 The isogram of land settlements in Tianjin

subsidence was first found out at some bench marks of Tianjin. In 1966 a large funnel formed. Up to now there have been three centers more than 2 meters of maximum settlements(Fig. 2). High rates of settlements and fast expansion of areas are its basic characters. In 1977 the range of more than 1 meter of settlement only covered 42 km², but had expanded up to 135 km² by 1981. In other cities, such as Wenzhou, Guangzhou and Shantou, although the land subsidence isn't remarkable recently, the contents exploited from groundwater must be effectively controlled because suitable geological environments to land subsidence exist in these cities.

The land subsidence not only brings about the construction foundations subsiding but also make the seawater intruding or submerge cities. Respectively in 1976 and 1979, the typhoon with violent tides hit Ningbo city . Seawater poured into wharfs, factories and residential areas along the Yong River so that the factories and schools had to be closed. The local government had to spend millions of RMB Yuan to heighten damp proof walls.

In order to control the land subsidence, almost all of these cities have issued the decree for groundwater administration, and taken measures of artificial recharging with a good result. In China, Shanghai is the earliest city in which the theories and methods of land subsidence control were studied. Since 1965 the subsidence has basically been controlled in urban district, but still continues in suburbs. Recently the environmental benefits brought with the artificial recharging is queried whether the recharging water (seawater) will polluted the aquifers or not. This is a contentious question for which a deep research should be done as soon as possible.

3.4 The regional crustal stability

Chinese coastal cities are adjacent to the Pacific Seismic Belts. Both in northern cities and in southern cities exist the active faults (Fig.3), so these cities' crustal stability must be on no account ignored. The country-famous active fault ,Tanlu Fault Belts, had induced a number of great earthquakes since Quaternary. Dis-

Active fault • Earthquak

Fig.3 Active fault & earthquak in eastern China

strous Tangshan earthquake occurred just near this belt. Along this belt Dalian, Qinhuangdao, Tianjin , Qingdao and Yantai scattered. Although their crustal stability have been attached great importance, close attention must be paid research works, such as identification of activity of faults, earthquake risk zoning and so on, to make provisions before troubles occurring . In Fuzhou, Xiamen, Guangzhou and Shenzhen, while watching closely the influence of Taiwan seismic belts, we must strengthen to keep watch on some of large and middle active faults. Raising the quantitative precision should be the main target for evaluation of regional crustal stability.

3.5 The stability of foundations on special rocks and soils

The stability of foundation is a common problem in all the coastal cities, on which some special rocks and soils exerts a tremendous influence.

Soft clay is widely spread over the coastal cities of China (Table 4). The origins of these

Table 4. The thickness and origin of soft clay in the coastal cities

City	Facies	Thickness (m)
Shanghai	littoral	>25
Wenzhou	lagnoon	44 – 47
Fuzhou	liman	<10
Tianjin	littoral	7 – 20
Lianyungang	littoral	5 – 10
Ningbo	lagnoon	10 – 25
Guangzhou	delta	3 – 9
Xiamen	littoral	15 – 25
Zhanjiang	lake bog	< 20
Nantong	littoral	10 – 15

soils are mostly littoral facies or lagoon facies. Generally their lithological characters are mucky clays or mucky loams with great moisture contents (over liquid limit) and high compressibility. Besides, they are often in a state of flowing or very high plasticity, showing bad engineering characters . They are often looked as harmful foundations to projects. Among twenty-six flats, built in 1974, of Tianjing Port, almost all of three-floor or four-floor flats had subsided by 20 to 50 centimeters before 1976 because they were on soft clay. The breakwater of another port were based on soft clay in thickness of 12 meters, which faint to be found in advance. In 1955, its embankments, after a high tide, suddenly collapsed with a maximum settlements of 3.3 meters. On the base of field investigation in 1970, the method of lateral sand cushion loading berms were used for the consolidation of embankments. Up to now

they have been in steady state. Therefore engineering investigation is very important in coastal cities. Close attentions must be paid to the distribution of soft clays . Although the thickness of soft clay in Xiamen and Zhanjiang isn't great, the outstanding relief of bedrock, as well as variable thickness and facies sometimes make natural foundations a unequal settlement, to which we must maintain sharp vigilance.

Jurassic to Cretaceous granites widely dispersed over the eastern coastal cities have undergone a high weathering. The thickness of weathered crust ranges from 30 to 40 meters, and sometimes the maximum will be up to 60 or 70 meters. The mantle rocks may reduce the tip resistance and increase the depth of foundation. On granite hills the landslide and soil erosion occurs from time to time. The bearing capacity of eluvial soil has not been made full use up to now, which is being studied by the IHEG.

3.6 Harbor outwashing and clogging

Harbor is one of the most important projects in coastal cities. Whether it can be under normal condition or not will influence the economic development of these cities to a great extent. The characters of outwashing and clogging of harbors are determined by the conditions of rocks, soils and landforms of coast. The harbor of such cities on the border of Huabei Plain, Yangzi Delta, Zhujiang Delta and Mingjiang Delta(Fig. 4), as Tianjin, Shanghai, Nantong, Guangzhou and Fuzhou chiefly show clogging at debauches all the year around. The clogging substances are mostly sands brought with rivers. Others, such as the ports of Yantai, Qinhuangdao , Lianyungang and Zhanjiang,belong to rocky coasts, are mainly effected by outwashing from sea waves unless the terrain slopes gently. These kinds of outwashing and

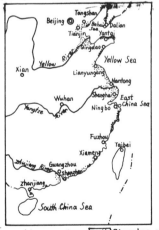

Outwashing coast ☐ Clogging coast
Scale: 1,35,000,000

Fig. 4 The types of outwashing & clogging of Chinese coasts

clogging originated from natural outcome are difficult to harness, but it is possible to reduce the degree of outwashing and clogging in the way of protecting ecological environment of drainage area, solidifying the coast of port or dredging the river course. On the contrary, those outwashing and clogging induced by artificial events not only should be averted or eradicated but also can be done. Main artificial events include destroying the ecological environment of drainage areas (e.g. felling woods and injuring slopes), changing the normal hydraulic conditions and so on. A ten-thousand-tonnage cargo ship, at the time of high tide, could pass through the channel of the Yong River in Ningbo, but after building a dam at the mouth of Yao River, the upper reaches of Yong River, has lowed the water level of channel so that it was at the arrival of tides that a three-thousand-tonnage ship could go through. A embankment built at the mouse of Yong River near Zhaobao Hill changed the course of flowing sands and identify clogging of river with the result that the spade ship have to be used for keeping a normal depth of channel. Although this kind of clogging is easy to treat, terrible costs had to be prepared. Therefore the construction of city must be planed in advance, and a close attention should be paid to their effect on environments.

4 CONCLUSION

Both the change models and the main problems of environmental geology in coastal cities of China are discussed above. It is evident that the occurrence and developments of these problems are not only due to morals but also to unreasonable usage of the environment. Developing and protecting city's environment is the central issue of the environmental geology in city from start to finish. Exploiting water resources and construction foundation are the inevitable requirements of city's development, but land subsidence, water pollution, outwashing and clogging of port mete out severe punishment to unreasonable usage of environment. We should have full understanding of relationship between them, and effectively treat it in practice. Only by doing well city plan as well as engineering investigation and research work can all of problems be avoided or solved.

REFERENCE (omitted)

The reactivation mechanism of Fushun old underground excavation area acted by tectonic fault

Le mécanisme de réactivation d'une ancienne exploitation souterraine à Fushun sous l'influence de l'activité de la faille tectonique

Nie Lei
Engineering Geology Department, Changchun Geological College, Changchun, Jilin, People's Republic of China

Sun guangzhong
Institute of Geology, Chinese Academy of Science, Beijing, People's Republic of China

ABSTRACT: According to traditional ground subsidence theory, when underground excavation take place, groundsurface deformation last only 3 to 5 years then it will be stable. But if there is a tectonic active fault passed near the old underground excavation area, a stable old excavation area can reactivate. The activity of fault can cause stable underground excavation reactive and makes the stable excavation roof collapse again, It makes the ground surface subsidence again. Tectonic fault cause old unfull excavation reactive mainly by two reason: 1. By the tectonic stress released. 2. By small earthquake. It has happened in some regions where underground mining has stopped at least 20 years and cause a huge number buildings damaged seriously. A typical example, Fushun coal mine, is given in this paper. According to the reactive mechanism, the groundsurface deformation of Fushun coal mine has been predicated.

RESUME: Selon la théorie traditionnelle de la subsidence de terrain, quand des exploitation souterraine ont lieu, la déformation de terrain dure seulement 3--5 ans, puis stable. Quand il y a une faille active téctonique qui passe près du lieu de l'exploitation ancienne, toutefois, cette ancienne exploitation stable peut être réactivé. L'activité de faille peut aboutir à l'activation de l'ancienne exploitation souterraine, et le toit de l'exploitation stable s'effendre de nouveau, cela provoque encore la subsidence de terrain. Il y a dewx raisons principleux pour expliquer ce phénomene: 1. Quand des failles sont actives, la contrainte téctonique près d'elles ect libérée dans une certaine mesure. 2. L'activité de faille provoque le petit tremblement de terre, qui fait aussi l'effoudrement et la déformation. Certaines déformations de terrain ont été prévus d'après ce mécanisme de reativation. L'activation s'est produit dans certains régions où des mines souterraines s'étaient arrêtées depuis 20 ans au moins, et a gravement abîmé des batiments énormes. Un exemple typique, la mine houille Fushun, est donné dans ce texte.

1 THE ENGINEERING GEOLOGICAL CONDITION OF FUSHUN COAL MINE

Fushun coal mine openpit which has been cut about 70years is located in northeast of china, Liaolin province. Now a huge openpit about 6.6km long, 2.2km wide and 300 metres high has been formed. Under the northern wall ground, there are many underground excavations of Shenli mine and Shenbujin mine. On the surface near northern wall there are some important factories, such as Fushun No.1 oil refine factory, power plant and parts of Fushun city.

A huge amount of deformation on the northerm wall of Fushun openpit has been happened for long time. It has made serious damage to a lot of buildings of Fushun No. 1 oil refine factory and the transportation of the northern wall. finding out the northern wall deformation main reason is the key to prevent the deformation and harness the slope. In order to find out the northern wall deformation reason, it is important to investigate the geological condition detail.

1.1 Common geological conditions

The initial landform of Fushun openpit region is plain. The surface slowly incline to Hun river which is about 2 km away from the north of the openpit. The openpit top average elevation is about 80 metres, botowm is -220 metres and the northern wall angle is about 25°. Each step of the openpit is about 10 to 20 metres high.

The main stratum from new to old is: 1. Quaternary Period: loess, muddy sand and gravel on surface about 20 metres high. 2. Eocene epoch: green shale(120--530m), oil shale (80--157m) and coal (30--145m). The

stratum crop out on the northern wall and
the bottom of the openpit. 3. Paleocene
epozh: mixed collour tuff and basalt, crop
out on the southern wall and between fault
F1 and F1a.

The openpit is in the southern side of
Fushun syncline coal basin. The stratum of
Tertiary period in its northern side have
mostly disappeared because of the cutting of
Hun river major fault F1. The attitude of
the major fault F1 through from northern
wall is NW350<55° and the secondary fault
F1a pass through No.1 oil refine factory and
power plant is NW350° <70° and 200 metres
away from F1. During Cretaecous period and
Eogene period, Hun river fault showed
dextral so that the synclinorium in the west
part of northern wall was formed. The
engineering geological conditions about
Fushun coal basin and west openpit shows in
fig.1.

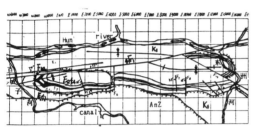

Fig. 1 The engineering geological map of
fushun coal basin

1.2 The projects of surface, openpit mining and underground excavations

The underground excavations are bounded by a
hundred metre coal pillar (EWO pillar). In
its east there is shenli mine which has
fulled by wast shale. The excavations of
No. 501 to No.510 were mined in 1960's and
the elevation of its bottom is about 428 to
475m. The excavation from No.532 to No. 537
were mined in 1970's and the elevation of
its bottom is about-480 to-525m. The
excavations of No.537 to No.577 and No. 615
were mined in 1971 --1978.The No. 447 was
mined in May 1985---June 1986. ALL of these
Shenli mine underground excavations have
stopped mining for above ten years except
No.447 excavation.

The west of EWO coal pillar is Shenbujin
underground excavations which are mined
during 1952 to 1977 have stopped mining for
at least ten years.Compared with Shunli mine
there are two unfulled underground
excavations which are 10000M² and 1600M² and
10.5M to 27.4M high in W143 to W267, N795 to
N913 and W321 to W363, N844 to N870. The
others also have been fulled with wast shale.

The openpit mining project is mainly in
the western of E1000 . The mathod of
electspade cutting and train transportation
are used, so coal and striping-substance are
cutting on the same horizontal steps which
is about 8--12M high.In the past few years
some steps arrive to 20 M high by
combination steps. The elevation of the
bottom of openpit now is about-220M. The
angle of the whole slope is only 10 or
more. The ground surface had reached the
limited boundery before 1980. The slope
angle above No.12 step has arrived at the
biggest angle and the variation of the slope
angle is well-distributed along the slope
strike. That is to say the affection of
openpit mining on the deformation is equial
along the strike of the northern wall.

Fushun No. 1 oil refine factory between
E800--W1500 is corresponding to the two
unfulled excavations and the EWO coal pillar.
Power plant (E800-E1500) is corresponding to
the No. 447 excavation which has stopped
mining no long before. All of the
factories,undrground excavations and openpit
mining project show in fig.2.

Fig. 2 The projects distribution of Fushun
openpit

1.3 The research of the tectonic active characteristics of Hun river fault(F1)

F1 which pass through the upper northern
wall is the major fault of Hun river fault
and it is the northern part of Tanlon Fault
which is a very important active fault in
the east of china.Its activity is relative
to the motion of the chinese plate, Indian
plate and the pacific ocean plate. The
chinese plate mainly moved to the south
before Cretaceous period and the action of
Tethys ocean crust in the west and the
Pacific ocean crust in the east is much weak.
In the east of china the sinistral and
transtational motion is predominant as to
Tanlou fault the strike of which is NNE.
From Cretaceous period to Eosene period, the
violent collid of Indian plate on the
chinese plate made the fault dextral
translation and cause the whole tectonic

forms in Fushun openpit. From Neogene period
to Quaternary period the collid of Indian
plate to the Asia plate ended which cause
the Indian continent motion direction
changed. This motion form last untill today.
   From the satellite picture, Hun river
looks to be a continous line and strictly
controls the land form in both sides of the
fault. All of those are correspondig to the
characteristics of land form observed in
field. There is the phenomenon of synchro
nous corner in some distributary of the Hun
river. At the same time there are at least
two-period striae in fault zone, one is
dextral thrust striae from Cretreous to
Eogeno period, the other is sinintral
traslation striae.
   The gravitational field along the Hun
river fault shows appareently form gradient
zone. The value of gravitatious on the both
sides of the fault is various and the
equivalent line of the height of earthcrust
is not continuous. All those show that the
Moho plain has been cut by the fault. The
fault cut through lithosphere. The content
of Rn apparently so up in F1 and F1a zone.
Is also one of the characteristic of the
active fault.
   There is Pleistocene epoch basalt in the
Chao county on Hun river fault zone and it
is distribution along the Fushun--Mishan
fault system. Except the magma activity in
Quaternary period, there is a
weak-earthquake zone along the Hun river
fault. These make clearly that the fault is
a tectonic active fault. From the earthquake
frequency diagram (Fig.3) we know the
earthquake activity increase period is 11
year or so. Besides there is a small
increase period every three or five years.
   There is a high tectonic stress in both
sides of the fault. The maximun horizontal
tectonic stress is 96--144 kg/cm^2. Its
direction is NE65° --84° . The smallest
tectonic stress is 45 kg/cm^2 or so.

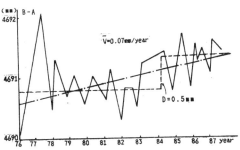

Fig.3 Level measurement curve of Hun river
       fault

   From the level measurement curve (fig.3)

the velocity (0.07 mm/year) of the fault
motion in the perpendicular direction can be
getten and the slide velocity can estimate
between 0.1 to 1 mm/year. At the same time
based on the curve the fault active period
is 11 yeat can be made sure and there is a
small increacement period every 3--5 years.
The result is completely correspondent to
the result of the earthquake frequency
diagram (fig.4).

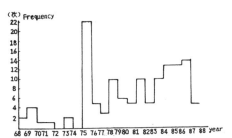

Fig.4 Earthquake frequency map

2 ROCK MASS MOVEMENT CHARACTERISTICS

2.1 Rockmass movement characteristics at
    the underground excavating time

Based on a large number of in-situ rock
movement measurement, the typical
deformation curve (fig.5) was given. From
this curve. The deformation and the wave
crest are mainly in the demountain direction.
So the underground excavating mainly
affection the upper part of the northern
wall and the groundsurface. It hardly has an
affection on the middle northern wall and
the southern wall. The value of the biggest
horizontal displacement appear on the
northern side of the biggest subsidence
value. To the northern boundary of the
underground excavation area the horizontal
displacement value is nearly zero. The
boundary angle in the demountain direction
is very small only about 37° --40° , so the
deformation of underground excavation effect
on groundsurface to a very far distance.

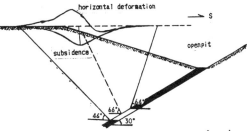

Fig.5 The typical deformation curve of under-
      ground excavation in Fushun region

2.2 The deformation charactors of the openpit northern wall now

There are severy measurement curves distributite on the northern wall and groundsurface. The measurement curve parallar to the northern wall are 510 curve, Shinpin road curve, route 28, step 10 curve and step 12 curve. The deformation for each curve have similar charactoristics. The deformation amount is closely related with different full types of underground excavations. Near the position of W200 and W300, there are two unfulled excavations, so the deformation value is very large on the upper part of slope and groundsurface. Near the coal pillar and total fulled excauation areas, the affection suddenly decrease. Near the position E1000, by the affection of No. 447 excavation area which excavated in recently year, the deformation also very large. At the position of middle slope, there the affection of underground excavtion is very small, on the whole measurement curve of step 12 the deformation is very small. This kind deformaton charactors show that the deformation of the northern wall and groundsurface is controled by the full types of underground excavtions.

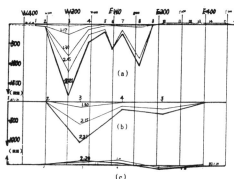

Fig. 6 Deformation curve of route 28, step 10 and step12

From the deformation of No.1 oil refine factory, it can be seen that at the position near W150 and W350, there are two subsidence basin. Its position and scale are very closely relative to the two unfulled underground excavations. This relationship shows the unfulled underground excavations still is the major factor controled the ground deformation. Figure 7 shows the deformation of the No.1 refine factory.

In the slope perpendicular direction, deformation curve form basin shape. The maximun deformation on the upper part of northern wall. But to the step 12 the deformation reduce seriously. The deformation near groundsurface is also small.

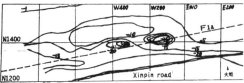

Fig. 7 Deformation curve of No.1 oil refine factory (from 7 to 10 month, 1987)

The whole deformation basin posite on the upper part of the slope. The characteristics of deformation curves (Fig. 8) is same as the deformation curve characteristics of underground excavation (figure 5). It indicate that the unfulled excavation is still active now and it is the main reason of northern wall and groundsurface deformation, eventhrough the underground excavating has been stopped for at least 20 years.

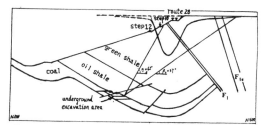

Fig. 8 The deformation characteristics on the slope perpendicular direction

By the finite element analysis and similar material model test, it can be seen that when underground mining, near the middle of deformation basin, the major deformaton is subsidence. When openpit mining, on the slope cause elastic rebound, on the groundsurface it cause a little subsidence. The deformation characteristics of the northern wall are similar to the underground mining deformation characteristics. it demonstrates that the major deformation of the northern wall is caused by underground excavation activity.

To the total fulled underground excavations, the deformation is much less. It demonstrates that the total fulled underground excavations only have a little effect to the northern wall deformation.

The openpit mining amount is same along the northern wall. In the perpendicular deriction the deformation don't form subsidence basin. Its deformation begin when the mining begin and the deformaton end very fast. The whole deformation process is much shorter than that of underground mining. Those deformation characteristics also are determined in finite element analysis and in similar material model test. The deformation characteristics of northern wall isn't corresponde to that of openpit mining. It demonstrates that the openpit mining is not the major deformation reason of the northern wall and groundsurface, and of cause not the major reason of buildings damagement.

Other factors, such as explosion, draining ground water, from the projects characteristics are not correspond to the deformation characteristics, their effects are excluded. On the basis of excluding the effects of openpit mining, explosion, draining groundwater, the deformation major reason, that is unfulled underground excavation reactivity, is futher determind.

3 OLD UNDERGROUND EXCAVATION REACTIVATION MECHANISM

3.1 The reactivation mechanism of Fushun excavation area acted by tectonic Hun river fault

The deformation on Fushun openpit northern wall was concentrated in some years, but in others the deformation decrease. Figure 9 shows those charateristics. The damagememt of buildings also shows the same characteristics. The slope deformation and buildings damagement appear periodly and its period is same as the Hun river fault active period. It demonstrates that the fault activity makes the old underground stable excavation reactivity. It is the main reason of Fushun coal mine northern wall deformation and buildings damagement. The fault active period getten from earthquake data and fault deformation measurement and the northern wall deformation and buildings sierious damagement period show in table 1.

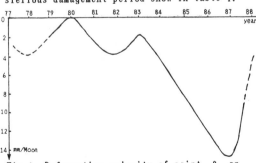

Fig.9 Deformation velocity of point 8 on No.1 measurement curve

Table 1. The relationship among F1 active periods, building damage periods and ground deformation times

fault measurement active period	1974-1975	1977	1980	1985-1986
earthquake increase period	1975	1978	1981	1985-1986
buildings damage concentrated period	1975-1976	—	—	1985-1987
deformation velocity peak period	1975-1976	—	1981	1985-1987

One reason of old excavation reactivation is that when the fault active the stress concentrated near the fault released some. This cause the friction force between joint decrease even become tensile, so the stable roof with high angle joint collapse again. This process has been shown in friction model test (Fig.11 to Fig.14). In finite element analysis when stress decreased a huge number tensile stress areas are formed (figure 10) . It shows that when stress decrease (which caused by fault active) the roof of the old excavation will colapse again. This process is called reactivation of old excavation.

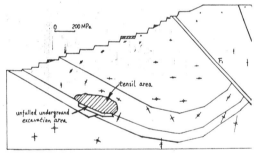

Fig.10 A large number of tensile area appear in the roof of unfulled excavtion when stress decrease

Fig.11 When excavating fissure formed and roof begain collapse

Fig.12 The limited stable roof formed

Fig.13 Stress decrease cause fissure
continue development

Fig.14 Roof collapse again(old excavation
reactivation)

Another reason caused old excavation reactivation when it effect by active fault is that when fault activity increase the miniearthquake increase. When acted by the quake force, the roof will be collapsed again and cause the old excavation roof reactivation. This action process also be proviced by similar material model test.

The old excavation reactivation process makes the collapsed area and fissure area developement to up side, so the excavation reactivation makes the deformation even more serious than the deformation at excavating time. The development process of collapsement and fissure area shows in figure 11. The curve 1 and 'a' present the position of collapsed area and fissure area

when excavating. The number 2, 3 and ' a, b' are the position of new collapsed area and new fissure area.

For a totally fulled old excavation, the material fulled in excavation area can suport the roof and the roof can't be collapsed. This is why the deformation in the position of total fulled excavation areas are much less.

According to this reactivation mechanism of old underground excavation area, if a unfulled excavation area is near a tectonic fault, by the affect of the tectonic fault active, the excavation area stable time will not be only 3 to 5 years as the traditional theory. it can be collapse for a long time and its reactive period will be the same as the tectonic fault active period.

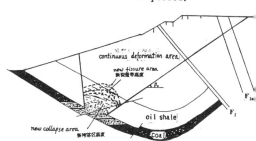

Fig.15 Collapse area and fissure area
development when old excavation
reactivation

3.2 The Fushun openpit northern wall and groundsurface deformation pridication

According to the mechanism of the northern wall deformation, that is tectonic Hun river fault active cause old excavation area reactivation and cause the deformation and according to the fault active period (11 year) , the future deformation of the northern wall can be predicated. At the position of unfulled excavations, the next seriouse deformation will be taken place between 1995 to 1998 which is the Hun river next seriouse active period. At that time the deformation of northern wall near W200 and the No. 1 oil refine factory groundsruface will be seriouse again. If the old excavations have not been fulled, the huge deformation will be taken place no matter the openpit is mining or not.

REFERENCES
E.Hoek & E.T.Brown 1980. Unerground Excavation in Rock. Canada.
Nie Lei 1988. The active charactoristics of Hun river tectonic fault. Beijing.
Sun Guangahon 1988. Rock mass Constructure mechanics. Beijing.

6th International IAEG Congres / 6ème Congrès International de AIGI, © 1990 Balkema, Rotterdam. ISBN 90 6191 130 3

Environmental impact due to tunnelling in crystalline rocks

L'impact sur l'environenment de la construction d'un tunnel dans la roche cristalline

B.Olofsson
Royal Institute of Technology, Stockholm, Sweden

ABSTRACT: Studies of groundwater levels around the Bolmen tunnel in southern Sweden have shown that the drawdown in the vicinity of a rock tunnel is a complex function of many factors, including the distance from the underground construction, the hydrogeological conditions both in the rock mass and in the overburden and the tectonical conditions. The lowering of groundwater levels in the overburden is supported by major tectonic zones having good hydraulic contact with the overburden. Two different linear regression methods for the estimation of the drawdown have been used. A minor average lowering of the groundwater levels (<0.1 m) can sometimes give rise to water supply problems in domestic dug wells since the drawdown is often confined to the summer season only.

RESUME: Des études sur des niveaux hydrostatique entourant le tunnel de Bolmen au sud de la Suède montrent que l'abaissement autour d'un tunnel construit dans la roche dépend de nombreux paramètres, entre autre, de sa profondeur, des conditions hydrogéologiques de la masse rocheuse ainsi que des couches de terre et des conditions tectoniques. Le rabattement de la nappe aquifère est favorisé par de vastes zones tectoniques en contact hydraulique avec les couches de terre. L'abaissement a été calculée avec deux méthodes de regression linéaires differentes. Parfois un petit abaissement (<0.1 m) peut causer des problèmes de ressources d'eau dans des puits domestiques piochés parce que la dépression de la nappe s'accentue souvent pendant l'été.

1 INTRODUCTION

Excavations of tunnels and rock caverns generally cause many problems leading to increased costs for support requirements. A considerable part of these costs for rock support in hard rock tunnelling is related to inflow of water (Helfrich et al, 1979). The lowering of the groundwater table causes environmental problems, especially in populated areas where ground subsidence can cause damage to buildings, sewers and roads (Morfeldt & Hultsjö, 1973, Bjurström, 1977, Straskraba, 1983, Carter & Shirlaw, 1985, Yao et al, 1987). Dug wells used for water supply may dry up, leading to claims for compensation (Nordberg, 1983). The drawdown is sometimes related to changes in the groundwater chemistry, which may be an irreversible process (Lindskoug & Nilsson, 1974, Olofsson & Ericsson, 1985).

On the other hand, a pressure tunnel constructed for water transportation can result in an outflow of water, raising the groundwater table and causing instability problems, slides or water-logging on the ground surface (Sandegren, 1972, Straskraba, 1983, Nordberg, 1983).

Impact on groundwater conditions related to tunnelling and excavation of rock tunnels has been described in many reports from projects in Sweden, e.g. from oil storage caverns (Larsson et al, 1977), power plant stations (Carlsson & Olsson, 1978) and from several tunnels (Ahlberg & Lundgren, 1977, Sund et al, 1977). The drawdown in tubes, wells and boreholes adjacent to the underground construction have often shown an anisotropic and heterogeneous hydraulic character. A systematic study of hydrogeological factors and their significance for the influence on the groundwater conditions during tunnelling i crystalline rocks, has not previously been carried out in Sweden. However, the construction of an 80 km long rock tunnel in southern Sweden, the Bolmen tunnel, enabled a long period study of groundwater levels and groundwater chemistry in the vicinity of the tunnel. The targets were to identify and quantify the changes in the groundwater conditions caused by the construction of the tunnel and to relate the changes to the prevailing hydrogeological, geological and tectonical conditions in the tunnel and its surroundings. The project was conducted on both a regional scale along the whole length of the Bolmen tunnel and on a detailed scale, in an area, Staverhult, where the tunnel intersects a tectonic zone.

2 SITE DESCRIPTION

2.1 The Bolmen tunnel

The Bolmen tunnel, is an 80 km long water supply tunnel running from Lake Bolmen to Perstorp in Scania. It was constructed during the years 1975-1986 and is located at a depth of between 30 and 100 m below ground surface. It has a cross-sectional area of approximately 8m², *figure 1*.

SMÅLAND

SCANIA

Figure 1 The Bolmen tunnel

2.2 Geology and tectonics

The bedrock along the Bolmen tunnel is dominated by different types of grey and red, often banded, gneisses, interlayered with more basic rocks such as amphibolite and amphibolite gneiss. Minor parts of the tunnel consist of a more massive and homogeneous gneiss-granite. In the southern part many dolerite dikes was found, varying in thickness from 1 to 80 m.

Most of the rock has been folded along a flat-lying axis mostly trending WNW. The northern part of the tunnel is dominated by fractures and fault zones trending NNE, whereas the southern part is characterized by a NW-WNW fault system (Stanfors, 1987).

The overburden mainly consists of sandy-silty till, usually 4-8 m thick, minor parts of glaciofluvial material and a widespread distribution of peatland.

3 METHODS

3.1 Investigation programme

Measurements of the groundwater level in the overburden were carried out 4-12 times per year from 1969 to 1986 in around 400 dug wells and tubes along the Bolmen tunnel. In addition, groundwater pressure levels in the bedrock were measured in 21 boreholes. Two reference areas consisting of approximately 20 tubes, far away from the tunnel, were also included in the long period measurement programme. Apart from this, 48 additional points were measured on a weekly basis from 1984 to 1986 within the detail studied area, Staverhult. During this period the tunnelling intersected three major parallel tectonic zones running through the area. Among the measurement sites in Staverhult were 10 boreholes in bedrock, 15 tubes in the overburden and in the superficial bedrock and 23 dug wells, *figure 2*.

The comprehensive investigation programme in Staverhult also included measurements of leakage into the tunnel, geophysical investigations in boreholes, geological and tectonical mapping in the tunnel and on the ground surface, tracer experiments, water chemical investigations, hydraulic tests in boreholes and tubes and rock stress measurements.

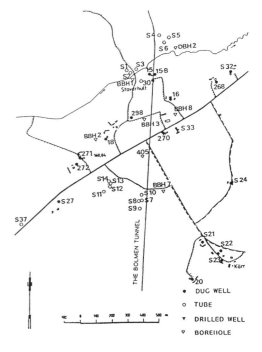

Figure 2 Measurement sites in the Staverhult area

3.2 Estimation of drawdown

One of the main problem of the study was to separate the influence of the tunnelling on the groundwater conditions from the normal seasonal fluctuations of the groundwater levels, which in this area ranges from 0.5 to 2 m in the overburden. Since the groundwater level measurements were carried out rather irregularly, time-series analysis could not be applied. Instead, two different linear regression analysis methods were used. Method 1 was based on a comparison between the best equations for the fitted lines, before and after tunnelling through the studied area:

$$y_i = a'+b' \cdot x_i \qquad (1)$$

$$y_j = a''+b''x_j \qquad (2)$$

where i and j represent groundwater level values before and after tunnelling respectively. The difference between the two straight lines is an expression of the drawdown which can be calculated from the equation:

$$s = \frac{\sum_{j=1}^{N}(a'+b' \cdot x_j)-(a''+b'' \cdot x_j)}{N} \qquad (3)$$

where s is the drawdown, $x_{j=1,2..N}$ are the values from the independent variable after the tunnelling period and a',b',a'',b'' are the constants for the fitted lines.

The point of time when the drawdown started was calculated using stepwise linear regression analysis. If the different values of the coefficient of determination (r^2) is plotted versus time, a marked reduction of the r^2-value gives a rough estimation of the time of influence, *figure 3*.

Method 2 derives from the Double Mass method, developed for the analysis of precipitation in USA and described by Merriam(1937). The method has been modified by Ehlert(1972) and is presented by Svensson(1988). It is based on the calculations of accumulated differences from the average groundwater level in two studied measurement sites according to:

$$x_i = \sum_{j=1}^{i}(dx_j-1) \qquad i=1,2,..N \qquad (4)$$

$$y_i = \sum_{j=1}^{i}(dy_j-1) \qquad i=1,2,..N \qquad (5)$$

where dx_j and dy_j are the differences from the average groundwater level in the two sites respectively. The relationship between the variables is then given by:

$$z_i = x_i - y_i \qquad i=1,2,..N \qquad (6)$$

A perfect linear relationship between the variables gives a horizontally orientated line if z_i is plotted versus the number of measurements (i) and hence versus time, *figure 4*. Significant differences from the horizontal

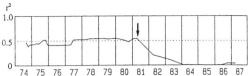

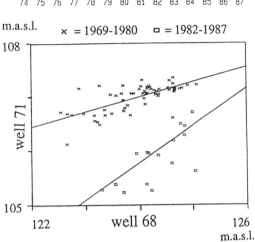

Figure 3   Estimation of drawdown and point of time for the influence in well 71 using linear regression and stepwise linear regression versus well 68.

line are an expression of inhomogeneity in one of the two groundwater level series from which the amount of change and the point of time can be calculated.

Estimation of the drawdown using these methods was carried out using a microcomputer. Method 1 was applied to all measured sites along the Bolmen tunnel and in the detail studied area Staverhult, whereas method 2 required more manual work and was therefore used only as a control method for drawdown in about 10% of the wells. The accuracy using these methods is estimated to ± 5 cm.

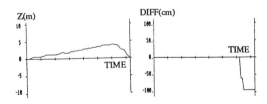

Figure 4   Estimation of drawdown and point of time for the influence in well 270 versus well 14, using a modified Double Mass method

4 RESULTS

4.1 Drawdown

More than half of all dug wells along the Bolmen tunnel were affected by the tunnelling. *Figure 5* shows the drawdown divided into classes. No simple mathematical relationship between drawdown and distance from the Bolmen tunnel has been found from the long groundwater level series in the dug wells around the whole length of the tunnel. However, divided into distance classes, obvious class differences can be identified which show that the drawdown decreases with increasing distance from the tunnel, *figure 6*. An estimation of the relationship between the slope of the fitted regression lines before and after tunnelling has shown that the construction of the tunnel was often followed by an increasing slope of the line.

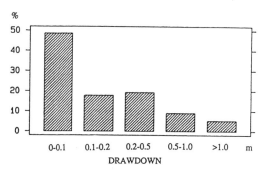

Figure 5   *Drawdown in dug wells around the Bolmen tunnel*

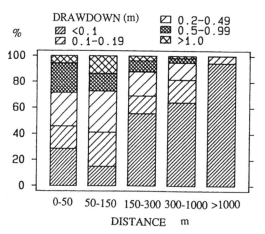

Figure 6   *Drawdown in dug wells in relation to distance from the Bolmen tunnel*

In the detailed area studied in Staverhult, the drawdown seems to decrease with the logarithmic value of the distance. A drawdown amounting to 0.3 m was measured up to 800 m from the tunnel along one of the main parallel tectonic zones running NE-SW through the area, *figure 7*.

4.2 Water leakage into the tunnel

Quantitative measurements of water leakage into the tunnel within the Staverhult area gave a leakage of around 6-7.5 l/s,km and no increased inflow was measured during tunnelling through the tectonic zones. Detail geological mapping showed that approximately 16% of the fractures in Staverhult were connected with water inflow. The most important leakages were restricted to two different strikes of fractures, N60-70W and N40-50E, which were often represented by minor fracture zones filled with rock fragments and sometimes partly filled with clay minerals. Short fractures in the hard rocks were of little interest from a hydrogeological point of view. The total porosity in the bedrock estimated from fracture data, is 0.26%. However, water flow can occur in only some of these fractures and more or less along channels on

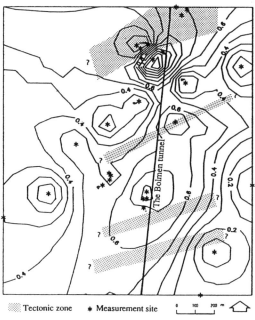

Figure 7   *Estimated drawdown in the Staverhult area. Equidistance 0.1 m.*

the fracture plane (Abelin, 1986, Hakami, 1988). Hence, the estimation of kinematic porosity from values of fracture frequency, fracture aperture and fracture pattern according to Carlsson & Olsson(1981) gives a value of 0.1%. The value estimated in Staverhult is in accordance to other crystalline rock areas (Freeze & Cherry, 1979, Carlsson & Olsson, 1981)

A rough estimation of the change of the groundwater reservoir due to tunnelling was carried out using porosity and drawdown values. The total leakage into the Bolmen tunnel in Staverhult during 1984-1986 was 20-30 times greater than the decrease of groundwater storage in the bedrock and in the overburden due to the drawdown. Hence, the estimated groundwater recharge in this area, approximately 300-400 mm/year, has wholly compensated the leakage into the tunnel.

4.3 Hydraulic properties

The hydraulic tests carried out in the Staverhult area included elementary recovery tests, single borehole tests and interference long period pumping tests in boreholes (Gustafson & Wallman, 1989). Additional recovery tests in tubes and dug wells were also carried out (Olofsson et al, 1988). The investigations gave hydraulic conductivity values in the order of $1 \cdot 10^{-7}$ m/s in the rock mass and $1-5 \cdot 10^{-7}$ m/s in the glacial till, which agree well with other areas in Sweden (Carlsson & Olsson, 1981, Lundin, 1982, Espeby, 1988). Slightly higher hydraulic conductivity values (10^{-5}-10^{-6} m/s) were obtained from tubes inserted in sandy-silty sediments and in a heavily fractured superficial rock masses in the northern part of Staverhult.

4.3 Groundwater chemistry

Groundwater chemical analyses of leakage water in the tunnel showed different types of water coming from adjacent fractures. This supports a heterogeneous and anisotropic hydraulic character in the rock mass.

Analyses of groundwater chemistry from the boreholes and tubes under influence in the Staverhult area have shown a drainage of water from the overburden into the bedrock as a result of the tunnelling. Geochemical simulations using the programme PHREEQE described by Parkhurst et al(1980), have been carried out. Within the programme, solutions from the overburden having a typical groundwater composition in this area have been mixed in different proportions with typical bedrock solutions. The simulations continued until the final solution reached a composition close to the analysed chemical composition in some boreholes and tubes which were obviously affected by tunnelling. The simulations showed that the rock mass contained 5-75% of drainage water from the overburden, as a consequence of tunnelling. The highest values were found in heavily fractured superficial bedrock situated in a topographical depression and covered by sandy-silty till and well sorted sandy sediments.

5 DISCUSSION AND CONCLUSIONS

The project shows that drawdown in the overburden around a tunnel is a function of many factors such as the distance from the tunnel, the tunnel depth, the geological conditions in the rock mass and in the overburden, the tectonics and the size of the groundwater reservoirs. Of particular interest is the major tectonic zones, the composition of the overburden, especially well sorted sandy sediments and a good hydraulic contact between the overburden and a heavily fractured superficial bedrock. The overburden thus acts as a water reservoir situated on top of the hard bedrock. The hydraulic conductors in the bedrock are fractures and fracture zones which can transport water down to the tunnel. This heterogeneity of the rock mass is one of the main explanations for the chemical differences between leakage water collected from different fractures. Inleakage of surface water directly into tunnels has previously been shown by Shimada & Ishii(1986).

Hence, during planning of tunnels in hard rocks, attention must not only be shown to the conditions in the rock mass. The leakage into the tunnel also depends on the hydrogeological conditions of the overburden. Since the kinematic porosity and thus the groundwater reservoir in the bedrock is small compared to the overburden, the lowering of the groundwater pressure level in the bedrock increases the infiltration of groundwater from the overburden into the rock mass. This infiltration is favoured by a fractured superficial bedrock, shown by the chemical mixing between groundwater from the overburden and from the bedrock.

The cone of depression in the overburden around a tunnel or rock cavern in hard rock will usually then have a fairly heterogeneous appearance. Local drawdowns will appear along the conductors in the bedrock, such as tectonic zones which are in good hydraulic contact with the overburden. Thus the tunnel can sometimes affect the groundwater levels in the overburden at a distance of 10-30 times the tunnel depth, whereas in other areas no influence at all can be measured in separate aquifers right above the tunnel (Olofsson, 1990).

The small lowering of the groundwater table (0.1-0.2 m), measured in many of the dug wells around the Bolmen tunnel is rather small compared to the natural seasonal groundwater fluctuations (0.5-2m) in glacial till in this district and will usually not affect the water supply for well users. Many of the most strongly affected dug wells were replaced with drilled wells by the tunnel construction company. However, the increase of slope in the fitted regression lines means that the seasonal amplitude of the groundwater level has increased, usually giving a marked drawdown during the dry seasons, whereas the influence is much smaller during the wet seasons. Therefore, a very small annual average drawdown (<0.1 m) may sometimes give rise to water supply problems if the lowering of the groundwater table is accentuated during dry periods. Long time series of natural groundwater level fluctuations, such as the Groundwater Net in Sweden (Nordberg & Persson, 1974), are valuable when analysing outside impact on the groundwater conditions. The linear regression methods used in this project for distinguishing the difference between natural and unnatural declines of the groundwater level is particularly useful for irregularly spaced time series.

6 REFERENCES

Abelin, H., 1986: Migration in a single fracture. An in situ experiment in a natural fracture. Doctoral thesis, Department of Chemical Engineering, Royal Institute of Technology, Stockholm, 169 pp.

Ahlberg, P., Lundgren, T, 1977: Groundwater lowering as a consequence of tunnel blasting (Grundvattensänkning till följd av tunnelsprängning). Swedish Geotechnical Institute, Report No. 1, Linköping, 60 pp. (In Swedish, English summary)

Bjurström, G., 1977: Changes in the groundwater level, geotechnical consequences. (Grundvattenytans nivåförändringar, konsekvenser från geoteknisk synpunkt). Swedish Council for Building Research, T2:1977, Stockholm, 148 pp. (In Swedish)

Carlsson, A., Olsson, T., 1978: Hydrogeological aspects of ground water inflow in the Juktan tunnels, Sweden. International Symposium on Water in Mining and Underground Works, Proceedings of SIAMOS-78, Vol. 1, Granada, pp. 373-389.

Carter, R.W., Shirlaw, J.N., 1985: Settlements due to tunnelling in New York. Tunnels & Tunnelling Vol. 17, No. 10, pp. 25-28.

Ehlert, K., 1972: Control of homogeneity in hydrological time series. (Homogenitetskontroll av hydrologiska tidsserier). Nordic Hydrological Conference in Sandefjord 1972, pp. 47-59. (In Swedish)

Espeby, B., 1989: Water Flow in a Forested Till Slope. Doctoral thesis. Department of Land and Water Resources, Report Trita-Kut No. 1052, Royal Institute of Technology, Stockholm, 175 pp.

Freeze, A., Cherry, J., 1979: Groundwater. Prentice-Hall Inc., 604 pp.

Gustafson, G., Wallman, S., 1989: The Bolmen Project. Hydraulic properties of the rock mass. Swedish Rock Engineering Research Foundation, BeFo 160:3/89, Stockholm, 51 pp. (In Swedish, English summary)

Hakami, E., 1988: Water flow in single rock joints. Licentiate thesis 1988:11L, Luleå University of Technology, 92 pp.

Helfrich., Bergman, M., Carlsson, A., Franzén, T., Granlund, N., Nord, G., Palmqvist, K., Stanfors, R., 1979: Optimum Site Investigation. Swedish Rock Engineering Research Foundation, BeFo 18:2/79, Stockholm, 226 pp. (In Swedish)

Larsson, I., Flexer, A., Rosén, B., 1977: Effects on ground water by excavation of rock store caverns. Engineering Geology Vol. 11, pp. 279-294.

Lindskoug, N-E., Nilsson, L-Y., 1974: Groundwater and Urban Planning. Report from STEGA 1966-73, Swedish Council for Building Research, Report R20:1974, 163 pp. (In Swedish, English summary)

Lundin, L., 1982: Soil moisture and ground water in till soil and the significance of soil type for runoff. Uppsala University, Department of Physical Geography, UNGI Report 56, 216 pp. (In Swedish, English summary)

Merriam, C.F., 1937: A comprehensive study of the rainfall on the Susquehanna Valley. Transactions. American Geophysical Union, pp. 471-476.

Morfeldt, C-O., Hultsjö, S., 1973: Small drops often tip big cities. (Litet dropp stjälper ofta stor stad). Väg-och vattenbyggaren No. 2, pp. 601-604. (In Swedish)

Nordberg, L., 1983: Some impact on groundwater from hydro-electric power development in Sweden. International Conference on Groundwater and Man 1983-12-05. Australian Water Resources Council Conference Ser. 1983, Vol. 8, pp. 307-316.

Nordberg, L., Persson, G., 1974: The national groundwater network of Sweden. Geological Survey of Sweden, Ser. Ca 48, 160 pp.

Olofsson, B., 1990: Groundwater conditions when tunnelling in hard rocks - a study of water flow and water chemistry at Staverhult, the Bolmen tunnel. Swedish Rock Engineering Research Foundation, BeFo, Report, 175 pp. (In preparation)

Olofsson, B., Bjarnason, B., Gustafson, G., Leijon, B., Stanfors, R., Wallman, S., 1988: The Bolmen Tunnel Research Project - Final Report. Swedish Rock Engineering Research Foundation, BeFo 160:2/88, Stockholm, 100 pp.

Olofsson, B., Ericsson, L-O., 1985: Environmental effects due to heat extraction from bedrock and groundwater. (Miljöförändringar vid värmeutvinning ur berg och grundvatten). Swedish Council for Building Research, BFR Report R149:1985, 142 pp. (In Swedish)

Parkhurst, D.L., Thorstenson, D.C., Plummer, L.N., 1980: PHREEQE - A computer program for geochemical calculations. U.S. Geological Survey Water-Resources Investigations 80-96, 210 pp.

Sandegren, E., 1972: Effects of water leakage, a study of some sections of the Kramfors-Ytterlännäs water main federation tunnel. (Följder av vattenläckage, en studie över vissa avsnitt av Kramfors-Ytterlännäs Vattenledningsförbunds tunnel). Geotechnical Department, Swedish State Railways, note 30, Stockholm, 24 pp. (In Swedish)

Shimada, J., Ishii, T., 1986: The influence on water chemistry in the crystalline bedrock groundwater by the Tsukuba tunnel construction. Colorado State University 8th Geohydrological Waste Management Symposium, Fort Collins Feb. 5-7 1986. Publ: A.A. Balkema, pp. 467-476.

Stanfors, R., 1987: The Bolmen tunnel project. Evaluation of geophysical site investigation methods, Swedish Nuclear Fuel and Waste Management Company, SKB, Technical report 87-25, Stockholm, 52 pp.

Straskraba, V., 1983: Ground-water as a Nuisance. GeoJournal No. 75, pp. 445-452.

Sund, B., Roosaar, H., Bergman, G., 1977: Water leakage in rock tunnels - effects and influence areas. (Vattenläckage i bergtunnlar - dess verkan och influensområde). Swedish Council for Building Research, BFR, Report R36:1977, Stockholm, 64 pp. (In Swedish)

Svensson, C., 1988: Analysis of affected groundwater levels. (Analys av påverkade grundvattennivåer). Geohydrological Research Group, Chalmers University of Technology, note 84, 35 pp. (In Swedish)

Yao, C.L., Chong, M.K., Broms, B.B., 1987: Subsidence from groundwater lowering by tunnels and impounding ponds. International Conference on Foundations and Tunnels 1987-03-24, London, GBR. Publ: Edingbury Engineering Technical Press 1987, Vol. 1, pp. 7-14.

6th International IAEG Congres / 6ème Congrès International de AIGI, © 1990 Balkema, Rotterdam. ISBN 90 6191 130 3

Deep seated rockslide initiated by ore mining

Déformation profonde provoquée par l'extraction de minerais

J. Rybář & B. Košťák
Institute of Geotechnics, Czechoslovak Academy of Sciences, Prague, Czechoslovakia

M. Uher
Rudné doly, Příbram, Czechoslovakia

ABSTRACT: A deep seated rockslide in Western Bohemia initiated by the mining of a tin ore body is analyzed. In order to protect mine plant buildings monitoring of deformations have been organized and alarming limits determined.

RESUME: On évalue un glissement profond dans la Boheme de l'Ouest provoqué par l'extraction d'un gisement de minerais d'étain. Pour protégér des objects miniers exposés on a introduit les méthodes d'auscultation des déformation et determiné les valeurs des criteres d'avertissement.

1 THE MINE

Gravitational deformations which developed during mining of an ore body in the rock massif in West Bohemia are analyzed. Mineralization of tin and wolfram is found at a contact between a stock of late Variscan granites with the mantle of highly tectonically fractured biotitic gneisses and migmatites (Fig.1). In the outcrop the ore was repeatedly mined in the past by underground methods. The uncontrolled mining was most intensive in the 16th and 17th centuries, when the original structure of the massif was largely disturbed by caving to the depth of about 150 m. In the 80ties an open pit mine was opened here to excavate caved stocks. After closing the open pit by the end of 1987 the mining was transferred to underground level 50 to 150 m under the pit bottom. The main hoist came from the levels 500 and 475 m a.s.l. Level 500 extraction was kept uniform at 50 % output of blasted rock. Discharging of caved stock sometimes makes cut through hanging wall and overburden so that sink holes develop on the surface. In the level 475 room and pillar process is applied, however, also abandonned old cave- in stock is hoisted under the 50 % condition like from the level 500. Level 425 is mined by room and pillar with quickly solidifying filling so that failure of the hanging wall is effectively limited.

2 DEFORMATIONS

Even old geodetical surveying had indicated sliding of the pit slopes, it was therefore natural to accept a more comprehensive system of monitoring when underground excavations started. Succesfully, from the very beginning by early months of the year 1988 the activity of movements was confirmed not only on the pit slopes but also in the wide foreground. By the transition of 1988/1989 a tensile crack appeared not very far from the head frame and the other important plant buildings. Under this circumstances alarming limits were determined and accepted as warning criteria against the risk in the mine operation.

3 MONITORING

The monitoring system is shown in Fig. 2, however only elements noted in this work are marked by numbers. Two boreholes A1 and A5 for precize inclinometry are considered to be the main elements of the system to track the massif in the depth. The boreholes are regularly checked with Glötzl inclinometers. Other boreholes are equipped with brittle cables (e.g. A3). Two profiles of survey pillars (P1 to P4; P5 to P6) are checked by tape extensometers and precize levelling on the surface. Trilateral sign system is used to check the main tensile crack (points MP3, MP5,

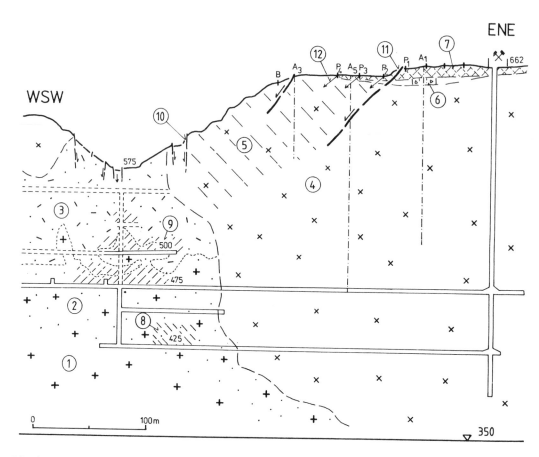

Fig.1 Cross section 1-1' through the ore mine in Western Bohemia.
1 - granit; 2 - ore bearing granit (greisen); 3 - worked out areas; 4 - biotic gneiss
and migmatites; 5 - the same as 4, rock shattered by sliding; 6 - soils and debris; 7 -
old mining wastes; 8 - room and pillar mining with quickly solidifying filling; 9 -
advance with continuous caving; 10 - sinkholes; 11 - sliding surface; 12 - survey pillars and displacement orientations.

MP7 etc.). Besides, levelling is extended
to check foundations of plant buildings
and nearby points. Other points close to
the pit margin are checked as to position
and level movements (points A to E etc.).
Fig.1 shows displacement orientations of
zone points, too.

4 OBSERVATIONS

Results of the monitoring on the surface
of the area are presented graphically in
Fig.3. Here, displacement rates of characteristic points are related to the output

from levels 500 and 475.
    Even the first deformation measurements
which took place in early months of 1986
when underground operations at the level
475 began, reported displacement rates
close to 20 mm per month. Clearly, this
represented an aftermath of the open pit
operations. Hauling from the level 500
started later in march 1988. The rates of
most investigated points linearly increase with the increasing output from this
level. It is evident, especially in the
second half of the year 1988. By the end
of 1988 displacements between points P1
and P2 reach the rate of 46 mm per month.
The pillar P2 sinks down by 32 mm/month.
Then, from the start of 1989 a tensile

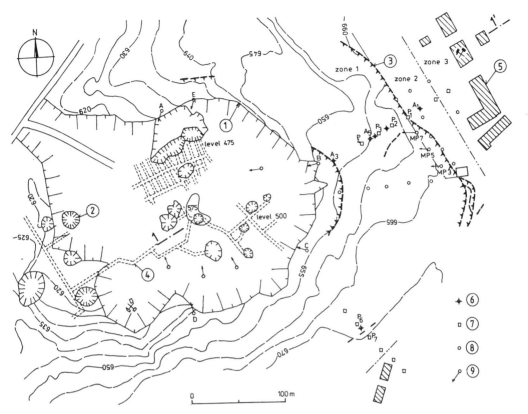

Fig.2 Situation of the ore mine in Western Bohemia.
1 - edge of the old pit; 2 - sinkholes; 3 - main crack with the evidence of sinking;
4 - levels 475 and 500 m a.s.; 5 - plant buildings threatened; 6 - drilling holes for
inclinometry and brittle cables; 7 - pillars for tape extensometry; 8 - geodetical survey points; 9 - the same as 8, displacement orientations.

crack is found between P1 and P2, which
quickly reaches a length of about 250 m.
The crack in the plan looks as a multi-
fold line composed of several straight
sections oriented to NNW-SSE and NW-SE.
This is the orientation of tectonically
weakened planes and zones tilted to SW,
as verified in drifts, in the shaft, as
well as in boreholes.

Following the occurance of the main
crack in the terrain, in the middle of January 1989 hauling from the level 500 was
stoped. However, it was increased from
the level 475 at the same time. The halt
at the level 500 has been traced in the
course of terrain deformations. The reaction is delayed and restraint. Rates of
movements slow down with the exception of
the section P1-P2.

After releasing the output restrictions,
since April 1989, the course of deformations in all investigated profiles has be-

come stable. There seems to be delayed
fluctuation in displacement rates in correlation with output rates. The maximum
in horizontal displacement between P1 and
P2 reaches 43 mm/month (September 1989),
the vertical movement of P2 being 30 mm/
month at the same time. Another result has
been observed with points located close to
the edge of the original pit. In the 2nd
half of the year 1989 the rate here gets
increasingly higher for horizontal as well
as vertical movements. The average in horizontal displacement for the point C comes to more than 65 mm/month.

It can be seen from Fig.3, that the slope area of the old pit had been continuos-
ly in slow movement, since the start of the
underground extraction, and present underground extraction did not initiate movements, only accelerated the old ones evidently. Fluctuations in the displacement
rates can be related to the output fluc-

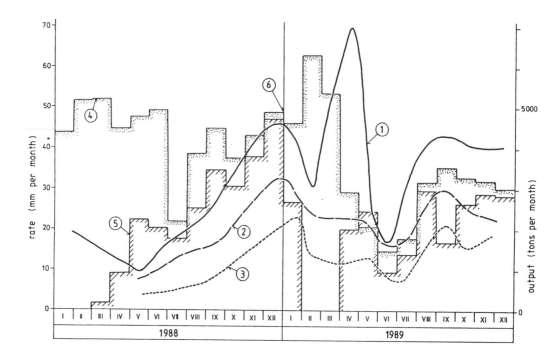

Fig.3 Relation between the underground output and displacement rates.
Displacement rates: 1 - increase of the distance between pillars P1 and P2; 2 - sinking of the pillar P2; 3 - increase of the distance between the pillars P6 and P7. Output rates: 4 - level 475; 5 - level 500; 6 - appearance of the main surface crack.

tuations. Reaction to the level 500 output is beyond any doubt, the level 475 output on the other hand cannot be correlated with slope movements. The halt in the level 500 extraction results in damping down of the slope movements, not complete stop. For that a halt of a length of many years would be needed, necessarily.

To evaluate the character of the mass movement underground, inclinometric measurements in the borehole A5 became essential. Observation resulted promptly in an indication of a barrel shift at the depth of 34,5-36,0 m, recognized as shearing along a plane.

Other underground observations are performed in the mine and can be considered as part of the monitoring system. Let us mention overcoring tensometry, microseismic radiation and convergence measurements at the level 500. Moreover, rock wall inspection is a routine. However, underground measurements have not proved any stress changes or other instability signals that could be understood as unfavou-

rable result of slope movements of the original pit mine.

5 ZONES INSPECTED

A decision to end the halt in production from the level 500 was conditioned by mining inspection with a demand to establish alarming deformation limits in respect to safe operation of the main shaft. For that reason four different zones were recognized, to be treated separately (Fig.2).

Zone 1 is the area of very active movements of the pit slope region. It is limited by the prominent tensile crack in the foreground.

Zone 2 is behind this crack. This is an area in close contact with the sliding area. Activation of movements in it will signal slide propagation towards the plant buildings and the shaft.

The zone has been limited by a line that cuts distance between the crack and plant buildings at the middle approximately.

Zone 3 is behind that line. It is the area with the main frame which is to be protected.

It is evident that austerity increases as the investigated zones get closer to the plant and the alarming limits must follow the same tendency.

Zone 4 covers the depth, the underground massif, where informations are scarce. The process here involves solid rock, and propagation of movements may signal fatal development of events.

Monitoring is a system of observation allowing for quick reactions to changes that occur in the area. Ideally,,it should provide an automatic feedback system with a capacity to intercept precursory signals and give forcast. As a matter of fact the system is far from ideal and it is the alarming limits that must tender the forcast at a certain level of information. So, we consider alarming limits as an alarming forcast that the situation is going to be changed and new decision will be needed. The new anticipated situation is evidently progresive failure in which reactions of the massif may get out of control and risk may definitely increase. According to this view corrections of alarming limits may be accepted when new information is obtained. The set of limits is therefore temporary, open to improvements.

Clearly, the most important zone of observation will be the interior of the massif (zone 4), with all possible information about failure propagation and deformation changes. On the other hand slope movements will not be considered dangerous, if correlate with mining advance and do not show any progressive tendency on slopes (zone 1).

6 ALARMING LIMITS

Alarming limits were proposed as follows.

Zone 1. Sliding area. Continuous sliding is accepted as not dangerous. The acceleration of displacement beyond the level 1,5 times of the present average is considered alarming.

Zone 2. Contact area. At present the zone does not show active movements. As alarming limit is considered:
2a subsidence higher than 72 mm;
2b new cracks and shearing displacements (horizontal or vertical) on cracks, higher than 25 mm.

Zone 3. Protected area. It can be assumed that failure may advance only through the zone 2 to reach the zone 3. Higher sa-

fety required to avoid damage.

As alarming limit is considered:
3a shearing subsidence observable as terrain tilt between observed points by 0,5 mm per 1 m;
3b appearance of cracks (open or covered) showing deformations as large as: 2 mm in subsidence; 8 mm in crack opening and/or horizontal shear.

Zone 4. Underground zone. Deep zones of hard rock massif with a danger of deep crack propagation.

As alarming limit is considered:
4a shear over 25 mm in an inclinometric borehole;
4b shear strain over 80" in the massif indicated in an inclinometric borehole, if calculated for the full lenght of the borehole;
4c shear in a borehole indicated by brittle cables.

Any alarming limit must be verified in double i.e. either in two points, by two methods, or by two independent measurements taken at one point in regular intervals of observation. This is to avoid mistakes and instrumentation errors. One month interval is generaly accepted for observation.

The above limits were in general accepted to protect the described mine and approved by mine inspection.

7 SLIDING SURFACE VERIFIED

As reported before, inclinometric measurements indicated shear at the depth of 34.5 - 36,0 m (borehole A5), and the alarming limit 4a was exceeded in January 1990. This is the anticipated situation which calls for a new decision in mining advance, safety measures, or possibly for a change of alarming limits. A group of specialists met then and considered the new situation. There was no doubt that the indicated shear represented sliding along the very same plane, which cuts the surface by the main crack. The situation does not represent therefore an increase of risk, and only the actual depth of the sliding plane has been verified by the observation. It was decided therefore to accept alarming limit conditions of zone 1 for the whole wedge of rock above the sliding plane and the prescribed conditions of zone 4 for deeper zones only. Moreover, the mechanism of the slide in the depth can be now better understood.

2685

Engineering geological survey and monitoring lead to a conclusion that slow slope movements in the massif of overburden gneiss result in slow and continuous pressure built against not compact, highly heterogenous, fractured and compressible body of old caved mines. The pressure builds up slowly. Slope movements are actived by mining and, at present, cannot be any more regulated by output limitation. The volume resulting from slope movements of rocks is essentially less than the volume deficit resulting from extraction. The deficit is not therefore compensated satisfactorily by slope movements and by sinkhole drops from the toe area of the pit. The space under the sinking volume is therefore showing porosity increase by gapping in the body. With that compressibility increases, so that pressure increase due to slope movements is less than pressure decrease due to mining. In this situation a conclusion may be drawn that the slide itself, i.e. neither the kinetic energy of gneiss, nor the weight shift due to accumulation of it, can induce a sudden destruction of mine openings in the solid massif of the ore under the present methods of extraction.

Inclinometric observations from the borehole A5 as well as observations on the terrain lead to a suggestion, that the broken wedge in the slope of the old pit move down continuously as one unit separated from solid beds by a predisposed plane. The sliding surface, in lower depth at least, is continuous, and almost planar in accord with tectonic structure. A suggestion that failure originated here during one the older episodes of mining, cannot be avoided. In greater depths a more complex spatial structure of planar discontinuities which can make movements possible, is anticipated. Here probably a single shear plane has not cut through, yet. In all horizonts, movements should be understood as shifts in space, which follow the shape of pyramidal horn of the old pit, at sides and front slope, especially. The shape of horn has oriented also the main sinkholes, centered close to its bottom.

The assumption about the planar movement in the heart of the slide supports our conviction that the slope deformation will stay (if the present extraction methods not changed) for a considerable period of time limited to the space of presently known and verified cracks shown in Fig.1.

6th International IAEG Congres / 6ème Congrès International de AIGI, © 1990 Balkema, Rotterdam. ISBN 90 6191 130 3

Dissolution of salt and influence of dissolution to the town of Tuzla

Dissolution du sel et effets de la dissolution sur la ville de Tuzla

I. Vrkljan & N. Sapunar
Faculty of Civil Engineering, University of Zagreb, Zagreb, Yugoslavia

M. Stević
Faculty of Geology and Mining Engineering, University of Tuzla, Tuzla, Yugoslavia

ABSTRACT: Exploitation of the salt deposit of Tuzla is performed through pumping of the saline ground water. The uncontrolled dissolution of this deposit provokes the uncontrolled settlement of the ground surface at the location of the old town of Tuzla. Until the present time, the town's center has settled for more than 13 m. Geological and hydrogeological properties of the salt deposit and the method of repairing significant historical monuments in the zone of great settlements are presented in this paper.

RESUME: Le gisement salifère de Tuzla est exploité par pompage de l'eau souterraine salée. Le procédé incontrôlé de dissolution provoque le tassement anormal de la surface du terrain sur lequel la vieille ville de Tuzla est située. Jusqu'à présent, le centre-ville de Tuzla a subi le tassement de plus de 13 m. L'article décrit les propriétés géologiques et hydrogéologiques du gisement salifère et la manière de réparation des bâtiments historiques importants situés dans la zone des grands tassements.

1. IDENTIFICATION OF THE PROBLEM

Until recently, the salt deposit in Tuzla was the only known salt deposit in Yugoslavia. First written documents on the salt exploitation date back to 1477. Archeological excavations have shown that the salt from Tuzla was exploited already in prehistorical times. The sources of saline water existed in these times. Until 1886, the salt was produced by evaporation of saline water that sprung to the surface or was obtained by digging shallow wells. In 1886, the mass exploitation of salt started by pumping saline water from the drilled wells. This type of exploitation is still being applied.

The method of salt exploitation by dissolution of salt deposit is widely applied in the mining practice. In such cases, the controlled quantity of fresh water is supplied from the surface to the salt deposit level. The process of salt dissolution is controlled by appropriate procedures which means that the position, shape and volume of the dissolved salt can be defined in advance. By appropriate distribution of these areas their stability is obtained so that there are no deformations of the ground surface.

However, in the town of Tuzla, the salt is obtained from the ground water which dilutes the salt in the course of its flow. Since hydrogeological conditions of this deposit are quite complex, the process of salt dissolution takes place without control. For that reason, this method of salt exploitation is called "method of uncontrolled salt dissolution". This method of exploitation is unique in the world.

The uncontrolled dissolution also provokes the uncontrolled process of deformation of the massifs that are located above the saline series. The massif deformations can be noted all the way to the ground surface.

Since 1967, the salt is also exploited by classical mining method. Therefore, the deposit is divided into two exploitation fields (Fig. 1). In the field "A", the salt is exploited by pumping the underground saline water using the method of the uncontrolled dissolution of salt. In the field "B", the salt is exploited by classical mining method and the explosives are used for excavation. Between these two fields, there is a protecting pillar approx. 200 m wide whose role is to protect the salt mine from the effects of exploitation in the field "A".

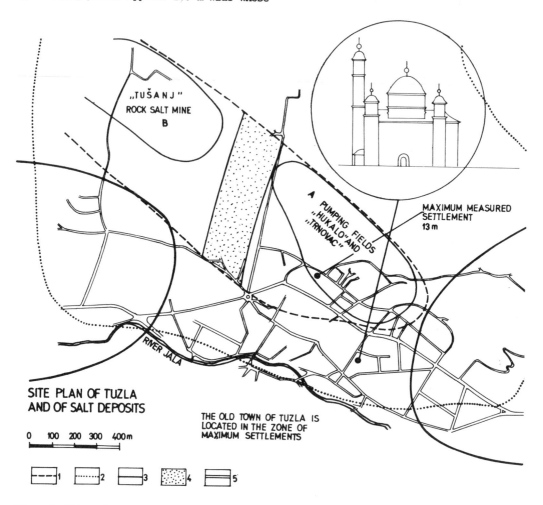

SITE PLAN OF TUZLA
AND OF SALT DEPOSITS

0   100   200   300   400m

THE OLD TOWN OF TUZLA IS LOCATED IN THE ZONE OF MAXIMUM SETTLEMENTS

Fig. 1 Site plan of Tuzla. 1-deposit boundary, 2-boundary of measured surface displacement, 3-Zone of current urbanization, 4-protecting pillar, 5-streets.

Typical cross section through the salt deposit is presented in Fig. 2. The wells are drilled within the triangular network and are spaced at 30 m intervals. 177 wells have been drilled up to the present time. 20 wells are currently being used.

Due to the massif deformation, the wells become damaged or completely destroyed. In the beginning of exploitation, wells usually lasted up to 40 years whereas today they last only about 2 years.

This unique procedure of exploitation is limited by the following factors: small ground-water inflow, small percentage of the deposit exploitation, brief period of the well exploitation, great damages to the town provoked by the settlement of the ground.

Due to the mass deficit in the underground, the ground surface settles, the inclinations change and specific deformations occur in the form of shortening or lengthening of the distance   between two points.

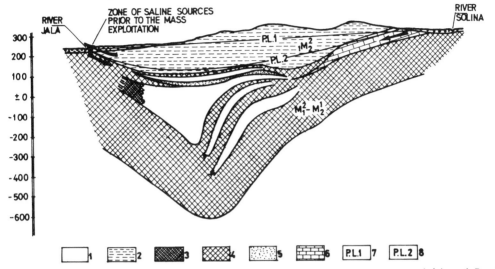

Fig. 2   Cross-section of the salt deposit.   1–Salt, 2–Flysch, 3–Anhydride, 4–Banded
series, 5–Pelite, 6–Thin–layered limestones of Tuzla, 7 and 8–Piezometric level

First geodetic surveys of the ground settlement were made in 1914. Since 1956 and up to the present time, systematic geodetic surveys are being performed. An area of approx. 8 km² is covered by geodetic networks. Up to this date, the settlement has been observed on the area of approx. 7 km². Maximum vertical settlements amount to approx. 13 m. The total measured volume of subsidence amounts to approx. 5 million m³. It should be noted that this settlement process is continuous, i.e. it does not happen suddenly nor in the form of individual failures.

All this information about the ground settlement would not be of particular interest if this area were not completely urbanized. The old part of the town of Tuzla completely lies on the ground that is affected by this settlement. A great number of structures had to be demolished since their repair proved impossible. Among the damaged buildings there are some historically significant structures from the 16th and 17th century. The old town slowly dies and gradually becomes only a link between the new parts of the town that are being formed outside of the settlement zone (Fig. 1).

In recent times, some measures have been taken to save at least those structures that are considered to be of significant importance. Successful repair works performed on two church structures that are located only 400 m away from the maximum settlement point are described in the following section of this paper.

2. GEOLOGICAL AND HYDROGEOLOGICAL PROPERTIES OF THE DEPOSIT

The rock–salt deposit in Tuzla is a sedimentary deposit created during the process of chemical sedimentation by evaporation of sea water in the lagoon-type basin. The deposit is of complex and heterogeneous lithological structure which is due to the frequent oscillations in intensity and to interruptions in the process of sedimentation. The deposit is located within the syncline "Trnovac" – "Tušanj". In the plan view, the deposit has a form of asymmetrical ellipse. The longer axis of the ellipse amounts to approx. 2.5 km whereas the shorter amounts to approx. 900 m (without anhydride which occurs at the wing of the syncline, approx. 600 m). The axis of the syncline is inclined so that the salt occurs at depths ranging from 150 to 600 m.

If we consider the stratigraphy of the ground, the most significant is the so called *banded series* which contains all salt formations. This series is mainly formed by the banded and laminated marls that pass into anhydritic, dahamitic and bituminous marls. It is estimated that the influence of the banded series spreads over the interval of approx. 500 m. This series has been classified as the Upper Burdigalian – Helvetian (P. Stevanović – M. Eremija, 1960), or Helvetian (I. Soklić, 1964).

The floor of the banded series is formed by the lake-lagoon type sediments of the Lower Burdigalian. It is made of marls, claystones, sandstones, aleurolites and conglomerates of red color and is therefore called the *red series*.

The roof of the banded series is formed of the Lower Tortonian sediments of the open sea, developed in the form of thin-layered flysch formations. From the lithological aspect, flysch is represented by alternations of clays, laminated and banked, gray and whitish clayey marls with intercalations of clayey fine-grained sandstone. The flysch spreading varies from 100 to 250 m.

The deposit itself is formed by the beds of rock salt, Glauber's salt and anhydrides. Salt formations are the most widely spread in the central part of the Trnovac – Tušanj syncline. Towards the wings of the syncline, the salt series gradually becomes less dominant and it disappears quite abruptly at the boundary of the deposit. The greatest part of anhydrides is found at the south wing of the syncline. After the boundary zone, where it is mixed with halite, the anhydride passes into the lateral continuation of an adequate series (II and III A series). It is often tectonically integrated into the breccioid mass of the south boundary of the deposit.

Pelite of variable spreading is found at the floor of the shallowest salt series. In the petrographic sense, this is the tuffitic – pyroclastic rock. Smaller quantities of pelite have also been found in the deeper parts of the deposit.

Marls belonging to the banded series, first salt series and pelites are laterally linked with the thin-layered limestones of Tuzla.

The structure and morphology of the deposit is much more complex when compared to that of the surrounding sediments, although the salt series were submitted to the same tectonic influence during the orogenic movements. This can be explained by the differences in rheological properties of the salt and other sediments.

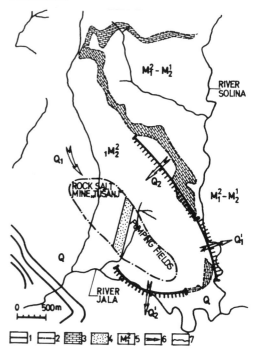

Fig.3 Geological map of the wider area of the rock-salt deposit in Tuzla according to the geological map of P. Stevanović (1:10.000). 1-Coal beds, 2-deposit boundary, 3-Thin-layered limestones of Tuzla, 4-Protecting pillar, 5-Designation of lithological members, 6-Direction of inflow and draining of the deposit at the time of the natural water flow (Đurić, N., 1986), 7-River courses.

Internal tectonics of salt formations is very complex. This is a specific aspect of the salt tectonics which occurs as a consequence of both different mechanical properties of individual salt types (with regard to their structure) and of different properties of salt and other secondary sediments. Internal tectonics of salt formations is manifested through abundant and variable micro and macrostructures such as: complex rugose folds, interbedded folds, small-scale overthrusts, microdiapirs, fractures, microfaults, breccia zones, salt filled joints, schistosity etc. Particular attention should be paid to the folded and fissured intercalations of the secondary sediments which, under the influence of complex internal tectonic processes, gradually lost their continuity and were separated into independent pieces and blocks of free spatial orientation and, for that reason, they now "fluctuate" in the salt mass. Such structure of saline series is the cause of the great frequency and abundance of unnecessary intercalations which amounts to 40% of the total mass.

Hydrogeological properties of the deposit

Hydrogeology of the deposit and of the surrounding sediments is quite complex and has not as yet been fully explored. If we consider hydrogeological properties of individual lithological series, then we must know that these properties change in time. Hydrogeological properties of a certain series depend on the degree of modification of the initial state of the massif under the influence of salt dissolution and cave-in processes. We therefore have to distinguish properties of the undisturbed massif from the properties which massif obtains some time after the beginning of the exploitation.

Banded series with salt deposits represents a complex formed of sediments that have different hydrogeological properties. If we take into account experiences from the salt mine "Tušanj", then these saline series can be considered impervious. Water has not been encountered throughout the exploitation period. Layers with the greatest ground-water quantity have been determined through the exploratory boring, boring of exploitation wells and through experience obtained during the shaft excavation at the salt mine "Tušanj". The greatest ground-water inflow comes from the banded marls which form the immediate roof of the first salt series (first primary water-bearing horizon). The water appears as the artesian water in both water-bearing horizons. The water also circulates through other parts of the banded series but in somewhat smaller quantities.

Hydrogeological properties of flysch formations have been better defined than those of the banded and red series. Flysch sediments are quite disturbed as can be seen in the different orientation of layers and in frequent loss of mud during drilling. Fracture porosity is dominant in these sediments and the interstitial porosity is also quite frequent. Flysch sediments can be considered as a poorly permeable environment.

Since the exploitation method is based on the salt exploitation, through pumping of the saturated ground water, it is quite evident that the natural balance of stress states and of the ground-water regime has been disturbed.

Sediments of the banded series, which form the immediate roof of the salt series together with its dissolution product, break and hence their porosity and permeability increases considerably.

The flysch sediments which form the roof of the first salt series have also been partly influenced by the cave-in and deformation processes, which greatly increases the porosity and permeability of this series. Thus the depth of the first water-bearing horizon increases in the area where the first salt series were dissolved.

Water is supplied to the deposit through the end parts of the banded series, by infiltration of the surface water and from the rivers Jala and Solina.

Intensive saline water exploitation caused the fall in piezometric levels of individual water-bearing layers. Fall of piezometric level in turn provoked the change in the ground-water flow. Saline sources have dried up and today we have at such places supply of water from the surface instead of the draining. Thus, the ground-water flow has completely changed.

Saline-water exploitation persists to this date and the inhabitants of Tuzla still struggle against the effects of settlement of their town.

3. REPAIR OF DAMAGED STRUCTURES

Structural damages are due to the uneven settlement and to the horizontal deformation of the ground.

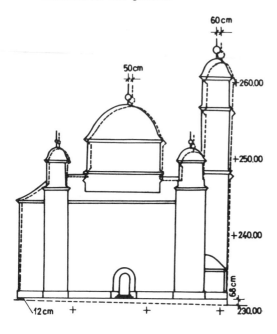

Fig. 4 Uneven settlement of the church.

Damaged structures of smaller importance are usually being demolished. Nevertheless, the structures of historic significance are being repaired so that they can still be used despite the great and uneven settlement values. We will present bellow the way in which one church 25 m high and the church-tower 32 m high have been repaired.

Due to the uneven settlement, the cross at the top of the church deviated 60 cm from the vertical (cf. Fig. 4). Designers were faced with the following problems:

- The church and its tower had to be brought into the vertical position. This means that one end of the church had to be lifted for 12 cm and the other for 60 cm.
- Such system of lifting had to be applied that would enable normal use of the church in the future despite the fact that the underlying soil will continue to settle and that the need for

horizontal correction of the church position will always be present.

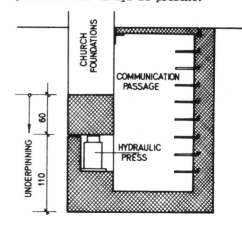

Fig. 5 Underpinning of the church foundations

Today, two years after the church has been repaired, we can state that all problems have successfully been solved. The system that was installed in order to lift the church can be used at any moment and thus, negative effects of the salt exploitation can be annulled.

Position of one of 42 presses is presented in Fig. 5. The empty space that remained between the reinforced concrete girders after the lifting of the structure was filled with concrete.

Fig. 6 Church after repair works

Total weight of the church that was lifted amounts to 36000 kN. The church was lifted using 42 hydraulic jacks. Jacks were divided into 6 systems. All systems were controlled from a single point. The entire lifting process lasted 340 minutes.

4. CONCLUSION

The uncontrolled salt exploitation performed by the saline water pumping causes the uncontrolled settlement of the ground surface. The old town of Tuzla is located in the zone of maximum settlements. A considerable number of structures had to be demolished since the maintenance costs in the conditions of constant settlements are extremely high. The structures of historical significance are being repaired in such a way that they can be brought to the stable state even in the conditions of continued irregular settlement of the ground.

REFERENCES

Vrkljan, I. (1981). Protection of the urban area of Tuzla against the consequences of rock salt exploitation by leaching, Mining and metallurgy quarterly, Vol. 28, no. 2-3, pp. 223-238, Ljubljana, Yugoslavia.

Stević, M., Jašarević, J., Ramiz, F. (1979). Arching in hanging walls over leached deposits of rock salt, Proc. International Congress on Rock Mechanics, Vol. 1, pp. 745-752, Montreaux, Suisse.

Sapunar, N., Planinc, R. (1989). Horizontal levelling of structures of an Orthodox church in Tuzla, Građevinar, Vol. 41, no. 1, pp. 31-37, Zagreb, Yugoslavia.

Trend forecast of land subsidence cycle

Prévision de tendances pour cycles de subsidence de terrain

Yan Tongzhen & Zhou Chuiying
China University of Geoscience, Wuhan, People's Republic of China

ABSTRACT: Land subsidence cycle is defined as its life-span. It is considered that subsidences may be widely occurred due to their different origins and varieties. However, the only ones of them mainly caused by artificial pumping of underground water are concerned in the paper. Xi'an and Ningbo cities are cited as typical examples. The emphasis is focussed on time trend forecast of their cycles. Geological background are discussed and mathematical models of Poisson cycle, established by the dynamic, monitored data. Fundamentals is discussed. Basic assumptions are suggested. By calculation a series of forecasted data are gained; and dynamic curves are plotted. Land subsidence cycle of both Xi'an and Ningbo would be 125 and 80 years separately.

Résumé: Le cycle de subsidence de terrain est défini comme la vie de subsidence. Tl est évident que la subsidence de terrain peut exister largement en raison de leurs differentes origines et variétés. Mais dans cet article il s'agit seulement de la subsidence de terrain causee principalement par artificiel pompage d'eau souterraine. Les deux villes, Xi'an et Ningbo sont presentées comme les typiques exemples. Cet article met l'accent sur la prévision de trend sur le temps des cycles de subsidence, et établit les modèles mathématiques du Poisson cycle par les dynamiques données de monitoring, et discute le contexte geologique et le principe fondamental, et pose les fondamentales conditions supposées, Par calcul, une serie de données prévisionnelles sont gagnées, et les dynamiques courbes sont tirées. Les cycle de subsidence de terrain pour Xi'an et Ningbo sont de 125 et 80 années séparément.

1 THINKING ABOUT THE TREND FORECAST

In the paper land subsidence is considered to be limited system. The system in nature might be regarded as organism. Generally the once cycle or life-span must be developed with common ecological characteristics including several evolutional stages: initiation, growth, maturation, decline and death stages. Among them the maturation one is emphasized so as to find the occurring or thriving date. An once-through cycle of land subsidence may be fitted and predicted well by developed, mathematic model of Poisson cycle. Due to back-analysis of landslides happened already being studied well with the same model, (Yan, 1989), so positive analysis of subsidences may be also researched by the principle of Poisson cycle. Basic assumption conditions for the trend forecast are as follows:

1) life-span of once-through cycle of a land subsidence is limited, and theoretically it may be fitted well with lower, absolute mean error rate by Poisson cycle;

2) future trend subsidence is determinated and forecasted on the basis of monitored, dynamic data series;

3) forecasted results must be controlled by subsided data monitored which reflect all effective factors acted upon subsidence.

For stage or trend forecast it is necessary to build the typical monitoring points, lines or even nets well. The monitored data may be of the velocity vectors or the displacements with horizontal or vertical components. If the horizontal components are zero or neglected, then a land slide event may be substantially transformed into a land subsidence. The monitored data series of the subsidence are the complicated reflections of its particla mechanism.

2 FUNDAMENTALS-POISSON CYCLE

Principle of Poisson cycle is originated from Poisson distribution which is one approached method for binomial distribution.Its theoretic basis is the Binomial theorem.However,both of the distributions are stochastic,dispersion methematic models.By means of theoretic derivation the determinated model may be derived

$$y_t = At^n e^{-t} \quad \text{(Weng,1984)}$$

By successive derivation it results

$$y_t' = \frac{dy_t}{dt} = y_t(\frac{n}{t} - 1) \quad \text{and}$$

$$y_t'' = y_t \frac{1}{t^2}((t-n)^2 - n); \text{let } t<n, \text{then } y_t' > 0;$$

$t=n$, $y_t'=0$; $t>n$, $y_t<0$; and let $y_t''=0$,then

$t=n+\sqrt{n}$.By limited integration $y_t dt = An! = \Sigma y_t$,

then $y_t/\Sigma y_t = t^n e^{-t}/n!$.From above characteristics it is known that the thriving and decline proccesses of an event may be to divide five stages:I.accelerating stage,t=0 to$(n-\sqrt{n})$;II.general ascending stage,t=$n-\sqrt{n}$ to n;III.top stage,t=n;IV.general descending stage,t=n to $(n+\sqrt{n})$;V.rather slow descending stage,t=$(n+\sqrt{n})$ to∞.These mean that a developing process of an event of a limited system y with time t is thriving directly propotional to t^n and decaying inversely propotional to e^{-t},then the model curve is firstly divergent and secondly convergent, and capably used to describe the life-span of limited system.The macro-prediction model of basic value of the products of world natural gas is

$$y = b + at^n e^{-t}, \quad (1)$$

$$n = ((t_2 - t_1) + Ln\frac{y_{t2}}{y_{t1}})/Ln(\frac{t_2}{t_1}) \quad (2)$$

3.TREND FORECAST OF LAND SUBSIDENCE—TWO EXAMPLES

3.1 Xi'an city subsidence

The typical district of the subsidence is situated at Xiaozhai,South Suburb of Xi'an City.It is also the location of ground fissure zone of Xiaozhai-Guanyinmiao with NEE strike.The origin of ground fissures may be regarded as topographic feature of recent active,tectonic faults under the ground (Zhu et al,198.) within Weihe Graben

up(7).According to the literatures and their figures showing echelon fissures,it is clear that the active feature in plan must be belonging to anti-clock-wise type of shearing-gravity faults with dip direction to SSE.This is just the opposite case of base fault systems within Weihe Graben with dip to NNW and,somewhat alike the fault systems of the rear part of a graben-horst landslide where the sliding plane is the case of Lintong-Chang'an gravity fault of North slope of Qinling Mt..It is considered that main subsidence component may be caused by the wide pumping of ground water in deeper aquifer composed of silt sand beds intercalated with fossil soil beds as aquifuge and covered with thin loess of Quaternary System,while the minor subsidence is due to the relative,vertical displacement caused by ground fissures.The supperposed sum velocity of the both subsidence of Xi'an is faster than that of either Shanghai or Tianjin City.Until now it is hardly to determine that how fast of both velocities are respectively in Xi'an City.However,as a whole, we may have to do a trend forecast for Xiaozhai district,Xi'an City.

From monitored,subsided data before 1984 listed in Table 1 and by formulae (1) and (2) we may gain n=5.13,a=0.4078,b=6.0921. Thus we may build a forecasting equation as

$$v = 6.0921 + 0.4078t^{5.13}e^{-t} \quad (3)$$

By equation (3) we gain the fitted (t=0-3) and predicted (t=4-21) data series listed in Table 1.

From equation (3) and Table 1 the evolution stages of an once-through cycle of the subsidence,Xi'an may be divided as

I.Accelerated stage,t=0-$(n-\sqrt{n})$=0-2.865, i.e.from 1972-1983.5(a)

II.general ascending stage,t=$(n-\sqrt{n})$-n=2.865 -5.13,i.e. from 1983.5-1992.5(a)

III.top stage,t=5.13,i.e.the mid-year of 1992

IV.general descending stage,t=n-$(n+\sqrt{n})$= 5.13-7.395,i.e.from 1992.5-2001.6(a)

V.rather slow descending stage,t=$(n+\sqrt{n})$- ∞=7.395-21,i.e. from 2001.6-2056 (a) or even more.

The predicted results show that the peak value of subsiding velocity will occur at 1992 to 1996 ; decreasing velocity of subsidence may initiate at 2001 ; the least velocity (v = 0.0018 cm/a) may be to start at 2050 when the hazard of subsidence will end the threat for urban construction engineering. By data series listed in Table 1 the Figure 1 may be plotted to show the trend forecast of developing stages of the subsidence from an once-through cycle.

Table 1 Data series predicted for the subsidence of Xiaozhai district,Xi'an City

Per 4-year	1968	1972	1976	1980	1984	1988	1992	1996	2000	2004	...2056...
No.of time series monitored t_0	1	2	3	4	5						
Velocity of subsidence,v_0(cm/a)	3.0	3.65	6.46	11.0	12.0·						
Predicted v(cm/a)		6.09	6.24	8.03	11.78	15.25	16.75	16.01	14.14	11.96	...6.09...
No.of time t series	0	1	2	3	4	5	6	7	8		...21 ...
Residual differences (v_0-v)		-2.44	0.22	2.98	0.22						
Error rate= (v_0-v)/v_0		-0.67	0.03	0.27	0.02						

Remark:Totol absolute mean error rate $|\overline{ER}|$= 8.675 %

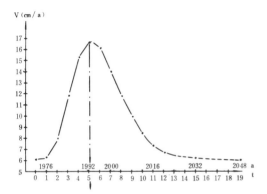

Fig. 1 Trend forecast of the subsiding velocity of Xiaozhai,Xi'an City(downmost arrow showing maturation stage of the subsidence.)

3.2 Ningbo City subsidence

The mining of ground water initiated from the beginning of 1950's.The land subsidence was discovered at the beginning of 1960's. Nowadays,the mining amount approaches to $10*10^6$ m³/a and the area of subsidence spreads out about 160 km². Ningbo City situates at the central part of Ningbo Basin with mountains around but the north part nearby the sea. The mean elevation of the land is about 2.1 m above sea level.The basin is made of tectonic fracture-graben of Late Creataneous Period.The loose sediments are composed of littoral and fluviolacustrine facies of Pleistocene and Holocene series.They belong to sedimental mode of littoral-continental alternations.The main sediments are composed of marine clay and sand-gravel and the later one is confined aquifer.The radius of descending funnel of ground water level is 12-15 km which reaches the basin boundaries. The subsiding depression coincides with the form of descending funnel of ground water level.The most thickness sum of Quaternary

strata approaches 100 m.Based on the monitored data performed by researchers (Shen, et.al 1989) the paper builds forecasting equation as follows

$$v=2.1414+0.0778t^5e^{-t} \qquad (4)$$

By equation (4) the calculated,predicted velocity data series are listed in Table 2. From the v data series of Table 2 we obtain Fig. 2 .

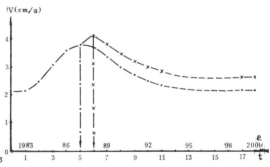

Fig.2 Trend forecast of the subsiding velocity of Ningbo City (downmost arrow showing maturation stage of the subsidence cycle,.normal predicted point;x predicted higher point superposed by 2Sp)

Table 2 and Fig. 2 denote that the maturation stage is reached at 1987 to 1989; while the death stage will be approached at 2050 with subsiding velocity (2.616+ $3.2253*10^{-16}$) (cm/a) . The evoluting stages of an once-through cycle of the subsidence , Ningbo may be divided by n value as being done for Xi'an City.

Table 2 Data series predicted for the subsidence of Ningbo City

Year	1981	1982	1983	1984	1985	1986	1987	1988	1989	1990	1991	1992	1993...2000...
No.of time series monitored t_0	1	2	3	4	5	6	7						
Subsided velocity v_0 (cm/a)	1,65	1.91	2.17	2.83	3.00	3.60	4.00						
No.of time series t	0	1	2	3	4	5	6	7	8	9	10	11...18...	
Subsiding velocity predicted v (cm/a)		2.14	2.17	2.48	3.08	3.60	3.79	3.64 ▲4.12	3.33 3.81	3.00 3.48	2.71 3.18	2.50 2.98	2.35...2.14... 2.83...2.62▲...
Residual differences (v_0-v)		-0.23	0	0.35	-0.08	0	0.21						
Error rate= $(v_0-v)/v_0$		-0.12	0	0.12	-0.027	0	0.053	Total mean $|\overline{ER}|$= 0.5 %					

Remark: 1. Residual standard dispersion $Sp=\{((-0.23)^2+0+(0.35)^2+(-0.08)^2+0+(0.21)^2)/(6-2))\}^{1/2}$
=0.2376; 2Sp=+0.4753;

2. Data series with ▲ are gained by adding 2Sp=0.4753

4 CONCLUSION

Land subsidence is a special case of landslide of which the horizontal displacement component is zero or neglected.The studied results of positive analysis of two urban subsidence by Poisson cycle fundamentals are basically identical either with that of verhulst'biological mathematic model (Ningbo City) or with the actual trend (Xi'an City).The fitted error rates of the two examples are absolutely 0.5 to 8.7 %; The further corollary should be suggested that the processes of limited system other than slide and subsidence such as reservoir seisms etc.might be predicted by the same fundamentals.

REFERENCES

Shen,X.Y.et.al.1989,Physical mathematic model and prediction on subsidence of Ningbo City. Earth Science-Journal of China University of Geoscience,Vol.14,No.2 p.135-144.

Weng,W.B.1984, Fundamentals of forecasting theory (in Chinese), Petroleum Industry Press. P.77-86.

Yan,T.Z.1989,Stage predictions of landslide and subsidence from an once-through cycle. Earth Science, foreign language edition. No.1. (in print)

Yan,T.Z.1989New mode for occurring time forecast of landslide hazards.Progress in geosciences of China (1985-1988). Papers to 28th IGC,Vol.1,Geological Publishing House, Beijing, China. P.137-140.

Yan,T.Z.1989, New mode of forecasting land subsidence regularity (in Chinese). Earth Science, Vol.14, No. 2

Zhu M.R.et.al.1986, Some environmental engineering geological problem encountered in the urban construction of the city of Xi'an. Proc. of 5th IAEG Vol. 3 P.6,7,7,--1967--1975.

6th International IAEG Congres / 6ème Congrès International de AIGI, © 1990 Balkema, Rotterdam. ISBN 90 6191 130 3

The problem of subsidence caused by coal cutting under the flood-bank

Problème de l'abaissement de la digue contre les inondations par l'extraction de la houille

H.C. Zhang

Hydraulic Engineering Administration of Yi-Shu-Si Rivers, Committee of Huaihe River, Ministry of Water Conservancy, People's Republic of China

ABSTRACT: Coal cuttings under the flood-banks of important rivers and lakes are related to the safety of flood prevention. Based on the test data and engineering records of coal cuttings under the Weishan Lake flood-bank in the east part of China, the subsiding lengths of the flood-banks in coal cuttings are estimated in this paper with the method of: (1) "Cut and Trial Method"; (2) "Geotechnic Mechanics Method"; (3) "Analysis of Structural Mechanics Model Method". A preliminary investigation on the mechanism of subsidence and crack are also conducted in this article. The author hopes that all of his works will be beneficial to the research of safety of flood prevention for coal cuttings under banks.

RESUME: On extrait de la houille sours les importantes digues contre les inotations, le fait concerne le probleme de la securite de la prevention des crues. Le present article, d'apres les documents d'essai et les ouvrates reels interesses de l'extraction de la houille sours la digue contre les inotations de larc de Wei Shan, a l'est de la Chine, en utilisant: (1) la methode de l'estraction d'essai; (2) la methode de la mecanique des sols et des roches; (3) la methode analytique de modele de la dynamique structurelle, a fait le calcul quantitatif de longueur de l'abaissement de la digue provoque par l'extraction et on a fait aussi la recherche preliminaire sur la mecanque produite de l'abaissement et de la faille. L'auteur espere que tout ce travail favoriserait la recherche sur le probleme de la securite de prevention des crues de l'extraction de la houille sous les digues.

Currently, coal cuttings under buildings, under railways and under water bodies (referred to as "three-under" for short) are developing rapidly both in China and at abroad and many successful experiences have been accumulated and very big advance has been acquired in the field of theoretical research. However, coal cuttings under flood protection banks of major rivers and lakes, especially under thick, loose and soft overlying stratum and relatively thinner roof rocks, coal cuttings by wholly stope-caving method of large span which will induce subsidence and crack and etc. to the flood protection banks, still have to be studied by gathering more data. The author investigated this topic by taking the coal cutting under flood protection bank of Weishan Lake in east China as an example.

1 HARM TO FLOOD PROTECTION FROM SUBSIDENCE CAUSED BY COAL CUTTINGS UNDER FLOOD BANK

The flood protection bank of Weishan Lake is a portion of flood protection bank at the west of Nan Si Lake. The said flood protection bank at the west of Nan Si Lake was built after the flood of the year 1957, with a total length of 131 km., which can control flood up to 5×10^9 cubic metres and protect farm land of 3.6 million Mu and inhabitants there of more than 3 million people. The bank is also protecting the safety from flood of two main railway lines, Long Hai railway and Jin Pu railway, and the industrial and mining districts of Xuzhou, Jining and Zaozhuang three cities. Certain mine conducted coal cutting under the bank at the west of Weishan Lake (stake no. 76K+436 -77K+076) to the depth of 230 - 460m. The thickness of coal seam being cut was 4 m. The situations of the flood protection bank of Weishan Lake above the stoped out workings (stake No. 75K+950- 77K+520) within the distance of 1533 m. were: in January of the year 1985 the maximum subsidence of bank above stope

face 7115 was 1.5 m and the length of
subsidence was 476 m. The subsided
portion of the bank had been improved by
raising higher before the flood of that
year. Before the flood of the year 1988
several lateral tensile cracks were found
on both ends of the bank above 4 stope
faces. Those cracks were caused by
subsidence which was induced by coal
cuttings. Those lateral cracks were in
rather large scale and extended relatively
far, across the bank from the top of the
bank to its foot, generally in the length
of 30 m up to 50 m and the longest one
was around 100 m which ran through the
bank from the beach behind the river to
the beach facing the river and then led
to the side of Jin Hang canal in Weishan
Lake. The width of the top opening of
the crack was 15-20 cm and the depth of
the crack was more than 5 m.

The subsidence of bank caused by coal
cuttings and the lateral crack arisen
therefrom were very disadvantageous to
the safety of flood protection. The
safety water level for the upper lake of
Nan Si Lake was in the elevation of 36.50
m and that of the lower lake, i.e.
Weishan Lake was in the elevation of
36.00 m. When the water level of flood
approached the safety water level of
flood protection, the flood may possibly
have concentrated seepage and seepage
flush along the lateral crack which ran
across the bank, or the bank above the
stoped out workings might subside
suddenly in large quantity which may
induce burst or collapse of the bank and
cause flood disasters. Therefore, the
study of subsidence of flood protection
bank caused by coal cuttings is very
significant.

2 STRATUM GEOLOGY AND SITUATION OF THE
STOPE OF COAL SEAM

The coal seam stoped by certain mine
under the flood bank of Weishan Lake was
No.7 coal seam of Shanxi group of Lower
Permian. The occurrence of rock layer
(coal seam) was N70°E, NW < 15°. Its
lithology was interlay of shallow sea
facies lime stone and shale stone. The
embedded depth of coal seam was 230 -
420 m. The thickness of Quaternary
alluvial lacustrine formation was 80 -
270 m. The thickness of No.7 coal seam
just under cutting at that time was 5 m.
The miner adopted "Method of Strike Long
Wall Wholly Caving" for stoping. The
stope direction was orthogonal to the
bank. The 4 stope faces under the bank
in the sequence of from south to north

were 7114, 7115, 7117 and 7119. The
width of each stope area, i.e. stope
face, was 150 m. Stope pillars of 10 m
were reserved between two stope faces.
No.7 coal seam had been stoped for the
depth of 4 m. A stoped out workings with
span of 640 m along the direction of the
bank had been formed (please refer to
Fig.1 for details).

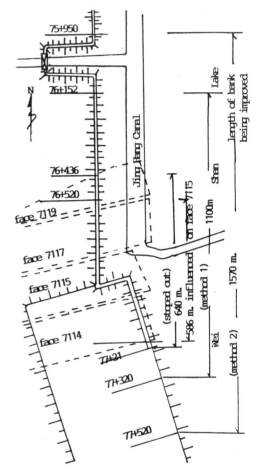

Fig.1 Coal-cutting positions under bank of
Weishan Lake

3 MECHANICS MECHANISM OF SUBSIDENCE AND
CRACK OF BANK CAUSED BY COAL CUTTING
UNDER IT

The subsidence of flood protection bank
due to coal cutting was the results of
deformation and rupture of rock and soil
bodies under the bank but above the roof
of the stoped out workings. Under the
action of sole weight stress, the rock
and soil bodies above the roof of the

2700

stoped out workings shifted downward toward the stoped out workings, then stress concentrated zones emerged at the upper corners of both the right side and the left side of the stoped out workings. The reaction of supports at the two corners (as shown in Fig.2) were indicated by the following expression:

$$R_A = R_B = \frac{1}{2} \gamma ZL \quad \ldots \ldots \ldots \quad (1)$$

where R_A, R_B are reaction of supports at point A and B respectively, KPa;

γ: unit weight of rock and soil, KN/m^3;
Z: depth of stope, m;
L: span of stope, m.

From the above expression it was learnt that the vertical concentrated stress at points A and B were $\frac{1}{2}L$ fold of sole weight stress at points not stoped yet of the same depth. If the depth of stope under the bank of that mine was taken as 420 m and its span to be 640 m, then the vertical concentrated stress at points A and B would be 320 folds of the sole weight stress at the point of the same depth before being stoped, that is,

$$\sigma_A = \sigma_B = 2.42 \times 10^3 \text{ MPa}$$

Points A and B at both ends of the stoped out workings were under such a big pressure, initial shear damage would be occurred to the rock, then the rock at the upper part of both ends of the stope out workings would trace the shear damage surface of points A and B and would expand upwards. The maximum principal stress also changed from vertical to horizontal because of the downward shift of rock above the roof of stoped out workings toward the stoped out workings. Tensile crack of certain depth would be happened on the surface of rock at the upper part. That was the process of subsidence and deformation of rock on the roof of stoped out workings.

The overlying soil layer of stoped out workings was situated below the water level of the lake and it was saturated soil. Dragged by the movement of underlying rock toward stoped out workings and the action of self weight, this kind of saturated soil would also have subsidence and deformation. Area of draw cracks would emerge at both ends of the subsidence area. The deformation of saturated soil by force was elasto-plastic deformation [1]. Under the condition that the level of stress maintained unchanged and deformation was going to occur continuously, then creep behaviour appeared. The consequence of creep was that the shear angle, or subsidence angle, of saturated soil became smaller and smaller. Reaction on the ground surface was the influence of stope, or to say, the length of subsidence of the bank, became longer and longer (please refer to Fig.3). This is the mechanics mechanism of subsidence and crack of the flood protection bank induced by underground coal cuttings. In Fig.3 it showed that generally the subsidence of the flood protection bank right above the stoped out workings was the maximum one, and it decreased gradually towards both ends in a symmetrical way and formed a "sink basin". It was just like an upward open parabola along the topographic line in the direction of axis of the bank. The deformation and rupture of rock and soil bodies overlying the stoped out workings were apparently in multiple zones, i.e. collapse zone of rock at the bottom, crack zone of rock surrounding outside the inverted cone of collapse zone, then the overlying rock and soil layers were bending sink zone. Draw crack (lateral crack of the bank) of certain depth appeared at both ends of bending sink basin of the surface of rock and soil layers. What the administrative department of hydraulic engineering concerned most were the length of subsidence of the bank and the scale of lateral crack.

4 CALCULATING THE LENGTH OF SUBSIDENCE OF FLOOD PROTECTION BANK DUE TO COAL CUTTINGS

The flood protection bank subsided due to the influence of coal cuttings under the bank. The subsiding length may be calculated by "Cut and Trial Method", "Geotechnic Mechanics Method", and "Analysis of Structural Mechanics Model Method".

(1) Cut and Trial Method

Generally, before the mine district be started overall cutting formally, one or several representative stope faces should be selected to carry out trial cuttings and to measure relevant data to be used as bases of calculation for the range of ground subsidence in cutting design [2]. The following formula may be used for calculation:

$$S = Z_1 ctg \gamma + Z_2 ctg \beta + 2Z_3 ctg \phi + L + S_k \quad \ldots \quad (2)$$

2701

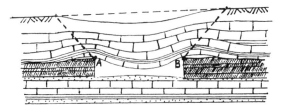

Fig.2 schematic drawing of rock and soil deformations above stoped out workings

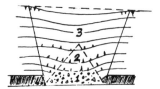

Fig.3 deformation rupture of rock above stopedout workings
1. collapse zone
2. crack zone
3. bending zone

where
S-subsiding length of flood-bank due to coal cutting, m;
Z_1-thickness of overlying rock at the rising end of stoped out workings, m;
Z_2- thickness of overlying rock at the descending end of stoped out workings, m;
Z_3- thickness of overlying soil on overlying rock of stoped out workings, m;
r-travel angle of overlying rock at the rising end of stoped out workings, degree;
β-travel angle of overlying rock at the descending end of stoped out workings, degree;
L-span of stoped out workings in the direction of the bank, m;
S_k-additional distance for safety protection, m.

(2) Geotechnic mechanics methods

According to Coulomb's strength theory, the vertical stress at both ends of the roof rock of stoped out workings were concentrated, similar to uniaxial compression test that the plane of rupture and the maximum principal stress, or the vertical stress plane were in the relationship of $\beta = 45°+\frac{1}{2}\phi$. Shear failure happened on the wall of rocks in the stoped out workings, which shifted downwards under the gravitational force of the overlying rock and expanded upwards following the shear crack plane $\beta=45°+\frac{1}{2}\phi$ [3]. Bending deformation was emerged on upper rock. Draw crack of certain depth would appear at both ends and the depth could be calculated by the following expression:

$$Z_o = (2C/r)tg(45°+\frac{1}{2}\phi)$$

Based on the analyses mentioned above, the length of subsidence of bank caused by coal cuttings may be calculated by taking $\beta=45°+\frac{1}{2}\phi$ as subsiding angle, or breaking angle, of rock and take internal friction angle of saturated soil ϕ_2 as travel angle of soil layer. The expression for calculation is:

$$S=(Z_1+Z_2)ctg(45°+\frac{1}{2}\phi_1)+Z_3ctg\phi_2+L \quad (3)$$

where ϕ_1-internal friction angle of shear resistance of overlying rock in stoped out workings (deg.); ϕ_2-internal friction angle of total stress of saturated consolidated undrained shear of overlying soil body in stoped out workings, degree; all the other symbols are the same as those used in formula (2).

(3) Analysis of structural mechanics model method

The key point of this method is that pursuant to the stratification of roof rock in stoped out workings and the splitting of rock mass by joints, fissures or faults, the roof rock in stoped out workings is considered as a multi-layered superposed beam of synchronous deformation and the first layer of rock mass above the roof is to be considered as a simple beam under even load from above. Such a mechanics model is set up to facilitate the study of the deformation of rock and soil mass on the roof in stoped out workings.

If the dip angle of rock above the roof in the stoped out workings is less than 20°, this kind of slowly inclined rock might be treated approximately as horizontal rock. To treat rock of certain thickness (such as with the thickness of 1 m) on the roof of stoped out workings as an beam placed on both ends of the stoped out workings and the pressures of rock and soil on the beam are distributed uniformly. As initial bending emerged from the roof, ends A and B would have the action of bearing and hinging (refer to Fig.2). So the deformation rupture of rock above the stoped out workings might be simplified and to be studied as the deformation

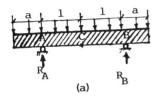

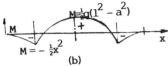

Fig.4 Structural Mechanics Model
(a) forces on beam
(b) bending moment of beam

rupture of a simple beam with equal cantilevers at both ends and being loaded evenly [4] (refer to Fig.4).

Now the deformations of rock and soil above the stoped out workings are calculated. From Fig.4 the bending moment between points A and B is obtained as:

$$M = \frac{1}{2}\gamma ZB(1^2 - a^2) \quad \ldots \ldots \quad (4)$$

where
1 - half of the span of the stoped out workings, (m);
a - length of cantilever beam beyond points A and B, (m);
γ - unit weight of rock and soil, (KN/m^3);
Z - thickness of rock and soil layers on beam, (m);
B - width of beam used for calculation, (m).

In the above expression, when a=1, the bending moment at the centre of the beam M=0, i.e. there will be no bending deformation of the beam because the actions of uniformly distributed loadings inside points A and B are offset by the actions of uniformly distributed loadings outside points A and B.

While a=0, the bending moment at the centre of the simple beam is of the maximum value, i.e. $M_{max} = \frac{1}{2}\gamma ZB1^2$.

From the aforesaid analysis it is learnt that the beam will have bending deformation when $0 < a < 1$, which means theoretically rock outside of points A and B with distances equal to or longer than half of the span, 1, there will be no bending deformation occurred. Hence the length of deformation (S) of the

entire beam of double support and hinging and equal cantilevers is $21 < S < 41$. As $1 = \frac{1}{2}L$, so $L < S < 2L$. This expression shows the total length of cutting influence caused by the fact that the roof rock in stoped out workings sunk and generated bending deformation and then the rocks outside the two end points of the stoped out workings were affected and generated bending deformation. Considering the bending of rock will draw the overlying saturated soil to be sunk or slided, then the length of cutting influence in soil will be longer than the length of cutting influence in rock. Therefore the subsidence of flood protection bank caused by coal cutting and the length of crack influence can generally be calculated by the following expression (when the depth of cutting is less than or equals to the span of stope area):

$$L < S < (2 - 2.5)L \quad \ldots \ldots \quad (5)$$

Three methods of calculation are mainly described above. Moreover, there are the Engineering Comparison Method, Probabilistic Integral Method, etc. which are also often used. The subsidence of the bank of Weishan Lake caused by coal cutting (cutting influence) was calculated by using the aforesaid 5 methods with their results results listed in Table 1.

5 CUTTING TEST AND VERIFICATION OF MONITORING IN PRODUCTION

As to how are the values in application of those methods of calculation mentioned above should be verified by cutting tests and monitorings on the spots during production.

(1) Verified by cutting tests

For the purpose of obtaining experience of coal cutting under flood protection bank and to provide necessary data for design of coal cutting, certain mine conducted trial cutting from October of 1980 till August of 1981 on coalface 7112 (not a stope area under flood protection bank). The observed data of June, 1982 are listed in Table 2. Its related subsidence and deformation curves on rock and soil masses are plotted in Fig.5.

It is learnt from verification of trial cuttings that during trial cutting the length of subsidence of ground surface S was near to the calculation results by using methods (1) to (4) mentioned in Table 1, and the value calculated by method (5) was on the small side, and the

Table 1. Calculated figures for extent of influence induced by coal cuttings in certain mine under the bank at the west of Weishan Lake

method used	formulas used for calculation	span of stoped out workings L (m)	thickness of cutting H (m)	depth of cutting (m) rise Z_1+Z_3	depth of cutting (m) descend Z_2+Z_3	thickness of soil Z_3 (m)	travel angle of rock (°) rise r	travel angle of rock (°) descend B	travel angle of soil (°)	extent of cutting influence, S	maximum subsidence of bank roof, W_0 (m)	S/L
(1) cut and trial method	$S=Z_1ctg\gamma + Z_2ctg\beta + Z_3ctg\phi +L+S_k$	640	4	230	420	150	75	59	38	1100	3.5	1.72
(2) geotecnic mechanics method	$S=(Z_1+Z_2)ctg(45°+\frac{1}{2}\phi)+Z_3ctg\phi_2+L$	"	"	"	"	"	58	58	24	1533	(0.7H)	2.40
(3) structural mechanics model method	$M=\frac{1}{2}\,rzB(1^2-a^2)$ $(0\leqq a\leqq1)$	"	"	"	"	"	"	"	"	$1.5L\leqq S\leqq(2-2.5)L$ $1280\leqq S\leqq1600$		2-2.5
(4) engineering comparison method	same with method (1)	"	"	"	"	"	70	58	35	1306		2.04
(5) probabilistic integral method	$W(x)=\dfrac{W_0}{r}\displaystyle\int_0^\infty e^{-(x/r)^2}\,dx$	"	"	"	"	"	"	"	"	980		1.53

2704

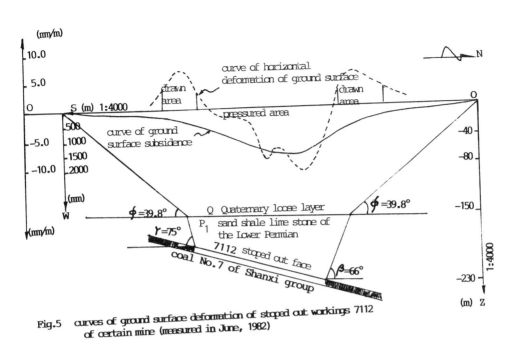

Fig.5 curves of ground surface deformation of stoped out workings 7112
of certain mine (measured in June, 1982)

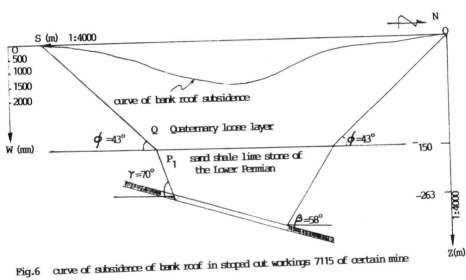

Fig.6 curve of subsidence of bank roof in stoped out workings 7115 of certain mine

Table 2. results of measurements on site and graphic calculation for ground surface deformation in trial cutting of stope area 7112

stope	L (m)	H (m)	depth of cutting (m) Z_1+Z_3	Z_2+Z_3	thickness of soil Z_3 (m)	travel angle of rock γ	β	travel angle of soil ϕ	maximum subsidence W_{max}(mm)	maximum horizontal deformation mm/m	S (m)	S/L
7112	220	2	190	230	150	75°	66°	39.8°	1600	+6 / −11	627	2.85

Table 3. Results of measurements and graphic calculation for deformation of bank roof in stope area 7115

stope area	span of stope L (m)	thick-ness of stope H (m)	depth of cutting (m)		thick-ness of soil Z_3 (m)	travel angle of rock (°)		travel angle of soil (°)	maximum subsi-dence W_{max} (mm)	extent of cutting influence S (m)	S/L
			Z_1+Z_3	Z_2+Z_3		γ	β				
7115	160	2	216	263	150	70	58	43	1500	586	3.66

value acquired by method (2) was on the safe side.

(2) Verified by field measurement of production in stope area

Coalface 7115 of certain mine was started to be cut in September of 1984 and ran through the flood protection bank of Weishan Lake and cut further inwards for 15 m towards the Lake. The bank subsided abruptly in January of 1985 (it was just within the active season of subsidence at that time). The portion of bank above the stoped out workings produced serious torsion and subsided then. Pavement on the bank top tilted towards the Lake about 30°, wire poles were slant, the beach was full of water. According to measurement the cutting thickness, or stope empty, was 2 metres, which induced the bank to be subsided for 1.5 metres and the length of influence of subsidence was 476 metres. In order to guarantee the safety of flood protection it was decided instantly at that time that the improvement of the bank should be carried out by putting on earth for the length of 696 metres [the direct linear distance was 586 metres (see Table 3)].

Fig.6 is the curve of subsidence of coalface 7115 by actual measurement.

The actually measured data and engineer-ing records stated above show that the length of influence of subsidence S is comparatively near to the results calculated with method (1) to method (4), the result calculated with method (5) is on the small side and the result calculat-ed with method (2) is on the safe side..

(3) Results of verification

Results of cutting tests and production monitorings show that the former three methods (those methods are the key topics of description in this paper) are of great value in application. The length of cutting influence (subsidence and crack) is:

$$1.5L \leqslant S \leqslant (2 - 2.5)L$$

6 CONCLUSIONS

(1) Coal cuttings under flood protection banks of important rivers and lakes are related to the safety of flood protection. Coal-cutting departments should bear the responsibility and duty of providing detailed plan of mining and program of exploitation to hydraulic engineering administrative departments and to submit information regularly on the situation of exploitation and subsidence monitoring.

(2) Measures of tunnel through the bank are recommended for stoping under water body for the projects of stoping under rivers and lakes as ratified by govern-ment authorities concerned, to minimize the stoping under the bank to the utmost. If it is really necessary to stope under flood protection bank, then safety measures should be taken. The cutting speed must be controlled to reduce abrupt subsidence and to protect the integrity of the bank. Stoping under the flood protection banks of important rivers and lakes should be stoped 3 months prior to the flood or during the flood season to avoid flood disasters induced by the abrupt subsidence of flood protection bank and lateral crack occurred.

(3) To strengthen the monitoring of deformation of the portion of flood protection bank which is affected by coal cuttings and to measure regularly the vertical displacement and horizontal displacement in order to control the development of subsidence and deformation of the bank

(4) It is necessary to predict the length of subsidence of the bank caused by coal cutting when it is conducted under flood protection banks of important rivers and lakes, in order to set up the improvement plan of the bank. Large span and wholly stope-caving method are adopted for coal cuttings in thick, loose soil layer and relatively thinner roof rock, when (depth of cutting) $Z \leqslant L$ (span of cutting), then the length of subsidence of flood protection bank caused by coal cutting could be determined within the range of $1.5L \leqslant S \leqslant (2-2.5)L$.

(5) "Geotechnic Mechanics Method" and "Analysis Method of Structural Mechanics Model" for the calculation of length of subsidence of flood protection bank caused by coal cuttings were only the author's preliminary research outcomes which are subject to further investigation whether they are well in conformity with the real situations of coal cuttings under the protection bank.

REFERENCES

[1] H.C. Zhang, "Research on seismic thixotropyin muck base", "Proceedings fifth international congress international association of engineering geology", A.A. Balkema/Rotterdam, Netherlands .Boston, U.S.A., 1986

[2] Z.Y. Wu, et al., "Handbook for Coal Minw Swaifn", Coal Industry Press, 1984, pp. 692-729

[3] H.C.Zhang,"Environmental condition and evaluation of high slope stability of Jiangyin nuclear power .,station", "Proceedings of the international symposium symposium on engineering geological environment in mountainous areas", Institute of Geology, Academia Sinica, Beijing, China, May 1987, pp.773-782

[4] D.F. Kers (Canada),Basic Mechanics of Rock (translated into Chinese by H.N. Lei et al., Publish House of Metallurgy Industry, 1978, pp.166-168

5.4 Underground storage of energy, liquids and waste
   Stockage souterrain de l'énergie, des liquides et des déchets

Engineering geological modelling of the underground waste disposal site in Loviisa, Finland

Modélisation de géologie de l'ingénieur du site de stockage souterrain de résidus à Loviisa en Finlande

P.T. Anttila
Imatran Voima Oy, Finland

ABSTRACT: Site investigations have been made in the Loviisa power plant area for assessing the suitability of the bedrock for the final disposal of low and medium-level nuclear wastes. The study area covers a small island on the coast of the Baltic Sea in southern Finland.

The bedrock consists of coarse grained Precambrian rapakivi granite. Typical for granitic rocks, two vertical and one horizontal fracture sets are present, the last mentioned of them being dominant.

A three-dimensional engineering geological model has been prepared by using computer aided design (CAD). The model covers a bedrock block, which is some 400 x 600 m wide and 200 m thick. Three gently dipping fracture zones divide the rock mass into zones of intact and broken rock. The model forms the basis for the groundwater flow modelling and siting of the repository.

RESUME: Les études géologiques ont été effecturées dans la région de la centrale nucléaire de Loviisa dans le but d'évaluer l'adaptation du socle rocheux pour le stockage final des résidus nucléaires à faible et à moyenne radioactivité. Le terrain de recherches couvre une petite île située sur la côte de la mer Baltique dans la Finlande méridionale.

Le socle rocheux consiste en du granite de rapakivi Précambrien. Typiquement aux roches granitiques, elle comprend deux séries de fissures verticales et une série horizontale. La série horizontale est dominante.

Une modélisation géologique tridimensionnelle a été élaborée avec l'aide de moyens informatiques (CAD). Le modèle couvre un bloc du socle rocheaux d'environ 400 x 600 m de côté sur 200 m d'épaisseur. Trois zones de fissures à inclinaison légère partagent la masse rocheuse en zones de roche intacte et de roche cassèe. Le modèle forme la base de la modélisation de la circulation des eaux souterraines et de l'aménagement du stockage.

1 INTRODUCTION

In Finland there are four nuclear power reactors in operation. Imatran Voima Oy (IVO) operates two 445 MWe units in Loviisa and the Industrial Power Company Ltd. (TVO) two 710 MWe units in Olkiluoto. These power companies are legally responsible for the conditioning of the nuclear waste, including its disposal.

Geological conditions of Finland favour the disposal of nuclear wastes deep into the hard Precambrian bedrock. The power companies have considered it most appropriate to construct separate repositories for their own plants for low and medium-level reactor wastes. The construction of the Olkiluoto repository started in 1988; as to Loviisa, the construction decision has not yet been made. TVO is preparing to dispose of the

spent fuel, too, in the Finnish bedrock. The spent fuel of the Loviisa power plant is returned to the Soviet Union.

This paper deals with the results of engineering geological studies and modelling of the bedrock in the Loviisa site. It is located on a small island of Hästholmen on the southern coast of Finland some 80 km east of Helsinki (Figs.1 and 2).

2 SITE INVESTIGATIONS

The main object of the investigations has been to clarify the bedrock conditions of the planned disposal site in order to assess the safety of the disposal. They also form the basis for the siting and rock mechanical design of the repository. The focus of the studies have been the most cru-

Fig. 1. Location map.

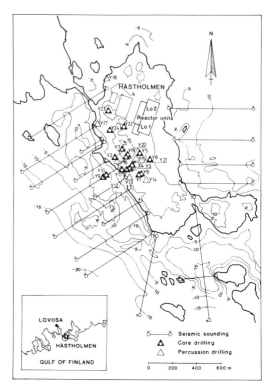

Fig. 2. Investigation map.

cial properties considering groundwater flow, i.e. fracturing and hydraulic conductivity of the bedrock. Investigations have discussed previously by e.g. Gardemeister & Rouhiainen (1984) and Anttila (1986, 1988).

Site investigations, carried out during 1980-88 have comprised the following activities:
- geological mapping of outcrops; focused especially on fractures
- exploratory drillings; mainly diamond core drilling (24 boreholes) with core logging
- geophysical surafce and borehole measurements; especially single-hole methods like dipmeter and in situ state of stress as well as seismic and electrical cross-hole methods have been very essential in structural modelling
- hydraulic measurements; permeability tests, pressure and tracer tests
- groundwater studies; GW-table monitoring, water sampling and chemical analyses
- laboratory tests; mainly petrological and rock mechanical studies

The location of the boreholes and the investigation lines are presented in Fig.2.

3 INVESTIGATION RESULTS

3.1 Bedrock

The topography of the rock surafce is gently undulating, being on the island generally below +10. In the sea area there are depressions, filled by glacial till and clay, where the maximum depth to the bedrock is some 70 m from the sea level.

The study site and its vicinity belong to the western part of Precambrian Wiborg rapakivi massif. Rapakivi granite is an anorogenic igneous rock, which intruded into the older bedrock in three phases c. 1,700 - 1,640 Ma ago (Simonen 1980). The rock type of the island has several varieties, which occur as gently sloping lenses and layers. The most common rapakivi type is pyterlite, which has typical rapakivi texture i.e. round potash feldspar ovoids (up to 80 mm in diameter) surrounded by a plagioclase mantle in a fine to medium-grained ground mass.

Rock material is generally fresh and strong. The uniaxial compressive strength of pyterlite averages 172 MPa and the tensile strength 11.8 MPa. The elasticity parameters, static Young's moduls is on an average 64.5 GPa and Poisson's ratio 0.19. The strength values are somewhat lower than the mean value for granite, but still rank high to very high according to ISRM classification.

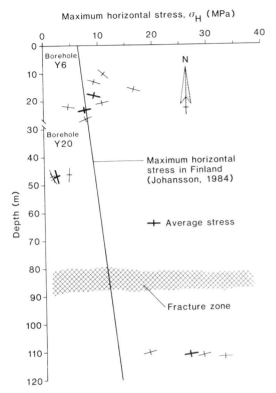

Fig. 3. Maximum horizontal stress field of the bedrock.

repository.

It seems like the gently dipping fracture zone at the depth of 80-90 m divides the rock mass in two parts with respect to stress conditions. Above the zone, the stress has been partly relieved, whereas below it a higher stress has been locked.

3.3 Fractures and fracture zones

The structural direction of the rapakivi ganite massif is typically NE-SW and NW-SE. The same main directions repeat themselves in the orientation of the depressions in the sea area, and also on the outcrops of the study site (Fig.4).

The fractures on the outcrops are almost vertical, the fracture frequency averaging 0.6 pcs/m. The fractures in the NW-SE direction seem to be longer than the other ones, also the widest fractures concentrate on this direction.

The third and most strongly developed fracture set is horizontal, or actually gently dipping to the east, as observed on the core samples. Thus the fracturing is nearly cubic, a common feature to granitic rocks. The fracture frequency of the horizontal fractures averages 1.3 pcs/m.

Some 7% of the total borehole length (2531 m) can be classified as fracture zones containing 40% of the total of 6,000 fractures. The majority of these are concentrated on two main fracture zones, so called upper and lower fracture zone. These zones are crushed and in part strongly weathered dipping gently (10-30°) to the east-northeast. Between the two zones there is a fracture zone some metres wide, so called intermadiate zone, dipping gently to the southeast. These three fracture zones may be caused by compression in the E-W direction and may represent low angle reverse faults.

3.2 Rock stress

In Finland, a relatively strong horizontal stress field has been measured in the upper part of the bedrock, generally being of the order of 5-15 MPa at a depth of 100 m (Johansson 1984). This is a characteristic feature of old shield areas. The stress is generally considered to be related to global plate tectonics.

The bedrock of the Hästholmen island has a strong horizontal stress field in the E-W direction, measured in boreholes Y6 and Y20. The values of 7.5-9.3 MPa in the uppermost part of the bedrock corresponds well to horizontal fields measured in Finland (Fig.3). The reliability of the low stress values at the depth of 40-50 m are questionable due to fractures at the measuring point. At the depth of 110 m, the average main stress of 26.7 MPa is nearly double that of the normal values. The minimum horizontal stress is 0.3 MPa and the vertical stress -2.5 MPa, i.e. it is probably a vertical tensile stress caused by horizontal compression. This must be taken carefully into consideration in rock mechanical design of the re-

3.4 Hydrogeological conditions

The groundwater table of Hästholmen is close to the sea level, the maximum height being some metres above it. Variations of the height of the sea level are reflected in the groundwater, at a time lag of one to three hours. It means that these two water phases are hydraulically connected.

The hydraulic conductivity of the rock mass varies between $k=10^{-9}$ and 10^{-5} m/s. The mean value for the intact rock (fracture frequency >10 pcs/m) is $k=4.8 \times 10^{-7}$ m/s and for the fracture zones $k= 2.2 \times 10^{-6}$ m/s. The upper fracture zone seems to be the most conductive zone and thus most crucial considering groundwater flow.

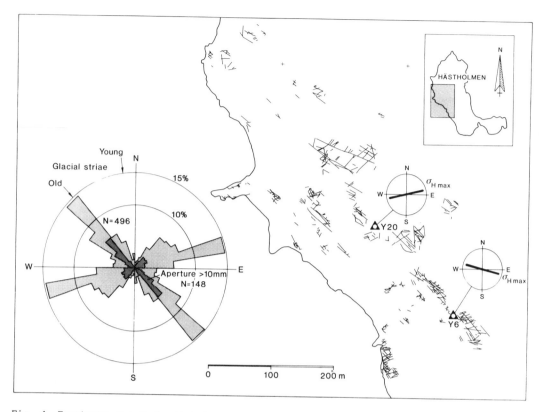

Fig. 4. Fractures and their distribution. Directions of glacial striae and maximum horizontal stress field are also shown.

The groundwater has two zones differing as to their characteristics. Rainwater has infiltrated into the bedrock and has created a fresh groundwater layer, the lower edge of which generally follows the upper fracture zone. Under this layer, the groundwater is saline, with a total salt content of 5-9 °/∞. The saline groundwater originates from the Baltic Sea and is some 4,400-10,000 years old (Kankainen 1986). Tritium determinations have shown that the saline groundwater is in a nearly static state.

4 ENGINEERING GEOLOGICAL MODELLING OF ROCK MASS

The three gently dipping fracture zones are the only ones that could be traced between the boreholes. Especially the electrical and seismic cross-hole methods have played an important role (Rouhiainen 1986). Structural interpretation of all the investigation data have been the basis for the three-dimensional modelling of the study site (Anttila 1986,1988, Pirhonen 1986).

Fig.5 shows the engineering geological model of the study site from two different directions, giving a very clear picture of the gently dipping, plate-like structure of the bedrock. The model was prepared using computed aided design (CAD) in the Intergraph Vax environment with Engineering Site Package (ESP) program, developed by Intergraph Corporation, and GEOREK program, developed by Imatran Voima Oy.

The model has been the basis for the groundwater flow model, where the upper and lower fracture zones form the primary flow routes in the rock mass of the study site. Concerning the siting of the repository, the zones of intact rock are all large enough in size. Since the saline groundwater has been assumed to be almost stagnant, the rock mass between the upper and intermediate zones (rock units II, IIIa, IIIb in Fig.6) forms the most suitable site for the repository in view of the safety of the disposal.

The latest lay-out for the repository facilities and their siting in the rock unit IIIa are shown in Fig.6 (Anttila 1988).

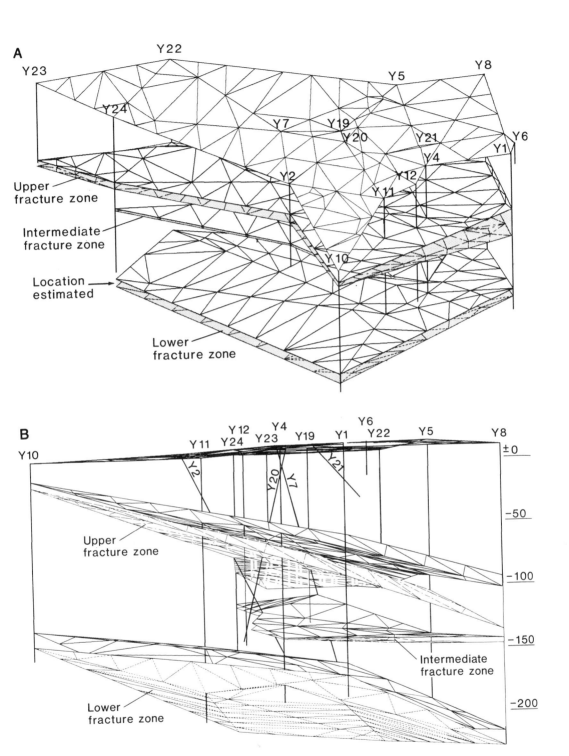

Fig. 5. Engineering geological model of the rock mass. A. Isometric view from SW to NE. B. View from SE to NW.

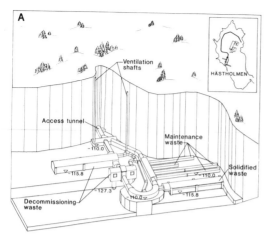

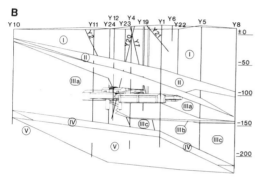

Fig. 6. A. General lay-out. B. Location of the repository facilities (Anttila 1988).

5 CONCLUSIONS

Engineering geological modelling of the bedrock in Loviisa is based on the interpretation of results gained by different investigation methods. Cross-hole geophysics has played an important role in correlating structural features between the boreholes. By CAD techniques the model can be inspected three-dimensionally in various directions and sections, and utilized more effectively in siting and design of the repository.

The model reveals gently dipping zones of intact and broken rock. The repository is planned to be built in intact rock mass, at the level of -90 to -130, separated by two main fracture zones with higher permeability. At the siting level the groundwater is saline and constitutes a nearly ideal environment for the disposal.

REFERENCES

Anttila, P. 1986. Loviisa power station, Final disposal of reactor waste, Geological and hydrogeological conditions of the island of Hästholmen. Nuclear Waste Commission of Finnish Power Companies, Report YJT-86-05, 66p.

Anttila, P. 1988. Engineering geological conditions of the Loviisa power plant area relating to the final disposal of reactor waste. Turun yliopiston julkaisuja, Annales Universitatis Turkuensis, Sarja-Ser. A. II. Biologica-Geographica-Geologica, 72, 131 p.

Gardemeister, R. & Rouhiainen, P. 1984. Investigations for final disposal of reactor waste at the Loviisa power plant site, Finland. Seminar on the Site Investigation Techniques and Assessment Methods for Underground Disposal of Radioactive Wastes, Sofia 1984, IAEA-SR-104/4, 19 p.

Johansson, E.J.W. 1984. Primary stress state of rock and stress measurements. Helsinki University of Technology, Laboratory of Mining Engineering, Licentiate thesis, 79 p.

Kankainen, T. 1986. Loviisa power station, Final disposal of reactor waste, On the age and origin of groundwater from the rapakivi granite on the island of Hästholmen. Nuclear Waste Commission of Finnish Power Companies, Report YJT-86-29, 56 p.

Pirhonen, V.O. 1986. Siting a rock cavern complex by means of 3D mapping of fracture structure of a granite block. Bull. Int. Ass. Eng. Geol. 34, 67-72.

Rouhiainen, P. 1986. Loviisa power station, Final disposal of reactor waste, Geophysical methods in structural study of the bedrock of Hästholmen. Nuclear Waste Commission of Finnish Power Companies, Report YJT-86-10, 37 p.

Simonen, A. 1980. The Precambrian in Finland. Geol. Surv. Finland, Bull. 304, 58 p.

Stockage souterrain dans un shale: caractérisation de la roche en paroi et en déblai

Underground storage in a shale: Characterization of intact and excavated rock

G. Ballivy, B. Benmokrane & J.-C. Colin
Laboratoire de mécanique des roches, Département de génie civil, Université de Sherbrooke, Sherbrooke, Québec, Canada

R. Poulin & R. Simard
Le Groupe SNC, Montréal, Québec, Canada

ABSTRACT:Ordovician shale and limestone sedimentary formations in the Lower St.Lawrence are suitable for liquid natural gas storage sites. This region highly industrialized has transactlantic port facilities and is serviced by a trans-Canada gas pipeline. This paper describes geomechanical investigations carried out at the Rock Mechanics Laboratory at the Université de Sherbrooke into the properties of in situ and excavated rock. The latter, totalling several hundreds of thousands of cubic meters, were examined to determine their suitability as fill for surface structures. Special studies focuses on the influence of saturation cycles (water and liquid propane gas) on the rock, the chemical stability of gas, and the mechanical properties of shale.

RESUME: Les formations sédimentaires des Basses-Terres du Saint-Laurent (Québec), calcaires et shales ordoviciens, constituent des sites favorables pour la réalisation de projets de stockage de gaz naturels liquéfiés. En effet, cette région très industrialisée est accessible par bateaux transatlantiques et elle est desservie par un gazoduc transcanadien. Il est décrit ici les investigations géomécaniques qui ont été conduites au Laboratoire de mécanique des roches de l'Université de Sherbrooke. Elles ont permis de définir les propriétés des roches en paroi et aussi d'examiner si les matériaux excavés, plusieurs centaines de milliers de mètres cubes, pourront être réutilisés comme matériau de remblai pour des aménagements de surface. Des études spéciales sur la roche intacte ont porté en particulier sur l'influence des cycles de saturation à l'eau puis au gaz liquide (du propane), sur la stabilité chimique du gaz et aussi sur les propriétés mécaniques du shale.

1 INTRODUCTION

Le stockage souterrain de gaz naturels liquéfiés (GNL) existe depuis plus de 40 ans et le premier stockage à caractère commercial fut une cavité excavée en 1953 dans l'Illinois (E.U.), pour l'entreposage de 42,000 m^3 de propane. Actuellement, ce procédé existe sur les cinq continents où l'on compte plus de 110 ouvrages de ce genre, d'âges variables.

Le principe du stockage souterrain de GNL, développé pour réduire l'impact environnemental et éviter les potentiels accidents des stockages conventionnels de surface (guerre, catastrophe naturelle, vandalisme...), est de confiner les GNL par l'eau. A une pression hydrostatique supérieure à la tension de vapeur du gaz stocké, la cavité devient étanche, si la recharge de la colonne d'eau est assurée.

Les cavités de stockage de GNL ont très majoritairement été réalisées aux Etats-Unis, dans des formations sédimentaires, et en particulier, le shale. Aussi, de par leur composition de shale et de calcaire ordoviciens, les Basses-Terres du St-Laurent (Québec) ont retenu l'attention des concepteurs pour la réalisation de trois cavernes souterraines de 100 000 m^3 chacune (Projet SOLIGAZ). Chaque caverne aura une section de 160 m^2, à une profondeur moyenne de 100 à 150 mètres, dans les shales de groupe de Lorraine. La température du massif se situera aux alentours de 15°C (pas de réfrigération nécessaire).

Le Laboratoire de mécanique des roches de l'Université de Sherbrooke a conduit les investigations géomécaniques sur le shale (la roche encaissante) afin de déterminer les propriétés géotechniques de la roche en paroi et d'examiner les caractéristiques géomécaniques des matériaux excavés, dans le but d'une utilisation en remblai pour des aménagements portuaires. Les résultats de ces études en laboratoire sont décrits ici.

2 CARACTERISATION GEOLOGIQUE ET GEOCHIMIQUE DE LA ROCHE ENCAISSANTE

2.1 Géologie générale

Les shales du groupe de Lorraine (ou schistes argileux) qui font l'objet de la présente étude, proviennent d'une campagne de sondage menée en septembre 1988 sur le site de Varennes (Fig. 1a). Les échantillons ont été prélevés de deux forages entre approximativement 94 et 200 mètres de profondeur, dans la formation de Nicolet. Cette formation repose sur les shales d'Utica et les calcaires de Trenton comme le montre l'ancien forage réalisé à 6,5 km au NE du site de Varennes (Fig. 1b).

Le shale de Lorraine est caractérisé par une stratification subhorizontale et une lithologie alternante de strate de shale gris-noir, micacé, fissile, parfois très fossilifère, et de minces lits de grès fins ou de silstone calcaireux gris (moins de 3 cm). Ces minces lits, parfois aussi dolomitiques, montrent de fréquentes figures sédimentaires (stratifications obliques, granoclassement) et atteignent exceptionnellement un décimètre. On note la rare présence de dykes et filons-couches centimétriques à décimétriques, reliés aux proches intrusions montérégiennes du Crétacé. Deux lits minces de couches volcaniques vertes (porcellanite) ont été repérés, dont l'un est à la profondeur propice aux cavités de stockage (100 à 150 m).

2.2 Analyses minéralogiques et géochimiques

Les analyses du shale en lames minces par diffractométrie aux rayons-X ont mis en évidence (Tableau 1) la prédominance des minéraux argileux: chlorite, illite et interstratifiés représentent plus de 80% de la roche. On note aussi une faible proportion de quartz et de feldspath (moins de 10%). La texture, fine et ori-

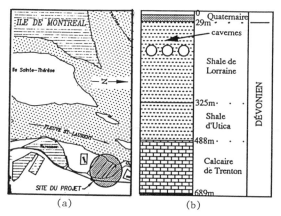

(a)   (b)

Fig. 1 – Situation géographique (a) et stratigraphique (b) du projet

entée dans le plan de fissilité, enchevêtre de l'apatite (1 à 2%) et de la pyrite (4 à 5%).

D'un point de vue chimique, le shale possède une forte proportion de silice (60%) et d'alumine (14%). Le cortège, chimique des minéraux argileux semble complet (Tableau 2) et la présence de phosphore (dans l'apatite), de titane, de dioxyde de carbone et de soufre (dans la pyrite) est notable.

Les minces lits de silstone ont une granulométrie grossière composée de quartz, d'argile, de micas et de minéraux opaques, liés par des argiles et cimentés par du calcaire ou par de la dolomie. Le filon-couche et les minces dykes se définissent comme un lamprophyre alcalin riche en augite, à quartz et chlorite. Les lits de cendre volcanique sont pauvres en chlorite, et ne possèdent ni quartz ni feldspath, le mica est de la muscovite.

D'une manière générale, le shale est caractérisé par la présence de traces de pyrite, fine et cubique, submicroscopique (0,5 à 7%). D'un point de vue géotechnique, la présence de ce minéral n'est pas un inconvénient ici: le gonflement, observé dans les matériaux de type shale, et attribué à l'oxydation de la pyrite (CMFE, 1985) n'est pas gênant pour l'utilisation du matériau excavé en remblai émergé ou immergé.

Par contre, si les conditions d'oxydation de la pyrite sont réunies (H_2O et O_2), la réaction va libérer de l'acide sulfurique H_2SO_4, dont une partie sera consommée pour former du sulfate ferreux

Analyse optique	(%)	Analyse par diffracto. RX
Minéraux argileux	85	Micas, chlorite, kaolinite, illite
Quartz + Feldspath	7	
Apatite	1	
Opaques	4	Pyrite
Divers	3	

Tableau 1. Analyse minéralogique moyenne du shale de Lorraine (gris-noir)

SiO_2	Al_2O_3	FeO	MgO	CaO	Na_2O	K_2O
%	%	%	%	%	%	%
59,2	14	5	3,7	3,8	1,5	2,9

TiO_2	P_2O_5	MnO	CO_2	S	H_2O	P.A.F
%	%	%	%	%	%	%
2,6	1,6	0,9	3,8	1,2	3,2	8

Tableau 2. Analyse chimique moyenne du shale en poids d'oxydes (3 échantillons)

$Fe_2 (SO_4)_3$, et dont l'autre partie sera transportée par les eaux de percolation. Ce processus d'acidification, décrit par Thomas et al., (1989), devrait être limité aux seuls remblais émergés et la présence de carbonates dans les fissures neutralisera la formation d'acide.

3. RESULTATS ET DISCUSSIONS DES ESSAIS DE CARACTERISATION GEOMECANIQUE REALISES EN LABORATOIRE

Le programme expérimental réalisé en laboratoire comprend deux étapes.

Les essais réalisés dans la première étape (étape I) ont porté sur l'identification des caractéristiques physiques, mécaniques et hydrauliques de la roche intacte ainsi que la compatibilité de cette roche (shale de Lorraine) avec le gaz naturel liquide, en l'occurence le propane. Ces essais furent réalisés sur des carottes de roche (diamètre d'environ 45 mm) provenant de deux forages verticaux effectués sur le site de construction projeté des cavernes ou réservoirs souterrains, pour l'entreposage de gaz naturel liquéfié (Fig. 1a). Il s'agit du forage No 1, identifié par VA-1 et du forage No 2, identifié par VA-6; la profondeur de prélèvement des échantillons utilisés dans les essais est comprise entre 94 m et 200 m de la surface du terrain naturel. L'objectif des travaux de cette étape I était, en effet, de définir les paramètres géotechniques de cette formation rocheuse afin de concevoir les réservoirs souterrains (creusement, géométrie, stabilité des parois). Aussi, le concept final a pu être défini par méthode numérique (éléments frontières); il s'agit de trois cavités de section 13 x 18 (m^2) et de 110 m de longueur. La distance entre les cavités est de 39 m, à une profondeur moyenne de 160 m.

La réalisation de ce projet de stockage engendrera une importante quantité de matériau excavé (quelque 400 000 m^3). Dans le but de son utilisation pour la construction de remblais ou jetées immergés (aménagements portuaires), il a été étudié dans la deuxième étape (étape II), les caractéristiques de ce shale sous une forme également broyée, obtenue par simple trempage à l'eau des éprouvettes (éclatement), condition que subira la roche à l'excavation. Les essais de laboratoire de cette étape II ont vu la réalisation d'essais de compactage, de gonflement, des cycles de mouillage et séchage et de compression triaxiale sur du matériau compacté.

3.1 Essais sur échantillons de roche intacte - Etudes de l'étape I

Le Shale de Lorraine est une roche sédimentaire argileuse finement litée qui se débite très facilement si on l'expose à l'air libre ce qui a nécessité l'élaboration de méthodes d'essais particulières. Ainsi, les échantillons utilisés dans les essais (entreposés dans des boîtes de 1,5 m de longueur depuis leur prélèvement) ne furent pas resaturés au laboratoire car, après un séchage à l'air, ils se débitent en disques dès qu'ils sont ré-immergés dans l'eau. En plus, de sévères précautions ont été prises lors de la préparation des éprouvettes (sciage, surfaçage, usinage) à cause de la fragilité de cette roche (Ballivy et Benmokrane, 1988).

3.1.1 Caractéristiques physiques

Le tableau 3 présente un résumé des caractéristiques physiques de la roche intacte déterminées sur des échantillons provenant de diverses profondeurs de prélèvement. Le séchage des échantillons dans le four a été réalisé par ventilation d'air chauffé à 45°C. Lors des pe-

W	W_sat	ρ_{dry}	ρ_{hum}	ρ_s	V_L
(%)	(%)	(g/cm³)	(g/cm³)	(g/cm³)	m/s
1,44	3,4	2,64	2,66	2,75	1533
[11]	[4]	[10]	[11]	[4]	[5]

W: teneur en eau (conditions du laboratoire);
ρ: masse volumique; V_L : vitesse des ondes longitudinales; [ ]: nombre d'essais

Tableau 3. Caractéristiques physiques
de la roche intacte

sés dans l'eau pour définir le volume,
les échantillons furent recouverts d'une
mince couche de paraffine et des corrections appropriées furent appliquées (Ballivy et Benmokrane, 1988). Les résultats
obtenus n'ont montré aucune évolution
significative des caractéristiques physiques, mesurées ici, selon la profondeur
de prélèvement (94 m - 200 m), ni de
différence marquée entre les deux sondages (forages VA-1 et VA-6).

On note une masse volumique moyenne des
grains élevée (ρ_s), représentative d'une
importante quantité de micas et d'oxydes.
La porosité efficace semble se situer
autour de 2%. Les vitesses d'ondes longitudinales mesurées au laboratoire sont
faibles (de 1027 m/s à 2585 m/s), et les
valeurs les plus basses sont obtenues
sur les éprouvettes à forte fissilité
potentielle. Ces essais ont été réalisés
sans l'application de charges de confinement sur les échantillons et il est
probable, pour ce type de roche, d'obtenir des vitesses in situ supérieures à
celles mesurées en laboratoire; à cause
de l'effet de compaction; la fissilité
ne peut se développer qu'à la sortie des
carottes du trou de forage. Le shale se
présente donc comme une roche à porosité
moyenne, mais dont la taille des pores
est extrêmement fine, ce qui est une caractéristique des roches argileuses. La
plus grande part de cette porosité se
situe dans les plans de fissilité.

3.1.2 Caractéristiques hydrauliques

Deux types d'essais de perméabilité ont
été conduits. Le premier, en écoulement
longitudinal classique, caractérise la
perméabilité dans un sens perpendiculaire aux feuillets, c'est-à-dire celle de
la matrice. C'est une perméabilité minimale d'environ 3.10^{-12} m/s (Tableau 4).

En écoulement radial convergent, par
contre, nous mettons en évidence la per-

méabilité inter-feuillets sous un état
de contrainte de compression (Ballivy
et Niemants, 1982). De plus la saturation
préalable aux essais de perméabilité a
développé la décohésion inter-feuillets.
Les échantillons étant confinés, les coefficients mesurés caractérisent alors
la perméabilité horizontale de la roche,
d'autant plus grande que la roche est
décomprimée (abords des excavations).
Avec une valeur moyenne de $4,6 \cdot 10^{-10}$
cm/s, un rapport d'anisotropie de perméabilité égal à environ 150 est mis en
évidence. C'est une valeur maximale pour
le shale due à la fissilité développée
des éprouvettes testées, une fois mises
en contact avec l'eau.

Pour les zones fracturées, les dykes
(ou filons-couches), et les minces lits
gréseux, les sondages ont permis de noter un fort colmatage de ces zones par
le matériau argileux issu du shale. Ce
shale semble donc être un excellent
aquiclude.

3.1.3 Caractéristiques mécaniques

Les résultats des essais de compression
uniaxiale (résistance en compression C_o)
et de compression diamétrale (résistance
en traction T_o) ont montré une faible
évolution entre 94 et 200 mètres de profondeur. Il semble (Fig. 2) que C_o
croisse légèrement alors que T_o décroît
sensiblement. Il est notable que l'essai
de compression diamétrale (ou essai brésilien) sollicite aussi la cohésion inter-feuillets: une explication est donc
proposée, en ce sens que la fissilité
augmente avec la profondeur (diminution

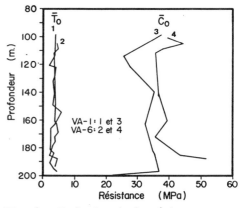

Fig. 2 - Evolution de la résistance en
compression (C_o) et de la résistance en
traction (T_o) en fonction de la profondeur pour les 2 sondages

Type d'écoulement	Identification de l'échantillon	Coefficient de perméabilité (K(cm/s))	Valeur moyenne K(cm/s)
Ecoulement radial	VA-6 (162,5)	$2,26 \times 10^{-8}$	$4,6 \times 10^{-8}$
	VA-6 (124,8)	$7,00 \times 10^{-8}$	
Ecoulement longitudinal	VA-6 (148,10)	$1,97 \times 10^{-10}$	$2,8 \times 10^{-10}$
	VA-6 (144,05)	$4,85 \times 10^{-10}$	
	VA-6 (144,15)	$1,83 \times 10^{-10}$	

( ) : profondeur de prélèvement de l'échantillon en mètre

Tableau 4. Résultats d'essais de perméabilité (roche intacte)

de T_o) de même que la compaction (augmentation de C_o).

Ces effets restent néanmoins peu sensibles, et il est plus notable que les valeurs obtenues à partir des éprouvettes du forage VA-6 sont en moyenne supérieures de près de 20% aux résultats des éprouvettes du forage VA-1.

Le tableau 5 résume les résultats moyens des essais de compression uniaxiale, de compression diamétrale, d'adhésion roche-coulis et d'essais de cisaillement direct. Ces derniers ont été réalisés à charge normale constante, sur des éprouvettes cylindriques de 45 mm de diamètre, dont les surfaces en contact (plan de cisaillement) furent polies. L'angle de frottement résiduel obtenu ainsi varie de 26 à 34°.

Les essais d'adhésion roche-coulis ont consisté en des essais d'arrachement de barres d'acier, scellées avec un coulis de ciment dans des éprouvettes de shale. Dans cette technique de modèles réduits d'ancrages pour l'évaluation en laboratoire (Ballivy et Dupuis, 1980) de la résistance à l'adhésion roche-coulis, on utilise des barres d'armature de diamètre égal à 9,5 mm et un coulis de ciment portland de type 10 (type I ASTM),

avec un rapport eau/ciment égal à 0,4. Après 7 jours de mûrissement (f'_c = 28 MPa: résistance en compression simple du coulis de scellement), la résistance au cisaillement ultime roche-coulis avoisinait les 2 MPa.

Des essais triaxiaux ont été menés sous une presse MTS à haute rigidité, sur des éprouvettes cylindriques (45 mm de diamètre) d'élancement sensiblement égal à 2. Le comportement du shale jusqu'à au moins 10 MPa de pression de confinement est de type élastique avec perte de résistance, après la contrainte de pic (écrouissage négatif) (Fig. 3). Les valeurs moyennes obtenues (11 échantillons VA-1, 15 échantillons VA-6) sont résumées dans le tableau 6.

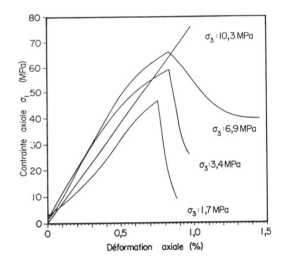

Fig. 3 - Courbes contrainte-déformation d'essais de compression triaxiale (roche intacte)

No du trou de forage	Résistance en compression uniaxiale C_o(MPa)	Résistance à la traction T_o(MPa)	Adhésion roche-coulis (MPa)	Module de Young E* (GPa)	Angle de frottement résiduel φ (°)
VA-1	31,55 ± 5,97 (60)	3,48 ± 1,14 (59)	2,2 ± 0,55 (6)	9,4 ± 3,0 (10)	18 à 33 (8)
VA-6	39,33 ± 4,2 (19)	4,08 ± 1,21 (45)	1,93± 0,36 (6)	13,1 ± 3,9 (4)	26 à 34 (6)

( ) : nombre d'essais; * : Module de Young tangent évalué à 50% de la charge de rupture en compression uniaxiale.

Tableau 5. Résultats d'essais mécaniques (roche intacte)

No du trou de forage	σ_3 (MPa)	σ_1 (MPa)	Module de Young E^* (GPa)	Paramètres de Mohr Coulomb ϕ(°)	c (MPa)
VA-1	3,45 à 10,35 (11)	23,2 à 76,1	6,86 ± 1,97	32 à 35	8
VA-6	1,73 à 10,35 (15)	38,6 à 84,2	8,50 ± 2,86	33 à 37	10,8

σ_3 : pression de confinement appliquée sur l'échantillon; σ_1 : contrainte verticale à la rupture de l'échantillon; * : module de Young tangent évalué à 50% de la charge de rupture; ( ) : nombre d'essais; ϕ : angle de frottement interne; c : cohésion.

Tableau 6. Résultats d'essais de compression triaxiale (roche intacte)

3.1.4 Reconnaissances géochimiques: compatibilité du shale et du propane

Pour caractériser la compatibilité physico-chimique du shale et du propane, il a été procédé à des essais de trempage de disques de roche par cycles de 8 heures: 4 h dans un bain de propane liquide puis 4 h dans le propane gazeux. Après 40 cycles complets (environ 13 jours), les chromatographies en phase gazeuse du propane n'ont pas montré de différences significatives avec celles réalisées avant trempage (Tableau 7). Par contre, lors d'un essai de vérification, le propane liquide s'est coloré de rose au contact du shale, ce que nous attribuons à la formation d'un complexe métallique. Ce phénomène n'a été constaté qu'une fois, mais si l'oxydation de la pyrite devient possible lors de l'excavation, il est probable qu'une faible quantité de soufre et de fer passe dans le propane si les conditions d'oxygénation sont favorables (Thomas et al., 1989).

Nature	Propane après 20 h		Propane après 580 h		
Chromatographie	214	215	217	219	221
CH	0,6	0,7	0,9	0,9	0,7
C_2HC	9,9	9,8	9,8	9,8	9,9
C_3H_8	89,0	89,0	88,7	88,7	88,8
C_4H_{10}	0,4	0,4	0,5	0,5	0,5
Divers	0,1	0,1	0,1	0,1	0,1
Total	100%	100%	100%	100%	100%

h : heure

Tableau 7. Composition du propane commercial au début et à la fin du trempage avec le shale

D'un point de vue physique, chaque échantillon, préalablement aux conditions ambiantes du laboratoire a absorbé plus de 0,2 à 0,3% de son poids en propane et la résistance à la traction des disques imprégnés (essai brésilien) a temporairement doublé de valeur (Tableau 8). Ce comportement est réversible: le shale supporte beaucoup mieux les cycles de saturation au propane que ceux à l'eau où, dans ce cas, il se débite très facilement.

En fait, saturer une roche au propane permet d'obtenir un comportement équivalent à celui qu'aurait la roche sèche. De par sa faible constante diélectrique ($\epsilon < 1$), le propane n'est pas capable de venir baisser l'énergie spécifique de surface développée lors du processus de fissuration (théorie de rupture de Griffith). Mais à long terme, la rupture du shale pourrait intervenir à des valeurs de chargement faible. Il ne semble donc pas prudent de retenir les résistances déterminées sur le shale saturé au propane.

Trou de forage	T_0 (H.R. = 0,4) (MPa)	T_0 (Sat. propane) (MPa)
VA-1	3,5 ± 1,1 (59)	7,5 ± 1,2 (5)
VA-6	4,1 ± 1,2 (45)	8,3 ± 1,2 (5)

T_0 : résistance à la traction indirecte (essai brésilien sur disques); H.R. : humidité relative; ( ) : nombre d'essais.

Tableau 8. Résistance à la traction pour des éprouvettes de shale intactes conservées à l'H.R. du labo (40%) puis saturées au propane

Le mouvement du propane dans la roche peut s'expliciter par la théorie de la double couche (Cases, 1968), amplement décrite dans la littérature ces vingt dernières années, et selon laquelle un fluide de basse constante diélectrique ne peut substituer un fluide de plus forte constante diélectrique (ε eau = 80), s'ils sont immiscibles. Ceci, confirmé par exemple par les essais de Fernandez & Quicley (1985), ou Storey et Peirce (1989) sur des argiles, permet de supposer que le shale, même imprégné superficiellement par le propane, conservera des caractéristiques mécaniques identiques à celles de la roche saturée à l'eau.

Pour s'en assurer, on a effectué des essais de traction indirecte sur des éprouvettes conservées à 98% d'humidité relative (la saturation à l'eau les aurait détruits) avant saturation au propane et il n'a pas été noté d'amélioration de T_o sur les cinq échantillons testés ($T_o \cong 4,8$ MPa).

3.2 Essais sur échantillons de roche broyée (ou excavée) - Etudes de l'étape II

Cette étape, qui a pour objectif d'évaluer la durabilité et le comportement mécanique du shale excavé, en vue de son utilisation pour la construction de remblais ou jetées immergés, a vue la réalisation des essais suivants en laboratoire:

- essais de compactage sur du matériau tamisé au tamis 20 mm (moules standard et CBR);
- essais de compression triaxiale sur du matériau, tamisé au tamis 2,5 mm, et compacté;
- essais de gonflement sur du matériau compacté dans un moule CBR ou sur des éprouvettes de roc intactes;
- essais de mouillage-séchage sur du matériau initialement constitué d'éprouvettes de roc intactes.

Tous les détails concernant la préparation des échantillons et l'application des différentes méthodes d'essais sont décrits ailleurs (Benmokrane et Ballivy, 1989).

Le matériau original, utilisé dans ces essais, était constitué de carottes de roche intactes (diamètre d'une carotte = 45 mm) provenant des deux forages décrits précédemment (VA-1 et VA-6).

Le shale de Lorraine est une roche sédimentaire argileuse qui se débite très facilement si on la trempe dans l'eau. De ce fait, le matériau granulaire utilisé dans les essais qui suivent a été obtenu par une simple immersion d'éprouvettes de roche dans de l'eau (Fig. 4).

3.2.1 Essais de compactage

Des essais de compactage dans des moules standards et CBR à base perforée ont été effectués, et pour chaque essai, on a déterminé la granulométrie du matériau avant et après compactage, comme le montre la figure 5. Celle-ci illustre les fuseaux de courbes granulométriques et pour l'ensemble des essais réalisés (6 essais de compactage).

Les résultats obtenus ont montré que le shale broyé et compacté présente des valeurs de masse volumique sèche (ρ_d) comprises entre 1,8 et 2,24 T/m^3, pour des teneurs en eau (W) de 5,9% à 8,2%, ce qui est une caractéristique favorable à la compaction. Toutefois, le shale s'est avéré être un matériau dont la granulométrie est sensible à la compaction: les chocs, aux essais standards ou CBR, ont déplacé le fuseau granulométrique (avant compactage) vers les fines (Fig. 5), ce qui peut poser un problème de rétention de nappe en cas de remblais émergés.

3.2.2 Essais de gonflement

Afin d'évaluer le potentiel au gonflement du shale, il a été réalisé des essais de gonflement qui consistaient à immerger dans l'eau potable des carottes de roche intactes, de longueur d'environ 10 cm, ou du matériau compacté à sec (dans le moule CBR), et à mesurer les variations de hauteur de l'échantillon en fonction du temps. Les gonflements obtenus sont de l'ordre de 3% (2,31 à 3,6%) pour les carottes intactes et ils correspondent à l'écartement des feuillets du shale et au gonflement des matériaux argileux (montmorillonite, illite, ...). Sur le matériau compacté à sec, le gonflement a atteint 1,73%, et il ne résulte que de l'expansion des argiles. Ces valeurs de gonflement maximum ont été atteintes après une durée de conservation dans l'eau d'environ 3 heures; au-delà de cette période, il y a stabilisation.

3.2.3 Essais de mouillage-séchage

Le degré de dégradation du shale intact, sous fluctuations hydrométriques, a été examiné en soumettant le matériau à des cycles de mouillage-séchage. L'essai consistait à soumettre le matériau, constitué initialement de carottes de shale intactes (de longueurs comprises entre 1 et 5 cm), à quatre cycles de mouillage-séchage, et à la fin de chaque cycle, le

(a)   (b)

(c)

Fig. 4 - Transformation du shale intact (a) par immersion dans l'eau et séchage à l'air (b, c)

matériau est tamisé pour déterminer sa granulométrie. Le mouillage fut effectué dans de l'eau potable (23°C), tandis qu'un four réglé à 110°C fut utilisé pour le séchage du matériau. Un exemple de répartition granulométrique du shale après quatre cycles de mouillage-séchage pour un essai, est illustré à la figure 6. Ces résultats montrent que le shale se détériore rapidement lorsqu'exposé à des cycles d'humidification-séchage, et que cette détérioration semble se stabiliser après le troisième cycle pour les conditions d'essais considérées ici. Aussi, la valeur du D_{50} (diamètre à 50%) passe d'environ 12 mm au premier cycle, à environ 3 mm au 3e cycle, ce qui montre que ce procédé a entraîné la production des fines et la modification de la structure granulaire.

3.2.4 Essais de compression triaxiale

Les propriétés mécaniques et hydrauliques du shale broyé (ou excavé), en vue de son utilisation pour la construction de remblais (stabilité, pente), ont été examinés en réalisant des essais de com-

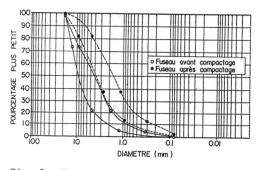

Fig. 5 - Fuseaux granulométriques obtenus avant et après compactage à l'essai Proctor sta ou CBR

Fig. 6 - Répartition granulométrique du shale après 4 cycles de mouillage à l'eau (23° C) et séchage au four (110° C)

pression triaxiale sur des éprouvettes cylindriques (diamètre - 50 mm, hauteur - 95 mm) constituées d'un matériau (diamètre maximum des grains - 2,5 mm) compacté à l'aide d'un pilon. Les trois éprouvettes utilisées ont été soumises à un essai triaxial du type CID (consolidé isotropiquement drainé) et en suivant la procédure d'essai conventionnelle appliquée pour les échantillons de sol. Un résumé des résultats obtenus est présenté dans le tableau 9. Celui-ci montre que la rupture du matériau intervient entre 0,8 et 1,8 MPa selon la pression de confinement (0,2 à 0,5 MPa), avec une cohésion de 43 kPa et un angle de frottement interne égal à 32°; cette valeur est analogue à celle obtenue pour l'angle de frottement interne du shale intact. Le coefficient de perméabilité minimale obtenu après consolidation varie de 10^{-5} à $7,8 \times 10^{-7}$ cm/s, ce qui confirme la présence des fines.

Essai No	W (%)	ρ_d (g/cm³)	σ_3 (kPa)	σ_{1f} (kPa)	K_c (cm/s)	E_o (MPa)
1	8,7	2,09	200	814	$7,77*10^{-7}$	32,5
2	10,5	2,04	350	1271	$4,30*10^{-6}$	25,0
3	8,2	2,20	500	1803	$1,04*10^{-5}$	29,0

W(%) : teneur en eau du matériau lors du compactage; ρ_d (g/cm³) : masse volumique sèche du matériau après compactage; σ_3 (kPa) : pression cellulaire appliquée sur l'échantillon; σ_{1f} (kPa) : pression axiale à la rupture; K_c (cm/s) : coefficient de perméabilité du matériau après consolidation; E_o (MPa) : module de déformation initiale.

Tableau 9. Résultats d'essais de compression triaxiale sur du shale broyé (passant tamis 2,5 mm) et compacté

4. CONCLUSIONS ET RECOMMANDATIONS

Ces essais de laboratoire réalisés sur des échantillons prélevés dans deux forages réalisés à Varennes (près de Montréal, Québec), pour un projet de stockage souterrain de gaz naturel liquéfié, confirme que la formation rocheuse examinée est homogène et qu'elle a les propriétés du shale de Lorraine, formation des Basses-Terres du Saint-Laurent (Québec, Canada).

Les résultats des essais réalisés sur les carottes de roche intactes, incluant les propriétés mécaniques et hydrauliques et aussi la compatibilité physico-chimique du shale avec un gaz naturel (le propane), montrent que cette formation rocheuse constitue un site favorable pour le stockage souterrain de gaz naturel liquéfié.

Les essais de trempage propane-shale ont montré qu'il y a compatibilité chimique entre ce gaz naturel et cette roche; en effet, les analyses de la composition chimique du propane, avant et après trempage, n'ont pas révélé de différences significatives, et aussi les propriétés mécaniques du shale après trempage ne se trouvent jamais réduites.

Par ailleurs, cette roche sera facile à excaver, ce qui n'est pas sans importance, connaissant la grande quantité de matériau à excaver (quelque 400 000 m³). En plus, la perméabilité matricielle de cette roche est considérée très faible (inférieure à 10^{-10} m/s). Pour les zones fracturées, les dykes (ou filons-couches) et les minces lits gréseux, les sondages ont permis de noter un fort colmatage de ces zones par le matériau argileux issu du shale. La stabilité des parois jugées critiques durant l'avancement en chantier peut être assurée en utilisant du béton projeté ou un renforcement par ancrages injectés.

Ces travaux de laboratoire sur des échantillons intacts (étude de l'étape I) seront complétés par des essais in situ, dans les trous de forages et aussi dans des galeries de reconnaissance (mesures des contraintes, propriétés mécaniques et hydrauliques,...). Ces travaux en chantier permettront d'optimiser la conception des excavations et auront aussi pour but d'examiner l'effet d'échelle; les échantillons prélevés durant les sondages ayant une section de 160×10^{-6} m² alors que les cavernes de stockage auront une section de l'ordre de 160 m².

Enfin, les études de l'étape II, comprenant les essais sur du shale broyé, montrent que ce matériau, une fois excavé, pourrait être utilisé pour la construction de remblais ou jetées immergés, même s'il se détériore rapidement lorsqu'exposé à des cycles d'humidification et de séchage. Toutefois, certaines précautions doivent être mises en oeuvre. Comme le shale de Lorraine, une fois mis en contact de l'eau, se débite avec formation de particules fines, celles-ci peuvent donc être entraînées par l'eau particulièrement dans la zone de marnage. Un des moyens de s'opposer efficacement à l'entraînement de ces fines est la mise en place d'un filtre dans le remblai. Aussi, pour protéger le remblai de l'érosion due aux vagues, on utilise un perré constitué de roches dures (calcaire, roches ignées,...). Dans ce cas, on peut aussi prévoir, en plus du filtre, un géotextile non tissé entre le remblai et le perré.

5. REMERCIEMENTS

Cette étude a été rendue possible grâce à une aide financière du consortium Soligaz inc. (Québec, Canada). Les travaux expérimentaux ont été réalisés grâce à la participation active de monsieur Georges Lalonde et monsieur Martin Lizotte, techniciens au Laboratoire de mécanique des roches de l'Université de Sherbrooke.

6. REFERENCES

Ballivy, G., Dupuis, M. (1980). Etudes en laboratoire de l'adhésion roc-coulis pour divers types de roc, Bulletin de l'AIGI, No 22, 241-243.

Ballivy, G., Niemants, P. (1982). New Laboratory Techniques to Test Grouted Fractured Rocks, Special conference on grouting in geotechnical Engineering, Session X, New Orleans, ASCE Publication. Ed. by W.H. Baker, 809-821.

Ballivy, G. et Benmokrane, B. (1988). Propriétés géotechniques des roches du site de Varennes - Projet Soligaz; rapport GR-88-12, Département de génie civil, Université de Sherbrooke, Québec, 70 p.

Benmokrane, B. et Ballivy, G. (1989). Essais sur le shale de Lorraine en vue de définir les paramètres de construction de remblais et jetées immergés - Projet Soligaz. Rapport GR-89-10-2, Département de génie civil, Université de Sherbrooke, Québec, 22 p.

Canadian Foundation Engineering Manual (CMFE) (1985). Canadian Geotechnical Society, 2nd Edition, 460 p.

Cases, J.M. (1968). Les phénomènes physico-chimiques à l'interface: Application au procédé de la flottation. Thèse Université de Nancy, Mémoires Sciences de la Terre, No 20, 120 p.

Fernandez, F., Quigley, R. (1985). Hydraulic Conductivity of Natural Clays Permeated with Simple Liquid Hydrocarbons, Canadian Geotechnical Journal, No 22, 205-214.

Storey, J.M.E. and Peirce, J.J. (1989). Influence of Changes in Methanol Concentration on Clay Particle Interactions. Canadian Geotechnical Journal, Vol. 26, 57-63.

Thomas, M.A., Kettle, R.J., Morton, J.A. (1989). The Oxidation of Pyrite in Cement Stabilized Colliery Shale, Quarterly Journal of Engineering Geology, London, Vol. 22, 207-218.

Some unique uses of underground space in Kansas City (USA)

Quelques usages uniques de l'espace souterrain à Kansas City (Etats-Unis)

Syed E. Hasan
Department of Geosciences, University of Missouri, Kansas, City, Missouri, USA

ABSTRACT: Located in the midwestern U.S.A., Kansas City with more than 1.4 million sq m of developed underground space, has been the world leader in conversion and secondary use of mined-out space for human occupancy. Presently, there are 29 underground sites that are scattered over an area of 9728 sq km in the metropolitan Kansas City. There are over 250 businesses, employing about 2500 persons, who spend a 40-hour work week in the underground spaces. Uses of the underground space include warehousing, retail businesses, offices, different types of manufacturing, photographic laboratory, sound recording studio, restaurant, etc. In recent years, however, some innovative and unconventional uses of the underground space in the area have occurred. These usage include: (i) mushroom farming, (ii) ice manufacturing, and (iii) higher education institution.

The paper outlines the geologic setting of the area and discusses the geotechnical characteristics of the rock formations encountered in the underground excavations. A brief review of the history of development of secondary space for human use and occupancy in Kansas city is also included. Detailed account of the geologic and related factors that led to the three unconventional uses of the underground space is presented in the paper. Prospects for future use of the mined-out space for other purposes are also included.

ABSTRAIT: Situé au midwest des Etats-Unis, Kansas City avec plus de 1,4 million de milles carrées d'espace souterrain, exploité est à la tête du monde en ce qui concerne la conversion et l'exploitation secondaire d'espace miné pour des buts variés. Actuellement il y a 29 installations souterraines qui sont dispersées sur un espace de 9 728 kilomètres carrés dans l'agglomération de Kansas City. Il y a plus de 250 entreprises commerciales, employant environ 2500 personnes, qui travaillent une quarantaine d'heures par semaine dans ces espaces souterrains. L'exploitation de l'espace souterrain comprend l'emmagasinage, le commerce de détail, bureaux, la fabrication, des laboratoires photographiques, des studios d'enregistrement du son, des restaurants, etc. Cependant ces dernières années des usages innovateurs et originaux des espaces souterrains ont eu lieu dans la région. Ces usages comprennent: (i) champignonnières, (ii) fabrication de glace, et (iii) établissements d'instruction supérieure.

L'article décrit l'endroit géotechnique de l'espace et discute des caracéristiques géotechniques des formations de rocher rencontrées dans les excavations souterraines. Il inclut également l'historie de développement de l'espace secondaire pour l'exploitation et l'habitation á Kansas City et présente un compte spécifié des facteurs géologiques qui ont abouti aux trois exploitations innovatrices de l'espace souterrain. L'étude expose aussi des perspectives pour l'exploitation future des espaces minés destinées à d'autres usages.

1 INTRODUCTION

Conversion of mined-out spaces for secondary use, involving human occupancy, has been going on in Greater Kansas City for nearly 50 years. Presently there are over 1.4 million sq m of developed underground space in the area. Additional space, resulting from limestone mining, is being created at the rate of 0.25 million

sq m annually. There are nearly 5.2
million sq m of potentially developable
space of which 0.51 million sq m can be
converted into secondary use within 90 days
(Ward, 1983). With about 2500 persons
working full-time at various businesses
that are located underground, Kansas City
has emerged as the world leader in the use
and human occupancy of the underground
space.

Greater Kansas City, with a population of
1.5 million (MARC, 1989), includes the
eight county metropolitan area extending
over the states of Missouri and Kansas.
The City lies in the midwestern United
States and because of its central location
(Figure 1), serves as a node for major
transportation lines.

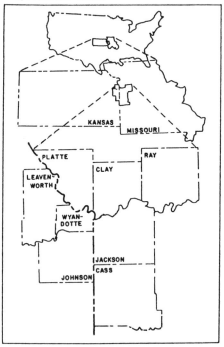

Figure 1. Map of country, states and the
eight-county area of Greater Kansas City
(from Hasan, et al., 1988)

Underground mining of limestone, that
began over a century ago, is still
continuing. Most of the developed
underground spaces, however, are located in
mines that were abandoned long time ago.
Figure 2 shows the distribution of
developed underground sites in Greater
Kansas City. Out of the 29 sites, 22 are
located in mines that were abandoned after
the completion of limestone extraction;
seven sites are located in active mines,
where limestone mining is going on

concurrently with conversion and secondary
use of part of the mined-out areas. These
underground sites are used for a variety of
purposes that include conventional
merchandise and frozen food warehousing,
precision instrument manufacturing, storage
of vital documents, retail and wholesale
businesses, sound recording studio,
restaurant, photographic laboratory,
philatelic sales and ordering division of
the U.S. Postal Service, automobile and
recreational vehicles storage, industrial
machineries and equipment, etc. (Figure 3).

Specialized use of underground space
started as early as 1954 when the Brunson
Instrument Company decided to locate its
precision instrument manufacturing
facilities in underground space created
after limestone mining. The navigational
and other precision instruments,
manufactured by the Brunson Instrument
Company, have been used in the NASA's space
exploration program and by the U.S. Defense
Department. Because of the high level of
accuracy and rigid specifications-tolerance
of the order of fifty-millionth of an inch-
Brunson had difficulty in calibrating these
sensitive instruments in his above-ground
facility, mainly due to vibrations caused
by moving traffic and wind. Lack of
adequate control of dust, temperature, and
humidity was another factor that led to
underground location. Mining of limestone
began in 1954 and the operations were moved
into the subsurface in 1961. The
underground environment enables calibration
of the instruments any time of the day or
night, something that was possible only
between the hours of 2:00 and 4:00 a.m. at
the above-ground factory.

Recently, however, three underground
sites were brought under use that can truly
be called unique. These uses include: (a)
mushroom farming, (b) ice manufacturing and
(c) underground location of an educational
institution. These operations are being
carried out at Atchison, Independence and
Parkville (sites no. 1, 7, and 29 in Figure
2). A discussion of these unique uses and
the advantages of locating these facilities
in the underground is presented, after a
brief review of history of underground
space development and the geotechnical
features of the rocks occurring in Greater
Kansas City.

2 HISTORY OF UNDERGROUND SPACE DEVELOPMENT
IN GREATER KANSAS CITY

Underground mining of limestone in the area
began in the 1880s to meet the requirements
for buildings and other constructions.

2728

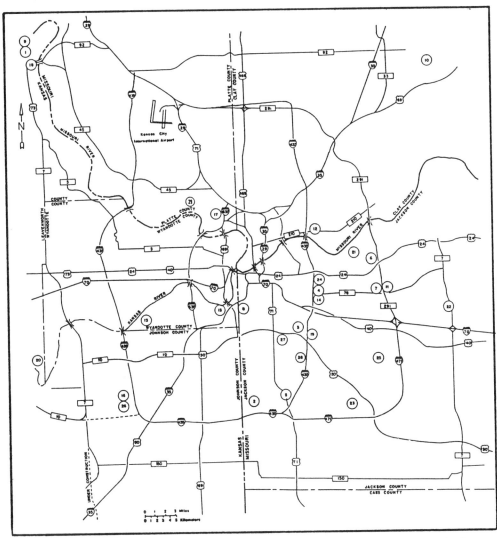

Figure 2. Underground sites in Greater Kansas City. 1-Atchison Underground; 2-Bannister & Holmes Business Park;3-Botsford Ready Mix;4-Brunson Instruments;5-Byers Crushed Rock;6-Carefree Innerspace;7-Commercial Distribution; 8-Dean Fairmont;9-DIPEC;10-Excelsior Springs;11-Geospace;12-Hunt Midwest Enterprises, Inc.;13-Inland Storage & Distribution Center;14-Interstate Underground;15-J.A. Tobin-Kansas City;16-J.A. Tobin-Lenexa;17-K.C. Quarries;18-Leavenworth Underground;19-Marley Research Center;20-Mid-continent Underground;21-Missouri Portland Cement;22-Pink Hill Quarries;23-Rock Acres;24-Solar Park;25-Union (Noland Road);26-Union (Renner Road);27-Vance Brothers;28-Winchester Center;29-Park College; (from Ward, 1983).

These mines were abandoned after the expiration of lease or completion of mining operations. Since the sole objective in such operations was to extract as much limestone as possible, the mine operators did not consider it necessary to ensure stability of the mine, beyond the immediate, short-term requirements. Large scale mining of limestone created a huge volume of void spaces, which progressively got into various stage of deterioration. These abandoned mines remained as waste lands in an otherwise scenic area until the 1940s, when some developers envisioned the possibility of converting these mines into usable secondary space. Thus it was in 1944 that one of the mines was converted into a "natural cooler" space for storage of perishable food at a cost of nearly $2 million. This site, located at Atchison, about 80 km from Kansas City, Missouri (number 9 in Figure 2), is being used for storage of machineries and industrial equipment since 1952.

Dean Fairmont Company (number 8 in Figure 2), acquired an abandoned mine in the 1950s and after additional mining, converted it into a storage facility located close to the business and industrial district in Kansas City, Missouri. Since then many mines have been converted into usable secondary space and several new underground mining ventures have been undertaken with the dual purpose of resource extraction and secondary use of the mined-out space.

3 GEOLOGIC SETTING

Sedimentary rocks, dominated by limestone and shale, and aggregating 274 m, occur in Greater Kansas City. A generalized stratigraphic column of these rocks is shown in Figure 4. The two limestone units that have been extensively mined are the Bethany Falls and the Argentine

Figure 3. Uses of underground space. A) U.S. Postal Service's Philatelic Sales Division; B) Frozen food storage (Photos courtesy; Hunt Midwest Enterprises, Inc.)

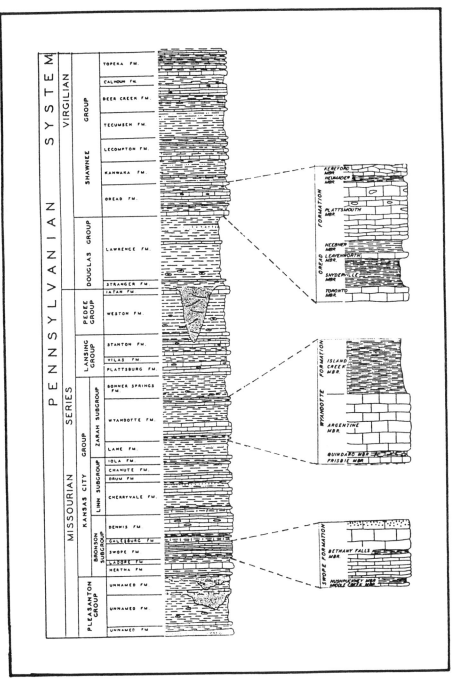

Figure 4. Generalized stratigraphic column for rocks occurring in Greater Kansas City with details of formations where underground sites are located, shown on the right (from Howe and Koenig, 1961).

Limestone. All underground spaces in Greater Kansas City, with the exception of the two sites at Atchison (numbers 1 and 9 in Figure 2) have been developed in these two limestone units. The Atchison underground sites have been excavated in

the Plattsmouth Limestone of the Oread
Formation (Figure 4) which occurs nearly 84
m above the Bethany Falls Limestone.

The rocks in the area are nearly
horizontal, with a general dip of 1/5° or
1.9-3.8 m/km in W-NW direction. The area
has not undergone major deformation; small
folds, faults, and flexures are locally
present and these may cause the beds to
assume steeper dips at places.

Jointing is common in the rock
formations. Two prominent sets strike NW-
SE and NW-SE with steep to near vertical
dips and spacing of 6 to 7.6 m.

The topography of Greater Kansas City is
characterized by gently rolling hills and
valleys with limestone escarpments. Relief
is 154 m and bluffs comprising limestone
provide convenient portals for driving the
tunnels and galleries. The level valley
floors facilitate extending the roads and
rail lines into the underground
developments. Weathering and erosion have
removed large thickness of the overlying
strata, especially in the vicinity of
Missouri River. Here the Bethany Falls
Limestone occurs at shallow depths (<10 m)
at some places. The Pleistocene glaciation

Table 1. Engineering properties of the
Bethany Falls Limestone (taken from Hasan,
et al; 1988).

Parameter	Value, Average	Number of tests
Specific Gravity		
Apparent	2.65	15
Porosity, Apparent, %	4.06	
Hardness, Shore	40.7	40
Absorption, %	1.5	10
Soundness, Freeze-thaw,%	0.97	
Los Angeles Abrasion		
Loss, Grading B, %	26.0	7
P-wave velocity, m/s	5500.0	
Tensile Strength, Direct, MPa	3.82	
Tensile Strength, Brazilian, MPa	3.54	
Modulus of Rupture, MPa	7.44	
Uniaxial Compressive Strength, MPa	63.9	33
Young's Modulus at 50% of ultimate strength, MPa	0.065×10^6	
Poisson's Ratio at 50% ultimate strength	0.266	
Point Load Index, MPa (perpendicular to bedding)	3.81	40
Rock Quality Designation	63.0	34

Table 2. Engineering properties of the
Argentine Limestone (taken from Hasan, et
al, 1988)

Specific Gravity, Apparent	2.64
Bulk	2.52
Bulk, SSD	2.58
Absorption, %	2.58
Los Angeles Absorption Loss, Grading A, %	33.4
Uniaxial Compressive Strength, MPa	53.0

has left behind a mantle of loess capping
the bedrock.

3.1 Geotechnical characteristics of mined
limestones

Some of the important geotechnical
characteristics of the Bethany Falls,
Argentine, and the Plattsmouth limestone
that have led to extensive exploitation of
these units include: adequate thickness,
6-8 m; negligible solutioning, wide joint
spacing, nearly horizontal dip, favorable
chemical and engineering properties. Table
1 gives the engineering properties of the
Bethany Falls Limestone and Table 2 those
of the Argentine.

The ground cover over the mines is of the
order of 10-37 m, and one or two layers of
nearly impermeable shales overlie the mined
limestones; these shales inhibit any
vertical movement of surface water into the
underground openings.

Presence of shale partings in limestone
provides the weak plane along which the
rock breaks during blasting operations,
thereby leaving smooth and level roofs and
floors.

4 UNIQUE USES OF UNDERGROUND SPACE

Recently three entirely different
operations were located in the underground
space in Greater Kansas City. They
represent uses that are unique in the sense
that these are the first such uses in the
area and include: mushroom farm, ice
manufacturing, and educational
institution. A detailed account of these
uses is presented below.

4.1 Toto mushroom farm

Part of the Atchison Underground (number 1

in Figure 2) was converted into a mushroom farm in the early 1987 by a mushroom growing family that was producing it at an above-ground farm in Pennsylvania. Because of the availability of raw materials locally—limestone, straw, brewer's grain—and the presence of underground site, they decided to relocate in the midwestern U.S.A. and selected the underground site for their operations. Currently 1.62 ha space is being used for mushroom growing, with anticipated addition of about 5-10 ha in future (the Atchison Underground has a total available space of about 13 ha).

Nearly 24 persons work in the mushroom producing operations. The main reasons for locating mushroom production in the underground are: (a) uniform humidity and darkness, (b) constant temperature of 12.2-15.6°C year-round, and (c) energy cost savings. Uniform humidity, at 50 percent, greatly cuts down the water usage for maintaining the desired moisture content of the growing medium. Except for light watering once a week, growing mushroom in the underground space does not require additional irrigation. The darkness helps in allowing the mushroom crop to attain its right color, size, and texture. Security is another important factor and controlled exit and entry points result in reduced insurance cost against losses due to theft and vandalism.

4.2 Ice manufacturing plant

Commercial Distribution Center (number 7 in Figure 2) is one of the larger underground space development in Greater Kansas City. It has a mined-out area of 0.64 million sq m of which 0.195 million sq m have been converted into secondary space which is being used for conventional and frozen food warehousing. In June, 1989, nearly 3100 sq m was converted for ice manufacturing. The plant utilizes anhydrous ammonia as the refrigerant, which is piped from outside to the ice-making plant, located in the underground.

The plant can manufacture about 73 metric tons of ice/24 hours. The production is currently sufficient to meet demand of local market, but during special occasions, when the demand exceeds production, ice manufactured earlier, is stored in the freezer rooms in the underground. Temperature in the freezer rooms is maintained between −23.3°C and −17.8°C.

Some of the advantage of locating the ice manufacturing plant underground includes: (a) energy costs saving—up to 70% as compared to above ground warehousing

(Stauffer, 1978), (b) lower production cost—nearly 20% cheaper than comparable above ground operation, and (c) little waste of ice during packaging and loading.

The ambient temperature in the underground space is about 14°C in unoccupied areas. In occupied areas, the temperature could be as high as 18.5°C, which is due to the heat generated by equipment and people, and from the movement of traffic. The incoming water from the city water supply system is at a temperature of 21.6°C; this water is led through pipes in the underground space over a distance of about 120 m to the ice making plant. This causes the water to lose heat and its temperature drops down to 10°C before it is led through a series of filtration devices to the refrigeration plant. This is in contrast to the surface ice making operations where the incoming water is at a temperature of 28.3°C and, therefore, must be chilled before being fed into the refrigeration plant. This is one of the major energy cost saving factor. The other factor is the elimination of waste resulting from the ice that gets partially or totally melted during packaging and loading operations. Again, the ambient temperature at the point where ice is packaged into 3.6-4.5 kg plastic bags, is 18.3°C, and because the loading docks are located within 30 m, and where the same 18.3°C temperature prevails, there is virtually no loss of ice due to thermal shock or melting.

The existing plant supplies nearly 1/3 of the total requirement of ice in Greater Kansas City. Plans are in place to double the capacity without significant modifications of existing facilities.

4.3 Park College underground facilities

Park College, the second oldest institution of higher education in Greater Kansas City, was founded in 1875. It is a private four-year liberal arts college and had an enrollment of nearly 500 in 1989, of which 42 percent lived on-campus. Future plans call for increasing the enrollment to 800 with 350 students living on-campus, during the next four years. Its 800-acre, wooded campus is located on a bluff north of the Missouri river that rises over 60 m above the river. Because of its need to expand and the commitment to preserve the natural quality of the surface land, coupled with favorable geologic setting, and the advantages of underground, Park College administration, in 1979, embarked upon an ambitious project of underground

development. Initial phase of the plan
called for location of its library,
computer center, classrooms, a large
meeting room; faculty, registrar's, and
student government offices; art gallery,
physical plant, and mail room in the
subsurface space specially created after
limestone mining. Additional space, not
used by the College, will be leased or
rented to commercial operators for storage,
etc. The College is also looking into the
possibility of installing energy-saving
equipment to exchange cool/warm air from
the underground to existing buildings on
the surface. It is of interest to note
that the underground facilities do not
require temperature control, despite the
wide temperature fluctuation in Greater
Kansas City-high of about 39°C during
summer to low of below 0°C during winter
months. This is due to the year-round
constant temperature of about 14°C in
unoccupied portions and 20°C in occupied
areas. This has resulted in great energy
cost savings for the College.

4.3.1 Method of excavation

Excavation of limestone began in 1981 and
the new facilities were occupied in 1988.
Limestone excavation is done using
conventional drilling and blasting
procedures. Mining is currently being done
in the upper limestone, the Argentine
(Figure 4). This limestone is nearly 6.1 m
thick, wavy, bioclastic, and thin to medium
bedded. Nearly 3.8 m of Argentine
Limestone is removed by excavation, leaving
1.8 m in the roof. The roof is reinforced
by installing resin bolts.
It is estimated that nearly 0.6 million
sq m of underground space can be developed
in the Argentine Limestone. Of this nearly
51000 sq m has already been excavated, and
11160 sq m has been developed for secondary
use as of December, 1989. Mining of the
Bethany Falls Limestone will create an
additional 2.1 million sq m of space during
the later phase of development. It is
planned to add 18600 sq m of new space
annually during the next 15-20 years.

4.3.2 McAfee Memorial Library

The McAfee Memorial Library occupies 3906
sq m of developed space. Current
collection of the library is about 100000
volumes and the dedicated library space can
hold 200000 volumes. Also located in this
development are the Mabee Learning Center
which houses the computer labs, classrooms,

meeting room, registrar's office, student
government office, the Board of Trustees
room, faculty offices, and the mail room.
Physical plant office, mechanical rooms,
and warehouse are also located underground
Entrance to the underground facilities is
by means of two vehicular and two
pedestrian tunnels; in addition a main
pedestrian entrance connects the library to
the surface by means of a terraced passage,
called the Millsap Foyer. This dome shaped
entrance, 17.1 m wide and 6.71 m high, was
ingeniously constructed by using a nylon
hemispherical balloon, the inside of which
was coated with vegetable oil to prevent
its bonding with ferrocrete. The balloon
was inflated using an air pump, and was
anchored to the ground and made air
tight. An air lock was constructed to
allow for entry and exit of workers and
equipment. The balloon was kept inflated
all the time by pumping air. Ferrocrete
and insulation material were then sprayed
on the inside surface of the inflated
balloon. Ferrocrete consisted of 2.5-4 cm
long, thin gauge, stainless steel needles
mixed with concrete. In each cubic meter
of concrete, 29.7 kg of stainless steel
needles, approximately 60000 pieces, were
used. The 0.31 m thick ferrocrete was
applied in "lifts"; each day 2.54 cm was
sprayed and was allowed to harden overnight
before the next layer was sprayed on.
After 10 cm ferrocrete had been applied, no
ferrocrete was sprayed for the next four
days to allow for installation of steel
ribs and electrical conduits. After this
the remaining 18 cm of ferrocrete were
sprayed, again in 2.54 cm lifts. The final
2.54 cm thick layer consisted of concrete
without the stainless steel needles.
Testing of ferrocrete gave a strength of
48.3 MPa.
After the inside of the dome was
completed, its outside was sprayed with
five coats of butyl rubber, each 0.8 mm
thick, to make the structure water proof.
Drain tiles, sky light and a covering of
native loessal soil and vegetation,
completed the construction of the dome.

5 FUTURE PROSPECTS

In order to meet the needs of growing
population in Greater Kansas City, more and
more structures and facilities will be
located underground. It is likely that
sports facilities, swimming pools, waste
disposal sites and other facilities will be
located in the underground space created by
limestone mining.

6 ACKNOWLEDGEMENTS

The author wishes to extend his thanks to staff of Toto Mushrooms, Atchison, Kansas; Empire Ice, Independence; and Park College, Parkville, Missouri, for facilitating visits to their developments. Special thanks are due to Jeanne Brown, Department of Geosciences, University of Missouri, for her help in typing the manuscript.

6 REFERENCES

Hasan, S.E.; Moberly, R.L.; and Caoile, J.A. 1988. Geology of Greater Kansas City, Missouri and Kansas, United States of America: Bulletin of the Association of Engineering Geologists, Vol. 25, No. 3, pp. 277-341.

Howe, W.B. and Koenig, J.W. 1961. The Stratigraphic Succession in Missouri: Missouri Geological Survey and Water Resources, Rolla, Missouri, 2nd series, Vol. 40, 185 p.

Mid America Regional Council (MARC), Kansas City, Missouri, 1989. Dataline, Third Quarter, p. 7.

Stauffer, T. Sr. 1978. Energy Use Effectiveness and Operating Costs Compared between Surface and Subsurface Facilities of Comparable Size, Structure and Enterprise Classification: Unpublished Report to the City of Kansas City, Missouri, 42 p.

Ward, D.M. 1983. Underground Space: Inventory and Prospects in Greater Kansas City: University of Missouri-Kansas City, Center for Underground Space Studies, Kansas City, Missouri, pp. 7-8.

Disposal in or beneath a thick sedimentary sequence

Arrangement dans ou sous une série de couches sédimentaires

R.J.Heystee & P.K.M.Rao
Ontario Hydro, Toronto, Canada

K.G.Raven
Intera Technologies Ltd, Ottawa, Canada

ABSTRACT: Engineered barriers in a disposal vault will have a finite life, and the ingress of water, leaching of nuclear fuel waste and the mobilization of long-lived radionuclides will be inevitable. Therefore site selection strategies must place emphasis on the containment properties of the geologic environment. Studies that have been completed in Ontario and elsewhere indicate that the hydrogeologic nature of a sedimentary sequence may allow reliable predictions of ground water flow and thus enhancing the safety of the disposal vault.

Thick sedimentary sequences in three large regions of the Province of Ontario are potential sites for the disposal of nuclear fuel waste. The disposal vault could be constructed within a limb of the sedimentary basin in a competent rock formation and beneath a shale formation. Alternatively the vault could be constructed in the Precambrian basement rock beneath a thick sedimentary blanket. Both options would take advantage of the potentially favourable hydrogeological properties of a sedimentary sequence. In two regions ground water discharge from potentially suitable disposal locations would likely be into a large salt water body.

RESUME: Dans un dépôt souterrain, aucune barrière technique ne peut avoir une durée de vie illimitée, et la pénétration d'eau, le lessivage des déchets de combustible nucléaire et la mobilisation des radionucléides de longue durée sont des phénomènes inévitables. On doit donc, lors du choix de l'emplacement d'un dépôt, mettre l'accent sur les propriétés de confinement du milieu géologique. Des études effectuées en Ontario et ailleurs dans le monde indiquent que la nature hydrogéologique d'une série de couches sédimentaires peut être un indicateur fiable de la circulation souterraine d'eau, ce qui permet d'accroître la sûreté du dépôt souterrain éventuel.

Les séries sédimentaires épaisses dans trois grandes régions de l'Ontario sont des sites possibles pour l'évacuation des déchets de combustible nucléaire. Il serait possible d'aménager le dépôt souterrain dans un flanc du bassin sédimentaire d'une formation rocheuse appropriée, sous une formation schisteuse. Il serait également possible de l'aménager dans la roche du sous-sol précambrien, sous une couche sédimentaire épaisse. Ces deux possibilités tireraient profit des propriétés hydrogéologiques uniques et potentiellement favorables d'une série sédimentaire. Dans deux des trois régions, la résurgence de l'eau souterraine se ferait tout probablement dans une grande nappe d'eau salée.

1 INTRODUCTION AND BACKGROUND

The concept of nuclear fuel waste disposal in or beneath a thick blanket of sedimentary rock would take advantage of the favourable hydrogeological properties of stratified sedimentary rock units. The sedimentary blanket, which would include a thick shale unit, could act as an overlying hydraulic and radionuclide migration barrier. A competent rock formation within the sedimentary sequence or beneath it would host the disposal vault (Figure 1). It has been proposed by several researchers that disposal in a mixed geologic environment will enhance the safety of a deep geologic disposal facility (Cherry and Gale, 1979; Bredehoeft and Maini, 1981; Russel and Gale, 1982; Fye et al, 1984; Chapman and McKinley, 1987; Heystee and Freire-Canosa, 1988).

The Ontario Hydro study on disposal in or beneath a sedimentary sequence is being performed in parallel with and is

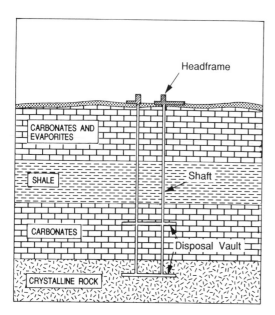

Figure 1   Disposal vault in or beneath
a thick sedimentary sequence

complementary to the more comprehensive
Atomic Energy of Canada Limited (AECL)
study on deep geologic disposal in plutonic
rock (Hancox, 1987). The broad
classification of plutonic rocks includes
all rocks crystallized from a molten state
deep within the earth's crust and would
include crystalline rocks beneath thick
sedimentary sequences. The AECL program is
evaluating all aspects of deep geologic
disposal in plutonic rock including
engineering design of the disposal vault,
geologic and hydrogeologic characteristics
of representative plutonic rock masses, and
the safety performance assessment of the
plutonic rock disposal concept. Ontario
Hydro is only studying the geologic and
hydrogeologic properties of thick
sedimentary sequences in Ontario and is not
duplicating any studies related to disposal
vault design and its assessment.

2 UNDERGROUND OPENINGS IN SEDIMENTARY ROCK

There are numerous underground openings in
sedimentary rock that have been constructed
for civil engineering and mining purposes.
These underground openings provide a unique
opportunity to observe ground water inflow
conditions from a large volume of rock.
The hydrogeologic nature of sedimentary
strata can be assessed at a scale that
would not be possible by using borehole

techniques. The underground opening
observations can be used to identify
sedimentary rock environments in Ontario
that are potentially suitable for nuclear
fuel waste disposal.

Raven et al, 1989 conducted a survey of
88 underground sedimentary rock openings
across North America. These openings are
constructed in shale, limestone, dolomite,
sandstone, or salt, and the opening depths
range from near surface to 700 m. The
findings of this study indicate that there
will likely be significant ground water
inflows to openings that are constructed in
dolomites, dolomitic limestones and
sandstones to depths of at least 250 m. In
contrast tunnels that would be constructed
in massive argillaceous limestones are
likely to be seepage-free (no visible
ground water inflow), particularly below
depths of at least 50 to 100 m. Although
shales are recognized as having low
hydraulic conductivity, few openings have
been constructed directly in shale at
depths below 50 to 100 m and ground water
inflow conditions could not be assessed.
Tunnels constructed within crystalline or
"clean" limestones would likely be
seepage-free if the openings are at depths
greater than 100 to 200 m and if low
permeability shale units overly and underly
the limestone formation. Inflow conditions
of underground openings in limestones and
to a lesser extent in shales may be wetter
if the openings are intersected by
permeable major structural discontinuities
such as transmissive vertical faults.

Figure 2 summarizes the interpreted range
of hydraulic conductivities for the
Palaeozoic rocks that were surveyed by
Raven et al, 1989 and for crystalline rocks
reported by Brace (1980,1984). The
Palaeozoic rock data includes borehole
hydraulic test results in the vicinity of
southwestern Ontario. The differences in
bulk hydraulic conductivity between certain
sedimentary rocks and crystalline rocks may
be attributed to the differences in
geomechanical properties and in-situ stress
conditions. Shales and shale-rich rocks
possess plastic deformation properties that
would promote self-healing of fractures
(Brace, 1980). Some of the limestone
formations that were surveyed by Raven et
al (1989) are subjected to high horizontal
in-situ stresses which would promote
fracture closure (Lee and White, 1986). On
the other hand crystalline rocks are more
brittle and thus more likely to maintain
permeable fractures. Thus it would be more
likely to construct a seepage-free
underground opening in fine-grained

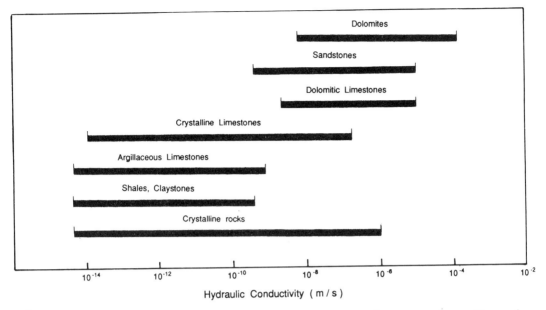

Figure 2   Interpreted range of hydraulic conductivity values for sedimentary and crystalline rocks (adapted from Brace 1980, 1984 and Intera Technologies Ltd. 1988)

limestones, shaley limestones and shales than it would be in crystalline rock.

3 HYDROGEOLOGICAL SUITABLE DISPOSAL ENVIRONMENTS

After more than a decade of research, there is now a consensus that the safe disposal of radioactive waste in certain deep geologic formations is possible. The Swedish and Swiss national disposal concepts have both undergone comprehensive reviews and have received conditional approval to proceed with their disposal plans (Chapman and McKinely, 1987; Nuclear Waste Bulletin, 1988). It is likely that other national programs will reach the same stage of development within the next decade and that repository construction will begin in several countries after the turn of the century.

There has been a great deal learnt about repository site selection and design through the collective efforts of these countries in their search for suitable disposal sites. For example at the start of many national disposal programs argillaceous formations, crystalline rocks and evaporites were selected as potential host formations because they were believed to be nearly impermeable to water. The original notion was that by constructing a repository in these types of geologic

formations the flow of water would be greatly limited in terms of both volume and velocity. This in turn would render the scenario of leaching and release of radionuclides to be an "unusual" event.

However now it is well accepted that engineered barriers will have a finite life and that ground water will eventually infiltrate the repository, leach the wastes and subsequently mobilize soluble long-lived radionuclides (Chapman and McKinley, 1987). This is supported by data from studies on engineered barriers and from field investigation programs of the geologic barrier. For example the long held belief that salt is dry and has no moving water has recently been questioned (Bredehoeft, 1988). Similarly in the Canadian program the field investigations at representative plutonic rock masses have demonstrated that these rocks are not "dry" as was originally thought (Boulton, 1978). Rather the plutonic rocks have been found to be usually saturated, the fracture fluid is saline and the rocks have measurable fractured-controlled permeability. With this realization the Canadian disposal program took a major new direction with siting research and hydrogeology becoming major components of the overall program (Cherry and Gale, 1979; Menely et al, 1982; Whitaker, 1987). Swedish, Swiss and British disposal programs recognize the importance the geologic barrier and have large siting research programs to collect the necessary data for repository design and performance assessments (Bjurstrom, 1988; Kowalski et al, 1988; Chapman et al, 1986).

The dominant role that hydrogeology will play in the long-term performance of a repository requires that it have an equally important role during the site selection process. Site selection factors which direct the search to hydrogeologically appropriate sites should be developed in conjunction with the traditional factors for tectonics and seismicity, minimum burial depth, constructability and socio-economic considerations.

The hydrogeological approach to siting emphasizes the containment properties of the overall geologic environment (rather than the intrinsic properties of the host rock). This approach to siting takes into consideration the fact that radionuclide releases to the geologic environment are likely to occur and must be controlled in a safe and environmentally acceptable manner. It identifies geologic environments that have the following attributes (Chapman et al, 1986):

(1) predictable ground water flow paths; preferably long and resulting in the

Figure 3   Geological regions of Ontario

progressive mixing with older waters or leading to discharge into a salt water body; and

(2) very slow local and regional ground water movements in an area with low regional hydraulic gradients.

Locations on limbs of three Ontario sedimentary basins (Figure 3) and beneath a thick shale formation would have these hydrogeologic attributes. In addition to having relatively low hydraulic conductivities, the argillaceous sedimentary rocks occur in predictable stratigraphic sequences that allow extrapolation of hydrogeologic conditions over large distances. Because ground water flow conditions are better understood in sedimentary sequences than in fractured crystalline rock masses the isolating capabilities of sedimentary rocks can be better assured. The potential down-dip flow of ground water along a basin limb and, in some cases discharge into a salt water body, are advantages to disposal in these geologic settings.

4 DISPOSAL IN OR BENEATH ONTARIO SEDIMENTARY BASINS

4.1 Great Lakes Lowland (Region 1)

The Palaeozoic rocks of the Great Lakes Lowland (Figure 3) are part of the Michigan and Appalachian Basins. In southwestern Ontario the maximum drilled thickness of the sedimentary sequence is 1440 m in the vicinity of Sarnia, Ontario (Brigham, 1971). The sedimentary rocks consist of a variety of carbonates, shales, evaporites and minor sandstones. Beneath the sedimentary sequence are the Precambrian Basement rocks which are a complex assemblage of metamorphosed igneous, sedimentary and volcanic rocks. The sedimentary rocks generally dip in a southwesterly direction with the oldest rock formations subcropping in the east and the youngest in the west (Figure 4).

On the basis of work done by Raven et al (1989) it is predicted that the Ordovician carbonate and shale formations of Region 1' are likely to be the most suitable for nuclear fuel waste disposal. These Ordovician formations comprise over 50 percent of the sedimentary basin volume in southwestern Ontario (Sanford, 1962). The Upper Ordovician shales thicken from 155 m in the north of Region 1 to 400 m in the south. There is no appreciable thickness of Upper Ordovician shales to the east of Toronto. The underlying Middle

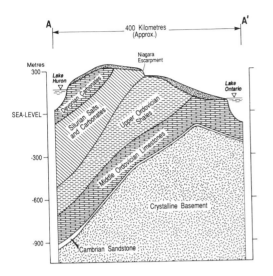

Figure 4   Great Lakes Lowland
Cross-section of Paleozoic rocks
(after Beards, 1967)

Ordovician carbonate rock formations vary
in thickness from 170 m near eastern
boundary of Region 1 to more than 300 m in
the west (Sanford, 1961; Winder and
Sanford, 1972). The repository could be
constructed within the thick sequence of
shales in a limestone formation (Russell
and Gale, 1982). Alternatively the
disposal vault could be constructed
underneath the Ordovician shales in
limestone or the Precambrian basement rock.

In addition to the presence of thick
shale formations, an important factor which
lead to the selection of Region 1 was the
apparent lack of oil and gas resources in
the Ordovician formations of this region.
The Ordovician formations produce 2 and
1 percent of the oil and gas in
southwestern Ontario. Of this production
the vast majority comes from three nearby
fields that are associated with dolomitized
limestone outside of Region 1 (Bailey
Geological Services Ltd and Cochrane,
1984a).

The underlying Cambrian formation in this
region is also not known to produce any
significant quantities of oil and gas
(Bailey Geological Services Ltd and
Cochrane, 1984b). Presumably there are few
abandoned and poorly sealed boreholes
within the Ordovician formations, and
future exploration for petroleum resources
will be limited in Ordovician formations of
Region 1.

Sanford et al (1985) has postulated that
Palaeozoic fault movements have occurred

from the late Cambrian to the late Devonian
times. The system of faults are believed
to transect the Palaeozoic rock strata and
at some locations rock blocks with sizes on
the order of kilometres are believed to
exist. With respect to deep geologic
disposal of nuclear fuel waste it will have
to established whether these faults exist
in Region 1 and their implications to
regional ground water flow will have to be
determined.

4.2 Hudson Platform (Regions 2 and 3)

The Hudson Platform lies in the central
part of the Canadian Shield and comprises
the remnants of the Moose River and Hudson
Bay sedimentary basins. The Precambrian
basement rocks are overlain by Palaeozoic
sedimentary rocks that consist of
limestones, dolostones, sandstones, shales
and evaporites. In the Hudson Bay Basin
the sedimentary rocks are up to 1000 m
thick beneath Hudson Bay. In the smaller
Moose River Basin the sedimentary rock
blanket is up to 760 m thick (Douglas,
1970). The two basins are separated by the
Cape Henrietta Maria Arch (Figure 5).
Since these sedimentary basins were once
connected to the above-mentioned
Appalachian and Michigan Basins, a close
correlation of geologic conditions has been
postulated (Sanford and Norris, 1975).

Limited drilling explorations during the
1960s and 1970s have found little evidence
of significant petroleum resources within
the Hudson Platform Palaeozoic rock
formations (Norris and Sanford, 1968;
Johnson, 1971). The corollary of this is
that there may be relatively little
exploration for petroleum resources in the
future. The harsh climatic conditions and
the rugged terrain will also inhibit the
exploration for and exploitation of any oil
and gas resources.

Within the Moose River and Hudson Bay
basins the only significant
(i.e., widespread and thick) shale unit is
the Upper Silurian Kenogami Formation
(Sanford, 1987). The argillaceous
component of this formation attains a
thickness of approximately 170 m and 800 m
in the Moose River and Hudson Bay Basins
respectively.

Within Region 2 of the Moose River Basin
it is assumed that ground water flow would
be down-dip towards the basin centre and
possibly discharging into the salt water of
James Bay. The topography of this region
is extremely flat and thus the hydraulic
gradients are predicted to be small. The
most suitable location for a disposal vault
would be beneath the Kenogami Formation

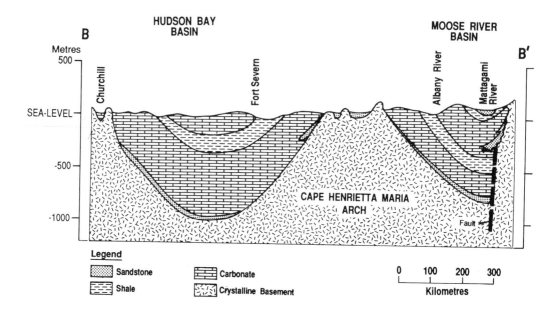

Figure 5   Hudson Platform : Cross-section of Phanerozoic rocks (after Douglas, 1970)

within or beneath the southern and western sedimentary limbs of the basin.

Region 3 is located on the shores of Hudson Bay and is centred on Fort Severn. Within this region it is expected that ground water flow will be down-dip and beneath the waters of Hudson Bay. It is proposed that the disposal vault should located along the shore of Hudson Bay and beneath the Kenogami shale formation. If the ground water from beneath the Kenogami formation discharges at the centre of Hudson Bay then it will have to do so through approximately 1 kilometre of sedimentary strata. The distance between this salt water discharge zone and the proposed disposal vault location is approximately 300 kilometres.

5 CONCLUSIONS

Deep geologic disposal within or beneath a thick sedimentary sequence appears to be an attractive alternative to the plutonic rock disposal concept. Large regions within three Ontario sedimentary basins have hydrogeologic conditions which may be potentially suitable for hosting a disposal vault. Common attributes of each region are that they have thick shale formations; each region has competent rock formations beneath the shale to host the disposal vault; there has been relatively little oil and gas exploration within potential host

formations in each region; and ground water is presumably down-dip towards the basin centre and in two cases the ground water most likely discharges into a large salt water body.

6 REFERENCES

Bailey Geological Services Ltd and R.O. Cochrane (1984a). Evaluation of the conventional and potential oil and gas reserves of the Ordovician of Ontario. Ontario Geological Survey, Toronto. Open file Report 5498.

Bailey Geological Services Ltd and R.O. Cochrane (1984b). Evaluation of the conventional and potential oil and gas reserves of the Cambrian of Ontario. Ontario Geological Survey, Toronto. Open file Report 5499.

Beards, R.J. (1967). Guide to the palaeozoic stratigraphy of southern Ontario. Ontario Department of Energy and Resources Management, Toronto Report 67-2.

Bjurstrom, S. (1988). Status of the Swedish nuclear waste management program. Waste Management '88, Tucson, Feb 28-Mar 3, 1988, V2, 11-18.

Boulton, J. (ed) (1978). Management of radioactive fuel wastes: the Canadian disposal program. Atomic Energy of Canada Ltd, Chalk River, Report 6314.

Brace, W.F. (1980). Permeability of crystalline and argillaceous rocks. Int J. of Rock Mech. and Mining Sci. 17, 241-251.

Brace, W.F. (1984). Permeability of crystalline rocks: new in-situ measurements. J. of Geoph. Res. 89, 4327-4330.

Brigham, R.J. (1971). Structural geology of southwestern Ontario and southeastern Michigan. Petroleum Resources Section, Mines and Northern Affairs, Toronto, Paper 71-2.

Bredehoeft, J.D. and T. Maini (1981). Strategy for radioactive waste disposal in crystalline rocks. Science 213, 293-296.

Bredehoeft, J.D. (1988). Will salt repositories be dry? EOS, Trans. Am. Geoph Union, March 1, 1988, p. 121, 131.

Chapman, N.A., T.J. McEwan and H. Beale (1986). Geologic environments for deep disposal of intermediate level wastes in the United Kingdom. Proc Int Symp on the Siting, Design and Construction of Underground repositories for Radioactive Waste, 311-328. Int. Atomic Energy Agency, Vienna.

Chapman, N.A. and I.G. McKinley (1987). The geologic disposal of nuclear waste. John Wiley and Sons, Chichester.

Cherry, J.A. and J.E. Gale (1979). The Canadian program for a high-level radioactive waste repository: hydrogeological perspective. In C.R. Barnes (ed), Disposal of high-level radioactive waste: the Canadian geoscience program, Paper 79-10, 35-44. Geological Survey of Canada, Ottawa.

Douglas, R.J.W. (ed) (1970). Geology and economic minerals of Canada. Economic Geology Report No. 1. Geological Survey of Canada, Ottawa.

Fye, W.F., V. Babuska, N.J. Price, E. Schmid, C.F. Tsang, S. Uyeda and B. Veloe (1984). The geology of nuclear waste disposal. Nature 310, 537-540.

Hancox, W.T. (1987). Safe, permanent disposal of nuclear fuel waste. Nucl. J. of Canada 1, 256-267.

Heystee, R.J. and J. Freire-Canosa (1988). Disposal beneath a thick sedimentary sequence in crystalline rock. Proc. Waste Management '88, Tucson, Feb 28-Mar 3, 1988, V2, 753-760.

Intera Technologies Ltd (1988). Inventory and assessment of hydrogeologic conditions of underground openings in sedimentary rock. Report to Ontario Hydro, Toronto.

Johnson, R.D. (1971). Petroleum of Hudson Bay Basin studied. Oilweek, May 10, 1971, p. 40, 43, 52.

Kowakski, E., C. McCombie and H. Issler (1988). Status of Swiss waste disposal projects. Waste Management '88, Tucson, Feb. 28-Mar 3, 1988, V2, 275-280.

Lee, C.F. (1981). In-situ stress measurements in southern Ontario Proc, 22nd US Symp on Rock Mechanics, Mass. Inst. Tech, Boston, Mass, June 28 - July 2, 1981, 435-442.

Lee, C.F. and O.L. White (1986). On some geomechanical aspects of high horizontal stresses in rock. Proc. Int. Symp. on Geomechanics, Beijing, Sept 4-10, 1986.

Meneley, W.A., R.A. Freeze, C.L. Lin, and F.J. Pearson Jr. (1982). Hydrogeology program review. Atomic Energy of Canada Ltd, Chalk River, Report TR-198.

Norris, A.W. and B.V. Sanford (1968). Operation Wimisk - an air-supported geological reconnaissance survey of Hudson Bay Lowlands. Ontario Petroleum Inst. 7, 1-33.

Nuclear Waste Bulletin (1988). Government approves Project Gewahr. Nuclear Energy Agency, Paris, 33-34.

Russell, D.J. and J.E. Gale (1982). Radioactive waste disposal in sedimentary rocks of southern Ontario. Geoscience Canada 9, 200-207.

Sanford, B.V (1961). Subsurface stratigraphy of Ordovician rocks in southwestern Ontario. Geological Survey of Canada, Ottawa Paper 60-26.

Sanford, B.V. (1962). Sources and occurrences of oil and gas in sedimentary basins of Ontario. Proc. Geol. Ass. of Canada 14, 59-89.

Sanford, B.V. and A.W. Norris (1975).
Devonian stratigraphy of Hudson Platform
- part 1: stratigraphy and economic
geology.   Geological Survey of Canada,
Ottawa. Memoir 379.

Sanford, B.V., T.J. Thompson, and
G.H. McFall (1985).   Plate tectonics - a
possible controlling mechanism in the
development of hydrocarbon traps in
southwestern Ontario.   Bull. of Can. Pet.
Geol. 33, 52-71.

Sanford, B.V. (1987).   Palaeozoic geology
of the Hudson Platform.   In C. Beaumont
and A.J. Tankard (eds), Sedimentary Basin
and Basin-Forming Mechanisms, 483-505.
Canadian Society of Petroleum Geology,
Memoir 12.

Raven, K.G., R.A. Sweezey, and
R.J. Heystee   (1989).   Hydrogeologic
conditions of underground openings in
sedimentary rocks.   Proc. Int. Cong. on
Progress and Innovation in Tunnelling,
Toronto, Sept. 9-14, 1989, 567-574.

Whitaker, S.H. (1987). Geoscience research
for the Canadian Nuclear Fuel Waste
Management Program.   Radioactive Waste
Management and the Fuel Cycle 8, 145-196.

Winder, C.G. and B.V. Sanford (1972).
Stratigraphy and paleontology of the
palaeozoic rocks of southern Ontario.
24th Int. Geol. Cong, Montreal, Excursion
A45-C45.

6th International IAEG Congres / 6ème Congrès International de AIGI, © 1990 Balkema, Rotterdam. ISBN 90 6191 130 3

Heated water storage in underground openings
Le stockage d'eau chauffée dans des cavités souterraines

Yoshinori Inada & Naoki Kinoshita
Department of Civil Engineering, Ehime University, Matsuyama, Japan

Hidehiko Nakazaki & Takao Ueda
Takenaka Komuten Co., Ltd, Tokyo, Japan

ABSTRACT: *Heated water produced by using surplus heat from sources such as garbage-burning plant will be utilized well in the near future for many purposes. To insure a stable supply, temporary storage in underground openings in rock mass may become an important·problem. In this study, leakage of heated water from openings were estimated by analysis using the physical properties of rock which were obtained by experiment.*

RESUME: *L'eau chauffé était produit par employant le chaleur surplus de brûler d'ordures peut être utilisé a beaucoup d'effets. Pour avoir une bonne réserve deau, ça devient important davoir l'entreposage temporaire dans les entrées souterains de rocher. Dan çette étude, la fuite d'eau chauffé des entrées étaient évalués par l'analyse qui utilise les propriétés physiques de rocher et de glace obtenant par l'experiencé.*

1. INTRODUCTION

For the purposes of improving the quality of life and saving energy, heated water produced by using surplus heat from garbage-burning plants etc.,will be utilized well in near future for many purposes, such as district heating and hot-water supply systems, melting of snow on roadways, heated swimming pools, fish farming and space heat for greenhouses.

As the utility quantity of heated water will change with a daily or seasonal cycle, a temporary storage in underground openings excavated in rock mass becomes an important problem to insure a stable supply(Inada etc 1988, 1989).

In this case, the leakage of heated water from the openings become an important problem.

Effect of high temperature and high pressure on permeability of rocks was investigated by experiment. And then using these results, leakage of heated water from openings was analyzed and discussed.

In this study, it is assumed that the opening has a diameter of 10m is excavated in fresh rock mass at a depth of 100 meters from the surface of the mountain, and heated water is stored directly.

2 PERMEABILITY OF ROCK AT HIGH TEMPERATURE AND PRESSURE

2.1 Specimen used for this experiment

The rock used for this experiment was granite (obtained in Oshima, Ehime, Japan). By measuring the elastic velocity of rock, we can know the rift plane and hardway plane. Specimens were prepared in the rift plane direction or the hardway plane direction agreeing with axial direction.

Physical properties of the specimen are shown in Table 1.

2.2 Experimental method and apparatus

The experiment was carried out at high temperature and high pressure. A schematic diagram of the apparatus for measuring permeability is shown in Fig.1.

Surrounding pressure, axial pressure and seepage pressure were devised to work independently by using separate hydraulic pressure.

Specimens were heated from room temperature to the required temperature in the cell. A heating rate of 0.5 °C/min was adopted for this experiment. In experiments, it is necessary to make the temperature in a specimen uniform. Specimens were put in the cell before heating and

Table 1 Physical properties of rocks.

	Wet			Specific gravity (Dry)	
Porosity (%)	Water content (%)	Degree of saturation (%)	True specific gravity	Bulk specific gravity
1.16	0.26	60.28	2.669	2.641

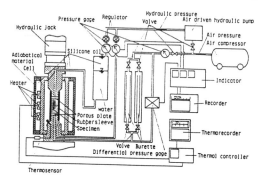

Fig.1 Schematic diagram of the test.

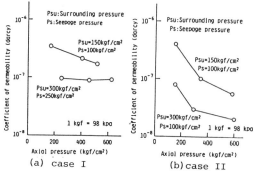

(a) case I　　　　　　(b) case II

Fig.2 Effect of axial pressure on permeability.

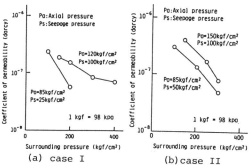

(a) case I　　　　　　(b) case II

Fig.3 Effect of surrounding pressure on permeability.

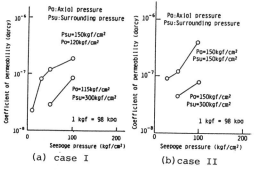

(a) case I　　　　　　(b) case II

Fig.4 Effect of seepage pressure on permeability.

after reaching the required temperature, specimens were kept at the required temperature for more than 1 hr, and then the permeability test was carried out. Darcy's law was adopted for estimating the coefficient of permeability.

Many mechanical investigations on anisotropy of granite have been done(Kudo etc. 1986). However, it is supposed that anisotropy should have some effects upon permeability.

2.3 Results and consideration

Results are shown in Fig.2(a),(b)-Fig.5 (a),(b). However, the experiment was performed under the following conditions:
(1) Case I :Saturated with water parallel to the rift plane.
(2) Case II:Saturated with water parallel to the hardway plane.

From Fig.2(a),(b), it is found that the coefficient of permeability decreases with the increase of axial pressure in both cases. This seems to be caused by the fact that granite is an aggregate of rock forming grains. That is, water flows through the microcracks which exist in mineral grains themselves and between mineral grains, but mineral grains deform with increasing axial pressure. As a result, water flow decreases.

Fig.3(a),(b) show the results of the case which changde the surrounding pressure. Likewise, the coefficient of permeability decreased with increased surrounding pressure, the same as the case which changde the axial pressure. However,

the gradient is larger in surrounding pressure than in axial pressure, so the surrounding pressure has a large influence on permeability. That is, the gradient is 6 times larger in case I and 2-3 times larger in case II.

The results in the case which changed seepage pressure is shown in Fig.4(a),(b). From these figures, it is found that the coefficient of permeability increases with the increase of seepage pressure. The

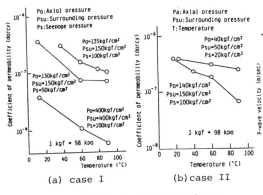

(a) case I   (b) case II

Fig.5 Effect of temperature
on permeability.

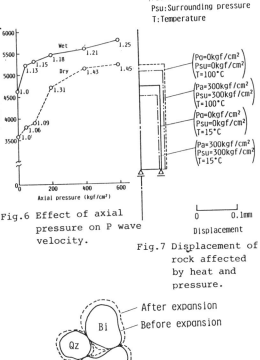

Fig.6 Effect of axial
pressure on P wave
velocity.

Fig.7 Displacement of
rock affected
by heat and
pressure.

Fig.8 Schematic diagram of expansion
of mineral grains.

reason seems to be that seepage pressure spreads or increases the microcracks existing in mineral grains themselves and between grains, allowing, water to flow easily.

Moreover, the results when keeping axial pressure, surrounding pressure and seepage pressure constant and changing the temperature of the specimen from room temperature to 90°C, is shown in Fig.5(a),(b). It is seen in both cases that the coefficient of permeability decreases with rising temperature. This reason seems to be that the volume of rock forming grains expands respective to rising temperature, but it also effects the inside by axial pressure and surrounding pressure. Therefore, water flow decreases and finally permeability decreases.

From the results mentioned above, it was found that permeability of granite is affected by surrounding pressure, seepage pressure and temperature largely, and very little by axial pressure.

The anisotropy effect upon permeability was seen only when axial pressure changed. From these facts, it is found that anisotropy does not have an effect upon permeability in the case where rock which is fresh and has few cracks.

Then, to make a fundamental consideration for the effects of heat and pressure on granite, the elastic wave velocity was measured and a fundamental analysis was made.

Specimens like these used for the permeability test were used for measuring the P wave velocity receiving axial pressure. The results are shown in Fig.6. From this figure, it is seen that the value increases with increasing axial pressure, For comparison specimens which were dried in a desiccator for a week, were measured for P wave velocity and the results are

shown in Fig.6. It is also seen that the value increases with increasing axial pressure. From these results, we can estimate that microcracks in rock will close and will disturb the flow.

Displacement of rock was analysed by FEM for following cases:

(1) the case in which the acting axial pressure and surrounding pressure are 300kgf/cm² respectively under normal temperature, (2) the case in which the surrounding temperature is 100°C, (3) the case in which acting axial pressure and surrounding pressure are 300kgf/cm² respectively under 100°C.

Results are shown in Fig.7. In the case of (2) and (3), it is seen that the rock expands by changing the surrounding temperature from 15°C to 100°C. However, in the case (3), as displacement is smaller than in case (2), it is found that axial pressure and surrounding pressure effect the internal structure of rock. In this analysis, as rock is supposed to be a homogeneous material, so displacement appears towards the outside. However, in the case of actual rock, as is shown in

Fig.8, rock forming mineral grains expand inwardly as adjacent grains become fixed to each other. As a result, microcracks will be closed. Therefore, actual displacement will be smaller than the results of Fig.8. From the results mentioned above, it seems that the flow will be disturbed with rising temperature and permeability will be decreased.

3 CHANGE OF PERMEABILITY OF ROCK AT HEATED WATER STORAGE

3.1 Simple equation for estimating coefficient of permeability

As temperature distribution and stress distribution change with time, the coefficient of permeability changes. That is, as leakage of heated water changes with time, simply estimating the coefficient of permeability corresponding to every state becomes important for analyzing leakage.

For this reason, a simple equation for estimating the coefficient of permeability was derived from the results of the experiment mentioned above. Equation (1) was obtained from the results of case I,

$$\log K=(-6340-0.56Pa-3.4Psu+3.4Ps-5.2T) \times 10^{-3} \qquad (1)$$

Equation (2) was obtained from the results of case II.

$$\log K=(-6100-1.6Pa-3.7Psu+4.7Ps-5.4T) \times 10^{-3} \qquad (2)$$

where, Pa:axial pressure (kgf/cm^2),
Psu:surrounding pressure (kgf/cm^2),
Ps:seepage pressure (kgf/cm^2),
T:temperature (°C), K:coefficient of permeability (darcy) ·

3.2 Estimation of leakage of heated water

Using Equation (1) and standardizing the coefficient of permeability before storage of heated water, the change of the permeability ratio from the start of storage to after 1 year is shown in Fig.9(a)-(d).

From Fig.9(a), it is seen that the coefficient of permeability decreases suddenly near the surface of the openings from the initial time and becomes the value of 1/30 of before storage after 1 day. The range of decreasing spreads with time and at near the surface of openings the value changes to 1/50 before storage after 1 year.

In fact,it is known that the viscosity

of water at 100°C becomes about 1/4 of the value at 15°C(Tokyo astronomical observatory,1988). However, if we consider these facts, it is thought that permeability near the surface of openings becomes about 1/10 of that before storage.

Also we must consider cases which need large quantities of heated water or which need multiple openings due to topography and geology. In this study, it is assumed that the diameter of openings is 10m, distance between surface of openings are 20m and 10m respectively, and the coefficient of permeability was compared. The results obtained are shown in Fig.9 (b),(c).

In the case where surface distance of the openings is 20m, it is seen that the coefficient of permeability decreases suddenly from early times in any area, between openings and outside the openings, and the value becomes 1/40 at the surface of the openings after 30 days. Conversely, the value becomes 1/25 as stress relaxation occurs after 1 year. This is caused by the fact that openings deform as shown in Fig.10(a) and compressive stress at tangential direction is relaxed. The value outside of the openings is same as for mono-opening.

In the case in which distance between openings is 10m, the value of coefficient of permeability near the surface shows a similar tendency. That is, the value is 1/15 between openings and 1/25 outside of openings. As is seen in Fig.10(a),(b), deformation of relaxation of stress is also larger in (b).

Moreover, Fig.9(d) shows the results when the 10 m diameter openings were arranged in a horizontal line. It is seen that the value after 1 day becomes 1/30 of that before storage, but becomes 1/3 after 1 year due to stress relaxation.

Using the coefficient of permeability obtained by Equation (1), we tried to estimate the leakage of water from the openings. In this study, total leakage of water is indicated as a ratio for initial storage volume of heated water. From the results, as is shown in Fig.11, total leakage of water after 1 year is 1.78% with mono-opening, 1.87% and 1.96% with two openings where distance between openings was 20m and 10m respectively. In the case where openings were arranged in a horizontal line, leakage of water was 4.23%. This volume 2.4 times larger than mono-opening. It is considered that this was caused by adjacent openings affecting each other and coefficient of permeability became larger in comparison with the case for mono-opening. In any case, as mentioned above, leakage of water is only 2-

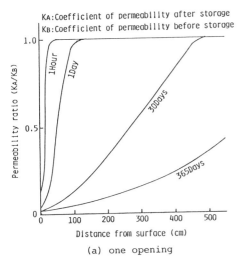

(a) one opening

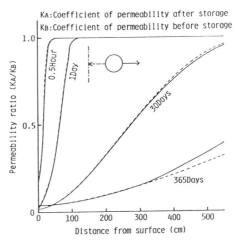

(b) two openings (distance between openings: 20m)

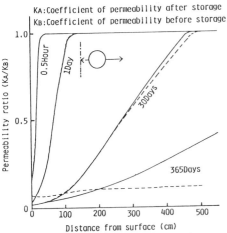

(c) two openings (distance between openings: 10m)

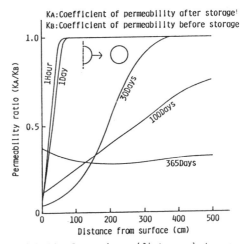

(d) lined openings (distance between openings: 10m)

Fig.9 Temporal change of permeability ratio along horizontal axis.

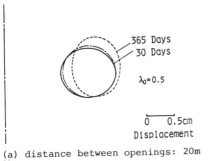

(a) distance between openings: 20m

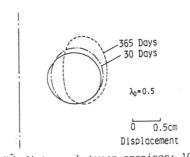

.(b) distance between openings: 10m

.Fig.10 Change of openings' shape.

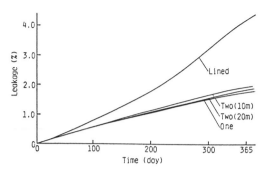

Fig.11 Leakage from the openings.

4%, so we can neglect it as insignificant.

4 CONCLUSION

The main results obtained in this study are as follows:

(1)   From the results of a permeability test using granite, it was found that permeability is effected by surrounding pressure, seepage pressure and temperature largely and only slightly receive axial pressure.

(2)   On the effect of anisotropy on permeability of granite, is was found for fresh rock used in this study that axial pressure has a slight influence, but surrounding pressure, seepage pressure and temperature have almost no influence.

(3)   To estimate permeability around openings which change with time, a simple equation for estimating the coefficient of permeability was derived from the results of the experiment.

(4)   From the results estimating the leakage of heated water from openings using the equation mentioned in (3), it was found that leakage of water is only 2-4%, which is neglegible.

REFERENCES

Inada,Y., Manabe,T., Ohashi,S. & Yoshikawa, M., 1988,"Stability of Underground Openings due to Storage of Heated Water", Proc.3rd Int.Conf. Underground Space and Sheltered Buildgs., pp.338-343.

Inada,Y. & Sterling,R.L.,1989,"Storage of Heated Water in Underground Openings", Proc.J.Geotech.Engrg., ASCE, 115, 5, pp. 597-614.

Inada,Y., Kinoshita,N., Nakazaki,H. & Ueda, T.,"Storage of Heated Water in Multiple Openings", Proc.Symp. on Utilization of Underground Space 1989, pp.77-82.

Kudo,Y., Hasimoto,K., Sano,O. & Nakagawa, K., 1986, "Relation between Physical Anisotropy and Microstructure of Granite", Proc.of Japan Soc.of Civ. Engrg., 370, 3-5, pp.189-198.

Tokyo astronomical observatory, 1988, "Science chronological table", Maruzen, pp.446-447.

Low temperature materials storage in underground rock mass
Le stockage de matières à température basse dans des masses rocheuses souterraines

Yoshinori Inada & Yuichi Kohmura
Department of Civil Engineering, Ehime University, Matsuyama, Japan

Shuntaro Ohashi
Applied Technology Co., Ltd, Osaka, Japan

Masaaki Yoshikawa
Kinki Geo-Engineering Center Co., Ltd, Kyoto, Japan

ABSTRACT: *From the view point of multiple-utilization of land and the reduction of heat loss from low temperature materials, storage of these materials in underground openings excavated in rock mass should be considered. In this case, leakage of liquid, gas and heat along the cracks becomes an important problem. One of the useful prevention technique is to enclose the gas with water, which we call a "water curtain system". This paper describes the physical properties of ice obtained by experiment at low temperatures. Finally, the behavior of rock and ice were analyzed theoretically and the effect on the water curtain system is discussed.*

RESUME: *Afin d'augmenter l'utilisation de la terre et de réduire la perte de chaleur, c'est important à examiner le question d'entreposage des matières ayant la temperature basse dans les entrées souterains qui sont excavés dans le rocher. Dans ce cas, le problème de la fuite de liquide, de gaz de chaleur devient important. La clôture de gaz dans l'eau, que nous appelons "la système de rideau d'eau", est une technique très utile. Çet article décrit les propriétés physiques de glace qui étaient obtenus par les expériénces à les temperatures basses. L'article analyse aussi théoriquement le fonctionnement de rocher et de glace et l'effet sur la système de rideau d eau.*

1 INTRODUCTION

In the future, storage of low temperature materials such as LNG, LPG and frozen food, in underground openings excavated in rock mass will be desirable for effective utilization of land. Especially in Japan, where a rapid increase of distribution of frozen food involving importing is anticipated, underground openings excavated in rock mass will be newly utilized as temporary refrigeration storage.

In a separate publication the authors have shown that such openings shrink towards the mountain where they are excavated and cracks occur radially advancing with time(Inada etc. 1986).

As leakage of gas and liquid from the cracks can be expected, a way of preventing this becomes an important consideration.

The authors had been proposed a water curtain system as one of the effective methods for preventing leakage of gas and liquid (Inada 1989).

In this study, the results of the investigation of the physical properties

of ice at low temperatures by experiment, simulation using large scale calculations are also described. Moreover, the effects of the water curtain system were investigated by experiment in the laboratory.

2. PHYSICAL PROPERTIES OF ICE AT LOW TEMPERATURES.

2.1 Strength and deformation characteristics of ice

Specimens used for this experiment were made as follows:

1. Distilled water was poured into the molding box.
2. It was kept in a vacuum desiccator for 1 hour with care that air bubbles in the molding box were removed from small holes which were set up on the surface of the molding box.
3. The molding box was put into the freezer used for microbic culture, as the temperature is controllable within the range of 1/10°C.
4. Cooling temperature was kept between -1∿-2°C for 2 days.

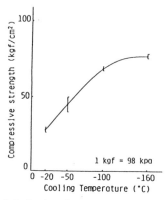

Fig.1 Relation between compressive
strength and cooling temperature.

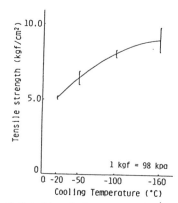

Fig.2 Relation between tensile
strength and cooling temperature.

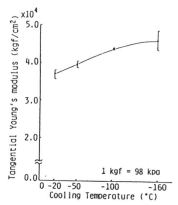

Fig.3 Relation between tangential Young's
modulus and cooling temperature.

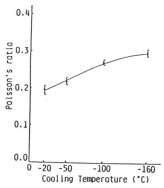

Fig.4 Relation between Poiison's ratio
and cooling temperature.

Through trial and error, it was found that if cooling temperatures were much lower, the ice froze away from the surroundings of the molding box, and one crack detectable by the naked eye, formed in the center of the specimen. Thus, a temperature which keeps the specimens uniform in a sherbet-like state was used.

Specimens were prepared as cylinders, 30mm in diameter and 60 mm in length for uniaxial compression tests, and as cylinders, 30 mm in diameter and 30 mm in length, for radial compression tests (Brazilian test) instead of the uniaxial tension tests.

Tangential Young's modulus and Poisson's ratio were obtained by a stress-strain curve obtained when compression tests were carried out.

Fig.1 and Fig.2 show compressive strength and tensile strength of ice at low temperatures, the values being an average of 5-6 measurements.

From these figures, it is seen that ice shrinks and its strength increase with falling temperatures. That is, the compressive strength of ice is 2.7 times larger at -160°C and tensile strength is 1.8 times larger at -20°C. Comparing this with strengths of granite which obtained by the authors (Inada 1980), 1.4 times for compressive strength and 1.3 times for tensile strength, the ratio of increase of ice is larger than that of rock. As the strain on ice at -160°C obtained by another experiment was found to be 3.9 times larger than that of rock, so the degree of shrinking and hardening of ice is larger than rock. These facts are also considered as factors that have an effect on rising strength.

Fig.3 and Fig.4 show tangential Young's modulus and Poisson's ratio at 30% of fracture stress respectively. The values increase with falling temperatures.

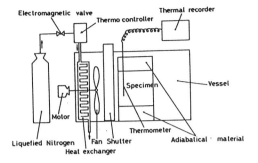

Fig.5 Schematic diagram of the test.

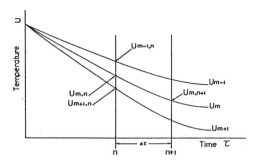

Fig.6 Schematic diagram of temperature
change with time.

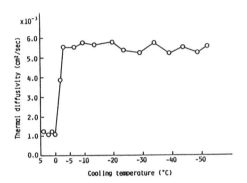

Fig.7 Thermal diffusivity at
low temperatures.

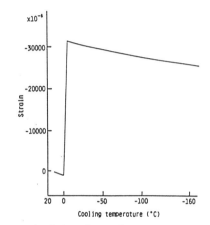

Fig.8 Strain at low temperature.

2.2 Thermal properties of ice at low temperatures

Thermal diffusivity of ice changing with temperatures has not previously been investigated. In this study, a combination of numerical (FDEM) (Inada etc. 1983) and experimental techniques was used.

This method has the following advantages:

1. This method is able to determine the thermal diffusivity of the rock at many different temperatures in one experiment while keeping the surface temperature constant.
2. The measurement time is short.
3. Fewer specimens are required.

The schematic diagram of the experimental apparatus is shown in Fig.5.

If a one-dimensional heat conduction equation is assumed, the temperature of element number m and time n is given by:

$$(\frac{\Delta x^2}{\kappa \Delta \tau} - 2)U_{m,n} + U_{m+1,n} + U_{m-1,n} = \frac{\Delta x^2}{\kappa \Delta \tau}U_{m,n+1} \quad (1)$$

where, $U_{m,n}$: the temperature in element m and in any time n

κ : thermal diffusivity

Δx : distance between elements

n : any time

n+1 : time later than n

$\Delta \tau$: time interval

If a series of temperature-time curves for closely spaced points within the specimen is obtained by experiment like Fig.6, the quantities of $U_{m,n}$, $U_{m+1,n}$, $U_{m-1,n}$ and $U_{m,n+1}$ can be determined for a particular time interval.

Then by using Eq.1, we can determine the value of   for an average temperature $(U_{m,n+1} + U_{m,n})/2$. The results obtained are shown in Fig.7.

It is seen that the heat conduction of water is lower but becomes higher rapidly at the time of changing to ice, and after changing to ice, the value of diffusivity becomes constant.

Thermal expansion of ice at low temperatures was measured by a comparison method using quartz glass which was proposed by the authors (Inada etc. 1971).

2753

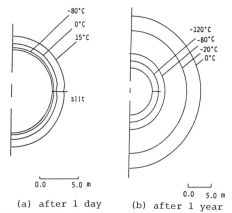

(a) after 1 day   (b) after 1 year
Fig.9 Temperature distribution around openings.

Fig.8 shows strain of water and ice at low temperatures. Sudden changes are seen at -5°C, but the gradient becomes lower after water changes to ice.

3. STORAGE WITH WATER CURTAIN SYSTEM

3.1 Temperature distribution considering freezing

As mentioned previously, in this study the water curtain system was adopted as one way to prevent leakage of gas, liquid and cooling air from cracks occurring around the opening.

The behavior of ice and rock mass were investigated by analysis. To facilitate understanding, slits which were larger than cracks were made and the behavior of ice in them was analyzed. That is, it is assumed that the 10 m-diameter opening is excavated at a depth of 100 m from the ground surface and two slits (3 cm in breadth, 2 m in length to the horizontal direction from the surface of the opening) are set up.

Fig.14(a),(b) show temperature distribution around the opening calculated by FDEM where LNG is stored. After 1 day, delay of heat conduction at the slit and its surroundings are seen in (a), however they are not observed after 1 year. It is considered that delay of heat conduction occurred due to existing water in the slit in early time. That is, as thermal diffusivity of water is about 1/10 that of rock, a delay of heat conduction occurred. But water changes to ice with time and as the value of diffusivity of ice is the same as rock, the delay of heat conduction disappears.

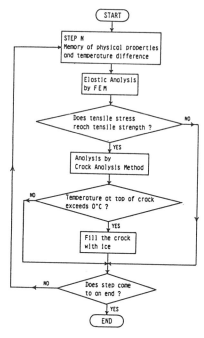

Fig.10 Flow chart of analysis.

3.2 Stress analysis considering freezing

In this study, Crack Analysis Method (CAM) (Inada etc. 1986) was adopted and the theory of Mohr's envelope was adopted as a failure criteria(Inada etc. 1987).

The time from start till 1 year when a semi-steady state was reached, was divided into 7 steps for analysis. When cracks occurred in ice in slit, it was supposed that the water entered into cracks from the top of the cracks momentary, if temperature of water over 0°C and change to ice momentary. The flow chart of the analysis is shown in Fig.10.

Fig.11(a)-(d), Fig.12(a)-(d) and Fig.13 (a)-(d) show the results of analysis of the behavior of ice in slits in the case in which LNG, LPG and frozen food were stored in the opening. Ice froze with a thickness of 78cm, 50cm and 58cm respectively from the surface of the opening after 1 day, but the ice was in an unstable state involving many cracks.

It is seen that cracks filled with new ice from the surface of the opening after 30 days. And the effect of the water curtain system was seen as cracks filled with new ice with a thickness of 50cm, 26cm and 31cm respectively after 1 year.

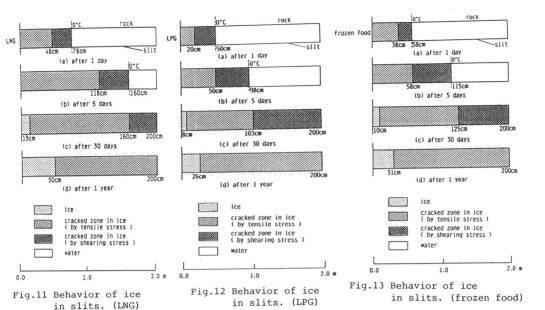

Fig.11 Behavior of ice in slits. (LNG)

Fig.12 Behavior of ice in slits. (LPG)

Fig.13 Behavior of ice in slits. (frozen food)

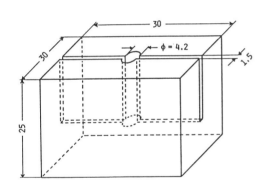

Fig.14 Schematic diagram of the test.

4. CONSIDERATION FOR STORAGE WITH A WATER CURTAIN SYSTEM

To confirm the effect of the water curtain system, experiments were carried out using a cement mortar specimen. A schematic diagram is shown in Fig.14.

In the experiment, the water level was made equal to the upper surface of the specimen, and liquefied nitrogen (-196°C) was poured into a central hole of the specimen. After water froze in the slit, the slit was enlarged by adding force and cracks was made in the ice in the slit. Leakage of liquefied nitrogen and ways to close the crack were observed.

Fig.15(a)-(c) shows the results of experiment. From(a), we can see the delay of heat conduction in the slit due to the frost state occurring on the surface of the specimen. This is due to the fact that heat conduction of water is lower as mentioned above.

Even if the crack was made as in (b), as water entered into cracks changed to ice momentary, crack filled with ice momentarily like (c) and leakage of liquefied nitrogen was not seen. From these facts, it is assumed that the water curtain system could be an effective way of preventing leakage.

7. CONCLUSION

The results obtained in this study are as follows:

(1) The strength of ice increases with falling temperatures. That is, the compressive strength of ice at -160°C is 2.7 times larger and tensile strength is 1.8 times larger than at -20°C.
(2) The value of tangential Young's modulus and Poisson's ratio of ice at low temperatures increases with falling temperature.
(3) Thermal diffusivity of water changing to ice at low temperature can be estimated easily by combination with FDEM and experiment.
(4) The strain of water changing to ice at low temperatures can be estimated easily by the Comparison Method using

(a) start of the test

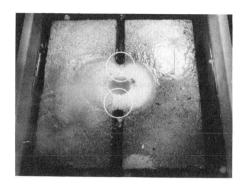

(b) occurrence of the cracks

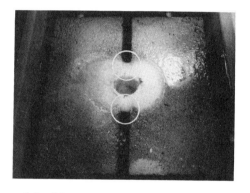

(c) effect on water curtain system

Fig.15 Test for the effect on
water curtain system.

quartz glass.

(5) To prevent the leakage of gas, liquid and cooling air from the cracks around the opening, the water curtain system was proposed.

(6) From the results of analysis and experiment, it was found that the water curtain system could be an effective way to prevent leakage.

REFERENCES

Inada,Y. Terada,M. and Ito,I., 1971, On the coefficient of thermal expansion of rocks, Suiyokwai-shi, Vol.17, No.5, pp.200-203.

Inada,Y. and Yagi,N., 1980, Thermal Properties of rocks at low temperatures, J. Soc.Mate.Sci.Japan, Vol.29, No.327, pp. 1228-1233.

Inada,Y. and Shigenobu,J., 1983, Temperature distribution around underground openings excavated in rock mass due to storage of liquefied natural gas, J.Min. Meta.Inst. Japan, Vol.99, No.1141, pp. 179-185.

Inada,Y. and Taniguchi,K., 1986, A theoretical analysis of the range of plastic zone around openings due to storage of LNG, U.S.Symp.Rock Mech., Proc.27th, pp. 782-788.

Inada,Y. and Taniguchi,K., 1987, Plastic zone around openings excavated in rock mass due to storage of liquefied natural gas, I.Min.Meta.Inst. Japan, Vol.103, No.1192, pp.365-372.

Inada,Y., 1989, Low temperature storage of gases in underground openings, Proc.Int. Conf.on Storage of Gases in Rock Caverns, pp.259-266.

Reservoir-induced earthquake and deep-well-injection earthquake
Séisme induit de réservoir et séisme dans l'eau injecté de puits profond

Chunshan Jin
Department of Civil Engineering, Dalian University of Technology, People's Republic of China

ABSTRACT: According to the fact that characteristics of local fluid earthquakes induced due to the impoundment to the reservoir and those induced due to deep-well-injection are very similar, it makes a possibility to trace an analogy between them. The investigation on mechanism of triggering the deep-well-injection earthquake is taken as basis of the study of reservoir-induced earthquake so as to achieve a way to predict it. The reservoir-induced earthquake is considered as a result of crustal dilantancy and fluid flow. Furthermore, the hydrodynamic mechanism of the induced earthquakes and regurality of change of critical pore-fluid pressure are also discussed in the paper.

RESUME: D'après le fait que le séisme de local fluid induit de l'eau retenue au reservoir se manifeste de même façon que celui induit de l'eau injectée au puit profond, ils sont donc comparables. L'étude sur le mécanisme du séisme induit de l'eau injectée au puit profond permet d'analyser le séisme de réservoir et de le prévoir. le séisme de réservoir est considéré comme um résultat de dilatation de l'écorce terrestre et de l'écoulement de fluid. Finalement, le mécanisme hydro-dynamique de séisme et la loi de l'évolution de pression critique de pore fluide ont été discutés.

1 INTRODUCTION

The concept of pore-fluid pressure, especially the concept of abnormal high pore-fluid prossure(Pa) and the effective stress, have great significance for almost all domaines of earth science. I believe that what the significance of $\tau = \tau_{\circ} + \mu(\sigma_n - p)$ in physical seismology is to the significance of $l_g N = a - b M$ in statistical seismology, which is very important for the research in the formation mechanism of Reservoir-Induced-Earthquake(RIE).

The stress increase and underground fluid transport are two big factors in leading to natural earthquake. RIE is the energy release in advance caused by the accumulation of tectonic stress field and hydrodynamic pressure field due to the engineering activities. It is the local phenomena in the development of earthquake caused by the two factors. RIE is mainly tectonic type. In contrast with natural earthquake, it has specific condition and background and is not directly caused by tectonic movement. The stress increase is limited, but water action is giant. Therefore, RIE is the result of crustal dila-

tancy and fluid flow. In China it happened a lot. We should know the fact that the increase of underground water seepage flow and Pa trigger effect is in the leading position. The paper is on the hydrodynamic mechanism in the formation of RIE, for stillwater pressure, as well as water load, is not strong enough to induce earthquake.

Analogy is basic study method of geology. In order to study waterdynamics mechanism in the formation of RIE, we should devote much attention to analogy between RIE and DWIE. RIE is a kind of fluid earthquake stimulated by instantaneous fluid pressure in the formation after the injection of high-pressure water. It has high probability, definite hydrodynamic characteristics, and clear mechanism. We can find out the RIE formation mechanism in analogy.

Lock of test research is one of the main reasons why there is no breakenough in the research on the RIE formation mechanism. Test is easy said than done, but range-limited DWIE offers idealer test basis. Although DWIE is much similar to RIE, it did not be paid more attention up till now.

DWIE is the best example and illustration testifed in practice when we go deeply into the research of RIE. Hydrodynamic pressure effect of DWIE is very large. Its specific conditions and backgrounds of earthquake occurring are:

1. Tectonic stress field is lower.
2. Loack-mass have fixed permeability and enclosed environment.
3. Water injection velocity and pressure can produce Pa within fixed underground range.

DWIE and RIE have the following identical characteristics:

1. Earthquake is concentrated at the local area around water injection well and in reservoir-dam area, focal volume is small, recurvence rate is high.
2. Deep well injection pressure reservoir water-level fluctuating speed and magnitude-frequency have very good corresponding relation. Earthquake focus is extremely shallow. Earthquake magnitude is low. It is a high b-value process.
3. Earthquakes occur at an aseismic or weak seismic area with general stress accumulation, it is nonlimited stress earthquake under the lower stress state, its array characteristic is similar. It is the energy release in advance.
4. The earthquake centre is usually located at the end of main seepage flow belt or in the joint of strong and week pervious bed. They all have hydrofracture character.

The identical character is the foundation of analogy, what must be paid attention to is a few key data in character: average depth of focus is 2-3km, magnituds is 2-4 Ms. tectonic stress field is lower, it is a high b-value (b≥1) process, etc. We can hever be careless with it. It is itself the combination of test theory and model because view and model without data support are meaningless.

We must note: more than 74 percent of the DWIE and RIE cases took place in a relatively stable aseismic or weak seismic area, quantitative grade of ground stress is limited. The earthquake process of high b-value means low stress state and high cracking (low strength) of medium at the focus.

One definite viewpoint is: when studing earthquake--triggering mechanism we should base our analysis on the large number of earthquake instance of < 5 Ms but > 2-4Ms (84.1%). The earthquake instance of > 5 Ms (15.9%) are not representative. This coincides with earthquake--triggering background and low magnitude characteristics under the low stress state. It also means there is no natural co-relation between

the reservoir paraneter and RIE.

There is a very good corresponding relation of seismicity with deep well water injection and seismicity with change rate of reservoir water level. High earthquake generating probability illustrates that there is medium of low confinent pressure and high Pa, that the effect of tectonic stress field is not direct and that it depends mainly on the Pa effect of partial area. Condition of initial geo--stress always fits Pa which cause faultactivity. Earthquake is induced in the way of the increase of Pa replacing the increase of shearing stress.

The Pa of DWIE is formed by man-made high pressure water injection, but the Pa of RIE is not formed by high pressure reservoir water injection and its water load is limit. We should analys what happened in the underground rock-mass of reservoir-dam area to know how Pa forms.

2 EFFECT OF WET DILATANCE AND UNDERGROUND CLOSED STRUCTURE

Gretener(1973) point out: "in order to form and maintain the abnormal fluid pressure, the fluid flow must be restricted. It means, that there are dosed conditions in both the lateral and the vertical direction." The earthquake-occured reservoir must have dilatance of earth crust--the tectonic environments of shallow crust, which enable the water in reservoir to take effect this is the geomechanical process of formation of earthquake.

The earthquake focus of DWIE in oil field distribute over the oil reservoir, the formative wet dilatance caused by hydrofracture after high pressure water injection is part closed environment (fig 1). It will make local stress field change, and the fault activity will be increased to occur earthquake too.

Characteristics of dilatant crustal deformation--DWIE and RIE distribute mainly over the low stress area within which strain energy is out big. The level of stress concentration in rock-mass is the state of dilatant stress, the prime deformation feature are dilatance of rock-mass, formation and development of enclosure area. But the major behavior of crustal deformation is effect of centre (surficial sivell and deformation) and effect of boundary(crustal tilt). The zone is the local swelling region situated in new geotectonic movement area, and action of effect of dilatance on inducing the fluid earthquake is a decisive.

Dilitant depth--under the action of tectonic stress, the range of lower rock strength in earth crust first occurs dilatance. It's general depth is less than 10km, the optimum depth is 3 km to 5km. Kemme(1972) shows, that all known depth of oil reservoir around is 3 km below surface on the whole.

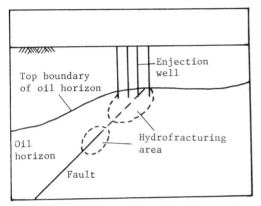

Fig. 1. Sketch map of earthquake occurring mechanism of deep-well-injection in oil field

The average depth of DWIE and RIE in earthquake source area coincides with it, again this is exactly the lower limit of primary circulative depth of underground water, so dilatant depth is the depth of earthquake source.
Dilatant stress--Pregnant earthquake region is located in a dilatant sector, the strain energy for dilatant phase is early released by earthquake. According to dilatancy theory and imitation test results, rock-mass will produce dilatance with the stress between 0.33 and 0.67 of the fracture strength upon the action of single axis pressure stress, this data approaches the golden section number-- 0.382 and 0.618. It is generally believed that the value of dilatant stress is about 300--600bar in an area, where porous fluid pressure equaling to static pressure of rock-mass, and the strength of crustal formation in tectonic area is estimated about 1×10^{6} Pa or so. This is approximate to the value of 400 ± 200 bars for crustal stress in a lower stress state, result of this investigation has been indicated by M.V. Gsouski.
The level of dilatance means the closure of dilatance, the formation and the development of enclosure region. The pregnant earthquake region of DWIE is situated in the better closed oil reservoir. Along

with the closed area of crack and crowed together--enclosure area is formed after hydrofracture, the high pressure is achieved too, and it brings on local stress field to change.
There are thousands injection wells in U.S.A. Many people think that rock-mass will generate rupture when they analyse the injection pressure and stress of these wells. But these injection wells don't induce earthquake. Considering their characteristics, the enclosure area isn't existed, and pressures are caused to break away.
The injection pressure of earth's surface and Pa aren't the same one concept, the development and formation of Pa is decided by enclosure area. Deep well injection induces 2 M earthquake in Chinese Wuhan Xiao Hong San after serious phenomenon of the well leak is occured, the depth of focus can be measured to give 1.5--3.0 km, and the well leak has good corrsponding relation with pump pressure, these prove the existence of enclosure area.
The occurence of RIE needs local closed environment too, not that it is formed by hydrofracture but that by these causes: nonuniformity of medium; the stress re-- adjustable action which making dilatance to evolve further and crowed together; the water level fluctuation of resevoir; the hydrostatic pressure causing by porous free flow; and both the development of enclosure area and the transport to dam.
The potential earthquake focus of RIE is distributed over the enclosure area, because the dilatance is repeatedly stable process that take place in the twinkle, and the porosity is rapidly expanded by propagation of unsteady impulse state of the dilatant stress for many times, within the enclosure area, the vacuum is generated, thus causes strong water absorbing effect.
The general trend of water content in rock is inversely proportional to depth. Because of dilatancy mechanism the anormlous high porosity is formed, so there will be unusual phenomenon in the reservoir area where earthquake may be occured. Pa is commonly associated with the anomalous high porosity, value b is stress characteristic factor of earthquake area, it is the function of stress. The high value be is related to the lower strenght of main shock and the lower stress state. The high value b (b ≥ 1) process of RIE and DWIE indirectly describes the highly fissured feature of medium in source area, it is the existent evidence for enclosure area.

3 HYDRODYNAMIC MECHANISM OF RESERVOIR-INDUCED EARTHQUAKE

The RIS and DWIE is the result of crustal dilantancy and fluid flow. Water is core component of earthquake source medium. Hydrodynamic mechnism deliberates the relationship between earthquake rupture and fluid activity.

3.1 Hydrodynamic pressure effect of DWIE

DWIE is the result of high-pressure water injection dynamic circulation. In order to produce Pa the fluid flow must be restricted, therefor Pa is the instant hydrodynamic pressure increase formed by high--pressure, high-speed fluid within the enclosed underground area.

High-pressure flow forces rock mass hydrofracturing and dilatancy, which forms enclosure area. High velocity flow forces violent increase of hydraulic grade, which produces violent water cycle. The occurrence of DWIE is the process of pressure reincrease of oil-bearing horizon. It is restricted by the Pa produced after water injection (table 1). The injection pressure is not necessary equal to the bottom pressure which is only meaningful for the formation of enclosure area due to hydrofracturing and further rock mass dilatancy. The bottom pressure is also not equal to the Pa of the formation within the enclosure area because the pressure will dissipate in the oil-bearing horizon if there is not an enclosed environment.

When the Pa increases to approach to general overlying load, the critical pressure to induce earthquake will create a kind of floation state which cause faultactivity. Therefore, the critical fluid pore pressure to form DWIE must be larger than the injection pressure (table 1).

The hydrodynamic process of DWIE is: high pressure water injection--hydrofracturing--formation of enclosed area--formation of Pa--floating state--faultactivity, earthquake induced. There is no water load effect in the system, but hydrodynamic pressure effect is clear. Similarly, the induced force sourse of RIE is also not the hydrostatic pressure but the hydrodynamic pressure which depends on the underground confining structure formed before earthquake and after impoundment.

3.2 Hydrodynamic characteristics of RIE

This is the analysis of underground confining pressure increase system within the reservoir-dam area. Here it must be stressed that what the significance of $\tau = \tau_s + \mu(\sigma_n - p)$ in physical seismology is to the significance of $\lg N = a - bM$ in statistical seismology. The seismic status coefficient b may predict seismic trand, but $(\sigma_n - P)$ may logically predict:

Pore-fluid pressure is strongly effected by the confinement pressure . It is only when the σ_n is lower that Pa is produced, which has the structural meaning.

In order to form and maintain the Pa, the fluid flow in the unit must be restricted, i.e. earthquake focus area must form a confined environment before earthquake. Even a very slow load increse can produce Pa in a confined rock formation.

In general, $(\sigma_n - Pa)$ reflects the confining degree within earthquake focus area and adaptability of primitive stress state for pore-fluid pressure.

Table 1. Examples Deep-Well-injection Earthquake

Zone	Rangeley Oil field	Ren Qiu Oil field	Da Gang Oil field	Denver	Matsushiro	Wu Chn Xiao Hvng San
Injection pressure (bar)	83	60–90	27–30	30–100	50	35–70
Earthquake magnitude (M)	4.0	2.0–3.0	1.5–3.0	3.0–5.0	4.0–5.0	2.0–2.2
Forcal depth (km)	2.0–4.0	3.0	2.0–3.0	4.0–5.0	2.0–5.0	1.5–3.0
Critical por-fluid pressure (bar)	245 –280	272 –450	331 –320.6	240 –415	360 –460	230

1. Pressure increase system in the earth-quake focus area.

Earthquake-occurring reservoir is usually situated in the dilatancy eara. When inclusion is connected to reservoir water through passage top-open-bottom-close uniclinal storage structure is formed (Fig 2).

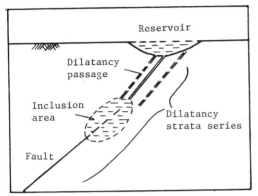

Fig. 2. Pressure increase system of uniclinal storage structure

This system has outstanding magnifying action to the pore-fluid pressure. The most active part is the stress inclusion area within which there exist mutual tran-smission and exchange of stress and fluid and the two alternating processes: one is reservoir water injection the other is that the inclusion absorbs water quickly. The former is the hydrostatic pressure caused by the free pore flow which can not force the overlying rock-mass to be in the floating state. The latter is the unsa-turated state, caused by dilatancy velocity of inclusion which is quicker than water velocity, forces inclusion to be in the vacuum state. When the inclusion is con-nected to reservoir water, it absorbs water quickly, which causes hydrostatic pressure to become hydrodynamic pressure.

The water velocity is restricted at the moment when inclusion is full of water, which forms water-hammer pressurization and produce the giant water-driving pres-sure Pa. When Pa is nearly equal to σ_n , the faultactivity occurs. The range and vacuum level of inclusion control the Pa.

The formation of RIE is the special combi-nation of stress and hydrogeologic con-dition. It has no obvious interrelation with the height of dam and capacity of reservoir.

2. Critical pore-fluid pressure
Unclinal storage structure formed be-fore earthquake causes the abnormal change of pore-fluid pressure. But whether ear-thquake occurs or not will depend on the Pa. Data from oil fields shows that Pa is produced within about 3 km depth, which coincides with the average earthquake source depth of DWIE and RIE. Usually, Pa is over 330 bar.

At 3 km underground depth, the general overlying load (geostatic pressure) is about 700 bar, the normal pore-fluid pre-ssure is usually 300-330 bar, and effe-ctive overlying load pressure is 330-460 bar. This means that, under the normal pore-fluid pressure, the effective stress on fault plane is still very high. There-fore, it is difficult to cause quick shear motion. Only when the pore-fluid pressure reaches 3/4 of the confining pressure (300-500bar) is the abnormally low effe-ctive stress produced. We can see from the cases in table 1 that the over limit of critical pore-fluid pressure to induce earthquake is 250 bar and upper limit 460 bar, i.e. when Pa=250--460 bar, a kind of floation state is created. Therefore, when Pa approaches 1/3-2/3 of geostatic pressure, the function of underground fluid in the process of rock-mass defor-mation is obviously increased and cause the change of local stress field, which expedite the faultactivity.

3. The abnormal golden section point of pore-fluid pressure
Pa/σ_n is close to the golden section point (0.382 and 0.618). What is inte-resting is that when the rock-mass is di-latating, the rate of stress and damaging stress is also close to the point. There-fore, we can believe that the dilatancy mechanism of rock-mass in the pragnant earthquake area and formation rule of Pa is controled by the golden section point. When the optimum depth of dilatancy area is know, it is possible to estimate the range of Pa in advance.

If the critical values at the 3 km un-derground are 267.4-432.6 bar, or:
700 bar x 0.382 = 267.4 bar
700 bar x 0.618 = 432.6 bar
This coincides with the statistical cri-tical valul of Pa (250 bar--420 bar) in the table 1. Therefore, similar to DWIE, the critical pore-fluid pressure to induce reservoir earthquake should close to that pressure which can be used for the predic-tion of RIE.

4. Water-absorbing effect of RIE
Pa is the power source to induce earth-quake. Its formation depends on water-absorbing level in enclosed area. So hydrodynamic effect of RIE mainly means water-absorbing effect of the enclosed

underground area and fluid pressure redistribution (water hammer pressurization)
caused by the block of high velocity flow
due to the water--absorbing effect towards
the enclosed area.

The natural earthquake occurs under the
condition of increasing high stress. Its
hydrodynamic condition is relatively stable, but tectonic stress is active. RIE
and DWIE occurs under the condition of
lower stress. Its stress field is relatively stable, but hydrodynamic condition
is always increasing so hydrodynamic condition is active. DWIE is high pressure
injection effect, but RIE is vaccuum
water--absorbing effect. In analogical
analysis, injection effect of DWIE is used
for the study on the water-absorbing effect of RIE because water-absorbing effect
causes the water-hammer pressurization in
the inclusion to produce Pa, finally it
induces earthquake.

4 CONCLUSION

DWIE and RIE distribute over dilatancy area.
They all belongs to fluid earthquake and
have similar characteristics. Both of them
can be analogized. DWIE has definiter
mechanism of formation. It can be used
as the basis of analysing mechanism of
formation of RIE.

RIE is considered as the result of
crustal dilatance and fluid flow. Its
physical mechanism includes the geotechnical process and hydrodynamic process.
The geotechnical process means the closed structure (inclusion) formed in the
rock series of dilatancy under the state
of lower stress. It is uniclical storage
structure. The hydrodynamic process is
the strong water--absorbing effect caused
under the condition of vaccuum environment
occurrer within the inclusion. The inclusion absorbs a large sum of water which
causes water--hammer pressurization. This
leads to the redistribution of pore--fluid
pressure. Then Pa is produced. It creates a kind of floating state and makes
flautactivity. Finally, it induces earthquake.

The dilatancy stress of rock--mass in
pregnant earthquake area and the Pa is
controled by the golden section point.
The critical pore--fluid pressure to induce earthquake is about 250--460 bar.

REFERENCES

Earthquake Research Institute of National
Earthquake Office, (1984). Induced Ear-
thquake of China. Publishing House of
Earthquake. Beijing,China.
Gretener, P.E. (1979). Pore Pressure:
Fundamentals, General Ramifications and
Lmolications for Structural Geology.
AAPG Continuing Education Course Note
Series 4.
Jin, Chunshan.(1986). A dilatancy and
water-hammer pattern of reservoir-induced
seism. Proc. 5th International IAEG
Congress. Buenos Aires.
Jin, Chunshan.(1988). Reservoir-induced
earthquake and anomalously porefluid
pressure. Porc. 3th nationwide engine-
ering geological congress. Chengdu,
China.
Institute of Geophysica, Academia
Sinica. (1976). Orerseas Seism Overseas
a survey of study of reservoir-induced
earthquake and deep-well-injection earthquake. Beijing, China.

A review: effects of (peri-)glacial processes on the stability of rock salt

Une revue: effets des processus (péri-)glaciaires sur la stabilité du sel gemme

A. F. B. Wildenborg, J. H. A. Bosch, E. F. J. de Mulder, R. Hillen, F. Schokking & K. van Gijssel

Geological Survey of the Netherlands, Netherlands

ABSTRACT: Since 1985 geological studies have been performed in the context of the Dutch research programme on disposal of radioactive waste in deep-seated salt bodies in The Netherlands. Geological literature and model studies were completed recently, which focused on the possible effects of (peri-)glacial processes on the geological stability of salt bodies during the next 100,000 years. Ground freezing, differential loading by an ice mass and glacial erosion may have marked effects on the stability of the salt bodies in the subsurface of The Netherlands. The conclusion of this study is that salt bodies of which the tops are at present located at a depth of less than 540 m below the surface, may be influenced by glacial erosion during the next 100,000 years.

RÉSUMÉ: Des études géologiques ont été entreprises dès 1985 dans le cadre du programme de recherche Néerlandais sur le stockage permanent des déchets nucléaires dans les masses de sel aux Pays-Bas. Récemment une étude bibliographique et des modélisations ont été faites à propos des effets potentiels des processus (péri-)glaciaires sur la stabilité géologique des masses de sel au cours des 100.000 ans à venir. La formation de pergélisol, la charge différentielle liée à la calotte glaciaire et l'érosion glaciaire ont pu affecté la stabilité des masses de sel dans le sous-sol Néerlandais. En conclusion, il a été montré que la stabilité des masses de sel, dont les sommets se trouvent à moins de 540 m sous la surface du sol, peu être affecté par d'érosion glaciaire au cours des 100.000 ans à venir.

1 INTRODUCTION

In 1984, a national research programme was formulated to investigate the feasibility of permanent disposal of radioactive waste in the salt bodies that are present in the deep subsurface of The Netherlands. As part of this programme a geological study was started in 1985 with an inventory of the salt bodies in the subsurface of The Netherlands and a characterization of these rocks and the overburden in (hydro-)geological terms. The Zechstein salt bodies are either undisturbed beds or accumulations (pillows or diapirs; Figure 1). The salt occurs at depths of several kilometers to about 100 m below the surface. In addition to this inventory, a literature study of limited extent was performed of the (hydro-)geological processes that may affect the geological stability of salt bodies (Geological Survey of The Netherlands, 1988).

The stability of salt bodies may be influenced directly by subrosion, erosion or halokinesis. Subrosion is subsurface dissolution of rock salt and halokinesis refers to salt flow as a result of gravitational forces. These processes, in turn, are influenced by a complex of other natural processes of exo- or endogenetic origin (Figure 2).

It is expected that during the next 100,000 years, (peri-)glacial processes in particular may have an influence on the stability of salt bodies in the subsurface of The Netherlands. As a follow-up of the geological investigation mentioned before, these effects were studied in greater detail in the literature and by performing simulations in analytical and numerical models (Geological Survey of The Netherlands, 1990). The modelling study was carried out in close cooperation with the Netherlands Energy Research Foundation (ECN). The principal question in the research was: To what extent can (peri-)glacial processes during the next 100,000 years have an effect on the stability of the salt bodies in the subsurface of The Netherlands? The study of the (peri-)glacial effects was approached from a very conservative point of view. This paper presents a review of the essential results of that study.

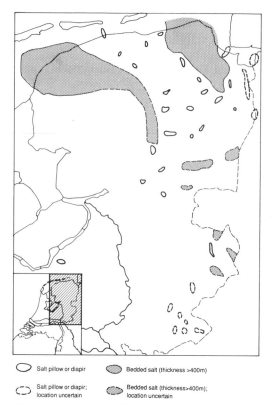

Salt pillow or diapir
Salt pillow or diapir; location uncertain
Bedded salt (thickness >400m)
Bedded salt (thickness>400m); location uncertain

Fig. 1: Locality map of Zechstein salt bodies in the subsurface of The Netherlands (after Geological Survey of The Netherlands, 1988).

2 PERIODICITY OF GLACIAL PERIODS

During the Quaternary (the last 2.3 Ma) glacial periods occurred intermittently, which also left traces in The Netherlands. From the palynological record of the past 900,000 years in The Netherlands eight glacial periods were identified (Zagwijn, 1985). During these cold phases large ice sheets spread over northwestern Europe and the northern American continent. Distinct changes in vegetation took place at the middle latitudes: forests were replaced by tundras or by polar deserts.

The periodic character of the interglacial-glacial periods in the Quaternary has been the subject of widespread debate since the mid 19th century. A well-known theory is the astronomical or Milankovitch theory, which relates long-term cyclic changes in summer insolation at high northern latitudes with assumed interglacial-glacial cycles. These time-dependent changes in intensity of solar radiation that enters the atmosphere at specific latitudes

are caused by alterations in the orbit of the earth around the sun. Milankovitch (1941), a Jugoslavian physicist, provided an important contribution to this theory by deriving a mathematical relationship between these variations in insolation and several orbital variables, viz.:

1 The obliquity of the earth's axis which varies with a periodicity of 41,000 years. It influences the intensity of the seasonal changes in both hemispheres.

2 The precession of the equinoxes with periods of 19,000 and 23,000 years. This precession is brought about by the spinning of the earth's axis around the normal to the ecliptic. Depending on the eccentricity of the earth's orbit, this precession leads to antagonistic changes in intensity of seasonal changes in both hemispheres.

3 The eccentricity of the earth's orbit with periods of 100,000 and 413,000 years. An increase in eccentricity gives rise to an amplification of the precession effect and a small decrease in global insolation.

Since the mid nineteen seventies detailed geological data sets from deep-sea cores have become

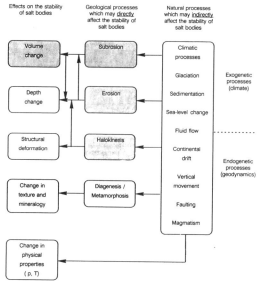

Fig. 2: Diagram of natural processes that may directly or indirectly influence the stability of salt bodies. Geological processes which cause the most relevant changes in the stability of salt bodies have been shaded. Note that only a few important relationships between the processes and the stability of salt bodies have been indicated by arrows (after Geological Survey of The Netherlands, 1990).

2764

available in support of the astronomical theory. Data for the past 800,000 years show cyclic changes in the stable-isotope record of oxygen, which are believed to be a function of the continental ice volume, correlate remarkably well with the orbital periods of 19,000, 23,000, 41,000 and 100,000 years (e.g.: Hays et al., 1976; Imbrie et al., 1984).

The paleoclimatological record for the last 800,000 years shows that the 100,000-year cycle is the most pronounced one, which seems to be in contradiction with the relatively small effect that the corresponding astronomical cycle has on the total changes in insolation. The reaction of the climate system of the earth to orbital forcing is apparently non-linear. One explanation for this discrepancy may be found in the modulating effect of the glacio-isostatic subsidence of the lithosphere (see, for example, Covey, 1984).

On the basis of the stable-isotope record and the calculated insolation changes, Berger (1981) predicted climatic changes for the next 100,000 years. He concluded that, apart from the long-term climatic effects of the emission of greenhouse gases, a major expansion of the ice sheets on the northern hemisphere can be expected about 60,000 years After Present (A.P.), which will be comparable to the magnitude of the Weichselian glaciation. This is confirmed by theoretical simulations of future changes in ice volume using a calibrated ice-sheet model. These simulations revealed a distinct increase in ice volume for the period between 50,000 and 110,000 years A.P. (Oerlemans & Van der Veen, 1984; Figure 3).

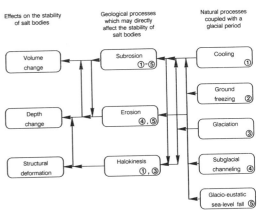

Fig. 4: Diagram of (peri-)glacial processes that may influence the future stability of salt bodies. Note that only a few important relationships between the processes and the stability of the salt bodies have been indicated by arrows (after Geological Survey of The Netherlands, 1990).

Leaving the climatic effects of the emission of greenhouse gases aside, it can reasonably be assumed that a glaciation will occur in northwestern Europe in the second half of the next 100,000 years, which may also affect The Netherlands. The effects of such a potential glaciation have to be taken into account in the long-term safety assessment of salt bodies in the subsurface of The Netherlands as potential disposal sites for radioactive waste.

3 (PERI-)GLACIAL PROCESSES AND THE STABILITY OF SALT BODIES

(Peri-)glacial processes may influence the stability of salt bodies because of the increase in erosion, subrosion and halokinetic rise of the salt bodies (Figure 4). During a glacial period the mean annual temperature will probably decrease to -8 °C at the latitude of The Netherlands, which is about 20 °C below the maximum mean annual temperature for this zone during an interglacial (Eissmann, 1981). As a result of cooling of the atmosphere, the temperature in the subsurface will also decrease. Under these conditions plastic deformation of rock salt occurs at lower rates and dissolution of rock salt takes place at a reduce rate because of its lowered solubility and the decreased flow rate of groundwater (Section 3.1).

Furthermore, climatic cooling during a glacial period may result in the formation of discontinuous or continuous permafrost (permanently frozen ground; Section 3.1). The presence of permafrost can cause a reduction in the thickness of shallow

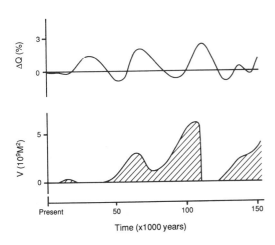

Fig. 3: Above: Prediction of the future variation in insolation at 65° N.L. Below: Simulation of concurrent changes in ice volume (after Oerlemans & Van der Veen, 1984).

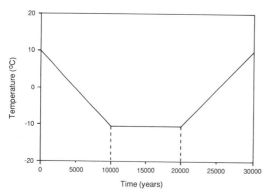

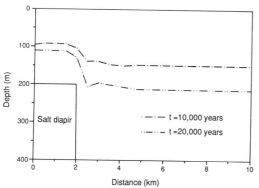

Fig. 5: Above: Assumed temperature change at the earth surface for a numerical simulation of thermal effects in the subsurface. Below: Calculated depth of the 0 °C-isotherm at t = 10,000 years and t = 20,000 years (after Prij & Benneker, 1989).

aquifers and possibly cut off the recharge by meteoric water. These changes will probably also reduce the subrosion rate of rock salt. On the other hand, the subrosion rate might increase because the pore-water pressure in a confined aquifer below the permafrost may increase.

As a consequence of the accumulation of large ice masses on the continents, the sea level will fall by 100 to 200 m at most (see compilation study of Courbouleix, 1985). The related lowering of the base level of erosion will result in intensified denudation of the earth surface. During the Weichselian glacial period erosional valleys were formed in the region of The Netherlands, which reach to a depth of 50 m below the present ordnance level of The Netherlands. This value has been used as an estimate of the maximal depth of fluviatile erosion during the next 100,000 years (see Chapter 4).

Eventually the ice sheets may spread over part of The Netherlands. Recharge of groundwater by meltwater and an increase in the hydraulic gradient

below the ice sheet can step up the subrosion rate of rock salt. The rates at which the salt structures rise may increase, as a result of the differential loading exerted by the ice mass on the subsurface (Section 3.2).

Erosion below or directly in front of the ice sheet can create incisions of a depth of several hundreds of metres, which may have a marked impact on the stability of salt bodies. As a consequence of glacial erosion, meltwater can reach the salt because the impermeable strata above it are eroded (Section 3.3).

3.1 Ground freezing and groundwater flow

Ground freezing results eventually in the formation of permafrost, i.e. frozen rock or soil in which the temperature remains below 0 °C during at least two successive years. In Central Europe, evidence is present for a well-developed permafrost during the last three glacial periods (Eissmann, 1981). Eissmann reconstructed the minimum thicknesses of the permafrost layers for these periods in Eastern Germany as being 20 to 50 m. Continuous permafrost with a minimum thickness of 10 tot 20 m is suggested to have occurred during the last stadial of the Weichselian in The Netherlands (De Gans & Sohl, 1981).

Theoretical simulations of permafrost formation have been performed at the Netherlands Energy Research Foundation (ECN) to obtain a better understanding of the maximum possible thickness of permafrost that can be expected during a glacial period in The Netherlands (Prij & Benneker, 1989). Analytical calculations were carried out to determine the sensitivity of the permafrost thickness to variations in lithology, geothermal gradient and temperature drop for a fixed period of 50,000 years. The thickest permafrost (about 400 m) was calculated for rock salt (2 °C/100 m, ΔT = 20 °C) and the thinnest one (100 m) for clay (4 °C/100 m, ΔT = 15 °C).

Subsequently, numerical analyses were performed for a model of a shallow salt diapir with an axial symmetry, which is supposed to be representative for The Netherlands. In the model, the top of the rock salt is overlain successively by a cap rock of gypsum with a thickness of 50 m and by 150 m of clay and sand. A period of 30,000 years was simulated, for which a maximum drop in ground temperature of 20 °C was assumed (Figure 5). The most important conclusion from this simulation is that the permafrost does not penetrate into the salt diapir. After a period of 20,000 years the 0 °C-isotherm attains a depth of about 100 m above the diapir, while at the same time the 0 °C-isotherm lies at a depth of about 200 m in the country rock of the diapir (Figure 5).

However, permafrost thickness was probably overestimated, among other things because the heat convection in groundwater and the phase transition water ice were neglected and the temperature drop was maximised. The actual maximum thickness of permafrost during a glacial period in The Netherlands will probably not exceed 50 to 100 m.

A lower groundwater temperature will result in a lower dissolution rate of rock salt. Because of the higher density and viscosity of the water, its flow rate will decrease by an estimated value of up to 20% of the present interglacial rate. In addition, the longer residence time of groundwater results in a higher total content of dissolved matter (Williams & Van Everdingen, 1973). Together, these factors will possibly lead to a reduction in the subrosion rate of rock salt.

In (peri-)glacial conditions the pattern of groundwater flow is primarily determined by the position of the ice sheet and the impervious permafrost rather than by topography. Basal meltwater from the ice sheet instead of meteoric water recharges the shallow confined aquifers, as far as they are not frozen. The main direction of groundwater flow differs fundamentally from the present interglacial direction. During deglaciation the flow rate can attain its maximum value (Boulton, 1988: pers. comm.). These changes in groundwater flow may increase the subrosion rate of rock salt.

As yet, the net effects of (peri-)glacial conditions on groundwater flow and the related subrosion rate are not known. Geohydrological modelling is needed to quantify this relation.

3.2 Differential loading by an ice sheet

Halokinetic movement is primarily driven by a horizontal variation in the geostatic pressure on the rock salt, which causes deviatoric stresses in the salt. The differential loading exerted by an ice sheet on the surface of the earth, may increase the deviatoric stress in the salt and consequently may speed up the rise rate of salt structures that occurs in non-glaciated conditions.

The thickness of the marginal part of an ice sheet could increase by a maximum of about 100 m/km, which value drops to 1 m/km for the central part (Boulton & Jones, 1979; Boulton et al., 1984; Drewry, 1986). This means that the pressure gradient is greatest in the peripheral zone of an ice sheet. For this reason the possible effect on the rise of salt is expected to be the most pronounced in the salt structures that are located in the peripheral zone of

an ice sheet. The increase in the rise rate is counteracted by the decrease in temperature in the deep subsurface resulting from climatic cooling during a glacial period. At a depth of 2 km, the temperature may diminish by up to 4 °C.

Statements in the geological literature on the relation between differential loading by ice and the accelerated rise of salt diapirs are almost entirely hypothetical (Gripp, 1952; Illies, 1955; Picard, 1966). To obtain a better understanding of the existence of such a relation, theoretical simulations were carried out at the Netherlands Energy Research Foundation (Prij & Benneker, 1989).

The individual effect of differential loading by ice was calculated in a 2-D numerical model of a shallow salt diapir (Figure 6; see also Section 3.1). In the model the thickness of the ice sheet increases from zero above the diapir to 1200 m at a distance of 10 km laterally of the diapir, which is equivalent to an increase in vertical pressure of 12 MPa. The assumed increase in thickness of the ice is a very conservative approximation of the real situation during a glaciation. Initially, the deviatoric stress

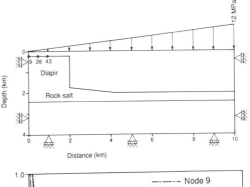

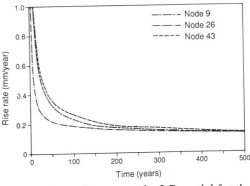

Fig. 6: Above: Geometry of a 2-D-model for the numerical simulation of changes in stress and strain related to differential loading by ice. Below: Rise rate of a salt diapir as a result of differential loading by ice (after Prij & Benneker, 1989).

resulting from the differential loading by ice will result in a large rise rate, which drops off very rapidly to a constant value of about 0.14 mm/year because of the release of deviatoric stress (Figure 6). In this calculation was not accounted for the negative effect of a decrease in temperature on the rise rate of the salt.

In subsequent analytical calculations the stress regime under non-glaciated conditions and the additional effect of an ice sheet on the stress in the subsurface were included. The geometric model describes a shallow salt diapir with an axial symmetry. The sensitivity of the rise rate was tested to variations in deviatoric stress and in temperature in the rock salt which can be expected under glacial conditions. The maximum additional rise rate which was calculated for a shallow diapir under the influence of differential loading by ice, is 0.08 mm/year. If the effect of the decrease in temperature is taken into account, the additional rise rate amounts to 0.06 mm/year. In both cases, the initial rise rate under non-glaciated conditions was assumed to be 0.07 mm/year. The order of magnitude of the additional rise rate that results from differential loading by ice, is well comparable to the outcome of the numerical calculation.

The various assumptions in the modelling exercise, which are not discussed here, are very conservative. This means that the additional rise rates calculated, have probably been overestimated.

The maximum average rise rate calculated for a shallow salt diapir in the subsurface of The Netherlands on the basis of Quaternary geological data, is 0.3 mm/year (Geological Survey of The Netherlands, 1990). For a period of 100,000 years this results in a rise of 30 m. The maximum period during which part of The Netherlands will be glaciated in the next 100,000 years, has been estimated to be about 40,000 years. This implies a maximum additional rise of a shallow salt diapir of about 6 m in the next 100,000 years (40,000 years * 0.14 mm/year). The additional rise rate of 0.14 mm/year is the result of the numerical calculation that has been discussed before. In the worst case, the total maximum rise of a shallow salt diapir in the next 100,000 years is not expected to exceed 40 m (≈ 30 + 6 m).

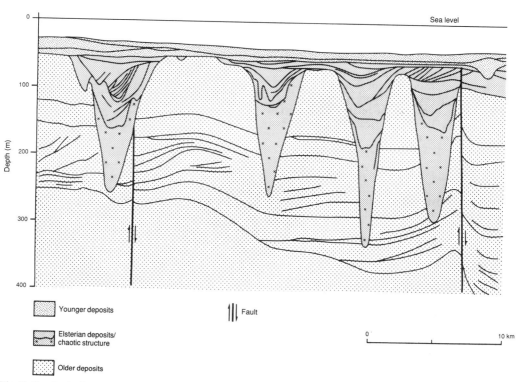

Fig. 7: Geological section across Elsterian channels in the present North Sea area (after British Geological Survey and Geological Survey of The Netherlands, 1984).

2768

3.3 Glacial erosion

The Elsterian, Saalian and Weichselian glaciations have left distinct erosive traces in large parts of northwestern Europe. Channels in the subsurface of glacial origin can be recognized at many places. A widespread net of glacial channels originated in the Elsterian. It forms a broad W-E band from the British Isles to Russia (see Ehlers et al., 1984).

Individual incisions created during the Elsterian reach a length of 6 to 20 km and a width of 2.5 to 6 km. A maximum depth of 450 m was described in the present North Sea area. The Elsterian channels are 9 to 13 km apart. Sills may be present in the channel floors. Figure 7 shows a section across Elsterian channels in the present North Sea area.

Various theories exist to explain the origin of the Elsterian channels. One theory states subglacial piping as formation mechanism of the channels (Boulton & Hindmarsh, 1987). During phases of high meltwater production and high porewater pressures, part of the meltwater is discharged along the basis of the ice and forms tunnels. These tunnels deepen as a result of creep of the underlying sediments into the tunnel and headward erosion. The ice sheet subsides in the deepening channels.

A second theory is proposed by Wingfield (1990). The deep channels, or major incisions according to Wingfield, are formed during deglaciation as a result of catastrophic emptying of lakes in or on the marginal parts of the ice sheets. Initially, the water is discharged through subglacial tunnels, which widen and deepen very rapidly. Within a very short period, however, subaerial erosion takes place because of the ice above the tunnel collapses.

Somewhat similar phenomena are the jökulhlaups, known from Iceland. These are sudden subglacial discharges of water through subglacial tunnel valleys (Spring & Hutter, 1981). The depth of these tunnels (up to 100 m) is less than the depth of the major incisions, referred to by Wingfield.

The formation of glacial channels may have a distinct impact on the stability of salt bodies, as can be seen in the salt diapir of Gorleben in Western Germany (Duphorn, 1986; Bornemann & Fischbeck, 1986; Bundesanstalt für Geowissenschaften und Rohstoffe, 1987). In a few places the top of the rock salt is in direct contact with the coarse sands in the basal part of an Elsterian channel, which circumstances may accelerate the subrosion of salt. During glacial erosion in the Elsterian, K-Mg salts were dissolved selectively down to a depth of 90 m below the base of the cap rock.

4 CONCLUSIONS

Apart from the long-term effects of the emission of greenhouse gases into the atmosphere, a major glaciation of northwestern Europe can be expected during the second half of the next 100,000 years. The magnitude of this glaciation will probably be comparable to that of the last glacial period in the Quaternary. Such a glaciation will probably also affect the region of The Netherlands. In the long-term safety assessment for the permanent disposal of nuclear waste in rock salt, one has to consider the effects of (peri-)glacial processes on potential disposal sites for nuclear waste.

The worst-case glaciation scenario for the next 100,000 years states that the effects of (peri-)glacial and other geological processes may result in a 540 m reduction in thickness of the overburden above a salt body. This 540 m is the sum total of the maximal effects of:
1 Fluviatile erosion down to a depth of 50 m.
2 Diapiric rise of salt (40 m).
3 Glacial erosion down to 450 m.

The salt bodies in the subsurface of The Netherlands of which the tops are at present at a depth of less than 540 m below the earth surface, may therefore be affected by intensified subrosion and subglacial erosion during the next 100,000 years.

ACKNOWLEDGEMENTS

We acknowledge the constructive suggestions of M.W. van den Berg and the linguistic improvements of the French version of the abstract by Th.A.M. de Groot. The drawings were prepared by A. Koers and P. Marselje assisted with the lay-out of the text.

REFERENCES

Berger, A.L. (1981). The astronomical theory of paleoclimates. NATO Advanced Study Institutes Series C: 501-525.
Bornemann, O. & Fischbeck, R. (1986). Ablaugung und Hutgesteinsbildung am Salzstock Gorleben. Zeitschrift der Deutschen Geologischen Gesellschaft 137: 71-83.
Boulton, G.S. (1988). Personal communication.
Boulton, G.S. & Hindmarsh, R.C.A. (1987). Sediment deformation beneath glaciers: Rheology and geological consequences. Journal of Geophysical Research 92: 9059-9082.
Boulton, G.S. & Jones, A.S. (1979). Stability of temperate ice caps and ice sheets resting on beds of deformable sediments. Journal of Glaciology 24 90: 29-43.

Boulton, G.S., Smith, G.D. & Morland L.W. (1984). The reconstruction of former ice sheets and their mass balance characteristics, using a non-linear viscous flow model. Journal of Glaciology 30 105: 140-1152.

British Geological Survey & Geological Survey of The Netherlands (1984). Quaternary geological map. 1:250,000 Series: Indefatigable Sheet 53°N-02°E.

Bundesanstalt für Geowissenschaften und Rohstoffe (1987). Tätigkeitsbericht 1985/1986.

Courbouleix, S. (1985). Étude géoprospective d'un site de stockage. Vol. 1 Climatologie: évolution du climat et glaciations. Sciences et techniques nucléaires EUR 9866FR: 137 pp. Commis. Commun. eur.

Covey, C. (1984). The earth's orbit and the ice ages. Scientific American 250 2: 42-50.

De Gans, W. & Sohl, H. (1981). Weichselian pingo remnants and permafrost on the Drente Plateau (The Netherlands). Geologie en Mijnbouw 60: 447-452.

Drewry, D. (1986). Glacial geologic processes, 276 pp. Edward Arnold, London.

Duphorn, K. (1986). Das subrosive Sicherheitsrisiko bei der geplanten Endlagerung von radioaktiven Abfällen im Salzstock Gorleben aus quartär-geologischer Sicht. Zeitschrift der Deutschen Geologischen Gesellschaft 137: 105-120.

Ehlers, J., Meyer, K.-D. & Stephan, H.-J. (1984). The Pre-Weichselian glaciations of North-West Europe. Quaternary Science Reviews 3: 1-40.

Eissmann, L. (1981). Periglaziäre Prozesse und Permafroststrukturen aus sechs Kaltzeiten des Quartärs: ein Beitrag zur Periglazialgeologie aus der Sicht des Saale-Elbe-Gebietes. Altenburger Naturwissenschaftliche Forschungen 1: 171 pp.

Geological Survey of The Netherlands (1988). Geologische inventarisatie en ontstaansgeschiedenis van zoutvoorkomens in Noord- en Oost-Nederland. Report 10568.

Geological Survey of The Netherlands (1990). Nadere ordening van zoutstructuren ex fase 1 met de nadruk op (peri-)glaciale verschijnselen in de komende 100.000 jaar. Report 30010.

Gripp, K. (1952). Inlandeis und Salzaufstieg (Klima und Lagerstätten). Geologische Rundschau 40: 74-81.

Hays, J.D., Imbrie, J. & Shackleton, N.J. (1976). Variations in the earth's orbit: pacemaker of the ice ages. Science 194: 1121-1132.

Illies, H. (1955). Pleistozäne Salzstockbewegungen in Nord-Deutschland und ihre regionale Anordnung. Geologische Rundschau 43: 70-78.

Imbrie, J., Hays, J.D., Martinson, D.G., McIntyre A., Mix, A.C., Morley, J.J., Pisias, N.G., Prell W.L. & Shackleton, N.J. (1984). The orbital theory of Pleistocene climate: support from a revised chronology of the marine δ^{18}O record Milankovitch and climate 1: 269- 305.

Milankovitch, M. (1941). Canon of insolation and the Ice-Age problem, 484 pp. Königliche Serbische Akademie, Beograd.

Oerlemans, J. & Van der Veen, C.J. (1984). Ice sheets and climate, 231 pp. D. Reidel Publishing Company.

Picard, K. (1966). Der Einfluß der Tektonik auf das pleistozäne Geschehen in Schleswig-Holstein. Schr. Naturw. Ver. Schleswig-Holstein 35: 99-113.

Prij, J. & Benneker, P.B.J.M. (1989). Invloed van een tijdelijke temperatuurdaling en landijsbedekking op de gesteentetemperatuur en - druk: 122 pp. Report ECN-89-12, Netherlands Energy Research Foundation.

Spring, U. & Hutter, K. (1981). Numerical studies of jökulhlaups. Cold Reg. Sci. Tech. 4: 227-244.

Williams, J.R. & Van Everdingen, R.O. (1973). Groundwater investigations in permafrost regions of North America: a review. 2nd Intern. Conf. Permafrost, N. Am. Contr.: 435-446.

Wingfield, R. (1990). The origin of major incisions within the Pleistocene deposits of the North Sea. Marine Geology 91: 31-52.

Zagwijn, W.H. (1985). An outline of the Quaternary stratigraphy of The Netherlands. Geologie en Mijnbouw 64: 17-24.

6　Engineering geology of land and marine hydraulic structures
　　Géologie de l'ingénieur des structures hydrauliques sur terre et sur mer

6.1　Flood control and erosion protection; environmental impact
　　Contrôle de la marée et protection contre l'érosion; influences sur
　　l'environnement

6th International IAEG Congress / 6ème Congrès International de AIGI, © 1990 Balkema, Rotterdam. ISBN 90 6191 130 3

Keynote lecture: Some notes on the role of engineering geology in the design process of land and marine hydraulic structures

Conférence thématique: Quelques notes concernant la rôle de la géologie de l'ingénieur dans la processus de faire le plan des structures hydrauliques sur terre et sur mer

Willem J. Heijnen
Delft Geotechnics, Delft, Netherlands

ABSTRACT: Engineering geology expertise is not used efficiently in the complex design process of land and marine structures. For the solution of many problems connected to the design of foundations co-operation between civil engineers and geologists is of essential importance in order to obtain more certainty regarding uncertainties in the foundation soil.

RESUME: La géologie de l'ingénieur n'est pas utilisé à une manière efficace dans la processus de faire le plan des structures hydrauliques sur terre et sur mer. Pour la solution de beaucoup des problèmes géotechniques lié aux projets de fondation la coopération entre les ingénieurs et les géologues c'est d'importance essentielle pour d'obtenir plus certitude concernant des incertitudes le sol de fondation.

For the design op hydraulic structures, proper knowledge of the interaction processes between the structure at one side and the foundation soil and sometimes also the free water is needed for a reliable assessment of the failure and deformation mechanism. This is especially the case in the ground beneath the structure when predictions have to be made about the behaviour of the complex system of the granular skeleton of a soil in which the pores are saturated with water under complicated dynamic and cyclic actions.

Very often the importance of interaction processes is underestimated by the designer which could imply that dangerous aspects are not exposed and, consequently, are overlooked.

Especially marine hydraulic structures are exposed to large dynamic and cyclic loadings due to the wave action. But only in rare cases the response of the structure to these loadings influences the magnitude of these loadings significantly. This is because of the rather stiff behaviour of the structure in comparison with the persistent character of the wave pressures.

The structure transmits these forces to its foundation and from there to the foundation soil underneath. The processes involved in this transmission process are generally of a very complicated nature and, therefore, difficult to model properly. Consequently, quite an amount of engineering judgement has to be called in order to simplify the various aspects of the problem in such a way that decisive phenomena are still represented properly in the model.

It seems that the civil-engineering profession is good in this respect. Many large structeres have been erected already in deep seas in very hostile environments and they all show a behaviour which is in reasonable agreement with the predicted performance. At least it is seldom worse. Only in very special cases the civil engineer tries to create computation models in which all material properties and also the mechanical systems are simulated in detail. On the contrary, he mostly characterises the structure and the soil in a rather crude way, although based on his knowledge about the processes which dominate the behaviour of the system. The most critical part in this approach certainly is the characterisation of the ground. It concerns not only a proper simulation of the composition and stratification of the soil and the material properties of the various layers, but also a reliable estimation of possible variations.

It is extremely important to be as certain as possible about uncertainties.

It is in this part of the approach that the civil engineer can benefit from the skill and knowledge of the geologist. It is the field which we call nowadays 'engineering geology'. However, this name needs to be defined. There are, as the name says, two components: engineering and geology, which implies co-operation between engineers and geologists. Not competition! Geologists should offer their their skill and expertise to the engineer who has to create structures for the benefit of the community and the industry. On the other hand, the knowledge of the engineer about the behaviour of soils under the influence of complicated processes connected to the genesis of the ground can be helpful when analysing geological processes.

Land, marine and coastal structures have very often to be founded on difficult soils. In practice there is only very seldom room for a free selection of the location of the structure because of restraints concerning its use.

Of course, in case of a free choice the geology could certainly supply the civil engineer with adequate information about the general soil conditions of the possible locations. They can form a sound base for the site selection.

But also in cases where such a free choice not exists civil engineers and geologists should co-operate.

Because of technical and financial constraints prediction models often have to be problem-oriented, which means that the models must be focussed to the most adverse design situation in connection with the dominating aspects of the material behaviour of the various ground layers.

For hydraulic structures the effects of structure movement on the soil-water matrix is quite often the determining factor in the design of the foundation. This implies that soil characterisation for computation modelling, but also for physical modelling should simulate pore-water pressure generation and dissipation of excess pore-water pressures properly. In such a case correct and detailed modelling of the stress-strain behaviour of the soil particle skeleton is less important and may be simplified to semi-elastic and plastic models.

It will be clear that this process of simplifying the ground in such a way that it can be transformed into a computation model which fits the demands with regard to the performance, requires input from both disciplines, civil engineering and geology. This means that the civil engineer and the geologist should not understand only each others 'language', but also the most important phenomena of soil-structure behaviour and its ties with geological processes. On the basis of such an understanding they may be able to master the problem, which the design of the structure including its foundation, in order to satisfy the requirements with respect to safety, serviceability, quality and durability.

Nowadays terms of reference also normally include a request for quality assurance, which means that clearly described procedures, tests and methods should be followed in the whole of the design process.

Quality and quality assurance are not just slogans offering possibilities to shift responsibilities from one party involved in the realisation of the project to another party. In principal they should serve the purpose of attaining a higher degree of reliability and as a consequence a better insight in the risks. The strong relation with safety and serviceability requires a good understanding of those features of the processes in structure and ground which dominate the resistance of the hydraulic structure against the expected most severe attack by the hostile environment.

The question raised here is how optimal use can be made from the expertise of all parties involved, including the geology.

Experience has learned me that we are still far away from the ideal situation in this respect. Quite a lot of important work is done separately by experts from the two disciplines and there is no coherence at all between their studies. From the viewpoint of quality assurance this is not acceptable. We must try to improve the situation and a congress like this one could be of great help in this respect.

Even for a huge structure like the Storm-Surge Barrier in the Eastern Scheldt Estuary the results of the geological investigation were only used incidentally. Could more use have been made of them, is an important question. I do not think so. The engineers involved in the process of designing the barrier had already a rather good notion of the difficulties they might encounter. A large amount of information has been gathered by the engineers of Rijkswaterstaat about the behaviour of the soft marine deposits in the Eastern Scheldt Estuary.

Moreover, a considerable research programme on flow slide is carried out in co-operation between Rijkswaterstaat, Delft Hydraulics and Delft Geotechnics. In the past model tests in a large flume were done showing some of the features which govern the process. The work is proceeded with more theoretical studies.

Although we are still not able to predict the occurrence and the size of large flow slides on the basis of geological and soil mechanical data, we know that such events could not be excluded from the analysis of the foundation for the barrier which means that full-prove precautions had to be taken. However, I think that it is important, not only because of scientific curiosity, to improve our knowledge regarding the behaviour of large flow slides in sandy and silty soils. Such knowledge will also be valuable for the estimation of the risks connected with liquefaction problems in areas where serious earthquakes have to be anticipated.

It is, in my opinion, not sufficiently investigated whether the knowledge of geological experts about geological processes in coastal regions could be helpful or not.

But let us go back to the question if engineering geology could have been important for the design of the foundation of the Storm-Surge Barrier in the Eastern Scheldt. Of course, geological maps have been provided by geologists of the State Geological Department but they were only used as a guide for planning the soil-investigation programme.

This is not remarkable, as the alignment for the barrier was selected on the basis of the already existing infrastructure, the works already carried out for the originally planned earth dam, hydraulic aspects and further on geographical restrictions. Geotechnical aspects played only a minor rôle in this aspect.

Consequently, geotechnical activities were concentrated on the design of the foundation of the 66 piers on the bed of recent marine deposits in the deep channels Roompot, Schaar and Hammen. Therefore, information had to be gathered about the mechanical properties of these marine sediments which were used in rather complicated computation models. Large and small model tests were performed in order to calibrate these computation models. Complicated interaction processes were also investigated in large-scale model tests. First in a wave flume in Oregon, USA, and later in the large wave flume, the ´Delta-goot´, in the Netherlands.

As a result of all these investigations, it was decided to replace part of the existing sand and silty sand deposits by better sand, and to compact the top 10 to 15 metres of the foundation bed by vibrating ´needles´.

This process of site investigation, soil testing, studies and design took many years. Only in the part of the soil investigation on site which was, of course, extremely important and very substantial, some sort of co-operation between the geotechnical specialists and experts in the field of geology or engineering geology existed. In the other fields of activities engineering geology expertise was not used although some experts in this discipline were involved in some parts of the design process.

It is very difficult to estimate now if this must be seen as an omission. I do not think so. There were too many restraints with regard to the alignment of the barrier and the location of the piers. moreover the designers already had, before

they even started, a rather good notion of the soil condition and the problems to be solved. On the other hand, the solution of many other problems connected with the long-term safety aspects such as erosion, danger of nearby flow slides, displacement of shoals and coastal banks, etc. could certainly have benefitted from the knowledge and skill of experts in geology. And I doubt if this skill was used properly, and is used in the right way for many other land and marine hydraulic structures.

Too many times we see that considerable effort has been put in a geological description of the building site which only is of limited use for the designers.

As said before, these reports should be much more dedicated to the engineering aspects of the problems to be solved.

In the field of dredging we notice a rapid change. Engineering geologists are extensively used successfully for the planning of dredging activities all over the world.

But there are many other fields in geotechnical engineering where the contribution of engineering geology could be of paramount importance. Not separated from the designer but inside the project teams.

The 6th Congress of the International Association of Engineering may be a step further in this direction.

The Jucar river flood of October 20th, 1982
La crue de la rivière Jucar le 20 octobre 1982

M. Arenillas
Universidad Politécnica, Madrid, Spain

J. Botella, R. Cortés, J.A. Ferri & R. Martinez
Universidad Politécnica, Valencia, Spain

ABSTRACT: In this paper we resume a study on the Júcar river flood (October 20th 1982) which produced the collapse of the Tous dam. By means of this study it has been possible to establish a reasonable explanation of the phenomenon as a whole and to make a global analysis of the flood's catastrophic consequences. Basically we consider the behaviour of the Tous dam.

RESUME: Dans cette communication nons résumons un étude de la crue de la riviére Jucar, le 20 Octobre de 1982, qui a produit la déstruction du barrage de Tous. Au moyen de cet étude il a été possible d'établir une explication raisonnable du phénomène dans sa totalite, et de faire une analyse des conséquences catastrophiques de la crue. Nous avons considéré fondamentalment le comportement du barrage de Tous.

1 INTRODUCTION

The Júcar Basin is located on the eastern region of the Iberian Peninsula, between the Iberian mountain range and the Mediterranean sea (Fig. 1). It has a total surface area of 21542 km² which is mainly composed of mesozoic and tertiary sediments. The first of these is made up of carbonated rocks and the second of clay bearing materials. The last section of the basin is a large coastal plain. Both the Júcar as well as the other rivers of its network have a torrential character, with similar systems to those present in most of the Spanish Mediterranean rivers. This leads to important variations in the volume which it transports, both on a yearly and inter-year basis. On average, the Júcar Basin discharges 2100 hm³/year to the sea, whereas it receives approximately 520 mm/average precipitations per year; thus, the average run-off coefficient is less than 0.20.

Half way long the river, the Júcar descends from the Castillian plateau towards the Mediterranean by means of a very steep channel which forms numerous canyons. This favorable topography has determined the construction of different dams in the main river and in its affluents. The reservoirs' water is used for producing electric energy as well as providing irrigation for the highly productive agricultural lands located on the coastal plains.

The Tous dam was constructed in the last section of this area. This dam was the first phase of a project which was eventually to have a greater height. The structure which was built was made up of two concrete abutments and a central rock fill section with a clay vertical core. The dam top was at a 98.50 m above sea level, that is, 41.50 m above the river bed. The reservoir had a total volume of 51.5 hm³ to the highest control water level (84.00 m).

The surface spillway, and especially its capacity, has been one of the most troublesome subjects after the collapsing of the dam. It was made up of a concrete structure, located on the right side. It was a gate spillway, the sill having a datum level of 77.00 m, made up of three wagon gates 15.33 m wide and 10.50 m high; it had been tested with a reduced scale model which had been seen to function correctly. According to Spanish legislation on large dams, the spillway

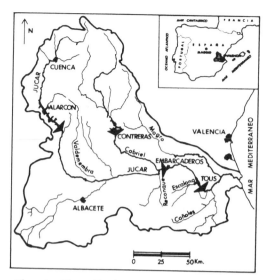

Fig. 1. Júcar Basin

2 THE FLOOD OF OCTOBER 1982

The Júcar Basin and in general all of the Spanish Mediterranean basins are subject to storm processes which are often very important, and especially from the end of the summer to the beginning of winter. This gives rise to intensive rain and the corresponding floods. In fact, historical references state that over 50 large floods occurred in the Júcar basin between 1328 (first report written on this subject) and the present. Although these reports do not give precise quantifications of the flow and volume, the information available states that the most important flood of the Júcar, before that of 1982, occurred on November 4th 1864. The studies carried out after that flood established a peak flood of 6000 m³/s in the roundabouts of the area where the Tous dam was later built, whilst at the mouth of the Júcar, approximately 60 km downstream, a flow of 12000 m³/s had been reached (that flood was precisely one of the elements which was used for the sizing of the Tous dam). Detailed studies carried out recently showed a maximum value of 5500 m³/s of the peak flood at the site of the destroyed Tous dam.

was projected for a 7000 m³/s capacity, this flow corresponding to the maximum value of the flood with a 500 year return period.

Thus, the peak flood of the 20th of November of 1982 was much larger than the previous figure. During that period of time, Spain was suffering an important drought which had already lasted various years, including 1983, despite intensive storms which took place in very specific areas, such as that which we are dealing with at present.

On the 19th of October of 1982, the Tous dam's personnel had not been notified of the possibility of heavy rain the next day, because the meteorological information available at the time did not foresee such a sudden and intense event. However, at 8 o'clock a.m. of the 20th of October, the Tous dam had received 19 hm³ of water, which, added to the water which the dam had stored the previous night, meant that the water level had surpassed the top level of the spillway gates by over one meter. The gates could not be opened because the heavy rain in the Júcar Basin which had begun, more or less, at midnight caused the almost immediate shortage of electric power and telephone lines, as well as the cutting off of nearly all roads leading to Tous.

The first waters of the flood which arrived at the reservoir were caused by the Escalona river, which meets with the Júcar near the dam. New stream flows arrived, transported by the Júcar. The water surpassed the dam top slightly after 4 p.m. (on October 20th), and thus, a slow degradation of the structure occurred, which accelerated after 7 p.m., coinciding with the almost total destruction of the central area (rock fill) of the Tous dam. This brought about a peak flood with a maximum value of approximately 15000 m³/s, which barely affected the land further down the river. This is due to the fact that the very heavy rain in this area, and the volumes of the affluents of the Júcar below the Tous dam (especially Sellent and Albaida) had made the people abandon the lower areas of the villages many hours before the dam had actually broken. This resulted due to the general flood situation present in the entire region during the morning of October 20th.

The storm which led to this catastrophical situation is probably the most important which this region has suffered during its history. This is at least what has been concluded from available written documents. The heavy rain extended from the sea to the

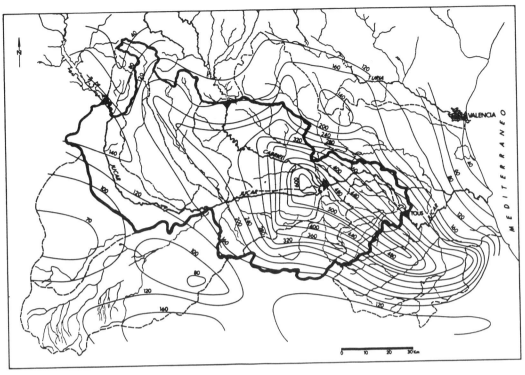

Fig. 2. Map of isohyets. Area affected

mountain ranges of the Castillian plateau, having an extension of 170 km, which is far more, for example, than that which occurred during the flood of November 4th, 1864. The isohyets elaborated by the National Meteorological Institute (INM) after the flood, have allowed us to quantify the rainfall in October 1982 in the area (Fig. 2). Referring to the Júcar Basin, located upstream of the Tous dam, we must keep in mind the presence of two large dams, one at the Júcar itself (Alarcón) and the other at its affluent Cabriel (Contreras) which held all of the runoffs which arrived at the reservoirs, and which prevented the water flow downstream. This reduces the area of the Júcar basin which leads its runoffs to the Tous dam. On the other hand, we must note that an extensive area of the basin, located in the mid region of the right side, is endorheic or semiendorheic, and even though it carried important volumes toward the Júcar during the flood, the truth of the fact is that these volumes arrived at the Tous dam very late, and thus, its influence on the main section of the inflow hydrogram in the Tous reservoir is quite limited. Due to this

fact, we have not considered its effects in our study.

Therefore, with the above mentioned reductions, the area of interest upstream of the Tous dam corresponds to a surface area of 6780 km². The said isohyets of the INM have us allowed to estimate a total rainfall of 1880 hm³ on this surface, which corresponds to an average rainfall of 277 mm. Downstream of the Tous dam, the affected area (Júcar basin and two small lateral streams which drain directly to the sea) has an extension of 4310 km², and it received a total rainfall of 976 hm³, which is equivalent to an average rainfall of 226 mm.

According to the data of the pluviograph of the Nuclear Power Plant of Cofrentes, located approximately 35 km from the Tous dam and which practically coincides with the most important nucleus of the rainfall, 80% of the total rain was concentrated during a time span of 9 hours (between 7 a.m. and 4 p.m. of October 20th) whilst during various hours the intensity per hour was above 100 mm (7 to 8 a.m. and 9 a.m. to midday). More than 500 mm of rainfall was noted in this maximum nucleus during the 24 hours of the 20th of October, which is known as the highest daily intensity recorded in Spain until that date.

Inmediatly after the collapsing of the Tous dam, we carried out a first study on the flood (Arenillas et al. 1985). At that time, we tried to establish the minimum reasonable values of the flows and volumes which defined the inflow hydrogram in the Tous dam, using very conservative criteria, in order to analyze the behaviour of the dam during the flood. This study allowed us to obtain two interesting conclusions:

1. The peak flood occurred after the collapsing of the dam, due to the importance of the flows which arrived from the furthest section of the basin (thus, the conclusion of other reports, which only studied the flood until the moment that the dam broke, were corrected).
2. The dam would have collapsed even if the gates of the spillway had been operated (which contradicted the opinion of those who stated that the dam would not have collapsed if the gates of the spillway had been opened).

After that date, the first author of this article had to act as an expert for the public prosecutor during the court case which was carried out due to the collapse of the dam. Thus, we considered that a new study should be carried out, whereby we could establish a new inflow hydrogram in the Tous dam, as close as possible to reality. By means of this hydrogram, we also wished to reanalyze the behavior of the dam during the flood. However, we were not able to present this study during the court case although many of its conclusions were used on that occasion.

Now that we are presenting this study, we have tried to concentrate exclusively on the scientific and technical aspects of the phenomenon which was studied.

3 METHODOLOGY

A working methodology including two main aspects was used during the realization of the study:

1. Simulation model of the generation of the flood, based on a simplified version of the well known unitary hydrogram theory.

We began with the pluviometric information available on the distribution of the total volume of rainfall (Fig. 2) for each point, as well as the temporary sequence deduced from the available pluviograms, especially that registered in the Nuclear Power Plant of Cofrentes. To determine the efficient rainfall, the theory of the unitary hydrogram was slightly modified, applying a criteria of simple proportionality with respect to the rainfall (in any case, given the very high volumes of rainfall, the rainfall thresholds need not be considered).

Afterwards, various hydrological parameters were defined for each of the 23 sub-basins into which the area was divided: surface, altitudinal development, length of flow, mean slope and concentration time, all of which were necessary for the application of the method.

Thus, the different hydrograms of the flood which were calculated in each sub-basin were concentrated in the different points of confluence, and were carried downstream by the existing river network. The end result was the general inflow hydrogram in the Tous dam (Fig. 3, 4 and 5).

The calibration of the model was based on the information given by those activities stated in section 2, which define a series of points (time-flow) being an obliged passing points for the hydrograms. An important aspect of the calibration process is that carried out in the reservoir itself, being that its precise behaviour is known until the moment that the dam collapsed. The same could be said for what occurred the following day, whereby the infrastructures which did not suffer as much damage allowed the quantification of the flows which corresponded to the descending branch of the hydrogram.

2. Detailed analysis of the hydraulic behavior of 28 different sections divided throughout the whole generating basin of the Tous flood and situated on its main channels (Table 1). For each of these sections, the analysis is based on the following points:

- Precise definition of the shape and dimensions of the maximum wet section during the flood, as well as any available information on the possible intermediate situations.
- Definition of the hydraulic parameters

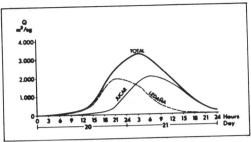

Fig. 3. Hydrograms concentrated in Júcar–Ledaña confluence

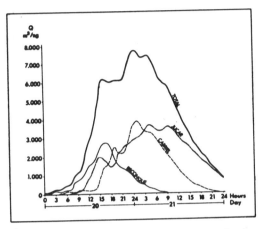

Fig. 4. Hydrograms concentrated in Embarcaderos

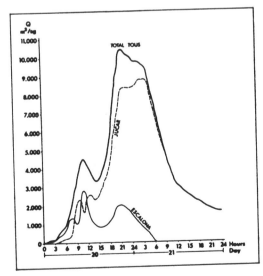

Fig. 5. Hydrograms concentrated in Tous

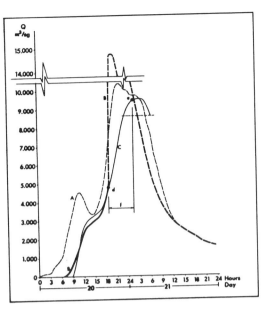

Figure 6. Inflow and outflow hydrograms in Tous. A. Inflow; B. Outflow in real situation: gates closed; C. Outflow in extreme hypothesis: gates open and reservoir initially empty; d. Real collapse of the dam; e. Hypothetical collapse of the dam; f. Maximum time span between collapses.

which determined the water movement in the neighbouring area associated to the section.
– Carrying out the absolute chronology of what occurred in the event, thus, defining the data which state the levels which had been obtained and the hours when they took place.
– Calculation of the flows by means of the Manning formula for the different depths. In general, and due to the lateral extensions of the flooded sections, these have been divided according to two or more sub-sections with similar roughness and depth coefficients.

The development of this methodology has required a large volume of work, both in the field and in the office. The data collection carried out in the field, which was essential for the realization of the activities stated in section 2., entailed a minute reconnaissance of the entire basin which was affected by the flood, in order to obtain sufficient information. Various interviews with on-the-spot witnesses of the phenomenon were carried out, as well as the checking of references and marks on the site itself which certify what was stated in the interviews. Afterwards, a measuring campaign of the main geometrical magnitudes in each of the sections of calibration was carried out.

According to what has been stated, the analysis allowed the definition of a series of partial hydrograms and, finally, the inflow hydrogram in the Tous reservoir (Table 2). The main characteristics of the flood were deduced from this hydrogram (Fig. 5):

- Duration of the main phase of the flood: 36 hours.
- Maximum inflow in the Tous dam: 10400 m^3/s.
- Total volume received by the reservoir: 864 hm^3.

Once the inflow hydrogram was defined, the behavior of the dam during the flood was studied. Thus, the outflow hydrogram was firstly calculated, where the abatement effect of the reservoir was considered until the moment of the definite collapsing of the dam. As previously stated, the gates of the spillway of the Tous dam could not be opened, thus spilling over the top of the gates and the dam, as the water height within the reservoir grew. Applying the corresponding discharge formulae, which were adjusted according to the height of water which was actually observed, we have been able to define the outflow hydrogram until 7:15 p.m. of October 20th. Afterwards, the dam collapsed, which motivated an important increase of the discharge volume. This section of the hydrogram was studied by the Centro de Estudios Hidrográficos (CEH) of the Spanish Ministry of Public Works (CEH 1983), which is a source we use to obtain the data to complete the outflow hydrogram which appears in Fig. 6.

Once the real situation had been established, we completed the study investigating the behaviour of the dam supposing that the gates could have been opened. Thus we have analyzed a first extreme hypothesis (which has not needed further enlargement), which consists in supposing that the reservoir was completely empty and the gates entirely

open when the flood took place. Using this criteria and applying the corresponding discharge laws (according to the water height in the reservoir, deduced from the inflow hydrogram), a new outflow hydrogram has been obtained (Fig. 6). As can be seen in the figure, in this extreme hypothesis the water would have reached the dam top at 11:00 p.m. on the 20th October. If the structure had resisted, the flow above the dam top would have lasted eight hours, and the height of water above the dam top would have surpassed 2.10 m. Being that the dam had actually collapsed approximately three hours after the dam's top had overflowed, with a water nappe similar to that previously indicated, it seems evident that this extreme hypothesis would have also led to the collapsing of the dam. (The test in a reduced scale model, carried out by the CEH to study the collapsing of the dam, proves that this occurred with a nappe which surpassed the dam's top by more than 1.50 m). According to our calculations, should this extreme hypothesis have occurred, the last phase of the collapsing of the dam would have started at 2:00 a.m. of October 21st. Thus, the CEH has deduced a peak flood of 18000 m^3/s (CEH 1989), that is, higher than that which actually occurred (15000 m^3/s), due to the volumes which entered the reservoir at that same time.

According to this analysis, it seems evident that any hypothesis between the real situation (closed gateways and reservoir with a water level reaching the spillway sill) and the extreme situation which was considered (opened gateways and empty reservoir) would have brought about the collapsing of the structure.

4 CONCLUSIONS

The study which we have carried out on the flood of the Júcar which occurred on October 20th, 1982, has allowed us to point out the following conclusions:

1. The peak inflow in the Tous reservoir reached a value of 10400 m^3/s and occurred between 10:00 and 11:00 p.m. on October 20th. In other words, three hours after the collapsing of the dam, which was the moment when the inflow was of approximately 7300 m^3/s.

Table 1. Control sections

Nº	NAME: RIVER/POINT	SURFACE AREA (m²)	HYDRAULIC RADII (m¹)	ROUGHNESS COEFFICIENT RIVER BED	REST	FLOW (m³/s)
1	Júcar/C.H. Millares	1208	12.01	0.050		9550
2	Júcar/Cª.Millares-Dos Aguas	1209	11.14	0.045		9400
3	Júcar/C.H. Cortes I	988	10.40	0.040		8900
4	Júcar/Presa Embarcaderos	(1)				(1)
5	Cabriel/Puente Hierro	990	11.30	0.045		3500
6	Cabriel/Casas del Río	723	6.18-1.95	0.035	0.060	2800
7	Júcar/Pte. Ac. C.N. Cofrentes	1050	6.30-1.18	0.040	0.060	4900
8	Reconque/Pte. Tran. Especiales	545	5.50-3.10-2.20	0.035	0.050	3000
9	Zarra/Zarra	360	4.90	0.045		1550
10	Zarra/Alpera	165	3.50-1.90	0.035	0.080	600
11	Júcar/Casa E. Piera	2003	7.26-5.23-2.66	0.040	0.070	4800
12	Júcar/Pte. Cª.Jalance-Cofrentes	1025	6.83	0.050		4800
13	Agua/Confluencia Júcar	(1)				(1)
14	Júcar/Embalse El Molinar	(2)				3500
15	Júcar/Pte. Nuevo Alcalá	685	6.40-2.90	0.030	0.050	3400
16	Júcar/Pte. Viejo Alcalá	(1)				(1)
17	Carcelén/Pte. Villavaliente	(3)				590
18	Júcar/La Recueja	896	4.80-4.95-2.50	0.030	0.050	3200
19	Ledaña/Molino Jorquera	653	5.35-2.10	0.035	0.080	2000
20	Ledaña/Galayar (4)	744	3.50	0.040		1800
21	Ledaña/Cenizate	445	2.48	0.030		1450
22	Júcar/Pte. Jorquera	450	9.45-7.10-3.10	0.030	0.050	2050
23	Júcar/Pte. Cubas	335	5.78	0.030		1800
24	Júcar/Puente Torres	486	6.40-4.00-3.35	0.030	0.050	1750
25	Júcar/Pte.Valdeganga	300	6.00-2.50	0.030		1600
26	Júcar/Pte. Cª Albacete-Casas Ibáñez	(3)				900
27	Valdemembra/Motilleja	1585	2.10	0.090	0.090	900
28	Mahora/Pte. Mahora	248	4.20	0.040		750

(1). Whitout flow calculation section; only passing times; (2) Calculated according to discharge sections of dam; (3) Calculated according to discharge sections; (4) There are two other contrast sections

Table 2. Main magnitudes in the area affected

Nº	NAME	SURFACE AREA (km²)	RAINFALL (Hm³)	(mm)	TOTAL RUNOFF (Hm³)
1	Júcar (Embarcaderos-Tous)	391	127.6	226	63.8
2	Escalona	540	241.6	447	120.8
	Suma 1 + 2	931	369.2	397	184.6
3	Cabriel (Contreras-Embarcaderos)	1364	499.1	366	224.6
4	Júcar (Jorquera-Embarcaderos)	809	236.2	292	106.3
5	Reconque	824	280.5	340	115.0
	Suma 3 + 4 + 5	2997	1015.8	339	445.9
6	Ledaña	894	218.7	245	105.0
7	Valdemembra	949	150.0	158	72.0
8	Júcar (Alarcón-Jorquera)	1009	125.8	125	56.6
	Suma 6 + 7 + 8	2852	494.5	173	233.6
	TOTAL	6780	1879.5	277	864.1

2. The Tous dam would have collapsed in any case during the flood, despite the situation of the reservoir and any feasible operation of the gateways. In a last extreme (gateways opened and reservoir empty), this collapsing would have occurred at 2:00 a.m. on the 21st of October, and thus, a peak outflow of 18000 m^3/s would have occurred.

3. The capacity of the spillway of the Tous dam was correctly defined according to the information which was available at that time (1973). The criteria which was used could even be considered as being conservative.

4. Despite this fact, the truth is that the Tous dam collapsed on October 20th 1982. This is explained by means of the extreme and sudden meteorological phenomenon, which could not be reasonably predicted. This is not only proved by all that has already been mentioned, but also by the fact that certain dams (and other installations which were projected with very strict criteria) near the Tous dam had not considered such extents, and thus, the original projects had to be redefined.

REFERENCES

Arenillas, M., Botella, J., Cortés, R., Esteban, V., Martínez, R. & Serra, J. (1985). La Avenida Extraordinaria del Júcar de 20 de Octubre de 1982. In Catástrofes Naturales, pp. 17-49. Publicaciones de la Cátedra de Geología aplicada a las obras públicas. ETSICCP. Universidad Politécnica de Valencia.

Centro de Estudios Hidrográficos (1983). Avance del Estudio hidrológico de la crecida del Júcar. M.O.P.U. (Unp.).

Centro de Estudios Hidrográficos (1989). Revisión del Estudio hidrológico de la crecida ocurrida en los días 20 y 21 de Octubre de 1982 en la Cuenca del Júcar. M.O.P.U. (Unp.).

Geotechnical aspect of endemic flood problem in the Lower Damodar basin (West Bengal/India)

Aspect géotechnique du problème des inondations au bassin inférieur de Damodar (Bengale de l'Ouest/Inde)

Barin Chatterjee
Geological Survey of India

ABSTRACT : Floods accompanied by bank failures are the annual twin events which constitute the greatest hazard in the lower Damodar basin.Towards south of the Damodar's elbow-bend,passing through constricted micro-relief zone, there is a series of aggraded channels from easterly direction to the present southerly direction.This area suffers very badly due to back-water flow from the trunk rivers Hooghly and Rupnarain on either side and also due to the released flood discharge from the upstream DVC reservoirs.As the eastern bank happens to comprise an industrial belt,it is well protected by embankments.The western side,the trans-Damodar area,comprises a dense network of channels in moribund stage.Climatic condition,geomorphological setting,underlying soil and groundwater condition,role of upstream dams and tidal influence all contribute to the run-off vis-a-vis rate of disposal problem.Thirteen flood sectors have been identified and remedial measures recommended.

Resume' :Le risque majeur du bassin inferieur de Damodar s'est marque', chaque anne'e,par des e'venements jumeau x quise presentent par des inondations accompagnée de de reptures du rivage.Au sud du meandre de la rivie're qui traverse le micro relief resserre'il existe une serie de canaux ã lits d'alluvions qui se remonte dans la direction allant du sud ã l'Est.Cette region se trouve essuyée fortement suite ã l'existance d'un courant d'eau du bras de decharge qui remonte des riveres Hooghly et Rupnarain se situant de part et d'autre ainsi que le debit de l'avant bec des bassins de retenue de Damodar.Comme la zone industrielle se trouvait sur la rive de l'Est elle se tient bien protegée par remblais. Quant ã la region d'enface,la rive occidentale comprend ur reseau resserre de cauaux en etat moribond.Des facteurs telsque la condition climatique,disposition gemorphologique, terrain ainsi que la condition de la nappe sousjascents, le role du bassin de retnue en amont et l'influence de la marée le tout contribue ã la repture vis a vis le taux de trasport.Apartirdes treize secteurs etablis les mesures riparatriees sontsuggerecs.

1.INTRODUCTION

The Damodar river, originating from the western plateau emerges south-east ward to the deltaic plain and meets the main river Hooghly at about 45 km downstream of Calcutta.Of the total basinal area of 23,700 sq km,an area of 19,700 sq km falls under the regulated system of the Damodar Valley Corporation (DVC) and the rest of the lower reaches of the basin comes under the purview of the present discussion.

The floods accompanied by bank failures are the periodic twin event which constitutes the greatest hazard in the middle portion of the Lower Damodar basin.The topography of the terrain,general storm direction and regression pattern impose a collective effect in causing the flood.

The storms have the peculiarity of swelling up earlier in the lower course before contributing in the upper catchment.The rain storms,originating from the bay head and moving in a north-westerly direction, come in close to the heel of one another, forming a chain of sequence,often hanging on to be reinforced by a subsequent one. The highest rainfall record in 48 hours' duration happens to be 708 mm (27-28 Sept, 1978).The average annual-rainfall in the catchment is about 1181 mm and ranges from a maximum of 1632 mm (1917) to a minimum of 696 mm (1966).The discharge in the river is very strongly correlatable to rainfall.In the pre-DVC period,15 % of the floods had the peak discharge below 5663 m^3/s,45% between 5663 and 8495,

m/^{3}s and 10 % above 12,742 m^3/s.In the
post DVC period,46% of the flood have the
peak discharge below 5663 m^3/s,36% between
5663 and 8495 m^3/s and 18% above 8495
m^3/s.The flood frequency analysis (Sharma,
1976)shows mean annual flood of 3,800 m^3/s
and most probable annual flood of 2650
m^3/s with a recurring interval of 11
years.In 1978,the all time record has
been exceeded by a flood of 33,330 m3/s
(DVC records).

II.GEOMORPHOLOGY

The principal geomorphological units are
(a) older deltaic plain and (b)younger
deltaic plain.The lower basin forms a
part of the younger plain which is subdivided into two sub-deltas,the apex of
both being common,located at the elbow
turning point of the Damodar river.While
the slope of the major subdelta is towards east,the secondary one slopes south
ward. Analysing contour disposition,it
is noted that the contours stretch out
wider apart generally from 36 m to 18 m,
still maintaining the upper level configuration pattern.Further below,the land
shows a gradual change of slope direction
from east through south-east to south,
rotating clock-wise about the common apex
of the subdeltas.The hinge line of the
changed direction of slope broadly lies
somewhere around 18 m in the upper level
and 8 m in the lower level.Further below,
the slope flattens out,where micro-relief
features can only be identified by slope
percentage analysis carried out with small
scale and closely spaced contour map.
Such a plan was prepared in connection
with the 'Diagnostic survey of the deltaic West Bengal' (Mukherjee,1979).Two such
micro-relief zones could be identified,
one coinciding with the zone of migration
of the river and the other with the zone
of endemic water logging towards upstream
(Plate-I).A third imperceptible one roughly demarcates the uppermost limit of the
tidal action.
Entering the lower reaches of the basin,
the Damodar river takes a sharp 'elbow'
bend and branches out into seven prominent
distributary channels (Plate I).From the
historical records available from 1550 AD
it has been observed that initially the
main flow was towards east-south-east
which gradually shifted towards south-
south west in a clock-wise direction and
at present the most active channel is the
Mundheswari river flowing towards SSW.On
the basis of the morphometric analysis,it
has been observed that the sinuosity value falls from the easterly flowing to the

southerly flowing branches,showing the
natural tendency of adjustment with the
steep gradient.Similarly,an increasing
width-depth relation indicates incised-
meandering associated with base level
lowering.
While the upland water moves down a
steep slope of initially 10°and then 5°,
the lower basin with less than 1° slope
has a poor disbursing capacity.After
bifurcation,the Mundheswari channel
carries about 90% of the flood discharge,
while its bankful carrying capacity
hardly exceeds 1982 m^3/s (Basu,1986).
The Mundheswari river meets the mighty
river Rupnarain about 10 km downstream
of the latter's confluence with another
major river,the Dwarkeswar,joining from
the opposite bank.Though the river Rupnarain is extremely dynamic in its tidal
regime and can effectively flush its
previous accretion,it fails to dispose
the flushy water column of the Dwarkeswar
before the Mundheswari brings down the
Damodar's and its own flood discharge.
The Rupnarain flows into the hooghly
further downstream of the confluence of
the Damodar.The area flanked by Rupnarain
to the west and the Hooghly to the east
falls within the realm of river micro-
tidal interaction, extending landward
from the low-tide mark to the upper limit
of the tidal influence at around 6 m
contour (Plate I).Here, the gradient of
the river is extremely low (0.08 m/km)
and the delta plain comprises detritus
brought down by the rivers Hooghly,Rup-
narain and Damodar,in an intermixed con-
dition.Adjacent to the active channels,
low-lying levees,backswamps and associated tidal creeks are the common geomorphic features.

III.GEOLOGY

The area is located over a zone of
transition of facies from continental
to marine in the direction from NW to
SE.Drilling for oil has revealed that
the basin is underlain by geosynclinal
trough and that there has been a number
of cycles of marine transgression and
regression(Biswas,1961).The stable shelf
is occupied by sediments of Mesozoic and
Tertiary age,striking NE-SW beneath a
cover of 90 m to 2700 m thick sediments
of Holocene and Recent age. While the
shelf area is tectonically undisturbed,
the marine portion is characterised by
subsurface domal structure of varying
dimensions (Sengupta,1972).Above the
Pleistocene sediments,a north-south trending sag has been reported to exist,which

incidentally coincides with the southerly course of the Damodar river.

IV. GEOTECHNICAL ASPECT

Analysis of the strata log reveals the prevalance of silt and clay down to 60 m depth in general and conspicuous sand layers of significant thickness at depth below 60 m, 150 m and 170 m from the ground level. Study of the near surface material by auger holes and pit has revealed the presence of several cycles of sand, silt and clay with plant remains in the southern and eastern part of the area. A black clay layer at a depth below present tide level bears special significance as it indicates eustatic changes of the sea level in the recent geological past.

Study of lithology and physico-mechanical properties of the deltaic sediments has led to infer that the grain size, in general, decreases down the slope, across the channel and with the increasing depth in each individual cycle of flood deposit. The Natural Moisture Content, Liquid limit and Plasticity index show a gradual trend of increase from upper delta to lower delta. There has been a consistent trend of reduction of voids down the delta, which retards the rate of infiltration. Localised steepening of the groundwater table implies that the speed of movement of the groundwater varies from place to place, depending on the rate of infiltration and is somewhat correlatable with the micro-relief zones. The range of water table fluctuation, flattening of water table in the microrelief zone and poor drainage condition are all interrelated in causing water logging problem.

The pre-consolidation pressure of the soil samples from and around the micro-relief zones show higher values to the tune of 3 kg /cm^2 over a background value of 0.9 kg/cm^2. This perhaps indicates initial depth of burial followed by subsequent squeezing up of sediments. Such deposition and uprising, basically contemporaneous process, are manifested as drape over reefs along the palaeoshore line (Morgan, 1968). Initiated by the rapid deposition of denser mouth-bar sand over less dense marine clay, the process continues as subsequent squeezing up of the differentially weighted clay in the form of diapiric type of intrusion. These squeezed up zones act as barriers to the subsurface path of infiltration and also as projected microreefs which

obstruct to the surface flow causing the acute water logging problem. The process of uprising is supposed to have its simultaneous effect of counter sagging, which is perhaps manifested by bowl-like depressions, subjected to endemic flooding.

V. FLOOD CARRYING CAPACITY

The old Damodar channel which at present carries only 10 % of the flood discharge, shows a progressive reduction of the channel's carrying capacity from 13 % in the upstream of the micro-relief zone, through 64 % - 72 % in the middle portion of it down to 3 % as the river emerges out of the constriction. The width of the channel gets abruptly reduced from more than 500 m to less than 100 m as it enters into the micro-relief zone. It maintains a uniform width of around 100 m through the entire zone and after emerging out, the width gradually increases to 180 m. Rate of aggradation is very high at the location of the Rail and Road bridges due to reduction of the river's carriage way. The flood profile of 1978 has indicated that the bridges acted as effective control in the lower reaches as the flood level has gone down by 1 m within a stretch of 100 m. The river Mundheswari has its own system of intricate drainage net work in a moribund stage while negotiating the same micro-relief zone in the trans-Damodar area to the west.

One more human interaction with the nature's delta building activity is the construction of embankments to save the thickly populated industrial belt located on the interfluve between the Hooghly and the Damodar to the east. This action has contributed to the natural tendency of pushing the flow more and more to the west and ultimately leading the main channel into the Mundheswari river.

The devastating flood which swept over a large part of West Bengal in 1978 had no parallel in the known history. The upper delta had suffered mainly from the overflow of the DVC canals and left bank spill above the elbow bend of the Damodar. The duration of flood in the upper reach was less comparative to the middle and the lower delta where the shallow channel of the interlocking abandoned rivers could not drain out due to their moribund condition and the then prevailing strong spring tide condition in the Hooghly river. In the trans-Damodar area the surging water from the Dwarkeswar and Mundheswari rivers swept over a vast land as a sheet of water and the bowl like depressions were subjected to the maximum distress for a considerable length of time.The hydraulic impact was so heavy that the left bank of Rupnarain river was breached open below its confluence with Mundheswari.Both the rivers Damodar and Mundheswari and their interlocking channels overflowed practically all along the length below their bifurcation and breaching was reported from innumerable locations.Massive flood water spread out in every direction and remained stagnant for weeks together.

Depending on the mathematically based hydraulic model for the simulation of flood propagation by using the system 11 FC of the Danish Hydraulic Institute, Basu(1986) has identified several flood sectors.Out of those,13 sectors falling in the trans-Damodar area have been further analysed on the basis of ground topography to estimate the volume of flood water in each individual sector as enumerated below,

Table 1. Volume of water in the flood of 1978,in flood sectors of trans-Damodar area.

Flood Plain No. *	Area (km^2)	Av.depth of water column (m)	Volume of water (m^3)
I.	58	0.05	2,900,000
II.	36	o.66	23,760,000
III.	46	0.005	230,000
IV	39	0.885	34,515,000
V	111	1.245	138,195,000
VI	89	0.9125	81,212,500
VII	64	0.9725	62,240,000
VIII	76	1.35	102,600,000
IX	49	0.925	45,325,000
X	58	0.6575	38,135,000
XI	13	0.58	7,540,000
XII	30	2.3525	70,575,000
XIII	47	0.005	235,000

* Flood plains,as shown in Plate I.

VI.ROLE OF DVC IN FLOOD MITIGATION

Voorduin (1945) had originally proposed a target for controlling a design flood of 28,320 m^3/s (1.5 times the maximum observed flood till then) and with a view to moderating this flood to 7080 m^3/s he suggested construction of seven dams, out of which only four dams have been constructed in the first phase.This can moderate a peak flow of 18,410 m^3/s to 7080 m^3/s and disregarding the condition of drainage congestion in the lower valley,these dams have to release the excess discharge. Subsequently the planned flood storage

has been further reduced due to lowering of the submergence level in case of two dams by 1.5 m and 6.1 m,respectively, because of land aquisition problems. With the existing provision, the theoretical exercise carried out by DVC shows that the 1978 flood with a three hourly peak of 21,917 m^3/s could be moderated to 5660 m^3/s,had the flood been received into empty pools.This too was far from adequate for the protection of the lower valley.Moreover,the proposition of receiving flood in the empty pools is now difficult as industrial water supply,as well as provision of hydroelectricity and irrigation components have become added responsibility of the DVC.At present the normal flood release by the DVC amounts to about 5700 m^3/s while the bankful carrying capacity of Mundheswari carrying 90 % of the flood is hardly 2000 m^3/s.The situation further worsens in the downstream reaches due to further reduction of carrying capacity.As a consequence to the DVC dams,the floods have been somewhat moderated but the duration is prolonged.

VII BANK FAILURE PROBLEM

Bank erosion and failures are common phenomena associated with floods.The fine grained soils get oversaturated resulting in reduction of both cohesion and frictional forces and consequently lowering the shearing resistance of the slope. During high energy flow,the foot of the bank where it is sandy, gets rapidly eroded away resulting clayey overhang to collapse subsequently.Underseepages are critical consideration when a hydrostatic head is developed on the river side face of the levee and water seeps out through pervious material.Thus,an artesian head and a hydraulic gradient built up in the sandy substratum towards landward face of the levees,with the ultimate formation of sand boils.The most susceptible areas for such bank failures are the locations where the top soil constitutes pervious point bar ridge,whereas, areas of thick clay in the backwamp and abandoned channels present little danger. Where point bars are located adjacent to the swales or abandoned channels and if these are parallel to the levees,the areas pose potential danger of underseepages leading to bank failures.With progressive increase of the incidences of bank failures there are chances of river migration,a common feature observed in the area.

VIII CONCLUSION

From the aforesaid discussion,it is evident that the climatic condition, geomorphological setting, topography of the terrain,underlying soil and ground water condition,role of DVC dams in releasing sudden influx of water and tidal influence,all contribute to the run-off vis-a-vis rate of disposal problem in the lower Damodar basin. Due to constricted passage through the micro-relief zone,which in turn is the manifestation of underlying geodynamic process,the intricate net work of moribund drainage fails to convey the flood water through the channel course. As a result,there is a tendency of the channels to overtop the banks and inundate the country side,particularly in the middle delta.As the left bank is protected by high embankment,the right bank is exposed to the vagaries of flood.Thirteen sectors of varying flood intensity have been identified in the trans Damodar area where bowl like depressions suffer from acute distress due to prolonged stagnation of water.

To contain such natural hazard,the following remedial measures have been suggested:
i)To increase flood cushion in the DVC reservoirs
ii)Dredging of the Rupnarain channel
iii)Construction of storage dam across the Dwarkeswar river
iv)Excavation of Mundheswari channel,Old Damodar channel and various N-S canals already existing for flood routing purpose (Plate I).
v)Excavation of a short-cut diversion channel (Plate I) connecting Old Damodar with Rupnarain,with provision of controlled discharge through Old Damodar course.
vi)Construction of embankments along left bank of Mundheswari between Rampur channel and the proposed short-cut.
vii)Re-excavation of easterly flowing abandoned courses for diverting a portion of the flood discharge directly to the Hooghly,as far as the ruling head of the latter permits.
viii)Controlling erosion in the upper catchment to reduce the silt discharge and providing silt traps.
ix)Recharging ground water through wells and ponds,to be located in the sandy horizons of the upper delta.
x)To avoid encroachment in the carriage way of the rivers,as far as practicable.

Acknowledgement

The work was carried out under the guidance of the Director, Environmental Geology Division,Geological Survey of India in collaboration with the Engineering Geology Division of GSI, by a team of Geologists comprising M/s B.B. Mullick, G.P. Sarkar, G. Ghosh, S. Lahiri and the author of the present paper.The author is solely responsible for the geotechnical interpretation of the present paper.The author is grateful to M/s D.P.Dhoundial, Director General and S.K.Banerjee,Deputy Director General,ER of the Geological Survey of India for their kind permission to present this paper to the VIth Congress of the International Association of Engineering Geology.

REFERENCES

Basu,A.N. 1986.Mathematical model for flood control of lower Damodar area Proc.Vol.II,53rd Annual Research & Development Session of Central Board of Irrigation & Power:129-147.

Biswas,B. 1961.Results of exploration for Petroleum in the western part of the Bengal basin,Standard Vacuum Oil Company's reports.

Danish hydraulic institute 1983. Mathematical modelling for flood forcasting and flood control,Interim report Vol.3 on flood control modelling for lower Damodar area.

Irrigation & Waterways Directorate, Govt. of West Bengal, 1978.Report on the floods in West Bengal, 1978.

I & W Directorate, Govt. of WB 1984. Lower Damodar drainage basin.

Morgan,J.P. 1968.Mudlump:Diapiric structures in Mississppi delta sediments.Mem.8,Am.Assocn.Pet.Geologist, Tulsa,Oklahama,145-161.

Mukherjee,K.N. 1979.Diagonostic survey of Bengal,Deptt.of Geography,Calcutta University.

Sengupta,S. 1972.Geological framework of Bhagirathi Hooghly-basin.Proc. Interdisciplinary symposium on Bhagirathi hooghly basin.

Sharma,V.K. 1976. Some hydrological characteristics of the Damodar river. Geographical review of India,Vol.38(4): 330-343.

Voorduin,W.L. 1945.Preliminary memorendum on the unified development of the Damodar river.Central Water & Power Board, Govt. of india.

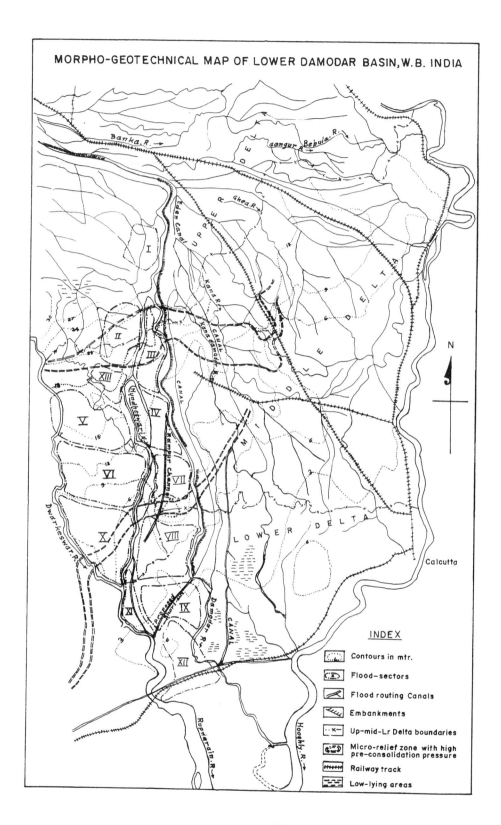

MORPHO-GEOTECHNICAL MAP OF LOWER DAMODAR BASIN, W.B. INDIA

INDEX

- Contours in mtr.
- Flood-sectors
- Flood routing Canals
- Embankments
- Up-mid-Lr Delta boundaries
- Micro-relief zone with high pre-consolidation pressure
- Railway track
- Low-lying areas

6th International IAEG Congress / 6ème Congrès International de AIGI, © 1990 Balkema, Rotterdam. ISBN 90 6191 130 3

Catchment analysis for sediment sources – a proposal
Analyse du bassin pour la détermination des sources du sédiment – une proposition

Yogendra Deva
Geological Survey of India, Lucknow, India

ABSTRACT: Reservoir siltation control measures depend upon recognition of potential sediment sources for fixing catchment management priorities. A technique for comprehensive catchment analysis, aimed at isolating critical sediment generating sectors, is suggested. It involves determination of 'sediment generation potential (pS)' of the catchment on a grid pattern, taking into account factors responsible for sediment generation, followed by calculation of share of each sub-basin in the overall sediment yield from the catchment. The sedimentation share, designated as 'sediment contribution (cS)', is obtained in pS km^2 and is used as the percentage of total sedimentation, rather than the absolute sediment yield.

Application of the technique for analysing Pench river catchment (4273 km^2) at Totladoh Dam in Maharashtra, India, has revealed that combined sub-basins of Ghatamali, Sukri and Mandhan nallas (569 km^2), with a 'cS' value of 189,477 pS km^2, account for 42.2% of total sedimentation, forming the most critical reach.

RESUME: Les mesure pour le controle des depots dans un bassin dependent au plupart sur la connaissance des zones qui apportent des sediments. C'est l'essential pour le maniement du bassin. Pour cela une technique pour l'analyse du bassin est suggeree. La 'potentialite du generer des sediments (pS)' dans le bassins est determine sur une grille, et puis les sediments apportee par chaqu'un du sub-bassin sont determines. Les sediments apportee sont designes sous le nom du 'contribution des depots (cS)' et ils sont tabules comme le pourcentage des depots totales en pS km^2.

l'Application de ce technique dans l'analyse du bassin de Pench (4273 km^2) au barrage de Totladoh dans l'etat de Maharashtra en Inde a montre que les bassins du Ghatamali, Sukri et Manshan (596 km^2) avec une valeur 'cS' de 189,477 pS km^2, apportes environs 42.2% des depots totales, et ainsi sont les plus importants.

INTRODUCTION

Reservoir sedimentation is a vitally important aspect of any river valley storage project. It governs the useful live storage of a scheme and forms the basis for fixing its life. In the early stages of the project planning itself, extent of sedimentation is assessed and additional storage (dead storage) is provided for its accumulation so that the project functions efficiently for a designed number of years (Indian Standard: 5477-1969).

The sedimentation assessment, as in practice today, , involves hydraulic studies of rivers encompassing measurement of sediment load carried by the river water. The data collected are averaged and projected for designing the storage. It is important to note that although codes exist for catchment management practices in order to control sedimentation of reservoirs (Indian Standard: 6518-1972), there is an absence of standard practices for identifying the critical sediment generating reaches for fixing management priorities. A systematic catchment analysis with reference to sedimentation, therefore, may help in identifying problem sectors for working out remedial measures which would not only enhance useful lives of reservoirs but would also help improve general environmental conditions.

In order to assess this aspect, a quantitative catchment analysis is proposed in this paper. The technique enables working out of relative sediment generation potential of various sub-basins in a catchment and also helps in identifying the contribution of these sub-basins in total sedimentation. Based on the analytical results, the critical sediment contributing reaches can be isolated and, depending upon the catchment characteristics, a decision can then be taken whether to trap the sediments at appropriate stages or to go in for remedial measures for arresting sediment generation at source.

MEASURING SEDIMENTATION

Sediment generation and its transportation form the two main steps in the process of sedimentation. Whereas terrain characteristics dicate potentiality of sediment generation, its transportation is controlled by the well defined, single largest measurable factor of surface runoff, other less responsible factors being gravitational force, wind, etc. Therefore, an adequate assessment of sedimentation lies in understanding and analysing the terrain characteristics in their proper perspective.

In agreement with general concepts on the matter, the potentiality of a terrain in terms of sediment generation is related to its status in the gemorphic cycle (i.e., its youthfulness), vulnerability to surface degradation and vegetative cover.

Status of a terrain in the geomorphic cycle has an important bearing over sediment generation. For obvious reasons, sediment generation will be directly proportional to the youthfulness of the terrain. If one considers the factors governing the terrain youthfulness, the representatives turn out to be relief and natural sculpturing. Both these factors are measurable, one as 'relative relief' and the other as 'drainage density'. Terrain youthfulness is proportional to both and has been represented by 'Ruggedness Number', expressed as the following relation (Verma et.al.,1976):

$$Nr = RR \times DD \qquad \dots\dots\dots\dots(1)$$

Where,

  Nr = Ruggedness Number
  RR = Relative Relief
  DD = Drainage Density

It is important to note that surface area available for sediment generation is proportional to both relief and drainage density. A higher Nr value, therefore, represents larger surface area.

Vulnerability to surface degradation ('Ground Erodibility') is a complicated variable dependent upon the intrinsic lithological and structural characteristics of the ground. However, for all practical purposes, these conditions can be adequately indexed depending upon the susceptibility of the ground to erosion. A five fold classification with corresponding index scale is proposed for this purpose (Table-I).

Table- 1. Classification of 'Ground Erodibility (Ge)'.

Character	Index Value
a) Loose soil, deeply weathered and disintegrated rock	5
b) Consolidated soil, deeply weathered rock	4
c) Weathered and highly jointed/ fractured rock, soft rock	3
d) Partially weathered and moderately jointed/fractured rock	2
e) Hard, sparsely jointed rock without weathering effects	1

Likewise, vegetative cover ('Vc'), if one goes into the details, also is a complicated factor. But, this has also been classified and indexed to suit the requirement (Table-2).

Table- 2. Classification of Vegetative Cover 'Vc'.

Character	Index Value
a) Grassy ground with deep rooted short trees	5
b) Mixed vegetation with partial grassy coverage	4
c) Mixed vegetation without grassy cover, cultivated land	3
d) Sporadic vegetation	2
e) Devoid of vegetation	1

Although the index values allotted to 'Ge' and 'Vc' are whole numbers, depending upon the combination of different classes of these factors, the effective index values are normally fractional numbers. For instance, a reach with 1/3rd area under category 5 of 'Ge', another 1/3rd under category

4, and remaining 1/3rd under category 1, will have an average 'Ge' value of 3.33.

With these factors, sediment generation is proportional to 'Ruggedness Number' (Nr), and 'Ground Erodibility' (Ge), and is inversely proportional to 'Vegetative Cover' (Vc). Therefore, the 'Sediment Generation Potential' ('pS') of an area can be expressed as the following relation:

$$pS = (Nr \times Ge) / Vc \qquad \ldots\ldots\ldots\ldots(2)$$

It may be seen that with uniform rainfall or surface runoff conditions in a catchment, actual sediment generation would depend upon 'pS'. It is therefore proposed to identify and zone the 'pS' conditions in a catchment for working out relative sediment generating potentiality.

In order to isolate the critical sediment generating reaches, each sub-basin in the catchment is to be analysed for its share in the total sediment yield from the catchment. It is likely that a sub-basin with higher sediment generation potentiality ('pS') may in fact be contributing less sediments due to smaller area it occupies in comparision to another sub-basin, which despite a much lesser 'pS' may be contributing greater sediment quantity, simply due to its much wider area. Therefore, since actual share in sedimentation from a sub-basin is proportional to its 'pS' and area 'A', its 'sediment contribution' ('cS') has been obtained as per the following relation:

$$cS = pS \times A \qquad \ldots\ldots\ldots\ldots\ldots(3)$$

The 'sediment contribution ' ('cS'), obtained in 'pS km2', is represented as the percentage of the combined 'cS' values of all the sub-basins.

It is very important to note that the whole concept of sedimentation measurement, as proposed here, is based on relative sediment contribution from various sub-basins in a given catchment and does not provide the absolute value of sediment yield. For this purpose, taking into account the surface runoff factor, it may be possible to calibrate 'sediment contribution' ('cS') against actual sediment yield by monitoring a few catchments. This aspect, however, falls outside the scope of present studies, and has not been dealt with.

COMPUTATIONAL PROCEDURE

The methodology adopted involves dividing the catchment into suitable grids followed by calculation of 'pS' for each grid. The data thus collected are represented as a 'pS contour map' highlighting the relative sediment generating potentiality of the catchment. Finally, each sub-basin in the catchment is analysed for its average 'pS' which, when multiplied by its area, provides its 'sediment contribution' ('cS') in pS km2, expressed as the percentage of the cumulative 'cS' of all the sub-basins. This therefore would help in isolating critical sediment generating reaches for subsequent consider-ations of adequate catchment management practices for controlling sediment flow into reservoirs. Some salient features of the technique adopted are given below.

The work has been basically carried out over Survey of India topographic sheets on 1:50,000 which incorporate vegetative cover details also. Reference of geological maps on 1:50,000, 1:250,000 & 1:2,000,000 and random geological mapping & field checks have helped in identifying 'ground erodibility' ('Ge') and 'vegetative cover' ('Vc') characteristics.

As the first step, the catchment is divided into grids of 9km2 area each with suitable index number. This is then followed by analysis of each grid for its 'pS' value.

Ruggedness Number (Nr)

The two components of Ruggedness Number, viz. Relative Relief and Drainage Density are measured independently.

a) Relative Relief:
Elevation sampling of the grid is carried out using '9 sample method', found to be most comprehensively representative. Eight sample locations lie on the perimeter of the grid at equal distances of 1.5 km, with the ninth location being at the centre of the grid. The perimeter samples, therefore, are shared with adjoining grids, the central sample being the only unique sample for the grid. The average value of 9 samples is taken as the representative general elevation of the grid, and the lowest elevation of the grid, when subtracted from this value, provides the 'relative relief' ('RR'). For data collection convenience, a transparent stiff PVC sheet with engraved grid boundary and sample locations has been used for this purpose (Fig-1).
b) Drainage Density:
This is a simple computation of measuring total drainage length in the grid followed

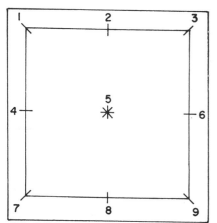

FIG.—I: ELEVATION SAMPLE FORMAT.

least count is fairly tolerable.

'GroundErodibility'('Ge')and 'Vegetative Cover' ('Vc'), for each grid, are determined and index values allotted as per the five-fold classification proposed in Tables-1 and 2 respectively.

The 'pS' values thus computed are suitably classified into five or more categories and then contoured to obtain the 'pS Contour Map' which illustrates comparative sediment generating potentiality of the catchment and also forms the basis for selecting critical sub-basins for assessment of individual sediment contribution. Finally, these sub-basins are analysed for their 'sediment contribution' ('cS') and their order of importance as sediment generating sources is worked out for subsequent remedial measures.

by its division by the grid area, which, excepting the catchment boundary region, happens to be $9 \times 10^6 m2$ (9 km2). The drainage length has been measured using a 'Map Measurer' (Rotameter) having a least count of 250m for a working sheet on 1:50,000. However, considering that the drainage length for a grid is normally more than 20,000m, this

CASE ANALYSIS : PENCH CATCHMENT

The concept of measuring sediment contri-bution, as discussed in previous paragraphs, has been applied to the Pench river catchment at Totladoh Dam (22° 39':79° 14';55 O/2) in Maharashtra, India (Fig-2). The 74.5m

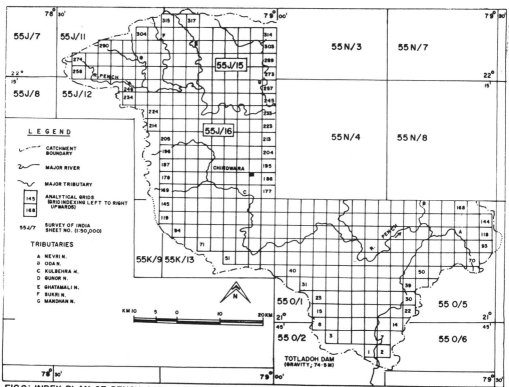

FIG.2: INDEX PLAN OF PENCH CATCHMENT AT TOTLADOH DAM, MAH.

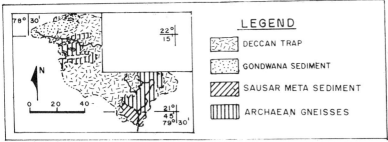

FIG.3 GEOLOGICAL INDEX PLAN OF PENCH CATCHMENT

high concrete dam across the river Pench constitutes a part of the 160 MW Pench Hydroelectric Project of the Maharashtra Irrigation Department, with an underground power house (2x80MW) and an 8.1km long tail race tunnel. The dam with TBL at EL 495.5m and river bed level at EL 421m, has been designed for a dead storage of 153 m.m^3 (5.4 TMC) against a gross storage of 1241 m.m^3 (43.8 TMC). The design is based on hydraulic data collected at site with projected sediment load figures. As per normal practice, the catchment characteristics have not been considered at any stage during designing.

The Pench catchment at Totladoh Dam occupies an area of 4273 km^2 with 79.8km^2 area coming under submergence as Totladoh Reservoir. The catchment attains its highest elevation of 1208m at Kala Pahar (55 J/11). The area is covered under Survey of India topographic sheets no. 55 J,K,N & O between lat 21° 40' to 22° 23' and long 78° 30' to 79° 30' (Fig-2). The Kulbehra N., the largest sub-basin in the catchment, constitutes the only major right bank tributary. The left flank of the river is sub-divided into several prominent sub-basins, viz., Nevri, Oda, Gunor, Ghatamali, Sukri and Mandhan nallas. Due to non-availability of topographic sheets under 55N, covering 27% of total area in its north-eastern sector, the work has been carried out for the remaining 73% area. The source of the river Pench south-east of Pachmarhi, the hill queen of the Satpura Ranges, however, is covered under the studies. The Pench catchment lies in geologically diverse formations (Fig-3). Whereas, a major portion of the catchment is occupied by the Deccan Traps (68%), unclassified granite gneisses occupy 18%, Lower Gondwana sedimentaries 9% and the Precambrian metasediments (Sausar) the remaining 5%. Large soil/alluvium covered reaches conceal these formations in part.

The catchment area under study has been divided into 317 grids of 9 km^2 area each (Fig-2). Only a few grids, located in the uppermost reaches of the catchment, are marginally smaller in area. Due to certain constraints, it has not been possible to analyse 46 grids, lying in the eastern sector, number of analysed grids, therefore, being 271 (2439 km^2). Each grid has been analysed for its 'pS' value and the 'pS contour map' has been brought out (Fig-4). For the purpose of analysis, 'pS' values have been divided into classes ranging from 0-25, 26-50, 51-75, and so on. The highest 'pS' grid is 262 with a value of 671, and the lowest 'pS' grid is 107 having a value of 11. Here, it may be mentioned that the 'pS' values obtained from the relationship worked out earlier in the text have been multiplied by 1000 for converting them into whole numbers. Since, in computations for different catchments this value may vary between 0.001 and 5.00, it would be advisable to multiply it with 1000 for the purpose of analyses. The studies reveal that (Fig-5) the predominant 'pS' class is 51 to 75 accounting for 19% area. However, significant features of 'pS' distribution are that whereas almost half the area analysed ranges between 'pS' values of 26 & 100, only about 6% lies above 400.

A study of catchment index plan (Fig-2) and the 'pS contour map' (Fig-4) indicates that whereas Kulbehra N. sub-basin deserves considerations due to its largest area (1035 km^2, 25%), the combined catchments of Ghatamali, Sukri and Mandhan N., lying in the upper reaches of the catchment, are important due to their high 'pS' values. The 'sediment contribution' ('cS') of these sub-basins has worked out to be (Table-3) 73,485 pS km^2 (16.4%) for Kulbehra and 189.477 pS km^2 (42.2%) for the combined sub-basins of Ghatamali, Sukri and Mandhan. Remaining sub-basins have a combined 'cS' value of 186,030 pS km^2 (41.4%).

It is therefore concluded that the critical sediment contributing sector of the Pench catchment at Totladoh Dam is the combined

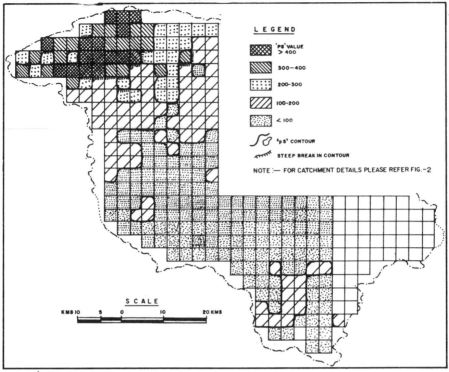

FIG.-4 'pS' CONTOUR MAP OF PENCH CATCHMENT.

Table- 3. Summarised 'sediment contribution' ('cS') of different sub-basins of Pench catchment.

S.No.	Sub-basin	Average 'pS'	Area 'A' km²	Sediment Contribution 'cS'	
				(pS km²)	%
1.	Kulbehra	71	1035	73,485	16.4
2.	Ghatamali, Sukri, Mandhan (combined)	333	569	189,477	42.2
3.	Remaining	159	1170	186,030	41.4

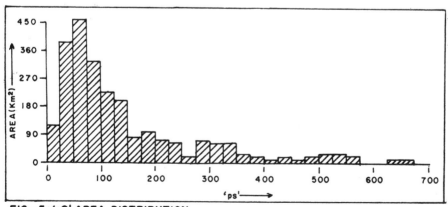

FIG.-5:'pS' AREA DISTRIBUTION.

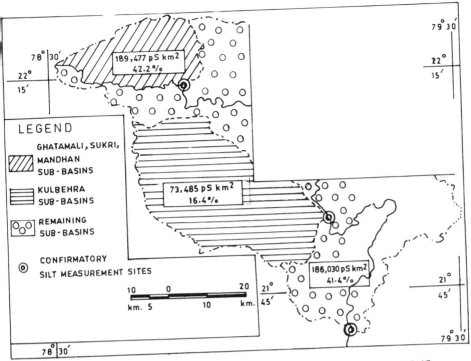

LEGEND

	GHATAMALI, SUKRI, MANDHAN SUB-BASINS
	KULBEHRA SUB-BASINS
	REMAINING SUB-BASINS
	CONFIRMATORY SILT MEASUREMENT SITES

189,477 pS km² 42.2%

73,485 pS km² 16.4%

186,030 pS km² 41.4%

FIG. 6-SEDIMENT CONTRIBUTION (cS)' MAP OF PENCH CATCHMENT.

sub-basins of Ghatamali, Sukri and Mandhan N. accounting for 42.2% of the total sediment yield from the catchment (Fig-6). The Kulbehra N., despite being the largest sub-basin, might be contributing only 16.4%. On the basis of these findings, in case the Pench catchment is ever considered for remedial measures against reservoir sedimentation, the combined catchments of Ghatamali, Sukri and Mandhan may be given priority.

The decision regarding actual remedial measures in the form of construction of check weirs, or improved agronomical or land-use practices, will have to be taken as per the physiographical, socio-economic and other important factors of the region under consideration. For this purpose the details of catchment management practices brought out by the Central Board of Irrigation and Power (1977) and also recommended by the Indian Standard (1972) may be referred. It may also be useful to consider these findings while contemplating development projects and environmental schemes in the catchment.

Finally, inferences drawn for sedimentation share from different sub-basins as above, would require confirmatory field checks for which sediment yield measurements may have to be conducted at locations indicated in Fig-6.

CONCLUDING REMARKS

The proposed analytical approach for measuring sediment contribution has been worked out within the limitations of scope for studies and is, therefore, open to modifications and improvements. The analysis has not taken into account the sediment source-deposition site distance factor, which would require some more field data. Further, it may once again be mentuioned that the 'sediment contribution' ('cS') has been used as a comparative factor for share in sedimentation from various sub-basins and does not provide an absolute value of sediment yield. A calibration of 'cS' (pS km2) against sediment yield, based on field data including the surface runoff factor, however, may help in establishing a useful relationship for assessing sediment yield through the proposed studies.

It may also be mentioned that although the work has been carried out over the toporaphic sheets and related geological and vegetative cover plans, it would be more convenient to make use of aerial photograph for these studies. This would enable instant assessment of all the three factors, viz., the ruggedness number, ground erodibility and the vegetative cover.

ACKNOWLEDGEMENTS

Grateful thanks are extended to the Director General, Geological Survey of India, for his kind permission to publish this paper.

REFERENCES

Central Board of Irrigation & Power (1977). Technical Paper No.20: Sedimentation Studies in Reservoirs. New Delhi.

Indian Standard Institution (1969). Indian Standard: 5477 (PartI)-1969: Methods for fixing the capacities of reservoirs. New Delhi.

Indian Standard Institution (1972). Indian Standard: 6518-1972: Code of practice for control of sediments in reservoirs. New Delhi.

Verma, V.K., Bhattacharya, G. and Deva, Y. (1976). Aesthetic evaluation of land-forms- a theoretical approach. Indian Geological Congress, University of Delhi, Vol of abstracts, Delhi, PP 37-38.

6th International IAEG Congress / 6ème Congrès International de AIGI, © 1990 Balkema, Rotterdam. ISBN 90 6191 130 3

Flood control programme in India

Programme de contrôle des crues en Inde

K.S. Murty
University Department of Geology, Nagpur, India

ABSTRACT: The flood-prone area has increased from 25 million hectares in 1953 to 34 million hectares in 1978 and is now around 40 million hectares. The average loss due to floods has worked out to Rs.4635.50 million in the last three decades, compared to just Rs.220 million in 1951. Over 5 million people are affected every year. The most frequently affected areas in India are in Assam, West Bengal and Bihar while Gujarat, Orissa and Andhra Pradesh suffer from this hazard to a lesser extent. A National Flood Control Programme was launched as early as in 1954 under which construction of embankments to the rivers, drainage channels, and raising of channels levels were undertaken. The Seventh Plan(1985-90) had a target of 2.5 million hectares to be brought under flood control at an investment of Rs.9473.9 million. One of the programmes is the preparation of a Master Plan for Flood Management and also of flood-zone maps. Soil conservation and afforestation programmes are proposed to be intensified, apart from new embankment schemes. Linking of rivers of North India to those of South India is a long-term programme to control floods.

1 INTRODUCTION

India is endowed with copious water resources. Its surface water alone amounts to 178 million hectare metres besides significant sub-surface storage(Dhanpat Rai 1983).

Table 1. Surface water resources.

Name of River Basin	Average annual run-off (Million Ha.M)	Utilisable flow
Indus	7.70	4.93
Ganges	51.00	18.50
Brahmaputra	54.00	1.23
West flowing rivers	28.80	6.92
East flowing rivers	34.80	33.80
Others	1.70	1.22
Total	178.00	66.60

The average rainfall in India is 110 cms and over 70 per cent of the total rainfall occurs between June and September, when spells of downpour cause a high-flood-generating water discharge in channels of rivers and streams. Either excess flow of water over the water carrying capacity of the river channel or reduction in the carrying capacity of the channel as a consequence of accumulation of sediments, can cause floods. Excess flow is sometimes due to heavy rainfall and cloudburst, melting of snows on a large scale with attendant bursting of dams, sudden and cexess release of impounded water behind dams, and bursting of man-made or landslide built dams. Cloudbursts have been reported in the Himalayas and Orissa and Central India(Abbi 1981). The Machu Dam II in Gujarat, and the Panshet and Khadakvasla dams in Maharashtra are examples of collapse of man-made structures.

Floods have frequented the past, prawl the present and hold the future. Numerous pre-historic settlements like the Mohenjodaro bear testimony to this. Though the precipitation may not have increased in the last hundred years, the frequency of floods and their intensity, however, have been increasing(Table 2).

Table 2. Large-scale floods and area affected from 1891-1979.

Year	Interval (Years)	Area affected (%)	Ranking
1892	-	33	6
1893	0	40	3
1894	0	44	2
1916	21	36	5
1917	0	38	4
1933	15	30	9
1936	2	27	11
1938	1	25	14
1942	3	30	8
1956	13	27	12
1959	2	28	10
1961	1	48	1
1975	13	31	7
1978	2	26	13

Of the 40 million hectares of fertile land affected by floods, 60 per cent lie in the Indo-Gangetic plains, where the population density is high. Rivers of the peninsular India, like the Godavari and Krishna, are also flood-prone(Rao 1975). The lofty barrier of the Himalayan mountains is instrumental in holding back the rain-clouds from moving northwards, thus causing their precipitation in torrents. Similarly, the Nilgiri-Anamudi in South India is responsible for heavy rainfall ón Kerala, and the Sahyadri range for excessive rain in the Western ghats. The desert areas in western India have an average annual rainfall of 15 cam, but they too in recent years have been ravaged by catastrophic floods. Several dams and embankments were destroyed and wells were filled with silt in Rajasthan(Seth et al 1982) in 1981. Tables 3 and 4 give details of major floods of rivers of India.

Table 3. Highest recorded levels of rivers of India.

River	Warning stage	Highest level	Date
	(Metres above S.L)		
Ravi	349.4	349.8	5.10.55
Beas	213.7	215.2	6.10.55
Satluj	210.3	211.2	6.10.55
Yamuna	382.2	385.9	1924
Ganga	198.7	199.0	17.06.63
Ramganga	210.9	211.8	6.08.69
Kosi	130.5	131.8	5.10.68
Teesta	86.1	87.2	5.10.68
Brahmaputra	104.2	105.5	17.07.69
Brahmani	20.1	22.8	1960
Mahanadi	21.0	27.9	1925
Godavari	16.7	17.7	16.08.53
Krishna	18.1	24.3	7.10.03
Kaveri	240.8	242.2	20.08.61
Tapi	29.0	31.11	6.08.68
Narmada	7.6	12.6	7.09.70
Mahi	100.5	104.9	15.09.59

Table 4. Frequency of floods between 1969 and 1981, more than 10.

River	State	Number of floods
Betwa	U.P.	16
Ghaghara	U.P.	19
Ganga	U.P.	10
Ganga	W.B.	14
Kosi	Bihar	13
Son	Bihar	12
Brahmaputra	Assam	11
Subansiri	Assam	25
Dhansri	Assam	21
Subarnarekha	Orissa	
Narmada	Gujarat	14
Godavari	A.P.	3(very severe)

(Dhar et al 1981)

2. DAMAGE AND LOSS DUE TO FLOODS

The average number of houses damaged during 1953-78 amounted to one million, while human lives lost were 1240 and cattle lost were about 80,000. The area affected in a single year was as high as 17.8 million hectares during 1976. In 1986, floods affected 252 districts, covering about 66,243 villages and 82 lakh(=8.2 million) hectares including 4.3 million hectares of cropped area and 52.2 million population were affected by floods. About 1180 human beings and 58,000 cattle were lost while 2 million houses were damaged. Table 5 gives details of flood damage and loss of crop and lives during the three decades between 1953-83.

Table 5. Flood damage, loss of crop and lives(1953-83).

Item	Average damage	Maximum in any one year
Area (m.ha)	8.66	3581(1983)
Population (mil.)	29.27	70.45(1978)
Cropped area (m.ha)	3.76	9.95(1978)
Value of crops (Rs.mil)	2548	11583(1983)
Houses	1.07	3.51(1978)
Value of damages to houses (Rs.mil.)	703	3839(1982)
Cattlehead lost(No.)	103621	618248(1979)
Human lives lost(No.)	1409	11316(1977)
Damage to public utility (Rs.mil)	1328	7841(1983)
Total damage (Rs.mil.)	4578	22922(1983)

(Padmanabhan 1984)

A close look at the trend brings out that the damage has risen disproportionately to the rise in the area affected. One reason may be higher inflation rate and the consequent rise in the assessed value of the loss of crop, houses and public utilities. Another reason could be the tendency on the part of the State Governments to report higher damage to get more Central assistance.

3. NATIONAL POLICY ON FLOODS

Floods necessitate massive rescue and relief operations. In 1985-86, expenditure on flood relief amounted to Rs.5121 million and in 1986-87 to Rs.3478.4 million. In the sixth five year plan, the total expenditure on relief operations came to Rs.12,000 million as against the investment of Rs.7800 million on the flood protection programme during the plan period(1980-85). The devastating floods of 1953 and 1954 in several parts of the country led to the announcement of a national policy on floods by the Government of India on 3 September 1954, which emphasised the need for taking up flood protection measures in a systematic and planned manner and outlined a timebound plan of action ranging from immediate phase(within two years) through short term(3rd to 7th year) to the long term(8th to 12th year). The outlay on flood protection measures has been increasing over the years(Table 6). Still it is felt that flood control has not received the priority it deserved, the outlay varying from 0.5 to 1.08 per cent of the total plan outlay.

Table 6. Outlay on flood control in the five year plans.

Plans	Period	Outlay on measures (million Rs.)
First	1951-56	133
Second	1956-61	492
Third	1961-66	866
Annual	1966-69	436
Fourth	1969-74	1718
Fifth	1974-80*	6049
Sixth	1980-85	7786
Seventh	1985-90	9474

* includes 1979-80.

(Kamta Prasad 1987)

The anticipated expenditure on the flood control measures during the seventh plan period was Rs.9700 million as against the approved outlay of Rs.10450 million.

The 1954 policy statement had underlined the importance of multipurpose reservoirs supplemented by measures like embankments, detention basins, channel improvements and other measures in moderating floods. The desirability of an a approach having a combination of several measures with due emphasis on multi-purpose reservoirs continued to be felt throughout the subsequent years as can be seen from the reports of several committees. However, the principal methods adopted in practice have been embankment and drainage channel supplemented by flood forecasting and warning.

4. ACHIEVEMENTS

Between 1954 and 1983, new embankments to a length of 12,905 km were constructed, 25,331 km of drainage channels were dug, 332 towns were protected and 47,000 villages were raised above the flood level. More than 12 million hectares, out of an estimated 32 million hectares of protectable area, were afforded a reasonable level of protection by these measures up to March 1984. It was envisaged in the seventh plan to provide flood protection to 2.5 million hectares additionally. So far 154 stations have been set up covering all the inter-State rivers for forecasting floods. Based on the data from upstream stations, the level of water at a particular place at a particular time can be calculated and information conveyed to local authorities 24 to 36 hours in advance of an impending flood. However, only two reservoirs have been constructed specifically designed to control floods and these are in operation over the Damodar in West Bengal and over the Rangali in Orissa. A pilot project has been taken up, with UNDP assistance, to improve forecasting on the Yamuna, while another scheme is coming up with Danish assistance for improving flood forecasting for the Damodar.

5. OTHER PROGRAMMES

The report of the National Commission on floods and other documents have cautioned against undue reliance on embankments and suggested a judicious combination of different types of measures such as soil conservation and afforestation, reservoirs, embankments, drainage channels, dredging, flood plan regulations, flood forecasting etc. Stress is laid on reservoirs as a means offlood control. Water conservation measures like watershed management, small, medium and large reservoirs, natural detention basins and groundwater storage have been recommended.

The concept of a "National Water Grid", envisaging Ganga-Cauvery link was put forward by K.L. Rao in the early seventies. He suggested that surplus waters of Ganga and Brahmaputra must be used to augment supplies to deficit basins so that semi-arid zones could be converted into food-yielding land. He also recommended flood water detention in reservoirs, particularly in the river basins of U.P. and Orissa, as also increased drainage facilities in West Bengal to save crops from floods. His plan (Rao 1975) has been modified into a "National Plan" which aims at multi-purpose and multi-objective development of the country's water resources, keeping in view the benefits of irrigation, land reclamation, flood control, hydro-power generation, navigation, pollution control, fisheries development etc. Flood control is envisaged as a major objective of this plan. The scheme outlines inter-basin and inter-State transfer of water, from the Ganges and Brahmputra which suffer from floods into the peninsular rivers. Diversion of west-flowing rivers in Kerala, Maharashtra, Gujarat and Tamil Nadu, diversion of surplus of the Mahanadi and Godavari waters to water short rivers like the Krishna, inter-linking of a few other west-flowing rifers north of Bombay and south of Tapi, and inter-linking of Ken with Chambal, are the main features. In this way, areas which are drought-prone, about one-third of the total area of the country, would also be getting water diverted to them. The National Water Development Agency (NWDA), set up in July 1982 gave shape to the outline of National Perspective for Water Resources Development, called the National Plan. (Fig.1) (Das, 1977 and Valdiya 1987)

A Master Plan for Flood Manage-

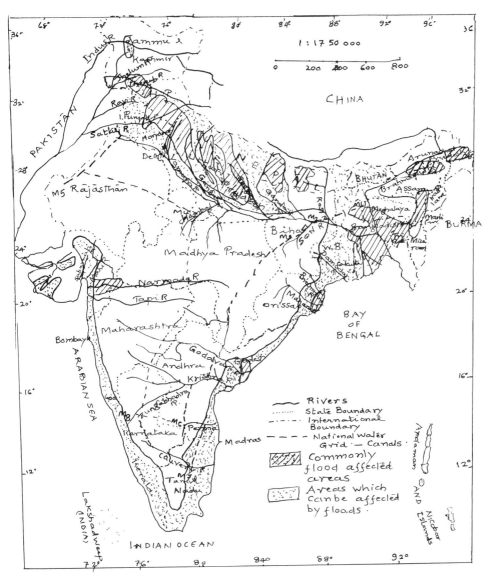

Fig-1. RIVERS OF INDIA AND FLOOD-AFFECTED AREAS.
(After D.C.Das, 1977 and K.S.Valdiya 1987)

ment is to be prepared after collecting hydrological and meteorological data and field investigations. Flood-zone maps need to be prepared as part of this plan on the scale of 1:50000. It is proposed to cover at least 25 million hectares by 2000 A.D. under the flood control programme. Experimental studies have been initiated by the Department of Science and Technology, under their scheme on Natural Resources Data Management System(NRDMS) making use of the remote sensing techniques. The Centre of Studies in Resource Engineering(CSRE) of the Indian Institute of Technology, Bombay has undertaken the study of flood plain zoning in the basins of Ghaghara and Gandak(Basu 1985).

The flood problem is of an inter-State nature and inter-State conflicts came in the way of tackling the menace on a permanent basis. An integrated view of the problem has to be taken and treated as a national problem. Inter-State river basin authorities may have to be set up, indpendent of the State control. Central financing for inter-State flood control works should become an integral part of the comprehensive approach to flood management.

REFERENCES

Abbi, S.D.S.,1981. Flash floods and their warning. Proc. Intl. Conf. Flood Disasters, Ind.Nat. Acad. Sci.,New Delhi,1:238-50

Das, D.C. 1977. Soil conservation practices and erosion control in India,- a case study. Soil Bulletin of F.A.O.,33:11-50.

Dhanpat Rai, 1983. Caring for India's millions - land and water aspects, Yojana, Dec.1-15, 4-8.

Dhar, O.N., G.C. Ghose and A.K. Kulkarni 1981. A catalogue of major devastating floods in India and their space-time distribution, Proc. Intl. Conf. Flood Hazards, Ind.Nat.Sci.Acad. New Delhi, 1:147-61.

Kamta Prasad 1987. Flood Control-Achievements and Challenges, Yojana Special 1987, 81-83.

Padmanabhan, B.S. 1984. Gaps in flood control. The Hindu, 10th August 1984.

Rao, K.L.1975. India's Water Wealth Orient Longmans, New Delhi, 255p.

Seth,S.L.,D.C. Das, and G.P. Gupta 1982. Floods in Arid and Semi-arid areas - Rajasthan. National Flood Commission Report, Vol.I, Chapter 10.

Sreelekha Basu 1985. Floods and their management, Yojana, April 1-15, p.13-17.

Valdiya, K.S. 1987. Environmental Geology, Indian Context, Tata-McGraw-Hill Pub.Co. Ltd., New Delhi, xii+583.

Genesis of river bank erosion in the Niger Delta System
Génèse de l'érosion des bords de rivière dans le delta nigérien

Chukwunonye Ozioma Owuama
IFERT, Rivers State University of Science & Technology, Port Harcourt, Nigeria

ABSTRACT: Niger delta system is a dynamic environment in which all of its towns and villages are situated at the banks of rivers and creeks constituting the distributaries of the river niger. Erosion at the outer bends of the banks are comparatively severe especially at inhabited locations. It is apparent that bank erosion depends upon soil type and river discharge but investigations on 28 settlements in the western part of this delta shows that the initiation of curvature and proportionate increase in the severity of erosion is caused by human activities. Problems of bank erosion in the niger delta cannot be solved by relocation of towns and villages but by constructing a series of shore protection measures.

1 INTRODUCTION

Niger delta represents a region of redistribution of waters of the River Niger into the Gulf of Guinea (Bights of Benin and Biafra). Many medium sized towns and villages are located at the banks of the various distributaries of the river. The Rivers Nun and Forcados form the major outlets. In this area rivers, creeks and canals are inter connected to form a network of fluvial system. Hydrographic observations using an automatic tidal gauge has shown that tidal variations extend as far as Sagbagrei Fig 1, and up to Odi during the dry periods of the year (December to May), NEDECO (1961). This could be due to a general low topography of the Niger Delta ecosystem. As a consequence, most of the region is submerged during peak flood. Vegetation is mainly mangrove or brackish-water type in the intertidal zones while in the fresh water districts rafia palms and other fresh water plants thrive. Most of the towns and villages rest on a 3m to 6m thick of silty organic clays underlain by fine to medium sand grains typical of alluvial deposits. There are however cases of very thick overburden of organic clays.

The sediments of the Niger Delta, in general, are predominantly of cenozoic age. The lithologies include shale (Imo Formation), mudstone, dark grey sandy shales (Akata Formation), sandstone, unconsolidated to partially consolidated sands and gravels (Benin Formation). The sub areal quarternary deposit consists of coarse layered sands and silts deposited in the river channels, point bars, levees and backswamps. Deposits of old beach ridge of cenozoic age, sands and interdunes, clays, muds and organic muds are also common.

An investigation on 28 towns and villages in the western part of the Niger Delta (Fig 1) reveals that the ecosystem is under constant threat of erosion, flood and siltation of navigation channels. It is the objective of this discussion to assess the genesis of the river bank erosion and propose a planning scheme for the containment of the hazard.

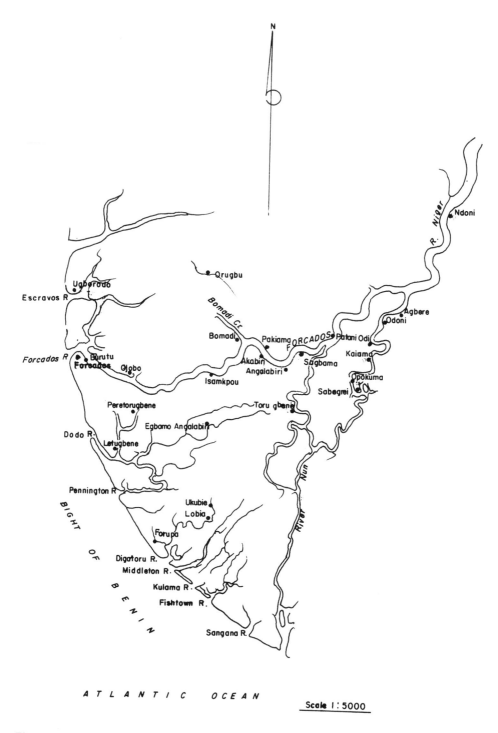

Fig.1: Map of Western part of Niger Delta

2 PROCESSES OF BANK EROSION

Erosion in this context is defined as a progressive disintegration and removal of the materials of the bank of a river channel through the force of river water. A body of water moving at a speed of V m/s has a velocity head of $\frac{\alpha v^2}{2g}$, where α is an energy coefficient and g the gravitational acceleration. Ideally $\alpha = 1$. The energy consequent on the head is dissipated as shear force when the flow comes into contact with the soils of the river bank. If the shear force exceeds the shear resistance much of the material will be removed by it. The limiting shear velocity u can be estimated from Shield's empirical relationship given by

$$U^2 = \psi \left(\frac{\rho_s - \rho}{\rho}\right)gd \quad \ldots \ldots \quad (1)$$

where ψ = coefficient of Shield's (=0.06), ρ_s = density of soil, ρ = density of water and d = mean particle diameter (d50). If v at the bank exceeds U erosion may take place.

Another process of bank erosion is the gradual disolution or removal of clay, silt and fine sand particles which could result to internal erosion and eventual collapse of the remaining soil skeleton. This phenomenon is assisted by the presence of tension cracks at the shoulder of the river bank. If the planes of weakness of these cracks are full of water the hydrostatic force that builds up causes instability of the bank. Furthermore, progressive undermining of the bank reinforced by secondary currents that move sediments at the bottom of the channel from outside of the bank in toward the point bar leads to steepening of the slope thereby facilitating failure.

All these phenomena singularly or collectively result to erosion of the river bank. However, another important factor in bank erosion is the relationship between the flow direction and geometry of the bank. When flow is directed towards the bank, the rate of erosion is high at the face. This is due to the disloca-tion of soil particles

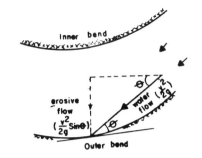

Fig 2 River flow and bank geometry

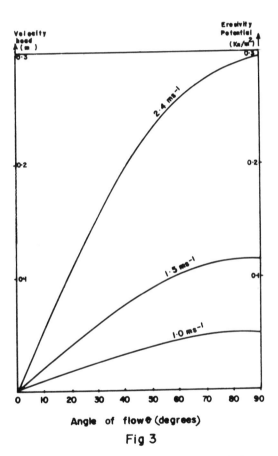

Fig 3

Relationship between relative erosivity potential and angle of flow θ

resulting from dissipation of energy whose magnitude is a function of $V^2/2g$. This energy (or head) is a maximum and most destructive when the flow is in a direction perpendicular to the bank, and decreases with decreasing flow angle θ, Fig 2. Fig 3 is a theoretical relationship between θ and the effective velocity head, h or relative erosivity potential p for three flow conditions.

h is defined by

$$h = (V^2/2g) \sin \theta \ . \ . \ . \ . \ (2a), \text{ and}$$

$$p = (\rho V^2/2g) \sin \theta \ , kn/m^2 \ . \ . \ (2b)$$

This is a maximum when θ is 90°. Hence the outer bends of river courses are areas of pronounced bank erosion where channel flow is mainly directed towards the bank.

3 CAUSES OF BANK EROSION IN THE NIGER DELTA

3.1 Human activities

Human activities are manifested in form of waves generated by speed boats berthing ashore; movement of people up or down the bank; anchoring of paddled conoes and clearing of protective cover along the bank.

3.1.1 Waves from speed boats

Waves initiated on the water by speed boats berthing at the shallow water fronts attacks the bank face with a wave celerity or velocity cw given by Daugherty et al (1977),

$$C_w = [g(y + \Delta y)]^{\frac{1}{2}} \ . \ . \ . \ . \ . \ . \ (3)$$

in which g is the gravitational acceleration, y the undisturbed depth and Δy the wave height. The direction of the wave is usually at an angle to the bank resulting to an energy breakdown at the boundary. This loss in energy is gained in the disintegration of soil particles. The loose soils are then dragged into the body of the water by retreating wave, secondary currents or gradually flowing water.

3.1.2 Movements on the bank

People entering or disembarking from boats, bathing in the river or fishing at the water front move up or down the river bank. This process weakens the soil structure thereby making it easily erodible by the river water or an overland flow that follows man-made pathways.

3.1.3 Piling for boat anchors and convenience apartments

In the process of installing anchors for paddled canoes or piles for convenience structures, the bank soil is weakened. This increases its erodibility potential.

3.1.4 Clearing of protective vegetation

Clearing of vegetal growths at the water fronts is commonly carried out by the natives. This leads to the exposure of the ground surface resulting to the development of tension cracks and gradual erosion of the overhanging soil.

3.2 Soil type

The nature of soil shouldering the river bank has an effect on the degree of degradation of its water front. The banks of rivers in the Niger Delta system are comprised of either clayey materials or alluvial sand, NEDECO (1961). Specifically at Agbere and Isamkpou, Fig 1, the banks are clayey while at Ndoni it is sandy. But resistant to erosion varies from one soil type to another, and the relative resistance to erosion along a canal after Wagner (1957) is shown in Table 1.

Table 1: Relative resistance of soils to erosion on canal (after Wagner, 1957)

Soil group	Group symbol	Erosion resistance
Inorganic silts, organic silts, very fine sands, organic clays, peat	ML,OL,MH, OH,Pt	0
Well graded gravels, gravels gravel sand mixture, little or no fine	GW	1
Poorly graded gravel sand mixture	GP	2
Clayey gravels, poorly graded gravel - sand clay mixtures	GC	3
Silty poorly graded sand- silt mixtures	GM	4
Clayey sand poorly graded sand clay mixtures	SC	5
Well graded sand, gravelly sands little or no fine	SW	6
Poorly graded gravelly sand, little or no fines	SP	7
Silty(gravelly) sand, poorly graded sand-silt mixtures	SM	8
Inorganic clays of medium to low plasticity, gravelly clays, sandy clays, silty clays, lean clays	CL	9
Inorganic clays of high plasticity, fat clays	CH	10

Generally, high plasticity inorganic clay is more resistant to erosion than fine sand or silt. It is therefore apparent that under the same set of environmental conditions an alluvial sand bank can more readily be eroded than a clay bank.

3.3 River discharge

The discharge Q of a river is a function of the mean flow velocity V and the cross-sectional area A. It is given by Bernouilli, (Webber, 1979),

$$Q = AV \dots \dots \dots (4)$$

From this equation the velocity (V) is higher when the size of the channel is reduced provided the discharge Q is constant. Similarly, an increase in Q for a given cross sectional area A leads to higher value of V. An increase in V then results to a greater erosivity potential of the river. In general, bank erosion in the Niger Delta is most pronounced during the yearly period of peak flood (October to November) within which time the rate of river discharge is at its peak. However, the effective cross-sectional area of a single channel does not change significantly in this ecosystem because of the presence of many creeks that accommodate excess discharge.

3.4 Curvature of river

Figure 3 shows a hypothetical relationship between the angle of flow and erosivity potential. A sharp curvature leads to high angle of flow and consequently high erosivity at the bank. This means that in a meandering river active erosion occurs at the outer bends while deposition occurs at the inner bends. The sharper the bend the greater is the erosion.

4 DISCUSSION

An investigation on 28 communities in the Niger Delta shows that about 80% of them are situated at the outer bends of rivers or creeks where bank erosion is severe, Table 2.

Table 2: Positional location of some small towns and villages in the Niger Delta system

Towns/ villages	Adjacent river/ creek	Location
Ndoni	R. Niger	Outer bend
Agbere	R. Nun	"
Odi	"	"
Kaiama	"	"
Sagbegrei	"	No bend
Opokuma	"	"
Patani	R. Forcados	O/bend
Isamkpou	"	"
Torugbene	"	"
Peretorugbene	"	"
Letugbene	"	"
Aleibiri	"	"
Pakiama	"	"
Angalabiri	"	"
Tororuwa	"	"
Sagbama	"	"
Forcados	"	Ill defined bend
Burutu	"	"
Angiama	"	"
Bomadi	Bomadi creek	Outer bend
Ojobo	"	"
Agbidiama	R. Dodo	No bend
Egbamo Angalabiri	Egbemo creek	Outer bend
Ukubie	Apoi creek	"
Lobia	Lobia creek	"
Foropa	Foropa canal	"
Orugbu	Orugbu creek	"
Ugborodo-Escravos	R. Escravos	"

A critical look at the origin of these bends indicates that:

-they are not entirely due to a preferential natural tendency of rivers to flow towards low hydraulic grade otherwise the settlements could have been located at the inner bends where erosion is not effective.

-soil type is not mainly responsible for locating these settlements at the outer bends. For example at Ndoni the river bank is a highly erosive silty to medium sand while at Agbere a few kilometers downstream it consists of a highly resistant medium plasticity inorganic clay. But both settlements suffer severe bank erosion.

-although river discharge in the Niger Delta varies in magnitude depending upon location the degree of erosion in these places is not directly related to it. For example the discharge of River Niger around Ndoni in its upper regime is about 9200 m^3/sec, while near Isamkpou Fig.1, River Forcados discharges at about 1800 m^3/sec (NEDECO, 1961). Inspite of these differences in discharge the severity of bank erosion at these places is of equal magnitude.

-a careful study has shown that human activities are the main causes of degradation of bank slopes in the Niger Delta ecosystem. The activities as highlighted in Section (3.1) above are precursors to the development of river bends.

These facts can be illustrated using the following examples:

(a) Ndoni is situated at the outer left bend of River Niger on what is now regarded as its third location. Previously the town was sited upstream at two different places and at different periods on either side of the Niger. At each time in the past the town was threatened by erosion due to increased curvature of the shore and then was relocated. At its present location on the outer bend of the river Ndoni is again seriously threatened by bank erosion.

(b) Isamkpou town situated at the outer left bend of River Forcados was originally located at the opposite bank which was an outer bend but due to severe bank erosion the site was abandoned. Again, Isamkpou at its present location is severely threatened by bank erosion.

5 CONCLUSION

It is apparent that the main cause of bank erosion in the Niger Delta is human activities. Hence, the most probable solution to this problem lies on:

-provision of a means of slowing down the rate of flow or diversion of the flow away from an inhabited area. This may be achieved by constructing a groyne field.

-protection of bank soil against direct contact with fluvial waters by means of sheet piles, retaining walls or reinforced synthetic bags.

-provision of a proper drainage to channel overland flow into the body of the river.

-education of the local inhabitants on mode of operation along protected water front so as to restrain damage to the structure.

6 REFERENCES

Dautherty, R.L. and J.B. Franzini 1977. Fluid mechanics with engineering applications. 7th edition. International Student edition. Kosaido Printing Coy. Tokyo. Japan.

IFERT Technical Report 1989. Reports on the preliminary investigations of Niger Delta. The Niger Delta Master plan project. Port Harcourt. Nigeria.

NEDECO Report 1961. The waters of the Niger Delta. Report on an investigation. Netherlands engineering consultants. The Hague.

Netherlands. 2nd revised edition.

Shields, A. 1936. Anwendung der Ahuilichkeitsmechanik and Turbulenzforschung auf die Geschiebewegung Mitteil PVWES Berlin No. 26.

Wagner, A.A. 1957. The use of the unified soil classification system by the bureau of reclamation. Proc. 4th Inter. Conf. Soil Mech. Foundation Engrg (London) Vol.1. P. 125.

Webber, N.B. 1979. Fluid Mechanics for Civil Engineers. S.I. Edition Chapman and Hall London 340 p.

An assessment of coastal erosion in parts of Akwa-Ibom State of Nigeria

Une répartition de l'érosion littorale dans quelques parties de l'état d'Akwa-Ibom au Nigéria

S. Clifford Teme
IFERT, Rivers State University of Science & Technology, Port Harcourt, Nigeria

ABSTRACT: The entire coastlines of Akwa-Ibom State in the southeastern portion of Nigeria are exposed to the influences of ocean waves and storm surges from the Atlantic Ocean that cause extensive erosion and destruction to both utilities and properties in these areas. Results of field investigations to assess the extent of erosion along the entire coastline of Akwa-Ibom State indicate that wave erosion had caused severe scours at the towns of Oron and Jamestown despite remedial protective measures. Also, it was observed that severe shore erosion and flood inundation have threatened a Federal Fisheries Terminal and residential buildings at the village of Uta-Ewa while a Mobil Oil Company Terminal at the village of Mkpanek is only several metres away from the ocean as a result of severe wave erosion.

The soils that occur along the coasts of the state vary from clays of low-to-moderate plasticity (CL) to non-plastic silty-sands (SM) with corresponding frictional angles ($\emptyset$) ranging from 7° to 32° and cohesive strength (Cu) from 165 KPa to 0.$\emptyset$ KPa. Thus, the coastal materials of Akwa-Ibom State can be regarded as having high to very high erodibility potentials.

RESUMÉ: Les littorals entirès d'Akwa-Ibom dans le sud du Nigéria sont exposés a l'influence des ondes l'ocean et houles de tempête qui viennent d'ocean Atlantique, ceci cause une érosion grave et détruit les propriétés et les utilités de ces places. Les resultats de l'enquête pour evaluer la limite d'érosion au long du littoral d'état entier d'Akwa-Ibom indique que les ondes d'érosion causés par la dégradation sévère des villes d'Oron et Jamestown malgre les réparateur mésure de protection. Aussi, c'était observé que l'érosion rivage et une inondation dé-bordé ont menacés les postes pêcherie du Government Federal et les maisons des indigiens du village d'Uta-Ewa, tandisque la poste d'usine de l'huilede Mobil au village de Mkpanek est quelque metres de coin de l'ocean acause de l'onde d'érosion sévère.

Les sols qui forment vers la côte de l'état se différènt de glaises de bas a plasticité modéré (CL) a l'ensables non-plastique (SM) avec les angles frictionals correspondents ($\emptyset$) de 7° a 32° et une force cohésive (Cu) de 165 KPa a 0 KPa. Alors, les matérials côtier d'Etat d'Akwa-Ibom peut être éstimé d'avoir haut a plus haut potential d'érodibilité.

1 INTRODUCTION

Akwa-Ibom State of Nigeria is situated in the southeastern portion of the country between longitudes 7°30' and 8°25' East of the Greenwich Meridian and latitudes 4°30' and 5°00' North of the Equator (Figure 1).

The entire southern boundary of the state, which is bordered by the Atlantic ocean, is subjected to the influences of ocean waves and storm surges that often cause extensive erosion of land and destruction to both utilities and properties within these areas. In order to assess the

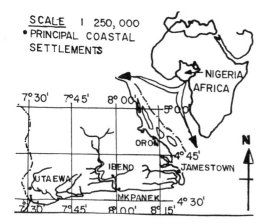

SCALE 1 250,000
• PRINCIPAL COASTAL SETTLEMENTS

NIGERIA
AFRICA

7°30' 7°45' 8°00'

ORON

IBENO   JAMESTOWN

UTAEWA

MKPANEK
4°30'

Figure 1
Location of study site

extent of coastal erosion along the
coastlines of Akwa-Ibom State, the
state government through a Task
Force on Landuse and Soils Surveys
commissioned a study for that purpose. This paper presents parts of
the findings of such a study.

2 METHOD OF STUDY

A field survey was undertaken to
assess the extent of coastal erosion
along the coastlines of the state.
During the field survey, field
measurements were made and soil
sampling carried out at specific
erosion sites to identify the soil
types and determine the engineering
properties of these coastal materials
with a view to evaluating their
potentials for erosion. Specifically
at each site visited, the following
parameters were obtained, observed,
measured or estimated as the case
may be:
   1. The shore length along settlements.
   2. Shore slope profiles and geometries.
   3. Estimate of extent of erosion
and amount of damage to houses,
utilities and other individual
properties.
   4. Estimate of erosion rate.
   5. Soil samples.
   6. Photographs of shore length,
shore slope, erosion damage and
other pertinent data necessary to
enhance both visual and quantitative

assessment of the problems at site.
   7. Presence or absence of erosion
protective measures and their performance levels.
   8. Modes of failures of erosion
protective structures, where such
failures have occurred.
   Item (8) above is the subject of
another paper (theme 6.3) in this
proceeding by this author.

3 ASSESSMENT OF COASTAL EROSION IN
   AKWA-IBOM STATE

An assessment of the erosion of any
given site must take into consideration the nature of the in-situ
materials and their geotechnical
characteristics in order to ascertain
their potentials for erodibility.
The causative factors responsible
for the erosion must also be identified to aid in the overall assessment of a particular location.
   The coastal regions of Akwa-Ibom
State are relatively low-lying with
occasional elevated areas which are
relics of cretaceous outcrops. These
elevated areas often serve as human
settlements along the coasts. The
vegetation along the coastline is
mainly mangrove forest with occasional rain-forest species occurring
slightly to the north. The major
rivers that drain the coastal
regions such as Qua Iboe and Imo
rivers are approximately perpendicular to the coastline and exhibit
a high degree of meandering before
finally draining into the Atlantic
Ocean.

3.1 Possible causative factors for
    the coastal erosion

Various possible factors and combination of factors are responsible
for the erosion along the Akwa-Ibom
State coastline. These include
hydraulics and hydrodynamics, local
geology, meteorology and hydrology
and, to an extent, human activities.

3.1.1 Hydraulics and hydrodynamics

Large quantities of fresh water from
the major rivers that seek entry into
the Atlantic Ocean usually come up
against tidal seawater surges upland-
bound. The resultant mixing of the
fresh and saline waters in these

coastal areas creates current patterns that help to scourge the erodible materials in these areas.

3.1.2 Local geology

The geology of the southern coastlines of Akwa-Ibom State comprises
Pleistocene-Oligocene coastal plains
sands; abandoned beach ridges;
relics of resistant cretaceous outcrops, and quaternary sediments.
These unconsolidated sediments have
low cohesion values and shear
strengths that make them very
susceptible to erosion by wave
activities.

3.1.3 Meteorology and hydrology

The annual mean rainfall in parts of
the coastal zones of Akwa-Ibom State
such as Ikot-Abasi, Eket and Jamestown reaches up to 4000mm with about
92% of the rainfall spread through
the nine months of March to
November. This rainfall pattern in
quantity and duration has been
observed to contribute immensely to
the flooding and erosion problems
in the coastal zones of southern
Nigeria (Fubara 1986).

3.1.4 Human activities

It has been observed during field
surveys that the practice of sand
mining by local residents is a
common phenomenon along parts of the
coastline. This activity has been
observed to contribute to the initiation of coastal erosion in certain
areas such as Ibeno along the estuary
of the Qua Iboe River

3.2 Physical characteristics of the coastlines at low and high tides

The shore profiles and geometries
that were observed to exist along
the coastlines are dependent on the
type of soils that occur in each
area. Generally, the shore profiles
along the sandy coasts have gentle
slopes as typified by the case at
Mkpanek. A schematic plan view of
the general conditions at Mkpanek
is shown in Figure 2.

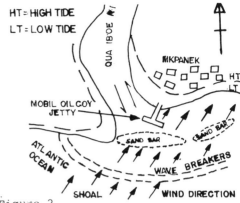

Figure 2
Plan view of general conditions at
Mkpanek

The inter-mixing of currents from
both the Qua Iboe River and the
ocean has created a complex hydrodynamic pattern that has resulted
in the creation of migratory shoals
and sand bars almost across the
mouth of the river. The general
shore slope geometry at Mkpanek is
shown in Figure 3.

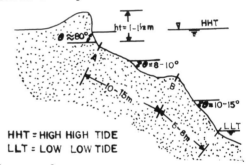

Figure 3
Shore slope geometry at Mkpanek

Generally, the uppermost portion of
the shore slope has a rather steep
($\theta \approx 80°$) slope face that is not more
than 1-1.50 metres high, followed
by a gentle slope with an angle of
about $\theta=8°-10°$ that stretches for a
distance of about 10 to 15 metres.
Usually but not always, a break-in-
slope (B in Figure 3) separates the
upper gentle slope from a lower one
that extends into the Atlantic Ocean
at low-low water levels.

The shore profiles along the
coast-lines with cohesive materials
are much steeper as shown by the
case at Uta-Ewa (Figure 4)

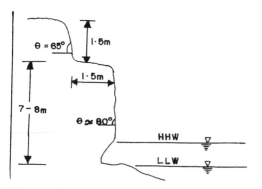

Figure 4
Shore profile at the Federal
Fisheries terminal, Uta-Ewa

The shore profile at Uta-Ewa comprises an upper portion that is
1.5Ø metres high with a slope angle
of about 65°. At the toe of this
section of the slope is a 1.50m
wide berm followed with a 7-8 metres
high slope with an average slope
angle of 80°. Evidence of severe
scouring occurs at the foot of the
slope between the low water and the
high water levels.

3.3 Erodibility potentials of coastal materials in Akwa-Ibom State.

Soil erodibility has been the
subject of numerous studies undertaken by various workers among whom
are Passwell (1973); Farmer (1973)
and Arulanadan et al (1975). In
general, an increase in the organic
content and clay size fraction of
a soil decreases erodibility. Also,
soil erodibility has been found
(Gray and Laiser 1982) to be low in
well-graded gravelly soils and high
in uniform silts and fine sands.
Erodibility also decreases with low
void ratios and high antecedent
moisture contents; increases with
an increase in sodium adsoption
ratios (SAR) and increasing ionic
strength of water. Field observations by the author have also shown
that coastal soil erodibility
increases with waves of high amplitudes and frequencies.
   A suggested hierarchy of erodibility of soils based on the
Unified Soil Classification System
(USCS) according to Gray and Leiser
(1982)is as follows:

Most erodible
ML>SM>SC>MH>OL>>CL>CH>GM>GP>GW
                              Least erodible
where the abbreviations, ML, SM etc
are soil types according to the
USCS nomenclature.

3.31 Some geotechnical properties of coastal materials in Akwa-Ibom State

Pertinent engineering properties of
coastal materials at localities
where erosion activities occur were
determined. Coastal material properties determined include soil consistency, particle size gradation
patterns, strength characteristics
and soil classification. The summary
of the geotechnical properties of
the coastal materials is presented
in Table 1. The coastal soil vary
from non-plastic, poorly-graded
silty sands and uniform fine sands,
to fairly plastic, gap-graded, clayey
sands, silty-sandy-clays and silty-
clays. The frictional angles ($\emptyset$)
range from 5°-16° for the silty-
clays to 25°-32° for the silty-sands
and uniform fine sands. The cohesive
strength (Cu) varies from 0.Ø KPa
for the silty-sands and uniform fine
sands to 45-195 KPa for the silty-
clays.

3.32 Erodibility risk zonation along Akwa-Ibom State coastline

Based on the geotechnical characterization of the coastal soils of
Akwa-Ibom State, in terms of classification types, consistency and soil
strength (Table 1), it is apparent
that most of these materials have
potentials for high to very high
erodibility. A case of severe coastal
erosion is shown in Figure 5.
An attempt has been made to zone the
state coastline into various degrees
of risk areas in terms of erodibility
potentials. Such a zonation is shown
in Figure 6.

4 CONCLUSIONS

The materials that constitute the
coastlines of Akwa-Ibom State in the
southeastern portion of Nigeria are
predominantly silty-sands, uniform
fine sands, silty-clays, clayey sands

Table 1
Summary of geotechnical properties of coastal materials along Akwa-Ibom State, Nigeria

Site	Soil type	Classi-fication (USCS)	Consistency W_L (%)	PI (%)	Strength Characteristics $\emptyset$ (degree)	Cu (KPa)	Remarks
Mkpanek	Silty sands, uniform sands	SM;SP	NP	NP	25-32	0.ØØ	Very highly erodible
Ibeno	Silty clays; clayey sands	CL;SC	14-18	8-12	5-14	75-195	Erodible
Uta-Ewa	Silty clays; silty sands	CL;SM	0-18	0-16	12-24	0-150	Erodible
Ikot-Abasi	Silty clays	CL	14-24	8-12	8-17	45-105	Erodible
Jamestown	Silty clays, clayey sands	CL;SC	18-24	14-19	5-8	85-145	Fairly erodible
Oron	Silty-sandy-clays;clayey-sands	SC; CL	9-13	4-9	5-16	45-180	Erodible

Figure 5
Severe coastal erosion at Mkpanek

and silty-sandy-clays. The frictional angles (Ø) of these materials range from 5°-16° for the silty-clays to 25°-32° for the silty-sands and uniform sands. However, the cohesive strength (Cu) varies from a low of 0.Ø KPa for the silty sands and uniform fine sands to a high of 45-195 KPa for the silty clays. The coastal materials thus have potentials for high to very high erodibility. An erodibility risk zonation along the state coastline seems to indicate that severe to very severe coastal erosion sites occur along the coast especially near to estuaries.

ACKNOWLEDGEMENT

The writer is grateful to the individuals from the various local government areas who helped in one way or the other during the field survey for this project. He is particularly grateful to Professor S.W. Petters, Chairman of the Task Force on Landuse and Soils Surveys and the Akwa-Ibom State Government for commissioning this study. His thanks also go to Mrs. F. Amadi for diligently typing of the manuscript.

REFERENCES

Arulanandan, K., 1975. Pore and eroding fluid influence on surface erosion. J. Geotech. Engr. Div.

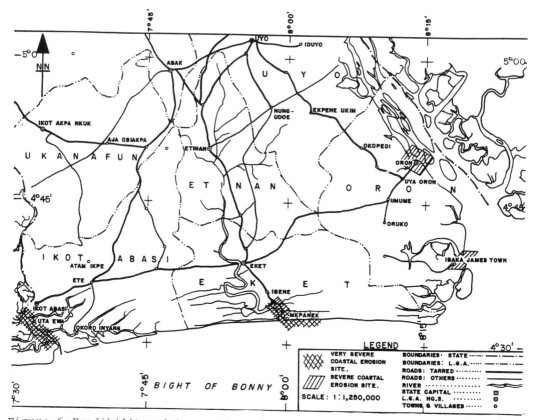

Figure 6 Erodibility risk zonation along Akwa-Ibom State coastline

ASCE, 101 (GT1), pp 51-61.

Farmer, E.E. 1973. Relative detachability of soil particles by simulated rainfall. Soil Science Soc. Amer. Proc. Vol. 37, No. 4, pp 629-633.

Fubara, D.M. 1986. Management and Conservation of Wetlands: The Niger Delta. Proc. National Workshop on Nigerian Wetlands. Port Harcourt, 27-29 August.

Gray, D.H. and A.T. Leiser, 1982. Biotechnical slope protection and erosion control. Van Nostrand Reinhold Company, New York, 271 pages.

Passwell,R.E. 1973. Causes and mechanisms of cohesive soil erosion; the State-of-the-art. HRB Special Report 135. Nas-NRC, Washington D.C. pp 52-76.

Engineering and geological aspects of bridge failures during the 1987/88 floods in Natal, South Africa

Les aspects de génie civil et aspects géologiques des défaillances des ponts pendant les crues de 1987/88 au Natal, Afrique du Sud

Marion A. Thomas & A. van Schalkwyk
Department of Geology, University of Pretoria, South Africa

ABSTRACT: Exceptionally heavy rains, resulting in severe flood damage in the coastal and inland regions of Natal, during September 1987 and February 1988 respectively, have severely damaged or destroyed a number of important road bridges.

In most cases the damage was the result of either foundation failure or because of a change in river course. Foundation failures took place mainly as a result of deep scour in alluvial deposits in which bridge piers or supporting piles have been located. Scour has also affected bridges founded on bedrock. Accumulation of trees and other debris on bridge piers and decks played a significant role in many of the failures.

Severe disruption of traffic and large repair costs as a result of bridge failures have highlighted the need for thorough investigations of geological and riverine environments before bridge construction.

RÉSUMÉ: Des pluies extrêmement fortes, ont causé des dégâts dans les régions côtières et intérieures du Natal, en septembre 1987 et en février 1988 respectivement, et ont endommagé sérieusement ou détruit un nombre important de ponts routiers.

Dans la plupart des cas, les dégâts ont été causés soit par la défaillance des fondations, soit par le changement du cours de la rivière. La défaillance des fondations s'est produite surtout à cause de d'érosion de dépôts d'alluvions où les piliers des ponts et les piles de support ont été installées. La force d'érosion a aussi affecté les ponts construits sur rocher. L'accumulation d'arbres et d'autres débris sur les piliers et le tablier des ponts a joué un rôle important dans un grand nombre de défaillances.

L'interruption de la circulation et les frais importants de réparation dûs à la défaillance des ponts ont mis en évidence le besoin d'études approfondies du milieu géologique et fluvial avant la construction d'un pont.

1 INTRODUCTION

Rainfall figures varying between 400 and more than 800 per cent of the mean, were recorded in Natal during the month of September 1987 (Van Heerden, 1988). During the four day period 26th till 29th September, falls of between 100 and 900 mm were recorded (Figure 1). The resulting floods were, in terms of loss of life and damage, probably the worst flood disaster ever experienced in South Africa. Loss of life amounted to 380 and damages were estimated at between R600 million* (Bell, 1987) and R1 000 million (Edwards, 1987). Most of the flooding took place in coastal regions where the highest rainfall was recorded. Above average rainfall, amounting to more than 300 mm during February 1988, resulted

* 1 SA Rand = 0,4 US$ (approx.)

in renewed flooding and damage to structures located farther away from the coast.

Widespread publicity was given to bridge failures which occurred during the floods and this led to claims that local technology was not up to standard. Problems at river crossings included damage to 28 bridges, 130 bridge approaches and 40 causeways. In many cases the damage to bridges was of a temporary nature, such as clogging with debris, damage to handrails or minor erosion. Structural failure to bridges amounted to 12, of which one was due to a deck floating off, 8 due to foundation failures and 3 due to a change in the course of the river.

This paper deals with bridges where geological factors contributed to the cause of damage or failure and describes the engineering solutions for the repair and prevention of future damage.

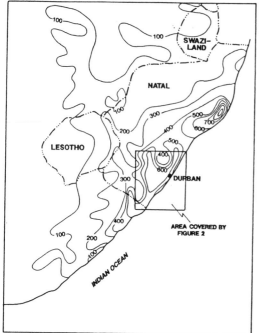

Figure 1. Map of Natal, showing total rainfall in mm for the period 26 to 29 September 1987. (After Van Heerden, 1988.)

2 WEATHER CONDITIONS AND FLOODING

Exceptionally high rainfall during September 1987 was caused by a sub-tropical cut-off upper air low pressure belt surrounding the southern hemisphere and an Atlantic high pressure cell ridging south-east of South Africa. In combination, these two weather systems led to more or less continuous heavy rains in Natal from the 26th till the 29th of September 1987 (Bell, 1987).

During most of February 1988, warm, sub-tropical air was in circulation over southern Africa. A cold front in the south on the 6th, followed by a high pressure system on the 7th, caused widespread heavy rains and local flooding in Natal. Similar conditions caused further heavy rains between the 23th and the 26th of February 1988 (Quinn, 1988).

Flood peaks at more than half of the river flow gauging stations, either destroyed the equipment, or exceeded the maximum calibration limits. Therefore, in many cases, indirect methods such as slope-area surveys had to be used to determine peak flows (Department of Water Affairs, 1989). Table 1 shows flood data for rivers at some of the damaged bridges.

Although some of the return periods are less than 50 years, it would seem, from the available records, that the 1987 flood peaks at the damaged bridges, were all higher than previously recorded events.

3 GEOLOGY AND PHYSIOGRAPHY

Basement rocks of Natal range in age from the c 3 300 Ma Archaean granite-greenstone rocks of the Kaapvaal Craton north of the Tugela River, through Proterozoic (c 1 000 Ma) granites and gneisses of the Natal Metamorphic Province, south of the Tugela River. Resting unconformably on the crystalline basement, are Ordovician-Silurian sandstone with sub-ordinate siltstone and shale of the Natal Group, which are unconformably overlain by thick deposits of Carboniferous-Jurassic diamictites, sandstones and mudrocks of the Karoo Sequence. Numerous sills and dykes of post-Karoo dolerite have intruded these rocks. Younger formations include Cretaceous calcareous sandstones and limestones and Quaternary calc-arenites, unconsolidated dune sands and dongha deposits.

Various cycles of uplift and erosion have resulted in an extremely rugged topography with deeply incised river valleys and steep river gradients between the high escarpment and the Indian Ocean. Major uplift towards

Table 1. Estimated peak discharge and return period for the September 1987 floods (Department of Water Affairs, 1989)

RIVER	BRIDGE (ROAD)	PEAK DISCHARGE (cub. m/sec.)	RETURN PERIOD (years)
Tugela	John Ross (N2)	9 400	35
Mpambanyoni	Mpambanyoni (N2)	1 200	25
Lovu	Illovo (N2)	1 800	35
Mdloti	Mdloti (N2)	1 960	90
Mvoti	Mvoti (N2)	5 400	180

the end of the Pliocene, followed by mono-clinal warping and seaward tilting of the coastal regions was responsible for thick accumulations of sediments in deeply incised river valleys along the east coast. Following eustatic lowering of the sea level during the late Pleistocene, a rolling landscape with rivers flowing on thick alluvial deposits developed along the east coast.

Economic activities in Natal are mainly agriculture, forestry, mining and tourism. Most of the people are concentrated in cities and rural settlements along the coastal areas. The road network consists of national roads from the Transvaal to Durban (N3) and along the coast (N2), and provincial and district roads connecting the various inland towns and settlements.

Because of the rugged topography, water absorbtion during heavy rains is limited, and floodwaters rush to the coastal areas where massive damage is done to the densely populated and most developed regions (Meineke, 1960).

4 BRIDGE FAILURES

Of the bridges which were damaged, the causes of damage can be classified into geological and non-geological factors (Thomas and Van Schalkwyk, 1988).

Non-geological causes include erosion of approach fills, clogging of the roadway by debris and the removal of a bridge deck by floating. These problems were mostly of a temporary nature and will not be dealt with in this paper.

The most seriously damaged bridges involved some geological problem, either failure of the foundation or failure caused by a change in the course of the river. A list of bridges which failed, and the main

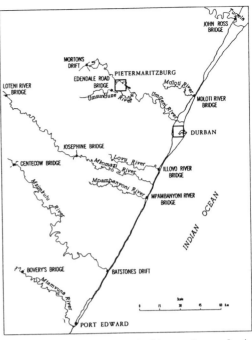

Figure 2. Location of bridges damaged in the 1987/88 Natal floods.

mechanisms of failure is presented in Table 2. Bridge locations are shown on Figure 2.

In order to illustrate the different mechanisms of failure, examples of each will be discussed.

4.1 Scour of rock foundation: John Ross bridge, Tugela River

The John Ross bridge, where the N2 road

Table 2. List of bridges and causes of failure.

RIVER	BRIDGE	MECHANISM OF FAILURE
Tugela	John Ross (N2)	Scour: rock foundation failure
Mtamvuna	Boverey's bridge (P59)	Scour: rock foundation failure
Mpambanyoni	Mpambanyoni (N2)	Scour: alluvium foundation failure
Lovu	Illovo (N2)	Scour: alluvium foundation failure
Mkomazi	Josephine (P5/2)	Scour: alluvium foundation failure
Loteni	Loteni (P27)	Debris: rock foundation failure
Mzimkulu	Batstone's Drift (P198)	Debris: alluvium foundation failure
Mzimkulu	Centecow (D65)	Debris: alluvium foundation failure
Mdloti	Mdloti (N2)	Change in river course
Mgeni	Morton's Drift (MR9)	Change in river course
Mzimduzi	Edendale Road	Change in river course

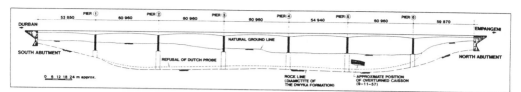

Figure 3. General drawing of the John Ross bridge. (Based on plans by Scott & De Waal Inc.)

crosses the Tugela River, was completed in 1959. The 412,5m long bridge deck was supported by six piers, with caissons founded on diamictite of the Dwyka Formation (Figure 3).

During construction, in November 1957, flooding caused one of the caissons to topple and it was buried under alluvium. In order to avoid the abandoned caisson, all the piers had to be re-sited and the span distances vary as shown on Figure 3. The design was also altered from a continuous deck to two decks with a joint over pier 5.

After three days of continuous rain over a large part of the Tugela basin, the John Ross bridge collapsed at 17:30 on the 28th September 1987 (Figure 4). Failure took place before the peak of the flood.

It was reported that the collapse started at pier 5. The bridge vibrated prior to collapse, which could have been caused by the turbulence and debris around piers and caissons exposed by scour. In a report by Knight and Dorren (1988), the following comments are made:

(1) Post-failure drilling has shown the diamictite at the sites of piers 5, 6 and possibly also 4, to be highly jointed at the founding levels, and the possibility of a fault near pier 5 was postulated.

(2) Pier 5 is at the centre of the permanent river channel and the estimated scour depth of 21,6m is below founding level of the pier. Undermining of the caisson was therefore possible.

(3) The caisson lost and left within the alluvium in 1957, probably added to the turbulence around pier 5.

(4) A rock berm between piers 1 and 4 and sugar cane planted out as far as pier 2, caused damming along the north bank and channelization at pier 5.

(5) A stormwater tributary joins the river between piers 5 and 6, adding to scour intensity there.

It was concluded that the load of water and debris on pier 5, which at that stage, was exposed to scouring over its full depth, led to failure within the jointed foundation rock. When pier 5 failed, the entire bridge became unstable and collapsed.

Figure 4. Photograph showing the failed John Ross bridge. (Photo: NPA)

It is interesting to note that, while the 28 year old John Ross bridge collapsed, the 60 year old steel road and railway bridges, located a few kilometres upstream, suffered no damage. The reason for this is that the sites for the old bridges were selected where the alluvium is much thinner and foundation rock is very good quality Natal Group sandstone. In both cases, however, marked deviations in the road and rail alignments were necessary.

The new John Ross bridge is located at its former position, but the new piers are

sited between the old caissons. The new
piers are made up of two circular columns,
connected on top by a capbeam and ending in
a rigid pile cap, founded on four or six
1350mm diameter permanently cased bored
piles of cast in situ reinforced concrete
(Anon, 1989). This type of pier configuration is well suited to skew river flow
and is hydraulically very efficient as the
circular piers do not collect large amounts
of debris.

The piles are up to 20m long, founded on
massive diamictite with average uniaxial
compressive strength of 150MPa and secured
by means of six or seven 40mm diameter
high-tensile steel dowels grouted 3m deep
into the rock. The alignment and rake for
the piles has been chosen such that they
will remain in compression under all load
conditions (Anon, 1989).

The rock berm along the north bank was
broken up to allow a more even flow of
water.

4.2 Scour in alluvium: Mpambanyoni River bridge

The Mpambanyoni River bridge carries the N2
road between Durban and Port Shepstone
across the Mpambanyoni River at Scottburgh.
The original bridge was completed in 1945
and consisted of seven spans, each 18,3m
long, supported on eight piers with friction piles driven 9m deep into alluvium
(Figure 5).

When the bridge was designed, the scour
depth was estimated to be between 1,0 and
1,5m. In 1985, one pier was washed out and
two deck spans failed. A new pier was
founded on 15m long friction piles.

During September 1987, a total of 437,9mm
of rain fell at Scottburgh. Of this total,
123,5mm was recorded on the 27th, and 168mm
on the 28th September (Weather Bureau,
1987). Two piers were washed away from the
southern side and three deck slabs collaps-

Figure 6. Photograph showing the damaged
Mpambanyoni bridge. (Photo: NPA)

ed (Figure 6). A third pier had to be demolished.

It was estimated that scour depth during
the floods was more than 6,4m which means
that the 9m long piles were left largely
unsupported (W. Martin, personal communication). The river water was seen to be
flowing at an angle of 37 degrees to the
piers and this would be a factor in increasing scour depth and water load on the
piers. It should be noted that the river
course had changed since the 1940's when
the bridge was originally sited.

The railway bridge, a short distance
downstream of the road, suffered no damage.
The reasons for this are that the southern
abutment of the railway bridge is founded
on good quality Natal Group sandstone, the
bridge deck is higher and to some degree,
the road bridge has aligned the flow of the
river to be parallel with the railway
bridge piers.

Repairs involved the construction of two
new piers founded on 24m long friction
piles. The new piers are aligned at 16,5

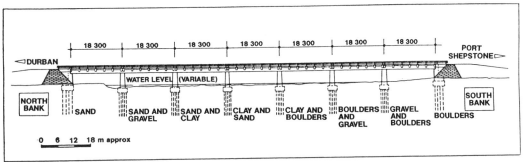

Figure 5. General drawing of the Mpambanyoni River bridge. (Based on plans provided by
NPA Bridge Office.)

degrees to the old ones in order to allow for changes in the river course.

4.3 Debris build-up causing failure: Loteni River bridge

The Loteni River bridge was located where the P27 road between Himeville and Nottingham Road crosses the Loteni River. It was constructed in 1925 and consisted of three, 10m long, deck slabs supported by two piers, spaced evenly between the abutments. The structure was founded on feldspatic sandstone with intercalations of closely laminated shale of the Tarkastad Sub-group, Beaufort Group, Karoo Sequence.

The bridge survived very high flood levels shortly after completion in 1925, and also stood up against the floods of September 1987 when the water level was just below the deck. As part of a bridge inspection programme on the 9th Febraury 1988, it was noted that bank protection works were in poor condition but otherwise the bridge was in fair to good condition.

A total of 409mm rain, which is 207 per cent of the mean for the month, fell at Himeville during February 1988 (Quinn, 1988). On the 24th February, the water level at the bridge rose to the top of the handrails and the bridge failed. Build-up of debris on the structure caused both piers and the southern abutment to be ripped from their foundations. The bridge was carried 200m downstream and was left lying in an inverted position on the river bed. Some of the founding rock was pulled out with the piers and could be seen stuck to

the bottoms of the overturned piers (Figure 7). The presence of weak mudstone interbeds in the foundation rock probably led to sliding or tensile failure within the foundation.

The piers for the new structure will have 4m long dowels penetrating two or three layers of the succession, and these will be bolted in place. In addition, the new bridge will be 20m longer and 1m higher, thereby doubling the area of waterway.

4.4 Change in the river course: Mdloti River bridge

The Mdloti River bridge crosses the river of that name on the N2 road, approximately 25km north of Durban. It was a 12-span bridge, completed in 1960. During the floods of September 1987 the bridge suffered damage to the north embankment, loss of the northernmost pier and the two adjacent deck spans. The rest of the bridge was only slightly damaged.

Before the N2 road was constructed, the Mdloti Lagoon had two main channels (Figure 8). Thereafter flow was restricted to the southern channel and the northern channel was silted up and overgrown by grass and reeds. When the river carried increased volumes of water, it tended to widen and straighten its course. With a subsequent increase in velocity it would then flow into the northern channel.

During the floods of 1987, water was diverted northwards where it was dammed against the approach fill and then diverted southwards to the nearest outlet which was the bridge. Increased turbulence around the

Figure 7. Photograph of the overturned Loteni River bridge with remnants of bedrock still attached to the piers. (Photo: NPA)

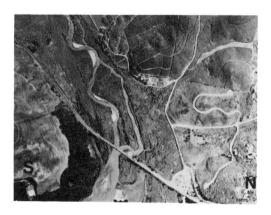

Figure 8. Air photo of the Mdloti River bridge showing the two river channels. (Photo: Department of Transport)

northernmost pier caused scour to an estimated depth of 10m and undermined the piles
driven to a depth of 9,10m. When the piles
moved, the pier toppled and the deck spans
collapsed. Rapid erosion of the sandy approach fill and northern bank of the river
provided a new course for the river and
scour at the reamining bridge piers was
reduced so that the bridge suffered no
further damage (Honey, personal communication, 1988).

To reinstate the bridge, a single span
replaced the lost spans and a new pier was
not needed. Reno mattresses were placed to
protect the approach fill and a rock berm
was constructed from the north bank to
divert water back into the main stream.
Although the waterway is slightly larger
than before, flood waters in the northern
channel will still be dammed by the rock
berm and the approach fill, and future damage to the north abutment is almost certain to occur.

5 CONCLUSIONS

The single most important geological factor
responsible for damage to most of the
bridges was the presence of thick alluvial
deposits subject to deep scour during times
of high flood.

In most instances scour depth was underestimated, and in the cases of older
bridges, pile driving technology at the
time of construction did not allow sufficient penetration or socketing into bedrock.

At times of high floods, rivers tend to
straighten their courses and revert to
previous channels which do not necessarily
correspond with the waterways provided by
the bridges.

Engineering factors responsible for failures include
(1) poor geometric layout, for example
siting of bridges on bends, asymmetric
alignment and skewness of piers;
(2) insufficient waterway due to inadequate hydrological design;
(3) poor foundation design due to inadequate site investigation and underestimat-
ion of scour depth and
(4) lack of piling technology.

Build-up of debris, especially when
bridge decks are overtopped, results in
excessive horizontal loading and was responsible for several foundation failures
and the removal of at least one bridge deck
from its piers.

6 RECOMMENDATIONS

For future bridges, it is recommended that
careful attention be given to geological
and environmental conditions in order to
design the structures in harmony with nature. Thorough engineering geological site
investigations for all river crossings are
considered essential.

Special attention must be given to foundation conditions and, where possible, all
bridges on deep alluvium should be founded
on large-diameter piles, socketed in, and
bolted to good quality bedrock.

In situations where the foundations cannot reach bedrock, piles should be founded
as deeply as possible and scour reduced by
correct design of piers and by increasing
the size of the waterway.

Overtopping of bridges should be avoided
as far as possible by adopting a design
flood with return period of at least 1:100
years. Consideration should be given to a
design capable of withstanding overtopping
during extreme floods.

ACKNOWLEDGEMENTS

The authors wish to thank the Geological
Survey of South Africa for the funding of a
research project on the engineering geological aspects of structures damaged during
the 1987/88 floods in South Africa, and the
Natal Provincial Administration and their
various consultants for information and
permission to present this paper.

REFERENCES

Anon 1989. John Ross Bridge over the Tugela
   River. A remarkable engineering achieve-
   ment. The Civil Engineer in South Africa,
   31(7):205-209.
Bell, D.A. 1987. Report to the insurance
   industry in respect of the Natal floods,
   September 1987. Unpublished report of the
   Commercial Union Assurance Co. of South
   Africa Ltd, Pietermaritzburg, 27 pp.
Department of Water Affairs 1989. Report by
   the Committee for the Study of Flood
   Damage in Natal in September 1987. Confi-
   dential report by the Department of Water
   Affairs, Pretoria, 72 pp.
Edwards, M. 1987. The Natal floods of September 1987, News Letter No. 462, Weather
   Bureau, Department of Environment Af-
   fairs, September 1987, 3 pp.
Knight, K. & D.I. Dorren 1988. Report on
   the failure of the John Ross Bridge over
   the Tugela River. Unpublished report to
   the Natal Provincial Administration, 11
   pp.

Meineke, E.N. 1960. Floods in the south-eastern coastal area, May 1959. Die Siviele Ingenieur in Suid-Afrika, September 1960, 191-199.

Quinn, S.J. 1988. The weather of February 1988. News Letter No. 467, Weather Bureau, Department of Environment Affairs, p. 20-25.

Thomas, Marion A. & A. van Schalkwyk 1988. Engineering geological aspects of bridge failures during the Natal Floods. Proc. Conf. on Floods in Perspective, SAICE/ CSIR, October 1988, Pretoria, 4.6.1-4.6.16.

Van Heerden, J. 1988. A season of floods over South Africa, July 1987 to June 1988. Proc. Conf. on Floods in Perspective, SAICE/CSIR, October 1988, Pretoria, 1.1.1-1.1.12.

Weather Bureau 1987. Daily rainfall from tape on 88-06-13. Weather Bureau, Department of Environment Affairs, MMF/596. 12.54.29.

Case studies of flooding and erosion phenomena in the Niger Delta and the effectiveness of their control measures

Etudes du phénomène d'inondation et d'érosion et l'effet de leurs mesures de contrôle dans le delta du Niger

P.O. Youdeowei
IFERT, Rivers State University of Science and Technology, Port Harcourt, Nigeria

ABSTRACT: The Niger Delta area of Nigeria with its characteristically low-lying terrain, high rainfall regime, and its overburden pedogenic nature of sandy/silty clay is seasonally and perennially subjected to flooding and erosion. During these episodes, hundreds of inhabitants are made homeless, and infrastructural and agricultural growth is greatly hampered. Over the years, attempts have been made to tackle these ecological hazards by various established bodies. The methods employed by these organisations have always been expensive and only recently have cheaper alternatives been considered, due to economic constraints. This paper examines the erosion and flooding situations in some communities noted as high risk areas and assesses the effectiveness of their present control measures (where there is). Recommendations for further efficacious, cheaper and more durable control methods are given for consideration.

RÉSUMÉ: La région du Delta de Nigeria avec sa terre caracteristicment peu élevé, avec sa pluie annuelle haute, est profondement subjecté a l'inundation et erosion. Pendant ces episodes d'inundation et erosion, les habitants restent sans maisons, l'infrastructure agricole gené. Pour plusieurs annees on a tanté de resoudre cet hasard ecologique. Les methodes employé par les organizations chargé de resouche ce problem ont etaint, couteux; mais récemment a cause de la crise economique ils ont commencé de mettre en consideration autres solutions methodique. Cet article examine les probléms d'inundation et erosion permi les communautés mentionné. Il fait aussi l'évaluation de l'efficacité des method du controle employait actuellement. L'article propose plusieurs recommandations.

1 INTRODUCTION

The Niger delta which occupies an area of over 36,000km^2, is situated in the coastal sedimentary basin area of southern Nigeria. Its formation has been attributed to the structural movement of the earth's crust and the physico-geographical processes of erosion and sedimentation leading to the establishment of an extensive sedimentary flat.

This plain of flats is traversed by several rivers and creeks of which there are two principal rivers, namely: Nun River and Forcados River systems. Through the action of these rivers and creeks, coupled with the physiographic nature of the region, the phenomena of flooding and erosion is well established as a periodic and perennial environmental hazard.

Consequent upon these flooding and erosion activities within the Niger delta, over 500,000ha of land which otherwise could have been used for agriculture and human settlement are rendered useless. In effect, no significant development in terms of infrastructure, agriculture and industrialization can be executed in the Niger delta, unless the problems of annual floods and erosion are controlled.

Some localities associated with acute problems of flooding and erosion within the Niger delta and regarded as high risk areas are

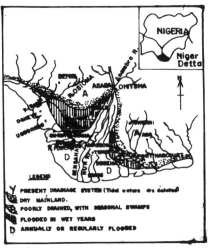

FIG·1; PROVISIONAL DRAINAGE ZONES OF THE
NIGER DELTA ⸱⸝(AFTER NEDECO, 1966)

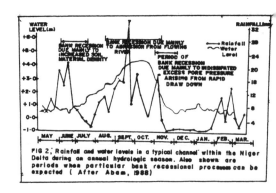

FIG 2, Rainfall and water levels in a typical channel within the Niger
Delta during an annual hydrologic season. Also shown are
periods when particular bank recessional processes can be
expected (After Abam, 1988)

highlighted in this paper with an
assessment of the efficacy of the
control methods presently adopted by
the inhabitants. Suggestions are made
whereby cheaper and locally available
materials may be utilised in some of
the localities as more durable control measures.

2 HYDRO-METEOROLOGICAL AND SOIL
  CHARACTERISTICS OF THE NIGER DELTA

The Niger delta is characterised by
two distinctive seasons, typical of
the tropical region: the dry season
(November-March) and wet season
(April-October). The dry season is
generally characterised by fairly
high daily temperatures and a monthly
rainfall usually less than 200mm.
The relative humidity is comparatively high (about 85%) even during
the dry season, and evaporation is
also high, reaching a monthly average
of about 300mm.

Since river discharge is derived
from precipitation, its development
is usually related to the rainfall
pattern (see Figure 1). Rainfall in
the Niger delta increases gradually
from march and reaches a peak in
July. In August, a slight fall is
experienced and a maximum peak is
attained in October. There is a
sharp fall in November as the dry
season is approached. Riverbank
failures frequently occur during this
period due to the rapid reduction in
the water level. A study of the

distribution of bank failures shows
that bank failure episodes are concentrated during the period of flood
recession (Figure 2)

The soil deposits of the Niger
delta comprises loose sands, silts,
fine clayey sands, clays, muds and
mangrove swamps. The soil profile
is mostly sandy clay underlain by
sands. This is irregular in some
cases due to the rivers changing
their courses.

The ages of these deposits range
from the Eocene to the Pliocene.
The older sedimentary series have
survived as outcrops on which the
human settlements have been built.

In the study of the geotechnical
characteristics of the Niger delta,
three main hydrometeorologic zones
have been identified, based on
hydrologic, vegetative and meteoro-
logic factors. These are:
 1. the freshwater zone
 2. the mangrove zone and
 3. the coastal zone as delineated
in Figure 3.

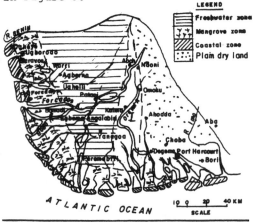

FIG:3, The major Hydrometeorologic zones of the
Niger Delta (modified after Teme 1988)

The freshwater zone is of a closer proximity to the hinterland. It is a triangular area that embraces parts of the River Niger and its major tributaries; the Rivers Forcados and Nun leading up to the boundary with the mangrove zone, southward. The low-lying areas within this zone experience yearly flooding, at which periods the two major rivers (Nun and Forcados) overflow their banks.

The soils of the freshwater zone are the most variable and have been found to consist of high plastic clays that usually occur in the backswamps linking river channels, to cohesive silty and clayey soils that are partly permeable.

The mangrove zone is identified as being adjacent and parallel to the coastal zone and occurs between this zone and the freshwater zone. This region is between 20-40km in width and is noted for its very poor drainage caused by its swampy nature and the high network of river channels and water courses that commonly traverse the terrain.

The soil deposits that make up the mangrove zone are mainly clays, shales and clayey shales (locally known as chicoco mud), silty clays and more commonly peat materials.

The coastal zone is the outermost region comprising beaches, estuaries, spits and bars linking the ocean with the Niger delta. This zone, like the mangrove zone is criss-crossed by multiple creeks and channels.

The soils of the coastal zone vary from mostly unconsolidated, non-plastic sands, gravels, silty-sands to occasional clayey sands and mud flats.

3 CASE HISTORY REPORTS ON EROSION AND FLOODING SITUATIONS

Flooding and erosion constitute the number one ecological problem in the country. While gully and rill erosion are common within the upland areas of the Niger delta (that is the freshwater zone), sheet erosion is prevalent both in the upland areas and within the coastal zones of the Niger delta. Bank erosion tend to be concentrated within the freshwater zones. In the coastal zone, the sea on its part scours, floods and breaks on the beaches causing severe loss of land.

Many low-lying areas in the freshwater zone and most places within the mangrove region are subjected to seasonal and perennial flooding, while the coastal or tidal areas is usually flooded daily.

In terms of agriculture, crop cultivation is greatly hampered due to the water-logged state of the soils in the inundated areas. Also the deposition of fine, silt materials at the shores or river banks of settlements creates a swampy environment which impedes the waterways and makes navigation and mobility difficult. As a result, most perennially flooded areas, especially in the coastal zone are sparsely populated.

Flooding in the Niger delta lasts for about 5 months (May-November), and as the flood recedes, soil erosion and river bank collapse occurs.

Three cases of areas afflicted with severe conditions of erosion and flooding, that are regarded as high risk situations are highlighted for consideration. It should be noted that there are at least twenty other serious cases like these within the Niger delta area.

Those communities in the Niger delta that are subjected to intense and perennial occurrences of flooding and erosion are more commonly located within the coastal and mangrove zones.

3.1 Ogheye-Ugogoegin communities

Ogheye **and** Ugogoegin are two proximate settlements on the western bank of the Benin river, about 5km to its firth into the Atlantic ocean.

These communities which are situated in the coastal zone exist on a low-lying, flat terrain. Their shore line is under tidal influence and hence is flooded daily. The landmass of the communities which varies from 40-50m from the shoreline to the backswamp is totally submerged at high tide.

The entire shore-length of the two communities is about 6km. the geometry of their shoreline is curvilinear to fairly straight, depending on the intensity of erosion, which . is more pronounced seaward as a result of the direct impact of tidal waves on the shoreline.

FIG:4; A typical deltaic community depicting a flooded environment.

The soil type in the area varies from mostly loose, fine sands at the riverbanks to silty sands and mud flats at the backswamp. There is a creek beyond the backswamp (Ogheye creek(, which makes the two communities exist as narrow, longitudinal islands. This further elucidates the dearth of land in the area.

3.1.1 Existing flood and/or erosion problems

Erosion is the key environmental problem in these settlements. This is in form of riverbank erosion aided by coastal erosion. Transversal flow of the river currents scours the banks, rendering them unstable. At ebb flow, the disaggregated sandy soils are swept off into the ocean. This is made considerably easy as a result of the low-lying nature of the settlements and the predominant sandy soil type. Coastal erosion is a consequence of the direct impact of tidal currents impinging upon the shoreline.

The rate of erosion at Ugogoegin is about 12m per year: it is reported that about 60m of land has been lost in 5 years. The situation is worse at Ogheye because of its nearness to the sea where about 2km of land has given way to water in 18 years, giving an average of over 200m of landloss in only one year!

Another matter of concern is the offshore oil exploration and exploitation activities by some oil companies close to these settlements. These activities though significant to the country's economic position, leads to minor subsidence of the nearby landmasses. This factor has

to be taken into consideration in the design of erosion protection measures for this locality.

3.2 Egbema Angalabiri community

This is a community situated in the mangrove zone of the niger Delta. The shoreline of this community is submerged under water in most places and extends for about 1km. The Dodo river which flows through the Egbema Angalabiri canal and is about 60m in width (with an average velocity of 0.8m/sec), together with water from the backswamp, totally inundates the settlement.

3.2.1 Existing flood and/or erosion problems

There is no evidence or record of land recession or erosion in this locality. The environmental problem is only that of severe flooding, which is at its peak in September when the rainfall is highest. There is virtually no dry land at any time of the year, especially since the land is of a flat topography, with no drainage facility.

This perennial water-logged condition is further heightened by the blockage of the canal with materials (mud) that were originally derived from the river bed during dredging activities.

3.3 Ugborodo community

Ugborodo is one of the few villages that make up Escravos in the coastal region of Bendel State. This settlement is located at the Escravos estuary into the Atlantic Ocean, and is adjacent to a GOCON (Gulf Oil Company, nigeria Limited) oil terminal.

The community presently exists as an island: being totally surrounded by water, with the Arunton creek flowing along the wing of the landmass close to the Gulf terminal and the other wing facing the Atlantic ocean. The shoreline, which is semicircular and curvilinear in shape has an overall length of approximately 1km. In all places, it slopes evenly (5-10°) towards the ambient waters, indicating a slightly higher hinterland. The soil type is mostly fine sands, which contributes to the high degree of erodibility of the area.

3.3.1 Existing flood and/or erosion problems

There is severe erosion due mostly to the tractive wave action on the shores. This is complimented by the construction of nearby canals such as the Arunton creek, which was the backswamp of the settlement but has now paced way for more water incursion into the community. Due to the extreme effect of erosion in this locality, Ugborodo was once declared a disaster area by the state government.

In very recent times, about 2km of what was once habitable land has been taken over by water. This leaves the community with less than 150m width of land, which is still being actively eroded away. At flow tide, the waves which rise as high as 2m, inundates the shore and as it recedes, carries away the incoherent sandy materials, thence further depleting the area of useful land.

There is also the problem of gradual land subsidence attributed to the nearby directional oil drilling activities by GOCON.

4 EXISTING METHODS OF FLOOD AND EROSION CONTROL AND THEIR EFFECTIVENESS

In the recent past, control methods of shore protection against erosion and flooding in the Niger delta included the use of sheet piling, retaining concrete wall and the use of synthetic mattresses. All of these methods though effective, were highly expensive in their implementation. Consequently, design proposals currently evolved for the control of these environmental hazards are now taking into consideration the use of less expensive and locally available materials.

Some of these methods already in use as pilot schemes are timber earth retaining structures (at Minima), timber groynes with cane-nets (Kaiama) and the used rubber tyre network system (Elem Bekinkiri). Even so, it is certain that the task of protecting all the areas affected by flood and erosion in the Niger

delta is extremely difficult in terms of time and money, so that preference should be given to high risk areas where human settlements occur, with places of oil terminal installations as other priority areas.

The three case studies cited in this paper are herein considered for their existing local control methods and the efficacy against flooding and erosion.

4.1 Ogheye-Ugogoegin communities

The inhabitants of these adjacent communities, being seriously threatened by the menace of erosion, build thatched houses on wooden piles as temporary abode. They keep shifting base inland as the land is further eroded away. It is obvious that this cannot continue for long as they will soon run out of land space to occupy.

There is no attempt by these communities to curb the problem of erosion apart from their intermittent retreat further inland.

4.2 Egbema Angalabiri community

As a result of the incessant water-logged situation in this locality, the community has built mounds of clayey shales (locally referred to as chicoco mud) around their houses to prevent water from flowing into their homes. They have also mounted timber logs along most of the shoreline, and Westminster Dredging Company during the canalization exercise, helped place heaps of mud along the entire shoreline. These measures have not proved effective against flooding and have only contributed to the swampy nature of the environment.

4.3 Ugborodo community

There is no flood or erosion control measure embarked upon by the inhabitants of this community, and even if there is, it is doubtful if it would be of any significance, since the situation is beyond the application of local conventional means of control. The inhabitants only move from one place to another in their bid to avoid the incessant invading waters.

5 DISCUSSIONS AND RECOMMENDATIONS

5.1 Ugborodo community

Presently, the rate of erosion is very high and alarming. From the havoc only recently done, it is estimated that if nothing is done to arrest the situation, in 5-10 years the present Ugborodo community could be completely effaced, due to the effects of both gradual land subsidence and the eroding sea waters.

Considering the gravity of the problem, it would be easier to evacuate the inhabitants from this locality, but experience has shown that to be a very difficult task as the inhabitants strongly object to it. They prefer to remain in the land of their fore-fathers even with the imminent threat of obliteration. Therefore, what may be recommended for now is the mitigation of the effect of scour and abrasion against the shore and the control of wave action. This may be done by the construction of revetments on the shore face and timber groynes along the bank traversed by the Arunton creek. The latter structure will reduce the current velocity by restraining and controlling the river flow. The types of revetments may include; the standard revetment with mattress, rip rap, gabions, wooden woven mattress, sack revetment, soil cement revetment, or the used tyre network.

Details of construction of these engineering structures will be deliberated upon further examination of the problem site.

5.2 Egbema Angalabiri community

Since the environmental problem in this community is that of severe flooding from the river Dodo, it is imperative that a proper river improvement work be conducted. This should involve the dredging of the river bed and its re-adjustment for free-flow.

The shores should be sandfilled above the flood level to a stable bank and a durable embankment constructed to prevent further flooding. The backswamp may be reclaimed for further extension of the landmass and there should also

be adequate drainage facilities for proper run-off of overland water.

5.3 Ogheye-Ugogoegin communities

These communities are subjected to similar dynamics of erosion as the Ugborodo community and as such the same protective measures as recommended for the latter community may be applied here. This should be followed by a well co-ordinated post-construction monitoring scheme to study the efficacy and durability of these semi-permanent engineering structures for future improvements.

6 CONCLUSION

It is realised that present control measures against flooding and erosion as embarked upon by most communities in the Niger delta is grossly inappropriate and in some cases futile. It is therefore hoped that further innovative and cost-effective means (with locally available materials) will be formulated and designed for the purpose of flood and erosion control in the Niger delta.

REFERENCES

Abam, T.K.S. (1988). Interim Report on Fieldwork to Zone 2 of Masterplan Project. IFERT Internal Report, RSUST, Port Harcourt, Nigeria.
ASCE Task Committee on channel stabilization works (1965). Channel stabilization of alluvial rivers; ASCE Journal of the Waterways and Harbours Division.
Carlson, E.J. and R.A. Dodge, (1962). Control of alluvial rivers by steel jetties; ASCE Journal of the Waterways and Harbours Division.
Chowdhury, R.N. (1979). Slope Analysis; Elservier, New York.
Fubara, D.M.J. (1983). Physical Conquest and Development of the Niger Delta, RSUST, Port Harcourt, Nigeria.
Gray, D.H. and A.T. Leiser (1982). Biotechnical slope protection and erosion control; Van Nortwand Reinhold Company, 271 pages.
Nedeco (1960). The waters of the Niger Delta; the Hague,

Netherlands.

Peck, R. (1967). Stability of Natural
   Slopes; J. Geotech. Eng. Div. ASCE,
   93(4): 437-451.

Petersen, M.S. (1986). River Engineering; Prentice-Hall, Eaglewood
   Cliffs, N.J.

Teme, S.C. (1988). Geotechnical
   Characteristics of the Niger
   Delta; Masterplan and Design of
   Flood and Erosion Control
   Measures in the Niger Delta.
   Progress Report No. 1, IFERT
   Technical Report, RSUST,
   Port Harcourt, Nigeria.

6.2 Offshore structures and seabed stability
Structures offshore et stabilité du fond de la mer

Characteristics of engineering geology in the Pearl River Mouth Basin
Caractère de la géologie constructive aux bouches du bassin de la Rivière des Perles

Chen Junren & Li Tinghuan
Guangzhou Marine Geological Survey, Guangzhou, People's Republic of China

ABSTRACT: The continental shelf in the Pearl River Mouth Basin is generally of good engineering conditions. However, the slope is a relatively unstable area caused by frequent earthquakes, intensive fractures, slides, turbidity currents, etc. Six strata with different geotechnical properties are divided within 120 meters below the seafloor in the basin. Two of them of low shear strength, strata 1 and 3, are unaccepted as bearing layers. Others of high bearing capacity are considered as good foundation for engineering facilities.

RESUME: Les conditions de la géologie constructive du plateau aux bouches bassin de la Rivière des Perles sont assez bien. La pente y est variable, souvent il y a des tremblements de terre, la rupture se développe intensivement avec des éboulements et des courants turbides. La zone à 120 mètres sous le fond de la mer pent être divisée en 6 strates dont la nature géotechnique est différente. Aux premier et au troisième strate, les forces contre cisaillement sont faibles, il n'est pas favorable à y utiliser comme strate de charge. Mais les autres strates dont la capacité de charge est relativement grande, sont de bonnes fondations pour des traveaux constructifs.

1 INTRODUCTION

Engineering demands associated with offshore oil development have made it necessary to survey the seafloor geological conditions in the Pearl River Mouth Basin that qualifies as an oil potential area. Guangzhou Marine Geological Survey has been carrying on the 1:200,000 regional survey since 1986. Based on integrated analysis of 15,300 km of seismic data, 124 gravity/piston cores and CPT results from 4 boreholes, the characteristics of engineering geology have been evaluated for first 4 international sheets which cover an area of 30,785 km² (Fig. 1).

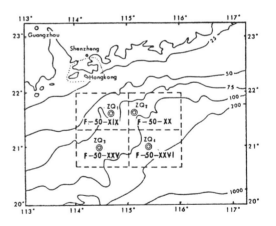

Fig. 1 Sketch map of studied area and locations of the boreholes

2 REGIONAL GEOLOGICAL SETTING

South China Sea has spread twice since it was formed during Late Mesozoic and Mid Cenozoic. Seafloor spreading resulted in a series of Cenozoic basins.

The Pearl River Mouth Basin, a divergent subsiding basin of 150,000 km² area, consists of six secondary structural units: northern fault terrance, Zhu-1 depression, Zhu-2 depression, Zhu-3 depression, Dongsha

uplift and Shenhu uplift (Jin, 1989). There are three groups of faults within the basin. Group 1 in orientation of ENE, 610 km long and 20 - 40 km wide, play a leading role, controlling the strike of the basin and the distribution of sediments. Group 2 in orientation of NE are located on middle of the basin, controlling the basement of the basin. Group 3 are younger faults and run across Group 1 and Group 2. Volcanic activities were frequent during Early Tertiary under the effect of faulting.

3 STRATIFICATION OF ENGINEERING GEOLOGY

Six geotechnical strata are divided within 120 m depth below the seafloor (Fig. 2). The geotechnical properties for each stratum are summarized in Table 1.

The results show that strat 1 and 3 are of low shear strength, especially in stratum 1 shear strength would reduce 75 percent when disturbed. So it is unsuitable to use these two strata as the bearing layers without foundation improvement. Strata 2, 4, 5 and 6 are the layers with high bearing capacity. Note not all 6 strata appear completely throughout the basin.

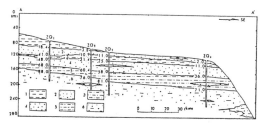

Legend: 1. cohesive soil 2. sandy soil w/ fine particles 3. silty clay 4. gravelly sand 5. silt & clayey silt 6. clayey sand & gravelly sand

Fig. 2 Geotechnical sequences revealed by boreholes

4 GEOHAZARDS AND CONSTRAINTS

Different types of geohazards and constraints have been observed in the basin (Li & Jin, 1989).

4.1 Buried channels

Buried channels with irregular geometry are common, especially in stratum 2. These channels incise the horizons to depths from several to tens of meters and their widths range from tens to hundreds of meters. The channel fills are mainly sandy clay, sand and gravelly sand that are distinct from nonchannel fills. Geotechnical properties are also different.

Table 1 Subbottom strata and their geotechnical properties

Strata	Thickness (m)	Major Properties
1	0 - 18.4	soft plastic, S_u < 10 kPa, w = 41.8 - 51.8 %, w > w_L, e > 1.
2	2 - 22.6	loose, w = 20 -30 %, ψ' = 38°, q_c = 20 - 40 MPa, u_w = -0.4 - 3.3 MPa, S_u = 61 ± kPa.
3	7 - 21.4	homogenous, OCR = 1.39, q_c = 10 kPa, c' = 0, S_u = 60 -70 kPa, ψ' = 20°.
4	20 - 32.4	well-graded & compact, w = 20 - 26 %, q_c = 71 MPa, u_w = -1.0 - 4.2 MPa, S_u = 120 - 280 kPa.
5	16 - 35	inhomogeneous, C_c = 0.34, w = 19.7 - 33.6 %, q_c = 1.35 - 17.22 MPa, u_w = 1.27 - 6.99 MPa, S_u = 110 - 140 kPa, ψ' = 24° - 27°, S_t = 7.5, c' = 0, OCR = 1.26 -1.54.
6	> 12	laminated, inhomogeneous & poor sorting, w = 21 - 25 %, OCR = 1, q_c = 0.03 - 68.8 MPa, u_w = -2.0 - 8.81 MPa, S_u = 610 - 770 kPa.

4.2 Storm sediments

Storm sediments are identified near surface in the basin. Storm sediments are mainly calcareous cemented fine sandrocks with medium hardness. Test results show their geotechnical properties (ψ' = 71° - 75° c' = 300 - 500 kPa) are distinct from those of surrounding soft sediments.

4.3 Buried hills

A few buried hills are located in the regions of the shelf and the slope. Fig. 3 shows a buried hill, assumed to be a Late Tertiary intrusive body, in the slope. The top of hill, 62 m below the seafloor, is 715 m wide and wall angle is 50° - 55°. Chaotic sediments exist on the foothill.

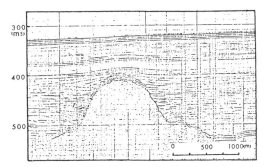

Fig. 3 Single-channel seismic profile showing a buried hill in the slope

4.4 Slides

Along the shelf break there is a significant sliding belt that is as long as 60 km and in orientation of ENE. The slides stretch approximately parallel to the contour lines and are considered to be resulted from earthquakes. In seismic profiles, obvious sliding planes and un- filled scarplets have been seen so they are assumed as younger slides.

Slide complex are found on the slopes of Weitan Banks (F-50-XXVI). Slide blocks im- bricate by ones, resulting in complexity of structures (Fig. 4).

Creeping slips appear in the basin with slope angles of less than 6°. Masses slip stratifiedly along the gentle creeping planes, presenting small failures in strata.

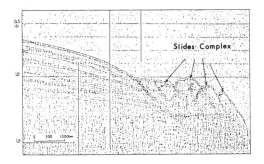

Fig. 4 Single-channel seismic profile showing slide complex

4.5 Shallow gas

Gas-charged sediments are identified in Sheets F-50-XIX and F-50-XXVI. Gas-charged sediments, inferred to be caused by the

presence of methane in tens of meters below the seafloor, are characteristic of dis- tinct seismic smear and reflection free zones in seismic profiles (Mullins & Nagel, 1983; Li & Jin, 1989).

Several "bright spots" related to high pressure gas pockets in deep strata are seen in digital seismic profiles. Such kinds of gas accumulation would make increase of pore water pressure and de- crease of effective normal stress of soil.

Seepage gas, presenting water-column ano- malies in seismic profiles, is observed in the basin (Fig. 5). Seepages are caused by the upward migrating of shallow gas through sediments and the subsequent seeping into oceanic water (Li & Jin, 1989).

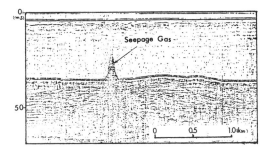

Fig. 5 Single-channel seismic profile showing gas seepage into the oceanic water (after Li & Jin, 1989)

4.6 Faults

Faults are concentrated in the northern slope fractures belt (Liu, 1981; Jin, 1989). Some of faults were formed in Yanshan Cycle and cut Mid and Late Pleis- tocene, keeping active during Quaternary (Fig. 6).

A series of shallow buried step faults are distributed on regions of the shelf and the slope. These faults are generally caused by continuous accumulation and dif- ferential compaction during depositional processes. Dip slips are only a few meters and the uppermost are quite near the sea- floor. So these synsedimentary faults are considered to be formed in the modern times.

4.7 Diapirs

Under the high pressure of overlying loads, locally underconsolidated soil has pene- trated into the overlying strata, even out of the seafloor (Li & Jin, 1989). Fig. 7

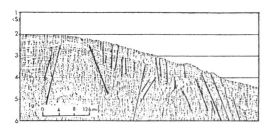

Fig. 6 Multi-channel seismic profile
showing fault structures in the slope
(after Jin, 1989)

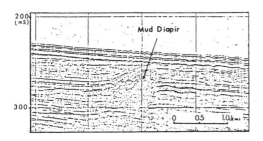

Fig. 7 Single-channel seismic profile
showing a mud diapir (after Li & Jin,
1989)

shows a typical mud diapir in Sheet F-50-XX.

4.8 Seafloor erosions

4.8.1 Active sandwaves

Sandy floor of the outer shelf are eroded
by strong bottom currents. So large area of
sandwaves exist in both Sheets F-50-XX and
F-50-XXVI between water depths of 130 -
250 m.

4.8.2 Turbidity currents

There are turbidity currents in the slope
area. Turbidity currents at high velocity
(up to 150 - 200 m/s) severely erode the
slope, resulting in the slope instability.

4.9 Earthquakes

In the South China Sea earthquakes occurred
frequently (Liu & Zhuo, 1984). Statistics
indicate since 1067 earthquakes with M > 4
have occurred 273 times, 64 percent of
which being with M > 5 and the maximum

being with M = 8. However, small earthquakes held sway in the past two decades
(Gu, 1983 & 1984).

5 ZONATION OF ENGINEERING GEOLOGY

According to the engineering geological
conditions, the Pearl River Mouth Basin are
principally divided into 2 areas, further
divided into 5 subzones (Chen, et al.,
1989) (Fig. 8).

5.1 Continental shelf area (I)

In the shelf area Quaternary is greatly
thick and loose. Shallow faults are common.
The slope ratio is $0.314 - 0.747 \times 10^{-3}$.
Calculation, based on the theory of infinit
slope, indicate the soil, whether cohesive
or sandy, would be stable under the effect
of gravity load. However, by the joint actions of mighty waves and earthquakes with
M > 5, sandy soil within 70 m below the
seafloor would liquefy and slide, and those
in depths of more than 70 m would keep
stable. No slide is observed in the shelf.
  Shelf area is further divided into 3
subzones.

5.1.1 Neritic accumulation plain (I_1)

This subzone is located in the northern
basin within water depths of 30 - 100 m,
where tidal action is strong. Deep faults
are common, resulting in disintegration of
bedrocks. Shallow faults are also common,
some of the uppermost being just only a few
meters below the seafloor.
  Sediments in Subzone I_1 are mainly grey
and olive green cohesive soil with loose
texture and an abundance of organism. The
particles with d < 0.1 mm are over 50 percent. Statistics of geotechnical properties
(w > 40 %, G_s = 2.7 g/cm³, e > 1, w > w_L)
present sediments are clayey soil that
would appear as slurry when disturbed and
replasticized. Test results indicate the
shear strength of soil would reduce when
little sand are mixed in, and increase
until sand content reaches a certain degree.
  Storm deposits and shallow gas exist in
Subzone I_1. Earthquakes with M > 5 occurred
sometimes in past 100 years. The seafloor
is relatively unstable.

5.1.2 Mixed accumulation plain (I_2)

This subzone borders on the south of Subzone I_1 in water depths of 60 - 120 m,
where tidal action is weak. The seafloor is

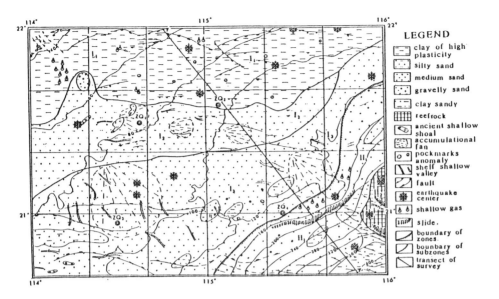

Fig. 8 Zonation map of engineering geology in the Pearl River Mouth Basin. Location of mapped area (dashed boxes) are shown in Fig. 1

generally flat with various microrelif. Shallow faults are common, but deep faults are not.

Sediments in Subzone I_2 are basically clayey sand and silty sand. Over 50 percent of particles are with d = 0.1 - 2.0 mm. The soil (w = 24 - 30 %, c_{cu} = 0.044 - 0.459 kPa, ψ_{cu} = 17° - 32°) is of high cohesion intercept.

Earthquakes in Subzone I_2 occurred seldom. Analysis of the soil from tens of cores indicate the seafloor of most region would be stable by the joint actions of the gravity load, waves and earthquakes.

5.1.3 Relic accumulation plain (I_3)

This subzone is located in the middle area in water depths of 80 - 140 m. There are variety of relic geomorphic features, such as ancient deltas, subaqueous shoals and scour valleys. Serious fragmentation of bedrocks is associated with deep faults, maximum dip slip reaches 2,000 m. Shallow faults exist locally. Evidence of neotectonism is insignificant.

Sediments in Subzone I_3 are well-graded, compact sand and gravel. The soil (w = 20 - 26 %, q_c = 71 MPa, G_s = 120 - 280 kPa, u_w = -1.0 - 4.2 MPa) is of high cohesion intercept.

Earthquakes in Subzone I_3, most being with M < 5, occurred frequently. The seafloor is relatively stable.

5.2 Continental slope area (II)

In the slope area geomorphy is very complex and slope ratio is 13 - 23 X 10^{-3}. Quaternary is generally thin and loose. Numerous faults have caused bedrocks fractured.

The slope is further divided into 2 subzones.

5.2.1 Slope accumulation subzone (II_1)

This subzone is located in the upper slope within water depths of 120 - 440 m. Erosions of bottom currents and turbidity currents have caused large area of sandwaves and valleys, the latter being with widths of 5 - 10 m and incisions of 1 - 2 m. Shallow faults are spread broadly.

Sediments in Subzone II_1 are mainly gravelly sand, clayey sand and clay with coarse debris. Analysis of stability show steep topography makes it possible for loose soil to slide along slipping planes under gravity load. For example, a potential slipping plane in coreing station 17-3, in water depth of 160 m, exists in 36 - 90 m below the seafloor and radius of arc is more than 90 m. In fact, various types of slides have been observed in this subzone during survey. Earthquakes occurred frequently.

2841

5.2.2 Slope reefrock subzone (II₂)

This subzone is distributed in southeastern edge of surveyed area. Reefs and bedrocks are exposed directly or covered with thin clayey sand. Topography changes roughly and steeply, water ranging from 60 - 320 m. Both deep and shallow faults are common.

Subzone II₂ is located in northern slope earthquake belt, so that earthquakes, some were with M > 6, occurred frequently, resulting in activities of slides and turbidity currents.

6 CONCLUSIONS

Based on the analysis of the available data, the following conclusions can be stated:

1. Within the surveyed area, the shelf is a generally stable area. Engineering geological conditions in Subzones I₂ and I₃ are better than those in Subzone I₁. The slope is a relatively unstable area with frequent earthquakes, faults, slides, turbidity currents, etc.

2. Of 6 strata within 120 m below the seafloor, strata 1 and 3 are unacceptable for bearing layers without foundation improvement. Others are considered as bearing foundation for engineering facilities.

3. Geohazards and constraints in the Pearl River Mouth Basin include buried channels, buried hills, storm sediments, shallow gas, slides, faults, diapirs, earthquakes, etc. These factors will endanger and/or restrain the operational facilities. Detailed surveys are recommended prior to specific engineering activities.

ACKNOWLEDGEMENTS

Financial and technological assistance for this project is provided by the United Nations Development Programme (CPR/85/044). The authors wish to thank Guangzhou Marine Geological Survey for their permission to publish this paper.

REFERENCES

Chen Junren, et al. (1989). The evaluation of engineering geological conditions in the Pearl River Mouth Basin. Proceedings of UN Meeting on Marine Engineering Geological Survey for Petroleum Exploration in Developing Countries. Guangzhou, Guangdong. (in press).

Gu Gongxu, et al., ed. (1983). Chinese earthquake catalogue (1831 - 1969). 894 pp. Beijing.

Gu Gongxu, et al., ed. (1984). Chinese earthquake catalogue (1970 - 1979). 334 pp. Beijing.

Jin Qinghuan, ed. (1989). Geology and oil-gas resources of South China Sea. 417 pp. Beijing.

Li Tinghuan & Jin Bo (1989). Seismic facies analysis of the seafloor instabilities in the Pearl River mouth region. Marine Geo-technology 8: 19 - 31.

Liu Yixuan (1981). Analysis of fracture systems in coastal region of South China. 120 pp. Beijing.

Liu Yixuan & Zhuo Jialun (1984). A discussion on the stability of geological environments in terms of the neotectonic features of the northern South China Sea. Tropic Oceanology 3: 55 - 62.

Mullins, H. T. & Nagel, D. K. (1983). High-frequency seismic data detect shallow hydrocarbons. World Oil 197: 133 - 138.

6th International IAEG Congres /6ème Congrès International de AIGI, © 1990 Balkema, Rotterdam. ISBN 90 6191 130 3

Force-controlled push sampling as a measure of in-situ shear strength and sample quality in offshore site investigations

Carottage à contrôle de force comme mesure de résistance au cisaillement in-situ et qualité d'échantillons en reconnaissance des terrains sous-marins

Ger de Lange & John ten Hoope
Fugro-McClelland Engineers, Leidschendam, Netherlands

ABSTRACT: In offshore soil investigations flush thin-walled shelby tubes are pushed hydraulically at a constant rate of penetration into the soil at the bottom of a borehole. Analysis of the recorded forces has enabled accurate definition of the disturbed zone directly below the drillbit as well as whether stones or gravel may have disturbed the sample during penetration of the tube.
Moreover, using a conventional pile capacity side wall friction relationship it has proved possible to construct in-situ (remoulded) shear strength vs depth profiles. Cross plots of the pushing force vs the laboratory shear strength show good correlation. This may indicate that analysis of the force data of the hydraulic push sampler is very suitable for in-situ shear strength evaluation, especially when the object is to obtain wall friction values for pile capacity calculations.

RESUME: En reconnaissances des terrains sous-marins, carottiers à paroi mince et surface unie sont poussés dans le sol au fond d'une forage par pression hydraulique et au vitesse constante. Analyse des forces enregistrées a mis en état de définer la zône remaniée directement au dessous de la couronne de forage aussi bien que l'observation des éléments grossiers qui auraient remaniés la échantillon pendent le pénétration du carottier. Par ailleurs, c'est possible de construire des profiles de résistance (remaniée) au cisaillement in-situ contre profondeur par appliquer un relation classique de frottement lateral de pieu. Le diagramme de la force poussante contre la résistance au cisaillement en laboratoire montre une bonne corrélation. Ça peut indiquer que l'analyse des forces enregistrées du carottiér hydraulique convient à évaluer la résistance au cisaillement in-situ, particulièrement pour obtenir le frottement lateral employé dans les calculations de capacité des pieux de fondation.

1 INTRODUCTION

For more than a decade it has been common practice in offshore site investigation to obtain push samples with thin walled "shelby"-type tubes. The hydraulically driven wireline push (WIP-sampler) and PISTON samplers in use at Fugro- McClelland (Figure 1) use such flush stainless steel tubes of 1000 mm length and 2 mm wall thickness with a sharp leading cutting edge. The tubes are pushed into the soil at a penetration rate of approximately 20 mm/sec, with variations depending on the resistance encountered. The oil pressure needed to advance the sample tube into the soil is recorded with penetration depth and stored on floppy disk. The use of this data to determine sample quality and as a measure of shear

strength of clays is explored
further in this paper.

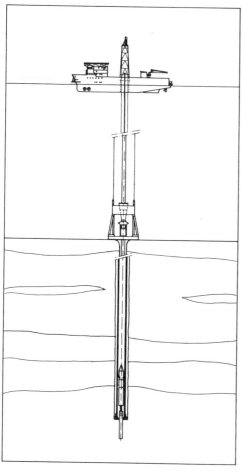

Fig.1 WIP-sampler with SEACLAM

2 METHOD

The method to obtain push samples
which has become a standard in
offshore site investigations, is by
using a wireline hydraulic push
sampler (Richards and Zuidberg,
1986). After drilling with an open
drill bit to the depth where a
sample has to be taken, the
wireline sampling tool is lowered
to the drill bit. The tool consists
of three principal elements:
1. the sample tube
2. the hydraulic ram
3. the remote measuring control
   unit.

At Fugro-McClelland two types of
samplers are used: the WIP sampler
(acronym for WIreline Push Sampler)
and the PISTON sampler. Both
utilize flush stainless steel tubes
with outer cutting edges and 72 mm
inner diameter. The tubes are 1000
mm long and provide a net sampling
length of 900 mm. The WIP sampler
can also accommodate a connecting
head for sample tubes of 50 mm
inner diameter. Sample tubes
with cutting shoes, often used to
reduce outer wall friction or to
retain a core catcher are not
discussed here because of their
irregular cross-section. The area
of the hydraulic ram is different
in both sample tools because of the
stationary piston of the PISTON
sampler: 38.3 cm^2 and 31.4 cm^2 for
the WIP and PISTON samplers
respectively. The remote measuring
control unit (RMC) contains, among
electronic circuitry necessary for
cone penetration testing with the
wireline cone penetrometer (WISON)
which can also be attached to it, a
volutronic, which measures the
amount of oil flowing into the
cylinder, and an oil pressure
transducer. The amount of oil
flowing through the volutronic is a
measure of the pushed length of the
hydraulic ram and in turn of the
penetration of the sample tube. The
electronic signals are transmitted
to the drill deck via an umbilical
cable containing the electronic
wires and the oil pressure hose.
When the tool reaches the drill
bit, it latches automatically under
it's own weight in a special bottom
hole assembly directly behind the
bit. Oil pressure is administrated
through the oil hose at a constant
rate, thereby ensuring an
approximately constant penetration
rate. Reaction force for the
pushing action may be provided by
the bit load. Frequently the
resistance of the soil exceeds the
safe bit load. Reaction is
therefore provided by a ballasted
seabed clamping unit.

3 THEORY

Analysis of the plots of oil
pressure against penetration depth
(Figure 2) shows that in clays a
fairly smooth relationship is

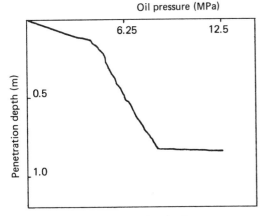

Fig.2 Oil pressure vs penetration depth

obtained. When used in more or less homogeneous clays, the resistance to penetration is governed by friction of the clay along the inner and outer surfaces of the sample tube. Thanks to the sharp cutting edge and the small wall thickness, the zone of influence of the end bearing pressure is negligible compared to a standard cone penetrometer. It is therefore proposed that the behaviour of a push sampler can be compared to the behaviour of driven steel pipe piles like those used for the foundations of offshore structures. Taking this reasoning one step further, the undrained shear strength (s_u) can be obtained by using a relationship between wall friction and s_u similar to the empirically derived relationships for foundation piles such as the ones recommended by the American Petroleum Institute (reference).

Considering the end bearing of the sample tube as an offset of the Oil Pressure vs Depth plot then the gradient of the plot represents the sum of the outer (F_o) plus inner (F_i) wall friction per unit of penetration (1). In terms of force this is represented by:

$$\frac{F_T}{l} = \frac{F_o + F_i}{l} \qquad (1)$$

By analogy with capacity, the outer wall friction can be related to undrained shear strength by:

$$F_o = \alpha * s_u * A_o \qquad (2)$$

where A_o: the outer wall area (m^2)
α : an empirical factor varying between 0.5 and 1.0 depending on the shear strength.
s_u: undrained shear strength (kN/m^2)

The inner friction is considered to be reduced by a factor of 0.8. The complete expression then becomes:

$$F_T = (\alpha * s_u * A_o) + (0.8 * \alpha * s_u * A_i) \qquad (3)$$

where A_i: the inner wall area (m^2)

For sample tubes with 76 mm outer diameter and 2 mm wall thickness the friction per length of penetration can be simplified to:

$$\frac{F_T}{l} = \alpha * s_u * 0.42 \qquad (4)$$

Then: $\alpha * s_u = 2.38 * \dfrac{F_T}{l} \qquad (5)$

For the PISTON sampler this expression becomes

$$\alpha * s_u = 7.5 * P_{oil}/m \qquad (6)$$

where P_{oil}/m: gradient of the Oil Pressure vs Depth plot in MPa/m
A cross plot of laboratory undrained shear strength measurements vs $\alpha * s_u$ obtained from the Oil Pressure vs Depth plots will then generate an α factor for the push sample tube (Figure 3).
Several considerations should be noted. Firstly, the reduction factor for inner friction may not be 0.8. Secondly, the α factor may vary with shear strength. Both factors may be related to in-situ soil pressures, which are dependent on the past history of the soil. It has been observed when sampling overconsolidated glacial clays that the sample diameter is smaller than the inner diameter of the tube for at least part of the sampled length. This effect can be seen on the Oil Pressure vs Depth plot as a convex curvature, indicating reduction of inner friction, and should be taken into account when deducing the shear strength. Because more research should be done to study these factors, and

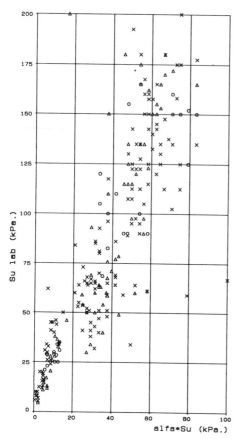

Fig.3 Alpha*s_u(tube) vs s_u(lab)

for reasons of simplicity the authors adhere to a simple α factor in this paper.
As with the bearing capacity factors, which are used to deduce shear strength from cone penetration resistance (N_C factors), the applicability of the α factor is localized by the type of clay. For other types of clay similar relationships should be derived.

4 PLUGGING

It has been demonstrated that, when the inner wall friction becomes greater than the end bearing, the sample will no longer enter the tube and the tube becomes plugged (Holt and Ims, 1985). This plugging is visible on the oil pressure vs depth curve as a steepening of the curve, due to the loss of inner friction build-up (Figure 4).

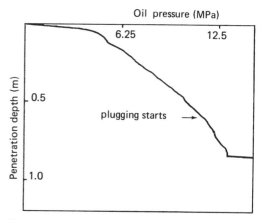

Fig.4 Plugging effect on oil pressure vs depth curve

Partial plugging i.e. the intermittent entering and plugging of the clay causes a similar effect. Other phenomena may also be deduced from the curve, such as disturbance of the clay at the bottom of the borehole (Figure 5), obstruction by cobbles and stones (Figure 6) or layer changes.

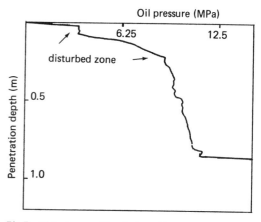

Fig.5 Disturbance effect below drill bit

It has been observed that when plugging occurs, the failure surface along which the clay shears has the shape of a cone below the sample tube. Such a cone has been recovered frequently and, as would be expected from pile capacity end bearing theory, has apex angles of 45° to 60°. As such, the end bearing of the plugging sample tube may be compared to the cone

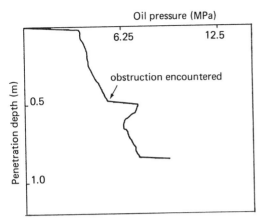

Fig.6 Increase in oil pressure due to obstruction of sampler

resistance (q_C) of the cone penetrometer. Plugging occurs when the end bearing is equal than or smaller than the inner friction of the sample tube, or:

$$F_i \geq Q_t \tag{7}$$

which is:

$$0.8*\alpha*s_u*\pi D_i*1 \geq q_C * \frac{\pi D_i}{4} \tag{8}$$

where: D_i: inner diameter of sample tube (m)

If q_C is expressed by the simplest correlation with undrained shear strength:

$$q_C = N_C * s_u \tag{9}$$

then the expression becomes:

$$0.8*\alpha*s_u*\pi D_i*1 \geq N_C * s_u * \frac{\pi D_i^2}{4} \tag{10}$$

Reduced this means that plugging will occur when the penetrated length obeys the following expression:

$$1 \geq \frac{N_C * D_i}{3.2 \, \alpha} \tag{11}$$

If the usually applied N_C factors (15-20) are substituted and the above found α factor of 0.3 to 0.55 is used then the expression reduces to:

$$1 \geq (7.8 \text{ to } 20.8) \, D_i \tag{12}$$

Taking the average, this means that the sample tubes will plug when the penetration length exceeds 14 times the inner diameter. For 72 mm inner diameter tubes, this means that on average, the maximum sample length is 1000 mm. It is therefore impractical to fabricate or use longer tubes of that diameter. Similarly, if 50 mm diameter tubes are used a maximum effective tube length of 700 mm would be sufficient. Premature plugging can thus be attributed to variations in N_C and α factors. The statement made by Holt and Ims (1985) that premature plugging can be predicted from measured cone resistance is valid. Whether the inner wall friction can be considered identical to the sleeve friction requires more research into the effects of the shearing surfaces in the clay when using the respective tools.

5 APPLICATION

The estimate of undrained shear strength obtained from the Oil Pressure vs Depth plots can be used as a preliminary value for the construction of shear strength profiles. When the sample tube is regarded as a model pile, the $\alpha*s_u$ value deduced from the plots, may find an application in the determination of design wall friction parameters in pile capacity calculations. It is noted that the mode of penetration lies somewhare between static behaviour and the behaviour during pile driving but it is the authors opinion that it compares best to the latter situation. Exploring these data further, an comparing them to cone penetration sleeve friction data will generate more confidence in the quantitative value of the sleeve friction data. It should be noted that diameter to wall thickness ratios are similar in foundation piles and sample tubes. This is important, because sample disturbance, due to the simple fact that the sample entering the tube may be subjected to post-peak strains, is dependent on this ratio (O'Reilly, 1987). Application of the foregoing to penetrations in sand has not yet

been researched. It is expected
that this situation is more
complex, due to a greater influence
of end bearing of the sample tube
edge and scale effects with respect
totexture and structure of granular
deposits.

6 CONCLUSIONS

It is concluded that in clays the
force controlled push sampler can
be used as a model pile. The
relationship between wall friction
of the push sampler and foundation
piles is the subject of ongoing
research. The penetration data of
the push sampler can also provide a
preliminary shear strength, without
extruding the sample.

REFERENCES

American Petroleum Institute 1989,
   1987, 1986. RP2A: Recommended
   practice for planning, designing
   and constructing fixed offshore
   platforms. Washington, API.
Holt, J.R. and Ims, B.W. 1985. The
   phenomenon of sample tube
   plugging. In R.C. Chaney and
   K.R. Demars (eds.), Strength
   testing of marine sediments:
   laboratory and in-situ
   measurements, p. 166-177.
   Philadelphia, ASTM.
O'Reilly, M.P. 1987. Site
   investigation, soil sampling and
   laboratory testing. Ground
   Engineering, Vol. 20, no. 8, p.
   2-8.
Richards, A.F. and Zuidberg, H.M.
   1986. Sampling and in-situ
   geotechnical investigations
   offshore. In R.C. Chaney and
   H.-Y. Fang (eds.), Marine
   geotechnology and nearshore/
   offshore structures, p. 51-73.
   Philadelphia, ASTM.

6th International IAEG Congres / 6ème Congrès International de AIGI, © 1990 Balkema, Rotterdam. ISBN 90 6191 130 3

Modélisation de la dynamique de glissements sous-marins

Modelling the dynamics of submarine slides

J. Locat
Groupe de Recherche en Géologie de l'Ingénieur, Département de Géologie, Université Laval, Québec, Canada

H. Norem & B. Schieldrop
Norges Geotekniske Institutt, Oslo, Taasen, Norway

RESUME: Un modèle visco-plastique à deux dimensions de simulation des avalanches de neige a été adapté à l'étude des glissements sous-marins et permet d'évaluer les distances de parcours de même que la vitesse du front tout en reproduisant la forme de la masse en mouvement tout au long de l'événement. Le modèle, quoiqu'incomplet, est calibré sur des cas connus ce qui permet de bien distinguer la coulée de débris du courant de turbidité subséquent.

ABSTRACT: A two-dimensional visco-plastic model developed for snow avalanches has been adapted to study the flow dynamics of submarine slides. It gives estimates of front velocity, run-out distance while reproducing the shape of the moving mass. Although incomplete, the model has been applied to known case histories with enough success to stimulate a discussion on the physic of submarine slides by clearly distinguishing between the dense flow (or debris flow) and the turbidity current generated by the initial slide.

1. INTRODUCTION

Le développement récent des ressources offshore a ravivé l'intérêt pour les études de glissements sous-marins, particulièrement ceux qui concernent les coulées de débris. Les glissements sous-marins ont une incidence directe sur au moins trois classes de problèmes: (1) l'érosion de structures enfouies (e.g. pipelines), (2) la conception de structures ancrées dans les fonds marins, (3) la génération de vagues dans des réservoirs naturels ou artificiels. On peut rappeler les conséquences néfastes de plusieurs cas récents dont celui de Nice en 1979 (Gennesseaux et al. 1980), et de Kitimat, Canada (Prior et al. 1982). Dans bien des cas, il semble que le phénomène est seulement décrit sur la base des bris de câbles sous-marins à différentes distances de la zone de glissement initial. Une telle approche est un peu risquée pour modéliser les coulées de débris qui souvent s'arrêtent bien en deçà de la distance maximale atteinte par le courant de turbidité. Dans la plupart des cas connus de glissements sous-marins, on reconnaît deux phases dynamiques. La phase I consiste en un écoulement initial de débris au front duquel se forme un "nuage" de particules en suspension. L'arrêt de la coulée de débris coïncide avec le début de la phase II qui est principalement dominée par le courant de turbidité amorcé lors de la phase I. La phase I est donc dominée par l'écoulement dense (dense flow) alors que le courant de turbidité de la phase II est constituée surtout par l'écoulement d'une suspension (suspension flow). Nous nous intéressons ici qu'à la phase I d'un glissement sous-marin.

L'article qui suit vise deux objectifs: revoir les développements récents des approches au calcul de la dynamique des glissements sous-marins, (2) présenter et appliquer une approche dérivée de l'étude des avalanches de neige (Norem et al. 1989, 1990), à quelques cas de glissements sous-marins.

2. APPROCHES A LA MODELISATION

Déjà, lors de la conférence de l'OTAN sur les glissements sous-marins (Saxov et Nieuwenhuis 1982), quelques scientifiques avaient abordé le problème de la dynamique de tels glissements: (1) la vitesse du front; (2) la distance de parcours. A ce jour, les approches se divisent en deux groupes. Le premier groupe utilise une approche basée sur la mobilisation de la friction de Coulomb (ϕ); le deuxième simule le problème avec un comportement visco-plastique pouvant inclure la friction de Coulomb.

2.1 Friction de Coulomb

Cette approche est habituelle en mécanique des sols (Hutchinson et Bhandary 1971, Hutchinson 1986 et Matos 1988). Dans un tel cas, le facteur principal de réduction de la résistance au cisaillement est l'excès de pression interstitielle dont la dissipation s'effectuerait selon la théorie de consolidation de Terzaghi:

$$T_v = \frac{c_v t}{h_d^2} \qquad (1)$$

où "T_v" est le facteur temps (time factor), c_v le coefficient de consolidation, "t" le temps, et "h_d" la demie distance de drainage. Cette approche offre l'intérêt de tenir compte de la variation d'épaisseur de la coulée durant le mouvement alors que les théories antérieures considéraient le sédiment incompressible. Ce modèle de Hutchinson et Bhandary (1971) néglige toute dissipation d'énergie dans la masse et considère donc que le profil de vitesse y est constant. Pour utiliser un tel modèle, il faut connaître la pression interstitielle (très difficile à évaluer ou à mesurer durant l'écoulement), le coefficient de friction, la cohésion (s'il y a lieu), la géométrie et les caractéristiques en consolidation. L'application de ce modèle est donc limitée aux cas où une résistance au cisaillement peut se développer à la base de la tranche en mouvement. Matos (1988) a réussi à calibrer ce type de modèle avec les données du glissement des Grands-Bancs de Terre-Neuve.

2.2 Approche visco-plastique

Les chercheurs intéressés par cette approche (Johnson 1970, Hampton 1970, Suhayada et Prior 1978, Karlsrud et Edgers 1982) ont utilisé une relation tenant compte à la fois des caractéristiques de friction du sédiment et ses propriétés rhéologiques:

$$\tau = c + \sigma' \tan\phi + \eta \frac{dv}{dy} \qquad (2)$$

où "τ" est la résistance au cisaillement, "c" la cohésion, "$\sigma' \tan\phi$" la friction de Coulomb, et "η" la viscosité. Cette approche tient compte du comportement visqueux du fluide, mais néglige l'interaction entre les particules durant le cisaillement. Pour tenir compte de cet aspect, Norem et al. (1989) ont introduit la pression de dispersion déjà formulée par Bagnold (1954, 1956) pour les écoulements granulaires. Norem et al. (1990) ont montré que pour les glissements sous-marins, l'écoulement est du type macro-visqueux de telle sorte qu'il y a linéarité entre le gradient de vitesse et la contrainte de cisaillement dans le fluide. Avec une telle approche, le modèle de Norem et al. (1989) s'adapte de la façon suivante pour un glissement sous-marin:

$$\tau_{xy} = c + \sigma' \tan\phi + m\bar{\rho} (dv/dy)^r \qquad (3)$$

où "m" est la viscosité de cisaillement, "ρ" est la densité du fluide et "r" un exposant égal à 1 pour un écoulement en régime macro-visqueux. Soulignons que ce modèle exige aussi une estimation des pressions interstitielles dans la masse en mouvement. Pour le modèle présenté et utilisé ci-après, la définition des contraintes dans la masse en écoulement est schématisée à la figure 1 alors que la géométrie de l'écoulement est illustrée à la figure 2.

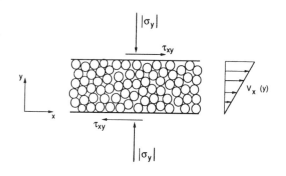

Figure 1. Définition des contraintes normales et de cisaillement et le profil de vitesses.

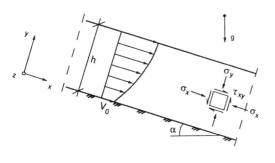

Figure 2. Définition de la géométrie en régime d'écoulement permanent.

Etant donné que le modèle est à deux dimensions, il faut tenir compte de la répartition des contraintes dans la masse en mouvement. Pour y arriver, on utilise l'approche par contraintes passives et actives ce qui permet, dès le départ, de suivre l'évolution de la forme de la masse (figure 3) un peu à la façon d'un glissement rétrogressif. A ce stade-ci, le modèle considère qu'il n'y a pas de perte ou de gain de masse durant l'événement et néglige la friction au sommet de la masse en mouvement. La solution de l'équation (3) est réalisée à l'aide d'un modèle numérique en différences finies mis au point par Irgens (1988).

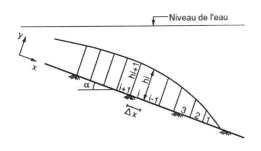

Figure 3. Disposition des éléments pour le calcul du démarrage du glissement sous-marin.

3. LE CHOIX DES PARAMETRES

De tous les paramètres physiques nécessaires, la pression d'eau est certes la plus difficile à évaluer. Le seuil d'écoulement est représenté par le terme "$c + \sigma' \tan\phi$". "c" est indépendant de la pression et correspond au seuil d'écoulement ou à la cohésion et est considéré nul pour les glissements sous-marins. Le terme de friction, bien que mobilisé dynamiquement, correspond, en première approximation, à l'angle de friction statique (Savage et Sayed 1984, Hungr et Morgenstern 1984). Quant à la contrainte effective, elle est fonction du poids de la colonne de sédiment (σ_y), de la pression interstitielle (u_w) et de la pression de dispersion (Bagnold 1954, Bagnold 1956, Norem et al. 1989).

La pression interstitielle est un paramètre très important, voire dominant dans un tel phénomène. Kummeneje et Eide (1961) ont observé que des sédiments liquéfiés pouvaient initialement avoir un excès de pression interstitielle de plus de 50% en sus de la pression hydrostatique. Sassa (1988) a aussi observé que la pression interstitielle de sables saturés augmentait de plus de 42% durant le cisaillement. Iverson et LaHusen (1989) ont aussi noté la même chose. D'autre part, Karlsrud et Edgers (1982) ont observé que l'angle moyen du plan de glissement d'avalanches sous-marines variait entre $0.3°$ et $5.7°$. Un tel angle reflète l'angle de friction maximum apparent maximum pour de tels phénomènes. Cruden et Hungr (1986) utilisent aussi une approche semblable pour évaluer l'équivalent d'un rapport de pression interstitielle (r_u) ayant pour effet de réduire la composante de friction à une valeur apparente observée. Par exemple, pour une valeur de r_u de 0.9 et ϕ de $30°$, on obtient un angle de friction apparent de seulement $3.3°$.

Quant à la viscosité, les travaux de Locat et Demers (1988), Wildemun et Williams (1985) de même que ceux de O'Brien et Julien (1986) permettent une première approximation de la viscosité dynamique des sédiments ayant un indice de liquidité compris entre 1.0 et 6.0.

Il reste à préciser les conditions géométriques. Dans le cas des analyses à rebours présentées ci-après, cet aspect demeure aussi important et difficile à cerner. Bien peu de cas permettent de connaître les conditions précises de l'état du sol et de la géométrie du terrain avant la rupture. Ces aspects sont précisés et discutés dans la section suivante.

4. ANALYSES A REBOURS DE CAS CONNUS

Trois cas sont présentés ci-après: Sokkelvik et Storglomvatnet en Norvège et celui des Grands-Bancs de Terre-Neuve au Canada. Chaque cas présente des aspects très différents. Celui de Sokkelvik est bien documenté en termes de morphologie avant et après le glissement, mais il n'y a pas eu de bris de câbles sous-marins et la vitesse n'a donc pu être mesurée. Celui de Storglonvatnet a été provoqué (Norem et al. 1990); les conditions avant et après rupture ont pu être estimées et les vitesses y ont été mesurées. Pour ce qui concerne celui des Grands Bancs de Terre-Neuve, il n'y a que les bris de câbles sous-marins comme données de la progression du mouvement. Dans ce dernier cas, le choix des paramètres géométriques initiaux est incertain étant donné la très grande étendue de la zone affectée par le tremblement de terre (Piper et al. 1986). Les paramètres généraux pour chaque cas sont fournis au tableau 1.

Tableau 1. Données paramétriques sur les glissements sous-marins étudiés

Paramètres	SOK	STOR	GB
cohésion, c (kPa)	0	0	0
Friction, ϕ (^{o})	30	35	30
γ (kN/m^3)	17	17	15
r_u	0.92	0.90	0.95
m (m^2)	10^{-5}	2×10^{-3}	10^{-4}
Epaisseur initiale (m)	10	6	100
Volume (10^6 m^3)	0.75	0.01	1000
Vitesse maximum (m/s)	23	21	66
Distance maximum (km)	2.0	1.3	80
Epaisseur finale (m)	7	2	27-57

SOK: Sokkelvik, STOR: Storglomvatnet, GB: Grands Bancs

4.1 Le glissement de Sokkelvik (Norvège)

Le glissement de Sokkelvik s'est produit en 1959 dans des argiles sensibles en bordure d'un fjord (Braend 1961); il a entraîné la destruction de quatre maisons et la perte de neuf vies humaines. Le volume total déplacé est de 0.75 million de mètres cubes de sédiments et l'épaisseur maximale du glissement a été estimée entre 15 et 20 m. Les dépôts créés par le glissement ont été cartographiés à l'aide de méthodes géophysiques. La figure 4 présente le profil avant et après la rupture.

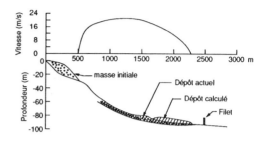

Figure 4. Le glissement de Sokkelvik, localisation des débris et calcul numérique de leur position, profil de vitesse de l'élément frontal.

La viscosité utilisée est de 10^{-5}m^2 (pour "m", ou η = 0.1 Pa.s). Pour ajuster le profil de déposition calculé à l'aide du modèle il a fallu utiliser un angle de friction apparent de 2.52° ce qui correspond à une valeur de r_u de 0.92 si on considère ϕ égal à 30°.

La figure 4 présente aussi le profil de vitesse de l'élément frontal en fonction de la distance. La vitesse maximale atteinte, un peu élevée, aurait été de 21.5 m/s. Toutefois, le modèle ne tient pas compte de la friction dans la partie supérieure de la masse en mouvement. Norem et al. (1990) ont indiqué que cette friction additionnelle pouvait augmenter la résistance au mouvement de presque 25%. Les calculs indiquent qu'après 60 secondes, l'élément frontal s'était déplacé de plus de 1567 m avant que le dernier élément ne commence à bouger. L'épaisseur maximale du glissement dans la zone de départ était de 18 m, mais elle est rapidement devenu comprise entre 3 et 6 m durant le mouvement avec une empilement terminal d'environ 7 m. La position des débris, calculée d'après le modèle, est légèrement en aval de la position réelle. Ceci peut aussi être dû, tel que mentionné plus haut, à la vitesse plus élevée permise par le modèle.

4.2 L'essai de Storglomvatnet (Norvège)

Le site de Sorglomvatnet (lac) est localisé au nord de la Norvège. Il a été utilisé comme site d'essai pour les techniques de liquéfaction par dynamitage

(Norem 1990). Pour cette expérience au lac Stroglomvatnet, environ 10 000 m³ de sédiments ont été liquéfiés sous l'eau. Les données préliminaires indiquent qu'à moins de 200 mètres de la zone de départ, la masse de

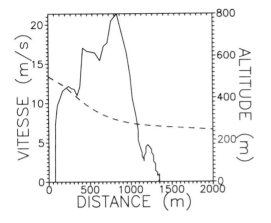

Figure 5. Résultats de calcul pour l'essai de Storglomvatnet (la ligne en tireté indique le profil du plan d'écoulement).

sédiments avait atteint une vitesse de 7 m/s et que la distance minimale parcourue était de 800 mètres. La dépression laissée dans la zone de départ a été estimée à environ 6 mètres, valeur utilisée dans les calculs. Le sédiment mobilisé est un sable silteux quartzitique, ce qui justifie un coefficient de friction légèrement plus élevé pour ce sédiment. A cause des conditions rencontrées lors de la réalisation de l'essai (Norem et al. 1990), les charges d'explosifs ont du être placées à la surface du sédiment ce qui a réduit l'efficacité du procédé de liquéfaction ce qui peut justifier un rapport de pression intertitielle plus faible pour cette expérience que pour les autres cas rapportés.

Les résultats du calcul numérique ont été réalisés avec les paramètres présentés au tableau 1 et sont présentés à la figure 5. L'ajustement des calculs aux observations révèle une oscillation des vitesses de déplacement assez typique des mouvements de pulsation de coulées de débris subaériennes. La coulée a une morphologie caractéristique avec un front bien prononcée et une queue étalée lors la phase terminale du mouvement. La vitesse maximale calculée est de 22 m/s pour une distance maximale de parcours de 1.3 km,

ce qui est compatible avec les observations préliminaires. Durant le mouvement, l'épaisseur des couches a variée de 6 m au départ jusqu'à des valeurs aussi faibles que 0.1 m (dans la queue).

4.3 Le glissement des Grands Bancs

Ce glissement est sûrement le plus connu (Heezen et Ewing 1952). Plusieurs chercheurs (Matos 1989, Kirwan et al. 1985) ont proposé des modèles qui permettaient d'interpréter les observations sur les temps de rupture des câbles sous-marins. Pour le modèle présenté ici, nous verrons que ces données sur les bris de câbles sous-marins sont à peu près inutiles si ce n'est de fournir une vitesse minimale pour la coulée de débris (dense).

Les données utilisées pour la simulation sont présentées au tableau 1 et les résultats sont illustrés à la figure 6. Etant donné l'importance du séisme qui a provoqué le glissement et de la quantité de sédiments impliqués, le rapport de pression interstitielle a été placé à 0.95 et la poids spécifique, légèrement plus faible, à 15 kN/m³. La vitesse maximale calculée est de 66 m/s et la distance de parcours de 80 km.

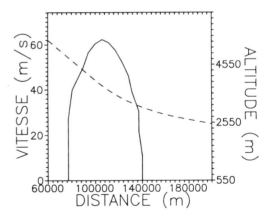

Figure 6. Résultats de calcul pour le glissement des Grands Bancs (la ligne en tireté indique le profil d'écoulement).

L'épaisseur de sédiment impliqué au départ est de 100 mètres. Il aurait été possible de faire progresser le sédiment jusqu'à des distances atteignant 800 km mais avec des vitesses irréalistes! En fait, les calculs indiquent clairement qu'avec des paramètres raisonnables, la masse de sédiments formant la coulée de débris se maintient à l'intérieur de la zone identifiée par Heezen et Ewing (1952) comme correspondant à la zone de glissement. Les résultats de calcul appuient donc fortement l'hypothèse que les câbles sous-marins ont été brisés par un courant de turbidité et non par une coulée de débris.

5. DISCUSSION

L'application du modèle introduit dans cet article à l'étude des glissements sous-marins permet de discuter tant du choix des paramètres que de la physique du mouvement.

Le modèle est encorte imparfait sous plusieurs aspects et sa plus grande faiblesse demeure la qualité et la quantité de données nécessaires pour le valider. Comme on peut le constater, pour arriver à décrire la physique du mouvement, il faut beaucoup d'informations tant sur la géométrie que sur les propriétés des sédiments de même que les conditions antérieures et postérieures au mouvement. La planification d'une expérience grandeur nature devient nécessaire (Locat 1987).

L'analyse du phénomène physique est, par contre, déjà possible. Pour le cas de Sokkelvik, le modèle fournit une bonne approximation tant de la position des débris que de leur épaisseur finale. Les vitesses calculées sont probablement trop élevées mais les distances de parcours sont acceptables. Le cas le plus nébuleux demeure celui des Grands-Bancs. En effet, on retrouve quelques contradictions dans la littérature récente portant sur cet événement. Kirwan et al. (1985), considérant l'événement comme un courant de turbidité, tentent de démontrer que l'origine du glissement serait environ 75 km en aval de la position proposée par Heezen et Ewing (1952). Si on utilise la position du câble correspondant au bris à 59 minutes, l'origine du mouvement selon Kirwan et al. (1985) serait environ 80 kilomètres en amont de ce point au lieu de 155 km tel que proposé initialement. Il est intéressant de noter que nos calculs suggèrent que la fin de la coulée se situerait environ 95 km an aval de la position du câble. Ceci permet de concilier les deux approches: une coulée

de débris qui se termine au pied du talus continental et qui amorce un courant de turbidité. Quant aux vitesses que nous avons calculées, elles sont quand même assez proches de celles estimées par Kuenen (1952) pour la portion du talus continental soit de 52 m/s, quoique qu'encore trop élevées (66 m/s).

6. CONCLUSIONS

L'application d'un modèle visco-plastique à quelques cas connus de glissements sous-marins a rendu possible la simulation raisonnable du mouvement de même que de la forme de la masse en mouvement. Bien que les valeurs absolues obtenues des différents calculs ne soient pas entièrement satisfaisantes les résultats permettent de mieux discuter et évaluer la physique du problème en distinguant nettement entre la coulée de débris (dense flow) du courant de turbidité (suspension flow).

7. REMERCIEMENTS

Les auteurs remercie l'OTAN pour le soutien financier à la collaboration entre la Canada et la Norvège. Nous remercions aussi le Conseil de Recherches en Sciences Naturelles et en Génie du Canada de même que le Conseil de recherche scientifique et industrielle de la Norvège pour les subventions accordées au projet ADFEX (Arctic Delta Failure Experiment). Nous remercions également nos collègues A. Hay (Memorial University) et J. Syvitski (Centre Géoscientifique de l'Atlantique, Canada) pour leurs commentaires durant la réalisation de cette étape du projet ADFEX. Nous remercions messieurs P. Therrien pour son aide à la modélisation et P. Gélinas pour la revision du manuscript.

8. BIBLIOGRAPHIE

Bagnold, R.A., 1954. Experiments on a gravity-free dispersion of large solid spheres in a Newtonian fluid under shear. Comptes Rendus de la Royal Society of London, Ser. A225:49-63.

Bagnold, R.A., 1956. The flow of cohesionless grains in fluids. Phil. Trans. Royal Soc. London, 249:235-297.

Braend, T., 1962. Skredkatastrofen i Sokkelvik i Nord-Reisa, 7 mai 1959. NGI Publi. No. 40.

Cruden, D.M., & Hungr, O., 1986. The debris of the Frank Slide and theories of rockslide-avalanche mobility. Jour. Can. Scie. Terre, 23:425-432.

Genesseaux, M., Mauffret, A., & Pautot, G., 1980. Les glissements sous-marins de la pente continentale niçoise et la rupture de câbles en mer Ligure (Méditerrannée occidentale). C.R. Acad. Sc. Paris, t. 290, série D:959-962.

Hampton, M.A., 1972. The role of subaqueous debris flow in generating turbidity currents. Jour. Sed. Petrol., 42:775-993.

Heezen, B.C., & Ewing, M., 1952. Turbidity currents and submarine slumps, and the 1929 Grand Banks earthquake. Am. Jour. Sci., 250:849-873.

Hungr, O., & Morgenstern, N.R., 1984. High velocity ring shear tests on sand. Géotechnique, 34:415-421.

Hutchinson, J.N., 1986. A sliding-consolidation model for flow slides. Can. Geotech. Jour., 23:115-126.

Hutchinson, J.N., & Bhandary, R.K., 1971. Undrained loading: a fundamental mechanism of mudflows and other mass movements. Géotechnique, 21:353-356.

Irgens, F., 1988. A continuum model of granular media and simulation of snow avalanche flow in run-out zones. Comptes Rendus de la XVII ICTAM, Grenoble, 1988.

Iverson, R.M., & LaHusen, R.G., 1989. Dynamic pore-pressure fluctuation in rapidly shearing granular materials. Science, 246:796-799.

Johnson, A.M., 1970. Physical process in geology. San Francisco Freeman et Cooper, Ed., 577 p.

Karlsrud, K., & Edgers, L., 1982. Some aspects of submarine slope stability. In: Marine slides and other mass movements, Saxov & Niewenhuis, éd., Plenum Press, pp.: 61-81.

Kirwan Jr, A.D., Doyle, L.J., Bowles, W.D., & Brooks, G.R., 1985. Time-dependent hydrodynamic models of turbidity currents analyzed with data from the Grand Banks and Orleansville events. Jour. Sed. Petrol., 56:379-386.

Kummeneje, O., & Eide, O., 1961. Investigation of loose sand deposits by blasting. Comptes Rendus de la 5e Conf. Int., Méc. Sol et Foundation, Paris, vol 1.

Locat, J., & Demers D., 1988. Viscosity, yield stress, remolded shear strength, and liquidity index relationships for sensitive clays. Can. Geotech. Jour., 25:799-806.

Locat, J., 1987. ADFEX: Arctic Delta Failure Experiment. Internal Report, Department of Geology, Université Laval, ADFEX Report 87-01, 26p.

Matos, M.-M., 1988. Mobility of soil and rock avalanches. Thèse de doctorat, Uni. Alberta, Canada, 360p.

Norem, H., 1990. Full scale experiment to study submarine slides. Preliminary field report from the pilot study in Storglomvatnet, Norway, Sept. 5, 1989. Memorendum, Norges Geotekniske Institutt, 3p.

Norem, H., Irgens, F., & Schieldrop, B., 1989. Simulation of snow-avalanche flow in run-out zones. Annals of Glaciology, 13:218-225.

Norem, H., Locat, J., & Schieldrop, B., 1990. An approach to the physics and the modelling of submarine slides. NGI Report 522090-2, 28 p., submitted to Marine Geotech.

O'Brien, J.S., & Julien, P.-Y., 1986. Rheology of non-newtonian fine sediment mixtures. Comptes Rendus, ASCE Conf., Minnesota, p:989-996.

Piper, J.W., Shor, A.N., Farre, J.A., O'Connel, S., & Jacobi, R., 1985. Sediment slides and turbidity currents on the Laurentian Fan: sidescan sonar investigations near the epicenter of the 1929 Grand Banks earthquake. Geology, 13:538-541.

Prior, D.B., Bornhold, B.D., Coleman, J.M., & Bryant, W.R., 1982. Morphology of a submarine slide, Kitimat Arm, British Columbia. Geology, 10(11):588-592.

Saxov, S. & Nieuwenhuis, J.K,, 1982. Marine slides and other mass movements. Plenum Press, 353p.

Suhayda, J.N., & Prior, D.B., 1978. Explanation of submarine landslide morphology by stability analysis and rheological models. Comptes Rendus de la 10th Offshore Tech. Conf., Houston, Texas, vol. 2, 346-360.

Wildemut, & Williams, 1985. A new interpretation of viscosity and yield stress in dense slurries. Rheol. Acta, 24:75-91.

6.3 Coastal protection and land reclamation
    Protection littorale et accroissement territoriale

The causes of morphological changes at the water edges of Orava-Reservoir in Slovakia

Les causes des variations morphologiques des rives du réservoir d'Orava en Slovaquie

Otto Horský
Geotest Brno, Czechoslovakia

ABSTRACT: The article deals with the evaluation of morphological shore changes of water reservoirs on the example of the Orava Dam. Considering the 30-year research of shore deformations, it was possible not only to judged objectively causes of these changes but also to suggest and to apply the methods of prognosis of shore deformations, that were proved by long repeated measurements

RESUME: Dans cet article le réservoir d'Orava est pris pour modèle d'évaluation des variations morphologiques des rives de réservoirs arti- ficiels. Suivant la prospection - durant plus que 30 années - des défor- mations de rives il était possible non seulement d'évaluer avec objecti- vité maximale les causes de ces variations mais en même temps de propo- ser et appliquer une méthode prognostique de déformations des rives, vé- rifiées par des mesures répétées.

Problems of rational utilization of artificial water reservoirs in Czechoslovakia connected with secu- ring their functions without any risk and in accordance with inte- rests of the conservation are beco- ming more and more important. Re- cently it's been shown very often that not only the reliably built dam but even a suitably situated and responsibly judged immersed area of the reservoir are the main pre-conditions for a safe and ra- tional operation.

Shoreline changes caused by swel- ling after filling the reservoir have been at the point of view in Czechoslovakia as early as from 1950s (Záruba, 1954). The rate of this attention was considerably in- fluenced by basic factors which had participated differently in shore- -reshaping of the reservoir (un- suitable geological and geomorpho- logical conditions, wave breaking, frequent and sudden decreases of water level at water systems, chan- ges of hydrometeorological condi- tions etc.). Orava-reservoir (fi- nished in 1953) is an example where

detailed studies of shore changes were necessary because of crashing stage of some parts of shore.

The Deformations of Shores at Orava Reservoir

With reference to immersed area and effective storage, Orava dam is one of the largest reservoirs in Cze- choslovakia. At its maximum capaci- ty the water covers an area of 35.08 km^2, the volume being 346 million m^3 of water and the shore- line measures over 90 km.

The reservoir has been in opera- tion for 37 years. The construc- tion of this dam had a favourable effect on industrial and agricultu- ral development of the whole region. Unfourable effects started to be visible on the shores of the reser- voir. The effects of water swelling (max. level is El 603 m over the sea) operational fluctuation of wa- ter level (a year operational cyc- le with winter decrease of as much as 15 m), wave breaking (max. height of wave is 3 m), as well as other factors (e.g. severe climat effects, frost -30°C) have the consequence that at some spots the banks receded by more than 45 m. The devastation of land increased as a result of landslides along the slope. Erosional cliffs up to 20 m high have been formed. The recession of bank caused danger to recreation units and agriculturally valuable land. In some places the recreation was limited.

The immersed area of the Orava Dam is situated in so called Orava basin mostly filled up with neogenic clayey sediments lying discordantly on paleogenic flysh rocks. Neogenic sediments are covered by quaternarial gravel and sandy fluvial accumulation and by slope loams and residual soils on slopes. Growing abrasion and landslides of shores hit mostly prequaternarial rocks.

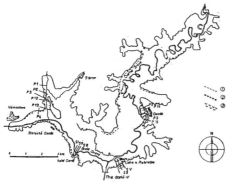

Fig. 1 Immersed line of the Orava Dam. The biggest abrasion and landslides occur in marked lokalities. 1 - Places damaged by abrasion, 2 - Places extremely damaged by abrasion, 3 - Places damaged by landslides

Paleogenic rocks are represented mostly by sandstones and consolidated claystones. They are hard to wash-out, claystones have small slaking and sandstones have no slaking, they resist to abrasion very well. It is presented in mantle slope loams and in little thick debrises mostly. But the recession of the bank did not reach over 5-7 m and there were no landslides. This type of rocks represents about 10 per cent of the shores in the surrounding of the reservoir.

The neogenic rocks are represented by clays, clayey sands, little consolidated claystones and sandstones with brown coal inserts. These rocks are slaked, washed-out and even simply washed out and not much resisting to abrasion. Sediments are mostly dipping into the reservoir (angle is 10-30°) what makes very favourable conditions for the creation of landslides. The lengthwise fault tectonics in neogenic basin edges form overdisponating areas for the creation of landslide scars. These depositing situation of neogenic rocks in combination with the intensive abrasion of shores as a result of wave breaking with the continuous washing-out and with transport of landslide material are important geodynamic processes which take a great part in the reshaping of shores. The deformation of shores is affected significiantly by geological situation in the edges of the neonic basin in a touch with paleogenic rocks. Neogenic clays and claystons form less pervious roof for water-bearing paleogenic rocks lying on.

The underground water acquires a pressure character in favourable depositing situation, relieves neogenic sediments and in this way actively affects during the creation of landslides. This factor combined with the change of hydrodynamic conditions after a sudden decrease of level and after the partial abrasion of the shores was the main cause of landslide under the Goral Hotel. (O. Horský, 1975).

After a drainage by horizontal boreholes the landslide stopped. Consecutive maintenance of the shore with stone piling with loading bench at the boot of the slope meant the end of the shore-reshaping process.

The cause of the pressure character of the underground water in neogenic sediments is even the limited circulation of the underground water as a result of the different perviousness of rocks. When beds are sloping downwards to the reservoir, water flows through sand-positions and positions of lignit beds. These positions are not continuous, therefore the impregnated positions sloping downwards to the reservoir are created. Especially during the sud-

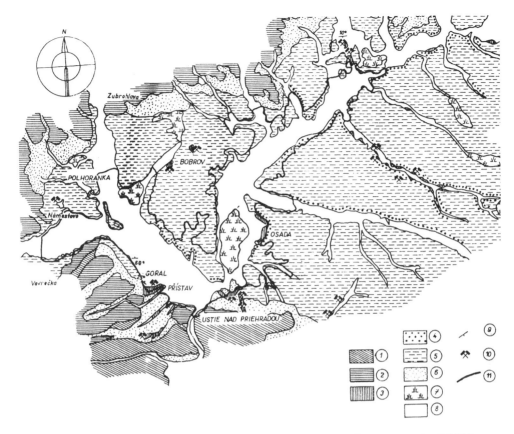

Fig. 2 The geological map of the Orava reservoir (O. Horský, 1969)

1 - claystones in flysch development (eocene), 2 - sandstones in flysch development (eocene), 3 - clays, sandy clays, weakly consolidated sandstones (neogene), 4 - gravels, 5 - loam with gravel, 6 - stony - loam debris, 7 - peat, 8 - recent sediment, 9 - the direction of the bed inclination, 10 - outcrops of coal beds in Neogene, 11 - the edge of backwater line

den decrease of level in the reservoir, water isn't able to flow out quickly and by the created overpressure protecting beds can snap. In winter during recurrently low level a protecting bed usually freezes over. During a sudden thaw under the accumulated pressure of the underground water its snapping and flowing out occur. Having suitable morphological conditions combined with the abrasion, the stability of a slope is broken and a landslide is created.

In outcrops of sandy positions

in formed erosional cliffs we can explore outflows of underground water. Freezing of the wall in winter causes a creation of a relatively nonpervious wall, which causes a hydrodinamic overpressure. During sudden thaw the wall snaps through and its partial collapse of dropping off apears (Osada).

The streaming of underground water through more pervious positions cause the large underground suffosion of sediments. Above erosional cliffs in Ustie nad Priehradou in the distance 10-30 m from the wall

in regular lines we can watch a
great deal of about 2-metre-deep
pseudo-sink holes, which during
rain are full of water. Water then
seeps through the boot of the cliff
and formed covern, even in the area
of flowing out, prove the intensive
underground suffosion. The process
of suffosion is more intensive especially during spring thaw, when
the level is incised by more than
10 m and slopes are impregnated by
water.

The main climatogenic factor is
a regime of precipitation, a régime of temperature and wind. Relatively low temperatures in winter
reaching 30 degrees below zero,
lead to deep freezing of slopes.
Water in voids and cracks enlarges
its volume what caused an intensive
softening and cracking of the edges
of erosional cliffs and exposed continuous slopes above the water area.
It leads to a negative change of
physico-mechanical properties of
clayey rocks and to an increase of
washing-out stage. Even running
down icy floes are important to modeling the shores in such softened
soils.

On contrary in summer when the
sunshine is intensive, shores de-
ssicate and contractional cracks
and to abraded shores,

Fig. 4 The extensive erosion of
exposed slopes and accumulations
under landslides and erosional
cliffs

Quick thaw in spring and intensive rain especially in summer
(when the precipitation is at maximum) affect rather unfourably, because they cause a large erosion of
exposed slopes and an accumulation
under the landslides and under erosional cliffs, which caused in some
cases the creation of large landslides.

The Prgonosis of Further Development of Shore Changes

Studying the size of shore deformation it was very important to de-
liminate the extent of height where
shore recession and washing-out
occur. It was proved experimentally
that the level with 2-8 per cent
frequency of the occurance according to the type of rock occurs
disturbingly in the Orava Dam. In
the extent of fluctuation of the
level, the lower edge of shore
changes was taken on El 594.30 m
and the upper edge on El 602.80 m.
The recession of the shores itself
means the section, the width of
which is determined at the bottom
by the line of the intersection of
the upper limit of shore changes
and the original terrain. At the
top it is determined by the surface
edge of the shore cliff or landslide scar.

From the survey from table L it

Fig. 3 An excesive lanslide of
neogenic sediments spreas gradually as deep as 170 m into the
slope

is evident that the shore recession above the elevation El 602.80 m ° reached in first fifteen years of the dam existence considerable values.

In the table 2 is the extent of shore reshaping, which considers even changes occuring under the level of the reservoir (Ust+Bz), which is considerably larger. The recession of shores profiled in 1983- after a 30-year operation of the water reservoir is evident from the table 3. To compare it better the recession of shores from 1967 is shown.

It is evident from table 3, that the shore recession of erosional type of shores in the Polhoranka locality reached 15-22.4 m till 1983 and the shore recession of a slidy type in the Ustie and Osada localities reached 24-56 m. As far as the spreading of landslides in Ustie upwards to the slope is concerned, the recession of shores was 153-177 till 1967 and till 1983 the parting edge removed by further 12 m.

Locality	The shore recession Ust/m	The shore recesion $m^3 . m^{-1}$
Osada	17,0 až 47,5	49,0 až 156,0
Polhoranka	10,4 až 22,0	27,0 až 106,0
Námestovo	7,4	40,0
Goral	28,6 až 45,8	160,0
Ústie n. Priehradou	153,3 až 166,0	

Table 1. The shore recession above the elevation El 602,8 m

Locality	The shore recession till 1967 Ust+Bz
Osada	45,0 až 68,5
Polhoranka	19,4 až 38,8
Námestovo	30,8
Goral	50,5 až 100,3
Ústie n. Priehradou	187,0 až 259,6

Table 2. The extend of shore reshaping including changes under the level of reservoir

Till 1983 the total shore-reshaping (Ust+Bz) in the Polhoranka locality reached values of 22.8-50.0 m and values of 56-87 m in the Osada locality.

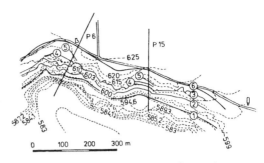

Fig. 5 The scheme showing the recession of the boot of a shore cliff in the Osada lokality in 1953-1968
1 - line of the lower limit of shoreline changes (El 594.30 m)
2 - line of the boot of erosional cliff in 1955, 3 - line of the boot of erosional cliff in 1962, 4 - in 1968, 5 - the upper limit of erosional cliff in 1968, 6 - - the prognosis of shore recession to final stage of reshaping.

In selecting the appropriate method for the prognosis of shoreline change, we considered the local geological and hydrogeological conditions. With a view to make the proper prognosis of shoreline recession it is important to determine both the lower and the upper edges of shoreline changes. This, however, is a function of water level fluctuation and the frequency of its occurance in direct relation to the local geological conditions. This disturbing phenomenon of water level takes places at the lower part of the reservoir at Orava Dam with the following frequency:
2% - in neogenic clay sediments El 595.50 m
3% - in sandy gravels El 597.75 m
5-8% - in stony-clayey paleogenic debris El 601.75 m

The water level at the upper part of the reservoir is found to occur in a disturbed state, with the frequency 2% which affects the loams in the slope, clayey alluvial loams, loess-loams and sandy loams or sands (figures 8).

The most frequent level in the first fifteen-year period of opera-

Profile	The shore recession Ust above the elevation El 602,8 m	Ust in	Prognosis of the assumed final recesion of shores above the elevation El 602,8 m	The shore changes under the elevation El 602,8 m	The total shore-reshaping till 1983	Assumption of final shore reshaping
	1967	1983	(Sa)	1983	Ust + Bz	Sa + Bz
Polhoranka						
P 1	15,2	15,2	15,2	7,6	22,8	22,8
P 2	22,4	22,4	23,0 (39,0)	3,5	25,9	26,5 (42,5)
P 3	10,6	18,0	20,7	10,4	28,4	31,1
P12	11,4	21,0	27,0	29,0	50,0	56,0
Osada						
P 5	21,2	25,0	38,0	62,0	87,0	100,0
P 6	47,5	56,0	76,0	29,0	85,0	105,0
P14	20,5	24,0	38,0	32,0	56,0	70,0
P15	17,0	34,0	48,0	38,0	72,0	86,0
Ústie n. Priehradou						
Cape A	32,0	52,0	77,0	82	134	159
Cape B	24,0	46,0	83,0	34	80	117

Table 3. The shore recession in profiles from 1983
The number in brackets means the change of the prognosis of shore recession for large erossion by accumulation

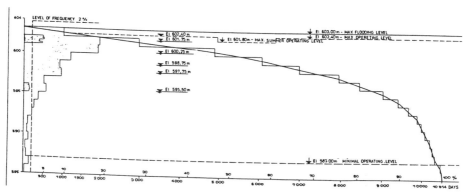

Fig. 6 The frequency curve of occurance of water stages and cumulative line of their getting over from 1954 to 1982 (the interval is 100 cm, in critical levels 25 cm)

tion is very important the El 601. 75 m level (8% of the occurances). In later years more equable and continuous distribution levels occured, so its share reduced during 30 years to 5%. This level affects the intensive abrasion of all the sediments except paleogenic rocks. In the range of its effect (600.55--602.15), strand plains are created, sliding towards the slope during the shore development.

It is necessary to add 1/2 of the wave height to the upper limit of the disturbing effect or the reservoir and 3/2 of the wave height to the lower limit in order to obtain the upper and lower limits of water edge changes.

The wave height 2h must be selected in each profile with reference to the depth of water in the reservoir, to the length of a wave and the speed of wind (fig. 7).

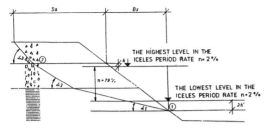

Fig. 7 The prognosis of shoreline recession at the Orava Dam for the erosional type of shores: 2h - the wave height for the minimal used level, α_1, α_2 - angles corresponding with the steady shore profile reaching the abrasion, α_3 - the stable inclination of an erosional cliff
Sa - the assumed shoreline recession above the maximum used level + h/m, Bz - the extend of shore changes under the max. used level + h/m

At the construction of shore shapes we started from the lower edge of the shore change in inclinations corresponding to the steady inclinations of the beach in pertinent soils, and from the part of the shore above the upper edge of the shore change corresponding to the steady slope depending on local geological conditions. The recession of water edge (Sa) is the horizontal distance of the point of intersection between the upper edge od the shore change and the original terrain and the edge of the shore cliff or the landslide scar.
In the graph showing the development of the shoreline recession these conclusions can be drawn:

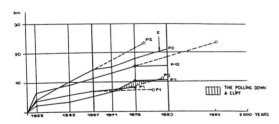

Fig. 8 The graph of recession of shores at the upper edge of shore changes at abrased type of shore

- - - according to the prognosis of development, ── reality ● finished abrasion

The effect of biotechnical measures occurs favourably at P1, P3 and P12 profiles. This effect meant the end of shore abrasion in 1976. From 1976 to 1983 there was no more shoreline recession. The solid line from 1976 to 1983 is moreless horizontal.
The development of P2 profile is disastrous. Here owing to an erosion of the Polhoranka river during lower levels of the reservoir, the falling down of the cliff after the sapping occured. That is why the prognosis line was belated (i.e. the course is lower than assumed).
Considering that measures against the erosion were done, the development of the abraded-erosional shore is in fact in the beginning again and the prognosis is not in force.
As far as the Osada locality is concerned, there is abrasion sliding type of shores in neogenic sediments. Profiles P5 and P14 are characteristic for the southern part of the locality. There had been shores more gradient before the reservoir was filled (15°-30°). The original channel of the river Čierna Orava was about 180-200 m far from present erosional cliffs. While in the southern part of the locality in the profiled area P6 and P15, the river aproached 60 m far from present erosional cliffs and before filling the reservoir as a result of erosion the river underwashed the slopes. Therefore before filling up, slopes were more sheer (30-50°) and locally even landslides occured.
Starting different morphological conditions qualify more gentle development of abrasion at the P5 and P14 profiles. The end of shore-reshaping at the upper edge of shore change we can expect in about 2000 (i.e. in 50 years the dam has existed.
The main developing stage of shore abrasion is finished in this area. In following years the plastic flow of shores will occur as well as the shore recession at the upper edge of shore changes by 2-3 m to the final phase of deformation.

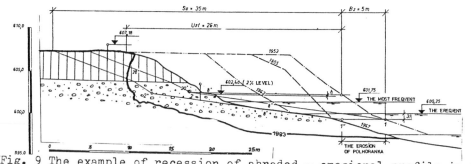

Fig. 9 The example of recession of abraded - erosional profile in the Polhoranka locality in gravels and loess-loams, evaluated in 1967. The thick line means the situation in 1983.
1 - the lowel edge of shoreline changes for the erosional type of shore, 2 - the upper edge of shoreline changes, 1´ - the lower edge of shoreline changes shifted by erosion, 2´ - the upper edge of shoreline changes shifted by erosion

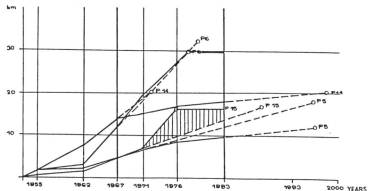

Fig. 10 The graph of shore recession at the upper edge of shore changes in Osada locality the abrasion - slide type of shore

As a result of recured landslide of erosional cliff at P6 and P15 profiles, consecutive landslide of the whole frontal shoreline occured. Essentially it meant the creation of 20 m high erosional cliff and the acceleration of abraded process at the upper edge of shore changes where this process is in fact in the end. (Wave breaking affects in slipped, simply washed-out soils. Slipped materials are gradually overgrown by vegetation and consecutive plastic flow of profile occurs at the upper edge of shore change. The plastic flow of 20 m high sheer erosional cliffs (landslide scar) will be long-termed and it will depend on physico-mechanical state of soils connecting the natural conditions around (fig. 15 a/b).

CONCLUSION

Relatively large research being realized in 1967/68 facilitated to limit endangered stages of shores at Orava Dam in details and with reference to prognosis of development of shore deformations to determine the extent of necessary maintenanced measures. The research from 1983 facilitated to verify the accuracy of prognosis from 1968 and to valuate the efficiency of realized measures.

The development of shore changes at Orava Dam shows that the process has lasted for 30-50 years since filling up the reservoir as far as washed-out and simply washed-out soils are concerned. The plastic flow of created sheer cliffs is long-termed and after the end of

Fig. 11 An example of the development of the abraded-erosional slide type in the Osada locality

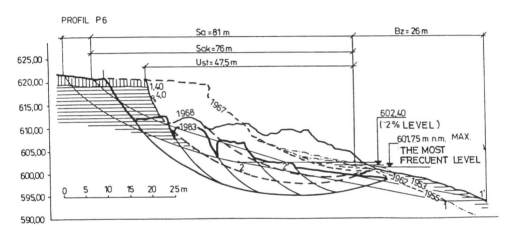

Fig. 12 An example of the development of the abraded-erosional slide type of shore at P6 profile in the Osada locality from 1968 and
the prognosis of its further development
1,2 - the lower edge of shore changes for abraded type of shore
1´, 2´- shifted upper and lower edges of shore changes for abraded-
        erosional type of shore
Sak - the prognosis of shore recession for abraded-erosional type of
      shores (thick line means the state from 1983)

the abrasion it doesn´t depend on
the influence of the reservoir (except possible deformations during
sudden inadmissible decrease or increase of water level in a reservoir, which can endanger the stability of a slope).
    With reference to erosional type
of shore the intensity of abrasion
is maximum in the first two or

three years, later it weakens, the relation has a straight line to parabolic character.

For erosionally slide type the intensity is considerable especially in the first 2-3 years. Later the effect is damped by formed high erosional cliffs. After about 8-10 years of the operation of the reservoir it leads to a greater extent to the disturbance of the stability of formed high erosional cliffs.

Slide soils don't resist to wave breaking, therefore they are quickly abraded what means in fact the quicker progress of abrasion towards the slope, repeated breaking of stability and repeated intensive abrasion.

This geodinamic process means higher speed of shore deformation and the quicker creation of the equilibration (after about 25-30 years at P6 and P15 profiles, and after 40-50 years at P5 and P14 with less sheer slopes).

The analysis of factors and conditions of shore deformation point out the dominating role of regional geological factors during the shore reshaping. The leading part represents an abrasion. But despite at any case it is not enough to valuate the erosional activity in order the total shore-reshaping has been judged and the prognosis of the shoreline recession has been determined.

The shore destruction in some localities affects very depressivly and continuous movement of materials and acute danger seem to exist there. It is necessary to point out that without thorough analysis of affecting factors and without studying the development in space and time we could make mistaken conclusions.

Achieved research is important both as far as the technic is concerned and even for its economic contribution as well (there is considerable saving of necessary means. It is clear that complex solving of these problems is necessary already in time of the research. Only by this way we will be able to procede great economic damage.

REFERENCES

Horský, O. 1969. The engineering--geological research for the suggestion of the sanation of shores of Orava Dam (in Czech), Roč. geol.s. (36-40), Geotest Brno.

Horský, O. 1971. The utilization of the aerial photogrammetry for searching of shoreline changes in immersed areas of dams (in Czech), Geol.průzk. 10/1971 300-302, Praha.

Horský, O. - Müller, K. Landslides on shores of Orava Dam. 1972, Sbor. HIG 10/1972, 59-71, Praha.

Horský, O. 1972.The influence of Orava Dam to the process of shore changes, Sb.PD 1972, 121-130, B. Bystrica.

Horský, O. - Woznica, L. 1973. Problems of prognosis of shore changes in régime of fluctuation of level in valley reservoirs PVE, Sb. PD 1973, 140-147, Ostrava.

Horský, O. 1974, The sanation of landslides at Orava Dam, Geol. průzk. 2/1975, 43-45.

Horský, O. - Muller, K. 1977. Investigation and landslide on the Orava reservoir shores, Sb. Geotest and landslides, 35-50,Brno.

Horský, O. 1974. The shore changes in immersed area of Orava Dam, Výzk.úk. 72-19-0744, Geotest Brno.

Horský, O. 1984. 30 years of Orava water reservoir and its effect to the shore deformation, Ing. st. 10/1984, 489-498, Praha.

Horský, O., Simeonova, G., Spanilá, T., The effect of exogenic processes to the shore reshaping in water reservoirs, Geol.průzk. 6/1984 163-166, Praha.

Linhart, J. 1963. Les réservoirs de barrages dans la Republique socialiste tchécoslovaque et l'etude géomorphologique de leurs rivages. Revue de Geomorphologie dynamique, Paris.

Woznica, L. 1967. The shore-reshaping of immersed areas of dams. The manuscript in Geotest Brno.

Engineering geological problems of the coast in Poland

Problèmes géologiques de l'ingénieur de la côte en Pologne

W. Subotowicz
Technical University of Gdańsk, Poland

ABSTRACT:The coast in Poland is about 500 km long. The two types of the coast have been distinguished. Those are: the cliff coast and dune coast. The geological structure of the first type is principally connected with glacial and fluvioglacial Pleistocene sedimentation, while the structure of the second type is the consequence of the Holocene sand accumulation. The maximum abrasion rate is 1,5 m/year. The abrasion coast in Poland to be sure is prevailing but the sections of relative stabilization and accumulation occur, too. The process of the coast reconstructioncan last for some generations and is repeatable. That character of the coast evolution takes place in conditions of interference of man into the geological environment. The existing coast protection systems about 126 km long, participate in the process, especially, that their history lasts already for centuries. This participation both restrains the rate of the degradation of protected coast and - unfortunately -enlarges the abrasion effects on neighbouring coast sections. The coast protection systems became the integral factor modelling the geological environment of the coast.

RESUME: La côte en Pologne est 500 km long. On a distingué deux types de la côte. C'est sont: le type de la falaise et le type de la dune. La structure géologique de premier type est unie pour la plupart avec la sedimentation glaciale et fluvioglaciale de Pleistocene et de second type - est la consequence de l'aacumulation Holocene de sable. La maximale rapidite de l'abrasion de la côte est 1,5 m/an. Abrasion de la côte prevalue, mais les sections de la stabilisation relative ou de l'accumulation peuvent apparaître aussi. Le procès de la côte peût durer par quelque generations et il est répétable. L'évolution de la côte parcours dans les conditions de l'ingerence de l'homme dans l'environment géologique. Les systèmes de la protection de la côte sur la distance de 126 km participent dans le procès, especielment, que leur histoire dure par les siècles. Cette participation d'un côté arrête la rapidité de la degradation de la côte, mais malheureusement agrandit l'abrasion des sections voisines. Les systèmes de protection sont devenus un facteurs integral modellant l'environment géologique de la côte.

The coast in Poland is about 500 km long. The two types of the coast have been distinguished. Those are: the cliff coast about altogether 100 km long and dune coast 400 km long (fig.1).

The cliff coast is structured of Pleistocene formations, namely: moraine clays and intermoraine loams, sands and gravels. The dune coast is mainly built of Holocene sand formations. The height of the cliffs and seashore dunes oscillates from several to tens meters. Actually their abrasion is observed. The maximum abrasion rate is 1,5 m/year (fig.1).

The abrasion causes the landslide phenomena especially at the cliffs. The two geodynamical types of the development of o-vercoasts have been distinguished , namely: the fallen ground type, the fall down type and the landslide type (Subotowicz 1981, 1982). Independently on hydrodynamioal factors, which are an immidiate cause of the geodynamical development of the overcoasts, land factors - especially in the case of cliffs - play a large role. It regards geological predispositions and water conditions of the underground and unuseful man interference during exploitation of the overcoasts. The landslides reach tens meters into back cliff, and in the case of the intensive storms - even 100 and more meters. The probability of such storms

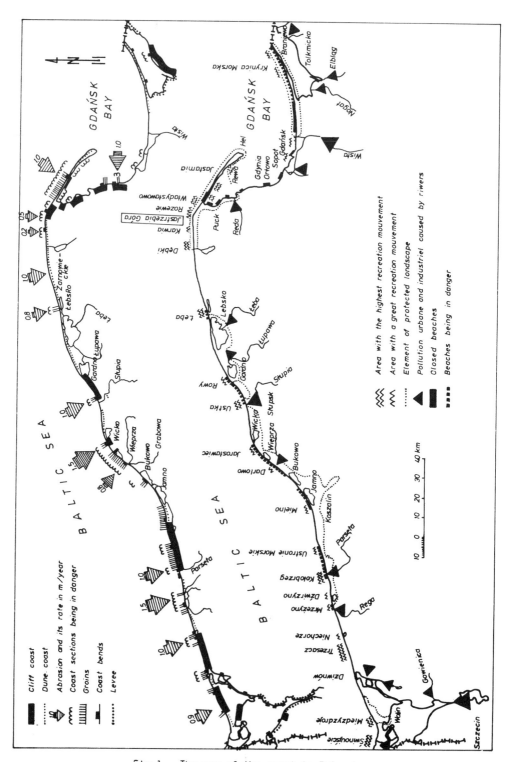

Fig.1   The map of the coast in Poland

occurs once in tens or a hundred years. So one should foresee a maximum single destruction of the overcoast and localize the objects of seaside settlements outside the dangerous zone. Actually the researches relating the notation of such a zone are carried out. The width of this zone should not be less than 100 to 150 m from cliff edge. In this connection the researches have been concentrated on the construction of a physical and mathematical model of the cliff coast transformation in Poland (Subotowicz 1988,1989). Among other things it matters the delimitation of a prognostic development of landslide phenomena at a cliff. To this aim one of the sections of cliff coast with the most inconvenient predispositions for formations of landslide phenomena has been chosen (fig.2).

Paralelly the measurements of basic dynamical parameters are carried out on other sections of the cliff coast in Poland, too. The results of the measurements will be a base of verification of the worked out lithodynamical model.

The results of the researches obtained for the cliff coast will be possible to refer to dune coasts, on which the mechanism of occuring geodynamical processes is simpler. In many cases the substrate on which a dune had been formed, is analogical to cliffs. In Polish conditions the substrate is always glacial and fluvioglacial formations.

Above-mentioned researches relating the delimitation of a zone of safe investment concern mainly lithodynamical conditioning. You must remember, that the final width of such a zone should also depend on economical conditioning taking into consideration the value of designed objects and systems of coast protection.

The rate and intensivity of the landslide processes occuring at the cliff or dune overcoast depends on the geodynamical type of undercoast (Subotowicz 1982). Especially important is the quantity of sand material which appears here. This quantity is deciding for bathymetry of bottom and for width of beach and, besides, for the quantity of energy unloading on a cliff or dune. At the same time that quantity decides on the rate of the sea abrasion. However it exists a differentiation in intensivity of coast abrasion depending on the type of geological substrate in which it has been formed. It appears when an active cliff is performed into a passive cliff and the abraded dune progressively stabilizises and even rebuilds (Subotowicz 1981, 1982).

The above process is variable in time and space, i.e. the natural stabilization of the coast in Poland on some its sections

of the cliff or dune coast becomes active one again. It results probably from the trangressive character of the Baltic Sea in regard to its south shore. The opinions, as to which of the factors: isostatical or eustatical is predominating, are divided.

So, the mentioned variability of evolution of coast line is characterized by cyclicity and its shape in plan is characterized by occuring of bay forms and sections protruded into the sea(Subotowicz 1982).

Different systems of the protection of the coast, accompany this development. Already from over 100 years attempts of preventing the abrasion processes are made. Groins, light and heavy bands, prefabricates and rock filling are built. Lately a beach nourishment on some sections of coast has begun. One of the elements of protection system are breakwaters localized at an entrance to small harbours.

Actually hydrotechnical protections of coast comprise 22% of all coast lenght i.e. 126 km and are concentrated in 20 regions (fig.1). Those objects perform their aim only locally and for a very short time. The evolution proecess of coast, as for its common destruction, is farther observed. The different hydrotechnical objects, placed on the coast, are a permanent element of the environment and participate in this transformation.

The mostly unfavourable influence of these objects on the coast results from the fact, t hat they disturb the existing sand material stream directed from West to East, the volume of which is about 200.000 m³/year (Subotowicz 1982). Accumulated sand material westwards - windward of port breakwaters, causes its negative balance eastwards - leeward. The similar phenomenon appears on a leeward of groins group. Consequently, decresing of sand material of coast zone leads to its intensive abrasion and deepening or making new bay forms along the coast line.

In last century especially on the coast sections where its relative stabilization appeared, many housing and resting objects had been built on the back overcoast. Besides, the border of a safe investment was not observed as it was not known in which distance from the edge of cliffs and dunes it goes. Those objects are actually placed in danger zone. This danger is caused not only by abrasion activity from the sea, but also from the very object. The object localizated near the cliff or dune edge, accelerates deprivation of overcoast stability. Besides, forming of landslide, especially at cliffs, is caused additionally by improper exploitation of the given object.

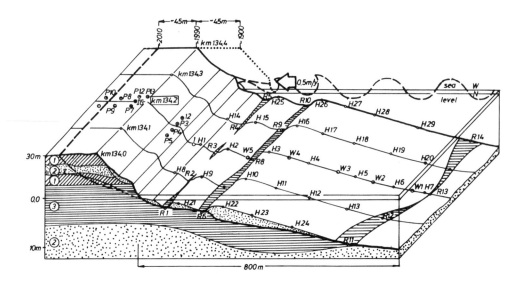

Fig.2 The cliff coast at Jastrzębia Góra. 1,I2, I6, P3-P13- structure drillings with
       piezometers and inclinometers; W1-W6- structure drillings from the sea level;
       H_1, H_2...- hydrosonders; R1, R2...- abrasion datum points; 1-moraine clay;
       2- intermoraine sand; 3- varved clay.

So, the coast in Poland is a special place of interest and engineering geological
researches. The artificial interference into the coast both from the sea and from the
land is so great that a natural process of
its evolution does not exist. This codepen-
dence of geological environment of coast,
with the hydrotechnical and building objects
on it, is an instance of the process being
a subject of engineering geological researches.

REFERENCES

Subotowicz, W.1981. Geologie und Dynamik
  an der polnischen Steilküste. Zeitschrift für Geologische Wissenchaften. 1:63-72.
Subotowicz, W.1982.Litodynamika brzegów
  klifowych wybrzeża Polski. Gdańsk. Osso-
lineum.
Subotowicz, W.1988. Litodynamiczny model
  brzegu klifowego w Polsce. Inżynieria
Morska. 2: 65-68.
Subotowicz, W.: 1989. Lithodynamical Model
  of Cliff Coast in Poland. Washington.
Abstracts 28 th International Geological
Congress. 3: 503-504.

Observed modes of failure of existing shore protection measures in parts of Akwa-Ibom State of Nigeria

Les différents échecs des mesures présentes de protection de rivage observés dans quelques parties de l'état d'Akwa-Ibom au Nigéria

S.Clifford Teme

Institute of Flood, Erosion, Reclamation and Transportation (IFERT), Rivers State University of Science and Technology, Port Harcourt, Nigeria

ABSTRACT: Major settlements along the coast of Akwa-Ibom State in Nigeria are experiencing extensive wave erosion along their fore-shores. However, the Nigerian Government had erected protective structures along the shores of three of these settlements namely, Jamestown, Oron and Ibeno. At Jamestown, the shores have been protected against wave erosion by a concrete overlay while at Ibeno and Oron, the shore protection measures consist of a combination of geotextile sandbags, rock gabions and a bituminous overlay. Recent field investigations have shown that extensive failures have occurred in the protective measures adopted at Oron and to a lesser extent at Jamestown and Ibeno.

The observed modes of failure of the shore protective structures of these sites consist of (a) tensile cracking perpendicular and parallel to the shoreline, (b) boundary failures, (c) separation of rock gabions, (d) tensile gashes in geotextile bags, (e) failure of gabion materials, (f) movement of anchor supports, (g) piping from back-swamps and (h) loss of boulders through toe of bank slope.

RESUMÉ: La pluspart des habitants de long de rivage dans l'etat d'Akwa-Ibom au Nigéria experiencent une vaste érosion de long de pré rivages. Néanmoins, le gouvernement du Nigéria a construit des structures protectifs de long de côtier des trois habitants qui sont: Jamestown, Oron et Ibeno. À Jamestown, les rivages sont protectés contre le flot d'érosion par une combinaison de geotextile sabre sac, des gabions roches et une concrete tandis que 'a Ibeno et Oron, les mésures protectifs de rivage consiste une combinaison de geotextile sabresacs, des gabions roches et une bitumen. Les enquêtes recentes ont montrés qu'il y a une échêc grave dans les mésures protectif adopté à Oron et dans une limite façon 'a Jamestown.

Les modes d'échêc observé d'une structure protectif . de rivage dans ces locations consistent, (a) le craquement tensile perpendiculaire et parallèle au rivage (b) une échêc de frontière (c) une séparation des gabions roches (d) les rupture tensile dans les sacs gabions (e) une échêc des materials gabion (f) mouvements des surportables de l'anchor (g) pépier de dos marais et (h) perte des roches par le doigt dabord penché.

1 INTRODUCTION

Many coastal settlements in Nigeria are exposed to various stages of wave-induced coastal erosion and occasional flood inundations. In the coastal state of Akwa-Ibom, situated in the southeastern portion of Nigeria, three of these settlements namely, Oron, Jamestown and Ibeno have, at various times, had their shores protected against erosion and possible flooding.

Oron is situated at the right-bank of the mouth of Cross River at approximately 8°15' longitude East of the Greenwich Meridian and at latitude 4°50' North of the Equator (Figure 1). Jamestown is located further south of Oron and at the estuary of the Jamestown River, along the coastline. It is approximately along longitude 8°19' East and along latitude 4°35' North. On the other hand, Ibeno is situated at the estuary of the Qua Iboe River, about two kilometres from the national coastline, with co-ordinates of 7°58' East and 4°35' North.

Field examinations have shown that structural failures have occurred within some of the protective measures used for erosion and flood control at these selected localities. This paper presents some of the observed modes of failure that occurred in these protective measures with some possible explanations for their occurrences.

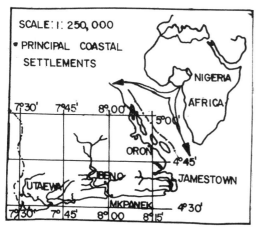

Figure 1
Location of study sites

2 DESCRIPTION OF EXISTING SHORE PROTECTION SITES IN AKWA-IBOM STATE

In each of the three localities where studies were carried out, brief descriptions of the topography, geology and vegetation patterns are made. Also, the type of protective measure adopted at each site is illustrated.

2.1 Jamestown shore protection site

The topography at Jamestown is rela-tively flat-lying but slopes gently towards the Atlantic Ocean. The lithology is characterised by the presence of friable sands and silty-sands overlying clays of low-to moderate plasticity in places. This sequence is further underlain by

sands and silty-clayey sands to great depths, (Federal Ministry of Works, 1981). The vegetation around the shores of Jamestown comprise mangrove trees and some brackish water varieties such as nippa palms with luxuriant undergrowths. Coastal erosion here is caused predominantly by action of waves.

The length of shores protected at this locality is 1.0 kilometre with an average protection width of about 25 metres. The shore protection works at Jamestown consist of sand-filled synthetic mattresses which form the base, overlain by rock-filled wire gabions. The gabions are further overlain by about 15cm thick concrete layer. No evidence of drain holes were observed on the concrete over-lay between the high and low water levels. A clay dyke protects the landward side of the compacted earth-fill that forms the major protection measure. A modified protection measure that incorporates drainage pipes for the Jamestown site is suggested in Figure 2.

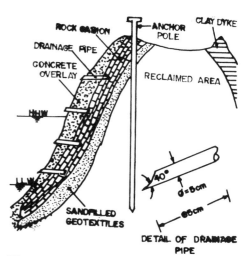

Figure 2
Suggested shore protection measures at Jamestown

2.2 Ibeno shore protection site

The topography at Ibeno is generally flat-lying with occasional depres-sions within the fresh-to-brackish water backswamps. The soil type is basically greyish to reddish-brown, lateritic silty-sands and clays which are low to moderately plastic in consistency. This sequence is underlain by sands, sandy-clays and shales to depths in excess of about 50-60 metres (Shell B.P. 1962; Federal ministry of Works 1981). The vegetation pattern is a mixture of mangrove trees at the shorelines and rafia palms with luxuriant undergrowth inland. Coastal erosion at Ibeno is caused primarily by waves induced by passing vessels on the Qua Iboe River and by sand-mining along the shores. The shore protection works at Ibeno are similar to those at Jamestown described in Section 2.1 above but have a bituminous overlay instead of concrete overlay. A schematic representation of the shore protection works at Ibeno is shown in Figure 3.

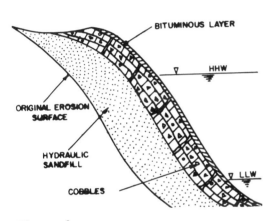

Figure 3
Schematic details of shore protection measures at Ibeno

2.3 Oron shore protection site

The topography at Oron is undulatory and attaining a height of approximately 80-90 metres above sea level at the highest point. The vegetation is mangrove at the shoreline and freshwater type inland. The lithology consists of reddish-brown, lateritic silty-sands and clays down to a depth of approximately 60-70 metres. This is underlain by sands, sandy gravels and silty-clays to depths in excess of 100 metres (Dar Al-Handasah 1981; Teme 1988). The shore protection measures at Oron consist of

hydraulic sandfill along the entire shoreline. Placed on top of this reclaimed material are sandfilled geotextile bags that extend well below the low-low-water level limits. These bags are overlain by rock-filled wire gabions which are covered by a bituminous overlay. No drainage holes were observed on the face of the protection measures at Oron during the field survey by the writer. A schematic cross-section of the protection measures is given in Figure 4.

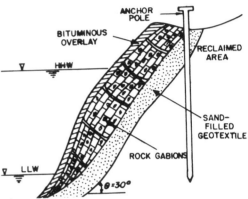

Figure 4
Schematic cross-section of shore protection measures at Oron

3.0 OBSERVED MODES OF FAILURE OF SHORE PROTECTIVE MEASURES

Evidence of failures of adapted shore protective measures were observed at some sites, notably Jamestown and Oron, during the field investigations by the writer. The different modes of failure of shore protection measures that were observed and studied include tensile cracking of overlays, boundary failures, separation of rock gabions from each other, tensile failures of geotextile bags, failures of wire gabions, movement of anchor supports, sand piping from backswamps and loss of boulders through toe of bank slope.

3.1 Tensile cracking of overlay

This was the most common type of distress sign observed on the surface of the protective measures. Tensile cracking occurred in two principal directions namely, perpendicular and parallel to the shorelines. Perpendicular tensile cracks were observed along major joints in concrete overlays (Figure 5) while parallel tensile cracks were observed in bituminous overlays, especially where breaks-in-slope were formed as a result of slope movements (Figure 6).

Figure 5
Tensile cracks in concrete overlay, perpendicular to shoreline of Jamestown

Figure 6
Tensile cracks in bituninous overlay, parallel to shoreline at Oron.

Tensile cracking was observed to be the precursor failure mechanism in all cases of protective measure failures at all sites examined. The tensile cracks were probably caused by (1) build-up of excess porewater pressures behind the protective measures and (2) disruption of equilibrium due to over-steepening of shore slopes through toe material erosion.

3.2 Boundary failures

This failure mode was found to be associated with boundaries between concrete overlays and bituminous overlays such as occur near locations of jetties or similar structures. An example is shown in Figure 7.

Figure 7
Boundary mode of failure at Oron

3.3 Separation of rock gabions from each other

Individual rock-filled wire gabions were often observed to be detached from adjacent ones. This detachment of rock-filled gabions was caused primarily by the rupture of strings connecting adjacent gabions. The phenomenon could be aggravated by steepening of shore slopes due to toe material erosion. The predominant direction of gabion separation was parallel to the shoreline (FIgure 8).

Figure 8
Separation of rock gabions from
each other at Oron waterfront.

3.4 Tensile failure of geotextile bags

This mode of failure was observed to
be caused by the displacement of the
rock gabions due to changes in slope
angle and geometry. In some of the
cases, the tears in the geotextile
bags were initiated by human activities such as pinning of sticks. Sand
loss from the torn geotextile bags
often led to deterioration of bank
slope protective measures (Figure 9)

Figure 9
Tensile failure of geotextile bag
at Oron.

3.5 Failure of wire gabions

Breakage of wire materials used in
construction of the gabions was
observed in several places along the
shores of Oron. These breakages were
caused primarily by human activities
such as the use of the gabions as
anchor supports for medium to large
sea-going vessels. Loss of rock
boulders from these broken gabions
were observed to lead to deterioration of bank protection measures at
Oron, (Figure 10).

Figure 10
Failure of wire gabions at Oron
waterfront.

3.6 Movement of anchor supports

Movement of anchor supports along the
crest line of the shore slopes
(Figure 11) was observed to be an
initial stage in the commencement of
massive bank slope failures at the
northern limit of the protected
shoreline at Oron. This phenomenon
is possibly caused by insufficient
embedment of the anchor supports.

3.7 Sand piping from backswamps

Sand piping from reclaimed areas
behind shore protection works was
one of the ways in which the stability of the system was observed to
be undermined along the waterfronts
of Oron. At a particular locality
in front of the National Museum,
extensive sand loss was observed to
have taken place (Figure 12).

Figure 11
Movement of anchor supports along
shore protection works at Oron.

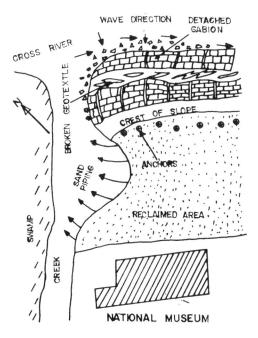

Figure 12
Schematic representation of erosion
and sandpiping near the National
Museum at Oron.

Immediate remedial measures in form
of placement of geotextile filters
and a drained overlay are necessary
to arrest the loss of sand from this
site.

3.8 Loss of boulders through toe of
bank slope

During low water levels, it was
observed that there were substantial
losses of boulders through broken
gabion bags at the toe of the bank
slope. This phenomenon (Figure 13)
is a result of failures of wire
gabions described in section 3.5
above.

Figure 13
Loss of boulders through toe of
bank slope at Oron.

4 DISCUSSION

The durability of shore protection
measures along the selected three
coastal settlements in Southeastern
Nigeria depends on the type of protection, provision of drainage facilities in the protective works,
effect of human activities and the
maintenance of these structures.
Design of shore protective measures
at localities should be site speci-

fic depending on site conditions and appropriateness of structures to these conditions. Whatever design option adopted for each site should make provisions for drainage facilities along the face of these structures. For instance, installation of weepholes would reduce the build-up of excess pore-water pressures behind concrete or bituminous overlays.

Interference of human activities with shore protection measures sited in settlements could lead to an early deterioration of the protective structures. Efforts should be made to reduce the interference of human activities with shore protection measures.

A maintenance programme for shore protection measures should be strictly adhered to, in order to prolong the life expectancy of the structures. This is an aspect that is most times absent from shore protection programmes especially in the developing world.

5 CONCLUSIONS

The shorelines of three settlements in the southeastern State of Akwa-Ibom in Nigeria have been protected against wave-induced erosion and flooding. One of these shore protection works at **Ibeno was** relatively new at the time of this study and had little or no deteriorations. However, extensive tensile cracking was observed in the shore protection works at Jamestown while severe deteriorations ranging from boundary failures, to loss of rock boulders from toe of bank slope were observed at the protection works at Oron.

ACKNOWLEDGEMENT

The writer expresses his appreciation to Professor S.W. Petters, Chairman of the Task Force on Land Use and Soils Surveys of the Akwa-Ibom State Government, for commissioning of this study and for permission to present some of the findings from it. He is also grateful to Mrs. F. Amadi of the Rivers State University of Science and Technology, for typing of the manuscript.

REFERENCES

Dar Al-Handasah Consultants 1981. Surveys and engineering design for erosion control of Ibeno, Ibekwe/Okorete and Jamestown, Mkpong/Utong, Okobo towns in Cross-River State, Nigeria. Technical Report No. 2, 47 pages.

Federal Ministry of Works 1981. Surveys and final engineering design for erosion control of Ibeno, Oron, Uta-Ewa and Jamestown, Mkpong/Utong towns in Cross River State, Lagos, Nigeria, 173 pages.

Shell Petroleum development Company of Nigeria Limited, 1962 Geological Series, Sheet No. 84 and 85 (Port Harcourt and Calabar).

Teme, S.C. 1988. Assessment of coastal erosion in Akwa-Ibom State of Nigeria. Technical Report submitted to Akwa-Ibom State Government Task Force on Land Use and Soils Surveys, 74 pages.

Comportement de remblais et de vasières liés à l'aménagement d'un site portuaire
The behaviour of embankments and muddy deposits induced by harbour installations

P. Thomas & R. Dupain
Laboratoire de Génie Civil, IUT St-Nazaire, ENSM Nantes, France

B. Gallenne
Port Autonome de Nantes, St-Nazaire, France

RESUME : Dans un site estuarien, les travaux d'aménagement portuaire conduisent à une évolution des vasières latérales et à la constitution de dépôts de dragage sous-marins. La recherche du comportement de ces sites remaniés nécessite une approche pluridisciplinaire intégrant les aspects cycliques et aléatoires de la dynamique estuarienne.
Dans cette étude appliquée à un estuaire aménagé (Loire-France) nous proposons une démarche scientifique comportant les phases suivantes :
- identification des caractéristiques géologiques, hydrauliques et morphologiques de la région
- détermination de la nature, de la genèse des sols et de leurs propriétés méca niques
- pour le dépôt sous-marin : interaction hydraulique maritime - géotechnique marine, comparaison des estimations avec les résultats de mesures in-situ par sondages ou traceurs radioactifs
- pour les vasières latérales au fleuve : prise en compte d'analyses chimiques dans l'estimation des phénomènes de rupture.
Cette méthodologie peut être appliquée à d'autres sites portuaires estuariens.

ABSTRACT : In an estuary, harbour installations lead to the evolution of lateral muddy deposits and to the accretion of submarine dredging deposits. Research into the behaviour of these sites requires a multidisciplinary approach, taking into account cyclical and random estuarine dynamics. In a case study of the Loire-estuary (France), we have taken the following scientific approach :
- identification of the geological, hydraulic and morphological characteristics of the area
- determination of the nature and the formation of soils and their mechanical properties
- for the submarine deposits : interaction between hydraulics and marine geotechnics, comparaisons of estimations and in-situ measurements
- for lateral muddy deposits : we have taken into account chemical analysis of interstitial waters in the estimation of failure phenomena
The proposed methodology may be applied to other harbour sites in estuarine environments.

Les aménagements portuaires estuariens engendrent fréquemment la constitution de remblais sous-marins et/ou terrestres. La recherche du comportement de ces sites remaniés est d'un grand intérêt à la fois économique (en raison des conséquences d'une éventuelle mauvaise tenue) et scientifique (interaction de contraintes hydrauliques et géotechniques s'exerçant sur des matériaux généralement dotés d'une forte cohésion).
Le site d'application de cette étude est l'estuaire de la Loire (France) zone portuaire marquée par des travaux d'aménagement liés à l'implantation de divers terminaux maritimes. La profondeur du chenal d'accès a été portée de -6m en 1939 à -10m en 1969 puis à -14m en 1986, à la suite de travaux de dragage et de déroctage effectués sur une vingtaine de kilomètres de long et 300m de large.
L'ensemble de ces travaux a conduit au déplacement de 5.10^7 m^3 de déblais partiellement utilisés pour le remblaiement

de zones industrielles ou transportés au large pour constituer un remblai sous-marin. Le dépôt sous-marin est situé à une vingtaine de kilomètres de l'embouchure de la Loire, à une profondeur de l'ordre de 20m.

Le site portuaire remblayé sur les bords de l'estuaire est situé dans une zone dont les sols initiaux avaient pour origine le remplissage de la paléovallée de la Loire (lors de la dernière transgression postglaciaire) ainsi que le colmatage des vasières latérales.

La recherche du comportement de ces deux sites remaniés placés dans deux zones hydrauliquement différentes (partie externe et interne de l'estuaire) nécessite une approche pluridisciplinaire intégrant les aspects cycliques et aléatoires de la dynamique estuarienne, et prenant en compte les différences d'origine et de sollicitations des sols terrestres et/ou sous-marins.

Après avoir caractérisé ces deux types de remblai, nous effectuerons une identification géotechnique des matériaux en place suivie d'une analyse de leurs comportements.

1 COMPORTEMENT D'UN REMBLAI PORTUAIRE SOUS-MARIN

1.1 Caractérisation du site

L'estuaire externe est caractérisé par l'existence d'un système de dépôt sédimentaire, modelant la morphologie de la région par son orientation Nord-Est Sud-Ouest qui s'enfonce dans un environnement de formations calcaires se retrouvant au large pour constituer des plateaux partiellement émergés qui modifient la déformation des vagues lors de leur propagation vers le rivage.

L'estuaire de la Loire est caractérisé par une meilleure protection aux contraintes océaniques que les estuaires proches de la Gironde ou de la Seine comme le montrent Allen (1982) et Thomas (1987).

Le dépôt de dragage est situé au large du chenal d'accès (fig.1) dans un environnement de zones sableuses ou rocheuses.

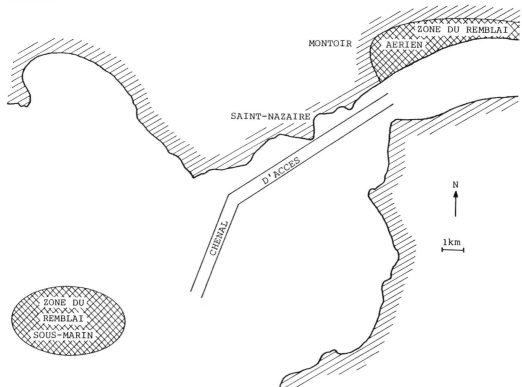

fig. 1 : l'estuaire externe de la Loire (France)

1.2 Identification géotechnique des matériaux déposés

La sédimentologie des matériaux dragués dépend du contexte géodynamique de la formation des dépôts (triage, apports grossiers, lévigation). La courbe enveloppe de la granulométrie des sédiments (fig.2) met en évidence la faible proportion de particules ayant une dimension supérieure à 80 µm. Dans les matériaux dragués, peut apparaître un sédiment typique de la région : "la jalle" vase compacte, ayant une cohésion élevée et une forte teneur en eau, mélange de limons argileux et de sables fins, pouvant se déliter en minces couches de quelques centimètres en raison de la présence de fines particules de micas et de limons. Les résultats des limites d'ATTERBERG permettent de considérer les vases draguées comme proches d'argiles très plastiques, en raison des ordres de grandeur des limites de liquidité W_L, de plasticité W_P et de l'indice de plasticité I_P :

$$80 < W_L < 120$$
$$35 < W_P < 60$$
$$45 < I_P < 70$$

Le remblai sous-marin a pour origine des rejets de matériaux extraits dans le chenal par des dragues aspiratrices en marche (densité des vases : 1,15 à 1,30).

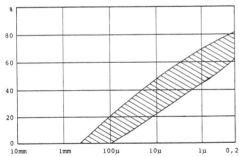

fig. 2 : distribution granulométrique

1.3 Estimation des contraintes océaniques

En raison de la position géographique et de la profondeur, les contraintes océaniques exercées sur les sols marins de la zone concernée sont engendrées essentiellement par la houle, phénomène aléatoire à la fois dans l'espace et dans le temps. A partir de l'exploitation de plusieurs années de mesures, que nous avons effectuées dans une zone peu éloignée au moyen d'un capteur à accéléromètre, il est pos-

sible de proposer une répartition probabiliste des amplitudes des vagues dans la région (fig.3) : l'amplitude des houles

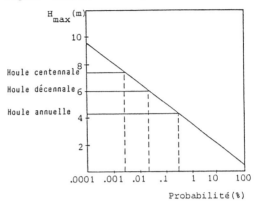

fig. 3 : estimations des amplitudes de houle

annuelles, décennales et centennales étant respectivement de 4,3m - 6m - 7,6m. D'autre part, les périodes les plus fréquentes sont proches de 7s, les tempêtes ayant une périodicité de l'ordre de 12s. Par ailleurs, en l'absence de capteurs directionnels de vagues dans la région, nous avons exploité le spectre des mesures de directions de houle effectuées dans le secteur Nord du Golfe de Gascogne par des navires océanographiques ou météorologiques au cours des 20 dernières années. La répartition (fig.4) met en évidence la prédominance des directions Ouest et Sud-Ouest. La connaissance de la distribution des amplitudes, des périodes et des directions de ce phénomène aléatoire que constitue la houle est indispensable pour connaître la probabilité d'occurrence des vagues ayant une énergie suffisante pour engendrer une instabilité ou une érosion du dépôt de dragage. En effet, les houles se déforment en fonction de leur angle d'incidence, par rapport aux lignes bathymétriques, et de leur période. C'est pourquoi nous avons appliqué le code de calcul RED (Khalifa (1987)) permettant d'exploiter le modèle mathématique de RADDER qui prend en compte la réfraction, la diffraction et les réflexions sur ce site marqué, en particulier, par la présence d'ilôts rocheux :

$$\frac{\partial \Psi}{\partial x} + [\frac{1}{2KCC_g} \frac{\partial(KCC_g)}{\partial x} - iK] \Psi - \frac{1}{2KCC_g} \frac{\partial}{\partial y} (CC_g \frac{\partial \Psi}{\partial y}) = 0$$

ϕ : potentiel bidimensionnel
C : célérité
K : nombre d'onde
C_g : célérité de groupe

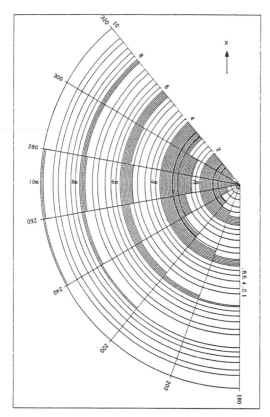

fig. 4 : distribution des directions de
houle

1.4 Comportement induit du dépôt de dragage

L'évolution d'un dépôt de dragage dépend
de la nature des matériaux rejetés (sables, vases dont la concentration varie
de 200g/l à 700 g/l) de leur évolution en
fonction des contraintes océaniques et de
la périodicité et de l'importance des
chargements anthropiques. La difficulté
de prélèvements d'échantillons non remaniés en raison du lavage subi lors de
leurs remontée, ajoutée à l'absence de
modélisation de l'action des houles et
des courants sur ce type de sol ne
rendent pas simples l'analyse de son évolution.
A partir de simulations expérimentales
sur modèle physique, Migniot (1981) propose de relier la vitesse critique d'érosion u_{xe} à la concentration C_S des
dépôts :

$$u_{xe} = 3,2 \times 10^{-5}(C_s)^{1,175} \text{ si } C_s < 240 \text{ g/l}$$

$$u_{xe} = 5,06 \times 10^{-8}(C_s)^{2,35} \text{ si } C_s > 240 \text{ g/l}$$

En raison de la faiblesse des courants
dans la zone du dépôt et en tenant compte
de la distribution spatiotemporelle des
houles mise en évidence précédemment,
l'application de ce modèle permet de cal-
culer la probabilité P_v d'occurrence
des vagues susceptibles de remanier des
vases de concentration 200 g/l présents
dans le remblai sous-marin. Les valeurs
de P_v ainsi obtenues sont généralement
inférieures à 0,1 sur la zone concernée.
Une simulation effectuée pour des sédiments vaseux de concentration 400 g/l indique une absence de remaniement annuel.
Cette estimation de la stabilité du remblai sous-marin est confirmée :
- par des mesures hydrographiques effectuées pour le calcul des cubatures en intégrant l'effet de la consolidation. Le
volume en place représente en effet 83%
du volume théorique des dépôts
- par une expérimentation par traceurs
radioactifs (opération réalisée avec le
concours du C.E.A.) consistant en un mar-
quage des matériaux dragués par de l'Or
198 au cours de leur immersion puis en un
suivi du nuage radioactif en suspension
après clapage des produits déposés sur le
fond.
Ces résultats confirment ainsi la validi-
té du choix de cette zone de création
d'un remblai sous-marin, qui avait été é-
tabli dans le but d'en rechercher la stabilité.

2 COMPORTEMENT D'UN REMBLAI EDIFIE SUR DES VASIERES LITTORALES

2.1 Originalité du site

Le matériaux sableux dragués lors des
travaux d'approfondissement du chenal à
la côte -13,5 m CM ont été réutilisés
pour remblayer hydrauliquement les vasières latérales apparues à l'arrière de la
digue de Montoir qui calibre le chenal
entre Donges et St-Nazaire (fig.5).

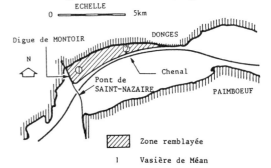

fig. 5 : zone aérienne remblayée

Les vases et argiles présentes dans ces vasières ont deux origines, l'une récente au sens géologique (-18000 ans) qui correspond au remplissage post glaciaire, l'autre récente au sens historique liée au colmatage latéral qui s'est réalisé de 1939 à 1980 lors de l'édification par étapes de la digue de calibration. Ces deux formations se caractérisent par des cohésions faibles : Cu = 35kPa pour les vases profondes les plus "anciennes" et 10 à 15 kPa pour les vases molles supérieures les plus récentes.

2.2 Les ruptures observées et leur interprétation à partir d'essais en place

Deux cas de rupture ont été analysés.
L'un concerne le remblai hydraulique déposé sur la vasière de Méan dont l'épaisseur prévue de 3m impose = 60 kPa environ. La rupture est franche et les mesures en place de Cu intact et Cu remanié permettent de montrer que celle-ci affecte les vases supérieures les plus récentes issues du colmatage latéral.
L'autre se situe sur une zone de stockage de matériaux (phosphates et potasse) imposant une surcharge de 100 kPa sur un secteur déjà en cours de consolidation sous l'action d'un remblai hydraulique plus ancien. Dans ce cas, on observe plus des grandes déformations qu'une rupture franche et rapide.
L'identification des sols, leur teneur en eau, leur cohésion non drainée mesurée en place au scissomètre de chantier, que ce soit sur des zones chargées ou en secteur vierge, montrent que la limite entre les deux horizons de vase se situe à -2 m CM (Dupain (1982), (1989)).

2.3 Prise en compte des analyses chimiques des eaux interstitielles

L'originalité des sols fins estuariens est qu'ils piègent dans leur structure une eau interstitielle chargée en ions qui, au départ, ont une composition liée à la nature et à la composition de l'eau dans laquelle ils se sont déposés. Les proportions relatives des différents composants sont celles de l'eau de mer, ceux-ci étant simplement dilués comme l'est l'eau saumâtre du milieu estuarien (environ 15 g/l dans ce secteur). De ce fait, des écarts dans ces proportions révèlent un apport extérieur.
Nos analyses chimiques ont permis de mettre en relief deux traceurs chimiques :
- Ca^{++} apporté en grande quantité par le dragage hydraulique du sable de Loire lors du rèmblaiement de la vasière de Méan.

La mixture pompée contenant de la crème de vase riche en Ca^{++} (Gallenne, 1986), cet ion se retrouve en quantité notable dans les vases remaniées par la rupture.
- Ca^{++} apportés en grande quantité par les stockages de phosphates et de potasse. Les eaux de pluie et les variations de niveau de nappe phréatique provoquent la percolation de ces traceurs chimiques vers les terrains inférieurs.
Nous avons pu distinguer les sels dissous d'origine marine, et datant de l'époque du dépôt du sol, de ceux issus de la pollution chimique en mesurant les concentrations en Ca^{++} et K^+ dans les eaux interstitielles, à différents niveaux.
La comparaison entre ces analyses et les concentrations des ions homologues de l'eau de mer ainsi que leur proportion relative par rapport à l'ion Cl permet ainsi de situer les ruptures de sol à -1,50 m CM sur la vasière de Méan et à -2,50m CM sous les stocks de phosphates et potasse.
Les analyses chimiques des eaux interstitielles contenues dans les vases marines permettent de confirmer que les sols atteints par les ruptures concernent les vases supérieures les plus récentes déposées par colmatage latéral.

CONCLUSION

L'étude du comportement de remblais portuaires sous-marins ou aériens issus de dragages, est rendue complexe par le caractère aléatoire de la nature et de l'évolution des matériaux dragués et de la distribution des contraintes d'origine océanique.
La méthodologie proposée dans cette étude permet de prendre en compte la variabilité des paramètres hydrauliques et géotechniques et d'obtenir ainsi une estimation du comportement de ces sites modifiés par les aménagements portuaires.
Pour le dépôt sous-marin, l'estimation des contraintes engendrées par les houles et les courants a permis d'obtenir des estimations de la tenue de cette zone.
Appliquée à posteriori sur un dépôt existant, les résultats de cette méthode ont été confirmés par des mesures in-situ effectuées par sondages ou par traceurs radioactifs.
Pour les vasières latérales au fleuve, l'analyse chimique des eaux interstitielles a contribué à l'estimation des phénomènes de rupture dans un site comportant un remblai portuaire et une zone de stockage.
L'ensemble de cette méthodologie peut

être appliquée à d'autres sites portuaires estuariens pour des études d'impact conduisant au choix, a priori, de sites d'immersion ou d'émersion de matériaux de dragage.

REFERENCES BIBLIOGRAPHIQUES

Allen, G.P. (1982). Mesures de houle en différents sites du littoral français. 424 pp. LNH, Paris.

Dupain,R. (1982). Liaisons entre les travaux d'aménagement et la géotechnique dans l'estuaire de la Loire : la concavité de Montoir, 197 pp. Thèse Doctorat Université de Nantes.

Dupain, R. (1989). Rupture de remblais en site estuarien. Apport de la géochimie Bull. AIGI 39, 91-97.

Gallenne, B. (1986). Nature et composition de vases draguées dans l'estuaire de la Loire, 18 pp. Rapport interne PANSN, Nantes.

Khalifa, M.A. (1987). Calcul automatique des déformations de la houle, 147 pp. Thèse Doctorat Université de Nantes.

O.R.I.S. (1984). Sédimentologie de l'estuaire externe de la Loire. Rapport CEA-PANSN.

Migniot, C. (1981). La pratique des sols et des fondation, 627-732. Moniteur des Travaux Publics, Paris.

Thomas, P. Estimations des cportement des interfaces air-eau et eau-sol dans un estuaire externe aménagé, 183 pp. Thèse Doctorat d'Etat, Université de Nantes.

Thomas, P., Gallenne, B. (1986). Estimation des phénomènes d'érosion et de sédimentation dans un estuaire externe aménagé : application à l'embouchure de la Loire. Proc. 5th Int. Cong. IAEG. 1603-1609, Balkema, Rotterdam

P.A.N.S.N. (1987). Etude par traceurs radiocactifs de l'évolution des produits de dragage dans l'estuaire externe de la Loire, document interne.

6.4 Harbours, causeways and breakwaters
Ports, levées et brise-lames

Research activities on mechanical rock cutting and dredging at Delft Hydraulics

Recherches de l'abattage mécanisé et dragage de la roche à Delft Hydraulics

P.M.C.B.M.Cools
Delft Hydraulics, Industrial Hydrodynamics and Dredging Technology Division, Delft, Netherlands

ABSTRACT: The paper presents a broad view of the laboratory research activities at DELFT HYDRAULICS on mechanical cutting and dredging of rock, performed over the last decade. The paper focusses on the description of the various experimental facilities, instrumentation equipment and experimental techniques used in the research programs, rather than the presentation of specific results.

RESUME: L'exposé présente une vue générale des activités recherches dans laboratoire à DELFT HYDRAULICS de l'abattage mécanisé et dragage de la roche, réalisé pendant les années passé. L'exposé se concentre sur la description des facilités expérimentales, l'équipement des instruments et les techniques experimentales appliqué dans les programmes recherches, plus que la présentation des resultats spécifiques.

1 INTRODUCTION

This paper presents a broad view of the laboratory research activities on mechanical cutting and dredging of rock, performed over the last decade mainly under commission of the Dredging Research Association, in which the major Dutch dredging companies and Rijkswaterstaat (Netherlands Public Works Department) participate. The rock cutting program is part of an extensive list of research programs on dredging technology, which involve not only the cutting of rock, sand, clay and intermediate soils, but also research on the suction process of a draghead, slope behaviour, hydraulic fill and erosion and environmental impact of dredging. In Steeghs et al. (1989) the history of these research programs is described.

In the following chapters especially research on rock dredging will be highlighted. The paper focusses on the description of the various experimental facilities, instrumentation equipment and experimental techniques used in the research projects, rather than the presentation of specific results. These projects involve soil mechanical tests, linear cutting tests, rotating cutting tests and numerical simulations. Each individual project aims at studying a specific aspect of the cutting process by looking at the phenomena or by measuring a specific parameter, like the stress distribution on a chisel. In a synthesis the results of all these partial processes are used to create a numerical model, which simulates the cutting process momentaneously.

2 ROCK PROPERTIES

Rock dredging, which often takes place during capital dredging projects, can be executed directly by means of a cutter suction dredger, or indirectly by blasting, ripping or breaking. Usually the uniaxial compressive strength (UCS) of rock is applied to indicate which method will be used; a cutter suction dredger is used on very weak up to moderately strong rock (UCS up to 50 MPa). Within this range the porosity increases rapidly (approx. linear) with decreasing UCS. Especially the weak rocks are highly porous (up to approx. 50 %) and the pores are assumed to be fully saturated with water.

Most of the rocks tested in our laboratory are limestones, which UCS vary between 2 and 35 MPa, but also other rocktypes like sandstone, gypsum and artificial rocks are examined. Before testing the blocks are saturated with water under vacuum.

3 RESEARCH APPROACH

Initially research aimed at predicting the required power consumption and production rate of a cutter suction dredger for a specific project during tendering. For this purpose a dredging flume has been built, in which cutter models of different scales have been tested, with diameters up to 0.55 m. The behaviour of individual chisels on a cutter has been examined in the cutting rig, where the rotating movement is schematized to a linear movement. In this rig not only chisels of the same scale of the cutter models have been tested, but also on prototype scale, in order to establish the influence of the length-scale.

In this way a large database of approx. 800 linear cutting tests has been generated.

This emperical approach however gives us just a poor understanding of the physics of the cutting process. Extrapolation of the curve-fits to rocks or working conditions not tested before is a hazardous enterprise. The alternative is the execution of a large number of additional tests, to expand the parameter-space. To meet these disadvantages the initial research aims have been extended with the gaining of more physical insight in the cutting process. For this purpose several individual research projects have been executed such as special linear cutting tests, soil mechanical tests and numerical simulations. In the following chapters this physical approach is further illustrated.

4 RESEARCH FACILITIES

In the former chapter two major facilities are mentioned: the dredging flume and the cutting rig. In this chapter their main features will be described.

4.1 Dredging flume

In this facility (photo 1) the processes of a cutter suction dredger (but also of a trailing suction hopper dredger or e.g. a trenching plough) can be studied on different scales. The dimensions of the flume are adjustable up to 50 m length, 9 m wide and 2.5 m deep. The trailing velocity is adjustable between 0.05 and 2.5 m/s. The rotation speed of the cutterhead is adjustable between 10 and 100 rpm. The pumps for mixtures are adjustable up to

photo 2

0.25 m^3/s. The cutting forces can be measured in three dimensions up to approx. 17 kN; the torque can be measured up to 5 kNm.

4.2 Cutting rig

In the cutting rig (photo 2) the cutting process of one or more chisels of a cutterhead can be studied in more detail. The rig itself is located inside a tank which can be pressurized up to 400 kPa, in order to simulate various waterdepths. The transducers of the measuring system can measure forces up to 100 kN. The cutting velocity is adjustable between 0.01 and 5 m/s. The maximum size of a sample amounts 0.7 x 0.9 x 6 m^3.

5 RESEARCH PROJECTS

The individual research projects mentioned in chapter 3 can be grouped in four research fields:

1. rotating cutting tests
2. special linear cutting tests
3. soil mechanical tests
4. numerical simulations

Hereafter the projects within each field will be described in more detail.

5.1 Rotating cutting tests

The behaviour of the entire cutterhead is studied in the dredging flume during several cutting projects. The influence of the length-scale is studied by using modelcutters with different diameters up to 0.55 m (photo 3). During the projects different rocktypes are examined and the operational parameters, especially the hauling and rotating speed, have been varied.

5.2 Special linear cutting tests

The purpose of the initial linear cutting tests was to create an extensive database of

photo 3

photo 5

photo 4

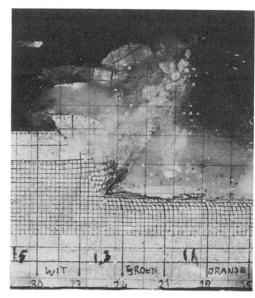

photo 6

testresults over a wide range of the inputparameters:
- chiselgeometry parameters: e.g. rake angle, width, bluntness
- operational parameters: e.g. cutting

velocity, cutting depth, waterdepth rockparameters: e.g. rocktype, UCS, degree of saturation, permeability

During these tests the cutting forces in three directions and the production was measured. From these data the specific energy can be calculated, which tells us more about the required power consumption during dredging. Apart from these tests with a solitary chisel on a flat surface (photo 4), an extensive project has been executed to measure the effect of cutting in grooves created by earlier chisels, as is the case with a rotating cutting head (photo 5).

After the change in the philosophy of research approach (mentioned in chapter 3) several projects have been developed and executed which involve special linear cutting tests. The projects aimed at studying a specific aspect of the cutting process.

During so-called looking-experiments (photo 6) the development of a crushed zone and the generation of fractures in front of the face and below the worn surface of the chisel is examined behind a glass window. With a chisel instrumented with strain gages the actual chisel forces, not corrupted by eigen-frequencies of the measuring system, are measured (photo 7). Two chisels equipped with piezo-electric pressure gages give insight in the stress distribution over the height and the width of a chisel (photo 8 and 9).

photo 7

photo 8

5.3 Soil mechanical tests

To study the material behaviour of rock during cutting conditions special material tests, apart from standard material tests like UCS and Brazilian test, have been executed which study specific aspects. These tests involve triaxial tests to establish a.o. the brittle-ductile transition and point- load, tensile and compressive tests under various loading rates and saturation degrees, done by DELFT GEOTECHNICS and Terra Tek Inc.

The point-load tests (photo 10) are of special interest because the failure mechanisms are very simular to those during rock cutting. For this reason additional tests have been executed

photo 9

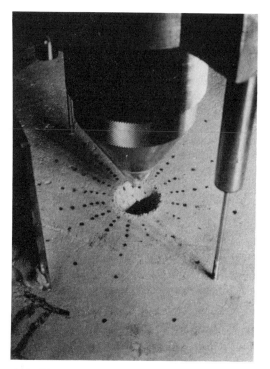

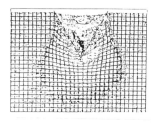

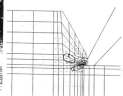

photo 12

photo 10

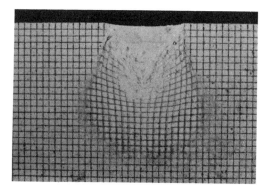

photo 11

on rock samples which were split into two parts in advance and supplied with a grid (photo 11).

5.4 Numerical simulations

To study the stress and strain paths which the rock follows during cutting, a computer program called DIEKA, developped by University of Twente (Huetink, 1986), has been applied, which can cope with large strains. The material- model initially built for steel has been modified for soft, porous rocks and provided with data from the soil mechanical tests of the former paragraph. With the modified program, runs have been executed to simulate axial point- load tests and cutting tests (photo 12).

The results of all former research activities have been implemented in a numerical model, called ROCKSIM, which simulates the cutting process under various input conditions. Because of the stochastic nature of the cutting process, ROCKSIM simulates the production of chips momentaneously and keeps a record of all geometrical data. Given this geometry and the material- parameters the stresses and forces can be calculated momentaneously. ROCKSIM can be installed on personal computer.

5.5 Future developments

In the near future more effort will be put in connecting the cutting model of rock,

ROCKSIM, with that of sand (van Os et al., 1987). This regards especially the weakly cemented soils, where the transition between both models still is not fully understood.

Another item under study refers to the conditions under which cracks propagate in a saturated, porous material under changing stress rates and to the development of a brittleness parameter e.g. the fracture toughness parameter or tensile strength, which gives a proper characterizion of fracture initiation of rock under these conditions.

The developments in numerical modelling of fracture processes are followed with great interest. Special mention should be made of the program DIANA (de Borst, 1986; van Mier, 1987; Rots, 1988) with a "smeared crack concept" and the program FRANC (Ingraffea, 1987; Wawrzynek, 1987), dealing with discrete cracks. These programs require a mainframe or workstation to run. With regard to the use of personal computers the program PLAXIS (Vermeer et al., 1984) can be mentioned.

Although these developments are very promising, the simulation of the cutting process in saturated, porous rock, requires a program which deals with a combination of an elasto-plastic materialmodel with softening behaviour, three-dimensional elements, rezoning because of the large strains, consolidation effects and crack initiation, propagation and possible arrest, under mixed-mode conditions.

Eventually, the program ROCKSIM and the model of groove interaction between individual chisels, will be implemented into a cutter model which simulates the behaviour of the entire rotating cutter. The use of a suitable CAD system may be helpful in describing the cutter-geometry and the tracks which all individual chisels follow through the rock. Such systems are applied successfully for roadheaders (Knissel et al., 1984; Morris, 1985).

6 APPLICATIONS

The results of the abovementioned research projects may be applied on a wide range of activities:

1. dredging
2. tunnelling
3. mining
4. drilling
5. waterjet assisted cutting
6. wear

6.1 Dredging

The program ROCKSIM is used by the dredging companies to estimate production and power consumption of rock dredging projects. When specific problems occur during the execution of a project the acquired physical understanding of the cutting process helps to find solutions.

The behaviour of a seagoing cutter suction dredger can be simulated by the program DREDMO, developped by Delft University of Technology and DELFT HYDRAULICS (Miedema, 1989). Eventually the cutter model which simulates the behaviour of the entire rotating cutter could be implemented in DREDMO.

Many aspects of the research of rock cutting are explored in other dredging research projects such as the U.S. Dredging Research Program (McNair, 1989).

6.2 Tunnelling

In tunnelling projects the use of roadheaders or tunnel boring machines (TBM's) are widely used. Apart from cutting under water instead of in air, the similarity between a cutterhead and a roadheader is striking (Hignett, 1984)

photo 13

A TBM may be equipped with chisels, roller discs or both in the same time, depending on the conditions (Ozdemir et al., 1987). At DELFT HYDRAULICS also roller disc experiments have been executed (photo 13). During cutting the same failure mechanisms as with chisels are observed; the geometry of the disc however allows a higher rock strength.

Many no-dig (micro-tunnelling) projects cope with the same problems as occur in dredging.

6.3 Mining

In mining operations use is made of mining machines or shearers which cut the rock or coal with adjacent chisels or even discs (Roxborough, 1988).

6.4 Drilling

Under certain conditions a number of phenomena observed during cutting are similar with drilling. Characteristic conditions of drilling are the small cutting depths and the large surrounding stresses in the rock and the drilling mud. During dredging these small cutting depths also occur when a chisel on the cutter enters the rock. Also the importance of the presence of water above the rock and the waterpressure has been demonstrated.

6.5 Waterjet assisted cutting

To reduce dust, wear and forces in tunnelling projects cutting is assisted by waterjets (Fowell and al., 1988; Hood etal. 1987; Summers, 1985). At DELFT HYDRAULICS dredging research on jet-assisted cutting has been examined (photo 14).

6.6 Wear

Wear of the applied tools plays an important part in all fields of application mentioned above. In the description of the wear process the level and distribution of the pressure and temperature on the tool are two main

modelparameters. For this purpose not only stresses are measured on the tool (see paragraph 5.2) but also the temperature distribution on the worn surface and inside the chisel is measured by means of four thermocouples. Metallurgical examination may show structural changes in the metal (photo 15). The research on wear is done in co-operation with Delft University of Technology (Verhoef, 1988).

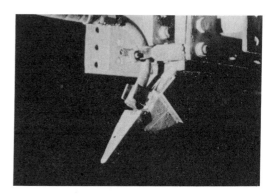

photo 14

photo 15

2896

7 CONCLUSIONS

In engineering practice classification systems of soil and rock are widely used (PIANC, 1984; Smith, 1987; Verhoeven et al., 1988; Spigolon et al., 1989). Often these classification systems are applied to determine the "dredgeability" of the soil. Research on rock cutting however has learned us that the assessment of the "dredgeability" or "cuttability" of rock by taking into account of the materialparameters of rock, only will address to a part of the problem. Also rock-independant parameters such as the chisel-parameters and the operational conditions will influence the decision whether a specific rock can be dredged easily or hardly.

ACKNOWLEDGEMENTS

The laboratory research activities presented in this paper are mainly performed under commission of the Dredging Research Association, in which the major Dutch dredging companies: Ballast Nedam Dredging, Royal Boskalis Westminster, HAM-Dredging, Van Oord-Utrecht, Volker Stevin Dredging and Rijkswaterstaat (Netherlands Public Works Department) participate.

REFERENCES

de Borst, R. (1986). Non-linear analysis of frictional materials. PhD thesis, Delft University of Technology.

Fowell, R.J. (1988). High pressure water jet assisted drag tool cutting. CARE '88, Conference on Applied Rock Engineering. 61-69. Dotesios Printers, Bradford-on-Avon, U.K.

Hignett, H.J. (1984). The current state of the art of rock cutting and dredging. US Army Corps of Engineers. Miscellaneous paper GL-84-17.

Hood, M., Geier, J.E., Xu, J. (1987). The influence of water jets on the cutting behaviour of drag bits. Proc. 6th International Congres on Rock Mechanics, 649-654. Balhema, Rotterdam.

Huetink, J. (1986). On the Simulation of thermo-mechanical forming processes. PhD thesis, University Twente, The Netherlands.

Ingraffea, A.R. (1987). Theory of crack initiation and propagation in rock. In: Fracture mechanics of rock. B.K. Atkinson (ed.), London, 71-110.

Knissel, W., Mertens, V., Kleinert, H.W., Mittmann, M. (1984). Verfahren zur Auslegung und Optimierung der Schneidköpfe von Teilschnitt-Vortriebsmaschinen. Glückauf, 120, nr. 23, 1534-1539.

McNair, C. (1989). Dredging Research Program of th U.S. Army Corps of Engineers. Proc. 12th WODCON, 43-50. Western Dredging Association, Fairfax, Virginia, USA.

Miedema, S.A. (1989). On the cutting forces in saturated sand of a seagoing cutter suction dredger. Proc. 12th WODCON, 331-352. Western Dredging Association, Fairfax, Virginia, USA.

van Mier, J.G.M. (1987). Examples of non-linear analysis of reinforced concrete structures with DIANA. Heron, Vol. 32, no. 3, Delft.

Morris, A.H. (1985). Practical results of cutting harder rock with picks in United Kingdom coal mine tunnels. Proceedings of the fourth international symposium, Tunnelling '85, 173-177. The Institution of Mining and Metallurgy.

van Os, A.G. & van Leussen, W. (1987). Basic research on cutting forces in saturated sand. Journal of Geotechnical Engineering, Vol. 113, No. 12, December 1987, 1501-1516.

Ozdemir, L. & Dollinger (1987). Laboratory studies of high speed tunnel boring. Proc. Rapid Excavation and Tunnelling Conference RETC, June 1987.

PIANC (1984). Classification of soils & rocks to be dredged. Report of a working group of the permanent technical committee II, Supplement to bulletin no. 47. PIANC, Brussels, Belgium.

Rots, J.G. (1988). Computational modeling of concrete fracture. PhD thesis, Delft University of Technology.

Roxborough, F.F. (1988). Multiple pass sub-interactive rock cutting with picks and discs. CARE '88, Conference on Applied Rock Engineering. 183-191. Dotesios Printers, Bradford-on-Avon, U.K.

Spigolon, S.J. & Fowler, J. (1989). Geotechnical descriptors for soils to be dredged. Proc. 12th WODCON, 697-708. Western Dredging Association, Fairfax, Virginia, USA.

Smith, H.J. (1987). Estimating the mechanical dredgeability of rock. Proc. 28th US Symposium on Rock Mechanics, 945-952. Balkema, Rotterdam/Boston.

Steeghs, H.J.M.G., de Koning, J. & Lubking, P. (1989). 25 years of dredging research in The Netherlands: Physics as a basis for innovations. Proc. 12th WODCON, 15-32. Western Dredging Association, Fairfax, Virginia, USA.

Summers, D.A. (1985). A review of waterjet excavation research. Proc. 26th US Symposium on Rock Mechanics, 895-903. Balkema, Rotterdam.

Verhoef, P.N.W. (1988). Towards a prediction of abrasive wear of cutting tools in rock dredging. Delft Progress Report, vol. 13, 307-320. Delft University of Technology, the Netherlands.

Verhoeven, F.A., de Jong, A.J. & Lubking, P. (1988). The essence of soil properties in today's dredging technology. Proc. Hydraulic Fill Structures, 1033-1064. Geotechnical Special Publication no.21, ASCE, New York.

Vermeer, P.A. & de Borst, R. (1984). Non-associated plasticity for soils, concrete and rock. Heron, Vol. 29, no. 3, Delft.

Wawrzynek, P.A. (1987). Interactive finite element analysis of fracture processes: an integrated approach. Masters thesis, Cornell University, Ithaca, NY, January.

6th International IAEG Congres / 6ème Congrès International de AIGI, © 1990 Balkema, Rotterdam. ISBN 90 6191 130 3

Geotechnical properties and problems of the Pleistocene Alluvium of Macau
Propriétés et problèmes géotechniques des alluvions pleistocéniques de Macau

F.M.S.F.Marques & M.O.Silva
Center of Geology, Department of Geology, Lisbon University, Portugal

ABSTRACT: Pleistocene Alluvium occurs in the offshore of Macau. It is covered by very soft to soft holocene fluvio-marine clayey silts (up to 25 m thick), which are, in reclamation areas, overlain by landfill. The basement of the unit consists of residual soils and colluvium of granitic origin.
Pleistocene Alluvium has been referred in Hong Kong and in some site investigation reports in Macau.
This paper presents geotechnical properties of this unit based on the analysis of large site investigation data in Macau. Variations in geotechnical behaviour related to the geomorphological context are also commented.
It is pointed out the presence of an interbedded level of soft to very soft pleistocene fluvio-marine deposit, with unexpected properties: the top of Pleistocene Alluvium is frequently overconsolidated by dessication and consists mainly of firm to stiff muddy sands, with ferruginous concretions; the base is composed of medium dense to dense silty sands, with a composition more similar to the substract. The presence of the interbedded fluvio-marine level with extreme values of N(SPT)=0 (under 29 m of sediment), R_p (CPT)=0,45 MN/m^2 (under 14 m of sediment) and with s_u/p close to 0.3 can have large engineering relevance.

RESUMÉ: Les Alluvions Pleistocéniques occeurent dans le offshore de Macau. Ils sont couverts par des siltes argilleux, très mous à mous, d'origine fluvio-marine holocenique (Épaisseur max. 25m), lequels sont, en certains endroits couverts par des remblais. Cette unité repose sur des sols residuels et des colluvions d'origine granitique. Les Alluvions Pleistocéniques ont eté dejá refereés à Hong Kong et sur quelques rapports de sondages à Macau. Ce travail présente des propriétes geotechniques de cette unité, baseés dans l'analise de large nombre de rapports de sondages de Macau. Des variations de comportement geotechnique, relationeés avec le contexte geomorphologique, sont aussi commenteés. On signale la presence d'un niveau interstratifié, trés mou à mou, d'origine fluvio-marine avec des propriétés inespereés: la partie supérieure des Alluvions Pleistocéniques est souvent surconsolidé par dessication et se compose, en general, de sables silto-argileuses mi-consistentes à consistantes, avec des concretions ferrugineuses; la base se compose de sables silteuses, moyennement denses à denses, avec une composition plus proche du substract. La prèsence du niveau fluvio-marin avec valeurs extremes de N(SPT)=0 (à 29m de profondeur), R_p(CPT)=0,45MN/m^2 (à 14m de profondeur) et des relations s_u/p proches de 0,3 peu avoir grande importance dans le point de vue de l'ingenièrie.

1 INTRODUCTION

The territory of Macau is located in South East Asia, in the southern littoral of the chinese province of Kwangtung, about 100Km south of Canton and 60Km West of Hong Kong (fig.1 a).
The territory is composed by the peninsula of Macau (area of 5.5km^2) and the islands of Taipa (5.5Km^2) and Coloane (7Km^2).
Since the sixties, extensive site investigation has been undertaren as result of an increased urban development, particulary in reclamation areas surrounding the peninsula of Macau.
The Pleistocene Alluvium of Macau (PA) is a complex sedimentary unit that fills the lower part of the depressions of the granitic basement existing in

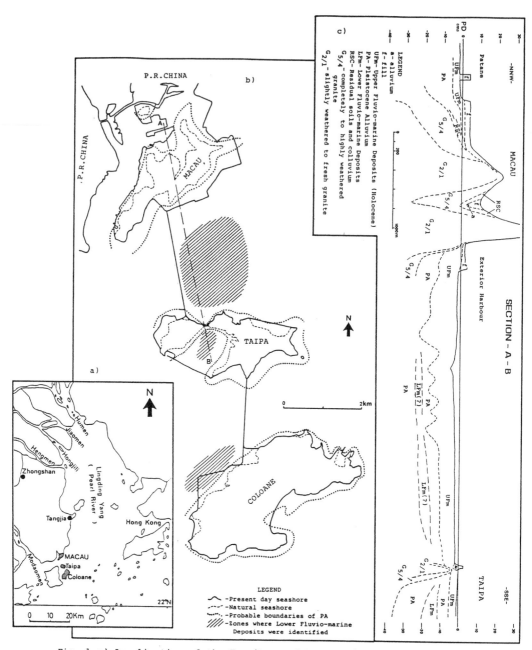

Fig. 1.a) Localisation of the Territory of Macau. b) Probable geological
boundaries of the Pleistocene Alluvium. c) Typical geological
section.

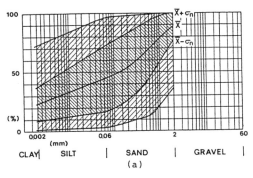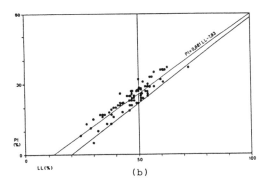

Fig. 2. Grain size (a) and plasticity (b) of the PA.

the region. The PA does not outcrop, and its knowledge is exclusively based on borehole data. In Hong Kong, correlative deposits have been studied providing a reference for the study of Macau terranes.

Geotechnical data used in the present work was obtained in several site investigation reports (59 reports,with 564 boreholes) carried out in Macau. This fact is limiting, because information lacks the desirable uniformity, soil description was made with variable criteria and laboratory tests often do not allow detailed analysis.

In spite of these shortcomings the data now obtained can provide considerable guidance to future site investigation in offshore areas of Macau.

This paper presents a part of a Msc thesis (Marques,1988), completed by new data.

2 GEOLOGY

The bedrock in Macau is composed of granite, probably upper jurassic in age, with minor occurence of andesite sills. This basement presents extensive weathering, up to 40m thick. Weathering zone is almost exclusively composed of completely to highly weathered granite.

Residual soils and colluvium occur over this irregular basement, whose main geomorphological features were formed in latter Pliocene or lower Pleistocene.

PA is the lower part of the sedimentary infilling of the pleistocene valleys, actually submerged. The unit is completely covered by soft holocene muds and in the nearshore, by beach sands.

The most complete successions of PA include three sub units, from top to bottom:

1. Upper PA, probably with age between 24.000 and 16.000 years BP, mainly clayey, frequently fissurated and mottled, with ferruginous concretions and usually with light colours: light red, yellow, light brown, greyish brown. These features indicate emersion, dessication and pedogenetic alteration. This sub-unit was probably formed under alluvial conditions with a sea level lower than the present.

2. Lower Fluvio-marine Deposits probably with age between 28.000 and 30.000 years BP, mainly composed of silty clay and clayey silts, sandy,with grey,dark grey and black colours. These colours suggest significant organic matter content.

Plant remains are frequent and less frequent the presence of shell debris.

3. Lower PA, probably with ages in excess of 36.000 years BP, mainly composed of silty-clayey sands, frequently with cobbles, with light colours (light brown, reddish brown, orange, white, light grey and pink). Towards the base, distinction from residual soils is usually difficult.

The lower fluvio-marine deposits were only positively identified in limited number of locations, mainly by two reasons: unsufficient soil description in borehole logs and irregular distribution of the sub-unit. Deposited in conditions defined by a high sea level, but probably lower than the present, it is likely to presume that its formation was restricted to immersed channels existing at that time. By consequence, specially in zones closer to the actual natural seashore, these deposits may be not present.

Comparison between these sub units and correlative deposits in Hong Kong (Yim,1983; Yim & Li,1983; Yim,1984) was discussed in a previous work (Marques,1988).

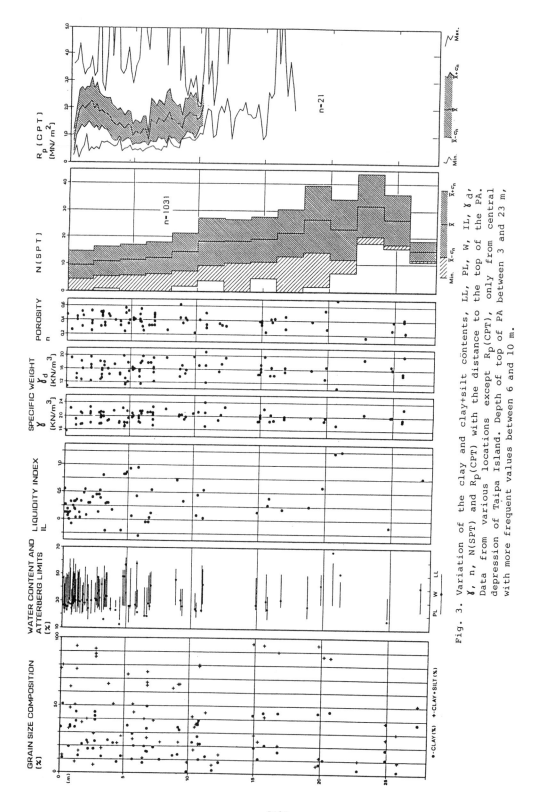

Fig. 3. Variation of the clay and clay+silt contents, LL, PL, W, IL, γ, γ_d, n, N(SPT) and R_p(CPT) with the distance to the top of the PA. Data from various locations except R_p(CPT), only from central depression of Taipa Island. Depth of top of PA between 3 and 23 m, with more frequent values between 6 and 10 m.

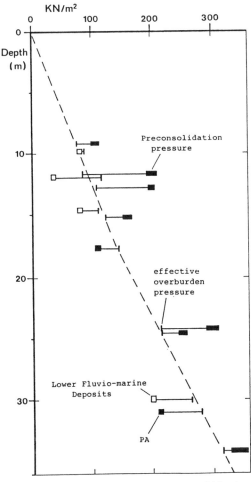

Fig. 4. Relation between preconsolidation pressure (from oedometric tests) and effective overburden pressure.

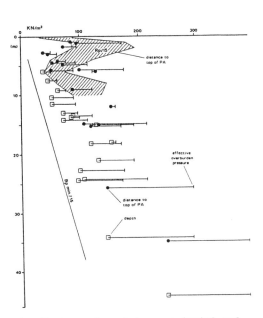

Fig. 5. Comparison between triaxial and CPT deducted undrained strength.

3 GEOTECHNICAL PROPERTIES

In terms of engineering geological mapping (Anon., 1976; Anon., 1981) the PA corresponds to a geotechnical complex. Because the interbedded fluvio-marine deposit was only positively identified in some locations, and it's presence in others was not detected, or is merely supposed, geotechnical data is presented in bulk. Anomalous results, positively related to the presence of that deposit will be separately discussed.

PA is quite heterogeneous. Grain size distribution (Figs.2 and 3) indicate a quite variable clay content. Available data indicate the dominance of sands, silty sands and silty clays. Finer soils correspond to inorganic soils of low to high plasticity (Fig.2).

Activity varied between inactive to active soils, suggesting variable clay mineral composition.

Variation with the distance to the top of the PA of index and strength properties are displayed in Fig. 4.

4 DISCUSSION

In general terms, available data seems consistent, in bulk. Dispersion of values of Atterberg limits, water content (w), specifif weight (γ and γ_d) and porosity (n) can be, in large part, assigned to grain size variability, which is easily understood considering that soils analysed correspond to flood plain deposits and to fluvio-marine deposits.

N (SPT) values in Macau cover a wider range than those indicated, for correlative deposits in Hong Kong, by Yim (1983), Liu & Gammon (1984) and Howat & Cater (1984), but with better agreement with those found by Willis & Shirlaw (1984).

Mean N (SPT) values (Fig.3) show a significant increase with distance to top of PA, until 20m below that reference. At larger distances there is an inversion of the referred mean trend due to the predominant number of test results

TABLE I - Summary of Properties of the Pleistocene Alluvium

Sub Unit	Thickness (m)	Dominant grain size	N (SPT)	R_p (CPT) (MN/m^2)	Remarks
Upper PA	2-10	Silts and sands, clayey	0-20	1-3	Fissurated, mottled and overconsolidated near the seashore
Lower Fm	0-19	Silty clay	0-4	0,5-1,5	Probably normaly consolidated. Presence of organic matter
Lower PA	4-38	Sands, Silty, clayey	10-30	2- 5	Almost regular increase of relative density with depth

obtained in boreholes located far away from actual seashore, where the PA is thicker. Actually, lower PA shows, in each location, a trend of increasing values of N(SPT) with depth. In this case there are superimposed results corresponding to different sub units (what means different composition), due to variable thickness of each one, what justifies the mentioned anomaly.

Oedometer tests gave preconsolidation pressures (Fig. 4) in some cases higher than effective overburden pressure, confirming the overconsolidation of the upper PA.

In situ (CPT, SPT) and laboratory (triaxial, CU) strength test results are consistent in a general way (Fig. 5).

However, some anomalous results deserve attention: γ lower than $17KN/m^2$; γd lower than $12KN/m^2$; $n > 0.5$; $LI > 0.75$; $N(SPT)=0$ up to a depth of 29m and tip resistance, $R_p(CPT)=0.45MN/m^2$ up to a depth of 14m.

The low strength values reported and results of oedometer tests, giving preconsolidation pressure lower than effective overburden pressure, are positively related wiht the lower fluvio-marine deposits.

Lower strength values deducted from field tests ($s_u=R_p(CPT)/15$; $R_p(CPT)=n$ N(SPT), with n=0.2 to 0.3) gave values for s_u/p around 0.3. Considering, for example the Skempton equation (Skempton,1954), with values of PI in the range 25-35 (range of values for positively identified soil tests in the lower fluvio-marine deposits), results obtained for s_u/p are within the range of 0,2 TO 0,24. Although should be noted that equations relating Atterberg limits with strength parameters for normally consolidatd soils, are empirical and questionable (Sridharan & Rao,1973) and that the role of organic matter,

almost always present in the Lower Fluvio-marine deposits, is not clear due to the absence of related data, calculated values suggest that very soft normaly consolidated soils occur in these deposits.

R_p(CPT) values, in particular (Fig.3), were revelative in terms of strength variaton of the two upper sub units of the PA.

Data analysis allowed the detection of frequent low values of N(SPT), in the range 0 to 4, corresponding to the Upper PA, in boreholes more distant from the present day littoral. This fact suggests that overconsolidation detected near the present day littoral possibly is not a general feature, what can be explained by the morphology of the region in the latter Pleistocene. Probably, the zones corresponding to river channels existing at that time were always submerged and by consequence preserved from overconsolidation by dessication.

5 SUMMARY AND CONCLUSIONS

Although data used presented limitations, the results of its analysis are believed to be relevant for future site investigation work in offshore areas of Macau.

The range of strength values detected for the PA in Macau is wider than those reported in Hong Kong.

The subdivision of the PA into 3 sub units, Upper PA, Lower Fluvio-marine Deposits and Lower PA seems relevant in terms of geotechnical properties.

Available data suggests that overconsolidation of the Upper PA probably is not a general feature, being limited to zones nearer the present day seashore.

Were reported very soft to soft probably normaly consolidated soils in the Lower Fluvio-marine Deposits. The conjunction of these two features, lack of overconsolidation in Upper PA and very soft to soft soils in the Lower Fluvio-marine Deposits, coupled with an holocenic cover of soft muds, will possibly mare geotechnical conditions particulary unfavorable in the more distant offshore of Macau.

In terms of site investigation CPT gave a very good immage of this unit. It is thougth that in the case of Macau the use of CPT in conjunction with boreholes will provide a more economical and informative alternative in future site investigation.

More investigation is needed in order to clearly understant the geotechnical properties of the PA, and particulary to explain the presence of very soft probably normaly consolidated deposits, at depths up to 30m, and with a probable age of almost 30.000 years.

6. AKNOWLEDGEMENTS

The authors wish to thank to the Center of Geology of the University of Lisbon for the finnancial support of the investigation carried out, and to the Governement of the Territory of Macau for the provision of site investigation reports used.

REFERENCES

Anon. 1976. Guide pour la préparation des cartes géotechniques. AIGI-UNESCO, Ed. Les Presses de L'UNESCO.Paris.

Anon. 1981. Rock and soil description and classification for engineering geological mapping. Report by the IAEG com. eng. geol. mapping. IAGE.Bull.,Nº.24,pp.235-274.

Howat,M.D.& Cater,R.W. 1984. The use of engineering data for mapping alluvial features. Geol.Soc.Hong Kong Bull. nº.1,pp.161-168.

Liu,K.H. & Gammon,J.R.A. 1984. Quaternary geology,weathering and geomorphology of Hong Kong. Geol.Soc.Hong Kong Bull, Nº.1,pp.49-59.

Marques, F.M.S.F. 1988. Contribuição para o conhecimento geológico e geotécnico do território de Macau. MsC Thesis, Dep. of Geology Ed. Lisbon (in portuguese).

Skempton,A.W. 1954. Discussion on the structure of inorganic soil. Proc. A.S.C.E., vol.80,pp.19-22

Sridharan,A. & Rao, S.N. 1972. The relationship between undrained strength and plasticity index. Geotechnical Engineering, vol.4, pp.41-53.

Willis,A.J. & Shirlaw,J.N. 1984. Deep alluvial deposits beneath Victoria Park, Causeway Bay. Geol.Soc. Hong Kong Bull. Nº.1, pp.143-152.

Yim, W.W.-S. 1983. Evidence for quaternary environmental changes from sea-floor sediments in Hong Kong. Palaeoenvironment Conf. East Asia Mid-tertiary. Centre of Asian Studies. Un. Hong Kong. 17p.

Yim, W.W.-S. 1984. A sedimentological study of sea-floor sediments exposed during excavation of the East Dam Site. High Island. Sai Kung. Geol.Soc. Hong Kong Bull.,Nº.1, pp.131-142.

Yim, W.W.-S. & Li. Q.-Y. 1983. Sea level changes and sea-floor surficial deposits of Chek Lap Kok. In abstracts.Geol. Surf. Dep. Hong Kong. W.W.-S. Yim & A.D. Burnett (eds.). Geol. Soc. Hong Kong and Un. Hong Kong, 1, pp.48-59.

Caractérisation et lois de comportements de sols stratifiés d'un site portuaire estuarien

The characteristics of stratified soil behaviour in estuarine site

G. Moulin, M. Kismi, R. Dupain & P. Thomas
Laboratoire de Génie Civil, ENSM Nantes, IUT St-Nazaire, France

RESUME : Dans certaines zones estuariennes, les travaux maritimes engendrent l'étude des sols argileux stratifiés.
Dans l'estuaire de la Loire (France), on rencontre ces sédiments dans les zones industrielles créées sur d'anciennes vasières latérales remblayées avec du sable. Ces matériaux sont soumis à des chargements portuaires provoquant des ruptures du sous-sol stratifié.
L'origine géologique du matériau est liée au surcreusement du lit rocheux du fleuve suivi de son comblement par des sédiments récents lors de la dernière période glaciaire.
La caractérisation physicochimique et le comportement mécanique de ce sédiment, alternance de sables fins et d'argiles, sont déduits d'expérimentations in-situ et en Laboratoire.
Malgré les différences de structures liées à son hétérogénéité, ce matériau a un comportement mécanique proche de celui d'une argile homogène. En particulier les caractéristiques géotechniques de ce sédiment sont très voisines de celles des argiles varvées du Canada.

ABSTRACT : In certain estuarine zones, it is necessary for marine works to study stratified clayed soils.
In the Loire estuary (France), these sediments, made-up of interbedded fine sands and clays, can be in industrial zones, built on ancient mudflats which have been filled with sands. These materials are subjected to loading effects from wharfs which provoke failures of the underlying stratified sediments.
The geological origin of these sediments is linked to the erosion of the underlying riverbed, and its infilling during the last glaciation.
The physicochemical characteristics and the mechanical behaviour of this material have been obtained by in-situ and laboratory experiments.
Despite the sediment being heterogeneous in composition, this material behaves like an homogeneous clay. In particular, the geotechnical characteristics of this sediment are very similar to that of the varved clays of Canada.

1 - INTRODUCTION

Les estuaires sont des zones géographiques de transition entre le milieu marin et le milieu terrestre.
Leur situation privilégiée a été depuis longtemps, à l'origine d'importants travaux d'aménagements.
L'estuaire de la Loire constitue un exemple intéressant de site aménagé : Construction de remblais sur sols compressibles liés au développement économique de la région, réalisation de grandes infrastructures de transport : routes, autoroutes, lignes ferroviaires, pistes d'attérrissage et zones aéroportuaires, avantports et zones d'activité économique portuaire.
D'autre part des zones industrielles ont été créés au bord de l'eau : pétrochimie, raffinerie, terminal méthanier, stockage de pondéreux.
Ces travaux nécessitent une bonne connaissance de la morphologie du substratum et de l'évolution du remplissage quater-

naire : des expérimentations in-situ et en laboratoire permettent de déterminer les propriétés physico-chimiques et mécaniques de ces sédiments.
Dans ce but, des sondages avec un carottier à piston stationnaire de 96mm de diamètre et des essais in-situ au scissomètre ont été réalisés.
Ces sondages sont localisés dans une surface de 25 mètres de diamètre, au Nord de la concavité de Montoir-de-Bretagne, sur la rive droite de l'embouchure de la Loire. Il s'agit de 5 sondages avec prélèvement de carottes entre 1,20 m et 14,20 m de profondeur, de 10 sondages avec prélèvement de carottes entre 5,20 m et 6,20 m, et de 5 reconnaisances au scissomètre, avec une série de mesures - distantes de 0,5 m - jusqu'à une profondeur de 15 m.

2 - CONTEXTE GEOLOGIQUE DU SITE

Les travaux d'aménagement effectués sur l'estuaire de la Loire donnent une bonne connaissance de la géologie du paléolit de la Loire, et de l'évolution du remplissage quaternaire de celui-ci.
La morphologie actuelle est le résultat d'actions successives d'érosion puis de sédimentation.
Ce paléolit est composé essentiellement de sédiments meubles, avec localement des pointements rocheux du substratum sousjacent.
La caractéristique essentielle de l'estuaire de la Loire est la grande profondeur du bed-rock, résultat du creusement lors de la régression préflandrienne contemporaine de la dernière phase glaciaire.
Il s'ensuit une morphologie très complexe révélée par de nombreux sondages effectués dans la région (2000 sondages référencés par le B.R.G.M.).

2.1 - La morphologie du substratum

La plus grande partie du socle est constituée de roches métamorphiques : micaschistes, gneiss, leptynites ...
Les bancs redressés presque à la verticale ont généralement une orientation armoricaine (110 à 140° Nord).
De nombreux filons de quartz et quelques massifs granitiques recoupent ces séries.
Le substratum sain est composé de roches non altérées souvent très fracturées.
La figure 1 (sondages distants d'environ 5 m) met en évidence la variabilité de cette structure qui présente une altération, fonction de la nature de la roche, de son obliquité, de l'ampleur de la

fracturation et des diaclases.
Le faciès résultant sera d'autant plus sableux que la roche-mère était riche en quartz, et d'autant plus argileux qu'elle contenait des feldspaths et des micas.
Suivant l'exemple de la figure 1, ceci implique au niveau des fondations d'une construction lourde, la réalisation de pieux avec des longueurs variant entre 3 mètres et 12 mètres.

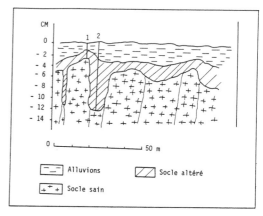

Fig. 1 - Schéma montrant l'altération en poche du socle.

Suivant F. OTTMANN et al (1968), le bedrock a été modelé et creusé par la Loire lors de la dernière période glaciaire.
Pendant cette période le niveau de la mer se serait abaissé de 110 à 160 m par rapport au niveau actuel. Au cours de cette régression le fleuve a creusé rapidement son lit, afin de rejoindre la mer sur la plate forme continentale.
Inversement, après la transgression flandrienne, le niveau de la mer est remonté pratiquement à son niveau actuel. Ainsi, par étapes successives, des sédiments fluviaux et marins ont remblayé la vallée creusée précédemment.
Le remplissage quaternaire de l'estuaire est donc le résultat de phases successives de régressions et de transgressions, dont les dernières, d'après L. BARBAROUX (1981), datent de l'époque gallo-romaine.

2.2 - Le remplissage sédimentaire de l'estuaire de la Loire

Lors de la remontée du niveau de la mer, un remplissage sédimentaire (postwürmien) du lit s'est opéré.
Il en est résulté une succession de sédiments constituée de bas en haut :
- de graviers et galets du remplissage post-glaciaire (sans doute lors d'une pé-

riode de régression),
- de sables gris à jaunes,
- de vases grises à noires à forte teneur
en eau,
- d'un complexe de sables fins et de vases dures appelé localement "Jalle" ; il
est caractérisé par une alternance de
lits de sables fins à très fins silteux
et d'argiles grises noirâtres, l'épaisseur de ces lits variant de quelques di-
xièmes de millimètres à quelques centimètres.
Ce faciès peut changer tant verticalement
qu'horizontalement créant une structure
varvée,
- de sables supérieurs appelés "sables de
Loire". (une partie de ceux-ci ont été
dragués dans l'estuaire et refoulés hydrauliquement sur la vasière de Méan pour
créer la zone industrielle de Montoir-de-
Bretagne),
- de vases molles, très compressibles
âgées de quelques dizaines d'années au
plus, et résultant du colmatage des vasières latérales.
Cette succession sédimentaire apparaît
nettement sur la coupe schématique présentée en figure 2.
Compte-tenu du relief tourmenté du paléolit, l'épaisseur des sédiments est très
variable et atteint 50m à certains endroits.
Dans ce contexte, l'installation de stockage de pondéreux induit des tassements
différentiels importants (sur une période
de 4 mois, des valeurs de 15 cm et plus
ont été observées).

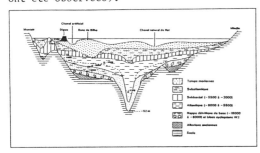

Fig. 2 - Coupe schématique du remplissage
de la Loire en fonction de l'âge
des terrains (datations
(d'après L.BARBAROUX, et al 1980)

3 - CARACTERISTIQUES PHYSICO- CHIMIQUES
DES SEDIMENTS.

Les sédiments de Montoir-de-Bretagne sont
essentiellement formés de "Jalle", dont
la caractérisation physique, représentée
figure 3, met en évidence une grande hétérogénéité pouvant s'expliquer par le

mode de formation.
Les teneurs en eau sont très variables et
sont comprises entre 47% et 110%.
Toutefois, ces déterminations sont ren-
dues difficiles par la grande hétérogénéité du matériau.
Ceci est montré par 40 mesures de teneur en eau effectuées sur une tranche de
carotte de 15cm de hauteur et de 9,6cm de
diamètre pour laquelle les teneurs en eau
varient de 65% à 100%.
Les limites d'Atterberg, mesurées selon
la méthode de Casagrande, montrent que la
teneur en eau naturelle W_o (moyenne de 4
à 8 valeurs) est généralement inférieure
ou égale à la limite de liquidité W_L.
A l'exception des 3 premiers mètres, l'é-
volution des caractéristiques granulomé-
trques est parfaitement corrélée avec les
évolutions du poids volumique moyen γ et
de la teneur en eau moyenne W_o.
Les valeurs extrêmes de ces caractéristiques en fonction de la teneur en argile C
(particules < 2 μm) sont les suivantes :
C = 25%, γ = 14,8 KN/m3, W_o = 105%
C = 10%, γ = 18,5 KN/m3, W_o = 55%

ce qui implique des indices des vides
plus élevés dans le premier cas que dans
le deuxième.
La caractérisation chimique des eaux interstitielles de cette région met en évidence l'importance de la concentration en
ions (Na^+ , K^+ , Ca^{++} , Mg^{++} , Cl^- ,
SO_4^{--}) et donc de la salinité (10 à 18
g/l).
Cette salinité peut être interprétée
comme caractéristique de l'origine marine des sédiments. Elle constitue également un moyen de reconnaissance des sols
atteints par des ruptures lors de la mise
en place de remblais (R. DUPAIN 1989).

4 - CARACTERISTIQUES GEOMECANIQUES DES
SEDIMENTS

La procédure expérimentale comprend des
essais oedométriques réalisés sur des
éprouvettes issues de carottes de 96mm de
diamètre prélevées lors de la campagne de
reconnaissance et des mesures de cisaillement au scissomètre (mesures in-situ et
en laboratoire).
Les essais triaxiaux consolidés non
drainés seront présentés dans une autre
étude.
Les essais oedométriques standards (58
essais) effectués sur des éprouvettes
(correspondant à des profondeurs de 1,60m
à 8,00m) permettent de caractériser
(fig.4) les indices de compression (Cc)
et de gonflement (Cs), la contrainte
effective de préconsolidation ($\sigma'p$),

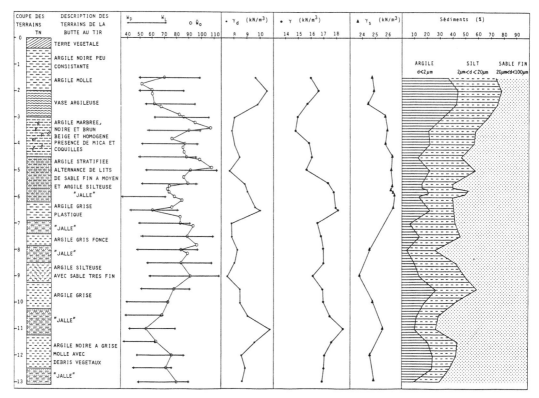

Fig. 3 - Argile de Montoir-de-Bretagne, caractéristiques physiques

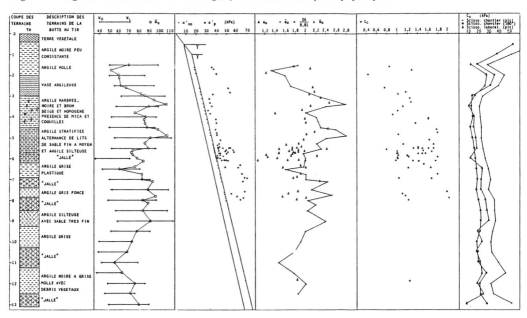

Fig. 4 - Argile de Montoir-de-Bretagne, caractéristiques géomécaniques

l'indice des vides en place (eo) du matériau.

L'évolution de σ'p fait, en particulier, apparaître une zone surconsolidée en surface, jusqu'à 2,80.mètres de profondeur, comme l'indique par ailleurs l'évolution de la cohésion non-drainée (cu) mesurée au scissomètre.

Pour des profondeurs de 2,80m à 8,00m le sol est légèrement surconsolidé, les diverses caractéristiques oedométriques du matériau - en particulier eo et Cc - varient dans de très grandes proportions.

L'indice des vides en place eo varie de 1,0 à 2,4 et l'indice de compression Cc de 0,4 à 2,1.

Compte-tenu de ces grandes variations et pour préciser celles-ci, 24 essais oedométriques ont été réalisés sur des éprouvettes correspondant à des prélèvements effectués à des profondeurs variant de 5,20m à 6,20m.

On observe des variations de eo et de Cc aussi importantes entre 5,20m et 6,20m qu'entre 2,80m et 8,00m.

Cette dispersion apparente des résultats dans cette zone est dûe aux changements rapides des faciès du matériau (alternance de lits sableux et de lits argileux).

Ce phénomène apparaît clairement sur la figure 5 représentant les courbes oedométriques de 4 éprouvettes voisines, découpées dans une même carotte pour laquelle la profondeur varie entre 5,55m et 5,85m. Ces courbes oedométriques font apparaître une variation de l'indice des vides initial $\Delta e = 0,70$, parallèlement l'indice de compression Cc varie de 1,19 à 1,69 (comme l'indique le schéma de la figure 5, les valeurs de Cc sont déterminées à partir de la pente intitiale de la courbe de compression).

Les valeurs respectives de la contrainte effective verticale en place (σ'vo = 38 à 39 kPa) et de la contrainte de préconsolidation (σ'p = 40 à 50 kPa) confirment la légère surconsolidation du matériau.

L'évolution rapide de ces caractéristiques a été mise en évidence en calculant eo d'après la teneur en eau moyenne Wo.

Pour une densité moyenne mesurée des grains de 2,65 et dans le cas d'un sol saturé, la courbe eo en fonction de Wo est tracée sur la figure 4.

Cette courbe confirme en partie les résultats obtenus à partir des essais oedométriques, et en particulier les évolutions très rapides de l'indice des vides en fonction de la profondeur.

Aussi, pour confronter ces résultats entre eux, le graphe Cc-eo a été établi, figure 6. Sur cette même figure sont représentées les argiles de Pornic et de Cordemais (G. MOULIN 1988) proches du site étudié, et les argiles de l'Est du Canada (S. LEROUEIL et al 1983).

Malgré la dispersion des points obtenus

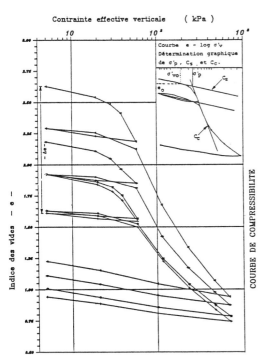

Fig. 5 - Courbes effort-déformation oedométriques de 5,55 à 5,85 mètres de profondeur

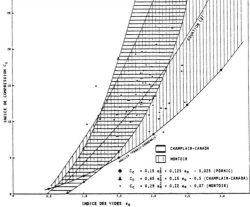

Fig. 6 - Relation entre l'indice de compression et l'indice des vides pour les argiles de Montoir-de-Bretagne.

dans ce graphe, une similitude d'évolution de Cc en fonction de eo est à remarquer.
Le fuseau obtenu pour l'argile de Montoir-de-Bretagne est limité en partie inférieure par la courbe caractérisant l'argile de Pornic, ayant pour équation :

$$\frac{Cc}{1+eo} = 0,405 \ \frac{wo}{100} - 0,025 \qquad (1)$$

et en partie supérieure par une limite voisine de celle des argiles de l'Est du Canada, dont l'équation moyenne est :

$$\frac{Cc}{1+eo} = 1,80 \ \frac{wo}{100} - 0,50 \qquad (2)$$

la relation moyenne pour l'argile de Montoir-de-Bretagne est

$$\frac{Cc}{1+eo} = 0,755 \ \frac{wo}{100} - 0,07 \qquad (3)$$

La comparaison de (1), (2) et (3) indique que, pour une même teneur en eau, l'argile de Montoir-de-Bretagne est plus compressible que celle de Pornic et généralement moins compressible que les argiles de l'Est du Canada.
D'autre part l'évolution de $\sigma'p$ en fonction de la profondeur suit celle de Cu de façon quasi linéaire :
à 3 m de profondeur $\sigma'p$ = 32 kPa et Cu $\simeq$ 20kPa
à 8 m de profondeur $\sigma'p$ = 65 kPa et Cu $\simeq$ 35kPa
Ainsi le quotient Cu/ $\sigma'p$ reste pratiquement constant, soit :

$$\frac{Cu}{\sigma'p} = 0,6$$

Ainsi, malgré une grande dispersion des indices des vides et de compression liée au caractère feuilleté donc hétérogène du matériau étudié, on arrive à une certaine constance de l'évolution de la cohésion non drainée avec la pression de préconsolidation.

5 - CONCLUSION

Dans cette région estuarienne, caractérisée par la continuité des faciès, les sédiments argileux présentent une hétérogénéité de structure qui est à l'origine d'une grande variabilité des caractéristiques géomécaniques.
Les variations horizontales et verticales du faciès qui sont engendrées par la configuration du paléolit affectent en particulier le comportement oedométrique du sédiment.

Cependant, malgré les changements rapides de la structure de ce matériau, nous avons pu montrer que celui-ci ne diffère guère mécaniquement d'une argile homogène.

REFERENCES BIBLIOGRAPHIQUES

DUPAIN, R., (1989). Rupture de remblais en site estuarien apport de la géochimie. Bull. AIGI n° 39, 91-97 Paris.
KISMI, M., (1986). Etude et caractérisation de sédiments estuariens proche de la concavité de Montoir. D.E.A. E.N.S.M. Nantes.
KISMI, M., DUPAIN, R., (1987). Un sédiment fin de l'estuaire de la Loire : "Jalle" nature et origine sédimentologique. 3èmes Journées Universitaires de Géotechnique St-Nazaire, 73-82.
LEROUIL, S., TAVENAS, F., LE BIHAN, J.P., (1983). Propriétés caractéristiques des argiles de l'Est de Canada. Can. Géotech. J.20 681-705
MOULIN, G., (1988) Etat limite d'une argile naturelle l'argile de Pornic. Thèse de doctorat Génie Civil Nantes, 121 pp
OTTMANN, F., ALIX. Y., LIMASSET, O., (1968). Sur le "lit ancien" de la Loire dans son cours inférieur. Bull. BRGM, Section 1, N° 2.

6th International IAEG Congres / 6ème Congrès International de AIGI, © 1990 Balkema, Rotterdam. ISBN 90 6191 130 3

Linear cutting tests in artificial sand-clay mixtures to study the influence of brittleness on abrasive wear of cutting tools

Des expérimentations de coupure linéaire dans des mélanges de sable et d'argile artificiels afin d'étudier l'influence de la friabilité sur l'usure de dents coupantes

M. W. Reinking
Delft University of Technology, Faculty of Mining and Petroleum Engineering, Section of Engineering Geology, Netherlands

ABSTRACT: Linear cutting tests on an artificial kaolinite clay were performed to distinquish different failure modes during the drying process. The influence of brittleness on abrasive wear of cutting tools was studied by drying an artificial sand-clay mixture. Three different failure patterns were examined: a ductile mode, at which the chip sticks to the cutting blade, a transitional mode and a brittle mode, at which chips were broken off. Brittleness is expected to influence the wear mechanism during cutting: the more brittle the rock or soil, the more three-body abrasion will occur. In ductile material two-body abrasion is expected to be of main importance.

RESUME: Des expériments de coupure linéaire ont été faits sur une argile kaolinite artificielle pour distinguer des modèles de cassure pendant la dessiccation. L'influence de la friabilité sur l'usure abrasive de dents coupantes a été examiné pendant la dessiccation d'un mélange de sable et d'argile artificielle. Trois modèles de cassure différent ont été examinés: Un modèle ductile, où la couche colle au tranchant, un modèle transistoire, et un modèle friable où la couche a été brisé. Il est supposé que la friabilité influence le méchanisme d'usure pendant la coupure: si la friabilité de la pierre ou du sol est plus grande, il y apparaîtra plus d'abrasion tri-corporelle. Quand aux matériaux ductiles, on suppose que l'abrasion bi-corporelle y soit essentielle.

1 INTRODUCTION

Abrasive wear of rock-cutter picks is one of the main problems occurring at many rock-cutting dredging projects. To develop reliable site-investigation techniques which can predict the severity of abrasive wear, one has to get a clear insight of all parameters influencing the abrasive wear process.

One aspect which is expected to influence this is the brittleness of rock [1]. According to Robberts [2], rock can fail in a brittle and a ductile manner during cutting. In purely brittle material, a rock chip would break off after penetration by a wedge-shaped bit and the bit would again penetrate the fresh rock. In a ductile rock the contact between the rock and wedge is continuous. From this point of view the abrasive wear is expected to be larger in rock which fail in a ductile manner [1].

In a study of the influence of failure mechanism on abrasive wear of cutting tools, the main problem is to obtain a test specimen which can deform in both brittle and ductile failure mode.

Furthermore, to compare the abrasive wear in both modes, the other properties of the specimen which influence the wear process have to be constant. This pilot study describes how a ductile and a brittle failure mode were obtained in different clay specimens.

2 LINEAR CUTTING TESTS WITH A DRYING KAOLINITE CLAY

2.1 The test procedure

For the experiments a French shaper, trademark Guillemin-Sergor et Pegard (GSP), was used. The cutting blade was of a simple chisel shape, 49 mm wide, 150 mm in length, a surface roughness of 0.2 μm and made of steel 37.

The clay used for the testing, was an

artificial kaolinite clay (table 1). The mixture has been chosen for its excellent drying properties. No shrinkage cracks appeared during the drying process.

Table 1   Properties of the kaolinite clay.

Property	Value
Atterberg Limits:	
Liquid Limit (LL) Plastic Limit (PL) Plasticity index (PI)	37.0% 16.5% 20.5%
Moisture Content	22.1%
Shear Strength (lab. vane test) peak residual	27.2 kPa 13.7 kPa

This kaolinite clay is an artificial mixture of 60% clay, mainly kaolinite, 30% quartz powder, and 10% calcite.

To prepare the samples for the cutting tests, specimens measuring 21 cm in length, 12 cm in width and 7 cm in thickness were cut from the clay blocks.
In order to observe the failure modes during the linear cutting, 7 samples with different moisture contents were tested. These different moisture contents were developed in the clay sample by allowing them to air-dry. The average moisture contents at which the specimens were tested were: 22.1% ; 20.2% ; 18.6% ; 16.7% ; 13.7% ; 11.6% and 7.7%. Within each specimen, two types of cutting were performed:
Type A: The cutting was performed at the center of the sample. The effects of the sides as well as the proceeding of the chip could be studied.
Type B: The cutting was performed at the side of the specimen. The cutting pattern at the cutting edge could be studied.

The following mechanical test parameters have been used:

- cutting angle: 30°
- cutting velocity: 6 m/min
- cutting depth: 5 mm
- stroke : 150 mm

The test arrangement is shown in figure 1. Only around the transition zone, a higher cutting velocity (50 m/min) was used to study the influence of the velocity on the deformation mode.

Figure 1. The test arrangement

2.2 Discussion and results

From the linear cutting tests, it was found that during the drying of the clay specimens three types of failure behaviour can be distinguished. Samples with a moisture content greater than 17% failed in a ductile manner. Samples with a moisture content lower than 16% showed a brittle failure mode and between these failure modes there exists a transition zone.
To characterise the different failure modes four parameters have been studied:

- the behaviour of the soil chips
- the sides of the cut
- the cutting surface
- the cutting pattern in front of the cutting edge

The ductile mode:
This failure mode has the same characteristics as the "flow" type of failure described by two Japanese investigators Hatamura and Chijiiwa [3]: a zone appears where shear deformation occurs continuously when the cutting blade proceeds. The soil chip sticks to the cutting blade along its whole length. In the direction of the cutting edge the thickness of the soil chip increases slightly (fig. 2). The cutting sides are straight as is shown in figure 2. The cutting surface is smooth.

The transitional mode:
As the cutting blade proceeds a chip develops on the cutting blade face as is

2914

the case during the ductile mode.
However, instead of sticking to the
blade, the chip curves away from it (fig.
3). The cutting side pattern and the
cutting surface shows the same
characteristics as developed during the
ductile mode, but the soil chip retains
the same thickness within its whole
length.

<u>The brittle mode:</u>
As the cutting blade proceeds, chips are
broken off in front of the cutting edge.
A slightly curved failure line appears as
is shown in figure 4. A "break out"
effect at both sides of the tool was
observed (fig. 4). A small contact area
(A) between the chip and the blade exist,
which decreases with decreasing cutting
depth (fig. 5). The cutting surface is
smooth.

With decreasing moisture content, the
soil chips were not able to stay intact
for there whole width. They tended to
crumble. Cutting with higher velocities
at the transition zone made no visible
difference with respect to the failure
mode.
    The break-out of rock at both sides of
the cutting tool is also observed by
Roxborough [4]. It is referred to as
"sidesplay" or "breakout".

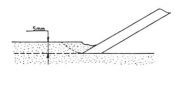

Figure 4. The brittle type of failure.

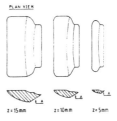

Figure 5. The chip dimensions.

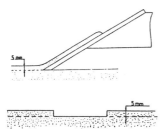

Figure 2. The ductile type of failure.

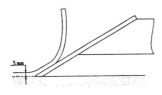

Figure 3. The transitional type of
failure.

3 CUTTING TESTS WITH A SAND-CLAY MIXTURE

3.1 The sand-clay mixture

To induce abrasive wear of the cutting
tool, the quartz content of the soil is
of main importance. An artificial sand-
clay mixture with 60 weight % quartz sand
was chosen. This mixture showed a nice
homogeneous structure. Also, when ovendried (110 °C, 24 hours), the quartz
grains were still held in place by the
clay matrix. Tests were done to
characterise the sand-clay mixture (table
2).

3.2 The test procedure

A special cutting test was developed to
compare the abrasive wear during cutting
in the ductile mode and the brittle mode.
The test consisted of cutting a groove 25
mm wide and 25 mm deep along the surface
length of the sand-clay mix core sample
parallel to its axis.

This was obtained by the use of the automatic heave of the shaper; automatically 50 cuts of 0.5 mm were made. With a sample test length of 200 mm this provided a total cutting length of 10 meter. Before and after each series of 50 strokes, the cutting blade was weighted and the loss of weight gave an indication of the abrasive wear.

Table 2. Properties of the sand-clay mixture.

Sand-clay mixture characteristics		
Property	Test	Value
Atterberg Limits:		
Liquid Limit (LL)	cone penetrometer	24.6%
Plastic Limit (PL)		12.75%
Plasticity Index (PI)		11.85%
Linear Shrinkage (LS)		5.29%
Moisture content (Mc)		14.5%
Shear strength:		
Peak	lab. vane test	31.5 kPa
Residual	lab. vane test	11.1 kPa
Brittleness index		1.84
Sensitivity		2.84
Cohesion	undrained triaxial test	30.0 kPa
angle of internal friction	undrained triaxial test	4.5°
Unconfined Compressive Strength (UCS):		
- dry (110 °C)	UCS test	2700 kPa
- wet (Mc = 14.5)	UCS test	37.3 kPa
Bulk density		2.17 g/cm³
Dry density		1.89 g/cm³

The cutting blade was of a simple chisel shape, 25 mm wide, 85 mm in length, a surface roughness of 0.2 μm, and made of steel 37.

The tests were performed with a cutting angle of 45 degrees, resulting in a front rake angle of 45 degrees and a clearance angle of 0 degrees. The cutting velocity was 8 m/min.

Cutting was done in both the ductile and the brittle mode. The brittle mode was obtained by drying the clay mixture in the oven at 110 °C for 24 hours. This was done to exclude linear variation of moisture contents throughout the sample. Before the actual test started the upper surface was levelled to obtain a straight surface.

3.3 Discussion and results

The results are shown in table 3.
Table 3. Weight loss of the cutting blade during ductile- and brittle failure.

Test No.	Weight loss (g)	
	ductile mode	brittle mode
1	0.001	0.026
2	0.000	0.022
3	0.000	0.030
4	0.002	0.024
5	0.000	0.020

In spite of the total cutting length of 10 meter and the high quartz content of the sample, the wear of the cutting blade was not measurable in the ductile mode. An attempt to obtain a measurable weight loss by increasing the cutting velocity to 54 m/min, failed. The weight loss stayed outside the precision of the balance. The average loss of weight of the tool during cutting in the brittle mode was 0.024 gram.

During cutting in the brittle mode the abrasion process seemed to be of a mixed type: both two-body* and three-body abrasion* appeared. The loss of weight of the cutting blade was caused by the quartz grains that sticked to the clay matrix and by quartz grains which were loosened by the cutting action. This was proven by examining the cutting surface after the last cut. A lot of loose particles were found on the cutting surface. Because of the large amount of the loose particles, the process of three body abrasion is expected to be the main cause of the wear. The occurrence of the crushing and trapping of loose particles between the intact rock and the tool is also observed by Roxborough [4].

The problem when cutting through a plastic clay is that no, or very little, abrasive wear of the cutting blade can occur as long as the quartz grains can be pushed into the clay matrix. This process happened during cutting in the plastic clay mixture that was used; the cutting surface was completely smooth.

From the previous observations it is expected that the main mechanism of the wear process in the ductile mode will be of the two-body type. During cutting in a ductile rock or soil, the loosening process of abrasive particles out of the matrix, is expected to be less, compared to brittle rock. Considering the observed types of abrasion, if the strength characteristics are similar, a higher abrasive wear would be suspected in the ductile mode.

* Two-body abrasion (sliding abrasion): wear caused by abrasive particles which are held fixed as in abrasive paper.

Three-body abrasion: involving loose particles between cutting tool and the cutting material, which may rotate as they contact the wearing surface [1].

4 CONCLUSION

Linear cutting tests in both a ductile and a brittle sand-clay mixture are promising for a study of the influence of brittleness on abrasive wear of cutting tools. During the drying process of such a mixture, a ductile and a brittle type of failure can be induced. Since parameters such as the mineral composition, the particle size and the angularity of the abrasive constituents are constant, only the brittleness and the strength of the test specimen vary. To exclude the influence of the mechanical strength parameters, the cutting forces have to be determined. If the cutting forces are known, factors can be introduced which make it possible to make realistic comparisons of the wear within both ductile and brittle modes of the test specimen.

Brittleness is expected to influence the wear mechanism during cutting: the more brittle the rock or soil, the more three-body abrasion will occur. In ductile rock or soil, two body abrasion is expected to be of main importance, provided that the quartz grains are fixed in the clay matrix.

More detailed investigation of the wear processes within the ductile mode is necessary to obtain a measurable quantity of wear. Since it is known that cohesive soils containing quartz can be the source of excessive wear, this is certainly an important topic for further investigation.

Acknowledgement

The work described in this paper forms a part of the research project (Abrasive wear of rock cutting tools by rock), which is supported by the Technology Foundation (STW).

REFERENCES

[1] Verhoef, P.N.W.: "Towards a prediction of abrasive wear of cutting tools in rock dredging", Delft Progr. Report 13, pp 307-320, 1988/1989
[2] Roberts, R.: "Applied Geotechnology", Permagon Press, 1981
[3] Hatamura, Y.; Chijiiwa, K.: "Analysis of the mechanism of soil cutting"
(1st Report), Bulletin of the JSME, Volume 18, No. 120, 1975
(2nd Report), Bulletin of the JSME, Volume 19, No. 131, 1976
(3rd Report), Bulletin of the JSME, Volume 19, No. 137, 1976
(4th Report), Bulletin of the JSME, Volume 20, No. 139, 1977
(5th Report), Bulletin of the JSME, Volume 20, No. 141, 1977
[4] Roxborough, F.F.: "Cutting rock with picks", The Mining Engineer, Vol. 132, pp 445-452, 1972

6th International IAEG Congres / 6ème Congrès International de AIGI, © 1990 Balkema, Rotterdam. ISBN 90 6191 130 3

Numerical modelling of transient flow in marine breakwaters
Modélisation numérique des écoulements transitoires dans les digues en mer

J. M. Usseglio-Polatera & A. Libaux
Laboratoire d'Hydraulique de France - LHF, Grenoble, France

ABSTRACT: The internal flow and pressure distribution within a deep breakwater at sea is fundamental for its stability under wave attack. A computer code accounting for non-Darcian flow equations and unsaturated zone has been developped and validated against in situ measurements at Calais, France.

RESUME: La répartition des écoulements et des pressions dans une digue en grande profondeur est fondamentale pour sa stabilité sous l'effet de la houle. Un code de calcul prenant en compte une formulation non darcienne de l'écoulement et la zone non saturée a été élaboré et validé sur des mesures in situ effectuées à Calais.

1. CONTEXT AND OBJECTIVES

Several accidental ruptures of deep rubble-mound breakwaters shortly after their construction have been reported in recent decades, namely Sines (Portugal) Arzew (Algeria), Tripoli (Lybia). Various explanations for the failure have been proposed: inadequate estimate of reference waves, local convergence of refracted waves, exceptional amplitude induced by trains of waves, weakness of blocks,.. Among these various reasons, the influence of the internal flow (hence pressure distribution) was very controversial and, particularly, after the Sines accident, the experts' interpretations varied from no-influence to major cause of the failure.

The French Ministry for Research and Technology decided to support further detailed investigations on the influence of internal flow variations on the stability, namely dynamics of flow and pressure distribution, heterogeneity of the breakwater structure, turbulent effects.

A series of research studies was carried out (references in Parisot, 1987) and it was decided to instrument a real breakwater (East Jetty at Calais, France) with pressure transducers in order to observe the propagation of waves within the breakwater.

This paper presents a further step of investigations, based on an advanced numerical model of flow, the main concepts of which have been validated against the data collected at Calais.

2. RELEVANCE OF NUMERICAL MODELLING

The application of a physical (scale) model to the problem requires a thorough reflection upon the relevant scale rules (Barends, 1988). They must not be founded on similitude only but on dynamic equilibrium of the actual forces in the model and in the reality. Actually, it seems almost impossible to meet all the requirements (scale rules for all phenomena involved) in a single model and there are significant risks of spurious results.

On the other hand, numerical modelling offers powerful tools although it may suffer from the incomplete formulation of the physical phenomena involved. In addition, the numerical model, when validated in depth, provides a predictive value in the framework of its basic assumptions.

Experiences of numerical modelling of flow and pressure variations within a breakwater under dynamic wave attack have already been conducted in the past few years (Barends, 1988).

The major originality of the present contribution is the introduction of an unsaturated zone which allows, among other advantages, a simultaneous computation of the flow velocities and of the water table.

3. FORMULATION AND CODING PRINCIPLES

LHF has recently developed an advanced software environment for the computation of unsteady flow and transport in heterogeneous porous media (saturated and unsaturated) including density effects (Usseglio-Polatera, 1989 and 1990). The framework of this software environment has been used and the formulation has been modified to account for non-linear flow equations (turbulent effects especially within the cover). A two-dimensional formulation in the vertical plane (2-DV) is implemented here.

3.1 Governing equations

Inertial forces are not negligible with respect to viscosity when the particles are coarse and a generalized Darcy's law must be applied. The Forchheimer formulation has been implemented in the form (Bear, 1979):

$$\vec{V}\,(\,\alpha + \beta\, \|\vec{V}\|) = -\overrightarrow{grad}\ h \qquad (1)$$

with: $\vec{V}$ Darcy's velocity vector
         h hydraulic head

Various formulations of the coefficients α and β have been proposed in the literature. The formulation by Ahmed-Sunada and Ward (Bear, 1987) has been selected since it relates α and β to only one parameter $\bar{d}$, the average dimension of the grain diameter through the following formulae:

$$\alpha = \frac{360.\nu}{g\bar{d}^2} \qquad \beta = \frac{0,55\sqrt{360}}{g\bar{d}} \qquad (2)$$

with:

g gravity
$\bar{d}$ grain diameter
ν fluid kinematic viscosity

The mass balance is formulated as follows:

$$\frac{\partial w}{\partial t} = -div\ (\vec{v}) \qquad (3)$$

w water content
t time

Variations of soil characteristics with respect to water content must be introduced to account for the unsaturated zone.

These formulations usually relate the hydraulic conductivity and the capillary suction to the water content in the Darcian framework. Therefore in non-Darcian formulation, an equivalent conductivity $\widetilde{K}$

$$(\widetilde{K} = \frac{1}{\alpha + \beta\ \|\vec{V}\|}\ )$$

must be defined and curves $\widetilde{K}(\theta)$ and $h(\theta)$ must be given for each type of soil.

Eq.(1) and (3) are thus combined following Richards' formulation based on the introduction of capillary capacity, as follows:

$$\gamma(h)\ \frac{\partial h}{\partial t} = div.\ (\widetilde{K}(\theta).\ \overrightarrow{grad}\ h) \qquad (4)$$

where

$$\gamma(h) = \frac{d\theta}{dh}$$

is the capillary capacity (calculated from $\theta(h)$).

Note that the capillary suction $\psi(\theta) = -h\ (\theta)$

3.2 Numerical scheme

An implicit finite difference scheme is used. The equivalent conductivity is discretised explicitly. A variant of ADI (Alternate Direction Implicit) method is used with an iterative procedure monotoring the level of accuracy. The capillary capacity is updated at each iteration since it must be expressed at intermediate time for the consistency of the discretisation. Within each iteration, two systems of tridiagonal matrices are solved through a Thomas double sweep algorithm. This scheme minimizes the computational time with good accuracy (Usseglio-Polatera, 1989).

3.3 Boundary conditions

Various types of boundary conditions are allowed by the code. For the case of breakwaters under wave attack, the piezometric head (water level) is prescribed at both faces of the dike.

At the seaward face, the formulation of N. Shuto (Shuto, 1972) is implemented; it gives a time-varying estimate of the run up/run down of waves, accounting for the embankment slope. The major inadequacy of this formulation is that it considers the dike as impervious hence ignores the influence on the run-up of flow through the dike cover. At the rear face, a constant water level (head) is prescribed. Above the water level on both faces, a seepage surface condition is implemented allowing the seepage of water along the emerged embankment.

3.4 Soil characteristics

The computational domain (breakwater) is divided into sub-zones of variable properties. Within each sub-zone, the relations between the equivalent hydraulic conductivity, the capillary suction (see 3.1) and the water content are provided. Typical curves are given on fig. 1. For each sub-zone, two couples of curves for two directions may be given to account for anisotropy.

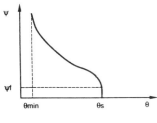

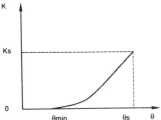

Fig. 1 - Typical curves for soil characteristics

4. APPLICATION TO CALAIS BREAKWATER

4.1 Calais breakwater

The East Jetty at Calais is 1100 m long and protected with ACCROPODE(R) blocks.
The instrumented section has a trapeziform cross-section (bottom width: 74 m, top width: 20 m, height: 20.25 m, embankment slope: 0.75).

The mean water depth is about 16 m at high water and 12 m at low water.
The core is made of 0-1000 kg blocks (thickness 2.00 m) and a layer of 6.3 m ACCROPODE(R) blocks (thickness 2.70 m). The rear face cover is made of a layer of 500-1000 kg blocks (thickness 2.40 m) and a layer of 4 m³ ACCROPODE(R) blocks (thickness 2.30 m).
The instrumented section includes five vibrating-wire pendulums (Telemac type) located on a same horizontal (6 m above the bottom) about every 7.5 m from the seaward side. The signals can be correlated with those transmitted by a Datawell buoy placed in front of the breakwater.

4.2 Results of measurements

An analysis of the real time measurements showed some filtration of the short periods and significant damping of the wave components through the cover (at 1st transducer), then lower damping through the core. No significant phase lag was observed. Table 1 shows the measured damping coefficients of wave components for a wave height of 4.0 m and a 10.0 sec. period.

4.3 The model

Fig. 2 shows the model of the dike (rectangular grid with $\Delta x = 1.0$ m and $\Delta z = 0.75$ m) and the location of the measurement points. In this case, three types of soils were defined, characterized by $\bar{d}$ (see 3.1)
$\bar{d} = 0.2$ m in the seaward face cover
$\bar{d} = 0.13$ m in the rear face cover
$\bar{d} = 0.03$ m in the core
So far, although it is allowed by the model, no differences have been made between the two layers in the cover.
The capillary fringe on the suction curve was considered as nil. A wave height of 4.0 m with a period of 10.0 s at high water was simulated. The time step is 0.05 sec. and 5 wave periods are simulated in succession.

4.4 Model results and investigations

Table 1 shows a comparison of computed wave damping coefficients against the measured one in the case described in 4.3.

Table 1. Computed and measured wave damping coefficients between H (incident wave), A1, A2, A3, A4, A5 (internal measurement points)
Period: 10.0 sec; wave height: 4.0 m

	H/A1	A1/A2	A1/A3	A1/A4	A1/A5
Measured (High water)	4.0	1.3	1.5	1.7	2.0
Computed (High water)	3.8	1.1	1.4	1.7	2.2

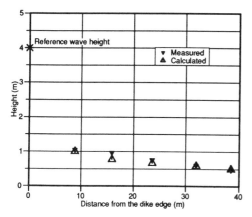

Fig. 2 - Computed and measured wave heights within the breakwater

Sensitivity analysis to the various parameters has been carried out and is still in progress.

Computations based on Darcy's linear law showed too small damping through the cover and too large one through the core.

The results are very sensitive to the respective values of $\bar{d}$.

The rear face boundary condition is influential and a constant head seems inadequate. Shifting the boundary, through an enlargement of the breakwater with a soil of high conductivity, significantly improved the results. The small discrepancies between measurements and computations must be related to the core heterogeneity.

Fine particles are progressively leached from upper levels and accumulate in the lower part of the core, leading to variable conductivities, hence variable $\bar{d}$. This point is currently under investigations. Figs. 4 to 6 show the distribution of pressure and velocity within the dike.

5. CONCLUSION

The developed modelling system, based upon Forchheimer's fomulation and accounting for the unsaturated zone, has shown very promising results, compared to in-situ measurements, with realistic soil parameters. Therefore, an application of the numerical tool for prediction of flow and pressure distribution within breakwaters at sea is envisaged.

Further investigations on the internal heterogeneity are planned as well as a second validation against other in-situ measurements.

The application of these results to the problem of stability will be considered subsequently.

6. ACKNOWLEDGEMENTS

This study was supported by a grant of the French Ministry for Research and Technology and conducted by Mr M. Couprie (SOLETANCHE) Mr P. Aristaghes (BCEOM), Mr. P. Galichon (SOGREAH), Mr J.M. Piau (LCPC) and Mr. A. Parisot (SPIE BATIGNOLLES).

REFERENCES

Aristaghes, P. 1986. Internal Flows in breakwater, Internal Report (in French) STCPMVN Compiègne, France.

Barends,F.B.J. & Holsher,P. 1988. Modelling interior processes in a breakwater, Proceedings of the Conference Breakwaters 1988, Thomas Telford Limited, London, 37-45.

Bear, J. 1979. Hydraulics of Groundwater, Mc Graw-Hill.

Bear, J. and Verruijt, A. 1987. Modelling Groundwater Flow and Pollution, D. Reidel Publishing Company.

Bonneton, M., Jardin, P., Lavedan, G., Moullard, P.Y. 1985. The MINOS Model, 2-D Representation of Subsurface Flow in Saturated and Unsaturated Zones, Proceedings of 21st IAHR Congress, Melbourne, Australia, August 1985.

Parisot, A., Aristaghes, P., Galichon & Cartier M. 1987. Wave induced pressures and flows in breakwaters: in situ experimentation and modelling (in French) - Actes du 3e colloque "Génie Civil et Recherche" Thème 4, 109-119.

Shuto, N. 1979. Standing waves in front of a sloping dike, Coastal Engineering Conference, Tokyo, Japan, Vol. 3.

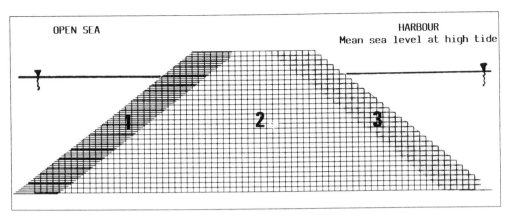

Fig. 3 - Layout of the Numerical Model of Calais breakwater

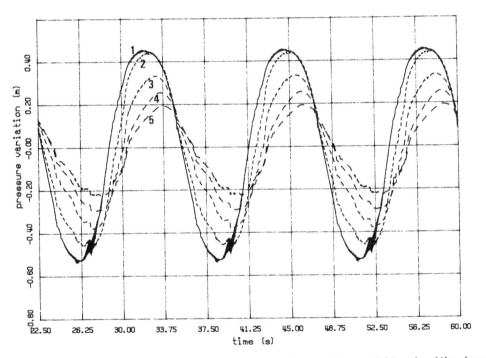

Fig. 4 - Timewise evolution of pressure at each transducer within the dike (pressures are given in meters of water)

Usseglio-Polatera, J.M. & Jardin, P. 1989. Software Environment for Transport and Dispersion of Contaminants within Saturated and Unsaturated Zones, Poster Paper to the International Symposium on Contaminant Transport in Groundwater, Stuttgart, FRG, April 1989.

Usseglio-Polatera, J.M. Aboujaoude, A., Molinaro, P., & Rangogni, R. 1990. 3-D Modelling of Coupled Groundwater Flow and Transport within Saturated and Unsaturated Zones, submitted to the Conference on Computational Methods in Water Resources, Venice, Italy, June 1990.

5.1 Maximum run-up

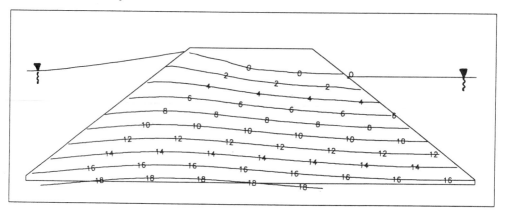

5.2 Mean level

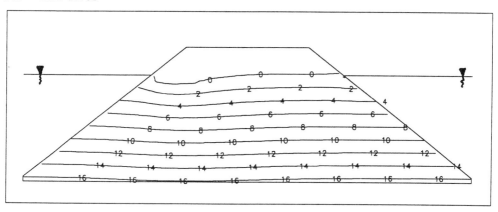

5.3 Maximum run down

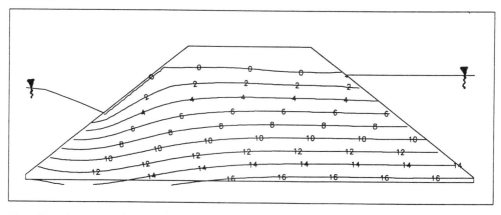

Fig. 5 – Pressure distribution within the breakwater at various times of the wave period. Wave height: 4.0 m; wave period: 10.0 sec. (Pressure are given in meters of water)

6.1 Maximum run-up

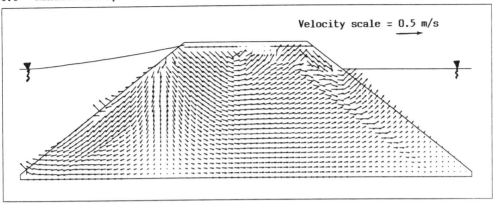

6.2 Mean level

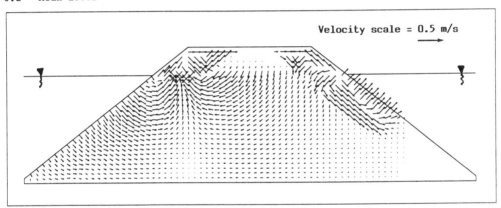

6.3 Maximum run down

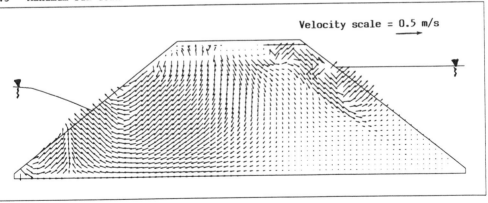

Fig. 6 - Velocity field within the breakwater at various times or the wave period.
Wave height 4.0 m, wave period: 10.0 sec.

Quarry yield and breakwater demand
Production de carrière et demande de brise-lames

J.K.Vrijling
Rijkswaterstaat, Directie Sluizen en Stuwen, Utrecht, Netherlands

A.H.Nooy v.d.Kolff
Delft Geotechnics, Delft, Netherlands

ABSTRACT: When operating a quarry, specially or temporary dedicated to the production of rock for a breakwater construction, economics usually urge a good match between quarry output and design requirements, as grading and quantity of the various functional elements of the structure. A simple method is presented to compare these two parameters.

RESUME: L'économie demande généralement de bonnes rélations de brise-lames entre carrière rapport et des projects demandés, comme granoclassement et grandeur des éléments différents fonctionnels de la structure et de faire usage d'une carrière, spéciale ou temporaire. Une méthode simple est présenté pour comparer les deux paramètres.

1 INTRODUCTION

All over the world many breakwaters and other coastal defence works are being constructed of quarried rock.

Usually such rubble mound constructions are built up of various functional elements (e.g. core, filter and armouring layers) consisting of specified gradings of rock to ensure satisfactory hydraulic and structural performance.

The required rock may be produced by specially dedicated, temporary dedicated or not dedicated quarries depending on geological conditions (availability of sufficient suitable rock), design requirements (gradings, quantities, etc.), economical reasons (costs, transport distances, etc.) and other considerations (environmental restrictions, political motives, etc.).

Engineering geological site investigations and trial blasting should give an indication of the most likely quarry yield that will be obtained. Such quarry output is usually presented as a weight distribution curve that has been estimated from a discontinuity survey or measured from results of trial blasting.

In particular when opening up a dedicated or temporary dedicated quarry it will be important to utilize as much of the produced materials as possible to avoid wastage and high costs. This may be achieved not only by choosing a favourable quarry location and applying suitable blasting techniques, but also by adapting the design of the construction to the expected production characteristics of the quarry without affecting the functional requirements imposed on the structure. This will require a design process consisting of a number of iterative stages incorporating not only structural and hydraulic considerations but also results of quarry yield analyses.

2 DESIGN REQUIREMENTS

The classical breakwater consists of various layers, each rock with its own function in the structure. A typical cross section is given in fig. 1.

The filter layer prevents the loss of the original bottom material via the coarse grained structure. The loss of material is caused by the vehement water movement inside the breakwater, when it is exposed to waves. To perform its function the grain size of the filter is designed in relation to the grain size of the original bottom material.

The body of the breakwater, the core, is constructed on top of the filter. This is generally a trapezoidal mound of quarry run with a loosely specified gradation curve.

The breakwater is defended against heavy wave attack by the armour layer. This layer is constructed of heavy rock elements that are stable under design wave conditions.
The stability of these rock elements is gouverned by the following parameter.

$$\frac{Hs}{\Delta\, Dn\, 50}$$

where Hs = significant wave height

$$\Delta = (\rho st - \rho w)/\rho w$$

$$Dn50 = (M50/\rho st)^{1/3}$$

The classical Hudson formula gives an upper value for this stability parameter. A more modern approach that also takes the wave period into account is proposed by Van der Meer [1]. Van der Meer's relation is given in fig. 2.

It is clear that the design wave height influences the size of the rock elements in the armour layer directly.
In many cases the size of the natural rock will not be sufficient and concrete elements have to be chosen instead.
On every interface between layers the problem of transport of material as described by the bottom filter layer has to be solved. Generally transport of material is avoided if the ratio between the diameters of adjacent layers is 6 - 10.
For severely attacked breakwaters this means a secondary layer between the core and the armour layer.

From the cross section of the breakwater and the required size of the rock for each layer, as follows from filter and stability relations, the material requirement can be specified as a function of the diameter of the rock (fig. 3).

Berm breakwaters use lighter rock relative to the design wave height than traditional rubble mound breakwaters. The lighter rock would have to be placed on a very flat slope to make it stable. Heavy equipment, i.e. cranes with long reach and high moment capacity, are necessary to construct such a breakwater.

Instead of placing the stones in two layers on a flat slope, the berm breakwater concept proposes that the rock is placed in a heap on the seaward face, requiring less heavy equipment. The rock mass is afterwards reshaped by wave action until a flat equilibrium slope is developed. Thus the main objective of the design of berm breakwaters is to determine the necessary size and extent of the heap of stones which will form the berm. It is of utmost importance to ensure that there is enough material in the berm to form the equilibrium slope.

Jensen and Sorensen [9] present some examples of berm breakwaters. In all their examples the stability parameters Hs/Δ Dn-50 equals about 3. With a design-significant wave height of 6.5 meters a figure of 7 tons follows for the M50 of the berm material. The equilibrium slope in the examples has a mean value of about 1:4.

Fig. 4 presents the cross section of the berm breakwater as developed in this paragraph.
It is easily understood that the material requirement of a berm breakwater as a function of the diameter of the rock differs considerably from the classical case.
From fig. 5 it appears that the amount of large rocks, is less in the case of the berm breakwater.

It is of course of utmost importance to equate the material demand of the breakwater to the output of the quarry dedicated to the project. For an economic design the cross section may have to be adapted to the quarry yield in an interactive design procedure.

3 QUARRY YIELD ANALYSIS

When opening up a dedicated or temporary dedicated quarry on behalf of the construction of a breakwater, quarry yield analyses should form an intrinsic part of the design process.
A quarry yield analysis will evaluate the results of (preliminary) engineering geological investigations, trial blasting and the subsequent fragmentation analyses in order to predict future production characteristics of the potential quarry.
An engineering geological investigation will provide valuable information about volume, quality and in-situ fracturing of the rock mass. A geological reconnaissance of the area will indicate the nature of

the rock, may in general terms assess the quantity of rock available and may select the most promising sites. Further study of these selected locations may include drilling, sampling and laboratory testing, geophysical investigations, discontinuity surveying, etc..

The properties of the rock may strongly influence the initial design and often the construction techniques, the capital costs, maintenance costs and life expectancy of the structure. The natural discontinuity pattern within the rock mass, to be studied from the borehole results and the discontinuity survey will limit the largest fragments that may be produced by the quarry.

Although literature suggests that fragmentation can be theoretically related to blasting parameters, rock mass properties and even to geological discontinuities within the rock mass; trial blasting will usually be carried out to establish the optimum blasting parameters in relation to the (required) fragmentation.
Blasting results depend upon many, often interrelated parameters as burden - spacing ratio, borehole geometry, specific charge, delay pattern and upon boundary conditions as geological discontinuity patterns and other rock mass properties. During trial blasting (and even during the production phase) these parameters should be changed systematically in order to find the best combination.

Evaluation of blasting results will demand the measurement of all rock fragments of the blasted volume. This may be done by calculating the weight of each individual block from its measured volume and unit weight or by weighing each block. An accurate measurement of the total blasted volume is required to estimate the percentage fines (i.e. smaller than 0.15 m^3), that has not been recorded. Results may be presented as histograms showing the number of blocks or percentage of total number of blocks in specified volume or weight ranges or as volume or weight distribution graphs (expressed as percentages of exceedence), sometimes referred to as expected quarry yield curves.

In contrary to common quarrying practice blasting for breakwater construction in a dedicated quarry is usually not aimed at obtaining good fragmentation, but is often directed solely to the production of large size fragments without much concern for the smaller fractions produced with it. Depending on the construction requirements this may imply considerable wastage and

consequently uneconomic quarry operations.

4 SUGGESTED EVALUATION TECHNIQUE

To facilitate the suggested interactive design process it will be important to enable a simple comparison between the expected quarry yield and the design requirements with respect to grading and quantity of the various elements of the breakwater.

The quarry yield, usually presented as a weight size distribution curve (expressed as percentages of exceedance), may be mathematically described by the Rosin Rammler equation, in its general form:

$$y = 1 - \exp\left(-\left(\frac{x}{x_c}\right)^n\right) \qquad (1)$$

in which:
y = cumulative weight in % finer than x
x = particle size (block size)
x_c = characteristic particle size (63% smaller than x_c)
n = index of uniformity

The characteristic particle size x_c (or x_{63}) can be determined from the quarry yield curve, while the index of uniformity n will follow from curve fitting. Experience indicates that the Rosin Rammler equation usually represents quarry yield curves reasonably well, although it should be realized that in its extremities (the fine and coarse fractions) it may deviate considerable from the actual situation.

By differentiating the Rosin Rammler equation a sieve density curve will be obtained that easily can be compared with the breakwater demand.

The Rosin Rammler equation has been fitted to the yield of various quarries.
In most cases a close fit is achieved.

	x_c	n
Diorite quarry Nicaragua	0.409	1.049
Grauwack quarry West Germany	0.20	1.20
Granite quarry (1) India (2)	0.95 0.90	0.80 0.80
Basalt quarry West Germany	0.73	0.97

As an illustration the characteristic particle size and the uniformity index are given for various quarries, studied or operated by Dutch engineers.

The sieve curve according Rosin Rammler and the accompagning sieve density curve are graphically presented for the diorite quarry and the basalt quarry in figs. 6 and 7.
As the sieve density curves are depicted on the same axes, as the breakwater demand curves, they may be compared directly.
The comparison given in fig. 8 and 9 shows that the berm breakwater fits the quarry yield far better than the rock breakwater. The demand for large quarry rock generated by the rock breakwater exceeds the yield considerably.

From the demand curve it can be seen that 25% of the total quantity of the rock breakwater is formed by the primary armour.
However, the best yield curve (Fig. 7) shows that only 15% of the rock produced, fits in this category. This means that after the core and the secondary armour have been built, considerably more rock would have to be blasted to produce the armour rock as only 15% meets the standard.

If the price of a ton of rock quarried is assumed to be constant, regardless of diameter and quantity, and equal to P, the average price of the armour rock per ton force is given by:

$$P_{armour} = \frac{yield}{requ.} * P + (1 - \frac{yield}{requ.}) * \frac{P}{yield} \quad (9)$$

where:
yield = % of quarry output x > d
requ. = % of demand x > d

If as in this example the yield is 15% and the demand 25% then the average price of armour rock equals 3.27 P, having assumed that the surplus rock has no value. The relative cost of the rock breakwater becomes 117%.
In that case the disadvantage of the berm breakwater caused by the larger material requirement would be easily compensated by the lower price of the armour rock as the required amount of 21% is equal to yield of 22% (D > 1.14 m compare Figs. 5 and 7)

The example shows the importance of the quarry yield analysis to select the breakwater alternative.

In figs. 10 and 11 the slightly adapted designs of the classical rock breakwater and the berm breakwater are confronted with the sieve density curves of the studied quarries.

The advantage of the berm breakwater, that requires lighter rock in the berm armour layer is clearly shown.
However only in the case of the granite quarry a close fit can be reached if secondary blasting or similar techniques is chosen to supplement rock for the secondary armour.

5 CONCLUSIONS

From the theory and the examples presented above a few conclusions can be drawn.
It is shown that in many cases the quarry yield curve is with sufficient accuracy described by the Rosin Rammler equation.
One of the conditions for an economical breakwater design is a close fit between the demand for rock of the breakwater and the quarry yield characteristics both expressed as a function of the rock diameter.
This is especially true for dedicated quarry operation.
For an easy comparison of rock demand and yield, the Rosin Rammler equation should be differentiated with respect to the rock diameter x. The resulting curve is called the sieve density curve.
In the case of a dedicated quarry and short hawking distance, the berm breakwater may be an economic solution although the amount of rock needed slightly exceeds the amount for a classical breakwater.
This is expected to yield insufficient heavy rock for the primary armour.

REFERENCES

Van der Meer, J.W., Rock Slopes and Gravel Beaches Under Wave Attack, thesis, Delft, 1988.
Jensen, O.J., Sorensen, T., Hydraulic Performance of Berm Breakwaters, in "Berm breakwater", ed's G.H. Wills, W.F. Baird, O.T. Magoon, Canada, 1987.
Langefors, U., Kilstrom, B., The Modern Technique of Rock Blasting, Wiley & Son, New York, 1978.
De Ruyter, T.F.M., The Influence of Discontinuities on Rock Blasting Operations, Dept. of Mining Engin. Delft, Univ. of Techn. Delft, 1985.

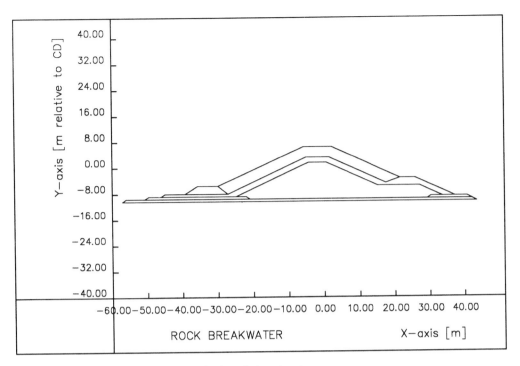

Fig.1. The cross section of a classical rock breakwater

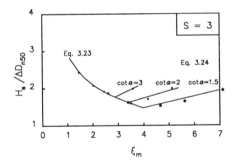

Fig.2. The stability parameter according to Van der Meer.

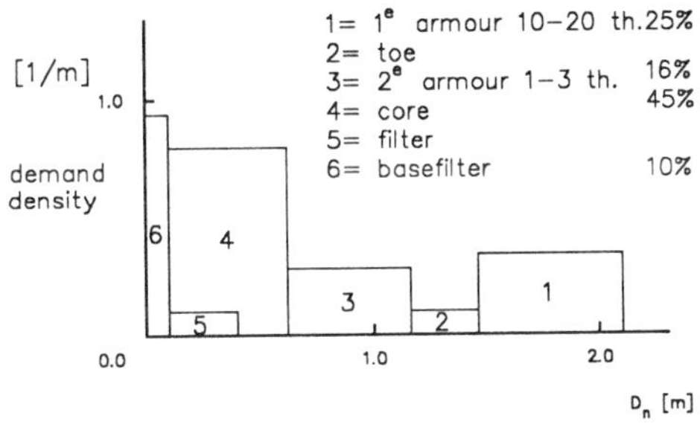

Fig.3. Demand curve rock breakwater.

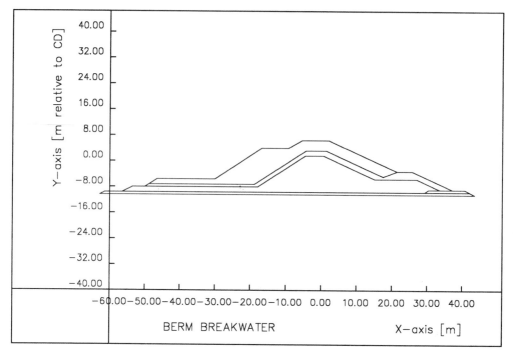

Fig.4. The cross section of a berm breakwater.

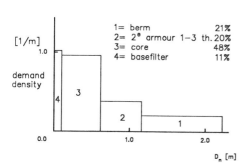

Fig.5. Demand curve berm breakwater.

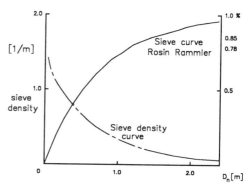

Fig.7. Quarry yield Rijkswaterstaat.

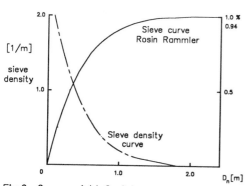

Fig.6. Quarry yield Corinto.

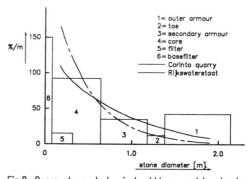

Fig.8. Quarry demand classical rubble mound breakwater.

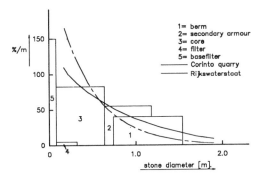

Fig.9.Quarry demand berm breakwater.

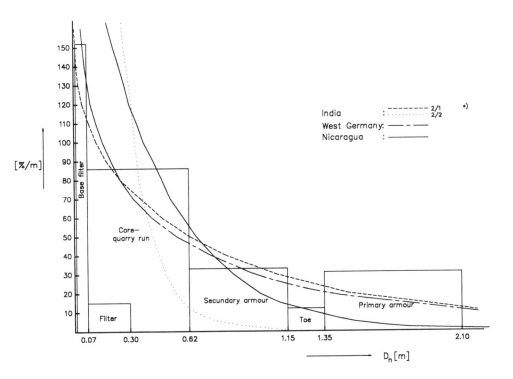

Fig.10. Classical rubble mound breakwater.

•) results of 2 testblasts with different blasting parameters.

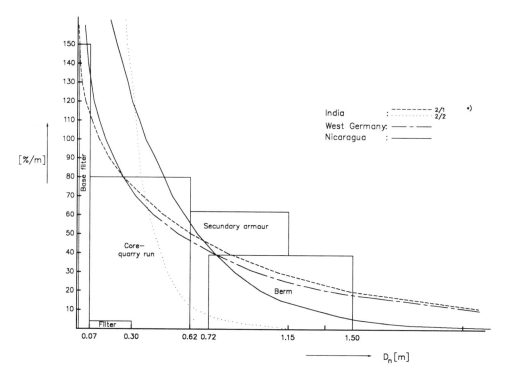

Fig.11. Berm breakwater.

*) results of 2 testblasts with different blasting parameters.

7 Construction materials
  Matériaux de construction
7.1 Exploration
  Exploration

Conférence thématique: La géologie de l'ingénieur et les matériaux de construction

Keynote lecture: Engineering geology and construction materials

Louis Primel

Laboratoire Central des Ponts et Chaussées, France

Résumé : Il devient de plus en plus nécessaire d'inclure les extractions de matériaux de construction dans une politique de l'utilisation des sols, tenant compte de critères économiques, techniques et d'environnement. Le géologue de Génie Civil est très bien placé pour jouer le rôle du fil directeur qui part des inventaires de ressources pour aller jusqu'au réaménagement des sites, en passant par l'étude détaillée du gisement et de son exploitation et l'utilisation des matériaux extraits.

Abstract : It is more and more necessary to include quarrying activities within a land planning policy, which takes into account economical, technical and environmental criteria. The engineering geologist is in a good position to act as the coordinator, starting from the inventories of resources, up to the site reclamation, through the detailed study of the deposit, the operating methods as well as the uses of the quarried materials.

Sous le vocable "Matériaux naturels" l'on regroupe habituellement tous les roches et minéraux industriels extraits du sol et du sous-sol, à l'exclusion des substances énergétiques (charbon, pétrole) et des minerais métalliques. Dans la pratique, on peut subdiviser ces matériaux en trois catégories :
- Matériaux de construction : granulats, argiles à tuiles et briques, matériaux pour chaux et ciment, gypse, ardoises, marbres et granites.
- Minéraux industriels : argiles céramiques, kaolins, bentonite, argiles fibreuses ; sables industriels (verrerie, fonderie) ; calcaires et dolomies pour charges ; feldspath, syénites ; diatomite, perlite, vermiculite, zéolite, talc, amiante, ...
- Matériaux pour l'agriculture : tourbes et amendements minéraux.

Ces substances occupent une place très importante dans l'économie. Leur exploitation engendre un chiffre d'affaires deux à trois fois supérieur à celui des substances métalliques (Fig. 1). Les prix de vente varient beaucoup selon les substances et les utilisations ; à titre d'exemple, en France :
- granulats, sables industriels : 30 à 100 F par tonne

- ciment, kaolin, attapulgite, bentonite, calcaires pour charges, perlite, vermiculite : de 500 à 1500 F par tonne
- ardoises, certains marbres et granites, certaines zéolites naturelles, quartz très purs de 3000 à 50000 F par tonne.

- Cette disparité au niveau des prix de vente implique des stratégies différentes selon les substances ; d'autre part les besoins dans les différentes substances varient énormément : pour certains matériaux, quelques kilogrammes, voire moins, par habitant et par an suffisent, alors que dans la plupart des pays développés chaque habitant consomme entre 6 et 8 tonnes de granulats par an. Ce qui implique également que si certains produits peuvent être transportés sur des milliers de kilomètres, il est par contre économiquement aberrant de dépasser (sauf exceptions) quelques dizaines de kilomètres pour des matériaux comme les granulats.

- L'ambition de l'auteur de ce rapport n'est pas d'étudier de façon exhaustive tous les problèmes liés à tous les matériaux naturels, un volume n'y suffirait pas ; elle est par contre d'essayer de montrer le rôle essentiel joué par le géologue à toutes les étapes, c'est-à-

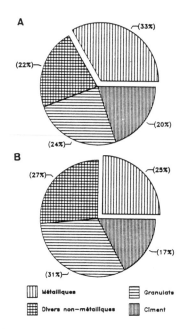

A

(33%)

(22%)

(20%)

(24%)

B

(25%)

(27%)

(17%)

(31%)

Métalliques   Granulats

Divers non-métalliques   Ciment

Fig. 1. - Estimation de la répartition du chiffre d'affaires pour les substances minérales non énergétiques.
A : en France en 1987. chiffre d'affaires total : 44 milliards de francs ; B : aux Etats-Unis en 1986. chiffre d'affaires total : 24 milliards de dollars.

(d'après P. Le Berre et J.F. Pasquet)

dire depuis la recherche des gisements jusqu'aux problèmes posés par l'utilisation des matériaux et leur comportement dans les ouvrages, en passant par les méthodes d'exploitation , les processus d'élaboration des matériaux dans les installations de traitement, et les moyens de remise en état des sites après exploitation. Pendant de nombreuses années, on a considéré que le rôle du géologue s'arrêtait à la mise en évidence des gisements ; nous montrerons à l'aide de nombreux exemples que sa compétence est en fait nécessaire à toutes les étapes citées ci-dessus. Compte-tenu de leur importance économique essentielle pour tous les pays la plupart des exemples que nous choisirons concernent les matériaux de construction, c'est-à-dire les granulats, les matériaux pour ciment, les argiles pour tuiles et briques, les pierres de construction et d'ornementation. Pour un certain nombre de ces matériaux, des statistiques de production au niveau mondial ont pu être établies : c'est ainsi que la production mondiale de ciments s'élève à 1 milliard 236 millions de tonnes en 1989, celle de chaux à 132 millions de tonnes, de gypse

à 105 millions de tonnes. Il est, par contre, très difficile d'établir des statistiques au niveau du globe pour les granulats ou les argiles pour briques et tuiles, les productions de nombreux pays étant inconnues. Citons simplement le chiffre de 7 milliards de tonnes de granulats produites par les dix plus gros pays producteurs. Malgré l'importance évoquée ci-dessus d'une conception globale des problèmes, nous traiterons pour des raisons de commodité de l'exposé, les différentes phases successivement, en insistant en permanence sur leur interdépendance.

LES INVENTAIRES DE RESSOURCES

Les problèmes se posent de façon différente selon les matériaux concernés. Pour certains produits, tels le ciment, un nombre relativement limité de carrières, en général très importantes, suffit à assurer la consommation. Dans les pays riches en calcaires, argiles et marnes, un inventaire complet des ressources n'est pas nécessaire, si ce n'est au niveau des régions dans lesquelles on souhaite pour des raisons économiques implanter les usines de production. Il n'en est pas de même pour les granulats, qu'il faut éviter le plus possible de transporter sur de longues distances et qui sont nécessaires en permanence dans toutes les régions de tous les pays. Pour beaucoup de pays, y compris les pays développés, l'usage de puiser les matériaux nécessaires aux ouvrages et bâtiments dans le sous-sol le plus adapté et le plus proche a été constant et longtemps sans aucune contrainte, mais le développement de la production, fréquemment multipliée par dix dans beaucoup de pays développés en l'espace de trente ans, la taille industrielle atteinte par les entreprises assurant les besoins des grands centres urbains et surtout l'effet cumulatif des surfaces affectées par l'extraction (25 à 30 kilomètres carrés par an en France), ont fait naître un problème sérieux.

Dans certains pays, en raison de la prise de conscience des impacts des extractions sur l'environnement naturel, on a même constaté des oppositions résolues à toute nouvelle extraction. C'est donc pour tenter de résoudre les quatre problèmes essentiels suivants qu'un certain nombre de pays développés se sont lancés dans des inventaires des ressources en matériaux, ces inventaires étant essentiellement centrés sur les granulats.
• L'épuisement des réserves, surtout en matériaux alluvionnaires, source tradi-

tionnelle des granulats pour bétons, en particulier autour des agglomérations grosses consommatrices (Londres, Paris, Sao-Paulo par exemple).

• Les conflits pour l'occupation du sol. Beaucoup de gisements se situent dans les vallées alluvionnaires, qui sont généralement des axes de communication, et ont donc une vocation aux concentrations urbaines ; les plaines alluviales sont par ailleurs des zones fertiles et dont l'hydrographie est favorable au développement de cultures spécialisées (maraîchage, pépinières) à proximité des agglomérations. Les développements urbain et agricole sont en concurrence directe avec l'ouverture de nouveaux gisements.

• L'accroissement très net des exigences dans le domaine de l'environnement, qui introduit des contraintes de plus en plus lourdes vis-à-vis de l'ouverture de nouveaux gisements : citons sans les détailler la protection des nappes aquifères, du régime hydrauliques des cours d'eau, des sites classés ou inscrits, la conservation des paysages, la prise en compte du bruit, des vibrations, des poussières et de la circulation des engins.

• Certaines techniques, en particulier dans le domaine routier, sont devenues au fil des années de plus en plus exigeantes quant aux caractéristiques des granulats utilisés, ce qui oblige donc à affiner les connaissances des réserves non seulement sur le plan quantitatif mais aussi sur le plan qualitatif.

L'accumulation des problèmes que nous venons d'énumérer tend tout naturellement à une limitation des autorisations d'exploiter. Comme il est hors de question d'interdire toute exploitation dans des régions entières, ce qui conduirait à accroître considérablement les coûts de transport, il est donc indispensable, en particulier dans les régions fortement urbanisées, de réaliser des enquêtes et études qui permettront aux différentes administrations concernées et aux exploitants de prendre des décisions cohérentes.

Il est très difficile de fixer de façon définitive un schéma standard d'étude, sachant que pour un secteur donné il faut prendre en compte successivement et parfois simultanément :
- l'état actuel des exploitations
- les contraintes de tous ordres
- les besoins à court, moyen et long terme (études économiques prospectives)
- le gisement potentiel
- les conséquences hydrauliques et hydro-

géologiques des futures exploitations
- la réutilisation des sites après exploitation.

C'est cet ensemble d'études et enquêtes qui permettra de bâtir un schéma coordonné d'exploitation de l'ensemble d'un secteur. Il s'agit là par définition d'un travail d'équipe, mais comme l'élément de base reste le gisement, tant sur le plan quantitatif que qualitatif, c'est le géologue de génie civil qui est en quelque sorte le fil conducteur de ce type d'études, y compris pour le volet économique : comment, en effet prévoir les futures consommations en granulats si l'on n'a aucune idée de l'évolution des techniques qui utilisent ces granulats ?

Bien que le géologue intervienne donc dans toutes les enquêtes et études, nous ne développerons ici que les aspects plus spécifiquement géologiques et géotechniques, c'est-à-dire les inventaires de ressources.

On peut regrouper ces derniers en deux grandes catégories :

1. **Inventaires à petite échelle** (1/50.000e -> 1.200.000e)

Ces inventaires sont réalisés à partir d'une documentation exhaustive et une étude géologique d'ensemble, c'est-à-dire d'une étude de terrain, appuyée sur l'examen des affleurements, des carrières en activité ou abandonnées ; les prélèvements réalisés lors de l'étude de terrain font l'objet d'une identification géotechnique complète. A cette échelle, n'interviennent en général ni la géophysique ni les sondages, si ce n'est au niveau de la documentation, qui prend bien entendu en compte les études ponctuelles antérieures. On peut même ajouter, compte tenu de l'expérience, que faire intervenir des mesures géophysiques ou des sondages avec une maille très lâche (1 point de mesure pour 1 ou plusieurs km²) peut être plus dangereux qu'utile.

Par quels documents se traduisent ces études, quelles sont leurs limites, leurs avantages, leurs inconvénients ?

a) Au niveau d'un bassin alluvionnaire, elles se traduisent par une estimation des cubatures par tronçons de vallées (avec une précision de l'ordre de 20 à 60 % suivant la densité des renseignements), ces tronçons ayant eux-mêmes une étendue variable (mais généralement de 10 à plusieurs dizaines de kilomètres), et par une appréciation globale des caractéristiques des matériaux (fuseau granulométrique, caractéristiques pétrographiques et mécaniques). Cette évaluation permet déjà de

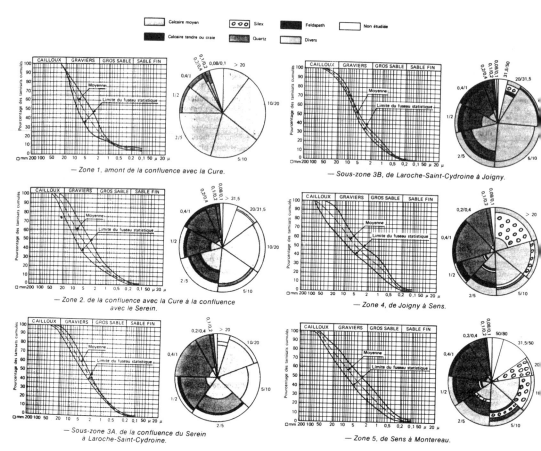

Fig. 2 — *Analyses granulométriques et petrographiques - Vallée de l'Yonne.*

prévoir quels problèmes plus précis devront être étudiés lors de la phase suivante : il faudra dans l'exemple de la figure 2 vérifier s'il n'y a pas de réactions de type liant-granulats entre les graviers de silex (très abondants dans deux secteurs de la vallée étudiée) et les ciments utilisés lors de la fabrication des bétons (Cf "Caractéristiques des matériaux - Conséquences pour leur utilisation").

b) Au niveau d'une région ou d'un département, les études englobent tous les matériaux, c'est-à-dire les matériaux alluvionnaires ainsi que les roches massives. Dans ce dernier cas, les estimations quantitatives n'ont plus aucune signification, seules les appréciations qualitatives concernant telle ou telle roche ont une réelle valeur (pétrographie + caractéristiques mécaniques + exploitabilité).

c) Etude d'un niveau particulier. Il est souvent intéressant dans les régions où les matériaux traditionnels alluvionnaires sont en voie d'épuisement, d'étudier un niveau particulier, actuellement peu ou pas utilisé, mais susceptible de devenir un matériau de substitution valable. La cartographie d'un niveau calcaire de la région de Paris (Fig 3) permet ainsi de délimiter les zones dans lesquelles ce calcaire est "dur" à la fois en surface (-> 5 m) et en profondeur (-> 20 mètres). C'est dans ces zones qu'il faudra implanter les études plus détaillées ultérieures. La notion de "dureté" est directement liée aux utilisations habituelles de ces matériaux pour les bétons et la viabilité et est donc directement dépendante des techniques utilisées et de leur évolution.
Elle est mesurée par un certain nombre d'essais normalisés qui sont supposés rendre compte de la résistance aux chocs, à l'usure, au polissage etc... (voir "caractéristiques et utilisations").

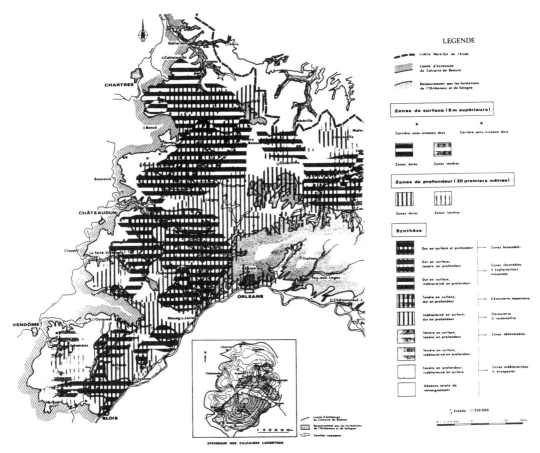

Fig. 3 - Cartographie des zones dures dans le calcaire de Beauce.

Avantages de ces inventaires généraux :
- coût peu élevé de l'étude
- pas de problème de confidentialité
- bon panorama des disponibilités, en particulier de l'éventail **qualitatif** (utile pour le maître d'oeuvre routier qui est amené à utiliser des matériaux de caractéristiques très variées, selon les types et les couches de chaussées concernés).

Inconvénients :
- Le principal inconvénient : ces études ne peuvent, en aucun cas, être utilisées pour définir une politique régionale des exploitations, et encore moins pour des décisions d'occupations des sols. Prenons un seul exemple : un tronçon de vallée de 10 km est défini dans un inventaire donné comme renfermant des sables et graviers 0-100 mm avec 50 % en moyenne de sable 0-5mm, et dont les gravillons ont un Los Angeles de 22 ; il arrive de trouver dans ce tronçon une

zone de 50 ou 100 hectares (c'est-à-dire environ 4 millions de m3 en vallée de Seine) renfermant des sables et graviers 0-20mm, avec 70% de sable 0-5 mm et un Los Angeles de 28. Le problème est encore plus évident en ce qui concerne les épaisseurs.

2. **Inventaires à échelle moyenne** (1/25.000e -> 1/5.000e)
Ces études concernent essentiellement des vallées alluvionnaires, dans la mesure où il s'agit des zones les plus sensibles et les plus convoitées par les différents utilisateurs de l'espace.

a) *Objectifs de l'étude* : il s'agit de connaître "avec une précision suffisante" les caractéristiques des gisements alluvionnaires, c'est-à-dire :
- l'épaisseur D de la découverte
- l'épaisseur G des alluvions
- les zonages correspondants
- la qualité des alluvions, qui détermine

leur utilisation dans les travaux publics et le bâtiment
- le volume des ressources alluvionnaires global ou par catégorie.

Le contenu de l'étude dépend de la signification de l'expression "avec une précision suffisante". Cette précision peut assez facilement être déduite de ce que l'on attend de l'étude : délimiter des zones auxquelles on affecte un coefficient "d'exploitabilité", la précision des limites étant compatible avec les plans d'urbanisme ; à savoir que toutes les études de ce type réalisées jusqu'à présent correspondent à des échelles variant entre le 1:10.000e et le 1:25.000e. Ces cartes d'orientation, dont nous verrons le principe d'établissement ci-dessous, permettent de réserver des zones à l'exploitation ou, au contraire, de les affecter à d'autres utilisations. Mais elles n'excluent cependant pas des hétérogénéités locales et ne doivent pas dispenser l'exploitant d'une étude beaucoup plus détaillée permettant de délimiter à l'intérieur de la zone "favorable" un gisement précis (études généralement pratiquées à des échelles de l'ordre du 1:1.000e, cf "Reconnaissance détaillée des gisements", ci-dessous).

b) Méthodologie et moyens mis en oeuvre :
il est indispensable de prévoir pour ce type d'étude une méthodologie et des moyens très souples, pour deux raisons essentielles, dont l'une est technique et l'autre conjoncturelle :
Les caractéristiques des gisements varient beaucoup d'une vallée à l'autre ;
Les moyens financiers consacrés à l'étude peuvent aussi varier et influer sur les méthodes retenues, et surtout sur les mailles de mesure. A ce point de vue, il est d'ailleurs essentiel de ne pas descendre au-dessous d'une certaine densité de mesures sous peine de risques importants d'erreurs.

La souplesse se traduira à la fois dans l'enchaînement des phases et dans le choix des méthodes. Les exemples que nous allons exposer concernent 4 ou 5 études importantes réalisées autour de grosses agglomérations françaises.

b1) Première phase d'intervention. Elle comprend :
- l'examen des cartes géologiques existantes (à l'échelle du 1:80.000e ou 1:50.000e selon les régions)
- la recherche et l'analyse de la documentation géologique locale (publications, code minier, archives des laboratoires,

etc.)
- l'examen des couvertures de photographies aériennes existantes afin de repérer tout indice favorable : localisation de graviers, zones remblayées, anciens lits mineurs, terrasses en bordures de vallée, etc.,
- une visite sur le terrain de tous les indices repérés, notamment les gravières avec notes sur
· l'épaisseur de la découverte
· éventuellement l'épaisseur du matériau exploitable
· la profondeur du niveau d'eau par rapport au terrain naturel
· les pollutions des ballastières après extraction.

Dès ce stade, des prélèvements d'échantillons sont effectués dans certaines exploitations pour analyse en laboratoire, (ce qui permet parfois d'utiliser les installations en activité pour le prélèvement des graves noyées).

Cette première phase permet de dégager un certain nombre d'éléments indispensables pour la poursuite de l'étude :
- mise en évidence de l'aspect évolutif du matériau, en relation avec la géologie des sites traversés
- aperçu de la nature pétrographique et de la granularité du matériau
- conditions de gisement du matériau, telles que
  · ordre de grandeur de l'épaisseur de la découverte et ses éventuelles variations d'épaisseur et de nature
  · appréciation sur les matériaux graveleux ou sablo-graveleux sous-jacents
- avis sur la nature du substratum
- renseignement sur la position et les fluctuations de la nappe phréatique et sur les relations entre cette nappe et les gisements
- avis sur l'intérêt éventuel d'élargir la prospection à d'autres niveaux, tels que les terrasses en dehors de la plaine alluviale.
- choix des méthodes de prospection à utiliser pour la phase suivante, compte tenu des données ci-dessus.

Elle permet souvent aussi à partir du suivi des exploitations existantes de définir des types de gisement et d'en tirer déjà des conclusions d'ordre pratique pour l'exploitation rationnelle des futurs gisements.
Il existe par exemple dans la Région de Paris deux types essentiels d'alluvionnement :
· vallée large et non meandriforme : où les gisements sont homogènes dans l'ensemble et hétérogènes dans le dé-

tail. On constate que quel que soit la
conduite d'exploitation, l'évolution de
la teneur en sable 0/5 mm à l'extraction
en fonction du temps se présente sous
forme d'une sinusoïde approximative, de
faible amplitude et de période courte ou
très courte (Fig. 4a).

. vallée meandriforme : gisements hétérogènes dans l'ensemble et homogènes dans
le détail. Quelle que soit la conduite
d'exploitation, l'évolution de la teneur
en sable à l'extraction se présentera
sous forme d'une sinusoïde approximative
de moyenne à forte amplitude et de
période moyenne à longue (Fig. 4b et
4c).

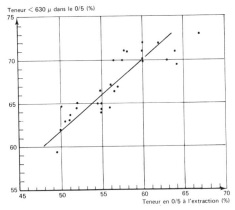

Fig. 5 — Corrélation entre la teneur en sable 0/5 à l'extraction et la
granularité de la fraction 0/5 (Bouafles).

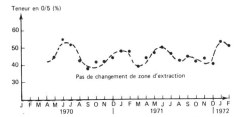

Fig.4a) — Evolution mensuelle des teneurs en sable à l'extraction.
Gisement de remblaiement — vallée large et non méandriforme.

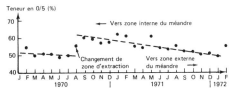

Fig.4b) — Evolution mensuelle des teneurs en sable à l'extraction.
Gisement de méandre (Bouafles - basse Seine).

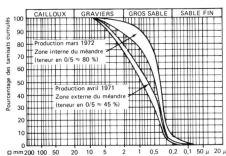

Fig.4c) — Evolution des courbes granulométriques - Prépondérance du
facteur gisement sur la production des fractions 0/d

Des suivis sur plusieurs années de carrières en activité ont de la même façon
permis de noter une très bonne corrélation
entre la granularité interne du sable
0/5 mm et la teneur en sable 0/5 mm dans
le matériau global extrait (Fig. 5).

Ces deux exemples montrent bien qu'une
étude de ressources ne doit jamais être
réalisée pour elle-même, mais que le géologue qui l'entreprend doit toujours avoir

en tête les phases ultérieures que sont
l'exploitation du gisement et l'utilisation du matériau.

b2) Seconde phase d'intervention

Le programme de cette seconde phase est
fonction des résultats obtenus au cours de
l'étude globale de la phase précédente. Si
les résultats obtenus sont très insuffi-
sants ou trop irrégulièrement répartis, il
n'est pas exclu d'intercaler une étude
préliminaire à cette seconde phase et dont
l'objet sera de compléter de façon suffisante la documentation.

Ce niveau d'étude atteint, on peut être
amené à éliminer certains secteurs de la
zone à prospecter, qui se révèlent dès ce
stade franchement défavorables.

Les techniques employées en 2ème phase
sont très généralement les suivantes :

- utilisation de la géophysique électrique
et plus particulièrement de la méthode
des traînés avec 2 longueurs de ligne,
l'une courte permettant de tester la
découverte, l'autre plus longue permettant d'apprécier l'ensemble découverte
matériaux sous-jacents. Il est également
possible d'utiliser à ce stade la
méthode magnétotellurique artificielle
(M.T.A.) que nous évoquerons plus en
détail dans le chapitre consacré à
l'étude détaillée des gisements.

A titre d'exemple, des résultats corrects
ont été obtenus sur plusieurs vallées du
bassin de la Seine en utilisant une maille
de 300 m longitudinalement à la vallée,
sur 150 m transversalement. Dans certains
cas, lorsque l'étude préliminaire a montré
que les épaisseurs de découverte étaient

faibles, ou bien lorsque le contraste entre la découverte et le matériau était mal tranché, il s'est avéré plus efficace de remplacer la longueur de ligne courte par des sondages rapides à la tarière à main : ils permettaient d'obtenir à la fois la nature et l'épaisseur de la découverte, et souvent le niveau de la nappe aquifère (découverte faible, presque toujours inférieure à 2 m).

- Etalonnage de la géophysique par des sondages rapides complétés par des diagraphies. Ces sondages revêtent déjà un aspect quantitatif (épaisseurs respectives de découverte et de matériau exploitable) ; éventuellement, aspect qualitatif (lit argileux ou pollution des matériaux mis en évidence par les diagraphies). Les sondages ne sont pas disposés selon une maille régulière, mais implantés en fonction des résultats de la géophysique. Leur nombre tourne autour de 2 par km².

Ces précisions sur le maillage montrent bien qu'il ne s'agit pas d'études très détaillées : la maille couramment utilisée en géophysique électrique (300 x 150 m) engendre des surfaces non prospectées de 4 hectares et demi. C'est-à-dire que si l'étude localise un gisement dispersé, il le sera dans la réalité ; par contre, si l'étude se révèle apparemment homogène, d'un seul tenant, rien ne prouve qu'au sein de ce gisement il n'y ait pas de zones impropres. C'est au futur exploitant de prévoir les études complémentaires lui permettant d'implanter son gisement avec précision.

- Enfin, cette deuxième phase se termine généralement par la réalisation de sondages avec prélèvements (méthodes variées selon le contexte). Toujours dans le cas des études évoquées ci-dessus, la densité moyenne des sondages est de l'ordre de 1 pour 2 km². Ils sont bien sûr presque tous implantés dans les zones définies comme étant les plus favorables par les méthodes précédentes (ce qui aboutit à un maillage plus serré).

Cette deuxième campagne, ainsi que les données documentaires permettent :

. de connaître l'épaisseur moyenne des matériaux alluvionnaires par section prospectée,

. d'établir une corrélation entre les valeurs de résistivité et l'épaisseur de la découverte et de confirmer le zonage établi à partir de la prospection géophysique.

. d'apprécier la qualité des matériaux exploitables, grâce aux essais réalisés sur les échantillons prélevés en ballastière et sur ceux ramenés par sondage. Il s'agit essentiellement de :

/ granulométries
/ analyses pétrographiques des différentes classes granulaires (par comptage)
/ mesures de la résistance mécanique (résistance aux chocs, résistance l'usure)
/ mesures permettant d'estimer la forme des granulats.

c) *Classement des disponibilités alluvionnaires*

Il est tentant pour des techniciens de s'en tenir à une expression cartographique des données brutes (épaisseurs, qualités, etc.). Mais, pour que les cartes soient utilisables par tout le monde, il est indispensable que la cartographie exprime un classement des gisements ; ce qui suppose le choix difficile de critères de classement. Ceux-ci peuvent être variés d'une vallée à l'autre. Ceux qui sont retenus dans l'exemple de la figure 6, correspondent au contexte de la Région Parisienne dans les années présentes (ces critères peuvent aussi varier dans le temps).

- Critères
Rapport du volume G du matériau exploitable sur le volume D de la découverte. Plus ce rapport est grand, plus le gisement est intéressant, à découverte égale. Il n'a été retenu dans le contexte technico-économique actuel comme "zone exploitable" que celles où G/D atteignait ou dépassait 2. (Dans des régions particulièrement pauvres en matériaux, il peut être envisagé de descendre à 1, si G est supérieur à 3 m).

D : épaisseur de la découverte. Plus D est grand, plus l'extraction nécessitera des moyens importants et plus l'épaisseur du matériau devra être importante pour les justifier. Dans la région considérée, la limite entre découverte "forte" et "faible" a été fixée à 2 m.

Qualité du matériau. Il s'agit essentiellement de la granularité. Ont été distinguées les zones sableuses et les zones

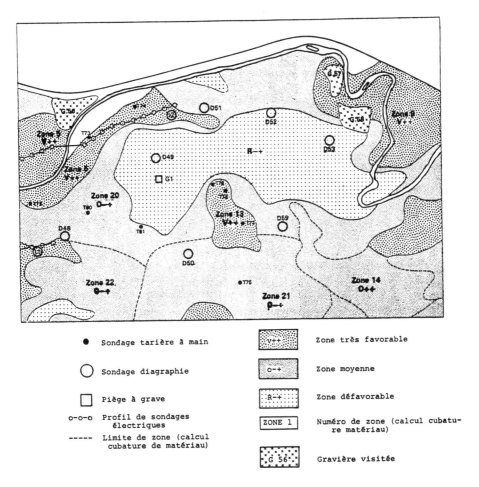

Fig. 6 — Exemple de carte d'orientation des potentialités d'extraction.

graveleuses, en considérant que, compte tenu du contexte géologique local, les secondes sont beaucoup moins répandues que les premières et sont donc plus "favorables".

- Présentation cartographique

La figure donne un exemple de carte d'orientation des possibilités d'extraction.

Les différentes zones sont affectées à la fois d'une couleur (ou d'une trame) et d'un indice (ici une lettre) qui sont définis par la prise en compte simultanée (tableau à 2 entrées) du rapport $\frac{G}{D}$ et de l'épaisseur de la découverte D.

Dans l'exemple donné, ce tableau conduit aux résultats suivants :

	$\frac{G}{D} > 3$	$2 < \frac{G}{D} < 3$	$\frac{G}{D} < 2$
D > 2	V -	O -	R -
D < 2	V +	O +	R +

V signifie zone très favorable, O zone moyenne et R défavorable. Chaque zone est en outre affectée d'un second signe + ou - selon qu'il s'agit d'un matériau graveleux (+) ou sableux (-).

2945

Un tel document est donc synthétique, clair et directement utilisable par l'aménageur. Il est cependant indispensable de disposer, sous forme d'annexes, de tous les documents de base. Celui qui a réalisé la synthèse est en effet amené à prendre des options qui seront peut-être différentes dans un autre contexte ou avec des normes d'utilisation autres. Il est nécessaire dans ce cas de pouvoir retourner aux sources des données pour réaliser une autre interprétation d'après les critères nouvellement choisis.

LA RECONNAISSANCE DETAILLEE DES GISEMENTS
Incidences sur les méthodes d'exploitation et d'élaboration

- Il s'agit là de la phase-charnière, celle qui conduit le géologue à utiliser toute la palette des disciplines des Sciences de la Terre. Il est évident que dans le domaine des matériaux de construction, les études de gisement varieront sensiblement selon qu'il s'agit de granulats, de matériaux pour cimenterie, d'argiles ou de pierres ornementales. Cependant la plupart des paramètres à étudier et la quasi-totalité des méthodes employées constituent en quelque sorte un tronc commun quel que soit le matériau produit. Dans un cas, les paramètres minéralogiques et chimiques seront primordiaux (argiles pour briques et tuiles ; argiles, marnes et calcaires pour les ciments), dans un autre ce seront les critères de fissuration du massif (pierres d'ornementation). De nombreux auteurs ont décrit les méthodes utilisées pour tel ou tel type de matériau, mais l'étude bibliographique montre à l'évidence que c'est le domaine des granulats qui a été le plus étudié ; la raison principale de la concentration des recherches sur ce domaine tient essentiellement au fait que les granulats sont des matériaux peu coûteux, et que les responsables des études ont été contraints de trouver des méthodes également peu coûteuses pour évaluer les gisements. On peut par exemple admettre de réaliser un très grand nombre de sondages carottés avant d'implanter une carrière destinée à alimenter une cimenterie, ce n'est pas le cas lorsqu'il s'agit de granulats. Le présent chapitre sera donc construit autour des gisements de granulats, tout en insistant à tel ou tel moment sur les paramètres plus particulièrement importants pour les autres matériaux de construction.

1. Objectifs de la reconnaissance détaillée

Les objectifs sont d'évaluer les quantités et qualités des matériaux commercialisables et des stériles, de préciser leur répartition et de recenser les contraintes particulières liées à l'exploitation et à l'environnement, de déboucher ainsi sur la définition d'un plan d'extraction et de remise en état du site et de contribuer à la conception rationnelle des installations ; permettre enfin la sélection qualitative des produits fabriqués tout en limitant autant que possible les investissements et les dépenses d'énergie.

2. Définition des paramètres à étudier

2.1. La morphologie et l'hydrographie

Ces paramètres sont à prendre en compte pour définir le plan d'exploitation et choisir l'emplacement des installations en vue de réduire les manutentions et les dépenses d'énergie et pour limiter les effets sur l'environnement (nuisances phoniques, impacts visuels, pollution de l'air et de l'eau,...). Une connaissance très précise de la morphologie du gisement permet en outre d'orienter la reconnaissance du sous-sol.

2.2. La découverte du gisement

On entend par découverte, la tranche de matériaux à extraire avant d'atteindre les couches exploitables. Longtemps considérée comme stérile, de plus en plus, la découverte est valorisée par sa réutilisation pour la protection de l'environnement (merlons anti bruits) ou la remise en état progressive du site. Dans l'étude de gisement, on déterminera donc le volume de la découverte, mais aussi les variations de son épaisseur et de sa nature en vue de définir les conditions optimales de son extraction et de sa réutilisation. On portera une attention particulière l'étude des horizons de surface (pédologie) en vue d'une bonne gestion de la terre végétale dans le cadre du réaménagement de la carrière.

2.3. Les matériaux exploitables

Il s'agit de définir le volume total de matériaux exploitables et sa répartition dans l'espace en sous-ensembles prenant en compte les caractéristiques du gisement et du matériau.
L'identification précise de chaque unité est très importante car elle conditionne

le plan d'extraction, la conception des installations et la qualité des granulats produits. Les principaux paramètres à prendre en compte sont :
- la nature pétrographique des matériaux,
- le comportement physique et mécanique des granulats,
- l'état de fragmentation du massif rocheux ou de la granularité d'un matériau alluvionnaire,
- les différentes pollutions ou hétérogénéités dans les horizons à exploiter.

Par exemple la fragilité, la fissuration, la composition minéralogique et la structure de la roche conditionnent son abrasivité, son aptitude à se fragmenter, la forme des granulats produits et la granularité du sable de concassage (teneur en éléments fins inférieurs à 100 microns surtout). La dureté du matériau permet d'orienter le choix des appareils soit vers un concasseur primaire à mâchoires ou giratoire soit vers un concasseur à percussion (matériaux non abrasifs, calcaires en général) pour autant que la production de ce genre d'appareil au point de vue granularité soit compatible avec les résultats recherchés. Comme nous le verrons plus loin, la notion de dureté évoquée ici est d'ailleurs difficile à saisir et à relier à une composition pétrographique précise.

- Autre exemple qu'il nous semble indispensable de détailler un peu, pour saisir les répercussions de la nature du gisement sur son mode d'exploitation : les différentes pollutions dans un gisement de roches massives :
. *pollution liée à la genèse du massif*, correspondant essentiellement :
  - pour les calcaires, à l'intercalation de lits argileux ou marneux ou de lentilles de nature différente du faciès d'ensemble. Il arrive que les deux types d'hétérogénéité liés aux conditions de sédimentation coexistent dans un même gisement :
  - pour les roches éruptives, à l'existence de filons qui recoupent le faciès principal et qui peuvent d'autant plus fréquemment être considérés comme des pollutions qu'ils sont souvent plus altérés que la roche encaissante.

Pollution « accidentelle » : la présence de zones très fracturées ou même broyées peut considérablement gêner une exploitation. Si l'étude préalable ne l'a pas décelée, il peut se produire que le front de taille attaque une telle zone

parallèlement à son allongement maximal. Comme ces zones correspondent très généralement à des matériaux très altérés impropres à l'utilisation, on peut être obligé de les traverser ou de les contourner et perdre ainsi plusieurs journées de production.

Fissures verticales et poches de dissolution argileuses : ce sont des phénomènes relativement fréquents dans certains massifs calcaires et ils influent directement sur la réussite des plans de tir et sur l'état du matériau après abattage, ce qui conditionne le mode de reprise et la maille de précriblage pour éliminer la fraction argileuse selon les conditions atmosphériques.

Une importante fissuration argileuse contraindra à effectuer des abattages de faible volume pour éviter que le matériau ne reste trop longtemps exposé l'humidité.

2.4. Les paramètres hydrogéologiques

Ils ont une importance primordiale seulement pour les gisements alluvionnaires, bien qu'il arrive que la présence de nappes dans les roches massives conduise à des débits importants. Il est indispensable de connaître :
- *les variations du niveau de la nappe* qui ont une incidence directe sur :
  • le déroulement même de l'exploitation (problème des crues),
  • le mode d'extraction de la découverte et même l'épaisseur tolérable de découverte qui peut varier du simple au double selon qu'elle est noyée ou hors d'eau,
  • le mode d'extraction des graves,
  • la propreté des matériaux,
  • le choix des installations pour les gisements terrestres, lorsque le lavage des matériaux est nécessaire.
- *la direction et le sens d'écoulement de la nappe, la perméabilité des alluvions* qui peuvent, dans certains cas (gisements pollués), avoir une influence directe sur la façon d'attaquer l'extraction, de manière à éviter un déplacement des fines vers le front de taille.

L'ensemble des paramètres hydrogéologiques conditionnent bien entendu aussi le choix du réaménagement.

3. Les méthodes et moyens d'études

Pour répondre aux objectifs précédemment définis, l'étude détaillée d'un gisement doit être réalisée à une échelle compatible avec la précision recherchée (1/2500 à 1/1000). Les études préliminaires, dans le cas d'un gisement neuf, ou l'étude préalable du front de taille, dans le cas d'une extension, permettent en général de définir les paramètres les plus sensibles du gisement et d'orienter le choix des moyens pour répondre aux problèmes posés. Il faut d'autre part rechercher les méthodes de reconnaissance les mieux adaptées au contexte local (occupation des sols, morphologie, géologie).

Ces différentes démarches permettent d'établir un plan de reconnaissance spécifique du gisement à étudier, en sélectionnant, parmi les méthodes qui vont être présentées, celles qui conviennent le mieux au projet.

3.1. Etude morphologique et géologique de surface

Toute étude de gisement commence par une étude morphologique et géologique englobant le site à exploiter et ses abords. Celle-ci a pu débuter lors d'études préalables visant à définir des secteurs favorables à l'exploitation de granulats à l'échelle du 1/25000 au 1/10000. A ce stade, les moyens suivants ont pu notamment être employés :
- Exploitation de la documentation géologique et géotechnique existante (cartes géologiques, hydrologiques, pédologiques,..., archives, dossiers géotechniques antérieurs, synthèses régionales,...) ;
- Analyse morphologique et géomorphologique par photointerprétation à partir de couvertures aériennes existantes ou de prises de vues spécifiquement réalisées pour mettre en évidence les phénomènes géologiques (émulsions spéciales, choix de la période et de l'angle de prise de vue,...).
- Levé géologique et structural précis, s'appuyant sur les fronts existants ou les affleurements naturels ;
- Echantillonnage des différents matériaux constituant le gisement, prélevés sur front de taille ou affleurements, ou en leur absence, par quelques sondages carottés ;
- Identification qualitative des échantillons-types par des études pétrographiques précises (analyse minéralogique quantitative, texture, altérations), des essais permettant d'évaluer la résistance mécanique des différents faciès et si possible des essais d'élaboration en vraie grandeur

permettant de définir les caractéristiques des granulats produits par type de roche (granulométrie, pourcentage de sables, de fines produites au concassage, qualité des fines,...).

Cependant, le niveau des études préalables étant extrêmement variable, il convient lorsque l'étude détaillée du gisement est abordée, d'apporter les compléments d'information géologique et géotechnique nécessaire à une bonne définition des objectifs et des moyens de la reconnaissance et de préciser à l'échelle de l'étude détaillée (1/1000) :
- la topographie de la surface du sol (nivellement très précis servant la base à l'évaluation des volumes) ;
- le levé géologique et structural, sachant que toute observation de surface est la moins coûteuse et qu'une bonne compréhension initiale de la structure du gisement permettra la meilleure organisation de la reconnaissance.

3.2. La géophysique de surface

En reconnaissance de gisements, les paramètres géophysiques discriminant le mieux la nature, l'altération ou la fissuration des terrains sont la résistivité électrique et la vitesse de propagation des ondes sismiques dans le sous-sol. Ces deux paramètres sont complémentaires, la résistivité étant sensible surtout aux variations de la quantité d'eau du sol qui dépend elle-même de sa nature, la vitesse de propagation des ondes sismiques variant en fonction des caractéristiques mécaniques du terrain de sa compacité et surtout de sa fissuration.

a. *Méthodes électriques à courant continu* (ou alternatif basse fréquence). Les sondages électriques donnent des indications sur la succession verticale des matériaux et sur leur résistivité. Ils permettent en outre de choisir les longueurs de lignes les plus appropriées pour la réalisation des cartes de résistivité et aident à leur interprétation. Les sondages électriques, lorsqu'ils peuvent être interprétés quantitativement, le sont soit grâce aux catalogues d'abaques quand le nombre de couches est inférieur ou égal à 3, soit au moyen de l'ordinateur.
Le traîné électrique et les cartes de résistivité permettent notamment de mettre en évidence les variations d'épaisseur d'une couverture conductrice sur un massif rocheux ou un gisement de graves. Ils peuvent aussi préciser la position d'un contact masqué entre deux formations à

résistivités différentes ou traduire les variations d'altération, de fissuration ou de pollution d'un gisement.

b. *Méthode magnétotellurique artificielle : M.T.A.* Cette méthode consiste à mesurer en continu à la surface du sol les composantes horizontales des champs magnétique H et électrique E créés par des émetteurs de radio-diffusion. On en déduit la résistivité apparente du sous-sol et la profondeur de pénétration de l'onde électromagnétique. On réalise donc ainsi un profil de résistivité du sous-sol en continu et beaucoup plus rapidement que par la méthode classique.

c. *La sismique réfraction*
On établit par cette méthode des coupes sismiques dans lesquelles les différentes couches sont caractérisées par une vitesse sismique variant en fonction de la compacité du terrain. Comme pour toute méthode géophysique la traduction de la coupe sismique en coupe géologique doit s'appuyer sur des sondages mécaniques. Des interprétations plus élaborées peuvent être envisagées par l'analyse du signal sismique.

3.3. Les sondages mécaniques

Les sondages mécaniques apportent des informations géologiques précises mais ponctuelles et coûteuses. Il importe donc de les implanter au mieux, sur la base des études géologiques et géophysiques préalables, mais aussi de choisir la technique la mieux adaptée au gisement et la moins coûteuse pour les objectifs recherchés qui sont à ce niveau :
- la production d'une coupe géologique précise ;
- le prélèvement d'échantillons représentatifs pour l'identification qualitative des matériaux ;
- la connaissance du niveau de la nappe (équipement des forages en piézomètres).

En reconnaissance de gisements alluvionnaires, les techniques de sondages sont très diversifiées en raison des particularités locales de chaque vallée. Pour définir la géométrie du gisement, on s'oriente généralement vers des sondages rapides avec prélèvement d'échantillons remaniés ; les matériels les plus utilisés sont : la tarière, simple pour les gisements hors d'eau, ou équipée d'un piège à graves en site noyé, la tarière continue, le vibrofonçage et le carottier battu.

Pour obtenir un échantillonnage représentatif, le choix des moyens dépend de l'hydrogéologie et du matériau :
- pour les gisements hors d'eau, on utilise le plus souvent la pelle mécanique ou la tarière simple en gros diamètre ;
- pour les graves noyées, le forage à la soupape et le vibrofonçage sont les techniques les plus performantes.
En reconnaissance de gisements rocheux, le sondage carotté apporte une information géologique sûre et un échantillonnage continu, qui permet d'étudier la fracturation et de réaliser des essais mécaniques. Cependant, les sondages carottés sont lents et onéreux, il convient d'en limiter le nombre aux seules fins d'étalonnage de moyens plus rapides lorsqu'il n'existe pas de coupes géologiques naturelles (affleurements) ou artificielles (front de taille) pouvant servir de référence.

La limitation en nombre des sondages carottés implique, en compensation, la mise en oeuvre de techniques de reconnaissance plus rapides et économiques, pour obtenir une information géologique suffisamment dense. C'est l'orientation qui a été prise en France, en particulier en développant la recon-naissance par forages destructifs au marteau perforateur, alliant la rapidité d'exécution à la possibilité d'utilisation du matériel de foration existant en carrière.
- L'exécution du forage est suivie par un géologue averti qui établit une coupe géologique dite "coupe sondeur" fondée sur l'examen systématique des sédiments remontés lors de la foration.
- En même temps, des échantillons sont prélevés par passes successives ou par horizon géologique en vue d'identifications plus précises en laboratoire (analyses pétrographiques, essais adaptés aux éclats,...).
- Pendant la foration, l'enregistrement continu de la vitesse d'avancement et éventuellement d'autres paramètres de forage (percussion réfléchie, pression du fluide d'injection ou pression sur l'outil,...) permet de préciser la "coupe sondeur" : ce sont des diagraphies instantanées.

3.4. Diagraphies différées

Ces diagraphies consistent à enregistrer dans un sondage, généralement d'une manière continue, en fonction de la profondeur, un ou plusieurs paramètres qui caractérisent le sol.

Les paramètres les plus fréquemment

enregistrés en reconnaissance de gisements sont les suivants :

a. *La radioactivité naturelle (R.A.N.), globale ou sélective*

Ce paramètre varie en fonction de la nature géologique des terrains. Certains bancs-repères, à radioactivité fortement contrastée, peuvent ainsi être suivis de sondage à sondage, permettant une reconnaissance structurale fine du gisement. La Sonde R.A.N. peut être mise en oeuvre indifféremment dans un forage nu ou tubé, vide ou plein d'eau et être utilisée aussi bien en site alluvionnaire que rocheux. (FIG. 7).

b. *Les diagraphies électriques*

Elles mesurent la résistivité des terrains. Les sondes sont de divers types, mais nécessitent pour la plupart un trou non tubé et plein d'eau (ou de boue), limitant ainsi les possibilités de mise en oeuvre. Ces diagraphies sont surtout intéressantes en site éruptif pour l'étude des variations d'altération et de pollution du massif rocheux. Elles complètent bien la R.A.N., peu sensible à ce type de variations.

c. *Les diagraphies soniques (ou micro-sismique)*

Le paramètre mesuré est la vitesse de propagation des ondes longitudinales dans le terrain au voisinage d'un sondage. Deux types de sondes existent : la sonde sonique discontinue utilisable en trous secs, la sonde sonique continue utilisable dans des trous remplis d'eau ou de boue. Le domaine d'application privilégié de cette diagraphie est l'étude structurale des massifs rocheux, la vitesse sonique

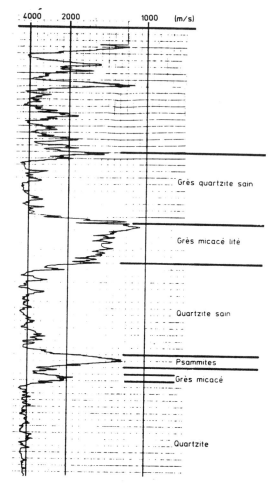

Fig. 8 - Exemple d'utilisation d'une sonde sonique.

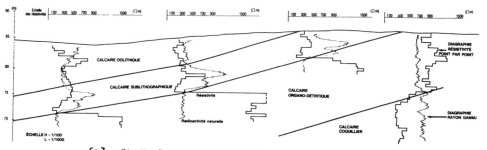

Fig. 7 — Géométrie d'un gisement déterminée grâce à un niveau marqueur de radioactivité.

variant, pour une roche donnée, en fonction de l'importance et de la densité des discontinuités qui affectent le massif. Elle sera utilisée pour prévoir la dimension des blocs extraits. Ces techniques peuvent être en particulier utilisées de façon très efficace, pour l'étude des gisements rocheux destinés à la fourniture d'enrochements, ou encore de pierres de taille ou d'ornementation (FIG. 8).

d. *Les diagraphies nucléaires*

Elles utilisent les phénomènes de diffusion des rayonnements gamma par la matière (sonde gamma-gamma) ou de ralentissement des neutrons rapides par l'hydrogène (sonde neutron-neutron). La diagraphie gamma-gamma permet d'atteindre la masse spécifique des terrains, tandis que la diagraphie neutron-neutron sert à mesurer leur teneur en eau.

e. *Association des diagraphies et interprétation*

Le coût des diagraphies étant faible par rapport à celui des forages, on peut

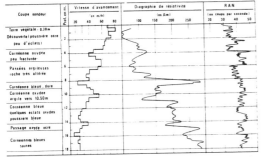

Mise en évidence de l'altération d'un massif de cornéennes par la vitesse d'avancement et la résistivité.

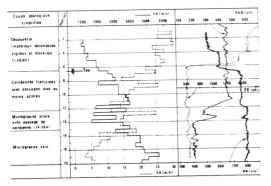

Fig. 9 – Reconnaissance d'un gisement de cornéennes et microgranite par association de diagraphies.

réaliser plusieurs diagraphies dans chaque sondage pour obtenir par recoupements successifs des données précises sur la coupe géologique. Dans le cas de la figure 9 par exemple, la radioactivité naturelle, et à un degré moindre les diagraphies NN et γ-γ (détecteurs éloignés), marquent bien les variations lithologiques (distinction cornéenne-granite) tandis que vitesses d'avancement et sonique permettent de déterminer l'épaisseur de découverte, les passages altérés et fissurés, la fracturation du massif.

L'interprétation est réalisée le plus souvent par recoupements graphiques, mais des méthodes de traitement des données par analyse discriminante ont été mises au point, en vue d'aboutir à une automaticité plus grande de la prospection.

3.5. **Identifications et essais**

Avant même d'engager l'investissement d'une étude détaillée de gisement, il convient de s'assurer que les matériaux à exploiter répondent aux critères de qualité exigés. Le problème se pose différemment s'il s'agit d'une extension de carrière ou de l'ouverture d'un nouveau gisement.

- dans le cas d'une extension de carrière, l'exploitation en cours ou antérieure a généralement permis de situer qualitativement les produits par rapport aux spécifications, en les classant suivant leur nature géologique par référence à la provenance sur le front de taille ;
- dans le cas où cette connaissance initiale de la qualité de la production en fonction de l'origine sur le gisement n'est pas connue, il est indispensable de l'acquérir. Pour cela, il est nécessaire de :
  • prélever des échantillons représentatifs des différentes unités géologiques du gisement, sur front de taille, affleurements ou par sondages non destructifs ;
  • réaliser une identification très complète de ces échantillons-types par étude pétrographique, mesure de la masse volumique, de la porosité, éventuellement analyse chimique ;
  • procéder, si nécessaire et si possible, à un test de concassage sur ces témoins, afin de prévoir les caractéristiques des granulats produits (granulométrie après concassage,...) et permettre la

réalisation des essais de comportement mécanique sur des matériaux se rapprochant de la production réelle ;
• évaluer, par des essais, la qualité de ces matériaux-témoins pour les situer par rapport aux diverses utilisations possibles et optimiser la conception des installations.

Les principales caractéristiques à déterminer sont les suivantes :
- la résistance aux chocs (essais Los Angeles ou après étalonnage essai de fragmentation dynamique) ;
- la résistance à l'attrition sous eau (essai microdeval en présence d'eau) ;
- le coefficient de polissage accéléré si une utilisation en couches supérieures de chaussée est envisagée ;
- la granularité des produits élaborés ou des matériaux bruts alluvionnaires ;
- la propreté, c'est-à-dire l'étude de la quantité et de la qualité des fines (sédimentométrie, équivalent de sable, essai au bleu de méthylène).

Suivant le matériau et les usages prévus, on peut ajouter des essais tels que la résistance à la compression, le module de déformation, la vitesse de propagation des ondes longitudinales, l'abrasivité, les essais d'altérabilité, l'essai de friabilité des sables...

Cette identification initiale des matériaux représentatifs des différentes unités géologiques du gisement va permettre de constituer une collection d'échantillons-témoins, auxquels on pourra se référer dans l'étude du gisement.

A ce stade on raisonnera essentiellement par analogie géologique à partir de la coupe de sondage et de l'identification des sédiments prélevés lors de la foration.

L'examen de ces prélèvements à la loupe binoculaire ou au besoin au microscope polarisant après confection de lames minces, constitue alors l'opération essentielle d'identification qui peut être complétée, si nécessaire, par des essais adaptés aux éclats et étalonnés par rapport aux essais classiques de granulats.

Plusieurs essais ont été étudiés en vue de cette étude géotechnique des éclats, notamment l'essai de fragmentation dynamique, l'essai de friabilité des sables et un autre essai intitulé "hauteur de sédimentation".

Enfin nous rappelons que les identifications doivent aussi porter sur la découverte du gisement. On pratiquera les essais de sols classiquement utilisés pour définir les conditions d'extraction et de réutilisation de ces matériaux dans le cadre du réaménagement de carrière.

4. Conclusions

L'éventail de techniques dont on dispose est donc très large. Aucune d'entre elles n'étant à elle seule suffisante pour résoudre tous les problèmes, il est indispensable de bien définir au préalable les objectifs - techniques, économiques et ceux liés à l'environnement - de la reconnaissance détaillée du gisement de granulats afin de marier au mieux les moyens d'investigation que l'on utilisera : géologie, géophysique de surface, sondages carottés et destructifs, diagraphies instantanées et différées, essais sur éclats et sur échantillons intacts. Il est évident que dans le cas d'autres matériaux, comme le ciment, la prospection fait intervenir d'autres paramètres. Il est de plus en plus courant que les cimenteries soient alimentées par deux carrières différentes de calcaire d'une part, d'argile d'autre part. Il arrive que l'un ou l'autre, ou les deux gisements soient hétérogènes. Il est donc indispensable pour assurer la régularité d'approvisionnement d'avoir une parfaite connaissance des gisements. Les reconnaissances géologiques par de très nombreux sondages carottés, à maille beaucoup plus serrée que pour les gisements de granulats, permettent dans un premier temps de dresser les variations géochimiques tridimensionnelles. La gestion des variations constatées peut ensuite se faire à travers une modélisation du gisement qui reproduit la variabilité de tel ou tel élément dans l'espace (silice dans le calcaire par exemple) en vue de simuler la gestion des ressources.

CARACTERISTIQUES DES MATERIAUX - CONSEQUENCES POUR LEUR UTILISATION

Toutes les roches ont été utilisées depuis fort longtemps par l'homme ; en fait depuis qu'il a été capable de saisir un objet dans sa main, c'est-à-dire il y a environ six millions d'années. Il s'est mis à tailler ces pierres il y a environ trois millions d'années et il faut attendre 6000 avant Jésus-Christ avant qu'il en fasse un autre usage qu'un simple projectile : c'est l'empierrement des huttes et parfois leurs abords immédiats dont on a des traces certaines. L'extension de l'empierrement aux chemins est sans doute liée à l'apparition de la roue et était limitée aux passages difficiles (sols plastiques, 3000 ans avant J.C. en Egypte). Peu à peu, à travers les siècles, les utilisations se sont diversifiées, les liants sont apparus, d'abord rustiques, puis de plus en plus élaborés. Il est tout à fait évident que les exigences des utilisateurs - que l'on appelle souvent en France des spécifications - varient beaucoup selon les utilisations. Dans le domaine de la construction, la plupart de ces spécifications reposent sur une longue et constante pratique des chantiers, en un mot sur l'expérience, parfois complétée, mais pas toujours, par des données scientifiques volontairement acquises. Il est vrai que ce manque de fondement scientifique peut s'expliquer : le suivi de l'évolution de matériaux dans une chaussée de route par exemple est une opération très longue qui doit être poursuivie pour être valable pendant plusieurs années. Par ailleurs, plus les techniques sont complexes et plus l'on est confronté à un nombre élevé de cas de figures. Si l'on prend l'exemple d'une chaussée, pour fixer une spécification donnée pour un type de granulat donné, on devra tenir compte d'un grand nombre de paramètres : nature du liant, structure de la chaussée, épaisseur des couches, importance du trafic, vitesse des véhicules. De la même façon, si l'on considère simplement deux des caractéristiques essentielles d'un béton, la durabilité et la résistance, on s'aperçoit qu'elles sont influencées directement ou indirectement par au moins vingt caractéristiques propres au granulat. Le but de cette communication n'est donc pas, et ne pourrait pas être, un panorama même partiel de toutes les caractéristiques des matériaux de construction, de la façon dont on les mesure - les essais - ni des limites que l'on fixe pour chacune de ces caractéristiques (les spécifications). Nous nous efforcerons simplement à l'aide de quelques exemples de montrer comment le géologue spécialisé dans les problèmes de matériaux peut contribuer de façon décisive à faire évoluer la compréhension des phénomènes complexes qui interviennent lors de la fabrication et de l'utilisation de certains matériaux très courants.

1. L'évolution des idées sur la propreté des sables utilisés pour les bétons de construction ou les matériaux de chaussées

Il y a quelques dizaines d'années, vers 1900, les ingénieurs demandaient que le sable soit "bien criant à la main", ou encore qu'il ne soit pas "gras". Ces expressions, qui peuvent maintenant faire sourire, correspondaient en fait à un souci tout à fait pratique : il fallait quel que soit l'usage, éviter d'avoir trop d'argiles dans les sables utilisés.

Les argiles suspectées depuis longtemps et ayant des dimensions très petites ne pouvaient être éliminées que d'une façon *le lavage*. Cette méthode enlevant à l'époque tout ce qui était inférieur à 50 ou 100 µm, il était facile de spécifier la limitation de la teneur en fines à un taux très bas. Ainsi, s'il y a peu ou pas de fines dans un sable, il y a peu ou pas d'argile dans ce sable.

Cette règle a subsisté très longtemps pour les sables à bétons et il a fallu attendre les années 60 pour s'apercevoir qu'enlever des fines inertes à un sable revenait à diminuer la résistance du béton.

Pour les techniques routières, cette prise de conscience s'est faite plus tôt, d'une part parce que l'on ne recherche pas de résistances aussi élevées que pour les bétons hydrauliques et, d'autre part, parce que le lavage est une opération coûteuse, surtout dans les gisements de roches massives, rarement pourvus en eau courante « naturelle ».

Il est donc vite apparu nécessaire de garder ces fines et cela d'autant plus que perdre de la résistance ou de la stabilité, quand on n'en a pas beaucoup, présente un certain risque.

1.1. L'équivalent de sable

L'utilisation de l'essai dit d'équivalent de sable (E_S), mis au point aux Etats-Unis résolvait assez bien ce problème pour les bétons hydrauliques, tant que leur teneur en fines restait voisine de 1 à 2 %. Mais, au-delà de cette teneur en fines, on s'est aperçu que cet essai présente de très sérieux défauts :

a) forte sensibilité aux fines inertes pour peu qu'elles soient trop fines. Par exemple, 4 % de fines siliceuses exemptes d'argiles et inférieures à 20 µm conduisent à un E_S de 4 ;

b) piégeage dans le floculat argileux d'une partie des minéraux inertes fins (calcite, quartz, feldspaths), ce qui les assimile à des argiles. Ces particules piégées peuvent atteindre 200 μm. Tant qu'il n'y a que peu de fines, ce défaut n'est jamais bien gênant (cas des sables à bétons pauvres en fines), mais pour les sables destinés aux assises de chaussées et aujourd'hui même pour des sables à bétons dont les teneurs en fines dépassent souvent 10 %, on constate une chute de l'équivalent de sable, parfois très forte, lorsque ces fines, bien qu'inertes, sont très fines (Fig. 10).

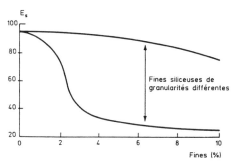

Fig. 10 — Influence de fines inertes sur l'équivalent de sable E_s

c) importance que prend le floculat avec l'argile la moins nocive qu'est la kaolinite. Or, toutes les études sur matériaux composites ont montré qu'on peut tolérer plus de kaolinite que d'autres argiles - Mais plus de kaolinite signifie une valeur de l'E_s très basse, située pratiquement toujours en dessous des valeurs admises dans les spécifications.

d) Le dernier défaut est lié à la dispersion des argiles, qui augmente avec la teneur en eau et donc souvent avec la durée du stockage non protégé et les diverses manutentions, dont le transport.

Il était donc à peu près indispensable de mettre au point un essai qui permette de mieux prendre en compte les fines nocives et en particulier les plus répandues d'entre elles, les argiles.

1.2 L'essai au bleu de méthylène

Les éléments argileux possèdent en présence d'eau un ensemble de propriétés spécifiques liées à leur « activité » ; ce sont notamment la plasticité, la cohésion, l'adhésivité, le caractère gonflant,

l'affinité pour l'eau, etc. Or, ces propriétés, qui confèrent à ces éléments un rôle fondamental et souvent nocif dans le comportement d'un matériau, sont toutes des propriétés de surface liées principalement :

- à l'extrême finesse des particules se traduisant par une surface spécifique externe élevée (de 5 à 80 m²/g pour la montmorillonite) ;
- à la structure particulière en feuillets des particules argileuses, qui leur permet de développer, pour les espèces les plus actives, une surface spécifique interne considérable, s'ajoutant à la surface spécifique externe précédente (80 m²/g pour la montmorillonite) ;
- Enfin, à l'état de charge de cette surface, où s'exercent des forces électriques intenses qui sont, pour une large part, à l'origine du phénomène d'activité.

Le problème était donc de trouver un essai rapide - condition supplémentaire et indispensable dans le domaine des contrôles de fournitures - résumant dans la valeur mesurée les trois aspects mentionnés ci-dessus. Après de nombreux essais, il est apparu que la capacité d'adsorption en bleu de méthylène de l'argile convenait bien, pour les raisons suivantes :

- l'adsorption est un phénomène de surface par excellence. Elle est utilisée couramment pour mesurer la surface spécifique externe des poudres comme dans la méthode classique de Brunauer, Elmet, Teller (BET)

- l'adsorption rend compte également, quand elle se produit sur des particules argileuses dispersées dans l'eau, de la surface interne devenue accessible,

- l'adsorption du bleu de méthylène dépend fortement des liens chimiques que les molécules de bleu peuvent contracter avec les charges de surface et, par suite, tient compte de l'état électrique de cette surface.

L'essai proposé consiste donc à mesurer la capacité d'adsorption en bleu de méthylène qui est la quantité de ce colorant nécessaire pour recouvrir d'une couche monomoléculaire les surfaces externe et interne de toutes les particules argileuses, présentes dans 100 grammes de granulat. Cette quantité est appelée « valeur de bleu » de ce granulat, exprimée en grammes de bleu par 100 grammes de matériau.

Cet essai - dit essai à la tache - présentait donc l'avantage de prendre en compte de façon très spécifique les argiles et donc :

- d'éviter de pénaliser des sables fins à fines inertes, riches en particules inférieures à 10 µm,
- d'éviter de pénaliser les sables à pollution kaolinique : cette dernière ayant une faible surface adsorbe assez peu de bleu.

Ce problème étant résolu, il était séduisant de s'en tenir à cet essai simple, rapide et peu coûteux. Cependant, de nombreuses études effectuées sur les matériaux composites (granulats traités aux liants hydrauliques ou hydrocarbonés) ont vite montré que dans ces derniers les argiles étaient généralement modifiées par les liants. Les observations actuelles semblent faire ressortir que :

• à long terme, c'est-à-dire pour la durée normale de service d'un ouvrage, c'est la surface totale (externe et interne) des argiles qui traduit le mieux le risque maximal encouru. C'est donc effectivement l'essai à la tache qui serait le plus significatif de ce point de vue.

• à court terme, seule la surface externe compte, car la partie interne, déjà par nature plus difficilement accessible à l'eau, est rendue quasiment inaccessible, par la présence des liants hydrocarbonés ou, dans une moindre mesure, hydrauliques. C'est aussi à court terme que l'on mesure la nocivité des argiles, lorsqu'on la fonde sur des essais mécaniques réalisés après une immersion limitée à 7 jours sur des éprouvettes de matériaux composites. Il est donc normal que l'on ait constaté que la surface externe est la mieux corrélée avec le comportement des matériaux composites. Ainsi les nocivités respectives d'une montmorillonite et d'une kaolinite sont dans un rapport de 3 à 4 (comme le rapport des surfaces externes) alors que celui des valeurs de bleu à la tache est de 15 à 20.

La première conséquence de ces observations, encore en cours, a été la mise au point d'une "variante" de l'essai au bleu à la tache, l'essai au bleu optique, dit turbidimétrique, qui mesure une valeur aussi proche que possible de la surface externe.

Si nous avons tenu à développer un peu longuement cet exemple, c'est qu'il nous semble significatif d'une évolution encore à peine entamée. En effet, les mécanismes des transferts, des équilibres et des changements de phases dans tous les matériaux composites, que le milieu soit carbonaté, silicaté, dolomitique ou mixte doivent encore faire l'objet de nombreuses recherches pluridisciplinaires (pétrographiques, minéralogiques, physico-chimiques, etc...) avant que l'on puisse déboucher sur des conclusions utilisables dans la pratique quotidienne des travaux.

- Un autre exemple que nous exposerons plus brièvement, illustre bien cette nécessité de parvenir à la compréhension des phénomènes fondamentaux si l'on veut déboucher sur des solutions économiquement viables, c'est celui des phénomènes connus sous le nom de réactions alcalis-granulats.

2. Les réactions granulats-liants

Des années de recherches menées par des laboratoires de divers pays, ont abouti à un certain nombre de constatations, dont les plus importantes sont les suivantes :

• les dégradations des ouvrages d'art et bâtiments en béton dues à des réactions entre granulats et liants apparaissent dans des délais variant de quelques mois quelques années.

• les désordres s'échelonnent depuis une fissuration insignifiante non évolutive et peu dense, jusqu'à une fissuration importante, dense et évolutive, conduisant à la ruine de l'ouvrage.

• les réactions conduisant aux désordres ne peuvent se produire qu'en présence d'eau ou dans une atmosphère dépassant 85 % d'humidité relative.

• S'il est vrai que les températures élevées favorisent les réactions, on en connaît aussi sous climat froid.

• La teneur basse en alcalins des ciments est un facteur de moindre risque, mais ne constitue en aucun cas une garantie totale de suppression des réactions.
On a même observé des dégradations du même type dans des ouvrages où le liant constitué de chaux ne contenait a priori aucun alcalin.
Il semble que parmi les ciments couramment utilisés en Génie Civil, les C.P.A. soient

plus volontiers à l'origine de la réaction que les ClK ou les ciments avec ajout de pouzzolane. Mais les observations ont montré aussi qu'il ne faut pas pour autant en conclure que laitier et pouzzolane sont des inhibiteurs de réaction.

• Le très grand nombre de cas observés permet d'affirmer que la presque totalité des roches est susceptible de donner naissance à une réaction (FIG. 11). Les minéraux les plus souvent mis en cause sont certains minéraux siliceux, comme la calcédoine et surtout l'opalé, mais l'on connaît de nombreux ouvrages sans désordres dans lesquels ces minéraux sont très abondants.

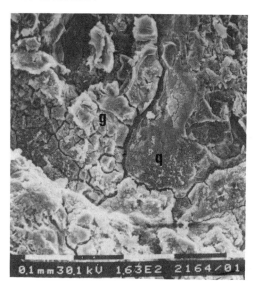

Fig. 11 - Gel de réaction alcalis granulats (g) à la périphérie d'un quartz (q).

Les quartz revêtus d'une pellicule de silice amorphe, ou encore ceux qui ont subi des contraintes tectoniques qui ont déformé leur réseau (extinction roulante) sont également réactifs. Il en est de même pour les feldspaths, les micas, en particulier lorsqu'ils sont altérés. On a aussi constaté le rôle important joué par les sulfures sédimentaires comme dévia- teurs des réactions de stabilisation lorsqu'on traite des sols, en particulier lorsqu'ils se présentent sous forme d'amas de framboïdes à surface spécifique élevée. En milieu alcalin, ils peuvent se détruire et donner naissance à de l'ettringite. Or les gels silico-calco-alcalins correspon- dant aux réactions alcali-granulat sont très souvent associés à d'importantes

cristallisations d'ettringite. Or ces sul- fures, dangereux en faibles quantités (moins de 1 %) peuvent se trouver dans de très nombreuses roches. On les rencontre par exemple, fréquemment associés à de la silice diffuse, dans des calcaires ou dolomies.

La structure et la texture des roches exploitées intervient bien entendu également, la réactivité étant conditionnée par l'accessibilité plus ou moins grande des minéraux réactifs aux solutions agressives.

De toutes ces constatations, il apparaît que les désordres n'ont pas pour seule origine les réactions alcali-granulats sensu stricto, mais aussi des réactions mixtes sulfato-alcali-granulats qui se développent lorsque sont présents simultanément les sulfures sédimentaires, la silice désordonnée et sans doute des argiles diffuses. Il est évident aussi que les réactions qui apparaissent avec un couple ciment-granulat donné ne se produisent pas lorsqu'un autre ciment - ou un autre granulat - est utilisé. Il s'agit donc d'un problème de couple ciment- granulat. En ce qui concerne les roches qui sont utilisées pour la fabrication des granulats, la première démarche est d'en faire l'étude pétrographique détaillée, c'est-à-dire :
- l'étude en lame mince au microscope polarisant
- l'étude en diffractométrie des RX et analyses chimiques
- l'étude par les méthodes thermiques ATD et ATG
- la microscopie électronique à balayage.

L'ensemble des observations faites permet de détecter les "roches à risque", même si l'on sait encore mal pour l'instant quantifier ce risque.

On peut donc dire que si le comportement macroscopique des matériaux est un sujet assez bien maîtrisé, en revanche l'analyse micro-structurale reste un champ immense de recherches qui pourraient bien bouleverser l'état des connaissances, sous l'angle de la durabilité et de la compatibilité des matériaux notamment. En dehors des matériaux les plus traditionnels évoqués ci-dessus (bétons- matériaux de chaussées) seront aussi concernées les utilisations les plus diverses :
- entretien et restauration de monuments historiques,
- fabrication de briques en terre stabilisée (sans cuisson),

- fabrication de bétons avec des sables et des ajouts divers,
- utilisation de tufs, encroûtements calcaires, de calcaires tendres, etc...

Cette analyse fine paraît le vecteur principal des évolutions à venir, le géologue se trouvant une fois de plus au coeur du problème.

Nous avons essayé de montrer à travers un certain nombre d'exemples à quel point le géologue était directement impliqué à toutes les étapes conduisant à l'utilisation des matériaux de construction, depuis les inventaires à l'échelle du pays jusqu'aux méthodes d'élaboration, en passant par l'étude détaillée des gisements. Il reste un domaine qui concerne directement toutes les exploitations à ciel ouvert, c'est celui de l'environnement.

CARRIERES - ENVIRONNEMENT ET GEOLOGIE

Dans la plupart des pays, même développés, les gisements ont été exploités pendant des siècles sans se soucier de l'utilisation ultérieure des sites. C'était seulement par hasard que de temps à autre tel ou tel site de carrière abandonnée était réutilisé.
Depuis quelques décades, voire seulement une dizaine d'années pour certains, la plupart des pays développés se sont dotés de législations sur l'environnement prenant en compte les exploitations à ciel ouvert. On peut dire pour simplifier que ce souci d'environnement se situe généralement à deux niveaux :

• dans les régions très urbanisées, le paramètre environnement est pris en compte tout à fait en amont, c'est-à-dire au stade des inventaires (cf. chapitre I).

En même temps qu'on inventorie les réserves en matériaux, on réalise des études socio-économiques, hydrauliques, hydro-géologiques, on passe en revue l'état actuel des exploitations, des contraintes de tous ordres, et on étudie les réutilisations possibles pour les futurs sites exploités. En fait, ces projets très globaux visent à intégrer les carrières dans un cycle d'utilisation de l'espace de façon à ce qu'elles ne soient plus qu'un instant dans l'utilisation de cet espace.

• pour tous les sites, qu'ils soient situés en zone urbaine ou rurale, les législations imposent en général, au moins

une remise en état en fin d'exploitation, souvent même un réaménagement plus ou moins complexe selon l'utilisation future prévue. A titre d'exemple, et sans naturellement essayer d'être exhaustifs, on peut résumer sous forme de tableau les principales possibilités d'aménagement des carrières après extraction des matériaux (Tab. 1).

Il est évident que pour la plupart - pour ne pas dire tous - ces réaménagements des études techniques préalables de tous ordres sont nécessaires. Certaines de ces études - en particulier les études hydrogéologiques pour les carrières en eau ou encore géotechniques pour les carrières à sec, pèsent d'un poids décisif dans la

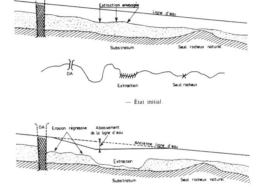

— État initial.

— Phase 1. L'extraction se fait à partir du matériau en place, des sédiments provenant de l'érosion régressive et du transport solide naturel.
Conséquences : abaissement localisé de la ligne d'eau, érosion régressive et augmentation de la pente amont, recoupement de méandres amont.

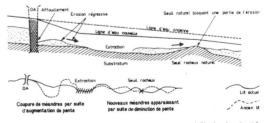

— Phase 2. L'extraction se fait à partir des sédiments provenant de l'érosion régressive et du transport solide naturel, de plus des sédiments sont entraînés en aval de l'extraction pour « nourrir » le fleuve.
Conséquences : abaissement général de la ligne d'eau, érosion régressive, augmentation de la pente amont, affouillement de l'ouvrage d'art, recoupement de méandres amont, érosion en aval de l'extraction, diminution de la pente en aval de l'extraction, méandrage en aval du fleuve.

Fig.12

Évolution du profil en long d'un cours d'eau suite à l'extraction de matériaux.

1 - CARRIERES EN EAU

Type et caractéristique		Critère d'environnement	Possibilités d'aménagement
Faible profondeur d'eau		Rural	Réserve ornithologique - Chasse du gibier d'eau Réserves d'eau Mise hors d'eau et réutilisation agricole ou sylviculture
		Péri-urbain et urbain	Coupure dans l'urbanisation Remblayage partiel ou total pour utilisation • zones vertes et de loisirs • zones constructibles
Profondeur d'eau moyenne ou forte		Rural	Pêche de loisir - Pisciculture - Baignade - Barque et canotage - Port de Plaisance - Bassin d'infiltration - Bassin de stockage d'eau
		Péri-urbain et urbain	Plan d'eau (lotissement au bord de l'eau) - Port industriel - Port de plaisance - Bases de loisirs polyvalentes

2 - CARRIERES A SEC

En fosse		Rural	Reconstitution du terrain - Reverdissement Agricole - Reboisement - Réserve naturelle
		Péri-urbain et urbain	Remblayage - Décharge contrôlée - Coupures vertes - Parc - Zone d'habitation - Zone industrielle - Lac artificiel
A flanc de relief	Parois meubles	Tous environnements	Mise en végétation
	Parois rocheuses	Vues éloignées Vues rapprochées	Confortement et traitement de la paroi Talu végétalisé
	Fond de carrière	Rural	Remise en végétation (prairie, agriculture, sylviculture) Réserve naturelle
		Urbain ou Péri-urbain	Parc de verdure - Parc de véhicules - Zone industrielle - Zone de loisirs

faisabilité de tel ou tel type de réaménagement. Comme nous l'avons fait pour le chapitre précédent, nous donnerons simplement quelques exemples qui nous semblent significatifs de l'importance de l'intervention du géologue.

1) **Extraction de matériaux dans le lit d'un cours d'eau.** L'exploitation dans un écoulement d'eau conduit toujours à des perturbations dont les répercussions se font sentir à des distances importantes. Elles consistent en :
- une augmentation de la section d'écoulement au droit de la zone d'extraction
- un abaissement de la ligne d'eau
- un accroissement de la pente, d'où augmentation des vitesses
- une érosion régressive
- un affouillement de certains ouvrages
- une attaque des berges
- un décolmatage des berges et du fond avec des modifications des échanges nappe-rivière
- des variations du niveau piezométrique.

Il s'agit là bien entendu d'un schéma de principe (Fig. 12). Il est évident que tous ces phénomènes ne se produisent pas dans tous les cas et que des études spécifiques à la fois hydrauliques, hydrogéologiques, sédimentologiques, des analyses de stabilité des ouvrages existants doivent être réalisés dans chaque cas précis.

2) De la même façon **les extractions sous le niveau hydrostatique ou dans la nappe alluviale** pour les exploitations du lit majeur entraînent toute une série de conséquences sur le gradient hydraulique en amont, sur le bilan hydrique de la nappe ou encore sur le coefficient d'emmagasinement, sans parler des colmatages éventuels à long terme du fond et des berges ou de la mise en communication plus ou moins directe des nappes superposées, etc... Les études menées au niveau d'un secteur de vallée peuvent ainsi conduire à une planification autoritaire des extractions pour tenir compte des conséquences immédiates et long terme.

3) **Les boues de lavage de carrière.** Il s'agit là d'un problème typique d'imbrication étroite des soucis techniques, économiques et d'environnement. Nous avons vu précédemment que la propreté des granulats commercialisés était un critère tellement important pour presque toutes les techniques que des recherches spécifiques ont été réalisées et des essais spéciaux imaginés pour en tenir compte. Ce qui signifie qu'il est souvent nécessaire de laver les produits. Ce qui suppose, après le lavage, une décantation des eaux de rejet avec, pour corollaire, une "production" inévitable de boues. On se trouve donc face à un problème technico-économique dû à la nécessité de respecter l'environnement. Dans un premier temps, le lavage est rendu nécessaire pour des raisons techniques, dans un second temps la teneur de matières minérales en suspension dans les eaux rejetées dans les cours d'eau et dans les nappes est limitée par la législation sur l'environnement et dans un troisième temps les boues récupérées doivent être stockées sans nuire à l'environnement ou réutilisées, ou alors, cas extrême, il faut éviter d'en produire. Ce problème a été étudié, il implique la mise en oeuvre d'études portant sur le matériau exploité, sur l'installation d'élaboration et sur le traitement des boues produites. On s'aperçoit ainsi qu'il existe quatre manières d'aborder le problème :

- la boue n'est plus produite : ce qui signifie qu'il faut supprimer le lavage et appliquer à la place un traitement adéquat agissant directement sur les fines argileuses, en installation de concassage. Les études réalisées jusqu'à présent ne sont pas convaincantes.

- la quantité de boues est réduite. La méthode consiste à abaisser la coupure d'élimination de la fraction polluante de 100 microns à 30 microns par exemple. Dans ce cas, il s'agit d'être certain que les fines comprises entre 30 et 100 microns ne sont pas nocives mais peuvent même au contraire être utiles. On se retrouve confrontés au vaste problème en cours d'études et de recherches évoqué précédemment sur les liaisons liants-granulats.

- la boue est réutilisée
  • soit sans traitement préalable en décharges contrôlées, en couches d'étanchéité, ou alors comme substrat ou composante d'un substrat en vue d'une revégétalisation, ce qui suppose, surtout dans ce dernier cas, des bilans minéralogique et pédologique très précis.

  • soit après traitement par un liant (briques ou blocs) : traitement aux liants hydrauliques supposant une composition minéralogique adéquate, en particulier celle de la phase argileuse.

  • soit après traitement thermique. Selon la nature de la boue peuvent être envisagées des fabrications de granulats expansés, de faïence calcaire, de ciment pouzzolanique, etc...

En fait, le problème qui ne peut être, une fois de plus, abordé que de façon pluridisciplinaire, aussi bien pour les techniques que pour les choix économiques, est ardu, car il consiste à trouver une ou plusieurs solutions permettant de réutiliser en grandes quantités (environ 6 millions de tonnes par an en France) des matériaux fortement hétérogènes, dispersés géographiquement (les 6 millions de tonnes sont produits par plus d'un millier de carrières), économiquement peu transportables et produits en quantités variant beaucoup dans le temps.

4) **Stabilité des fronts de taille et remise en état des carrières de roches massives.**

La remise en état des carrières de roches massives en vue d'une réutilisation du site doit tenir compte de nombreux éléments, le principal étant la présence des fronts de taille dont la disposition conditionne les utilisations possibles du site.

On peut distinguer :

- d'une part, les données liées au massif lui-même, c'est-à-dire la nature de la roche, la fracturation, l'altération, la structure...
- et d'autre part, celles qui dépendent des travaux exécutés, soit l'existence d'un front, sa hauteur, les problèmes de stabilité et donc de sécurité, l'insertion dans l'environnement.

L'étude doit tenir compte de tous ces éléments au cours de l'exploitation pour aboutir, le mieux possible, au résultat final sans travaux trop importants au niveau de la remise en état.

- Définition du risque

La définition du risque se fait en étudiant les paramètres qui contrôlent la stabilité du front de taille.

Description du massif rocheux et des roches

Le massif rocheux constituant le (ou les) front(s) doit avant tout faire l'objet d'une description géologique générale qui permettra la distinction de "zones" différentes en fonction des données lithologiques, pétrographiques et structurales (nature, faciès des roches, structures tectoniques...).

Conditions hydrogéologiques

Elles correspondent à la reconnaissance de l'origine du gisement et des alimentations des circulations d'eaux dans le massif. Localement, elles seront décrites par les valeurs du couple (Hi, Kj) de la charge hydraulique H et de la perméabilité K.

Dans le cadre d'un front de taille, elles seront simplement décrites par la seule valeur de la charge hydraulique, communément exprimée en mètres d'eau, au-dessus de la cote du carreau. Elle peut entraîner l'apparition de surpressions interstitielles et provoquer des instabilités.

Les paramètres géométriques

Les discontinuités du massif rocheux

Le terme correspond à toutes les interruptions physiques du massif rocheux ; elles se caractérisent par une faible résistance à la traction perpendiculairement à leur plan et par les conditions du cisaillement qu'elles permettent parallèlement à leur plan.

Leur présence a une influence notable sur la stabilité des terrains restant en place après exploitation.

Leur étude devra porter sur :
- leur densité (c'est-à-dire une quantification de l'état de fracturation du massif),
- leurs orientations et leur groupement en familles,
- leur comportement mécanique qui dépend largement de leur persistance, leur morphologie, leur ouverture et leurs remplissages.

Les principaux types de ruptures géométriquement possibles dépendent de l'orientation respective du front de taille et des familles de discontinuités.

On doit également tenir compte des possibilités d'insertion dans l'environnement : en particulier, on peut être conduit à terminer une exploitation par un front rectiligne ou sinueux, et dans d'autres cas, envisager plutôt de masquer ce front, ou de lui donner une pente et une conformation différente qui lui permette de mieux s'insérer dans l'environnement et en particulier d'être plus adapté à la remise en état envisagée.

Il est évident que beaucoup de ces éléments ont simplement à être complétés lorsque le gisement concerné a été correctement étudié lors de son ouverture, les méthodes de prospection détaillées (voir chapitre 2) permettant d'identifier la plupart de ces paramètres sur un site vierge.

- Recherche de moyens pour stabiliser les fronts de taille

Comment, à partir des données ci-dessus, peut-on aboutir au résultat final sans pénaliser l'exploitation de la carrière ?

Cette recherche fait appel à toutes les techniques :

• *A l'abattage d'abord :*

Lorsque le front de taille est proche de la limite d'exploitation, la méthode d'abattage doit être aménagée afin d'atteindre outre l'objectif "extraction de matériau", l'objectif supplémentaire "réalisation d'une face stable". Ce souci doit être prévu dès l'ouverture du gisement et au fur et à mesure de la progression des fronts en limites d'emprise.

Cette démarche permet de minimiser les instabilités et fournira le cas échéant des conditions saines pour une reprise éventuelle de l'exploitation ou un réaménagement spécifique.

Pour minimiser les dégâts infligés aux talus restant en place, la première méthode est de réaliser des tirs d'abattage de la masse correctement dimensionnés et soigneusement contrôlés, afin que le minimum d'énergie explosive soit envoyé dans les talus, en pure perte de rendement, et y provoque des dégâts.

Des techniques diverses, telles que l'abattage amorti, le pré-découpage, le post-découpage peuvent être utilisées, soit seules, soit en association. Par exemple :

• une surface bien pré-découpée sera efficacement protégée par un tir d'abattage amorti à l'approche de cette surface associé au tir de masse, lui-même dimensionné correctement.

• un post-abattage sera efficace si le massif dans lequel on le tente n'est pas déjà désorganisé par le tir de masse, d'où la nécessité d'un tir de masse bien contrôlé, associé le cas échéant à un abattage amorti à l'approche du talus provisoire, qui sera lui-même post-abattu.

• *A la remise du front de taille*

Les méthodes de confortation, d'adaptation, de parade, doivent être adaptées à chaque cas particulier.

Une étude géotechnique est nécessaire pour reconnaître les paramètres d'instabilité (cf. ci-dessus). On retrouve le plus souvent les problèmes habituels de stabilité de falaises pour lesquels est à disposition une série de parades variées.

On notera qu'un remodelage contrôlé à l'explosif selon les modalités décrites ci-dessus est toujours possible même en site sensible (proximité de l'habitat) et constitue souvent une solution efficace peu onéreuse. Ce remodelage permet d'obtenir une géométrie auto-stable.

Des méthodes de stabilisation par renforcement peuvent être mises en oeuvre lorsqu'un tel remodelage n'est pas possible ou ne permet pas d'obtenir cette stabilité (à cause de la structure défavorable du massif par exemple).

Il est évident que si toutes ces dispositions permettent en fin de compte de disposer de fronts de taille réutilisables, il reste encore à définir un certain nombre d'aménagements absolument nécessaires, tels que clôtures, accès signalisation et surtout l'entretien, qui sera la seule garantie de la stabilité des fronts dans le temps.

CONCLUSIONS

Comme nous venons de le voir, les carrières à ciel ouvert sont un des points sensibles de l'environnement. Il n'est plus possible, et cela est normal d'extraire n'importe où et n'importe comment. Mais, par ailleurs, comme nous l'avons indiqué en introduction, la consommation mondiale de matériaux de construction, et en particulier de granulats, est énorme.

Compte tenu des évolutions techniques prévisibles, cette consommation est appelée à croître encore beaucoup dans les vingt ou trente prochaines années. Il est donc absolument nécessaire de satisfaire cette demande tout en respectant l'environnement, ce qui suppose :

• de réaliser des bilans technico-économiques besoins-ressources, pour bien situer le volume et la nature des matériaux à extraire.

• d'optimiser l'emploi des matériaux disponibles, en dépassant les connaissances actuelles. Bien des matériaux non utilisés actuellement pourraient sans doute le devenir, permettant ainsi de ne pas trop concentrer les extractions dans les formations classiques, comme les alluvions.

Pour mener à bien ces tâches, des études pluridisciplinaires sont nécessaires, mais si l'on regarde l'ensemble du problème de près on s'aperçoit que seul le géologue peut jouer le rôle de fil conducteur. Cela suppose bien entendu de sa part non seulement une utilisation de toutes les disciplines et techniques classiques des sciences de la terre, mais aussi un suivi attentif des recherches nouvelles sur les méthodes de prospection, sur les matériels et techniques d'élaboration et d'exploitation, sur le comportement des matériaux et en particulier celui des matériaux composites.

C'est seulement en faisant progresser tous ces domaines que l'on assurera la pérennité des extractions de matériaux de construction, tout en respectant l'environnement.

Bibliographie

Archimbaud C ; Griveaux B ; Rat M (1977), Reconnaissance des gisements rocheux. Bull. Liaison L.P.C. n° spécial IV, p. 155-166

Archimbaud C ; Martin-Guillou Y ; (1977), Influence des caractéristiques des gisements sur la conception des installations, Bull. Liaison L.P.C., n° Spécial IV, p. 201-210

Archimbaud C ; Lassartesse J ; Mishellany A ; (1981), Exemple d'étude de gisement et d'impact d'une carrière de roches masives, Bull. Liaison L.P.C., n° 112, p. 262-270

Barisone G ; Bottino G ; Cardu M ; (1986), Italian peninsular aggregates and the alkali-silica reaction. Proceedings 5th International Congress of the IAEG, Buenos Aires, Balkema Publ., vol. 5, p. 1623-1632

Benaben J.P ; Nguyen Dac Chi ; Tourenq C ; (1989), Identification des argiles polluant les graves ciment par l'essai au bleu de méthylène ; une meilleure approche du problème de la propreté des sables, Bull. de Liaison des L.P.C., n° 164, p. 5-15

Bérubé M.A ; Fournier B ; (1986), Les produits de la réaction alcalis-silice dans le béton. Etude du cas de la région de Québec, Canadian Mineralogist, 24, 271-288

Collis L ; Fox R.A ; (1985), Aggregates : sand gravel and crushed rock aggregates for construction purposes - Geological Society of London - Engineering Geology Special Publication, n° 1

Demulder E.F.J. ; (1984), A geological approach to traditional and alternative aggregates in the Netherlands, Bull of the IAEG, n° 29, p. 49-57

Doridot M. ; Resende S ; (1977), Etude des gisements de matériaux alluvionnaires pour la préparation d'un SDAU - Bull. Liaison L.P.C., Numéro spécial IV, p. 141-153

Enrochements (ouvrages - caractéristiques et méthodes de détermination - exploitation des carrières et production d'enrochements naturels - recommandations de spécification pour enrochements et contrôles), 1989, Publication Laboratoire Central Ponts et Chaussées, Paris, 107 pages

Frouin L ; (1989), Etude expérimentale des intéractions surfaces minérales-bitume - Cas particulier des argiles, Publications L.C.P.C., 58, Bd Lefèbvre, Paris (15è), Rapport de Recherche n° 154, 67 p.

Granulats : Economie - Géologie - Prospection - Exploitation - Propriétés - Utilisations ; (1990), Publications de l'E.N.P.C. 28, rue des Sts Pères, Paris (6è), 450 pages, Figs., Tabl.

Héraud H. ; Puntous R. ; (1989), Les différents paramètres de la qualité du tir, de l'étude ou résultat, Industrie Minérale - Mines et Carrières, juin 1989, p. 57-61

Lagabrielle R. ; Chevassu G. ; (1984), Nouvelles méthodes géophysiques pour les gisements terrestres ou aquatiques, Bull. of the IAEG, n° 29, p. 111-116.

Lamond J.F ; (1984), Aggregates for concrete, general report 1st International Symposium on Aggregates Nice, France, Bull. of the IAEG, n° 30, p. 159-162.

Laviron F. ; Gand G. ; Longère P. Cognard E. ; (1981), Exemple d'étude d'impact pour l'exploitation d'un gisement alluvionnaire, Bull. liaison L.P.C., n° 112, p. 271-285.

Le Roux A. ; Cador C. ; (1984), Importance de la pétrographie dans l'approche des mécanismes de la réaction alcali-granulats. Bull. IAEG, n° 30, p. 255-258.

Livet M. ; Guillin J.C. ; (1981), Protection des berges et extraction de granulats dans le lit mineur d'une rivière. Bull. de liaison des L.P.C., n° 112, p. 122-130.

Mac Lellan A.G. ; (1984), Monitoring and Modelling progressive rehabilitation in aggregate mining ; a decade of Ontario expérience and a look at the future. Bull of the IAEG, n° 29, p. 279-284.

Mishellany A. ; (1981), Pollution des rivières par le lavage des matériaux en carrière. Bulletin de liaison des L.P.C., n° 112, p. 83-89.

Neeb P.R. ; Danielsen S. ; (1984), Evaluation of quaternary aggregate resources in Norway. Bull. of the IAEG, n° 29, p. 129-138.

Oberholster R.E. ; Krüger J.E. ; (1984), Investigation of alkali-reactive, sulphide-bearing and by-product aggregates. Bull of the IAEG, n° 30, p. 273-277.

Prax A. ; Primel L. ; (1977), Prospection des gisements alluvionnaires, Bull. Liaison L.P.C., n° Special IV, p. 211-226.

Primel L. ; (1969), Recherches sur l'évolution des propriétés des matériaux alluvionnaires dans un bassin et mise en évidence de quelques caractéristiques générales. Publications L.P.C., Rapport de recherche n° 1, 83 p.

Primel L. ; (1977), Les bilans de ressources en granulats. Bull. Liaison Labos Pts et Ch., Numéro spécial IV, p. 11-16.

Raynal J.J. ; Dampt F. ; (1989), Ciments français ; étude de la carrière de Beaucaire, Industrie Minérale - Mines et Carrières, Août Septembre 1989, p. 119-124.

Scott D.W., (1984) ; Aggregate resources inventory in Ontario, Canada, Bull. of the IAEG, n° 29, p. 163-168.

Shadmon A. ; (1989), Stone, an introduction, 140 p. Intermediate Technology Publications, 103 Southampton Row, London WC1B 4HH.

Shayan A. ; Quick G. ; (1989), Microstructure and composition of AAR products in conventional standard and new accelerated testing, 8th ICAAR, Kyoto, p. 475-482.

Tourenq C. ; Denis A. ; (1982), Les essais de granulats. Publications du L.C.P.C., Rapport de Recherche N° 114, 92 pages.

Tourenq C. ; Tran Ngoc Lan ; (1989), Mise en évidence des argiles par l'essai au bleu de méthylène. Application aux sols, roches et granulats ; Bull. Liaison L.P.C., 159, p. 79-92.

Unikowski Z. ; (1982), Influence des argiles sur les propriétés des mortiers de ciment. Publications L.P.C., Rapport de Recherche n° 110.

U.N.P.G. ; (1982), L'affectation des sols de carrières de granulats après exploitation. Collection technique n° 2, 79 p., 3, rue Alfred Roll, 75017 Paris.

U.N.P.G. ; (1983), Le tableau de bord de la production des granulats. Politique départementale. Collection technique n° 3, 53 p., 3 rue Alfred Roll, 75017 Paris.

The use of crushed sedimentary rocks for concrete production in Iraq
L'usage de la poudre des pierres sédimentaires dans la production de béton

M.A. Elizzi & B.G. Ikzer
Building Research Centre, Baghdad, Iraq

ABSTRACT: The southern part of Iraq suffers from shortages of suitable natural gravel to be used as coarse aggregate for concrete production. so, for the time being, gravel is being supplied from quarries outside this area, which obviously increases the cost of concrete produced. This research was carried out to study the possibility of the use of cushed sedimentary rocks as coarse aggregate, which are available in large quantities in southern part of Iraq, mainly as limestone. A geological surveying for rock quarries in Muthanna Governorate was made. It was found that the best location is "Al-Fadhwa" area, 45 km south west Samawa city.

Petrographic studies and physical and chemical analysis of this type of limestone, showed that this caorse aggregate satisfies the Iraqi Standard for concrete production.

Compressive strength tests of concrete specimen using crushed limestone as coarse aggregate showed that this type of rocks can be widely used for concrete production.

RESUME: La Partie Sud de l'Iraq souffre d'une manque de gravel naturelle convenable à utiliser comme agrégat dur dans la production de concrete, et pour cette raison ou doit l'apporter des autre parties du pays, ce qui augmonte le coût de concretes produits.

Le but de ce recherche est d'étudier la possibilité d'utiliser la poudre des pierres sedim-entaire comme agrégat dur, qui sont valables en grandes quantité's aen sud du pays, specialement comme limestone.

Plusieurs études géologique ont eté faites dans la ville de (Al-Muthanna), et on a trouvé que le meilleur endroit etait (Al-Fadhwa), situé à 45 kms dans le sud oest de la ville (Samawa).

Les études petrographique et les analyses physiques et chimiques éffectués pour ce type de limestone montrent que l'agrégat dur est satifusant pour la production de concrete selon le standard Iraquien, et les tests de resistence de compression du specimen de concrete montre que ce type de pierre peut-être utilisé largement dans la production de concrete.

1 INTRODUCTION

Natural gravel and silica sand are the main sources of aggregate used in concrete production in Iraq. During the last years the demand for aggregate sharply increased due to the large increase of concrete construction. The southern part of Iraq suffers from big shortage of suitable natural gravel to be used as coarse aggregate in concrete manufacturing, Mansour and Petrank (1980). Therefore, natural gravel is being supplied to the area from the middle part of the country, which obviously increases the cost of the constructions. This research was carried out to investigate the possibility of the use of a crushed limestone as coarse aggregate in concrete production, because the limestone reserves in the southern part of Iraq are enormous, Al-Komi (1983).

Geological , chemical and physical studies on crushed limestone of Al-Fadhwa quarry (the investigated area) were carried out to determine their suitability as coarse aggregate for concrete production. Alkali reactivity of limestone rock studies were also concluded on these aggregates. Finally compressive strength and shrinkage tests were also carried out on concrete specimens using this type of rock as coarse aggregate.

All tests showed acceptable results for the use of these crushed limestone rocks in concrete production.

to bottom; weathered brecciated chalky limestone (2.5 m); chalky fine grained recrystallized lime-stone (3 m); Dolomitic limestone (2 m); and chalky limestone with chert nodules (more than 8 m.), Elizzi and others (1985).

FIG (1) LOCATION OF AL-FADHWA AREA

2 GEOLOGICAL STUDY

Al-Fadhwa area was chosen to produce crushed limestone as coarse aggregate. Al-Fadhwa quarry is situated about 45 km south west of the city of Samawa, Fig.(1). The investigated area is underlain by the Euphrates formation (Middle Miocene). The layers are almost horizontal. The Euphrates limestone formation is composed of different types of limestone includes, from top

3 ROCK MATERIALS TESTS

3.1 Chemical analysis of limestone rock

Chemical analysis was carried out on several rock samples from different depths. The results of the analysis are given in Table (1)

3.2 Physical properties of the limestone

Core samples 50 mm diameter, 50 mm heights were tested to find their

Table 1. Chemical composition of the limestone rock

Oxide	CaO	MgO	SiO_2	Al_2O_3	SO_3	Fe_2O_3	L.O.I.	Total
Percent	33.6	17.1	2.22	0.77	0.45	0.16	45.21	99.51

compressive strength following ASTM
C170:1976 method. Abrasion tests of
rock samples showed that the average
value of the abrasion is 30%. Apparent
specific gravity and water absorption
were also measured according to BS812:
1967. Finally the impact value of the
limestone aggregate was found according
to the Iraqi Standard 47:1970. All the
physical properties results are listed
in Table (2).

Table 2. Physical properties of limestone aggregate

Test	Value
Apparent specific gravity	2.7
Total specific gravity	2.51
Water absorption	4.41%
Abrasion	30 %
Impact	11.33%
Compressive strength (N/mm^2)	32.47

3.3 Alkali reactivity

Potential alkali reactivity takes place
when the alkalies present in cement
paste (Na$_2$O, K$_2$O) are attacked by the
silica or calcium carbonate present in
aggregate. Insoluable complex compounds
are formed due to this reaction, and as
a result of that, cement paste is subjected to osmotic pressure and then
fractures, Hanson (1944).

To determine the potential alkali
reactivity of the aggregate used, the
rock cylinder method described in ASTM
C586:1975 was followed to measure the
length changes of rock specimens stored
in sodium hydroxide solution for three
and six months, the results are listed
in Table (3).

3.4 Grading of aggregate

Rock blocks were crushed by jaw crushers. The crushed materials then pased
on different vibrating screens to
separate the different sizes of aggregate produced. Iraqi standard No.45:1980
was followed to determine the required
coarse aggregate grades for concrete
production.

Table 3. Length changes of rock specimens

Item	Length change %	
	age 3 months	age 6 months
Measured value	0.024	0.057
ASTM limits	x < 0.05	x < 0.10

4 CONCRETE WORK

To determine the suitability of crushed
limestone as coarse aggregate for
concrete production, two different mixes
were selected, namely: 1:1.5:3 and 1:2:4
recommended by the Road Research Laboratory, which are commonly used in Iraq.
Crushed limestone was used as coarse
and fine aggregate, different w/c ratios
were tried or/and plasticizer was used
to improve the workability of the
concrete. Table No.4 gives all the
details of the concrete mixes.

4.1 Workability

Workability of concrete mixes was
determined by slump tests, according to
BS 1881-1970. All the results are
listed in Table No.4.

4.2 Compressive strength

Concrete specimens (100x100x100 mm) were
made of different mixes, and cured in
water until the time of testing.
Compressive strength of the specimens
were found at ages of 7, 28 and 90
days. Table 5 and Fig.(2,a,b) showes
the effect of aggregate type, w/c ratio
and the use of plasticizers on the
compressive strength of concrete at
different ages.

4.3 Linear changes (shrinkage) tests

Linear changes tests were carried out
according to BS1881-1970 requirements.
For each mix proportion, two prisms
(100x100x250 mm) of concrete were made.

Table 4. Details of concrete mixes

| Mix type | Mix ratios | | | | | Plasticizer % of cement | Work-ability (slump) mm | w/c |
| | cement | Fine Aggregate | | Coarse Aggregate | | | | |
		crushed rock	sand	crushed rock	gravel			
A	1	1.5	-	3	-	-	50	1.01
B	1	2	-	4	-	-	55	1.3
C	1	-	1.5	3	-	-	60	0.65
D	1	-	2	4	-	-	50	0.85
E	1	-	1.5	-	3	-	60	0.4
F	1	-	2	-	4	-	55	0.5
A1	1	1.5	-	3	-	1	50	1.0
B1	1	2	-	4	-	1	50	1.3
C1	1	-	1.5	3	-	1	50	0.5
D1	1	-	2	4	-	1	80	0.65
C2	1	-	1.5	3	-	0.5	60	0.55
D2	1	-	2	4	-	0.5	50	0.7

Table 5. Compressive strengths of concrete mixes at different ages, (N/mm^2)

Mix type		A	B	C	D	E	F	A1	B1	C1	D1	C2	D2
Concrete age (days)	7	7.5	4.1	25.4	16.9	41.4	33.4	8.9	5.7	45.8	42.0	38.4	33.0
	28	10.1	6.1	33.0	24.5	51.0	43.1	11.7	8.0	55.2	48.7	42.4	36.7
	90	12.6	8.0	38.6	26.4	57.5	51.4	14.4	10.0	59.0	52.1	51.2	43.2

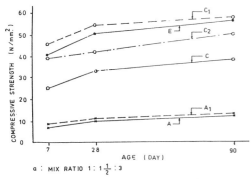

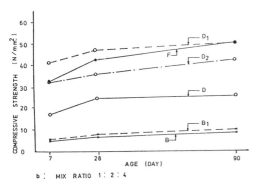

a : MIX RATIO 1 : 1½ : 3

NOTE : FOR MIX TYPE SEE TABLE NO.4

b : MIX RATIO 1 : 2 : 4

FIG (2 _ a & b) THE EFFECT OF AGGREGATE TYPE, W/C RATIO AND PLASTICIZER ON COMPRESSIVE STRENGTH OF CONCRETE .

The specimens were cured in water for 14 days then stored in air at room ambient. Linear changes were measured at different ages up to 180 days.

Fig.3 shows the effect of aggregate type, w/c ratio, mix proportions and the presence of plasticizer on the shrinkage of concrete.

5. RESULTS AND DISCUSSION

5.1 Physical properties of limestone

Table (2) shows that the abrasion of the tested rock is 30%, while according to the Iraqi Standard 47-1970, the abrasion of coarse aggregate suitable for concrete should not exceed 35%. The same standard requires impact resistance for the aggregate not more than 30%, while this resistance for the tested aggregate is only 11% as shown in Table (2).

5.2 Alkali reactivity

Table No.(3) shows that this type of limestone aggregate is suitable for concrete, as far as alkali reactivity property.

5.3 Workability of concrete

Table No.(4) shows that when using crushed rock as coarse or/and fine aggregate, w/c ratio was increased to maintain a good workability comparing with mixes using natural gravel and sand as coarse and fine aggregate. The reason for that is the higher water absorption, angularity, rougher surface texture of crushed rock aggregate. W/c ratio was reduced by using a small dosage of superplasticizer, (compare mixes C and D with C1 and D1 respectively, Table 4).

5.4 Compressive strength of concrete

Table 5 and Fig.2 show that using crushed limestone aggregate reduces the compressive strength of concrete comparing when using natural gravel. The reason of that is more water required to produce a workable concrete. In spite of this reduction in the strength, the concrete produced by using crushed limestone as coarse aggregate is suitable for the most concrete structures according to the compressive strength of concrete (at age of 28 days) required by CP 110:Part 1:1972, as follows:

5.4.1 Mixes of types C, C1, D1, C2 and D2 could be used for producing prestressed concrete (compressive strength according to CP110 is 30 N/mm^2).

5.4.2 Mixes of types C, D, C1, D1, C2 and D2 could be used for general reinforced concrete (compressive strength according to CP110 is 20 N/mm^2).

5.4.3 Beside the mixes mentioned in articles 5.4.1 and 5.4.2, mixes of types A, A1, and B1 could be used for plain concrete (compressive strength according to CP110 is 7 N/mm^2).

Using plasticizer improves the compressive strength of the concrete because w/c was reduced for nearly, the same workability.

5.5 Shrinkage

From Fig. No.3 it can be noted that the shrinkage of concrete produced by using crushed rock as coarse and fine aggregate (Mixes A and B) is higher than that of gravel and sand concrete (Mixes E and F), because the former mixes loss more water during drying process. On the other hand there was no significant difference in shrinkage values between gravel and sand concrete and concrete produced by using crushed rock as coarse aggregate, sand as fine aggregate (Mixes C and D).

6. CONCLUSIONS

6.1 Crushed limestone from Al-Fadhwa area can be used as coarse aggregate to produce good concrete.

6.2 Using crushed limestone as coarse and fine aggregate in the same mix is not recommended.

6.3 Using 1% of superplasticizer improves the compressive strength of the concrete.

6.4 Using crushed limestone from Al-Fadhwa area in concrete production solves most of the problem due to the shortage of coarse aggregates in the southern part of Iraq.

REFERENCES

Hanson, F.R.S., (1944). Cretaceous and tertiary reef formation and associated sediments in the Middle East. Am. Assoc., Petroleum Geologist, V.34, pp. 215-238.

Road Research Laboratories: Design of concrete mixes. Road notes No.4, London HMSO.

Mansour, J. and Petranek, J. (1980). Major occurances and deposits of limestone in Iraq. Report No.1037, D.G. of Geological Survey and Mineral Investigation.

Elizzi, M.A. and others (1985). Evaluation the use of crushed rocks in concrete production. Journal of Bldg. Research, Vol.4, No.1.

Al-Komi, M.A. (1983). Reserve of limestone and dolomatic limestone in Iraq: Special report, State Organization of Minerals, Baghdad.

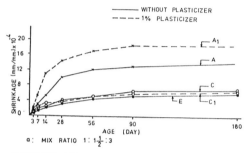

a: MIX RATIO 1 : 1½ : 3

NOTE: FOR MIX TYPE SEE TABLE NO. 4

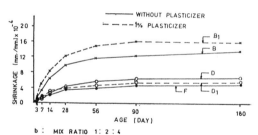

b: MIX RATIO 1 : 2 : 4

FIG. (3 - a & b) THE EFFECT OF PLASTICIZER, AGGREGATE TYPE AND W/C RATIO ON SHRINKAGE OF CONCRETE

6th International IAEG Congres / 6ème Congrès International de AIGI, © 1990 Balkema, Rotterdam. ISBN 90 6191 130 3

Le potentiel marbrier du Sud-Est de la France et ses contrôles géologiques
South-East France marble resources and its geological controls

R. Mazeran
C.R.G.M., Université de Nice-Sophia Antipolis, Nice, France

RESUME : Dans le Sud-Est de la France les roches et formations à vocation ornementale procèdent de contrôles géologiques variés parmi lesquels peuvent être identifiés :
- des contrôles sédimentologiques marins, pour notamment :
  • les calcaires subrécifaux à Rudistes de l'Urgonien et du Cénomanien de la Basse Provence et des chaînes subalpines septentrionales,
  • les calcaires siliceux hémipélagiques bordiers du bassin vocontien au Barrémo-bédoulien,
  • les calcaires bioturbés, de plateforme, du Dogger du promontoire jurassien de l'Ile-Crémieu,
  • les calcaires nodularisés mécaniquement et à faciès "ammonitico-rosso" du Malm briançonnais,
  • les calcaires nodularisés chimiquement du Tithonique ardéchois ;
- des contrôles sédimentologiques continentaux pour les brèches de pente et d'épendage contemporaines du début de la surrection des chaînons provençaux bordiers du bassin de l'Arc ou issues du démantèlement du pli jurassien de l'Epine, de même que pour les faciès d'encroûtement carbonatés d'âge vitrollien, des bords de l'étang de Berre ;
- des contrôles tectoniques pour les calcaires bréchoïdes (à stylolites et fentes de tension) des écailles frontales jurassico-crétacées du chevauchement sud-provençal ou des chevauchements subalpins méridionaux ;
- des contrôles sédimento-tectono-métamorphiques dans le cas des pélites du Permien de l'Argentera, des cipolins du socle pelvousien ou des marbres et cipolins de la couverture secondaire des unités alpines internes ;
- des contrôles magmato-tectono-métamorphiques pour les serpentinites du massif hercynien des Maures et pour les ophicalces et protogines du domaine alpin.

ABSTRACT : Decorative stones from South-East France can be reported to several controls with :
- marine sedimentary controls, particularly from :
  • urgonian and cenomanian rudistids reefs limestones from southern Provence and northern subalpine mountain ranges,
  • marginal hemipelagic siliceous limestones from the barremian-bedoulian vocontian basin,
  • mechanically nodularized limestones with "ammonitico-rosso" facies from the briançonnais uper jurassic,
  • chemically nodularized limestones from the ardechoise tithonian sequence,
  • bioturbated limestones from the Jura platform;
- continental sedimentary controls for mesozoic and cenozoic piedmont breccias in the Provence and Savoie areas and for montian carbonated incrustations in the Vitrolles area ;
- tectonic controls for the brecciated limestones from jurassic and cretaceous frontal scales of the south provençal and meridional subalpines thrusts ;
- all at once sedimentary, tectonic and metamorphic controls for permian marmorized ; pelites of the Argentera massive and variscan and alpines marbles ;
- all at once magmatic, tectonic and metamorphic controls for variscan serpentinous rocks and for alpine ophicalces and protogines.

1 INTRODUCTION

La valorisation des ressources naturelles du sous-sol français s'inscrit dans une réalité économique que nul ne saurait contester. En matière de roches ornementales, face à une demande croissante du marché, notamment au niveau du secteur bâtiment-décoration, la France se doit de relancer ses activités extractives aussi bien granitières que marbrières. Si le processus de relance est déjà bien amorcé en Bretagne, dans le Limousin, le Sud-Ouest et la Corse, il tarde à se déclencher dans le Sud-Est. Les raisons à ce retard incombent-elles à une géologie défavorable ou sont-elles à rechercher ailleurs ? C'est ce que nous avons essayé de savoir en entreprenant l'inventaire ainsi qu'un classement typologique des matériaux décoratifs de cette région. Le cadre de notre enquête concerne les départements suivants : Bouches-du-Rhône, Var, Alpes-Maritimes Alpes-de-Haute-Provence, Hautes-Alpes, Vaucluse, Gard, Ardèche, Drôme, Isère, Savoie et Haute-Savoie. Les grandes lignes de la géologie de l'espace considéré sont trop connues pour être rappelées ici.

Le recensement effectué à partir des multiples sources géologiques et géographiques existantes et surtout des documentations techniques spécialisées (Annales des Mines 1823, Mausolée 1976, Moniteur 1980) nous a permis de dénombrer dans le secteur étudié plus de cent carrières ouvertes à des fins ornementales. Une dizaine d'entre elles seulement continuent d'être actives de nos jours. La gamme des produits exploités, toutes époques confondues, comporte à la fois des calcaires marbriers, des brèches, des marbres et des cipolins, des serpentinites et des ophicalces, des granitoïdes, des onyx et des jaspes. La variété des environnements géologiques rencontrés explique la diversité des types de roches exploitées.

Différents contrôles géologiques apparaissent déterminants dans la prédisposition ornementale d'une roche à l'échelon régional :
- des contrôles sédimentogiques ; ils interviennent au premier chef, il va sans dire, pour toutes sortes de calcaires marbriers d'origine aussi bien marine que continentale et, à un moindre degré, pour des brèches d'origine tectonique ;
- ensuite, des contrôles tectonométamorphiques ; ils concernent à la fois des faciès paradérivés et orthodérivés ;
- enfin, très accessoirement, des contrôles magmatiques.

2 LES FACIES ORNEMENTAUX PROCEDANT D'UN CONTROLE SEDIMENTOLOGIQUE

2.1 Les contrôles sédimentologiques marins

Ils sont manifestes pour cinq sortes de roches exploitées actuellement, à savoir : des calcaires subrécifaux, des calcaires siliceux hémipélagiques, des calcaires pélagiques tachetés ou noduleux, des calcaires bioturbés. Des faciès à entroques et d'autres riches en matière organique, inexploités aujourd'hui, sont également à ranger dans ce groupe.

2.1.1 Les calcaires subrécifaux

Les plus intéressants d'entre eux, au point de vue marbrier, se rencontrent dans l'Urgonien et le Cénomanien supérieur de Basse-Provence (Bassin du Beausset), dans l'Urgonien des chaînes subalpines dauphinoises (Vercors, Chartreuse) et savoyardes (massif du Haut-Giffre), accessoirement dans le Portlandien du Vercors.

Les faciès à Rudistes (photo 1), ou à Rudistes et oncolites sont les faciès dominants dans ces complexes subrécifaux (Masse 1976, Philip 1974, Philip et Bilotte 1984). De véritables lumachelles à Ostréidés ou à Nérinées ainsi que des micrites à Préalvéolines s'intercalent parfois dans ces formations à Rudistes (Sainte-Anne d'Evenos). La richesse en Miliolidés de certains gisements est à souligner. Sur le plan de la dynamique sédimentaire, l'appartenance de ces dépôts à un domaine de plate-forme interne est aujourd'hui clairement établie (Arnaud-Vanneau et al. 1984).

Pour ce qui est des colorations, les teintes claires (crème, beige, jaune pâle) prédominent en Provence et dans le Dauphiné, alors que les teintes grises prévalent en Haute-Savoie. Des biosparites à fraction oolitique et des biopelmicrites noires sédimentées sur le bord externe de la plate-forme urgonienne sont même exploitées dans cette région (marbre de Sixt). Hormis ce cas particulier, la qualité esthétique des matériaux extraits apparaît selon les cas :
- liée à un contraste de couleur entre le fond de la roche et les fossiles (brun ambré sur crème, par exemple dans la pierre de Cassis) ;
- due aux contours capricieux des zones de décoloration des faciès gris, par oxydation superficielle (pierre du Mont Caume) ;
- ou induite par des ferruginisations secondaires, voire une stylolitisation qui introduisent des nuances d'ocre, de rose, d'orangé ou de vert (marbres de Sainte-Anne d'Evenos).

Lorsqu'elles se manifestent aléatoirement à travers le gisement, ces variations de teintes n'en constituent pas moins une gêne pour la conduite de l'extraction.

Sous exploités actuellement, les faciès subrécifaux de l'Urgonien et du Cénomanien du Sud-Est, dont il convient de souligner la qualité du poli, apparaissent comme un

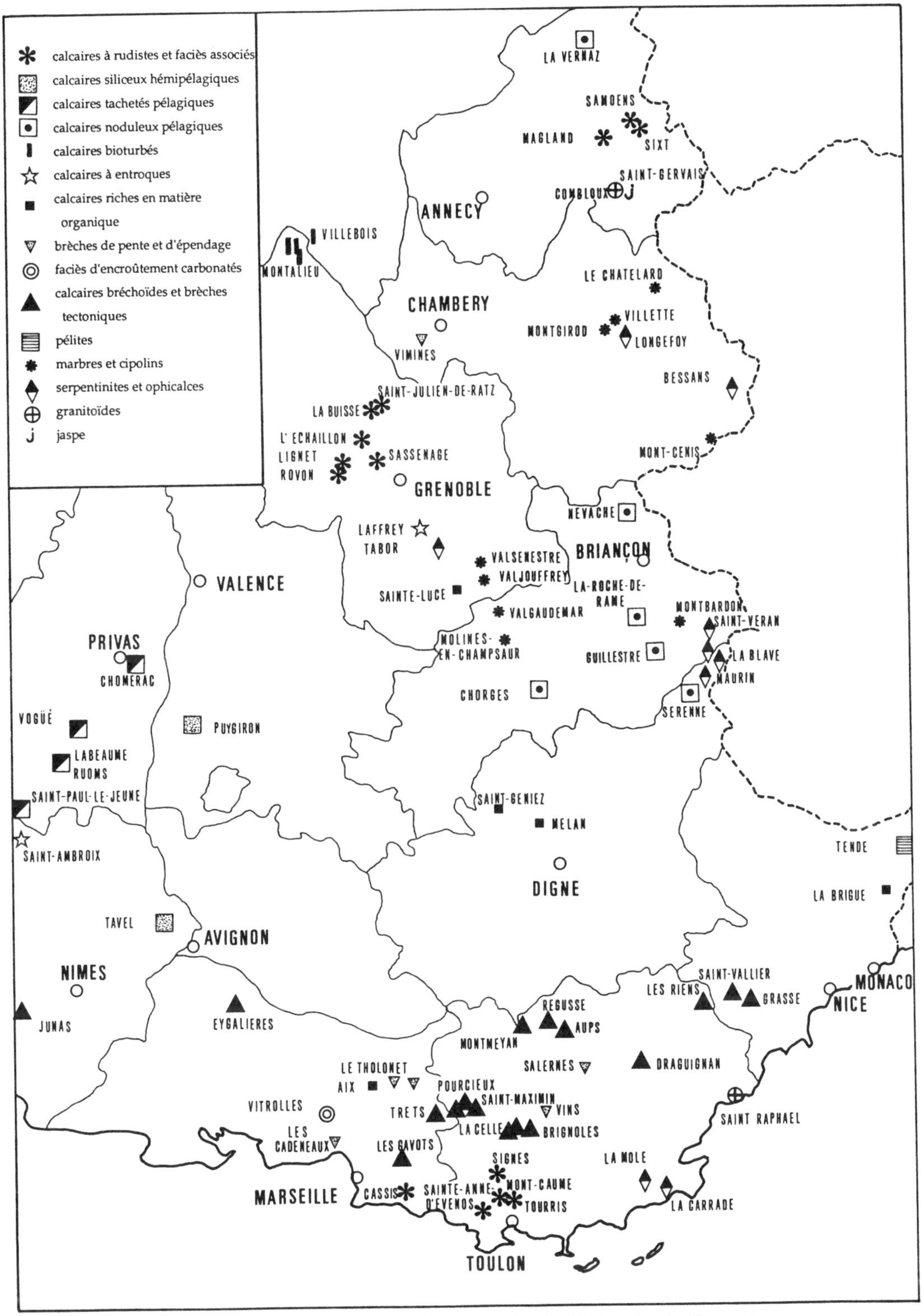

calcaires à rudistes et faciès associés
calcaires siliceux hémipélagiques
calcaires tachetés pélagiques
calcaires noduleux pélagiques
calcaires bioturbés
calcaires à entroques
calcaires riches en matière organique
brèches de pente et d'épendage
faciès d'encroûtement carbonatés
calcaires bréchoïdes et brèches tectoniques
pélites
marbres et cipolins
serpentinites et ophicalces
granitoïdes
jaspe
LA VERNAZ
SAMOENS
MAGLAND
SIXT
SAINT-GERVAIS
COMBLOUX
ANNECY
VILLEBOIS
MONTALIEU
LE CHATELARD
CHAMBERY
VILLETTE
MONTGIROD
LONGEFOY
VIMINES
BESSANS
SAINT-JULIEN-DE-RATZ
LA BUISSE
L' ECHAILLON
LIGNET
ROVON
SASSENAGE
MONT-CENIS
GRENOBLE
NEVACHE
LAFFREY
TABOR
VALSENESTRE
BRIANÇON
VALJOUFFREY
VALENCE
SAINTE-LUCE
LA-ROCHE-DE-RAME
MONTBARDON
VALGAUDEMAR
SAINT-VERAN
MOLINES-EN-CHAMPSAUR
PRIVAS
GUILLESTRE
LA BLAVE
CHOMERAC
MAURIN
CHORGES
SERENNE
VOGÜÉ
PUYGIRON
LABEAUME
RUOMS
SAINT-PAUL-LE-JEUNE
SAINT-GENIEZ
MELAN
SAINT-AMBROIX
TENDE
DIGNE
LA BRIGUE
TAVEL
AVIGNON
NIMES
SAINT-VALLIER
MONACO
LES RIENS
GRASSE
NICE
REGUSSE
JUNAS
EYGALIERES
AUPS
MONTMEYAN
LE THOLONET
SALERNES
DRAGUIGNAN
AIX
POURCIEUX
SAINT-MAXIMIN
VITROLLES
TRETS
VINS
SAINT RAPHAEL
LES CADENEAUX
LA CELLE
BRIGNOLES
LES GAVOTS
SIGNES
LA MOLE
MARSEILLE
CASSIS
SAINTE-ANNE-D'EVENOS
MONT-CAUME
TOURRIS
LA CARRADE
TOULON

Roches utilisées à des fins ornementales dans le Sud-Est de la France .

élément-clé de la relance de l'industrie marbrière régionale.

2.1.2. Les calcaires siliceux hémipélagiques

Ces faciès de transition entre les micrites pélagiques plus ou moins argileuses de la "fosse vocontienne", et les sédiments subrécifaux urgoniens précédemment décrits (Ferry 1984) présentent quelquefois des caractéristiques physico-mécaniques compatibles avec une utilisation marbrière. Tel est le cas des calcaires de Tavel, exploités dans le Gard, et des calcaires de Puygiron extraits dans la Drôme non loin de Montélimar. Il s'agit dans les deux cas de micrites ou de microsparites de ton uni, crème ou gris bleu, riches en silice (14-15%), celle-ci s'exprimant sous la forme de plages calcédonieuses à contour diffus de quelques centaines de micromètres. De grandes ammonites se rencontrent à l'occasion dans ces faciès datés du Barrémien. Relativement poreux (8% en moyenne), ces matériaux sont surtout employés localement dans le bâtiment. Les possibilités d'extension des carrières existantes paraissent limitées en raison de la forte inclinaison des bancs à Tavel (45°) et de la médiocre qualité des affleurements de Puygiron.

2.1.3. Les calcaires pélagiques tachetés

Ils sont l'apanage de la bordure ardéchoise de la "fosse vocontienne" au Tithonique et s'inscrivent dans un contexte de niveaux resédimentés bréchiques de même âge. Pétrographiquement ce sont des calcaires micritiques gris clair, à ammonites, parsemés de taches gris foncé et stylolitisés (photo 2). La différence de teinte entre les taches et le fond de la roche n'est pas imputable à une différence de lithologie mais à une plus ou moins grande richesse du sédiment en sulfure de fer, autrement dit à un phénomène de nodularisation chimique (Remane 1960). Ces matériaux sont extraits pour la construction et la décoration à Chomérac, près de Privas, et à Labeaume-Ruoms, non loin des célèbres gorges de l'Ardèche. En raison de leur grande dureté ils sont surtout recherchés pour les dallages. Leur utilisation reste avant tout régionale.Par le passé, des calcaires de même nature ont été extraits pour la construction dans la région de Chambéry.

2.1.4 Les calcaires pélagiques noduleux

Ils ont pour type le "marbre de Guillestre" du Malm briançonnais apparenté aux classiques "ammonitico-rosso" italiens. Comme eux, il est le résultat d'un phénomène de nodularisation mécanique sur des zones de hauts-fonds dans un environnement pélagique. Des phénomènes de dissolution-recristallisation, avec joints stylolitiques et fentes de tension, mis en valeur par un sciage à contrepasse, donnent parfois à cette roche (rouge ou verte selon l'état d'oxydation du fer) les caractères d'une tectonite. L'aspect le plus décoratif de ces calcaires demeure néanmoins attaché à la présence de grandes ammonites (photo 3). Une seule carrière exploite aujourd'hui, à Guillestre même, ce faciès autrefois très prisé aussi bien pour la construction que pour la décoration dans les Hautes et les Basses-Alpes. Les possibilités de développement des extractions semblent aléatoires, en raison de la concurrence italienne des "Rouges de Vérone" et autres faciès similaires.

2.1.5 Les calcaires bioturbés

Ils sont exploités au niveau de cette dépendance jurassienne que constitue le plateau de l'Ile-Crémieu (Isère), en rive gauche du Rhône ; ils font partie intégrante du grand bassin carrier de Villebois (Ain), nom sous lequel la pierre est commercialisée. Il s'agit de calcaires granulaires, un peu argileux, riches en traces de fouissage anastomosées, celles-ci étant d'autant plus apparentes que la roche est plus altérée (photo 4). La teinte de fond du matériau varie en conséquence du gris de fer au brun jaune. Des stylolites plus ou moins espacés se développent parallèlement à la stratification. L'association faunistique présente (Pectinidés, Brachiopodes) atteste de la faible bathymétrie du dépôt. De nombreuses carrières entaillent la formation datée du Bathonien moyen. L'aire de diffusion de ces calcaires marbriers dépasse très largement le cadre de la région Rhône-Alpes.

2.1.6 Les calcaires à entroques

Extraits plutôt comme pierres de taille par le passé, mais parfaitement aptes à prendre le poli, ils figurent de ce fait au nombre des roches répertoriées comme marbres dans la région Sud-Est, notamment dans le Dauphiné et les Cévennes. Ils correspondent respectivemet aux "calcaires de Laffrey" sédimentés sur le "haut fond" de la Mure, qui au Lias accidentait le Bassin subalpin (Sarrot-Renauld 1961) et aux calcaires de Saint-Ambroix (Gard) déposés sur la plate-forme bordière de ce même bassin au Bajocien. La nette différence de taille des entroques oppose ces deux faciès, de même que leur couleur (grise dans le premier cas, presque noire dans le second). Il est à noter que des indices de tectonique synsédimentaire sont quelquefois perceptibles au niveau des tranches polies sous la forme d'éléments bréchiques (Saint-Ambroix), ou de fentes comblées de micrite (Laffrey). Aucune reprise des extractions n'est envisageable sur les deux sites en question.

2.1.7 Les calcaires riches en matière organique

Comme les précédents ils ne sont mentionnés ici que pour mémoire, car ni les sites où ils furent extraits, ni leur qualité ne répondent aux normes actuelles d'une possible valorisation à l'échelon industriel. Les gisements répertoriés se rapportent dans leur majorité aux formations liasiques (Aix, Saint-Geniez-de-Dromont, Mélan, Sainte-Luce-en-Beaumont), plus rarement au Néocomien (La Brigue). D'autres niveaux furent très certainement sollicités à l'occasion. Dans les zones tectonisées, l'enchevêtrement des réseaux de veines calcitiques (calcite légèrement ferrifère) peut conduire à la réalisation du faciès "Portor", parfois signalé ici ou là.

2.2. Les faciès ornementaux procédant d'un contrôle sédimentologique continental

Ils sont représentés essentiellement par des brèches de pente et d'épendage polygéniques, des encroûtements carbonatés, accessoirement par des onyx calcaires et des jaspes.

2.2.1 Les brèches de pente et d'épendage polygéniques

Inexploitées aujourd'hui, elles n'en sont pas moins universellement connues en raison de leur très fréquente utilisation en ameublement (cf. les styles Louis XV et Louis XVI), ou pour la décoration d'édifices remarquables (église de la Madeleine à Paris, par exemple). Tel est le cas, tout au moins, des brèches de la région aixoise dénommées selon les cas brèches du Tholonet, brèches d'Alep, brèches Sainte-Victoire. C'est l'équilibre des nuances de jaunes, de roses, de rouges et de gris au niveau des éléments calcaires pluricentimétriques (parfois encroûtés de microcodiums) qui donne à ces roches leur qualité esthétique (photo 5). Des brèches de même nature, mais en général moins décoratives, furent aussi exploitées au Nord de Marseille (Les Cadeneaux) et dans le Var (Vins, Salernes). Datées du Bégudien ou du Paléocène, ces brèches sont le résultat du démantèlement, sous un climat rubéfiant, d'anticlinaux en cours de surrection lors de deux épisodes précoces des phases de plissement pyrénéo-provençales (Durand 1961, Durand et Tempier 1962, Chorowicz et Ruiz 1984). Le classement du site de la Sainte-Victoire élimine toute possibilité de reprise de l'extraction de ces faciès marbriers (les autres gisements n'étant guère à même de fournir des matériaux de qualité).

En Savoie, les brèches de Vimines formées au pied du chaînon de l'Epine, à l'Aquitanien, apparaissent fortement marquées par la pédogenèse. Le faciès réalisé est celui de glaebules laminaires mimant des oncolites (Truc 1975), à l'intérieur d'un encroûtement polyphasé (photo 6). Le site d'extraction depuis longtemps abandonné ne présente plus qu'un intérêt historique.

2.2.2 Le marbre de Vitrolles

Il s'agit là d'un calcaire extrêmement décoratif, exploité à des fins ornementales pendant près d'un siècle sur les bords de l'étang de Berre, mais relégué aujourd'hui à la production de granulats colorés. Sa teinte de fond rouge brique lui a valu le nom de "Rouge étrusque", et les nuances de jaune et de rose qui s'y ajoutent autour de géodes calcitiques, celle de "Rouge jaspé oriental". D'âge paléocène (vitrollien), ce faciès, longtemps considéré comme d'origine lacustre, est aujourd'hui interprété comme un faciès d'encroûtement carbonaté (Le Balleur et Triat 1985). La teinte rouge dominante est à relier à la présence d'environ 1% de Fe_2O_3 dans la trame micritique de la roche. La possibilité d'une reprise de l'extraction marbrière peut encore être envisagée en raison de l'originalité du matériau (photo 7).

2.2.3 Les onyx calcaires et les jaspes

Les faciès de concrétionnement calcitiques en milieu karstique ne sont ici signalés que pour mémoire. Leur exploitation s'est arrêtée avec celle des pierres de construction dans les carrières desquelles ils apparaissent sporadiquement, à l'image des "onyx de la Turbie" (Alpes-Maritimes). Quant aux jaspes richement colorés du "Verrucano" de Saint-Gervais-les-Bains, si leur renommée n'est plus à faire (Opéra de Paris), leur genèse reste mal élucidée.

3 LES FACIES MARBRIERS PROCEDANT D'UN CONTROLE TECTONIQUE

Ils sont représentés par des calcaires bréchoïdes et par des brèches tectoniques développées, pour la plupart, au niveau des écailles frontales du chevauchement sud-provençal et des chevauchements subalpins méridionaux. Le type même de ces "marbres" est le "Rose de Brignoles", mondialement connu, dont l'extraction est aujourd'hui suspendue si ce n'est pour la confection de granulats colorés. Il s'agit d'un calcaire sparitique micrograveleux, beige à crème, recoupé par un dense réseau stylolitique (rouges) et par de nombreuses fentes de tension calcitiques (photo 8). Le caractère massif de la formation urgonienne, située ici en position de lambeau de poussée (Aubouin et al. 1976), intervient favorablement sur les qualités physico-mécaniques du matériau. La valeur des produits marbriers fournis ailleurs par les écailles de Jurassique apparaît très inégale. Il est à noter aussi qu'une trop forte bréchification alliée à un enrichissement de la matrice en oxyde de fer peut nuire à la qualité esthétique du marbre (confusion possible avec des marbres reconstitués). Les zones de broyage associées à des failles verticales ou à des décrochements n'interviennent que très secondairement dans le potentiel ornemental régional. Il n'est guère que les brèches modifiées par la karstification qui retinrent jadis l'attention des exploitants en raison de leurs chaudes colorations (carrières des environs de Saint-Maximin).

4 LES FACIES ORNEMENTAUX PROCEDANT D'UN CONTROLE SEDIMENTO-TECTONO-METAMORPHIQUE

Ils comprennent d'une part des pélites vertes du Permien de l'Argentera et d'autre part, des marbres ou cipolins du socle pelvousien et de la couverture secondaire des unités alpines internes.

4.1 Les pélites de l'Argentera

Leur intérêt marbrier tient d'abord à leurs excellentes propriétés mécaniques qu'elles doivent conjointement à une teneur en silice élevée et aux effets d'un faible métamorphisme. La couleur verte, peu fréquente chez les roches utilisées en ornementation, est également un atout pour ce matériau communément désigné sous les noms de Vert de Tende, Vert de la Roya ou Vert du Levant. C'est à la présence de chlorite dans la phase phylliteuse (illites essentiellement) qu'est due cette coloration. Un microplissement synschisteux plus ou moins discret détermine l'apparition de nuances claires et foncées sur les tranches polies. Le matériau peut être utilisé dans tous les travaux du bâtiment, en intérieur comme en extérieur. L'inconvénient est qu'il ne prend pas le poli brillant. Une possibilité d'extension de la production est envisageable en raison de la demande régionale. Il est à noter que la tentative d'exploiter les pélites rouges de ce même Permien, pour l'ornementation, est restée sans lendemain eu égard au manque de dureté des produits testés et à leur fissilité.

4.2 Les cipolins du Pelvoux

Ils forment une suite d'affleurements discontinus au milieu des schistes cristallins, entre une ancienne formation volcanique et une ancienne série détritique (Le Fort 1970). L'épaisseur du niveau carbonaté, dont le rubanement est plus ou moins accusé, ne dépasse pas vingt mètres. La texture de ces cipolins est saccharoïde, leur teinte varie du blanc au vert pâle ou au rose saumon orangé. Le rubanement est déterminé par l'alternance des lits calcitiques avec des lits quartzo-micacés, ou quartzo-chloriteux, microplissés. Malgré des qualités esthétiques indéniables ces cipolins, difficilement accessibles ou trop fissurés, n'ont guère été exploités. L'insertion d'une grande partie des affleurements dans le Parc national des Ecrins élimine toute possibilité de valorisation.

4.3 Les marbres de la couverture secondaire des unités alpines internes

Ils appartiennent soit à l'anchizone du "domaine valaisan", comme le marbre de Vilette, soit à l'épizone du "domaine briançonnais interne", comme les marbres du Queyras et du Mont-Cenis.

Le marbre de Vilette, exploité dès l'époque romaine dans la haute vallée de l'Isère, est un calcaire à entroques faiblement recristallisé (entroques visibles à l'œil nu). Des stylolites et des fentes de tension calcitiques parcourent la roche dont la teinte oscille entre le blanc grisâtre et le gris bleuté. Le caractère massif des bancs permet l'extraction de blocs de grande dimension. Il est à noter qu'un faciès bréchique à fond violacé, recherché pour l'ameublement, fut autrefois extrait sur ce même site.

Dans le Queyras, près de la Chapelue dans les gorges du Guil, et à Montbardon, les calcaires cristallins de l'Anisien sont sporadiquepment extraits pour la construction et la décoration locales. Deux phases de déformation-recristallisation affectent ces calcaires, la seconde développant une "fibrosité" calcitique orthogonale par rapport à la schistosité antérieure qui est soulignée d'un pigment carboné.

Les marbres extraits à la carrière du Mont-Cenis se rattachent à la séquence carbonatée de couverture (Malm-Paléocène) du massif d'Ambin (Gay 1970). Ils comprennent tout à la fois d'anciennes brèches sédimentaires polygéniques et d'anciens calcaires argileux. Le faciès bréchique porte la trace d'une déformation cisaillante ductile. Le quartz est un élément accessoire de ces brèches dont la teinte générale oscille entre le blanc grisâtre et le gris bleuté. Des placages de séricite soulignent la schistosité alpine. Les marbres chloriteux associés montrent le classique rubanement des marbres néocrétacés de la Vanoise (Ellemberger 1958), avec alternances de bandes pluricentimétriques de calcite blanches et de minces lits chloriteux verts (photo 9). La paragenèse commune aux brèches marbrières et aux marbres rubanés est à quartz-albite-muscovite-chlorite-épidote-calcite. Les deux faciès prennent un excellent poli et se prêtent à l'extraction de blocs de grande dimension. Celle-ci doit cependant être interrompue pendant la saison hivernale en raison de l'altitude du gisement.

5 LES ROCHES ORNEMENTALES PROCEDANT DE CONTROLES MAGMATO-TECTONO-METAMORPHIQUES

Elles sont représentées par une plutonite acide "alpinisée", le granite du Mont-Blanc, par des serpentinites du socle et par des ophicalces alpines.

5.1 Le granite du Mont-Blanc

Matériau de base des constructions traditionnelles de la vallée de Chamonix, le granite du Mont-Blanc est aujourd'hui exploité pour l'ornementation à partir de blocs erratiques accumulés dans la vallée de l'Arve aux alentours de Combloux-Sallanches. Un traitement de surface variable est appliqué aux tranches sciées selon la qualité des blocs. Les faciès peu déformés sont affectés au polissage, les faciès schistosés soumis à un surfaçage thermique (photo 10). Pétrographiquement, ce granite, rappelons-le, est un granite porphyroïde à quartz, oligoclase, feldspath potassique et biotite (± chloritisée), contenant accessoirement de l'épidote. Les deux derniers minéraux donnent à la roche sa coloration verdâtre, acquise lors des évènements tectono-métamorphiques alpins.

5.2 Les serpentinites du socle varisque

Dans le massif des Maures, ces roches affleurent près de Cavalaire (la Carrade) et de la Molle où elles ont été extraites depuis des temps anciens, principalement pour la décoration des édifices religieux (Chartreuse de la Verne, etc.). La carrière de la Carrade connaît épisodiquement une petite activité extractive pour l'entretien de ce patrimoine. Macroscopiquement (photo 11), les serpentinites des deux gisements, dont la teinte de fond est vert sombre presque noir, apparaissent constellées de taches vert pomme rapportées à l'anthophyllite (Gueirard 1956). La teinte générale du matériau s'éclaircit au niveau des surfaces altérées, laissant apparaître à l'œil nu de fins alignements d'oxydes métalliques. Des fibres de chrysotile sont localement reconnaissables. L'antigorite est identifiable au microscope. La signification géodynamique exacte de ces deux pointements de serpentinites n'est pas connue.

Dans le massif du Tabor, au Sud de la Chaîne de Belledonne, une carrière a été également ouverte dans les serpentinites d'un complexe basique-ultrabasique, récemment considéré comme faisant partie d'une série ophiolitique d'âge cambro-ordovicien (Ménot 1984). Les roches en question, noires ou vertes, sont assez riches en magnétite et en chromite (Choubert 1936). Le gisement situé à 1850 m d'altitude, et difficile d'accès, apparaît actuellement sans intérêt.

5.3 Les ophicalces

Ces roches typiques des complexes ophiolitiques, et très appréciées d'une façon générale en marbrerie (les "Verts des Alpes"), ont été largement exploitées en Haute-Ubaye et dans le Queyras. Le faciès-type de la roche, à l'exemple du célèbre "Vert de Maurin", est celui d'une serpentinite bréchifiée dont les éléments anguleux, centimétriques, sont incorporés dans un lacis irrégulier de calcite, de talc et d'amiante (photo 12). L'opposition de teinte entre le veinage calcitique blanc à verdâtre et la trame serpentineuse vert foncé produit un effet très décoratif, renforcé par la qualité du poli. Des divers gisements autrefois exploités, seul celui de Maurin, apte à fournir des blocs de grande dimension, paraît susceptible d'être remis en exploitation.

Les serpentinites à pyroxène, à l'occasion recherchées dans le même contexte régional (Steen 1975) ou dans les ensembles ophiolitiques savoisiens, ne paraissent pas pouvoir être à nouveau extraites de façon rentable.

6 LES ROCHES ORNEMENTALES PROCEDANT D'UN CONTROLE MAGMATIQUE

Une seule mérite à cet égard d'être signalée pour des raisons historiques : l'estérellite. La roche en question a servi en effet dès l'époque romaine à confectionner des colonnes et divers éléments d'architecture (Bedon 1984). Dans les Temps Modernes, hors de simples emplois locaux en construction, le matériau fut utilisé pour décorer les cathédrales de Saint-Raphaël et de Monaco. Nous rappelerons qu'il s'agit d'une microdiorite quartzique porhyrique dans laquelle de grands cristaux de plagioclases zonés, de hornblende et de quartz se détachent sur le fond gris bleu de la pâte. La mauvaise aptitude des amphiboles à prendre le poli (tendance à l'arrachement) est la raison de la non utilisation actuelle du matériau en décoration.

7 CONCLUSION

Les points à souligner au terme de l'inventaire des matériaux de décoration du Sud-Est de la France sont :
  - une carence manifeste en granitoïdes,

- un potentiel de calcaires marbriers important et diversifié, sous-exploité actuellement, mais offrant des possibilités de valorisation certaines.

La région provençale nous paraît, en première analyse, la plus favorable à une relance des extractions par le biais des calcaires subrécifaux, des marbres de Vitrolles et de Brignoles. En région alpine, seul le gisement de Maurin nous semble encore présenter quelque intérêt.

Face à une concurrence étrangère très organisée, cette relance passe nécessairement par une redynamisation de la profession, impliquant elle-même le soutien des instances régionales et une plus étroite collaboration entre géologues, carriers, métiers du bâtiment et architectes. La création d'une "filière pierre Grand Sud-Est", où la Corse apporterait sa contribution pourrait en être le point de départ.

REFERENCES

Anonyme 1976. Essai de nomenclature des carrières françaises de roches de construction et de décoration, 254 p. Le Mausolée, Givors.

Arnaud-Vanneau, A., Arnaud, H.& Masse,J.P. 1984.Synthèse géologique du Sud-Est de la France.Faciès urgoniens. Mém. BRGM 125, 335-336.

Aubouin, J., Chorowicz, J.& Thiele, R. 1976.La terminaison orientale du chevauchement sud-provençal : de la Sainte-Baume à la Loube et au Candelon (Var). Bull. Soc. géol. Fr. 1, 179-190.

Bedon, R. 1984. Les carrières et les carriers de la Gaule romaine, 248 p.Picard, Paris.

Chorowicz, J. & Ruiz, R. 1984. La Sainte-Victoire (Provence) : observations et interprétations nouvelles. Géol. de la France 4, 41-55.

Choubert, G. 1936. La serpentine du Tabor et les roches qui l'accompagnent. Trav. lab. géol. Grenoble 18.

Durand, J.P. 1961. Quelques particularités des brèches de Saint-Antonin près d'Aix-en-Provence. C.R. Soc. géol. Fr. 71-72.

Durand, J.P. & Tempier, C. 1962. Etude tectonique des brèches du massif de Sainte-Victoire dans la région du Tholonet (Bouches-du-Rhône). Bull. Soc. géol. Fr. 6, 97-101.

Ferry , S. 1984. Synthèse géologique de la France, Crétacé inférieur.Mém. BRGM 125, 313-315.

Gay, M. 1970. Le massif d'Ambin et son cadre de Schistes lustrés (Alpes franco-italiennes). Evolution paléogéographique antéalpine. Bull. BRGM 1, 3, 5-81.

Gueirard, S.1956. La serpentine de la Carrade près Cavalaire (Var). Trav. lab. géol. Marseille 5, 85-101.

Héricart de Thury, M. 1823. Rapport sur l'état actuel des carrières de marbre de France. Ann. Mines, 8, 2-96.

Le Balleur, A. & Triat, J.M. 1985. Enquête sur les calcaires marbriers de la région PACA, 81 p.Antenne Ressources Minérales, Univ. Aix-Marseille III.

Le Fort, P. 1970. Les cipolins roses du Valgaudemar-Champsaur (massif du Pelvoux). Géol. Alpine 46, 111-116.

Masse, J.P. 1976. Les calcaires urgoniens deProvence.Valanginien-Aptien inférieur. Stratigraphie, paléontologie, les paléo-environnements et leur évolution. Thèse état, Aix-Marseille II.

Ménot, R.P. , Peucat, J.J., Piboule, M. & Scarenzi, D. 1984. Cambro-ordovician age for the ophiolitic complex of Chamrousse-Tabor (Belledonne massif, French external alpine domain). Ofioliti 10, 2-3, 527.

Philip, J. 1974. Les formations calcaires à rudistes du Crétacé supérieur provençal et rhodanien. Stratigraphie et paléogéographie. Bull. BRGM, 1, 3, 107-151.

Philip , J. & Bilotte, M. 1984. Synthèse géologique du Sud-Est de la France. Plates-formes carbonatées et bio-constructions à rudistes. Mém. BRGM 125,378-379.

Remane, J. 1960. Les formations bréchiques dans le Tithonique du Sud-Est de la France. Trav. lab. géol. Grenoble 36, 75-114.

Sarrot-Reynauld, J. 1961. Etude géologique du dôme de la Mûre (Isère) et des régions annexes, 207 p. Trav. lab. géol. Grenoble.

Steen, D.M. 1975. Géologie et métamorphisme du complexe ophiolitique de la Haute-Ubaye (Basses-Alpes,France). Bull. suisse Minéral. Pétrogr. 55, 523-566.

Truc, G. 1975. Les encroûtements carbonatés liés à la pédogenèse. IXme C.I. Sédimentologie, Nice. Livret-guide excursion A2, 47-51.

Documentation française du bâtiment 1980. Les pierres de France, 120 p. Le Moniteur , Paris.

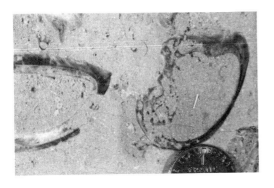

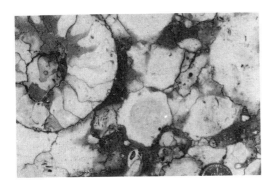

Photo 1 : calcaire à Rudistes, Cassis.

Photo 2 : calcaire tacheté, Chomérac.

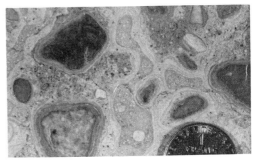

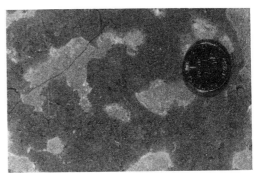

Photo 3 : calcaire noduleux, Guillestre.

Photo 4 : calcaire bioturbé, Montalieu.

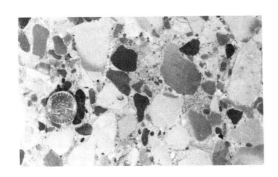

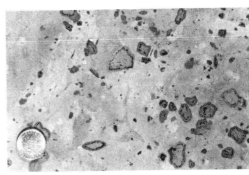

Photo 5 : brèche du Tholonet.

Photo 6 : brèche de Vimines

Photo 7 : "marbre" de Vitrolles (x 1/3).

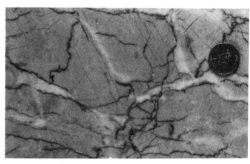

Photo 8 : calcaire bréchoïde, Brignoles.

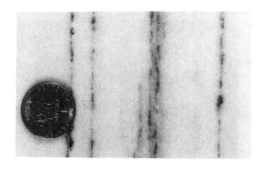

Photo 9 : marbre chloriteux, Mont-Cenis.

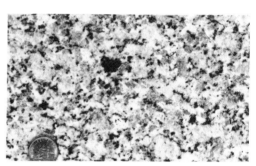

Photo 10 : granite "alpinisé", Combloux.

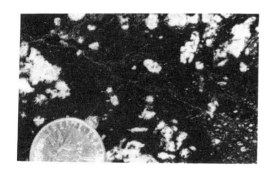

Photo 11 : serpentinite, la Carrade.

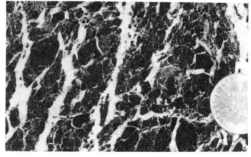

Photo 12 : ophicalce, Maurin.

Delineation and evaluation of rock types for construction material in an arduous terrain – A case study from Central India

Dessins et évaluations de type de rochers pour des matériaux de construction dans un terrain d'ardoise – une étude pratique dans le centre de l'Inde

R. Nagarajan
Indian Institute of Technology, Bombay, India

ABSTRACT: Satellite and aerial photographs were used in the delineation of various lithological and soil units and their utility in various civil engineering projects were assessed based on the physico-mechanical properties determined on the representative samples collected from the field, in the laboratory. Environmental parameters that are to be considered while removing this aggregates are also discussed.

INTRODUCTION

Rock and soil are used as aggregate in various civil engineering projects. Physico-mechanical properties and presence or absence of minerals evaluate their suitability for a particular type of activity Viz. concrete mix, building stone, communication (road/railway). Removal of these aggregates and the related uncontrolled developments would lead to environmental degradation on a regional scale. Hence, it is essential to have regional resources inventory to facilitate the optimal usage of these resources while preserving the environment.

This study highlights the preparation of a regional inventory on rock and soil as aggregates for various developmental projects including their areal distribution using various mapping methods aiding in regional resources management.

The study area is located in the Central India having rugged topography covered by deciduous forest growth and poor communication facilities. (Fig.1) It is pertinent to utilise all the available mapping techniques and prepare an inventory on natural resources using remote sensing data.

METHODOLOGY

Landsat MSS of February 1981 and Aerial photographs of 1973 were used in the delineation of areas exposed of various lithological units, soil cover and landuse and land cover pattern. Select field visits have been undertaken to collect the information on the rock mass weathering characteristics, fractures and also representative sample collection for laboratory studies.

GEOLOGY AND SOIL

The various rock types that are exposed (Fig.2) in the area include granite gneiss, meta sedimentary rocks like Quartzite, slaty shale and siliceous limestone (Pakhal group) and sedimentary rocks such as ferruginous sandstone, conglomerate, shale and shaley limestone of Lower and Upper Gondwana super group of rocks. Quartzite and Quartzite interlayered with shale occupies the structural landform such as ridges and inselbergs, while granite gneiss and siliceous limestone exposed on the undulating plains. Ferruginous conglomerate occupies ridges and ferruginous sandstone, shaly limestone, sandstone and shale exposed on the undulating ground.

In-situ soil formation over limestone and sandstone conglomerate is about 15' while Quartzite interlayered with shale is devoid of soil formation. Brown to reddish brown non-calcareous coarse loamy soil over sandstone, reddish brown fine loamy non calcareous soil on granite gneiss, dark greyish brown clayey, greyish brown clayey to silty clayey deep to very deep soil over the alluvium are some of the major soil type.

Vegetation growth in the study area include tropical southern moist deciduous type. its growth over limestone and sandstone conglomerate regions are dense, moderate over granite gneiss and sparse over rest of the terrain. Slaty shale do not support any vegetational growth. Representative rock and soil collected from the field were analysed for their physical, mechanical and mineralogical properties of rocks in Table 1 (a) and Geotechnical properties of in-situ soil in Table 1(b) are given. Areal extent of the various lithological units were measured using digital Planimeter Planix-7 and shown in Table 2.

UTILITY OF ROCKS

Considering the physico mechanical properties of rocks and the desired characteristics of rock aggregates for various purposes, the suitability (ISI 1963) of rocks are discussed. Granite gneiss,quartzite,siliceous sandstone having low water absorption moderate to high uniaxial compressive strength,durability and void index are best suited for the building stone. Coarse aggregates used in the concrete mix need to possess high durability and devoid of clay and gypsum coatings as that of granite gneiss and quartzite. However the presence of cherty bands in siliceous limestone (Pakhal) would impart a deleterious effect on the mix, hence not preferred. Quartzite having high silica content (94%) and hydrophilic nature could not be used as aggregate for bitumin mix for road, while hydrophobic granite gneiss is preferred. Shaley limestone and siliceous sandstone could be used as sub-base course alongwith granite gneiss and quartzite as base course in the road construction

ENVIRONMENTAL CONSIDERATIONS

While considering the various uses of rocks,it is essential to consider the various factors that would initiate the environmental degradations during the removal of them. It is anticipated that the removal of vegetation cover over the ferruginous conglomerate and sandstone would initiate large scale soil erosion. High frequency of fracture shown by quartzite and slaty shale would lead to rock fall. It is suggested a proper method of extraction to be followed in removing these aggregates.

A cursory look at the table reveals the spatial distribution of rocks. Volume of rock mass available for various uses were calculated based on the relative relief shown by landforms,and lithology. Relative relief has been calculated from the topographic sheets. Assuming the relief variations within the landform exposed of particular rock type is minimum the volume of rock has been calculated and shown in Table 2. The above calculated volume of rock is of regional assessment in the absence of any bore hole/sub surface data and are based only on the relief and the observed relief variation within the landform.However, similar exercise based on the bore hole/sub surface data and definite project of interest could calculated. Further studies on a small scale is in progress using the sub surface data. Spatial distribution of rocks,physico-mechanical properties and volume of rock,the aggregate requirement for large scale projects/regional developmental Projects could managed efficiently.

It can be summarised that the remotely sensed data with minimum ground data collection and analysis of rocks could be used for efficent management of the aggregates. Studies on similar nature could further minimise the transportation of these aggregates from one place to another. Further work on the storage of these information on Geographic Information System and retrieval of the same is in progress.

ACKNOWLEDGEMENT

Support and encouragement extended by Prof. C.Natarajan Head, CSRE and Director, IIT Bombay is highly appreciated.

REFERENCE

Indian Standard Institute 1963. Specification for coarse and fine aggregates from natural resources.IS:383,Calcutta.

Krynine DP & Judd WR 1957. Principles of engineering geology and geotechniques New York: McGraw Hill book co.

Nagarajan R 1984.Remote sensing and engineering geological studies around Sironcha,Gadchiroli Dt.Ph.d Thesis,IIT Bombay, pp 169.

TABLE 1 a   Physical, Mechanical and Mineralogical properties of Rocks

Mineralogy	1	2	3	4	5	6	7
Medium grained,angular to sub-angular quartz,felspar,microcline,biotite and opagues GRANITE GNEISS	2.79-2.81	0.48-1.11	0.2-0.3	2000-2300	0.2-0.22	15	Very strong stiff
Fine to medium grained saccroidal texture secondary quartz vugs & Veins QUARTZITE (Pakhal Group)	2.61-2.69	4.96-13.30	0.02-0.72	620-1400	0.8-1.2	18	Strong very stiff
Angular to sub-angular,fine grained micrite dolomicrite veins of quartz,calcite and chert SILICEOUS LIMESTONE (Pakhal Group)	2.70-2.81	4.51-14.14	0.03-0.18	730-1040	0.2-3.23	20	Moderate stiff
Fragmentary thin bands,kaolinite,illite SHALE (Pakhal group)	1.61-2.36		2.2-13.57	20-165			
Moderate sorting,medium to fine grained quartz,felspar and opaquesminerals in siliceous matrix SANDSTONE (Lower Gondwana)	2.58-2.60	10.8-12.0	0.19=1.09	470-485	7-10	10	Weak,stiff
Compact,kaolinite & illite SHALE (lower: Gondwana)	2,58-2.62	10.87-16.23	0.25-0.4	105-260	19.64-41.93	29	very weak Moderate stiff
Fine grained clay & aragonite bands,calcite and quartz crystals LIMESTONE (Upper Gondwana)	2.60-2.73	7.47-14.6	0.93-1.2	230-470	2.47-7.3	25-44	Very weak stiff
Medium to fine grained,angular to sub-angular grains,quartz,felspar in ferruginous matrix SANDSTONE (Upper Gondwana)	2.62-2.65	14-20	2.0-2.4	520-780	68-69	55-56	Very weak High Yielding
Angular to sub rounded chert,limestone shale and quartz in ferruginous matrix CONGLOMERATE (Upper Gondwana)	2.63-2.70	13-16	2.0-2.4	30-50	13-15	50-55	Weak highly yielding

1.True Sp-gravity   2. Porosity (%)   3. Water absorption (%)   4. Uniaxial compressive strength (Kg/cm^2)
5. Soundness test (MgSO4) % loss in 25 cycles   6. Los Angele's abrasion test   7. Deere & Miller's Engg.Classification

Table 1b Geotechnical properties of in-situ soil

Rock type	Soil classification		Texture	LL	PL	PI
	Unified	AASHO		%	%	%
Limestone	SP-SC	A-2-4	Sandy loam	23.0	13.6	9.4
(Pakhal)	SW-SC	A-2-4	Sandy loam	19.5	15.1	4.4
Sandstone (Upper Gondwana)	SP	A-3	Sand			
Conglomerate (Upper gondwana)	SW-SC	A-2-7	Sandy loam	23.5	12.6	10.9
Shale	GM-GC	A-2-6(3)	Sandy	30.0	18.0	12.0
(Pakhal)	SC	A-6(2)	Sandy loam	30.0	18.0	12.0

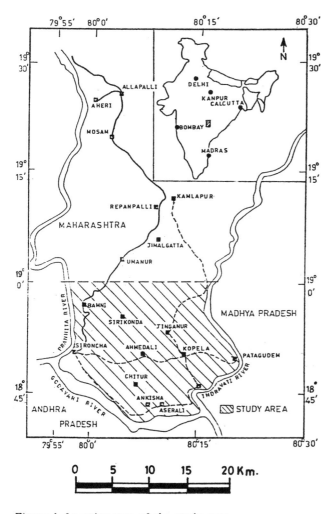

Figure 1. Location map of the study area

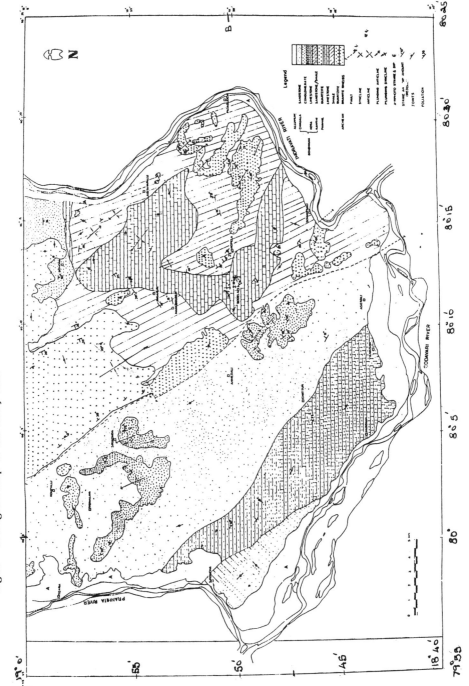

Fig.2.   Geological map of the study area

Table 2. Areal extent of various rock units in the study area.

Lithology	Area(sq.kms)	Relief(in m.)*	Volume of rock (m³)
Granite Gneiss	2.3	10	23.0
Quartzite	19.6	150	294.0
Quartzite with shale	5.0	125	625.0
Siliceous limestone	30.0	10	300.0
Slaty shale	22.2	5	111.0
Sandstone and shale	5.6	5	28
Shaly limestone	21.1	5	105.5
Ferruginous sandstone	44.0	10	440.0
Ferruginous conglomerate	11.0	20	220.0
Older alluvium	5.2	2	10.4
Alluvium	10.3	2	20.6

*
Relief considered for calculation of volume of rock incorporating the variations observed in the landform.

6th International IAEG Congres / 6ème Congrès International de AIGI, © 1990 Balkema, Rotterdam. ISBN 90 6191 130 3

Evaluation of some dimension stone occurrences in Saudi Arabia
Estimation de gisements de pierres de taille en Arabie Saoudite

W. M. Shehata
King Fahd University of Petroleum and Minerals, Dhahran, Saudi Arabia

M. A. El Mukhtar
Consoil, Jeddah, Saudi Arabia

M. Y. Badiuzaman
King Abdulaziz University, Jeddah, Saudi Arabia

ABSTRACT: Saudi Arabia possesses great varieties of geological formations that can be considered as important sources of dimension stones. Marble is one of these formations and is the most popular ornamental stone that is locally used. Granite and eruptive rocks are used to a much lesser extent. Large quantities of finished and unfinished ornamental stones are imported every year particularly from Italy. The purpose of this research is to assess some of the local occurrences through investigating their geomechanical properties and evaluating their potential commercial use. The assessment includes the marble deposits at Wadi Turabah, Jabal Khanugah and Jabal Al Khawar, and the granite masses at Jabal Jabalah, Jabal Al Bakri and Jabal Dheraya. A comparative study was also made between the geomechanical properties of these rock units and those of Biano Unito de Carrara marble and Rosa Baveno granite. With improved finish, some of the investigated dimension stones can compete with the imported ones and should be commercially used.

RESUME: l'Arabie Saoudite possède de grandes variétés de formations géologiques. Ces formations peuvent être considérées comme des réservoirs importants de pierres de taille. Le marbre qui en fait partie est la pierre ornementale la plus populaire employée localement. Le granit et les roches éruptives sont moins utilisées. Chaque année, une grande quantité de pierres ornementales finies et non finies est importee, particulièrement d'Italie. Le propos de cette recherche est d'estimer quelques uns des gisements locaux en cherchant leurs propriétés géomécaniques et en évaluant leur potentiel commercial. L'estimation englobe les dépôts de marbre de Wadi Turabah, Jabal Khanugah, et Jabal Al Khawar, les masses de granit de Jabal Jabalah, Jabal Al Bakri, et Jabal Dheraya. Une étude comparative fut également éffectuée entre les propriétés géomecaniques des gisements cités ci-dessus et ceux de Biano Unito de Carrara pour le marbre, Rosa Baveno pour le granit. Avec un usinage amélioré, quelques unes des pierres de taille examinees sont competitives avec celles importées et devraient être utilisées commercialement.

1 INTRODUCTION

A large number of marble, granite and other dimension stone occurrences have been located in the Arabian Shield (Figure 1). Many of these occurrences are either uneconomical, far from large cities or practically isolated.

A brief description of most of these occurrences and in particular the marble deposits is given by Laurent (1972, 1973) and an illustrative catalog of the Saudi Arabian stones is produced by D.G.M.R. (1401H).

This paper presents the engineering properties and assessment of several marble deposits existing in Jabal Al Khawar (Herein referred to as Khawar) and Jabal Al Khanugah (herein referred to as Khanugah) in Afif-Dawadmi area, and in Wadi Turabah (herein referred to as Turabah) south west of Taif (Figure 1). It also presents the engineering properties and

Dheraya (from now on termed Dheraya) in Afif-Dawadmi area.

Comparisons have been made between the engineering properties of the studied stones and those of some local and imported ones.

2 MARBLE TYPES

The marble occurrences in Saudi Arabia could be classified into 4 zones (Laurent, 1972). The investigated marbles represent two of these four zones namely Afif and Turabah zones. These two zones lie within areas that were geologically studied by Brown et al. (1962), Al Shanti (1976), Green and Gonzalez (1980) and Delfour et al. (1982). The marbles were previously investigated by Laurent (1970 & 1973), Dehlavi (1971), El Mukhtar (1985) and Badiuzaman (1985).

2.1 Stones description

The studied marble types have different decorative appeals, colors and grain sizes. Coarse grained white marble (mottled with gray) occurs north of Khanugah (Zone Ia) with yellowish marble existing east of it (Zone Ib). Light gray marble occurs as medium grained in the central section of Khanugah (Zone II) and as fine to medium grained in Turabah. Pink marble occurs as medium grained also in Turabah. Dark gray to black marbles are fine to medium grained in the southern section of Khanugah (Zone III) and medium grained in Khawar.

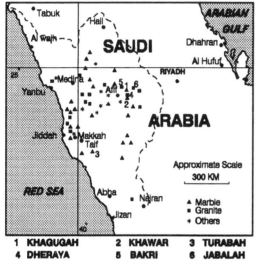

| 1 KHAQUGAH | 2 KHAWAR | 3 TURABAH |
| 4 DHERAYA | 5 BAKRI | 6 JABALAH |

Figure 1: Dimension stones distribution map in Saudi Arabia.

The dominant mineral constituent of the different marble types is calcite with traces of iron oxides and quartz in Turabah, dolomite, corundum and brucite in Khawar, and graphite, periclase and corundum in Khanugah.

Most of the studied marble types show attractive colors, although they can still accept more polishing. A reflectivity study conducted on laboratory polished marble slabs showed an improvement of 10% to 100% than the industrially prepared samples. Khawar black marble was one of those needed the highest polishing efforts in order to bring up its vivid color. Khanugah II and III practically required no further polishing.

Both Afif-Dawadmi and Turabah areas are intersected by several joint, faults and other dicontinuities. The marbles in the top 3 m at Khawar and Khanugah and the top 4-6 m at Turabah are relatively highly fractured as detected by the shallow seismic refraction surveys. This zone should be avoided in marble

slabs production and utilized only as crushed aggregate or reconstituted marble. This fractured zone is followed downward by a slightly fractured zone with block sizes ranging between 3-10 m^3 that can be dislocated making use of the present discontinuities. Below this zone the rock is massive and a suitable splitting method is required to produce the marble blocks.

The quantities of marble available in each of the studied occurrences are of economical size. The resources of the different marble types vary between 18 and 41 million cubic meters.

2.2 Engineering properties

Figure 2 shows the engineering properties of the studied marble as well as the properties of Jabal Ghuslan (Laurent, 1973) representing the local marbles and Carrara (Remez and Murrell, 1964; Laurent, 1973) representing the imported marbles. It also shows the ASTM (1978) standard specifications for exterior marble.

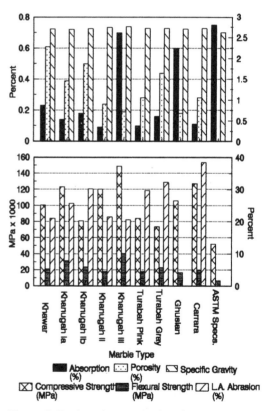

Figure 2: Engineering properties of marble

The specific gravity of all the studied marbles ranges between 2.66 and 2.74 with the highest values recorded for Khanugah II (2.72) and Khanugah III (2.74). These values are higher than that of Ghuslan (2.68) and the ASTM requirement (2.60). Although the porosity value is the lowest for Khanugah III (0.19%), its water absorption value is the highest (0.70%) but is still lower than the ASTM requirement (0.75%). Khanugah II and Turabah pink show the best combinations of specific gravity, porosity and water absorption. They are comparable with Carrara and better than ASTM specifications.

The highest compressive strength values are obtained for Khanugah III (149 MPa), Khanugah Ia (124 MPa) and Khanugah II (120 MPa). These values are comparable with Carrara (127 MPa) and also much better than the ASTM requirement (52 MPa). The flexural strength values, on the other hand, are the best for Khanugah III (41 MPa), Khanugah Ia (37 MPa) and Turabah gray (23 MPa). These values are relatively better than Guslan (17 MPa), Carrara (20 MPa) and ASTM requirement (7 MPa). The Los Angeles abrasion losses are relatively low for Khanugah III (20%) and Khawar (21%).

3 GRANITE TYPES

The investigated granite masses occur in the form of batholiths intruding the metamorphic rock units of the Arabian Shield. These masses were geologically studied by Delfour et al. (1982). Other than the brief examination of Baghdady (1982, personal communication), no evaluation was done for these masses for dimension stone production.

3.1 Stone description

The studied granitic masses are generally medium to coarse grained. The Jabalah granite is sodic in composition and light gray in color. The Bakri granite is also alkali in composition and pale gray with dark mica spots. The Dheraya granite, on the other hand, is pale pink to pale gray and locally porphyritic.

Both Jabalah and Bakri granites have attractive colors, although the study showed an increase in reflectivity of the laboratory polished slabs than the industrially prepared samples. An increase of only 1-2% was obtained for Bakri and Dheraya granites and 20% for Jabalah granite.

Some granitic masses are massive or very widely spaced jointed as Bakri and Dheraya masses or widely spaced jointed and faulted as Jabalah mass. Thick sheeting is also present in all the masses to form blocks as large as 18 m^3 at the foot hills.

The estimated reserves of the studied granites ranges between 22-38 million cubic meters for each type.

3.2 Engineering properties

Figure 3 shows the engineering properties of the three studied granite masses, in addition to the properties of Jamum granite (Laurent, 1975) representing a local stone type and Rosa Baveno (DGMR, 1401H) representing an imported stone type. It also shows the specifications requires by ASTM (1978) for structural granite.

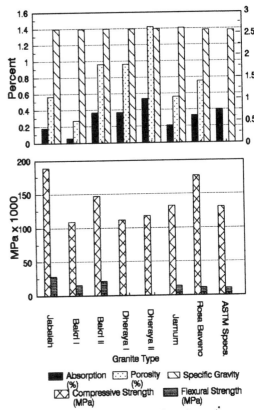

Figure 3: Engineering properties of granite.

The specific gravity of all the studied granite types ranges between 2.59 and 2.62 with the highest value recorded for Jabalah granite (2.62). These values are comparable to Jamum granite (2.62), better than Rosa Baveno granite (2.57) and exceeding the ASTM specifications (2.56). The water absorption for Jabalah (0.18%) and Bakri (0.06%) are comparable to Jamum (0.21%) and Rosa Baveno (0.33%) and are acceptable by ASTM (0.40%). The absorption value of Dheraya (0.55%), on the other hand, is not acceptable by ASTM. The porosity values for both Jabalah (0.57%) and Bakri (0.27-0.96%) are reasonable and comparable to Jamum (0.56%) and Rosa Baveno (0.75%) while that of Dheraya is relatively high (0.96-1.42%).

The uniaxial compressive strength values of Jabalah (189 MPa) and the western section of

Bakri (Bakri II) (147 MPa) are acceptable by ASTM (131 MPa) and are comparable to Jamum (132 MPa) and Rosa Baveno (177 MPa). Both Dheraya granites (112-118 MPa) and the western section of Bakri (Bakri I) (109 MPa) are not acceptable by ASTM. The flexural strength for both Jabalah (29 MPa) and Bakri (16-21 MPa) are higher than Jamum (14 MPa), Rosa Baveno (11 MPa) and ASTM requirement (10.34 MPa). The Los Angeles abrasion test was performed only on Bakri granite giving a relatively high percentage loss of 34%.

4 MICROFRACTURING

The severe Saudi Arabian weather conditions enhance the process of mechanical weathering which deteriorates the rock material through the formation of microcracks.

The degree of microfracturing is a difficult property to measure. However, it can be detected by the porosity value and the quality index (Figure 4) (Fourmaintraux, 1976). The quality index is defined by Tourenq et al.(1971)as:

$$IQ = 100*[V_m/V_c]$$

where V_m is the measured longitudinal velocity.
   V_c is   the   calculated   longitudinal velocity = 6.7 km/sec for marble.

The of the stress-strain curves for the rock material under cyclic loading may also give another way to evaluate the degree of microfracturing. A microcracked rock sample will behave more plastically at the first stages of loading due to the closing up of microcracks as the loading starts. Moreover, the cyclic loading of a microfractured sample shows continuous steepening in the stress-strain curves and therefore a continuous increase in the modulus of elasticity values. In other words, the microfractured rock gets stiffer with cyclic loading until all the microcracks are practically closed.   Figure   5   shows   the   stress-strain relationships of a massive unfractured sample from Khanugah II marble showing minor stiffening, and a microfractured sample from Dheraya granite showing a trend of continuous stiffening.

Figure 6 illustrates both the magnitude of the tangent modulus of elasticity of the studied stone types, and the rate of its increase with cyclic loading.

The modulus of elasticity value can also be used to predict the degree of microfracturing if it is measured both dynamically (E_d) and statically (E_s) (Figure 7). The higher the dynamic modulus of elasticity with respect to the static modulus, the more the intensity of microfracturing (Lama et al., 1978).

From the discussion above it can be concluded that the marble is less microfractured than the

granite. Khawar and Khanugah marbles are the least microfractured although Khanugah Ib and Khanugah   III   show   indications   of microfracturing.

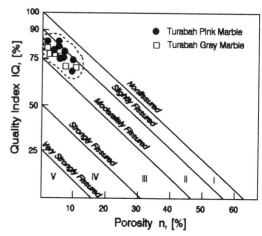

Figure 4: Degree of fissuring according to Fourmaintraux (1976).

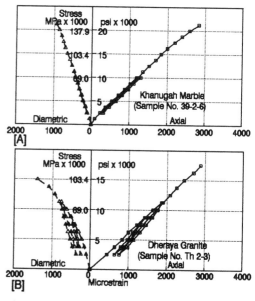

Figure 5: Stress-strain relationships of; [A] Khanugah III marble, and [B] Dheraya granite.

Turabah marble is generally moderately microfractured. Dheraya granite shows the maximum degree of microfracturing. Bakri granite is locally microfractured, whereas, Jabalah granite is relatively nonfractured.

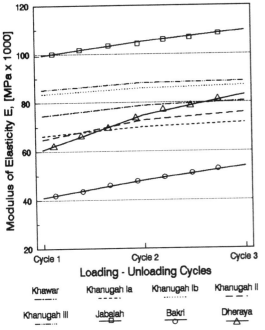

Figure 6: Increase in modulus of elasticity with cyclic loading.

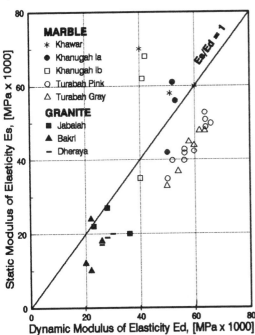

Figure 7: Correlation between dynamic and static moduli of elasticity.

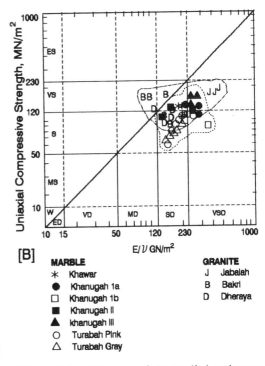

Figure 8: Classification of the studied rock types according to; [A] Deere and Miller (1966), and [B] Necdet and Dearman (1983).

5 ENGINEERING CLASSIFICATION OF THE STONES

The intact rock classification can also give an indication of the quality of the stones for construction purposes. The studied types of marble and granites were classified both according to Deere and Miller (1966) (Figure 8a), and Necdet and Dearman (1983) (Figure 8b).

Figure 8a indicates that the studied marble types are generally medium to high strength-high modulus ratio with Khanugah Ib, Turabah gray, Khanugah II and Khanugah III showing better stones. It also indicates that the studied granites range between high strength-high modulus ratio and high strength-average modulus ratio with the Jabalah granite better than the other types of granite.

Figure 8b shows that the marble is generally strong to very strong-slightly deformable to very slightly deformable with Khanugah III in the lead. The granites are very strong but its deformability varies between moderately deformable to very slightly deformable with the best stone being the Jabalah granite.

6 CONCLUSIONS AND RECOMMENDATIONS

Based on the stone description, engineering properties, degree of microfracturing and stone classification, the following conclusions are derived:

1. Khawar, Khanugah II and Khanugah III marbles, and Jabalah granite satisfies all the ASTM requirements. They are rated as excellent stones and are recommended to be exploited.

2. Turabah gray and Turabah pink marbles also satisfies the ASTM requirements and are rated as very good. Although Khanugah I is also very good, its inhomogetiety makes it noneconomical.

3. Bakri granite satisfies the ASTM requirements but its properties are not consistently satisfactory. The stone is attractive and can only be used in limited purposes.

4. Dheraya granite does not satisfy all the ASTM requirements. It is not suitable for the production of dimension stones although very large blocks can be obtained.

7 ACKNOWLEDGEMENT

The authors wish to acknowledge the support of the Faculty of Earth Sciences, King Abdulaziz University and the Research Institute of King Fahd University of Petroleum and Minerals.

8 REFERENCES

Al Shanti, A.M. 1976. Geology of Ad Dawadmi district, Saudi Arabia. DGMR Min. Res. Bull. 13, 57 pp.

ASTM 1978. Standard specifications for exterior marble. Amer. Soc. Test. Mater. Designation C 503.

ASTM 1978. Standard specifications for structural granite. Amer. Soc. Test. Mater. Designation C 615.

Badiuzaman, M.Y. 1985. Engineering geological aspects of Wadi Turabah marble occurrences. M.Sc. Thesis. FES. KAU. Jeddah.

Brown, G.B., R.O. Jackson, R.G. Bogue and W.H. Maclean 1962. Geology of the Southern Hijaz quadrangle. Saudi Arabia. USGS Misc. Geol. Inv. Map 1-210-A.

Dehlavi, M.R. 1971. Wadi Bidah-Wadi Turabah marble. BRGM. Open File Report 71-JED-22, 27 pp.

Delfour, J., R. Dhellemmes, P. Elasass, P. Vaslet, J. Brasse, Y. Indre and O. Dottin 1982. Geologic map of the Ad Dawadmi quadrangle. Saudi Arabia. DGMR Map GM-60A.

DGMR. 1401H. Catalog of Saudi Arabian Stones.Special Publication SP-1.

El Mukhtar, M.A. 1985. Geomechanical assessment of dimension stones at Ad Dawadmi-Afif area. M.Sc. Thesis. FES. K.A.U. Jeddah.

Fourmaintraux, D. 1976. Charecterization of rocks- laboratory tests. in La Mechanique des roches appliquee aux ouvrage du genie civil. M. Panet (ed.). Ecole Nationale de Ponts et Chaussees, Paris.

Green, R.C. and L. Gonzalez 1980. Explanatory notes to the reconnaissance geologic map of the Wadi Shuqup quadrangle. Sheet 20/41 A. Saudi Arabia.

Lama, R.N. & V.S. Vutukuri 1978. Handbook on mechanical properties of rocks. II. Trans Laurent, D. 1970. Prospecting for marble in the Afif area. BRGM. Open File Report 70-JED-21, 32 pp.

Laurent, D. 1972. Prospecting for marble in Saudi Arabia. BRGM. Open File Report 72-JED-19, 58 pp.

Laurent, D. 1973. Physical characteristics of some Saudi Arabian marble. BRGM. Open File Report 73- JED-23, 19 pp.

Laurent, D. 1975. Dimension stone in the Jeddah region-Al Jamum granite. BRGM. Open File Report 75-JED-3.

Remez, M.R.H. & S.A.F. Murrell 1964. A petrographic analysis of Carrara marble. Int. J. Rock Mech. Mining Sci. 1:217- 229.

Tourenq, C., D. Fourmaintraux & A. Denis 1971. Propagation des ondes discontinuites des roches. Proc. Int. Symp. Rock Fracture. Nancy.

7.2 Exploitation – Methods and environmental impact
Méthodes de l'exploitation et influences sur l'environnement

Engineering geological studies for the individuation of extractive basins in anthropized areas

Etudes pour l'individuation de bassins d'extraction dans des zones anthropizées

G. Barisone & G. Bottino
Dipartimento Georisorse e Territorio, Politecnico di Torino, Italy

ABSTRACT: The paper illustrates the methodology used in a wide study carried out on an hilly territory in the South-West of Piedmont (Northern Italy), study aiming to individuate the possibility to open large extractive basins in an highly anthropized country. This in order to permit a minimum environmental impact and a correct territorial planning, with a concentration of quarry activity in limited areas. All the data collected by in situ surveys and laboratory tests were synthetized in two thematic maps, showing in the first the potential interest of recovered deposits, in the second the constraints to exploitation posed by natural and human factors. The zoning criteria adopted were studied in order to permit an easy automatic elaboration of a third map, in which the real value of deposits is synthetized.

RÉSUMÉ: On expose ici la méthode suivie et les résultats obtenus pendant une étude conduite sur un territoire collinaire étendu dans le Sud-Ouest du Piémont (Italie du Nord); cette étude vise à focaliser la possibilité d'ouvrir des grands bassins d'extraction au milieu d'une zone fortement anthropizée, tout en minimisant l'impact sur le territoire. Les données acquises (in situ et en laboratoire) sont synthétizées dans deux cartes thématiques, illustrant l'une l'intérêt potentiel des dépôts individués, l'autre les constrictions posées par les paramètres (naturels ou artificiels) qui peuvent interférer avec l'activité extractive. Les critères suivis pour la rédaction de ces cartes permettent une facile élaboration automatique de la carte finale, qui synthétize la valeur réelle - à ce moment - des dépôts etudies.

1. INTRODUCTION

The research summarized in this paper was carried out, on behalf of the Quarry Office of Regione Piemonte, on a hilly territory about 650 km2 wide and located in the South-West of Piedmont region (Northern Italy).

In this area the growing human impact, particularly marked in the last decade, created a progressive increasing of the problems connected to the quarry excavation processes.

The main purpose of this study was to individuate large extractive basins, in order to permit the concentration of quarry activity, thus minimizing the environmental impact in an highly inhabited country and providing an adequate support (mainly based on thematic maps) to a modern and correct territorial planning.

2. GEOLOGICAL DATA

The geology of the country is characterized by the almost exclusive presence of sedimentary rocks, prevailingly connected to the last sedimentary trasgressive cycle of the Pliocenic period; in Fig.1 the most typical stratigraphic sequences are shown.

Taking into consideration only the two mainly quarried rocks (clays and sands), for each of them two sequences can be distinguished in the studied area, on the basis of lithological and technological characteristics.

2.1 Clays

Pliocenic sequence: mainly composed by marly-clayey marine sediments with a high constancy of attitude and sedimentological characteristics; for a quarry activity, the most interesting areas correspond to the

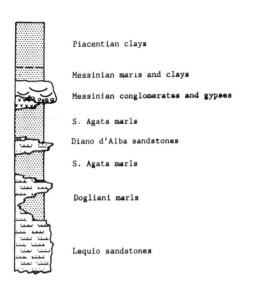

Piacentian clays

Messinian marls and clays

Messinian conglomerates and gypses

S. Agata marls

Diano d'Alba sandstones

S. Agata marls

Dogliani marls

Lequio sandstones

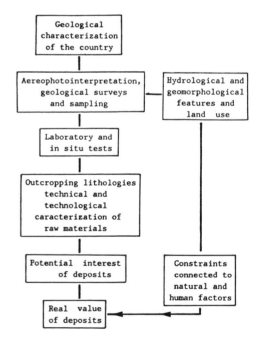

Lacustrine clays

Coarse sandstones (river channels)

Pelitic sediments (tidal planes)

Medium sands with crossed
stratification (tidal channels)

Fig. 1

Stratigraphic scheme of Langhian mio-pliocenic
sequence

- Stratigraphic scheme of Villafranchian
  sequence

layers with a lower carbonate content.

Villafranchian sequence: composed by
sediments corresponding to a transition
facies, with clays mainly of tidal plane or
lacustrine origin; the marked facies
hetereogenity is a serious limit for
extraction activities, despite the normally
good tecnological characteristics of these
clays.

2.2 Sands

Pliocenic sequence: near shore deposits,
prevailingly composed by sands and silty
sands; these deposits are marked by a
considerable granulometric homogeneity
along horizontal planes, whilst in the
vertical direction the granulometry of the
deposits increases from the bottom to the
top of the sequence.

Villafranchian sequence: characterized by
a transition facies, where the granulometry
is very variable in both the horizontal and
the vertical directions; the upper part of
the series is the most rich in coarse
deposits (gravels and sands).

3. METHODOLOGY

The methodology followed for the study is
schematized in Fig.2.

Fig. 2 -Flow Diagram of operational sequence

3.1 Geological-technical survey

In synthesis, after a first evaluation, made by mean of aereophotointerpretation, of geologic and anthropic aspects, a particularly careful and detailed survey was carried out, to better define the limits of outcropping formations and to evaluate the technically exploitable cubages (taking chiefly in consideration morphological, hydrological and anthropic factors).

During the surveys, significant samples were also collected for laboratory tests, tests mainly centered on a technical and technological evaluation of raw materials, to appraise their aptitude for actual industrial uses.

So defined the quantity and the quality (i.e. the economic value) of raw materials, the real value of exploitable recovered deposits was appraised by crossing this parameter with the exploitation constraints due to natural and human factors.

3.2 Laboratory analyses

To evaluate the quality of raw materials in the studied area 180 samples were collected and analyzed, the main purpose of laboratory tests being the evidentiation of the best possible industrial uses, without considering the possible actual utilization of the materials.

The study of technological characters of clays was based on the determination of the following data: mineralogical and granulometric composition, dilatometric and sintering curves, geotechnical characteristics. In Tabb.1,2 the results of mineralogical analyses and of geotechnical tests are summarized; granulometric and sintering curves (referred to Pliocenic and Villafranchian clays) are represented in Figg.3,4,5,6. The above cited tests evidentiate a uniformity of granulometric and mineralogic composition greater for Pliocenic clays than for Villafranchian ones; this greater uniformity, however, joins rather bad technological characteristics, mainly due to the constant presence of calcite, whilst Villafranchian clays - although very disuniform - have locally a better firing behaviour.

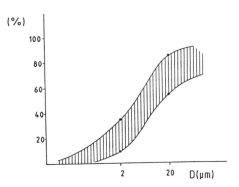

Fig. 3 - Granulometric distribution of Pliocenic clays

Table 1. Mineralogical composition of main lithotipes.

Geologic sequence	Main minerals	Subordin. minerals
Villafranchian clays	Illite,Smectite, Quartz	Kaolinite,Chlorite, Interstratified, Calcite,Feldspars
Pliocenic sands (Astian formation)	Quartz,Feldspars	Calcite,Dolomite, Clay minerals,Iron oxides
Pliocenic clays	Illite,Kaolinite, Quartz,Calcite	Chlorite,Smectite, Feldspars,Dolomite

Table 2. Geotechnical characteristics of clays.

Geologic sequence	w (%)	p (%)	wL (%)	wP (%)	IP	IA	Co (kPa)
Villafran. clays	28.2	1.97	55.9	34.2	21.7	10.4	150
Pliocenic clays	32.4	1.79	56.5	37.6	18.7	14.2	450

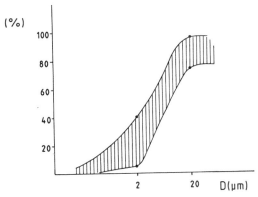

Fig. 4 - Granulometric distribution of Villafranchian clays

2997

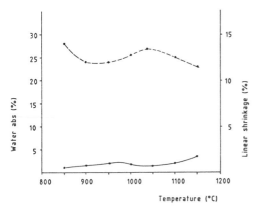

Fig. 5 - Sintering curves of Pliocenic clays

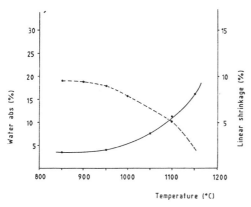

Fig. 6 - Sintering curves of Villafranchian clays

The sands outcropping in the studied area were evaluated mainly on the basis of mineralogical and granulometric characters; in Fig.7 the granulometric distribution of collected samples is pointed out. The sandy levels show a very high granulometric variability, with the fraction 200-500um varying between 20% and 90%; coarser sediments gather in the central part of Pliocenic sequence (Astian formation), while silty sands and silts are prevailing in the upper and lower parts of this sequence. In the Villafranchian sequence, silts are mainly found as intercalations in clayey levels, while the top of the sequence is characterized by the presence of coarse sands and sandy gravels. The mineralogical composition, summarized in

Tab.1 for the more homogeneous formation (the Astian sands), is constant enough in both the sequences, with quartz clearly prevailing (chiefly in Villafranchian sands) and a calcite content constantly low.

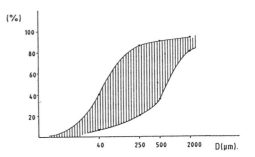

Fig. 7 - Granulometric distribution of sands

4. THEMATIC MAPS

All data collected by means of in situ surveys and laboratory analyses are synthetized in two thematic maps, the first (Deposits quality map) showing the potential value of recovered deposits, the second (Quarrying aptitude map) centered on constraints connected to geo-morphological and anthropic parameters; a sketch of these maps is shown in Figg.8-9.

The "Deposits quality map" was elaborated by considering both lithological and technological parameters (e.g., plasticity for clays or granulometry for sands); the "Quarrying aptitude map", that shows the less or more aptitude of the land to the opening of quarries, was obtained taking into consideration the following parameters: morphology of slopes; presence of slides or of erosion phenomena; presence of water courses, roads and railways or other important structures; extension of urban areas.

In the purpose of making the final map by means of computer aided calculation and design, each parameter considered for evaluating the real value (i.e. the possibility of exploitation) of recovered deposits was divided in three schematic classes, following the criteria synthetized in Tabb.3,4,5,6,7.

These parameters are: raw material quality; economically recoverable reserves; morphology; anthropization degree of exploitable areas. In this last case a fourth class was added, the "nul" class,

2998

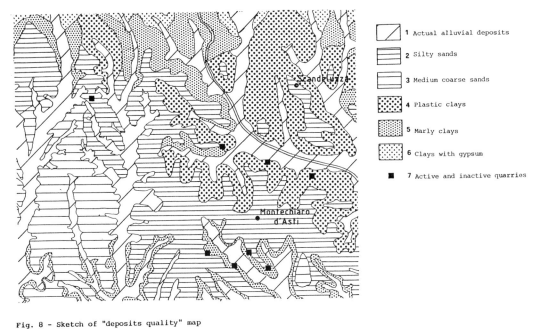

1	Actual alluvial deposits
2	Silty sands
3	Medium coarse sands
4	Plastic clays
5	Marly clays
6	Clays with gypsum
7	Active and inactive quarries

Fig. 8 - Sketch of "deposits quality" map

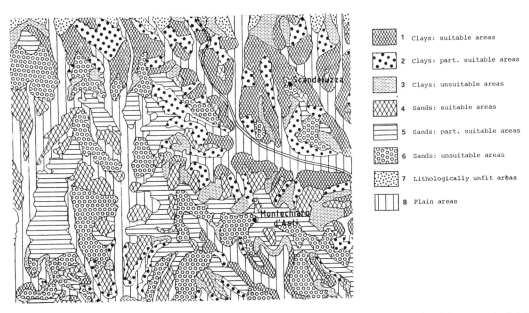

1	Clays: suitable areas
2	Clays: part. suitable areas
3	Clays: unsuitable areas
4	Sands: suitable areas
5	Sands: part. suitable areas
6	Sands: unsuitable areas
7	Lithologically unfit areas
8	Plain areas

Fig. 9 - Sketch of the "quarrying aptitude" map (extraction suitability in order to natural and human constraints)

Table 3. Clays quality classes.

Features	Quality classes		
	I	II	III
Mineralogy *	I,K,C,(S)	I,K,C,V,(S)	I,K,C,V,S
Granulometry (%)**	50A,40L,10S	40A,40L,20S	25A,75L+S
CaCO3 content	< 20 %	10-15 %	< 30 %
Sulphates content	traces	< 1 %	< 2 %
Baking temperature	950-1100 C	950-1000 C	900-950 C
Porosity on baked	5-20 %	8-12 %	15-25 %
Colour	clear	any	any
Kind of bricks	High fired	Valuable	Normal

* I = illite; K = kaolinite; C = chlorite; S = smectite;
V = vermiculite.
** A = clayey fraction,<2um; L = silty fraction,2-40um;
S = sandy fraction,>40um.

Table 4. Sands quality classes.

Granulometric features	Quality class
More than 50% of grains > 0.4mm (A1 of CNR-UNI classification)	I
More than 65% of grains > 0.075mm (A2-A3 of CNR-UNI classification)	II
More than 35% of grains < 0.075mm (A4,A5,A6,A7 of CNR-UNI classification)	III

Table 5. Deposits features and related classes.

Features	Classes		
	I	II	III
Deposit thickn	> 20m	20-10m	< 10m
Cover/deposit thickness	0-1/4	1/4-1/2	> 1/2
Homogeneity, areal extension *	H,H-M	M,H-M H,L	L,H-L M,L

* H = high; M = medium; L = low.

Table 6. Geomorphological features and related classes.

Features	Class
Top of the hills or regular and stable slopes	I
Planes or irregular or unstable slopes	II
Inondable areas, sliding areas, concave slopes	III

Table 7. Anthropization degrees and related classes.

Anthropization	Class
No settlements, roads, electric lines, etc.	I
Isolated houses or other structures of little importance	II
Inhabited area, with many houses or important structures (highways, railways,etc.)	III
Urban areas and neighbouring zones	Nul

corresponding to an urbanization degree involving a total exclusion of quarry activity, independently from the value of other parameters.

In Fig.10 a sketch of the final map ("Deposits real value map") is shown. This map was obtained crossing two intermediate parameters: the theoric value of the deposit and the constraints for the land use. The first parameter results from the pondered evaluation of raw materials quality and of exploitable reserves; the second from the crossing between morphological conditions and anthropization degree.

REFERENCES

Barisone,G., Bottino,G. (1981). Criteri di reperibilita' degli inerti. I materiali, i bacini estrattivi. Boll. Ass. Min. Subalp. 18.

Barisone,G., Bottino,G. (1987). Engineering geological mapping for the planning of the extraction processes in the middle Susa valley (Northern Italy). Proc. Int. Symp. on Eng. Geol. Env. in Mountainous areas, Beijing.

Bottino,G., Gola,G., Robiglio,G., Vigna,B. (1985). Cartografia geologico-tecnica per la valutazione ed il razionale sfruttamento delle argille del Bacino Terziario Piemontese. Boll. Ass. Min. Subalp. 22.

Bottino,G., Grassi,G., Stafferi,L. (1988). Le argille del Bacino Terziario Piemontese - Aspetti genetici e caratteristiche tecnologiche. Boll. Ass. Min. Subalp. 25

Francavilla,F., Bertolani Marchetti,D., Tomadin,L. (1970). Ricerche stratigrafiche, sedimentologiche e palinologiche sul Villafranchiano tipo. Giorn. Geol. 36.

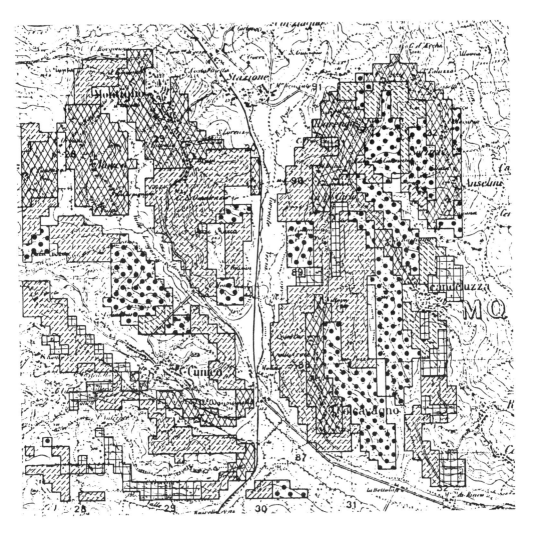

	Deposits of low interest
	Deposits of medium interest
	Deposits of high interest
	Anthropized areas

Fig. 10 - Deposits "real value" map

Functional requirements and alternative materials for embankments
Conditions fonctionnelles et matériaux alternatifs pour remblai

E.J.de Jong
Netherlands Concrete Society, Gouda, Netherlands

ABSTRACT: By formulating the functional requirements in relative material properties a judgement of the suitability of materials can be made. Therefore embankments are classified in parts. This gives a quick judgment after defining only the relevant properties of the materials is possible. The motive for this action is the growing lack of natural resources and growing offer of alternative materials. For waste products environmental laws must be formulated.

RESUME: Formulation les conditions fonctionnel en caracteristiques des materiaux d'importance on peu juger le modération des materiaux. Les remblais sont classifier en parties. Une methode rapide s'est possible apres qualification seulement les caracteristiques relevantes. Le motif pour cette action s'est le defaut progressif des materiaux naturels et l'offre progressive des materiaux alternatives. Pour les produits résiduaires le législation de milieu est necesaire.

1 INTRODUCTION

In the Netherlands annually about 30 millions tonnes of sand are used for embankments. Sand is everywhere for very small costs available and from a technical point of view very suitable. All expierence and standards are made for clean sand. It must be mentionned that other materials like clay were used in the past and even with good results. Dikes along the rivers were often built of clay and even heavy traffic passed the roads upon these contructions. The present generation of engineers is educated with a standard construction of sand and the slopes covered by clay to prevent uitspoeling. The material specifications for embankments are clearly made for sand and are restricted. The meaning of the standard is that if an engineer wants to use sand the sand should meet these standards. If the engineer wishes to use other materials he must proof that this material is fit for the purpose. In practice the formulated standards are used as general requirements for materials used for embankments.

At the end of the seventies the resistance against exploration was growing due to the green belt movement. Meanwhile the amount of industrial by-products was growing. Embankments were pointed as an ideal solution to solve these two problems in one. An ambitious program of investigation was started to find out the technical and economical possibilities of alternative materials for sand with regard to the environmental aspects.

2 RESEARCH PROGRAM

In 1982 the "research centre for road construction" (the present Centre R·O·W) organized a symposium about alternative materials in road constructions. He came to the principal question: What are the functional requirements materials used for embankments must meet?

The conclusion of the symposium was the need for a research program co-ordinated by the Centre R·O·W. The centre installed a steering group and five working groups to organise the program. One of these groups was the one named "Special materials for embankments". This group reported their results in 1988. Some of their results are presented in this paper. The working group was composed by experts of all kind of disciplines. The working

group formulated a research program which has to answer two questions:
- What are the functional requirements of embankments?
- What are the qualifications of potential materials?

In the end an opinion of the application can be given by comparing both results.

The general aproach was followed: research of (inter)national literature, laboratory-tests and verification in practice. Also research in the files with respect to selected materials and regions was foreseen.

3 REQUIREMENTS OF EMBANKMENTS

At the start of the working group only general descriptions of requirements were available. No stiff figures were present in the form of maximum or minimum angle of internal friction, consolidation, settlement or permeability in relation to the part or purpose of an embankment.

So at the start of all the activitities the purposes of embankments were considerated. Very soon it came clear that embankments can be devided in two types; one type with and one type without a constructive purpose.

Examples of embankments without a constructive purpose are sound barriers between roads and residential quarters. A constructive purpose can be a road construction. Both types can be presented in a scheme as presented in figure 1.

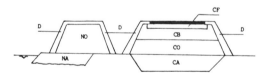

figure 1
Classification of embankments

Explanation figure 1:

GENERAL:
D : Cover to protect agaist erosion, base for the growing of weed, prevention of penetration of water and possible leaching.

A. NOT CONSTRUCTIVE EMBANKMENTS (without traffic)
NA : Fill up of excavation, ditches et cetera above or below groundwater.
NO : Raising in behalf of building sites, sound barriers etc.

B. CONSTRUCTIVE EMBANKMENTS (with traffic)
CA : Fill up excavation, ditches et cetera above or below ground water.
CO : Raising in behalf of roads (exclusive base or sub-base)
CB : Sub-base
CF : Base

The base is in the Netherlands considered as part of the road construction and not the embankment. Part CF will not be cited anymore.

To all of the now indicated parts of an embankment functional requirements can be formulated. These requirements can be devided in aspects of:
- civil egineering
- environment
- economics
- organization

For all parts and functional aspects the relevant elements are investigated. So the elements of civil engineering are listed after formulating the models of behaviour like collapse mechanism during construction and use. For part D the resistance against erosion is an important parameter. The instability in relation to the angle of the slope is of interest for the parts CB, CO and NO, while a push back due to high water tension counts for CB and CO.

4 REQUIREMENTS OF MATERIALS

After the distinction of embankments in elements and formulating the functional requirements of the relative elements the measure of importancy of material properties can be indicated. In table 1 some of the results are presented.

Table 1 : Value-indication of material properties related to the part of the construction.

Property	Part of the construction					
	CO	NO	CA	NA	D	CB
Density	1	3	2	3	1	1
Stability	1	1	2	2	2	1
Bearing capacity	2	2	2	-	2	1
Connexion	3	3	3	3	1	3
Frost behaviour	2	2	2	3	3	1
Swell and shrink	1	2	1	3	1	1
Persistancy to erosion	3	1	3	2	1	3
Fertility	3	2	3	3	1	3
etc.						

All the mentionned properties can be found by well-kown tests. For a special construction-part the value of the property related to the conditions of the embankment can be formulated.
Also the working-up in practice is of interest and so more. All of these elements are listed and investigated.

5 DESCRIPTION OF MATERIALS

As alternative of sand some groups of materials are considered:
Minerals of surface exploration : sand, sandy-clay, silt and dock-silt.
Stoney rubbish : minestone and sand of crushers.
Industrial by-products : gypsum, slag of incineration plants and bottom-ashes.

Of all these materials the possible relevant properties are determined by information of literature, file consultancy, laboratory tests and information of practice. It is not possible to present all the results in this paper.
For the materials in the groups of stoney rubbish and industrial by-products also environmental aspects are of great importance. Till this moment (spring 1990) there are no stiff figures in the Netherlands or abroad to distinguish a justified and not justified application. The polluting of the sub-soil is a problem every-one in the Netherlands is concerned about. So a lot of materials cannot be used without special conditions. Special conditions with regard to their use are formulated by the Ministery of Environmental. Because the obscurity of environmental laws in the judgement of the materials a summary of possible applications is given without considering the environmental consequences.

The working group reported about all the mentionned materials in the same way:
    definition
    available quantities
    materialproperties (as listed in table 1)
    general environmental aspects
    possible applications (only civil engineering).
The possible applications are summarised in table 2.

6 ENVIRONMENTAL ASPECTS

The environmental aspects of using alternative materials like furnace bottom ashe or slag of incineration plants are dominated by the fear of leaching of elements which can pollute the sub-soil and after-

Table 2 : Possible application of materials related to the construction part in figure 1 (only civil engineering).

Material	Part of the construction					
	CO	NO	CA	NA	D	CB
Sand	++	++	++	++	0	+
Clay (marine)	*	++	*	*	++	0
Clay (river)	++	++	*	*	++	-
Sandy-clay	++	++	++	++	++	-
Loam	+	+	*	*	-	0
Silt	+	+	*	*	-	0
Dock-silt	*	++	*	*	+	-
Sand of crushers	+	+	+	+	-	*
Minestone	+	+	+	+	-	*
Gypsum	*	*	*	*	-	*
Bottom-ashes	++	++	++	++	-	*
Powder fired fly-ash	++	++	+	+	-	+
Slag of incineration plants (MSW)	++	++	+	+	-	++

Explanation:
++ Suitible
+ Suitible under conditions
* Further research necessary
- Not suitible
0 Out of practice

words the drinking water. A test method is devellopped to determine the flow of elements in a previous fixed period. This period can meet hundreds of years.
To prevent the flow of elements which can pollute the sub-soil it is possible to immobilise these elements by binder (bitumen, cement, lime), screening by membranes or find a location where the flux cannot penetrate the sub-soil (f.e. a location on clay). It is till now not possible to indicate the limits where eco systems do not meet harm. Also because of the complicated combination of elements it is not to expect these limits can be formulated in general standards.
In all cases a balance must be made of environmental aspects (dumping of rubbish, the use of a dump belt as road base, the exploitation of natural resources, the influences on eco- systems).

7 CONCLUSIONS

Not only sand is suitable for embankments.

Local situations (place and moment) are important fot the choice of alternatives of natural kind.

There are positive and negative environmental aspects in the choice of materials. Local situations, economical, constructive an material conditions are of influence.

8 REFERENCES

Centre R·O·W, Publikatie 16, Resten zijn géén afval (meer), Bijzondere ophoogmaterialen; (Rests are no longer waste (anymore), special materials for embankments), Ede, the Netherlands, Octobre 1988.

Centre R·O·W, Publikatie 17, Resten zijn géén afval (meer), Primair gebruik van secundaire materialen; (Rests are no longer waste (anymore), primary use of secunairy materials), Ede the Netherlands, Septembre 1989.

Engineering geological study of Maadi sand deposits
Etude de la géotechnique des dépôts de sable de Méadi

S. Ossama Mazen
General Organization for Housing, Building and Planning Research, Dokki, Cairo, Egypt

ABSTRACT : Sand needed for foundry works are found in Egypt in arid zones necessitating either the transport of water to the site of operation or the transport of the sand to the water source. Consequently the capital investment and running cost of the separation plant required would be much higher than that of plant based on dry classification. This led to the studies into the possibilities of dry treatment of Maadi sand deposits, located South East of Cairo, by the use of screening for the course to medium size sand and the use of pneumatic appliances for the treatment of fine grained sands.
The present work, as a part of integrated investigation, included geological and topographical studies, exploratory borings and drilling and laboratory testing . It was found that the workable sand would represent no difficulty in quarrying and its characteristics indicate amiability to dry classification such as screening, air separation and washability.

RÉSUMÉ : Sable requis pour les travaux de fonderie est trouvé en Egypte dans les zone arides nécessitant soit le transport de l'eau Jusqu'a l'emplacement de l'operation, ou le transport du sable jusqu'à la source de l'eau. Par conséquent, l'investissment du capital et le cout du fonctionnement de l'usine de triage require seront de loin superieurs à ceux d'une usine basée sur une classification sèche. Ceci a abouti à des études sur les possibilités des traitment sec des dépôts de sable de Meadi situés au sud du Caire par l'emploi du criblage pour le sable à gros grains jusqu'au sable à grains de dimension moyenne, ainsi que l'emploi des appareils pneumatiques pour le traitement du sable au grain fin.
Le présent travail, en tant que faisant partie d'un examen intégré, comprend des édudes geologiques et topographiques, perforations et forages, et des analyses de laboratoire. Il a été établi que le sable, réalisable, ne représenterait pas de difficulté dans l'exploitation en carrière. Ses caracteristiques indiquent la nature clémente a la classification séche, tel que le criblage, la séparation de l'air et la possibilité du lavage.

1 INTRODUCTION

The sand needed for foundry work must be classified according to grain size and in most cases washed to eliminate the harmful ingradient. Such treatment is usually carried out wet; the cubic meter of sand requiring an average of twelve cubic meters of water per hour. This high requirement necessitates the availability of good and permanent source of water within reasonable distance from the deposit to enable its economical exploitation.

Although most sand deposits in Egypt are found in arid zones with no water source in their vicinity, Maadi sand, being the nearst to the River Nile, has been considered the most promising one. It is located approximately fifteen kilometers in a straight line South East of Cairo or eighteen kilometer from Maadi suburb. The site is easy to access; it lies on the road joining Maadi suburb to the Red Sea coast which is commonly called El-Kattamiya Road (Figure1).

From figure 1, it is evidensed that wet treatment of Maadi sand would necessitate either the transport of sand to a location on the Nile or to transport the water to the site of operation. In both cases, the capital outlay and working cost become prohibitive to economical exploitation of

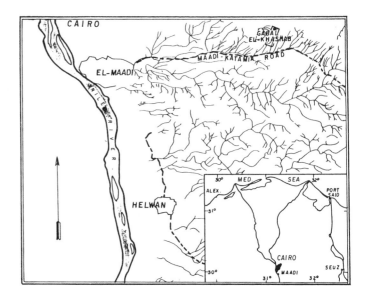

Figure 1
Guide map for the site and
area of study

the deposites. This led to the studies into the possibilities of dry treatment by the use of screening for the course grained particles and pneumatic appliances for the fine grained material .

The present work is a part of a comprehensive study that included planning, production, plants, buildings, economics, .. etc. It is concerned with two principal phases of the subject. The first is the geology and soil condition of the area of study relevant to quarrying, and the second is the amiability of Maadi sand to dry classification treatment.

The work includes geological and topographical studies, exploratory borings and drilling and laboratory testing of the samples extracted during drilling of bore-holes.

2 GEOLOGY AND TOPOGRAPHY

The outcrops in the area of study are mainly composed of yellow sand intercalated with red to gray clay belonging to the Oligocene formation of the region which has been lately termed Gabal El-Khashab Formation(Gabal El-Khashab means mountain of the wood or petrified forest). This term was given because of the occurrence of silicified wood as a distinct characteristic feature of the formation.

A number of geologists and research workers have passed through Gabal El-Khashab, studied and described the nature of the local geology of the area (e.g. Cuvillier, 1924; Shukri, 1953; Ghorab

and Ismaiel,1957; Farag and Ismaiel,1959; Said,1962; Said and Yosri, 1964; Said, 1975). From their study, the stratigraphic sequence can be summarised in the following; the youngest bed at top :

| Recent | —Wind blown sand and Wadi deposites (Sand) |

| Oligocene formation (Gabal El-Khashab formation) | —Upper yellow sand (Sand)
—Red clay bed (Clay)
—Sand intercalated with gray clay (Sand& clay)
—White sand (Sand)
—Yellow white sand (Sand)
—Lower yellow sand (Sand) |

| Upper Eocene (Maadi formation) | —Hard brown lime-stone (Lime-stone)
—Alternating (Shale) yellowish sandy and greyish shale beds
—Sandy limestone (Lime-marl stone & marl) |

The deposits lie, in general, in a flat terrain with very few topographic features.

3 SITE INVESTIGATION

As a first stage site investigation, thirteen shallow pits of an average two meters per pit were put down and sampled. The thirteen pits showed that the deposit was relatively uniform in size as far as

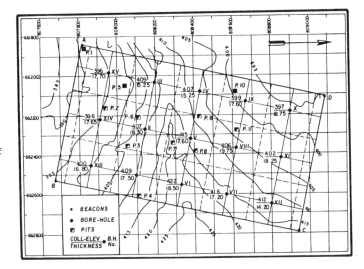

Figure 2
Contours of the area of
study and location of
boreholes

the vertical distribution of the grain was concerned. It was, therefore, decided to site the boreholes at the intersections of a grid every 240m both ways (Figure 2). The grid was planned to cover the area of study with the peripheral bore-holes at a distance of 120 meters from the boundary.

3.1 Drilling technique

Preliminary investigation showed that the sand layers were not compacted and with negligible bands of clay. Also, power drilling, previously experienced in a near by site, proved to be expensive. This made hand drilling, more favoured.

Hand drilling proceeded by the advance of a casing which is provided by a steel shoe with bevelled cutting edge. This advance into depth is accomplished by dead weight, placed on a detachable platform attached to the top of the casing which is in turn manually rotated.

Advancing in this way, the hand-drill may be said to make a core which, if necessary, the churning tools loosen up before the bailer is used to bring the sample to surface. As the hole deepens, the platform is temporarily removed and additional lengths of casing is added at the top. . The platform is then re-attached to the newly added casing, and drilling, churning and bailing are resumed.

3.2 Sampling

Sampling was carried out by a bailer fitted with a flap valve. Stones, when met, larger than the opening of the valve were broken down by churning tools and the fragments bailed out with the sand sample.

The bailed samples were discharged in pans and allowed to dry before being transferred into canvas sampling bags. All samples taken were analysed mechanically for grain size and chemically for iron contents.

4 LABORATORY WORK

Laboratory testing included the chemical analysis of sand, grain size distribution, iron content determination and washability tests.

Samples of the sand extracted during drilling of the bore-holes were sent to the laboratory for sieve analysis of each meter penetrated and for the determination of the iron contents of each sublayer of the geological section. Complete chemical analysis was done on intact samples. Washability tests were carried out on selected samples to assess the iron content reduction by simple washing.

5 DETAILED GEOLOGY

As a result of present drilling and pitting, different members forming the stratigraphic succession of Recent and Oligocene formation of Gabal El-Khashab area were encountered (see figure 3). A brief description of these members can be givin as follows:

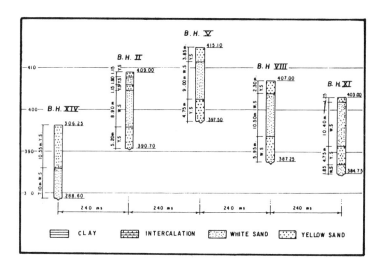

Figure 3
Borehole logs shown
across a longtudinal
section in the middle of
the area of study

5.1 Wind blown sands and Wadi deposits

These are superficial deposits consisting mainly of whitish yellow to yellowish brown drift sand with fragments of silicified wood, gravel pebbles and variable amounts of clay. It cover all the area with an average thickness of about two meters.

5.2 Upper yellow sand

This forms the upper member of Gabal El-Khashab formation. It consists mainly of whitish yellow, yellow and brownish sand of medium and course grains and occasionally of fine grains. The average thickness of this bed is about 8 meters.

5.3 The clays

These consist mainly of reddish bed intercalated with sands and grey clay. They underline the upper yellow sand and are followed downward by white sand intercalated with greyish clay of variable thickness. The series as a whole measures about 6 meters.

5.4 The white sand

This underlies the grey clay and overlies the lower yellow sand. The white sand may be divided into two horizons namely pure white sand constituting the main bed, which lies immediately under the clays, and a relatively thinner horizon of pale yellowish white sand underlying the main white sand. Streaks of iron bearing sand varying in thickness from one millimeter to two centimeters have been observed within this member.

The thickness of the white sand as revealed from drilling varies along the general trend of the deposit as well as across it. In some bore-holes it attained a thickness of over 20 meters while in other bore-holes the thickness was not more than 5 meters.

5.5 The lower yellow sand

Underlying the white sand and resting on the Upper Eocene series lies the lower yellow sand. It varies in colour from whitish yellow to reddish white. The sand is loose, fine to coarse grained. They owe their colouration to stains of limonitic iron oxide which form thin films around the grains. In certain horizons, the iron oxides cause cementation of the loose sand resulting in hard reddish streaks varying in thickness from few millimeters to two centimeters.

The total thickness of the lower sand has not been determined in present drilling where the thicknesses penetrated varied fron 7.6 to 13.80 meters. It is, however, reported to attain over thirty meters.

6 CHARACTERISTICS OF MAADI SAND

As a result of field investigation, sample examination and laboratory testing, the characteristics of Maadi sand deposits may be summerized in the following :

6.1 Chemical composition

Complete chemical analysis (Table 1) shows that the main constituent is silica which accounts for approximately 90% of the sand grains. Other constituents are alumina (three percent) and calcium oxide (one percent). The remaining one percent is made up of magnesia, iron oxide, soda (sodium hydroxide NaOH)and potash (calcium hydroxide Ca $(OH)_2$).

Table1: Chemical composition of Maadi sands

Constituents	White Sand %	Yellow Sand %
Silica SiO_2	95.56	94.99
Alumina Al_2O_3	2.11	2.81
Lime CaO	0.66	1.01
Magnesia MgO	0.58	0.49
Iron Fe_2O_3	0.26	0.31
Loss on Ignition	0.90	1.12
Total	100.07	100.73

6.2 Grain colour

The variation in colour of Maadi sand deposits from pure glossy white to yellowish and reddish white was identified by chemical analysis. Determination of the iron contents of soil samples indicated that:

1. Quartz grains representing the yellowish sand acquired this colour because of the stains of limonitic iron oxide which forms films, usually contineous around grains.

2. The reddish sand owe their colour to hard reddish streaks of cemented sands and iron oxide.

6.3 Grain size

Curves based on the results computed from average sieve analysis carried out on samples obtained from drilling are shown in Figure 4. They indicate that Maadi sand may be classed as coarse and medium to fine sand with very small amounts of superfines.

6.4 Grain roundness & sphericity

The study of the shape of the course and medium grain sand fraction resulted in the following :

1. The majority of grains of the coarse sand are subrounded with an average frequency of 66.7% for the yellow sand and 57.7% for the white sand. However, the white coarse sand grains are better rounded than those of yellow coarse sand.

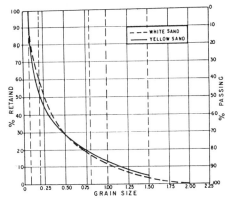

Figure 4   Grading curves of tested soil

2. The majority of grains of the medium sand are subrounded with an average frequency of 49.7% for yellow sand and 45.7% for white sand. However, the white medium sand shows better rounding.

3. The roundness study also shows that the coarse sand grains, in general, are better rounded than the medium.

4. Sphericity studies indicate that the majority of grains of both coarse and medium sand sizes belong to the medium and high sphericity classes. Grains possesing low sphericity classes are represented by a minor amount and may be completely absent in many of the investigated samples.

5. The study also showed that roundness and sphericity are inter-related; where the increase in roundness has been assosiated with an increase in sphericity.

7 AMIABILITY TO QUARRYING

By reason of the topography, the quarry can be of the subsurface or pit type operation. This constitutes of stripping the overburden followed by excavating the sand in three different horizons. Development, therefore, may proceed by excavating inclines from the surface to depth which would allow the establishment of the intended three horizons.   The sand beds, being unconsolidated and of almost horizontal layers, would represent no difficulty in quarrying.

8 AMIABILITY TO DRY CLASSIFICATION TREATMENT

8.1 Screening

The shape of the particles, their grain size distribution and the negligible amount of fine grained material present in the sand would enable the use of classical screens.

8. 2 Air separation

The separation of the small amount of silt and clay content would present no difficulty particularly when cyclons are introduced in the circuit.

8.3 Washability

As previously mentioned, stains of limonitic oxide form continous films around the particles of whitish yellow sand. It was found that such colouration can be easily removed by washing, provided that the appliances used allow for agitation and attrition.

Washability tests also proved that iron contents could be reduced to 0.08 Fe_2O_3% by simple washing with water.

9 CONCLUSION

1. The outcrops in the area investigated are mainly composed of sands belonging to the Oligocene formation of Gabal El-khashab region which attain a thickness of about 75 meters. Studies suggest that the deposits are of fluvio- marine depositional environment.

2. The sequence of the workable sands in the area from top to base is as follows:

Overburden: Made up of wind blown sand with fragments of silicified wood, gravel pebbles and variable amount of clay. Its average thickness is about 2 meters.

Upper yellow sand: Loose pale yellow sands with reddish streaks of concentrated iron-stained sand grains. The average thickness of this bed is about 8 meters.

Middle white sand: White unconsolidated sands free of silicified wood fragments and contains negligible amount of clay. Its average thickness is 10 meters.

Lower yellow sand: Yellow sand with patches of reddish colouration of undetermined thickness. It contains no silicified wood fragments and no clays.

3. Quartz account for approximately 96% of the sand grains. Other constituents are alumina(3%)and calcium oxide(1%). The remaining one percent is made up of magnesia, potash and iron oxide.

4. The workable sand, being unconsolidated and of almost horizontal layers will represent no difficulty in quarrying .

5. By reason of the topography, the quarry would be of the sub-surface or pit type operation.

6. The characteristics of Maadi sands indicate amiability to treatments relevant to dry classification such as screening, air seperation and washability.

REFERENCES

Cuvillier, J. 1924. Contribution a l'etrude geologique du Mokkatam. Bull. Inst. Egypt. 6: 93 -108.

Farag, I.A.M. & M.M. Ismail 1959. Contribution to the stratigraphy of the Wadi Hof area (North -East of Helwan). Bull. Fac. Sci. Cairo Univ. 34;147-168.

Ghorab,M.A. & M.M. Ismail 1957.Amicrofacies study of the Eocene and Pilocene East of Helwan. Egypt. J. Geol. 1: 105- 125.

Said, R. 1962. The Geology of Egypt, Amsterdam: Elsvier.

Said,R. & F. Yosri 1964, Origin and Pleistocene history of River Nile near Cairo, Egypt. Bull. Inst.Egypte.45:1-30.

Said, R. 1975. Subsurface geology of Cairo area, Memoires de l'Institut d'Egypte. 60:1 - 54.

Shukri,N.M. 1953. The geology of the desert East of Cairo. Bull.inst. desert Egypt. 3 (2) : 8 -105.

6th International IAEG Congres / 6ème Congrès International de AIGI, © 1990 Balkema, Rotterdam. ISBN 90 6191 130 3

Evaluation of crushed granitic rock aggregates in Ghana

Estimation des agrégats granitiques concassés des roches au Ghana

K.E.N.Tsidzi

Geological Engineering Department, University of Science and Technology, Kumasi, Ghana

ABSTRACT: A significant expansion in the building and construction industry of Ghana in recent times has necessitated a greater demand for the use of rock aggregates in concrete and road works. It is essential, therefore, that more quarriable rock outcrops or natural gravel deposits be located and their geological and engineering properties evaluated in addition to updating information on the existing quarries. The present paper summarises information, obtained from a thorough literature search and field and laboratory studies, on the suitability of the various Ghanaian granitic rocks for use as concrete and road aggregates. The choice of granitic rocks for the study is explained by the fact that they are widespread (covering about 25 percent of the total land surface) and also form the bulk of the country's aggregate resources. An attempt has been made to establish functional relationships between various rock aggregate properties with a view to providing a cost-effective approach to the selection of design parameters. The potential for alkali-aggregate reactivity of these rock materials in concrete has also been assessed from a critical study of their petrofabrics. The study generally seeks to emphasise the need for thorough engineering geological investigations in the economic exploitation of aggregates which, when combined with field performance studies, will provide the necessary framework for a reliable specification of Ghanaian rock aggregates for concrete and road works.

RÉSUMÉ: Une expansion considérable dans les industries de construction ghanéennes d'ajourd'hui rend nécessaire un grand besoin de l'utilisation des agrégats de roches dans la fabrication du béton et dans des travaux routiers. Il importe donc d'identifier des carrières et des dépôts de gravier naturel et d'évaluer leur propriété geologique afin de mettre à jour les données de carrières existants. La présente étude resume l'information obtenue par des recherches au laboratoire ainsi que des études sur le terrain dont le but était de déterminer la convenance des roches granitiques variées dans la fabrication du béton et des agrégats routiers. Le choix des roches granitiques pour l'étude s'explique par le fait quils sont en abondance au Ghana, couvrant vers 25% de la surface du pays, et constituant ainsi la plus grande partie de l'ensemble de ressources de pays. On a essayé d'établir la relation fonctionnelle entre la propriété de plusieurs roches qui tendent à fournir un approche rentable pour la procuration des paramètres de project. Le potentiel quant a la réactivité d'agrégat alkalin d'une étude critique de leur qualité et de leu teneur pétrofabrique est aussi évalue. L'étude vise à souligner la nécessite d'une recherche géologique le choix économique d'agrégats, qui couplé à l'étude réelle de performance, fournira le charperte d'une spécification fiable d'agrégat des roches ghanéennes pour la fabrication du béton et dans des travaux routiers.

1 INTRODUCTION

The Government of Ghana has recently embarked on a major roads rehabilitation and construction programme besides the serious efforts that are being made to satisfy the housing needs of the growing population. Consequently, a significant expansion has been witnessed in the country's building and construction industry leading to a situation in which the demand levels of rock aggregates far outweigh the present supply capacity. It is essential, therefore, that more quarriable rock outcrops or natural gravel deposits be located and their geological and engineering characteristics properly evaluated in addition to updating information on the existing quarries.

Granitic rocks usually provide the main sources of the country's natural materials for engineering construction including the use of crushed rock aggregates for concrete and road works. Despite their economic importance, however, there seems to be no comprehensive and systematic study carried out on the engineering geology of these geological formations in this country. Project-specific studies are scanty with the results not always readily generally available to the industry. Moreover, much more emphasis is sometimes placed on the mechanical aspects of the study programmes at the expense of assessing the influence of geological factors. In quarrying, for instance, it is not always appreciated that 'pockets' of mica schist could be associated with the granitic rocks, and that a high incidence of jointing could result in high degrees of rock weathering thus rendering the winning of such materials uneconomic. An engineering geological plan of the quarry faces, delineating zones of such unsuitable materials, would maximise the effectiveness and economy of the quarry operations. Additionally, a detailed petrographic study of the quarried rock materials may facilitate a more meaningful interpretation and enhance the reliability of the mechanical test results.

A somewhat comprehensive coverage of the sources and evaluation of the country's rock aggregates for the use of the Ghana Highway Authority (GHA) and all other agencies connected with the design, construction, and maintenance of roads in the country has been provided by the Building and Road Research Institute, BRRI, (Anon. 1987a) following a 10-year World Bank-sponsored Highway Materials Research Programme. A significant achievement of the project is an attempt to develop an engineering geological framework for the selection and use of Ghanaian rock aggregates for building and road construction purposes.

The present research seeks to provide an in-depth study of the engineering geology of Ghanaian granitic rock aggregates. The main objectives are:
1. to evaluate the general suitability of the rocks for use as concrete and road aggregates, and
2. to correlate selected aggregate properties so as to provide an easy and quick means of evolving relevant design parameters.

2 METHODOLOGY ADOPTED IN THE STUDY

The study was conducted in three distinct but interrelated phases: a critical survey of the available literature was followed by field work and laboratory testing and analysis.

2.1 Literature survey

A review was made of the relevant literature on aspects of the geological and engineering properties of granitic rocks within the country and elsewhere in West Africa. Offices and libraries of various organisations connected with the building and construction industry were consulted for such data.

2.2 Field work

This was carried out at selected sites of granitic rock outcrops such as working quarries. These quarries which have modern plant and equipment are mostly established by such big domestic and foreign enterprises as A. Kassardjian Limited, Allegemeine Bau Union (ABU), Bank for Housing and Construction (BHC), Construction Pioneers (CP), Ghana Highway Authority (GHA), Kas Products Limited, Limex-Bau/ABK, Regional Development Corporations, Social Security Bank (SSB), and State Construction Corporation (SCC) either as a commercial venture or for their own use on various building and road construction projects. The quarries are, therefore, usually in the immediate vicinity of urban centres or trunk roads so as to offset high haulage costs. The quarries studied, with materials embracing the four main granitic formations in the country (see Section 3), are given in Table 1. Their locations are shown in Fig. 1.

The field work mainly involved sample collection, material identification, outcrop geometry and weathering grade determination, discontinuity measurement, and evaluation of the quarrying technique. The relevant sections of Anon. (1981) were adopted for the rock mass characterisation.

2.3 Laboratory testing

The laboratory tests performed on the rock materials from the respective quarries were:
1. a petrographic analysis of rock thin sections by using an optical microscopic technique, and
2. the determination of the relevant physical and mechanical rock aggregate properties.

Table 1. Quarries Studied

Name of Quarry	Location	Main Rock Type(s)
Q1 ABU Quarry	Yawkwei, km 202 Accra – Kumasi motor road	Granite, granodiorite
Q2 Buoho Stone Quarry (KAS Products Ltd)	Buoho, km 10 Kumasi – Offinso motor road	Granite, granodiorite
Q3 GHA Quarry	Adoagyiri – Nsawam, km 36 Accra – Kumasi motor road	Migmatite
Q4 Kassardjian Quarry	Ntesere, km 17 Kumasi – Sunyani motor road	Granite, granodiorite
Q5 SCC Quarry	Near Esupon Beach, 5km East of Sekondi	Gneiss
Q6 Shai Hills Quarry	Shai Hills, km 25 Tema – Akosombo motor road	Gneiss
Q7 Upper Quarry	Pwalugu, km 150 Tamale – Bolgatanga motor road	Granite, granodiorite

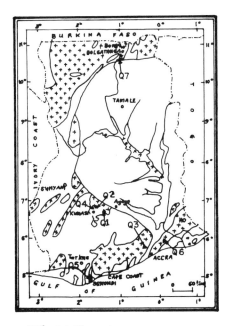

+ + Granitic rocks

Q1 ABU Quarry   Q5 SCC Quarry
Q2 Buoho Stone Quarry Q6 Shai Hills Quarry
Q3 GHA Quarry   Q7 Upper Quarry
Q4 Kassardjian Quarry

Fig. 1. Granitic rock areas of Ghana with locations of quarries studied (modified from Kesse 1985)

3 GEOLOGICAL ASSESSMENT

3.1 Rock outcrops

Outcrops of granitic rocks are widespread in Ghana, covering about 25% of the total land surface (Fig. 1). These rocks were believed to have been formed from post Birimian igneous and metamorphic activities during the Precambrian era (Kesse, 1985). They are usually grouped into Cape Coast granitoids, Dixcove granitoids, and Bongo granitoids, to which may be added a fourth group of Dahomeyan gneisses. This classification is of little significance for engineering purposes since all these rocks could be designated as 'granite' following the scheme of engineering group classification of the British Standards Institution, BSI, (Anon. 1975). However, distinctions made on the basis of their petrological peculiarities are deemed relevant to the interpretation of engineering test results.

Rock types include adamellites, aplites, gneisses, granites, granodiorites, migmatites, pegmatites, syenites, and tonalites which show a feldspar-quartz-mica-hornblende mineralogy and an interlocking fabric characteristic of true granites. The aplites and pegmatites usually occur as veins. Typical percentages of the main constituent minerals, as revealed by an optical microscopic modal analysis on several rock thin sections, are:

Feldspar   45 – 65%
Quartz   15 – 35%

Mica	5 – 30%
Hornblende	up to 20%
Accessories (garnet, magnetite, pyrite, etc.)	3 – 8%

Straining of quartz is observable in some rock types such as the Dahomeyan gneisses.

The rocks vary in grain size from fine to coarse and their colour is light grey, greenish grey, or pink. They may be massive, porphyritic or foliated. There are two major joints which are tight and widely spaced with average orientations of 298/70° NE and 060/25° SE respectively. However, there are certain portions of the rock outcrops which are closely jointed and highly weathered. These are believed to be fault zones. Materials from these zones and those from areas of high concentration of mica schist and of veins of aplite, pegmatite and quartz are particularly unsuitable for quarrying purposes. It is also observed that some of the outcrops are relatively too flat for the development of a quarry face while others are highly inaccessible on account of being remote from a road or exposed in a valley. In some cases, the outcrop frequency is so low to merit quarrying; rocks usually occur as scattered boulders separated by extremely weathered material.

3.2 Quarrying

The rock outcrops suitable for quarrying are dome-shaped, usually occurring as fairly flat-topped hillocks or inselbergs with little or no overburden. The quarries are generally sited at one flank of the outcrop and worked progressively towards the central portion so as to avoid, as much as possible, the winning of low quality moderately to highly weathered materials at the outcrop flanks. Volume of quarriable rock at most of the sites is generally between 2 and 5 million m³.

The economics of quarrying is enhanced by the outcrop geometry, virtual absence of overburden, widely spaced tight jointing with favourable orientation, absence of seeping water, low outcrop heights and relatively short haul distances. In most of these quarries, a blast pattern of burden-to-spacing ratio of 1:1 is adopted for an effective excavation of the outcrops, and the dual problems of flying rock fragments and ground vibrations could be eliminated or reduced by respectively inclining the drillholes at a small angle (up to 15°) towards the working face and using a series blasting technique. After blasting, the very large blocks are

further reduced by secondary blasting or by the use of large drop hammers to sizes which can be accommodated by the crusher jaws. Five different aggregate sizes (that is, 25mm, 19mm, 14mm, 10mm, and quarry dust) are normally produced after crushing and screening. The current selling price of the quarry products is up to ₵7,500 (about US $25) per m³.

4 ENGINEERING TESTS

4.1 Initial considerations

The quality of rock aggregates largely dictates their potential use for engineering construction purposes. In the course of any comprehensive evaluation of rock material quality, both geological investigations and engineering tests are deemed necessary. The engineering tests involve the determination of physical properties and mechanical behaviour of the rock aggregates. Physical properties, which are obtained without impairing the structural integrity of the aggregates, include shape and surface characteristics, relative density, and water absorption capabilities. Mechanical properties, on the other hand, are determined from the interactive effects of an applied load thus resulting in the destruction of the materials.

For crushed rocks, the two commonly used shape parameters as recommended in BS 812 (Anon. 1975) are elongation index and flakiness index. The test procedures for relative density and water absorption are also outlined in this standard.

It may be noted that strength tests such as uniaxial compression and point loading are grossly ineffective in assessing the engineering performance of rocks as roadstone and concrete aggregates because rock aggregate strength is essentially a measure of the resistance to pulverization or granulation. Nevertheless, serious attempts have been made by various institutions to devise certain tests that can reliably give a quantitative measure of the response of rock aggregates to mechanical force fields that may be operative under specific field conditions. The BSI, in particular, has recommended the aggregate crushing value (ACV) and aggregate impact value (AIV) as standard tests for assessing aggregate strength. The test procedures are detailed in Part 3 of BS 812 (Anon. 1975).

For weaker rock materials, the ACV and AIV were found to be relatively insensitive to strength variations because the materials become easily compacted at low applied loads thus suppressing further granulation at higher loads. In an attempt to overcome these shortcomings, the 10% fines load and modified AIV tests have been suggested. The 10% fines load is related to AIV (Anon. 1975) and ACV (Turk and Dearman 1989) by the following equation:

$$10\% \text{ Fines Load} = 4000 \times (\text{AIV or ACV})^{-1} \text{kN}.$$

The modified AIV test is very similar to the standard test, the only difference being that the number of hammer blows (normally 15) is limited to a number which will produce between 5 and 20 percent fines; the modified AIV is then obtained by multiplying the percentage fines by 15 and dividing the product by the number of blows (Hosking and Tubey 1969).

Another important mechanical property of aggregates, besides strength, is durability which includes both resistance to wear (or abrasion) and resistance to weathering or soundness. Abrasion resistance is particularly important when evaluating rock aggregates for use in heavily trafficked road surfacings or in concrete floors subjected to wear. Standard tests include the Los Angeles abrasion test of the American Society for Testing and Materials (ASTM Designation C131 - 76 (Anon. 1979)) and the Dorry abrasion test of BSI (Anon. 1975). The soundness test simulates the amount of mechanical weathering that rock aggregates may experience while in service - an aspect that normal mechanical tests may not be able to predict. The test is performed by subjecting the aggregates to a series of 24-hour cycles of immersion in a saturated solution of sodium or magnesium sulphate and oven-drying. Crystal growth in the voids within the aggregates exerts high pressures which cause disintegration. The test method is detailed in ASTM Designation C88-76 (Anon. 1979).

The retention of a bituminous film by an aggregate in the presence of water is determined by a stripping test according to the standard specification ASTM Designation 1664 (Anon. 1979).

4.2 Test Data

Results of engineering properties that appear to be mostly adopted for the aggregate specification requirements of the building and construction industry in Ghana are summarised in Table 2. These have been obtained following a critical literature search and tests performed on laboratory-crushed or crusher-run standard size aggregates in accordance with procedures specified in BS 812 (Anon. 1975) and the relevant ASTM Standards (Anon. 1979).

In the course of the literature search, it was necessary to exercise some judgement in assessing the available data in order to achieve some degree of reliability in the results presented in Table 2. Paramount among these was the rejection of results for rocks which were suspected to have undergone more than Grade II (that is, slight) weathering and also of results for which the test procedures were not specified.

5 SUITABILITY OF AGGREGATES FOR ENGINEERING PURPOSES

The suitability of rock aggregates for specific engineering construction works is more vital than their mere availability. Analysis of data presented in the previous sections indicates that the granitic rock aggregates satisfy the basic geological and engineering criteria for use as construction materials, and they are indeed being used on a variety of building and road construction projects within the country. For example, KAS Products Limited and Kassardjian Quarry provide over 80% of the crushed rock aggregates for building and road construction purposes in and around Kumasi; GHA Quarry at Nsawam is the main source of aggregates for the Nsawam-Koforidua area; Shai Hills Quarry provides the bulk of aggregates needed within Accra, Tema, Akosombo and Ho areas; SCC Quarry at Sekondi satisfies the aggregate requirements of most parts of the Central and Western Regions; and Upper Quarry is the sole supplier of aggregates within the Tamale-Bolgatanga area.

For road construction, the aggregates possess the requisite strength to withstand impact and crushing from the vehicle wheel loading as well as durability to resist abrasion and in-service weathering effects. They are, therefore, usually employed in both surface dressing and manufacture of bituminous mixtures, and occasionally as crushed rock base material. The mean aggregate impact and crushing values are 20% and 25% respectively, which are well within the specified tolerance limits and also compare very well with values obtained for granitic rocks elsewhere (see Ramsay 1967, Turk and Dearman 1988). Water absorption is generally low (0.5%) and the mean relative density is 2.68. These rocks are also found to be durable with mean abrasion value of 28% and sulphate soundness value of less than 1%. Their bitumen stripping value is generally less than 5% as a result of favourable shape and surface characteristics. The aggregates have fairly rough surfaces and their shape indices, elongation index and

Table 2: Summary of Results of Engineering Properties

Parameter	Unit	Mean Values (and Standard Deviation)	ASTM and BSI recommended value for good material
Relative Density (RD)	–	2.68 (0.10)	–
Water Absorption (WA)	%	0.5 (0.2)	≤ 1.0
Elongation Index (EI)	%	28 (9)	≤ 35
Flakiness Index (FI)	%	22 (6)	≤ 35
Aggregate Crushing Value (ACV)	%	25 (3)	≤ 30
Aggregate Impact Value (AIV)	%	20 (4)	≤ 30
Los Angeles Abrasion Value (LAAV)	%	28 (5)	≤ 40
Sulphate Soundness Value (SSV)	%	0.18 (0.06)	–
Bitumen Stripping Value (BSV)	%	2.6 (0.3)	≤ 5.0

flakiness index, are within tolerable limits averaging 28% and 22% respectively thus suggesting a generally cuboid shape. These indices are partially characteristic of the rock type but are also influenced by the crushing methods during aggregate production such as the nature of the crusher jaws and the reduction ratio. Even though no polishing stone value determinations were carried out in this study or reported in the literature, petrographic evidence coupled with experience indicates that these aggregates are resistant to polishing under traffic loading. Their light colour also offers night visibility especially when used for bituminous surface dressing.

Despite all these good qualities of Ghanaian crushed granitic rock aggregates, the roads engineer sometimes faces the problem of their inability to perform well while in service. There have been instances of failures on the country's roads and highways, the main failure mechanisms being stripping and pothole development, among others, which can be temporarily remedied by local patching but it becomes necessary in some cases to rehabilitate entire sections of the road. These failures may sometimes be attributed to poor performance of the wearing course resulting from excessive crushing of the rock aggregates under traffic loading or improper adhesion with bitumen. The rather poor aggregate performance that occasionally results can be explained by the fact that the variable nature of rock types in the majority of the quarries can give rise to the winning of a low quality material at a particular time. For instance, should there exist a high percentage of aplite, pegmatite, and vein quartz in the quarry product problems of loose bonding would be anticipated because of the hydrophilic nature of these materials. Also, excessive crushing may result when pegmatites or moderately weathered rocks predominate. Routine engineering geological investigations are, therefore, necessary to facilitate selective winning of rock in existing quarries.

As noted earlier, the crushed granitic rock aggregates covered by this study satisfy the basic requirements for use in concrete works; however, it is not certain if they always perform satisfactorily while in service. Indeed, there have been a few unexplained failures of concrete structures (bridges and houses) in the country. One of the most significant causes of deterioration of concrete structures all over the world, according to Swamy and Al-Asali (1988), is alkali-aggregate reaction especially alkali-silica reaction (ASR). Damage from ASR can occur when the hydroxyl ions present in the pore solution of a concrete react with

certain forms of silica (opal, flint, chal-
cedony, and highly strained quartz) in the
aggregate to form a gel which absorbs more
pore fluid, swells, and exerts a pressure
which can crack the concrete (Anon. 1987b).
The reaction can take place over a period
of time between a few months and several
years (Barisone 1984) and it is affected by
several factors such as the nature and
amount of reactive silica, amount of available alkali, and prevailing moisture conditions (Anon. 1987b , Swamy and Al-Asali
1988). ASR is commonly assessed in terms
of ASTM's chemical or mortar-bar methods
(Anon. 1979) and/or petrographic examination (Anon. 1987b).

Even though this phenomenon has for long
gained a worldwide recognition, it has been
known in Ghana as a mere bibliographic reference until recently when the author in
1988 investigated the potential ASR of aggregates from one of the major producing areas
- the Shai Hills Quarry. It was revealed
through petrographic studies that these
aggregates contain up to 30% of highly
strained quartz, but the risk of ASR could
be considered minimal on the basis of the
reasoning (Anon. 1987b) that damaging ASR is
possible only if the following three conditions of the existence of sufficient moisture in the concrete, a sufficiently high
alkalinity, and a critical amount of reactive silica in the aggregate are simultaneously satisfied. Nevertheless, this
evidence is not conclusive enough to debar
any possibilities of ASR taking place in
concrete structures within the country. A
more thorough and systematic research is,
therefore, being undertaken by the author
to specifically evaluate the ASR potential
of all the country's rock aggregates which,
hopefully, will provide a more satisfactory
opinion about this aspect of concrete deterioration.

6 CORRELATION BETWEEN SELECTED AGGREGATE
  PROPERTIES

Plots of selected engineering properties
such as WA, EI, FI, ACV, and LAAV have each
been made against AIV in an attempt to formulate an easy and quick estimation of
aggregate parameters. The choice of AIV as
the dependent variable is explained by the
fact that the test is simple and quick to
perform and also that it is a reliable
design parameter both in Ghana and elsewhere. Correlation coefficients and best
fit curves, where necessary, have been
computed.

There is no clearly defined relationship
between water absorption and aggregate

impact value within the rather narrow range
of water absorption values (0.2 - 0.9%) considered in the present study. This could
tend to suggest that water absorption is not
a useful parameter to quantify the mechanical
behaviour of Ghanaian crushed granitic rock
aggregates. Adoption of the relationship
established by Leverett (1970) between water
absorption and AIV for various British rock
aggregates including those from granitic
sources, for example, is not likely to give
any meaningful evaluation of the Ghanaian
granitic rocks and, hence, its use is not
recommended.

The shape factors (EI and FI), however,
seem to correlate somehow with AIV despite
the low coefficients of correlation. A plot
between FI and AIV is shown in Fig. 2. As
noted by Ramsay (1965) and Dhir et al (1971),
AIV is more sensitive to variations in the
proportion of flaky particles than of elon-
gate particles present in rock aggregates
and the flaky particles suffer most under
impact. Ramsay (1965), therefore, suggested
that a reasonable prediction of AIV can be
made for crushed granitic rock aggregates
from the proportion of flakes present, that
is, a value of 1 for every 10 units of flakiness index be added to 16 to achieve the
desired AIV - the value 16 representing the
petrographic character of granitic rocks.
A similar relation has been evolved in the
present study.

Strong Linear relationships have been
established between ACV and AIV (Fig. 3) and
between LAAV and AIV (Fig. 4). The sugges-

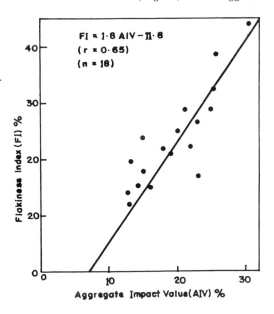

Fig. 2 Plot of FI versus AIV

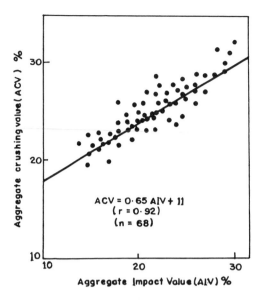

Fig. 3 Plot of ACV versus AIV

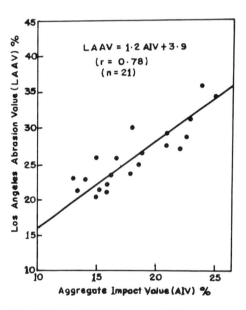

Fig. 4 Plot of LAAV versus AIV

tion, therefore, is that both ACV and LAAV can be estimated from a knowledge of the AIV. For the Ghanaian granitic rocks, the ACV and LAAV are respectively about 5 and 8 units greater than AIV. It is, however, worthy of note that similar studies elsewhere have shown ACV and AIV to be nearly the same (see Turk and Dearman 1988).

7 CONCLUDING REMARKS

The results of the study embodied in this paper indicate that granitic rocks form the bulk of Ghana's crushed rock aggregates which, by international standards, satisfy the basic requirements for use in both concrete and road works. However, their full exploitation to meet the growing demands is limited because of certain problems which are largely engineering geological and technical in nature. The engineering geological problems, in particular, could be solved by conducting regular checks on the quality of material being quarried - a sort of control measure of its general suitability for the intended engineering purpose. There is a need to specifically evaluate the potential alkali-silica reactivity of the country's rocks with a view to providing a more satisfactory opinion about this aspect of concrete deterioration which has hitherto been insufficiently addressed.

In an attempt to provide a cost-effective approach to the selection of design parameters, mathematical relations have been established between certain aggregate properties. It is recommended that the crushed rock aggregate specifications for road works, in particular, be reviewed to take account of the recent level of traffic increase within the country which has resulted in the gross underdesign of some roads and highways. A limiting AIV of 20% (corresponding to ACV of 25% and LAAV of 28%), for example, seems more reliable than the 30% being used presently.

REFERENCES

Anon. (1975). Methods for sampling and testing of mineral aggregates, sands, and fillers. Parts 1,2,&3, 60pp. British Standards Institution, BS 812, London.
Anon. (1979). Annual Book of ASTM Standards. Parts 14 & 15. Philadelphia, USA.
Anon. (1981). Rock and soil description for engineering geological mapping. Report by the Commission on Engineering Geological Mapping. Bull. Int. Assoc. Eng. Geol., 24: 235-274.
Anon. (1987a). Engineering geology and highway geotechnics of lateritic and saprolitic soils. Ghana Highway Authority Technical Paper HM/1,2,4,5. 89pp. Building and Road Res. Inst., Kumasi, Ghana.
Anon. (1987b). Alkali-silica reaction in minimising the risk of damage to concrete. Guide notes and model specification clauses, 34pp. Concrete Soc. Tech. Report 30.
Barisone, G. (1984). Petrographic analysis of aggregates related to alkali-silica

reaction. Proc. Int. Symp. Aggregates,
Nice (France). Bull. Int. Assoc. Eng.
Geol. 30: 177–181.

Dhir, R.K., Ramsay, D.M. & Balfour, N.
(1971). A study of the aggregate impact
and crushing value tests. J. Inst. Highway
Engrs., 17 – 27.

Hosking, J.R. & Tubey, L.W. (1969). Research
on low-grade and unsound aggregates, 30pp.
TRRL Report LR 293.

Kesse, G.O. (1985). The rock and mineral
resources of Ghana, 610pp. Balkema,
Rotterdam.

Leverett, I.J. (1970). Evaluation of rock
for use as aggregate and roadstone. Quarry
Manager's J., 197 – 204.

Ramsay, D.M. (1965). Factors influencing
aggregate impact value in rock aggregate.
Quarry Manager's J., 129 – 134.

Swamy, R.N. & Al-Asali, M.M. (1988). Alkali-
silica reaction – Sources of damage.
J. Inst. Highways and Transportation,
25 – 29.

Turk, N. & Dearman, W.R. (1988). An investigation into the influence of size on the
mechanical properties of aggregates. Bull.
Int. Assoc. Eng. Geol. 38: 143 – 149.

Turk, N. & Dearman, W.R. (1989). An investigation of the relation between ten percent
fines load and crushing value tests of
aggregates (UK). Bull. Int. Assoc. Eng.
Geol. 39: 145 – 154.

7.3 Testing and classification
Essais et classification

Effect of void characteristics and parent rock source on the strength of laterite gravel aggregates

L'effet du comportement des vides et de la source de la roche mère sur la résistance des agrégats de gravier de latérite

Enuvie G. Akpokodje & Peter P. Hudec
University of Windsor, Windsor, Canada

ABSTRACT: Few attempts have been made to quantify the relationship between porosity and mechanical strength of concretionary laterite gravels. A series of laboratory tests were performed on commonly used Nigerian laterite gravel aggregates to develop an empirical relationship for predicting strength of laterite gravels from porosity indexes.

The results show that the aggregate impact values of laterite gravels derived from similar parent materials, have significant linear correlations with water absorption, vacuum saturation, bulk dry density and specific gravity. No significant correlation exists between adsorption and aggregate impact value. This appears to indicate that macropores exert greater influence on the strength of concretionary laterite gravels than micropores.

An empirical relationship is presented for estimating the strength of laterite gravels from vacuum saturation, bulk dry density and specific gravity.

RESUME: Peu de tentnative ont été faites jusqu'ici pur quantifier le rapport entre la porosité et la résistance mécanique des graviers de laterité concrétionnés. On a donc fait une suite de tests au laboratoire sur des agrégats nigéruebs de graviér de latérite, d'usage courant, afin de développer un rapport empirique perrmettant de prédire la résistance des graviers de latérite à partir des indices de porosité.

Les résultats démontrent que les valeurs de choc des graviers de latérite dérivés de roches mères son en corrélation linéaire avec l'absorption de l'eau, la saturation sous'vide, la densité du matériau sec en vrac et les poids spécifique. Il n'existe pas de corrélation importante entre l'adsorption et la valeur de choc des agrégats. Ce fait semble indiquer que les macropores exercent une plus grande influence que les micropores sur la résistance des graviers de latérite concrétionnés.

On présente donc un rapport empirique pour l'estimation de la résistance des graviers de latérite à partir de la saturation sous vide, de la densité du matériau sec en vrac et du poids spécifique.

1 INTRODUCTION

Concretionary laterite gravels (or pisoliths) are used extensively in the tropics as coarse aggregates for concrete and black-top (bituminous) pavement constructions. They occur abundantly in all geologic terrains including those underlain by soft sedimentary rocks. They are cheaper to procure when hard rock aggregates are to be hauled from long distances. Field observations reveal that there are two major problems associated with the use of concretionary latrite gravels in road construction: (a) mechanical degradation during compaction and under repeated traffic loading and (b) high strength loss on wetting or saturation. These problems are generally attributed to low mechanical strength which is in turn mainly caused by relatively high porosity. However, only limited attempts have been made to quantify the relationship between void characteristics and strength of indurated residual materials.

Studies by De Graft-Johnson et. al.(1972) showed that water absorption and specific gravity are among the index properties which significantly correlate with strength of laterite gravels. However, laterites which are not genetically related, do not have any significant correlation (USAID,1971). Yaughan et. al. (1988) concluded from their study on residual soils that their engineering behaviour depends on the in-situ void ratio of the soil relative to that which it can achieve in the "de-structured" state. Results of studies on rock aggregates (Hudec,1980,1982) reveal that void characteristics and aggregate evaluation can be estimated from the results of simple tests such as water adsorption, absorption and vacuum saturation, bulk density and specific gravity. Once an empirical relationship is established between these properties and strength, then the latter can be accurately estimated from the results of simple and inexpensive tests. This is very important in the Third World tropical countries where both funds and facilities to carry out sophisticated aggregate evaluation tests are not always available.

This paper reports the results of a study carried out to determine the nature of the relationship between the void characteristics and strength of concretionary laterite gravels. This will enable an accurate prediction of the engineering behaviour and probable field performance of concretionary laterite gravels of various geological origins and weathering grades.

2 TEST SAMPLES AND PROCEDURES

Samples of concretionary laterite gravels (pisoliths) used for the study were collected from thirty-nine selected laterite profiles across Nigeria. These profiles are developed on three major rock types; granite/gneiss complexes (Precambrian), shales, and sandstones (Cretaceous-Tertiary). The shale units contain subordinate layers of sandstone and the reverse is also true for the sandstone units. The test samples were collected from laterite profiles exposed at the face of both abandoned and active borrow pits. Most of the borrow pits are sited at the middle/upper slopes of low residual hills.

The laboratory tests carried out included water adsorption, absorption and vacuum saturation, dry bulk density, specific gravity, abrasion and aggregate impact value (AIV). The samples were first washed thoroughly to remove all adhering fines and the fraction passing 16mm sieve and retained on 9.6mm sieve was used in all tests. Adsorption was determined at

approximately 95% relative humidity, using a humidity chamber saturated with $CuSO_4$ solution at 22 °C. Normal absorption was calculated from mass difference between oven dried and surface dry samples after 24hr. soaking. Vacuum saturation was determined in a similar manner but the oven dried samples were first boiled (totally submerged) for 30 minutes before 24hr. soaking. Both specific gravity of the sample powder and the bulk dry density were determined by water displacement method. Wide mouth density bottle was used for bulk dry density determination and the samples were left submerged for 24hr. before measuring the submerged weights. The normal absorption values were combined with the measured weights (oven dried and submerged weights) to calculate the actual bulk dry density. The abrasion values were obtained using the modified abrasion test procedures described in details by Hudec (1982). Basically, the test uses a modified Franklin Slaking durability apparatus, altered to provide a shelf which would lift and dump the sample during a single revolution. Steel balls approximately equivalent to the mass of the sample were inserted to act as an abrasive charge. The aggregate impact test was carried out following BS 812 (1975) procedures and using the standard equipment.

3 RESULTS AND DISCUSSIONS

The means and standard deviations of test results of the three groups of concretionary laterite gravels are presented in Table 1. Statistically significant differences (at 99% confidence level) exist between the means of laterite gravels derived from the Basement Complex and shales/sandstones (table 2). This indicates the significant influence of parent material on the physical and mechanical properties of concretionary laterite gravels. Lack of significant difference between the means of the shale and sandstone derived materials indicates similarity in genesis. This is evident from the intimate stratigraphic and lithologic relation (both units contain subsidiary interbeddings of each other) between these two groups.

The lowest mean values of strength (AIV), bulk dry density, and specific gravity occur in the samples developed on the Basement Complex rocks. This implies that concretionary laterite gravels developed from granite/gneiss complexes in Nigeria tend to be inferior in quality when compared to those formed from sandstones and shales. It should be emphasized that this generalization assumes that the samples tested are representative of these groups of rocks. The high values of absorption, vacuum saturation and apparent porosity exhibited by the shale derived gravels represent large volume of void spaces.

Table 1. Summary of means and standard deviations of test results; sorted according to parent material group.

	MEANS			STAND. DEVIATION		
TESTS	BC	SH	SS	BC	SH	SS
N	19	10	10	19	10	10
Ads	1.70	2.08	1.49	0.38	0.55	0.35
Abs	7.34	9.15	7.34	2.08	1.94	1.81
Vst	9.72	10.62	9.24	2.49	2.26	1.78
Sgy	3.15	3.40	3.16	0.20	0.12	0.26
Bdd	2.33	2.45	2.43	0.16	0.18	0.29
Apo	26.20	28.00	23.40	3.10	3.40	3.20
AIV	41.52	38.41	38.52	5.08	6.02	11.04
Abr	17.29	14.68	12.93	6.06	5.82	6.63
Ads Vol	0.17	0.20	0.16	0.06	0.06	0.03

BC=Basement Complex; SH=shale; SS=sandstone; Ads=Adsorption; Abs=Absorption; Vst=Vacuum saturation; Sgy=Specific gravity; Bdd=Bulk dry density; Apo= Apparent porosity; AIV=Aggregate Impact Value; Abr=Abrasion; AdsVol=ratio of adsorption to vacuum saturation.

Table 2. Statistical significance of means of test results.

Test Significance of means(>.99)of test results between parent material groups

	BC-SH	BC-SS	SH-SS
Ads	Y	Y	Y
Abs	Y	N	Y
Vst	Y	N	Y
Sgy	Y	Y	N
Bdd	Y	Y	N
Apo	Y	Y	N
AIV	Y	Y	N
Abr	Y	Y	N

Y=statistical significant difference between means.
N=no statistical significance difference between means.

However, this appears to be inconsistent with their high strength and bulk dry density values. These latter two properties seem to be more influenced by specific gravity, which is highest in the shale derived gravels. High specific gravity of the sample powder indicates higher concentration of sesquioxides which has been shown (Millard,1962; Acroyd,1963,1967; De

Graft-Johnson et.al.,1972) to enhance strength of laterite gravels. The relatively higher mean adsorption values of the shale derived gravels is due to larger surface area of micropores which could in turn, be attributed to the fine texture of the parent material.

The linear correlation matrix performed on the test results is presented in table 3. The

Table 3   Linear correlation matrix

	Ads	Abs	Vst	Sgy	Bdd	Apo	AIV
(Basement Complex)							
Abs	.62	1					
Vst	.48	.92	1				
Sgy				1			
Bdd	-.30	-.44	-.59	.81	1		
Apo	.44	.74	.76		-.54	1	
AIV	.71	.86	-.72	-.70	.68	1	
Abr	.75	.70	-.48	-.78	.56	.80	
(Shale)							
Abs	1						
Vst	.97	1					
Sgy	-.72	-.66	1				
Bdd	-.81	-.78	.82	1			
Apo		.98			1		
AIV	.98	.97	-.66	-.82		1	
Abr	.82	.80	.65	-.67		.78	
(Sandstone)							
Abs	1						
Vst	.90	1					
Sgy	-.68	-.73	1				
Bdd	-.68	-.73	.99	1			
Apo	.74		-.69	-.85	1		
AIV	.90	.94	-.88	-.82	.64	1	
Abr	.87	.94	-.79	-.79	.72	.91	

correlation coefficients at 95% confidence level, are 0.389 for the Basement Complex gravels and 0.550 both for the shale and sandstone derived materials. Non-significant correlation coefficients are omitted. The correlation between adsorption and the other properties is generally not statistically significant. This implies that micropores of laterite gravels have minimal effect on the other properties. This could be attributed to the relatively low proportion (16.3-20.2%) of micropores. Van Ganse (1957) reported similar low proportions of micropores in laterite pisoliths. As can be expected, there are very high correlations between the pairs of absorption-vacuum saturation and bulk dry density-apparent porosity. These tests can therefore be substituted for each other in material specification for laterite gravels.

Table 4 presents the linear regression equations between strength (AIV) as the dependent variable and the other void characteristic indexes as independent variables. The slopes (a) of the linear regression equations table 4 reveals that unit changes in the in the values of the

Table 4. Linear regression equations between strength and the other properties measured; sorted according to parent material groups.

(Y)	(b)	(X)	(a)	(R)
(Basement Complex)				
AIV = [1.73	*	Abs +	28.80]	0.71
= [1.85	*	Vst +	23.33]	0.86
= [-24.24	*	Sgy +	97.88]	- 0.72
= [-95.76	*	Bdd +	16.45]	- 0.70
= [95.76	*	Apo +	16.45]	0.68
= [0.70	*	Abr +	29.96]	0.80
(Shale)				
AIV = [3.98	*	Abs +	1.97]	0.98
= [3.43	*	Vst +	2.04]	0.97
= [- 6.74	*	Sgy +	13.86]	- 0.66
= [-14.50	*	Bdd +	2.83]	- 0.82
= [145.12	*	Apo +	4.40]	0.52
= [0.81	*	Aba +	26.55]	0.78
(Sandstone)				
AIV = [5.50	*	Abs -	1.87]	0.90
= [5.85	*	Vst -	15.53]	0.94
= [-32.89	*	Sgy +	118.52]	-0.88
= [-32.89	*	Bdd +	118.52]	-0.82
= [182.27	*	Apo -	4.20]	0.64
= [1.51	*	Abr +	18.98]	0.91

Y=Dependent Variable; b=slope;
X=Independent Variable; a=Intercept on Y;
R=Correlation coefficient.

independent variables results in greatest strength change in laterite gravels developed from sandstones. Those derived from the Basement Complex exhibit the least strength changes. The explanation for this trend is not apparent from the results of this present study. However, it could be related to their microstructure and mineralogy.

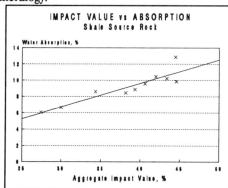

Figure 1 Relationship of water absorption of shale-derived laterites on Aggregate Impact Value

Some of the relationships are shown graphically in figures 1 to 5. Figure 1 shows the general non-significant relation between adsorption and strength. Figures 2 to 4 show the reduction in

strength (higher AIV) with increase in void spaces as indicated by higher values of absorption, vacuum saturation and apparent porosity. Figure 4 shows the negative correlation between specific gravity and aggregate impact value. Higher values of the former corresponds to greater amount of sesquioxides available for cementation which results in higher strength (lower AIV).

In developing an empirical relation between strength and the void characteristics of laterite gravels, adsorption can be ignored. Two independent variables are selected from the pairs of absorption or vacuum saturation and bulk dry density or apparent porosity. The two chosen

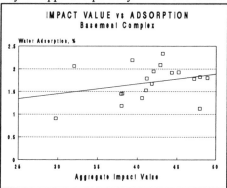

Figure 2 Effect of Adsorption on Impact Value

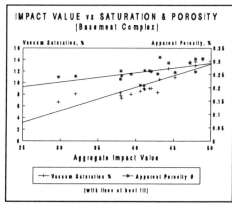

Figure 3 The effect of water saturation and porosity on the aggregate impact value.

properties are combined with specific gravity to constitute the three independent variables used in estimating the aggregate impact value. Multiple regression analysis performed using these three independent variables gives the following relations:

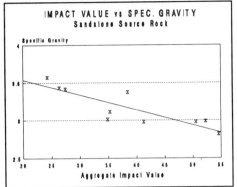

Figure 4 Relationship of Specific Gravity on the Aggregate Impact Value of a sandstone – derived laterite aggregate.

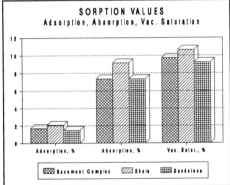

Figure 5 Effect of source rock on the sorption values of lateritic aggregates.

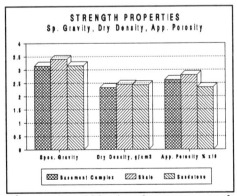

Figure 6 Effect of source rock on the porosity and density of lateritic aggregate.

BASEMENT COMPLEX LATERITES
Aggregate Impact Value =
[61.61 + 1.66Vst + - 8.6Sgy - 3.92Bdd]
R = .97
SHALE LATERITES
Aggregate Impact Value =
[141.25 + 1.34Vst - 21.41Sgy - 17.98Bdd]
R = .92
SANDSTONE LATERITES
Aggregate Impact Value =
[92.91 + 4.02Vst - 17.54Sgy - 14.86Bdd]
R = .87

Figures 5 and 6 give the means of the various properties measured classified according to the parent rock type. It can be seen that the source rock has a significant effect on these properties, and that the differences in the means follow a constant pattern for each set of the properties.

Figure 7 compares the means of experimental aggregate impact values with that calculated from the above equations. As can be seen, there is a very good agreement between the two. Statistical comparison of the means shows them to be the same - i.e., simple tests can be used to calculate the AIV (or the implied aggregate strength). The AIV equipment may not be available in many laboratories of the third world countries.

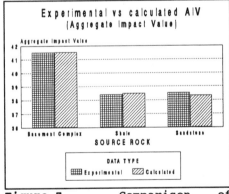

Figure 7 Comparison of experimental and calculated Aggregate Impact Values.

4 CONCLUSIONS

Voids in concretionary laterite gravels comprise mainly large pores which have greater influence on their strength than micropores. A unique and quantifiable relationship exits between void characteristics indexes and the strength of genetically related laterite gravels. Under broadly similar environmental and climatic conditions, laterite gravels developed from shales/shaly sandstones tend to possess higher strength than those derived from granite/gneiss complexes. This is primarily attributed to higher contents of sesquioxides.

5 ACKNOWLEDGEMENTS

The Natural Sciences and Engineering Research Council of Canada sponsored this research through a one-year Research Associate award received by the senior author.

6 REFERENCES

Acroyd, L.W., 1963. The correlation between engineering and pedological classification systems in Western Nigeria and its application. Proc. 3rd. Reg. Conf. Afri. Soil Mech. Found. Eng. Salisbury. 1: 115-118

------ 1967. Formation and properties of concretionary and non-concretionary soils of Western Nigeria. Proc. 4th. Reg. Conf. Afri. Soil Mech. Found. Eng. Cape Town. 1 : 47-52

British Standards Institute 812-1975. Methods for sampling and testing of mineral aggregates (in four parts).

De Graft-Johnson, J.W.S., Bhtia, H.S. and Hammond, A.A. 1972. Laterite gravel evaluation for road construction. J. Soil Mech. Found. Div. Am. Soc. Civ. Eng. 98 (SM 11) : 1245-1265

Hudec, P.P. 1980. Durability of carbonate rocks as a function of their thermal expansion, water absorption and mineralogy. ASTM Special Tech. Publ. 691 : 497-508.

------- 1982. Aggregate tests- Their relationships and significance. Durability of Building Materials. 1 : 275-300.

Millard, R.S. 1962. Road building in the tropics. J. App. Chem. 12: 342-357.

United States Agency for International Development(USID). Laterite and lateritic soils and other problem soils of Africa. An Engineering study report, AID/CSD-2164; 290p.

Van Ganse, R., 1957. Proprietes et applications des laterites au Congo belge. Acad. R. Sci. Colon., Mem. 8, t., fasc. 1: 46p.

Vanghan, P. R., Maccarini, M. and Mokhtar, S. M., 1988. Indexing the engineering properties of residual soils. Q. J. Eng. Geol. 21: 69-84.

6th International IAEG Congres / 6ème Congrès International de AIGI, © 1990 Balkema, Rotterdam. ISBN 90 6191 130 3

Preliminary research on the building materials used at the Peucetii Settlement of Monte Sannace in Apulia, Southern Italy

Recherches préliminaires sur les matériaux de construction utilisés dans l'habitat Peucetius de Monte Sannace, Pouilles, Italie du Sud

G. Baldassarre
Dipartimento di Geologia e Geofisica, Università di Bari, Italy

A. De Marco
Dipartimento Geomineralogico, Università di Bari, Italy

ABSTRACT: An archeological site, consisting of the remains of an urban settlement, is located at Monte Sannace in central Apulia. Built on the site of a group of Iron Age huts dating back from the 9th to the 8th century BC, the town developed with varying prosperity until the end of the third century BC, when it was destroyed during the wars between Rome and Carthage. The site stands on a hill formed of Cretaceous limestones and the building materials used are limestones and calcarenites.
The results of a geological field study were compared with those obtained from a laboratory investigation on the minero-petrographic, and in part physical, characteristics of the stones used on the sites. It was concluded that the limestones used on the site are identical to those which form the hill, while the calcarenites used on the site come mainly from outcrops at Santo Mola (a few Km from Monte Sannace) and partly, it is likely, from calcarenitic outcrops further afield.
Regarding certain volcanic rocks found on the site, it can be stated that they were brought from an entirely different region.

RESUME: Le site archéologique de Monte Sannace (Pouille centrale) garde les traces d'un habitat urbain qui, tout en partant d'une agglomération de cabanes du Premier Age du Fer (IX-VIII siècle a.J.C.) se développe jusqu'au III siècle a.J.C., quand il est détruit lors des guerres entre Rome et Carthage. L'ensemble se dresse sur un relief de calcaires du Crétacé et ce qui reste des matériaux de construction utilisés ce sont des calcaires et des calcarenites.
Si on compare les résultats des recherches géologiques sur place et ceux des recherches de laboratoire sur les caractéristiques minéro-pétrographiques et phisiques aussi, faites sur les roches des ruines, on peut conclure que les calcaires sont les mêmes de la colline, et les calcarenites viennent pour la plus grande partie des affleurements de la localité Santo Mola (quelques Km de Monte Sannace) et vraisemblablement d'affleurements plus éloignés.
Quant aux fragments laviques rétrouvés dans le site, on peut observer que leur provénance est sûrement extra-régionale.

1 INTRODUCTION AND AIMS OF THE RESEARCH

The Japigii peoples of Apulia were divided into three geographically distinct groups: the Messapii of the Salento region, the Peucetii of the Murge hills and the Daunii who inhabitied the area north of the R. Ofanto (6). The most important settlement of the Peucetii according to the literature (2) was at Monte Sannace, located on a hilltop a few Km NE of Gioia del Colle, dominating strategically the main roads of communications of the central Murge.

Today abundant evidence remains that an urban settlement once flourished at Monte

Sannace, on the site of an older group of early Iron Age huts. The urban complex was completely destroyed at the end of the third century BC during the Carthaginian wars (2).

The ruins of the city, which have been the subject of systematic archeological study for several decades, represent one of the most important records of the life and civilization of the Peucetii. It emerges from the explorations of the city up to now, that it developed in two distinct zones: an acropolis much stood on the flat area at the top of the hill, and the town extended across the flat area at the foot of the hill on the west side, from which there was the easiest access to the acropolis. Having thus given some background information regarding the settlement, we can now turn to the purpose of the research: the identification and analysis of the different types of stone used in the building of the settlement and the discovery of their places of origin. It has been possible to ascertain that: 1) the materials used were limestones and various types of calcarenites; 2) contrary to what has been previously hypothesized not all of the calcarenites came from the Santo Mola area (about 6 Km from Monte Sannace). At Santo Mola there was another Peucetii settlement, whose economy was based on the extraction and working of calcarenitic stone (2).

In this paper we investigate and compare the minero-petrographic characteristics, and also in part the physical properties, of the limestones and calcarenites outcropping in the area with those used in the construction of the settlement. It should be born in mind, however, that the results obtained suggest the need for further investigation regarding some of the calcarenites, in order to identify where precisely they were excavated and to gain information on the exchange network between these Peucetii communities and those which had no calcarenite available locally.

2 GEOLOGICAL AND MORPHOLOGICAL OUTLINE

In order to identify the areas from which the building materials used on the Monte Sannace site came, we considered the geological setting of the immediate surroundings, as well as that of areas located further away. The geological make-up is represented by a litho-stratigraphic succession (4)(5) composed of stratified Cretaceous limestone Formations (Calcare di Bari and Calcare di Altamura) with overlaying transgressive Quartenary calcarenites (Fig. 1). The latter include "Calcarenite di Gravina", "Calcareniti di Monte Castiglione" and "Depositi Marini Terrazzati".

In several places the limestones have been fractured and have undergone Karst phenomena. They also have an overburden of "terra rossa" which varies in thickness and extension. The calcarenites, known locally as "tufi", considered as a group, are for the most part massive or very little stratified, and softer than the limestones, so they can be worked easily, even with rough tools.

From a morphological point of view, the archeological site consists of an acropolis at the top of a hill and a settlement at the foot of the hill in a more or less flat area. The remaining sides of the hill slope steeply. More generally, both the limestone and the calcarenite areas develop in steps of moderate height alternating with undulating flat areas of varying size, with numerous erosional troughs ("lame"). It should be noted at this point that following prolonged heavy rainfall, the surface water which dug out these erosional troughs (and continues to do so today) must certainly have constituted one of the main sources of the inhabitants water supply (1).

3 ON SITE INVESTIGATIONS

The building stones found at Monte Sannace are mostly limestones and a variety of calcarenites. For the latter it was necessary to extend sampling to areas at some distance from the site. The limestones were mainly used for the lowest part (probably the foundations) of the perimetral walls of both the houses and parts of the city wall, as well as for the floors and load-bearing structures. The size and workmanship of the blocks varies considerably, ranging from c.10x 12x25 cm to c.150x50x50 cm, and from roughly-hewn to well-squared. However, the most common are plate-shaped blocks of an average size of 18x7x25 cm.

The following calcarenites amongst those used on the site were identified by means of macroanalysis:
- a very coarse-grained, well-compacted, yellowish calcarenite, used for the load-bearing structures of most of the city wall, and for the base of certain columns. The blocks of this calcarenite are well-formed and range from 65x65x25 cm to 100x 50x25 cm in size;
- a calcarenite of a similar colour to the previous one, but finer-grained and even more compact. It was used for the sarcophagi and columns on the acropolis and for some of the tombs on the plain. The sarcophagi are either monolithic, hollowed out of a single block, or simply constructed of stone plates lining a hole in the ground; in neither case were the lids of the sarcophagi made of plates of the same type of calcarenite. The sarcophagi are of notable size for the period, having a maximum length of 328 cm, a width of 180 cm, and a height of as much as 180 cm;
- a very well-compacted, fine-grained, pinkish yellow calcarenite, used for mortars, drinking troughs and grindstones.

Fig. 1 - Schematic

geological map.

1 - Quarternary terrains;

2 - Cretacic Formations.

Fragments of volcanic rock, which must have been brought from a completely different region, have also been found on the site.

It has been ascertained, by means of their stratigraphic position, that the limestones belong to the "Calcare di Altamura" formation. The entire hill is formed of layers of this limestone which vary in thickness down even to plate size, so too, in part is the flat area at the foot of the hill. There is a small stone quarry on the south side of the hill which was probably worked by the Peucetii themselves.

The calcarenites closest to the settlement are located a few hundred metres to the north, exposed along the "Canale Frassineto"; another calcarenite outcrop is located some Km away from the site, at Santo Mola, near Gioia del Colle; the furthermost outcrop is located between the modern towns of Acquaviva delle Fonti and Cassano delle Murge. The outcrops of all these calcarenites are discontinuous and their areal extent differs. It was observed that thin sandy and clayey silt layers overlie the calcarenites of the Canale Frassineto. The presence of the Cretaceous limestone rocks immediately beneath testifies that calcarenite layers are rather thin. They are medium grained, rather weak and can be referred to the "Calcarenite di Gravina", as can the calcarenites outcropping at Santo Mola. The latter, however, form layers of noticeably greater thickness, and their grain size, though variable, is prevalently coarse. Moreover, they can be pinkish or yellowish in colour and their degree of cement-

ation is variable. The same can be said of the calcarenites outcropping between Acquaviva delle Fonti and Cassano delle Murge, whose grain size, however, and degree of cementation vary greatly even within small area.

4 LABORATORY INVESTIGATIONS

The minero-petrographic characteristics of the limestones used on the site and of those which form the hill were tested, together with certain of their physical properties which are of interest for the purposes of the study. Similar tests were carried out on the calcarenites used on the site and on those of the various outcrops sampled.

The results show that the limestones used as building materials on the site are micrites with lenses of sparry calcite. Their allochemical components include intraclasts and bioclasts, which are scarce in some places and abundant in others and also partially recrystallized. Nevertheless the recrystallization process only reached the initial stage, as the internal cavities of the fossils are partially free. Some of these cavities, however, are the result of dissolution phenomena, so the limestones show partly primary and partly secondary porosity. In Duhnam's classification scheme (3), the lithotype to which they can be referred is "wackestone". Moreover, it should be stressed that there is very little difference between the unit weight of solid particles and the total unit weight of the limestones, therefore their degree of compactness must be very close to unity.

The limestones forming the hill show analogous characteristics, with certain insignificant differences concerning the quantity of allochemical components and of cavities caused by dissolution phenomena.

Identifying the calcarenites, by means of laboratory tests, proved somewhat more difficult. This was due firstly to the lithological differences between the three calcarenite formations mentioned above from which the Monte Sannace building materials may have come, and secondly to the considerable variety within each of the three formations. Therefore we concentrated our attention more specifically on the coarser calcarenites which were the most commonly used on the site and of which a larger number of sample fragments were available to us. Consequent to the lack of availability of fragments of the finer-grained calcarenites, testing of the latter has been restricted for the moment to further macroscopic and physical investigations. However, the necessary authorization to collect samples of the finer-grained calcarenites has already been requested. Tests will be carried out on these samples as soon as possible in order to complete the data for all of the calcarenites used on the site.

The coarse-grained calcarenites used at Monte Sannace, which could also be considered calcirudites, contain mostly rounded lithoclasts and bioclasts, while the moderately abundant non-carbonatic debris is mainly concentrated in the interstitial cavities. We observed a large quantity of psammitic insoluble residue consisting of quartz grains and scarce crystal chips of magnetite, flattered ilmenite crystals, tourmaline crystals, a few perfectly lengthened prismatic crystals of black pyroxene and corroded crystals of brown amphibole. Furthermore, their cementation, although variable, is generally quite good; their degree of compactness is not on the whole very high; while their imbibition coefficient is variable. As regards their total unit weight, they can be considered medium heavy rocks.

It should be noted that there is a certain degree of pollution affecting the clasts and the cement, which can be detected even by the naked eye in these calcarenites due to the presence of a pinkish and/or brownish pigment and or small superficial ballstones.

The calcarenites used for the sarcophagi are finer-grained than those described above, while those used for grindingstones and mortars, are finer still. Both, however, are well-compacted and seem more compact on the whole than the coarse-grained calcarenites described above.

Tests were carried out on all the calcarenite samples taken from outcrops except those collected in the Canale Frassineto. The latter were excluded since they could never have been worked due to their minimal thickness and their sandy and/or clayey silt overburden.

The samples collected at Santo Mola contain reasonably well-rounded limestone lithoclasts and bioclasts, which can be very large in size, and in some cases are covered with a pinkish yellow residual material. The packing of these calcarenites is not noticeably, while they give a large quantity of psammitic insoluble residue. The latter consists of quartz and feldspar grains, together with a small, though varying, percentage of heavy minerals (magnetite, ilmenite, tourmaline, amphiboles, pyroxenes) with similar characteristics to those observed in the calcarenites used on the site. Their degree of cementation and of compactness, moreover, are variable, although quite high in general.

The outcrops between Acquaviva delle Fonti and Cassano delle Murge have very variable minero-petrographic characteristics and thus correlations between them are not easy to establish; the same can be said of their physical properties (degree of cementation and compactness, imbibition coefficient).

It has not been possible to work out how the fragments of volcanic rock found on the site were used. These fragments are dark coloured and vesicular; in thin section they show a porphyritic structure with an intersertal groundmass. On the basis of the mineralogical composition of both the phenocrysts and the groundmass, they can be considered andesites.

5 CONCLUDING OBSERVATIONS

From the analysis of the existing data regarding the geological and morpholological conditions of the area of the central Murge, in which the Monte Sannace archeological site is located, combined with the results of the specific tests of this study, the following preliminary conclusions can be drawn:
- the stones used on the archeological site consist of limestones and calcarenites. The former were used for some parts of the city wall, for the lower part (probably the foundations) of the houses, for floors and for some load-

bearing structures. The calcarenites, instead, were utilized for long stretches of the city wall, the base of columns and for sarcophagi, grindstones and mortars;

- the limestones used have the same lithological and physical characteristics of the limestones which form the hill on which the settlement was located. Thus the limestone used was taken from the hill itself;

- amongst the calcarenite outcrops examined, that of Santo Mola is the one which almost certainly supplied the materials for the city walls, the columns and some of the sarcophagi: even though this outcrop is not the nearest to Monte Sannace. This confirms what had already been hypothesized (2). The hypothesis that the calcarenites used on the site came from the much nearer outcrops along the Canale Frassineto can be excluded since, apart from their lithological and physical characteristics, they are for the most part burried by a sandy and/or clayey silt cover. Furthermore they occur in such thin layers as not to allow the excavation of large blocks;

- regarding the calcarenite used for grindstones, mortars and some of the sarcophagi, it has not been possible to establish the origin of this stone with any degree of accuracy; this is not, after all, surprising as investigations into this subject have only just begun. However, it would seem reasonable to hypothesize that these materials could come from the "Calcareniti di Monte Castiglione" Formation, outcropping near Santeramo and Matera, and which were referred to above;

- finally, the presence of andesitic-type volcanic rock fragments, never registered before on the site, as far as we know, is of interest. Investigations are underway at the moment to discover their most probable place of origin, judging from the information given above on their petrographic characteristics; however, it is already possible to affirm that they were not brought from Mt. Vulture, the volcano nearest to the site.

REFERENCES

1 Baldassarre, G., Radina, B. & Radina, F. 1988. Pre-historic settlements in central Apulia (Italy) in relation to geological and morphological setting. In Géologie de l'ingénieur appliquée aux travaux anciens, monuments et sites historiques, p. 1623-1631. Rotterdam, Balkema.

2 Donvito, A. 1982. Monte Sannace. Archeologia e storia di un abitato peuceta. Bari, Schena.

3 Dunham, R. J. 1962. Classification of carbonate rocks according to depositional texture. Am. Ass. Petrol. Geol., 1, p. 125-178. Tulsa.

4 Servizio Geologico d'Italia 1966. Carta Geologica d'Italia al 100.000, Foglio 189 "Altamura", II Ed.. Roma.

5 Servizio Geologico d'Italia 1969. Carta Geologica d'Italia al 100.000, Foglio 190 "Monopoli", II Ed.. Roma.

6 Strabo (63 B.C.? - 25 A.D.). Geographica. In Les Belles lettres. Paris 1967, Ed. Lasserre.

6th International IAEG Congres / 6ème Congrès International de AIGI, © *1990 Balkema, Rotterdam. ISBN 90 6191 130 3*

A systematic approach to study chemical dissolution and concomitant physical disintegration of rock for construction

Approche systématique dans l'étude chimique de la dissolution et de la désintégration physique concommitante de pierres de construction

Christine Butenuth & M. H. de Freitas
Imperial College, Geology Department, London, UK

G. Butenuth
Labor Fuer Physikalische Chemie und Strukturchemie der Fachhochschule Aachen, Juelich, FR Germany

ABSTRACT: The strength of most rocks is reduced by the presence of water and an approach is described for investigating the cause of this weakening. Polymineralic and monomineralic rocks are shown to be unsuitable for such initial studies and it is reasoned that the surfaces of single crystals should be used to obtain an understanding of the fundamental processes that are involved. Results obtained from a study of the response of a cleavage surface of Iceland Spar to contact with water are reported and illustrate that events at the solid-liquid interface, as recorded by a measure of the pH of the liquid with time, can be described by an expression of kinetics of the first order with a time dependent rate constant. Measurements of acoustic surface wave velocity along the cleavage surface indicate that physical changes to the surface and its sub-strate can be linked to chemical changes recorded in the aqueous solution with which it is in contact.

RÉSUMÉ: La résistance de la plupart des roches est réduite en présence d'eau et une approche pour rechercher la cause de cet affaiblissement est décrite. On montre que les roches poly- et mono-minéraliques ne sont pas adéquates pour une étude initiale et aussi que les surfaces de cristaux individuels doivent être utilisées afin de comprendre les processus fondamentaux qui sont mis en jeux. Des résultats d'une étude des réponses de la surface de clivage du Spath d'Islande au contact de l'eau sont présentés: ils illustrent les évènements moléculaires à l'interface solide-liquide tels qu'ils sont enregistrés par la variation du pH en fonction du temps. Il est possible de décrire ces évènements par une expression de la cinétique du premier ordre avec une constante fonction du temps. Les mesures de la vitesse des ondes acoustiques sur une surface de clivage indiquent que les changements physiques de cette surface et de son substratum peuvent être liés aux changements chimiques enregistrés dans la solution aqueuse avec laquelle elle est en contact.

1 INTRODUCTION

The experimental work of Colback and Wiid (1965) quantitatively demonstrated the influence of water upon rock strength by showing the progressive loss of strength that occurs with increasing moisture content within the samples of rock tested. A quantitative description of the changes that occur on a mineral surface when this loss of strength occurs has yet to be achieved and it is to study this that the work described here has been undertaken.

It is difficult to accept that variations in rock strength arising from the presence or absence of aqueous solutions can be explained purely in terms of physical parameters, most notably the principle of effective stress, yet in practice this principle (which makes no reference to the quality of the pore fluids involved) has been used successfully for the purposes of construction. From this it can be concluded that the parameters measured in terms of effective stress, such as "cohesion", and those to which the value of effective stress is applied, such as "friction", carry the influence of the quality of the water involved; thus there is little advantage in further studying friction and cohesion to understand why aqueous solutions influence rock strength because attention has to be directed to the reactions that can occur on mineral surfaces.

When considering how best to study these phenomena it is important to appreciate the time over which such effects are to be studied. There will be those effects which occur once a mineral surface comes into contact with an aqueous solution and those which occur after repeated cycles of wetting and drying, and this distinction is worth noting because the latter need not be the summation of the former, even under identical environments. Should the environment change between successive periods of wetting, differences might be expected between the effects which occur. This means that research into the effects of weathering upon rock strength (such as is commonly conducted with building stones) has to be extraordinarily precise before it yields usable information of value to an understanding of the processes involved. Much of the work described at the recent conference in Athens (editors: Marinos and Koukis, 1988) bear witness to this.

Difficulties arise when a study of the effects of weathering (eg. variation of strength) is made via conventional studies of weathering and the magnitude of these difficulties can be appreciated by considering rock as a mechanical mixture of crystalline phases. The response of this mixture may be studied using either representative amounts of the mixture as a whole (as is the case when cores are tested) or representative samples of its individual phases. When considering the sample as a whole the effects observed will be influenced by the various minerals present: each may react differently and even different faces of the same mineral may not react with the same rate. Some averaging of effects therefore may occur from which it is difficult to disentangle the single effects that may lead to a loss in strength. For this reason it seems advantageous to use single crystals for detailed studies of the changes that can occur when minerals come into contact with water.

The work reported in this paper describes a systematic study of the microstructural changes which occur at a solid/liquid interface of a single crystal in contact with "water".

2 FACTORS TO BE CONSIDERED

The first factor to note is that the "water" which is used for such experiments should be considered as an aqueous solution. One reason for this is that most "water" contains gas in solution, but another very important reason is that when "water" comes into contact with mineral surfaces it changes its composition. When dealing with the response between aqueous solutions and mineral surfaces, it is apparent that adsorption, chemical reactions and other molecular processes, must be considered. In such systems the difference between chemisorption and chemical reaction may not be as readily distinguished as might be imagined. Eucken (1944) made the point that chemisorption processes which result in the formation of surface compounds may also be called surface reactions. For instance, in the case where water is chemisorpted at the surface of some solid oxides, the water molecules decompose to H- and OH- radicals which interact with the surface itself (Wedler, 1970).

Thus to begin the study it is necessary to obtain a quantitative description of a surface in both chemical and physical terms so that "chemical" changes can be compared with "physical" changes, where the latter should be of the sort that are related to parameters relevant to a measure of strength. This means a description of molecular change that may occur with time, once the aqueous solution is in contact with the solid surface, has to be obtained as well as a measure of the change which accompanies this but made using a physical parameter such as velocity of transmission. It has to be appreciated that at this level of investigation the differences between "chemistry" and "physics" become blurred. Nevertheless, by monitoring the interface with different techniques that describe its "chemical" and "physical" changes when in contact with an aqueous solution an attempt can be made to compare any molecular change that can be detected with changes that can be detected on a larger, non-molecular scale. To do this the experimental techniques which are chosen have to be able to quickly indicate the quantities required to be measured and this means that the rate of observation has to be greater than the rate of the process being observed.

Such an attempt to compare "chemical" and "physical" changes with one another necessitates that both types of experiments are performed under strictly comparable experimental conditions.

It is well known in material science that the behaviour of a material may be changed by changing its surface properties, eg. by using adsorption processes to change surface properties at a molecular scale. So for example, the Rehbinder effect (Meyer, 1968) explains the changed stress-strain behaviour of a material that is in contact with a surface active fluid solely by surface effects, such as adsorption processes. An understanding of the effects water may have on mineral and rock strength presents the prospect of being able to change rock behaviour, with time, in a controlled and desirable way, either by strengthening or weakening the material. To achieve this it is necessary to improve our understanding of two subjects:

(i) the qualitative and quantitative description of properties at a 'microscale', and of their change, and

(ii) the influence of these microscale properties upon material properties measured at a 'macroscale' and upon the change these properties can exhibit.

To study these kinds of effects and their implications a very good comparability and repeatability of experiments, which not only encompasses the experimental conditions but also the material used, has to be achieved. Therefore rocks do not seem to be a good choice of material to start with for, as mentioned previously, they usually contain different minerals, each of which will have different surface properties. The significance of this has recently been demonstrated by Butenuth (1988) who has shown that the dissolution kinetics of minerals are exceedingly difficult to understand; conventional analytical methods using powders are not sufficient for this work and for a systematic study to commence it is best to work initially with a pure mineral phase. This may be found in the form of either a mono-mineralic rock (eg. marble) or a single crystal of the rock forming mineral (eg. calcite) but care must be taken for different crystallographic planes through a mineral can be expected to exhibit different dissolution rates. Thus a surface cut through a mono-mineralic rock is unsuitable for this preliminary work as such a surface would contain not only the

random, or at least the different, orientations of the single crystals but also their interfacial boundaries. To avoid all these problems it seems best to start this kind of investigation on single crystals in which the orientations of the lattice on any face exposed is known. For this work single crystals of calcite have been used, as calcite is the building block of limestone, is a common mineral cement in many sedimentary rocks and occurs as an accessory mineral in many rocks.

3 EXPERIMENTS USING CALCITE

The solid surface used in these experiments was the $(10\bar{1}1)$ cleavage plane of calcite. Specimens of transparent cleavage rhombs of Iceland Spar from Mexico were used, the calcite being of hydrothermal origin and of considerable purity (see Butenuth and de Freitas 1989 for details). Microscopic examination revealed the rhombs to be of single crystals, ie. not a crystal mosaic, but to contain fluid inclusions with an approximate maximum length of 150×10^{-6} m. The edge length of the cleavage rhombs used was approximately 3×10^{-2} m in all cases.

Figure 1 illustrates the experimental arrangements used for measuring molecular change in the aqueous solution above the cleavage surface. For this a continuous measure of pH was obtained from a known and constant volume of solution sealed from the atmosphere by a layer of n-heptane: the area of solid surface so exposed was 3.87×10^{-4} m^2. In this way experiments using a constant volume of liquid phase over a constant area of interface could be closely followed in a closed system at room temperature and normal pressure.

An almost identical arrangement to that shown in figure 1 could be used for measuring the change occurring at a non-molecular scale and measured using the velocity of transmission in the plane of the cleavage surface: for this an acoustic microscope was used. The experimental arrangements can be described with reference to figure 1, where the shaft of the pH probe is replaced by the lens of the acoustic microscope: in this case the parameter continuously measured was a surface wave velocity.

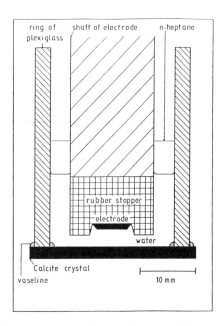

Figure 1. Experimental set up for pH measurements.

4 RESULTS

The concentration of hydrogen ions (C_H^+), measured by the experimental arrangement described above, changes with time such that when distilled water which is in equilibrium with the CO_2 of the environment (pH = 5.8) first comes into contact with the cleavage surface of calcite its value for C_H^+ decreases almost instantaneously, then increases for 5 or 10 minutes, passes through a maximum, and then decreases until no further change with time can be detected (at around pH = 8, which is the equilibrium value for calcite in contact with "water"). Repeating the experiments on the same surface but with fresh distilled water creates a similar response but eventually this apparent initial influx of H^+ ions can no longer be observed when fresh water is applied and only a decrease in C_H^+ with time is observed: details are given in Butenuth and de Freitas (1989). These observations with time permit this behaviour to be described in terms of formal kinetics as follows:

$$C_H^+ = (C_{Ho}^+ - C_{H\infty}^+)\, e^{-k(t)\,t} + C_{H\infty}^+ \qquad (1)$$

where

C_{Ho}^+ = hydrogen ion concentration at t_o and equal to the normal dissolution equilibrium of CO_2 in water (4.75 x 10^{-9} H^+ per 3 cm^3)

$C_{H\infty}^+$ = hydrogen ion concentration at t_∞ and equal to the normal dissolution equilibrium of $CaCO_3$ in water without contact with air (this is a small value compared with that of C_{Ho}^+ : 0.03 x 10^{-9} mol H^+ per 3 cm^3)

Equation (1) may be considered as an expression of kinetics of the first order but with a time dependent rate constant. Hence the new aspect of this expression is not that the dissolution kinetics can change with time (cf. Busenberg et al, 1976) but that a reaction of constant order can be expressed in terms of a continuous change of its rate. In practice it is not easy to distinguish from the experimental data whether a process is continuous or discontinuous (see Butenuth and de Freitas, 1989). In expressing k(t) in equation (1) as follows

$$k(t) = (k_o - k_\infty)\, e^{-\alpha t} + k_\infty \qquad (2)$$

the change in hydrogen ion concentration with time which was observed experimentally can be calculated. In equation (2) K_o is the rate constant at the beginning of the experiment, t_o, with a value of 100 x 10^{-2} per min and k_∞ is the rate constant at the time when the k value tends to its limit, for which the value is 2.35 x 10^{-2} per min.

Figure 2 illustrates the kinetic framework described by these results and shows how the concentration of H^+ ions (C_H^+) varies with time (t) and (α) : the open circles indicate points of inflexion which occur in the (C_H^+) - vs - (α) space and the silhouette represents the end member of the system seen in (C_H^+) - vs - (t) space, when (α) = 0. Similar results have been obtained by Butenuth (1989) for the dissolution kinetics of olivine in aqueous solutions.

At the moment, the work described does not provide evidence for the cause of such kinetic response but one important deduction can be made, viz. that within the volume of the solution no concentration gradient exists which can drive the process. Thus chemical reactions at the solid/liquid interface are determining the rate. This is to be expected because the solution contains chemical species other than those of the

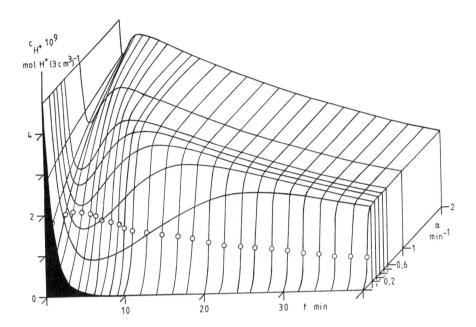

Figure 2. The hydrogen ion concentration (C_H^+) as a function of time (t) and (α), the following parameters being fixed. $C_{H_o}^+ = 4.75 \times 10^{-9}$ mol $H^+(3 \text{ cm}^3)^{-1}$

$C_{H_\infty}^+ = 0.03 \times 10^{-9}$ mol $H^+(3 \text{ cm}^3)^{-1}$ $k_0 = 100 \times 10^{-2}$ min^{-1} $k_\infty = 2.35 \times 10^{-2}$ min^{-1}

solid, corresponding to an irreversible chemical brutto equation, for example of the following type:

$$[CaCO_3]_{solid} + H_2O_{liquid} \rightarrow$$
$$[Ca^{2+} + HCO_3^- + OH^-]_{solution} \qquad (3)$$

The experiments conducted on calcite provided repeatable evidence to demonstrate that the total change in H^+ ion concentration per unit atomic lattice exposed to the solution on the cleavage surface is similar to the oxygen available for adsorption on that surface (Butenuth and de Freitas, 1989). Thus it is reasonable to visualise the molecular changes occurring in terms of H^+ ions becoming bound in some way to the surface with all possible positions to which H^+ ions can be attached being progressively filled.

All these lines of evidence suggest that the kinetics observed are caused by a shift of the molecular structure at the solid - liquid interface with time and such phenomena should be recognisable using appropriate methods for obtaining a measure of the physical condition of the surface.

As mentioned earlier identical surfaces have also been studied under identical environments using an acoustic microscope: details of this method are given by Briggs (1985). For these experiments a cylindrical lens was employed for its ability to generate acoustic waves in a known plane that would impinge as a line source of waves on the cleavage surface (also known as a "line-focus" lens: Kushibiki, 1985). This kind of lens was essential for the study of calcite, it being anisotropic ie. having quite different values of surface wave velocity in different directions. Figure 3 illustrates the range of velocities that were measured over the cleavage surface ($10\bar{1}1$) for one of the rhombs as a function of angle α. Figure 4 illustrates the change in surface wave velocity that occurs with time on a cleavage surface in contact with distilled water over a period of twenty four minutes. Such results are difficult to interpret being influenced by both the temperature of the aqueous solution (and in particular by any changes in its temperature during the period of measurement) and any change occurring in the physical form of the solid surface.

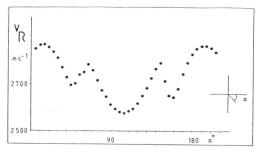

Figure 3. A graph showing the variation of surface wave velocity as measured by an acoustic microscope using a cylindrical (ie. line focus) lens at 200 MHz. The plot illustrates the various values of surface wave velocity with direction on the 10$\bar{1}$1 cleavage face as the face is rotated under the lens. The 10$\bar{1}$1 face is oriented horizontally and the axis of rotation is vertical to it, and co-axial with the lens. The result illustrates the anisotropy in this plane.

Consideration has also to be taken of the change in ionic composition of the solution with time and the effect this may have upon the acoustic properties of the solution. The combined influence of these parameters upon the experiments that have been conducted is not known in detail, however, the dependency of the velocity of acoustic wave propagation on temperature is well known so that a temperature correction for any measured velocity value can be calculated. Figure 4 incorporates a temperature correction. The dependence of acoustic wave velocity upon the ionic concentration of $CaCO_3$ in the aqueous solution is, to the knowledge of the authors, not known in detail, but an estimate based on the known dependency of the acoustic wave propagation on NaCl solutions (Bartel, 1954) led to the conclusion that the maximum velocity change for this system is likely to be in the third decimal place. The changes shown in figure 4 occur in the first and second decimal place, so if it is accepted that a genuine change in surface wave velocity is occurring with time then the cause for this change could reasonably be sought in a change of the solid and its interface.

Although the link between the molecular change and the larger "physical" change which accompanies it has yet to be defined, these results do provide an encouraging indication that the concomitant nature of such changes is now amenable to study.

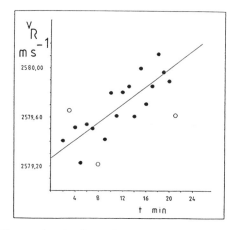

Figure 4. A plot of surface wave velocity on the cleavage face 10$\bar{1}$1, with time made using a cylindrical lens with wave propagation co-axial with the c-axis direction on the plane, at 200 MHz. This result has caused considerable interest in both fields of acoustic microscopy and crystallography.

5 CONCLUSIONS

5.1 A systematic approach to study chemical dissolution and concomitant physical disintegration of rock has been described in which attention is focused, by necessity, upon the behaviour of mineral surfaces.

5.2 The mineral surface of calcite cleavage has been chosen to begin this study so as to investigate not only the experimental methods required for such work but also the form of kinetics that the experiments reveal.

5.3 At a molecular level the processes seen to operate have a kinetic behaviour that is amenable to quantitative description. An interpretation of the meaning of these kinetics requires the completion of further work.

5.4 At a larger scale, measurements have been made of surface velocity changes in the solid portion of the solid-liquid interface under experimental circumstances and at a time scale that permits a comparison to be made between surface velocity changes and changes in the values of the hydrogen ion concentration.

5.5 These results show clearly that the relationship between "chemical" and "physical" changes in rock can be studied, but they also show that considerably more work has yet to be completed before a body of knowledge exists which will be sufficient to indicate the cause of strength variation in a mono- or polycrystalline material.

6 ACKNOWLEDGEMENTS

Work with acoustic microscopes was financially assisted by a grant from STATOIL through Professor J.S. Archer (Imperial College, London). Practical use of the microscopes has been facilitated by M. Issouckis (Leitz, Luton), Dr. A. Briggs (University of Oxford) and Dr. M. Somekh (University College, London). We are indebted to Professor Jun-Ichi Kushibiki (Tohoku University, Japan) for permitting use of his facilities for obtaining the data used to create Figure 4.

7 REFERENCES

Bartel, R. (1954). Sound velocity in some aqueous solutions as a function of concentration and temperature. The Journal of the Acoustic Society of America, Vol. 26, No. 2:227-230.

Briggs, A. (1985). An introduction to scanning acoustic microscopy, Royal Microscopical Society.

Busenberg, E & C.V. Clemency (1976). The dissolution kinetics of feldspars at $25^{o}C$ and 1 atm partial pressure. Geochim. et Cosmochim. Acta 40:41.

Butenuth, G. (1988). Aufloesungskinetik von mono- und oligo-mineralischen Koerpern in waessrigen Medien. Schwierigkeiten bei der Definition einer "einfachen Verwitterungsgeschwindigkeit", Teil 1, Wissenschaft und Umwelt, 3:135-150.

Butenuth, C. & M.H. de Freitas (1989). Studies of the influence of water upon calcite. Proc. of the International Chalk Symposium in Brighton, September.

Colback, P.S.B. & B.L. Wiid (1965). The influence of moisture content on the compressive strength of rocks. Proc. of the Rock Mechanics Symposium, University of Toronto, January 15-16.

Eucken, A. (1944). Lehrbuch der chemischen Physik. Vol. II, Part 2, Akademische Verlagsgesellschaft, Leipzig.

Kushibiki, Jun-Ichi, Noriyoshi Chubachi (1985). Material characterisation by line-focus-beam acoustic microscope IEEE Transactions on Sonics and Ultrasonics, Vol. SV-32, No. 2:189-212.

Marinos, P.G. & G.C. Koukis (1988). Proceedings of an International Symposium organized by the Greek national group of IAEG in Athens, 19-23 September 1988. The engineering geology of ancient works, monuments and historical sites, preservation and conservation.

Meyer, K. (1968). Physikalisch-chemische Kristallographie. VEB, Deutscher Verlag fuer Grundstoffindustrie.

Predicting the performance of crushed rock aggregates in highway wearing courses

La prédiction d'ouvrage des granulats en couches de roulement

D.C.Cawsey
School of Environmental Technology, Sunderland Polytechnic, UK

ABSTRACT: In service performance of highway wearing courses is dependent on many complex inter-related factors. Monitoring of road sites in the U.K. has demonstrated the influence of aggregate type and quality on mechanisms of deterioration. Overall performance of aggregates in bituminous wearing courses depends on: (a) soundness especially moisture-susceptibility and (b) adhesion with binder. Some 'sound' aggregates have been found to have relatively poor adhesion with binder. Although many tests have been developed to determine soundness and to assess adhesion, such tests are often carried out on single size aggregate and results are often difficult to relate to actual materials used. In this investigation laboratory performance tests have been carried out on mixes using different types of aggregates. These tests have revealed that differences in performance and modes of deterioration (stripping, fretting) are related to the different types of aggregate used. Such tests are leading to more realistic means of understanding and predicting in service behaviour of materials, and improved performance achieved by, for example, the use of additives. The investigations described have been supported by the U.K. Science and Engineering Research Council, Transport and Road Research Laboratory, Local Authorities and material producing companies.

RESUME: La comportement en place des revêtements bitumineux se repose sur des facteurs nombreux qui se rapportent. La surveillance des emplacements en chaussées en Royaume-Uni a montré l'influence d'espèce du granulat sur les mécanismes de détérioration. Le comportement des granulats en revêtements bitumineux dépend de: (a) résistance, surtont la susceptibilité à la moiture et (b) adhesivité bitume-granulat. On a trouvé l'adhésion mal en plusiers granulats 'résistants'. Bien qu'on a developé beaucoup d'essais pour déterminer la résistance et pour essayer l'adhésion, tels essais sont souvent comportés sur aggregats d'une dimension seul, et par consequence c'est difficile de correspondre les résultats avec les matériaux utilisés en actualité. En cette investigation, le comportement en laboratoire est examiné en utilisant les mélanges des divers granulats. Ces essais révélent que les differences du comportement et des modes de détérioration se sont rapportés aux espèces divers des granulats utilisés. Les essais amènent aux manières plus réalistes pour comprendre et predire le comportement en place des materiaux, et des progrés sont faits, par exemple, par l'utilisation des additifs. Les recherches décrits sont soutenus par le Science and Engineering Research Council du Royaume-Uni, le Transport and Road Research Laboratory, quelques Autorities Régionaux, et des Corporations qui produisent les materiaux.

INTRODUCTION

The in service performance of highway wearing courses is dependent on many complex interrelated factors. To facilitate improvements in the durability and performance of highway wearing courses requires investigation of these interrelating factors. The purpose of this is to consider the influence of aggregate type and quality on durability and performance.

DETERIORATION PROCESSES AND ASSESSMENT

In the United Kingdom premature deterioration of bituminous highway wearing courses has occurred with materials known to have complied with specifications and to have been correctly laid and compacted (Cawsey et al, 1987). Investigations involving the monitoring of particular road sites have demonstrated that two important processes leading to deterioration are: (1) disintegration of aggregate due to poor mechanical durability; and (2) loss of binder-aggregate adhesion.

Some rock types (e.g. gritstone) used as aggregates in highway wearing courses are susceptible to deterioration by weathering processes such as wetting-drying (Cawsey and Massey, 1988). Wetting-drying can cause swelling-shrinkage of moisture-susceptible minerals leading to volume changes which induce cracking in the rock. Progressive deterioration and removal of moisture-susceptible aggregate particles leads to the development of holes in the surface of the wearing course. Where many moisture-susceptible particles occur and deteriorate the resulting holes coalesce so forming "pot-holes" in the road surface. Trafficking assists the deterioration process.

Although the presence of moisture-susceptible aggregates can be detected by detailed mineralogical analysis (eg.XRD) these procedures are not normally used and such aggregates often give satisfactory results in British Standard aggregate tests. The use of a combination of test parameters, however, may permit detection of moisture-susceptible aggregates. In an investigation of four commonly used aggregates (three suspect and one control) from igneous rock sources strength was determined by the Aggregate Impact Ratio and soundness by the Magnesium Sulphate Soundness Test (Cawsey et al, 1987). Used separately test results were unable to properly diagnose the three suspect aggregates, but when results were used in combination an accurate diagnosis was possible (as shown in Figure 1).

The loss of binder-aggregate adhesion can result in physical separation of the binder and aggregate. This distress mechanism affecting the mixture is generally termed stripping. Observations of damaged highways have indicated that particular types of aggregates are more susceptible to stripping and that the process may be initiated from the road surface and/or from within the road structure (Cawsey and Gourley, 1989).

Scanning electron microscopy has been used to investigate the appearance of a range of aggregate types subjected to the stripping action of water in static and dynamic immersion and boiling tests. All the untested coated aggregates showed good adhesion and wetting, even the large flat quartz and feldspar crystals of the granites as shown in Figure 2a. Basalts and the slightly more porous limestones showed evidence of physiomechanical absorption of the bitumen as shown in Figure 2b. When tested most of the aggregates showed retraction of the binder into islands of binder. Figure 2c shows the appearance of a typical stripped area on a bituminous surface after testing in static immersion. Pits resulting from pulling of the bitumen up the side of air bubbles entrapped in the mix were also seen, as illustrated in Figure 2d. Areas showing peeling or displacement were more difficult to observe.

Close examination of the stripped surfaces indicated the presence of a second very thin coating associated with the aggregate surface. This second phase was observed on the flat crystal faces of the larger and fresher K-feldspar, plagioclase feldspar, quartz and vein calcite crystals, as shown in Figure 3a. Though difficult to observe, its presence was indicated by: (1) blistering under the electron beam; (2) the apparent adherence of fines on the surface; and (3) the detection of abundant sulphur on microprobe analysis. Examination of the area associated with the residual bitumen islands revealed that the thin film disappeared under the residual binder, as shown in Figure 3b. The photomicrograph showed a mirror image of the textured roll back on both the residual binder and the thin film, with the light fraction disappearing below the residual binder. It was therefore concluded that the thin film was a binder-aggregate interface phenomenon. Figure 4a shows the appearance of what seems to be an apparently stripped K-feldspar in a granite (magnification X330). When the magnification is increased to X8100, as in Figure 4b, the surface appears bumpy with small blisters evident in the thin film over the surface. The thin film was rarely observed on the smaller size and more porous aggregates. Further analysis of these stripped areas using a Laser Ionisation Mass Analyser (LIMA), showed that the surface was coated with a layer of organic material, of similar character to the original and residual binder.

It was inferred from this evidence that the film, originating as a separate lighter fraction from the binder, when in contact with an impermeable glassy

3046

mineral, e.g. a quartz or feldspar crystal, would set up a separate unabsorbed layer close to the interface, separating the interface from the whole binder mass. When the binder retracted this interfacial thin film remained attached to the mineral aggregate surface acting as a lubricating film over which the mass of binder could retract more easily. The film was rarely found on the more porous aggregates like basalt or limestone, probably due to absorption of the thin film into the surface of the aggregate. This postulated disbonding mechanism is illustrated in Figure 5. Such a mechanism explains why some 'sound' high strength aggregates have relatively poor adhesion with binder.

Although many tests have been devised, at present no generally accepted method exists, for recognition of aggregates liable to stripping. Different tests measure different characteristics and failure mechanisms and a combination of tests is preferable. The stripping potential index (SPI) developed by Cawsey and Gourley (1989) is based on three index test procedures.

$$\text{Stripping potential index} = \frac{SIT + 3TBT}{DIT}$$

where SIT = Static Immersion Test
             (HMSO 1962)
     TBT = Texas Boiling Test
             (Kennedy et al 1984)
     DIT = Dynamic Immersion Test
             (Cawsey and Gourley 1989)

A correlation of the index with performance testing by immersion wheel tracking is shown in Figure 6.

PERFORMANCE TESTING

Immersion Wheel Tracking Test

The Immersion Wheel Tracking test described by Mathews and Colwill (1962) was used to assess the performance of dense bitumen macadam wearing courses with a range of aggregate types. Three aggregate, types were chosen which were expected to be susceptible to stripping, viz. granite, quartzite and basalt. A fourth aggregate, gritstone, was chosen since it was not expected to be as susceptible to stripping and would provide a comparison with the other three aggregate types.

A previous investigation (Cawsey et al, 1988) into deterioration processes affecting dense bituminous materials in the wheel tracking apparatus considered the effect of sample density, immersion temperature, frequency of tracking, the applied loading to the sample and the effect of different solutions. From this work it was concluded that to produce a stripping failure of the sample, optimum test parameters were: temperature of immersion 30°C, tracking rate 30 cycles/min, and contact force on the sample of 1103N. One of the drawbacks of the Immersion Wheel Tracking Test is the length of time required to test a sample to failure but testing in a salt solution can be used to accelerate the deterioration of a test sample.

200mm diameter cores were taken from road sites and their performance assessed in the immersion wheel tracking apparatus. During testing the deformation of the material at the centre of the sample was recorded continuously. The sample was photographed at 3 hourly intervals. Stereo pairs of photographs were obtained. Longitudinal and cross-sectional profiles were also taken through the centre of the sample at 3 hourly intervals. A profile gauge consisting of a series of 0.5mm diameter wires clamped together vertically was used to measure the profile. Loose material was brushed off the surface of the sample to form a profile. The samples were assessed visually and the material lost from the surface of the sample was examined. Each sample was weighed at both the commencement and completion of the wheel tracking test. The weight after the test was taken several weeks after the completion of the test to allow the sample to dry out at room temperature.

Test Results

The immersion wheel tracking results are reported in Table 1 to the nearest hour. In these wheel tracking tests, failure time is defined as the time to produce a stripping failure along the full length of the rut formed in the sample. After a period of constant rate of rutting there was a sudden collapse of the samples tested. The quartzite, granite and basalt aggregates failed in this way. This type of sample collapse was also observed by Mathews and Colwill (1962) and it has been considered to be characteristic of stripping failure.

Apparent specific gravities within the range 2.2 to 2.4 do not affect the performance of the samples (Figure 7). Voids content also does not affect performance provided it falls within the range of 1 to 7%.

Figure 8 shows typical rut depth versus failure time curves for quartzite, granite, basalt and gritstone samples tested in salt solution. Initially, the sample recompacts and this may proceed for several hours. When extensive stripping has occurred in the sample, the rate of penetration of the sample increases suddenly and the sample fails. The gritstone samples recompact however and do not show a sudden increase in the rate of penetration.

Examination of the samples after failure showed that the collapsed rut contained loose coarse aggregate particles. Many smaller sized aggregate particles had been removed by the wheel tracking and were in the bath of the apparatus. Others had been taken out and collected. The smaller particles were the first particles to be plucked out. The upper surfaces of the particles of all four different aggregate types were stripped clean of bitumen by the abrasive action of the rubber tyre tracking over the sample. The under surfaces of the aggregate particles in both the granite and quartzite samples were frequently free from binder. The under surfaces of the basalt and gritstone aggregates usually had binder adhering to the surface after being plucked from the sample. If these particles were left in the rut long enough to abrade against each other, then this bitumen normally cleaned off the surfaces. Once the sample loses fines, its cohesional resistance is reduced and a good bond between the binder and aggregate can no longer exist.

During test, the fluid in the water bath became discoloured (brown) with a scum forming on the surface. This would indicate that spontaneous emulsification may be occurring. The quartzite, granite and basalt aggregate particles were more likely to become totally stripped of bitumen than the gritstone. The rut forming in the gritstone samples became very enriched with bitumen as the test progressed. Bitumen appear to migrate to the surface.

Observations of sections cut through the samples after testing showed gaps between particles. The void content of the sample would have increased and water would be able to move freely through the sample accelerating the deterioration which was occurring in the test. It is possible that pore pressure could develop causing further stripping of the aggregate. Some particles within the interior of the sample showed signs of binder loss.

Further investigations are being undertaken involving the use of additives incorporated into the mixes to improve binder-aggregate adhesion and inhibit the deterioration processes.

CONCLUSIONS

1. In service performance of highway wearing courses depends on complex inter-related factors. Moisture is a significant factor leading to mechanical deterioration of moisture-susceptible agregates and stripping caused by loss of binder-aggregate adhesion.

2. Although standard test procedures are not normally adequate for the prediction of in service performance of aggregates used in highway wearing courses, combinations of selected test parameters, e.g. the Stripping Potential Index, can indicate likely performance characteristics.

3. The observations in the immersion wheel tracking tests indicate that the mode of failure of aggregate types is not the same. The granite and quartzite samples appear to fail due to the loss of adhesion between binder and aggregate and this would initiate stripping failures. The loss of fines is followed by removal of coarse aggregate. This process would be accelerated by the pumping action of water in the growing volume of voids which occur as the internal structure is weakened by the loss of material from both the surface and interior of the sample. Samples that were cut into sections after testing showed loss of material internally.

4. The gritstone samples appear to start deteriorating by the loss of fines in a fretting type failure. Binder usually remains adhering to a substantial portion of the gritstone aggregate.

5. The failure of the basalt samples would appear to be initiated both by lack of adhesion between the binder and the coarse aggregate and by fretting. Some aggregate remains in adhesion with the binder whereas other aggregate is stripped.

6. For materials which are within specification, degree of compaction, voids content and density do not affect performance in the immersion wheel tracking test. Deterioration is mainly due to the loss of binder-fines followed by removal of coarse aggregate. Mechanisms involving

loss of binder-aggregate adhesion are considered to be important in determining durability and performance.

Acknowledgement

The research project was supported by the U.K. Science and Engineering Research Council and the Transport and Road Research Laboratory. The assistance is gratefully acknowledged of U.K. local authorities including Devon, Dorset, Nottinghamshire and Somerset CCS and U.K. material producing companies including E.C.C. Quarries Ltd, Redland Aggregates, Tarmac Roadstone and J. Wainwright.

REFERENCES

Cawsey, D.C., Bonner, D.C., Hargreaves, A.E., Moores, T.A. & Griffin, A. 1987. Effects of moisture on aggregate properties. Transport & Road Research Laboratory. Contractor Report 64, Crowthorne, U.K.

Cawsey, D.C. & Massey, S.W. 1988. In service deterioration of bituminous highway wearing courses due to moisture-susceptible aggregates. Engineering Geology 26:89-99.

Cawsey, D.C., Raymond-Williams, R.K. & Russell, W.A. 1988. Performance of dense bituminous macadam wearing course mixes tested under controlled laboratory conditions. Highways & Transportation, Vol.35, No.6:14-21.

Cawsey, D.C. & Gourley, C.S. 1989. Bitumen-aggregate adhesion: development of a stripping index. Proceedings 4th Eurobitume Symposium Volume I, 224-229.

HMSO. 1962. Bituminous Materials in Road Construction. Road Research Laboratory DSIR. Her Majesty's Stationery Office, London.

Kennedy, T.W., Roberts, F.T. & Lee, K.W. 1984. Evaluating the moisture-susceptibility of asphalt mixtures using the Texas boiling test. Transportation Research Record, 968:45-54.

Mathews, D.H. & Colwill, D.M. 1962. The Immersion Wheel Tracking Test. Journal of Applied Chemistry, 12:505-509.

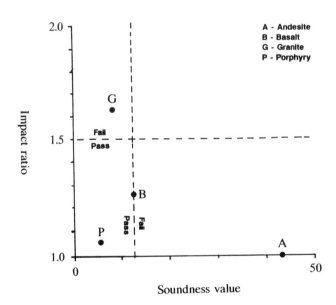

FIGURE 1   Impact ratio vs soundness value for igneous rock aggregates.

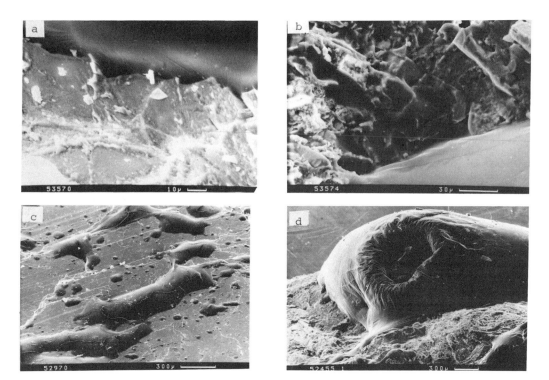

FIGURE 2   (a)   Good binder quartz adhesion in granite aggregate;
                (b)   Binder absorption into surface discontinuities on a basalt aggregate;
                (c)   Typical appearance of a stripped surface after static immersion;
                (d)   Retracted binder with pitting.

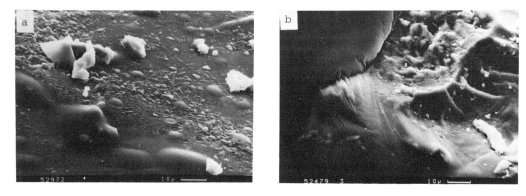

FIGURE 3   (a)   Thin film of binder on K-feldspar - note adherence of fines and the
                     presence of blisters on the mineral surface;
                (b)   Retraction of residual binder across the aggregate surface mirrored
                     in thin film.

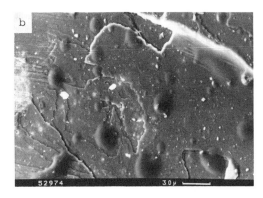

FIGURE 4   Appearance of apparently stripped K-feldspar crystal at
            (a) X330, and (b) X8100.   At the higher magnification the presence
            of the thin film on the mineral surface becomes apparent, with blistering
            and adherence of fines.

Non-absorptive   Absorptive

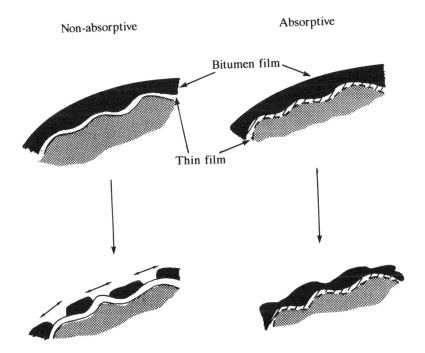

Bitumen film

Thin film

Stripping by slipping over   Very little slipping due
smooth non-absorbed thin   to absorption into surface
film surface.   of thin bitumen film, allowing
                                    texture on surface to retain
                                    the film.

FIGURE 5   Proposed mechanism to explain stripping.

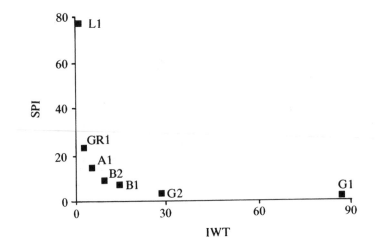

FIGURE 6   Correlation of Stripping Potential Index (SPI)
            and Immersion Wheel Tracking Test (IWT)

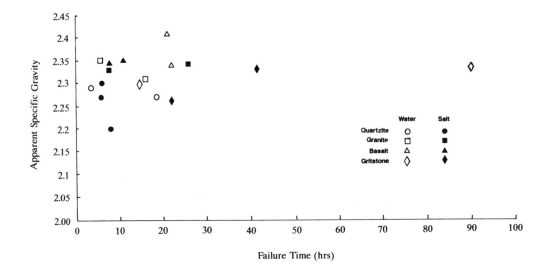

FIGURE 7   Influence of density on failure time in
            wheel tracking test

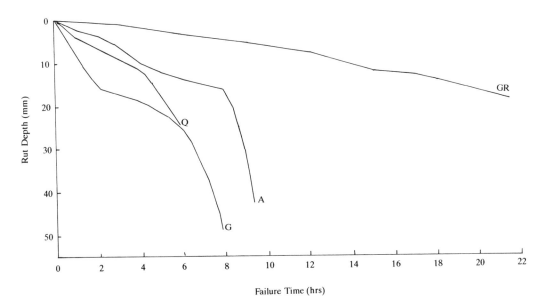

FIGURE 8   Typical failure curves from the immersion wheel tracking tests in salt solution.

Sample Reference	Aggregate Type	Apparent Specific Gravity	Estimated % Air Voids V_A	Wheel Tracking Test Failure Time (Hrs)	Length of failure time	Solution used	Thickness of Sample (mm)	% Weight Loss	Failure edge of length rut
QF2	Quartzite	2.27	6	19	Immediate	water	40	37	length rut
QF3	Quartzite	2.29	5	4	Short	water	30	12	End failed
QF1	Quartzite	2.27	6	6	Short	salt	40	2	length rut
QF5	Quartzite	2.30	5	6	Short	salt	30	19	length rut
A3052	Quartzite	2.20	5	8	Short	salt	90	8	length rut
BWP2	Granite	2.35	3	6	Short	water	50	17	End failed
BWP9	Granite	2.31	5	16	Intermediate	water	50	20	length rut
BWP1	Granite	2.34	4	27	Intermediate	salt	50	11	length rut
BWP11	Granite	2.33	4	8	Short	salt	50	33	End failed
LA1	Basalt	2.34	3	22	Intermediate to long	water	50	19	length rut
B1	Basalt	2.41	1	21	Intermediate to long	water	50	3	length rut
C1	Basalt	2.34	3	8	Short	salt	50	12	End failed
LA2	Basalt	2.35	3	11	Short	salt	50	17	length rut
BVG1	Gritstone	2.33	4	90	Long	water	50	8	length rut
BVG2	Gritstone	2.30	5	15	Intermediate	water	50	2	End failed
BVG3	Gritstone	2.33	4	41	Intermediate to long	salt	50	8	length rut
BVG4	Gritstone	2.26	7	22	Intermediate to long	salt	50	5	End failed

TABLE 1   Results obtained from the immersion wheel tracking test.

3053

6th International IAEG Congres / 6ème Congrès International de AIGI, © 1990 Balkema, Rotterdam. ISBN 90 6191 130 3

Assessment of rock durability through index properties

Evaluation de la durabilité des roches à travers des propriétés-indices

J. Delgado Rodrigues & F. Telmo Jeremias
Laboratório Nacional de Engenharia Civil, Lisbon, Portugal

ABSTRACT: Engineering geological studies of rock materials are very often concerned with durability assessment and with the task of defining acceptance/rejection criteria to be included in contractual documents. Accelerated ageing tests of several kinds are widely used for that purpose but their reliability may be questioned. In some circumstances, classification and numerical indices have been used, namely when rapidity is a determinant factor. This paper presents and discusses some of those ways of assessing durability and suggests a new index for this purpose. Porosity and swelling strain, the most important parameters of this index, are taken as indirect indicators of the rock materials susceptibility towards two of the most destructive weathering mechanisms: salt crystallization and the swelling of clay minerals. Some examples taken from carbonate rocks and greywackes support the usefulness of this way of predicting durability and the potential value of the proposed index.

RÉSUMÉ: Les études de géologie de l'ingénieur sur des matériaux rocheux sont souvent dirigées vers l'évaluation de la durabilité et vers la définition des critères d'acceptation/réjection pour leur inclusion dans des documents contractuels. On utilise parfois des indices numériques et des classifications, notamment si la rapidité est un facteur décisif. Cette communication fait la présentation et la discussion de quelques voix utilisées pour l'évaluation de la durabilité des roches et suggère un nouvel indice à cette fin. La porosité et l'expansion par imbibition d'eau sont les paramètres les plus importants de cet indice et ils sont considérés comme indicateurs de la sensibilité des matériaux à deux mécanismes d'altération particulièrement efficaces: la cristallisation des sels et l'expansion des minéraux argileux. Quelques exemples pris sur des roches carbonatées et sur des grauvaques montrent l'utilité potentielle de cet indice pour le but considéré.

1 GENERAL

Many circumstances in the practice of Engineering Geology call for a forecast of the behaviour of rock materials. This is particularly relevant for worked out construction materials, but it can also be found in studies concerning the in situ performance of rock bodies.

The assessment of rock durability has been tried by many authors, for many rock types and in very diversified applications. It is not our purpose to present one comprehensive analysis on the validity of all the tests and indices that have been used for this purpose, but only to summarise some leading ideas grasped when reading the literature consulted.

The soundness test (salt crystallization test) and the freeze-thaw test are among the best known ways of predicting the performance of materials. However, in spite of their interest, it seems that their utilizations have been pushed far beyond the limits of validity. In particular for most circumstances that use these methods for prediction, there is clearly a lack of parallelism between the decay mechanisms acting in situ and in the laboratory test and this fact may seriously impair the validity of the results for forecasting the real behaviour of materials.

Many other methods have been proposed and used more or less widely. The enumeration of some of them may be useful for understanding some basic assumptions underlying the choices of the proposed ways of prediction.

Farran et Thénoz (1965) suggest the use of permeability, specific surface and the water-rock reactivity for prediction of in situ performances of rock masses durability. Struillou (1969) recommends the use of hydrogen peroxide, envisaging assessment of the role of clay minerals. With a similar purpose, but using microscopy methods, were the ways followed by Scott (1965) and Weinert (1964, 1968). Franklin and Chandra (1972) and Gamble (1971) have made a wide use of the slake durability test and Delgado Rodrigues (1976) has adapted this same test to hard rocks by adding a pretreatment with ethylene-glycol and increasing the duration of the test. Swelling strain in conjunction with ultimate strength was suggested by Olivier (1976), Felix (1987a) made use of swelling strain in conjunction with the ratio between ultimate strength in wet and dry conditions and Delgado Rodrigues (1988) proposed swelling strain and porosity as basic parameters for the assessment of durability of carbonate rocks. The methylene blue value has been used by several authors (Ngoc Lan, 1980; Stapel and Verhoef, 1989).

Other authors have proposed more complex indices, two of which may be found in Smith et al. (1970) and Fookes et al. (1988).

The simulation in laboratory of climatic conditions, keeping the environment conditions close to those in force in the real cases is claimed by many authors as a good way to assess stone durability. However, the exact significance of the results is not always clear and the lengthness of tests clearly impair these methods of prediction as a routine way of work.

In the course of the studies carried out, mainly with carbonate rocks and greywackes, most indices were rejected for one or another reason, and an alternative way started to show promising results for the materials concerned.

The first approach was delineated for carbonate rocks and a very simple abacus for classification was proposed (Delgado Rodrigues, 1988). It uses swelling strain and porosity as the classification parameters (fig.1) and some applications so far available support the validity of this procedure. Both parameters are simple and accessible to most laboratories' testing facilities. These parameters evaluate two distinct rock properties, the importance of clay minerals and the amount of pore space. They react preferentially to two distinct decay mechanisms, which are changes in water content and salt crystallization. The roles of these parameters are not antagonistic, but a certain synergy may eventually exist when both parameters increase simultaneously.

Following the approach just outlined, another attempt was made, trying to introduce a certain degree of universality, by constructing one index that could be applied to any rock type without the need for auxiliary abacus. This attempt will be discussed in the next paragraphs.

2 INDEX OF ROCK DURABILITY

The underlying assumption of the forthcoming consideration is that swelling strain and porosity are the basic parameters for expressing the importance of the harmful components of the rock material. This is true for many (or even most) common situations, but this assumption needs a critical analysis whenever any specific practical application is to be made.

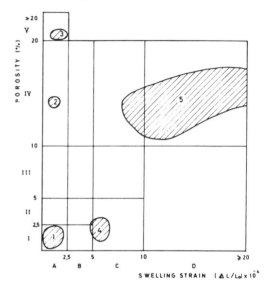

KEY

1 _ "LIOZ" limestone
2 _ "COIMBRA" dolostone
3 _ "ANÇA" limestone
4 _ "KEDDARA" glauconitic limestone
5 _ "MONDEGO" argillaceous limestones and marls

Fig. 1 Example of the utilization of a geotechnical classification abacus for carbonate rocks (Delgado Rodrigues, 1988)

In the course of the definition process, it came out that the index could benefit from inclusion of a strength parameter, in the assumption that two rocks with similar swelling strain and porosity would behave differently should they have different strengths.

For uniformity of future utilizations of the index, the following definitions should be taken into account:

a) Swelling strain (ε) - is the expansion (Δl) measured when the ovendried specimen is immersed in distilled water, divided by the initial length (L_o) of the specimen. Swelling can be monitored at regular time spans, 48h being the usual interval for computing the swelling strain. High accuracy is required in this measurement and, therefore, the equipment described in Nascimento et al. (1968) or similar is recommended.

b) Porosity (n) - is the ratio, expressed in percentage, between the volume of open (or accessible) pores and the bulk volume of the specimen. Volumes can be determined by resorting to hydrostatic weighings with vacuum-saturated specimens. The test procedure can be found in RILEM (1980), for instance.

c) Strength (R) - the ultimate compressive strength is the recommended parameter of strength. Unless specified otherwise, dry specimens should be used. In many current practical applications a compromise between simplicity, rapidity and accuracy may be accepted and compressive strength may be obtained indirectly by resorting to the determination of point load strength and using existing correlation formulas.

The basic expression of the index has the strength factor in the numerator and the summation of swelling strain and porosity in the denominator, and its mathematical expression is:

$$IRD = \frac{\dfrac{R}{Rt}}{n + 2a}$$

where,

IRD = index of rock durability
R   = ultimate compressive strength of the rock
Rt = strength parameter for comparative purposes
n   = porosity, expressed in percentage
a   = is the mantissa of the swelling strain (ε) when expressed as: $a \times 10^{-4}$

For keeping the index adimensional, Rt= 1 MPa when R is expressed in MPa. Rt should be transformed into the appropriate coefficient if different units are used for expressing strength. In order that the index might keep its comparative value, the swelling strain only can be expressed as mentioned, which means that the value used corresponds to the strain measured (ε) multiplied by 10^4. In this way, for most common rock materials, the weights that correspond to swelling and porosity in the proposed formula fall within the same range (between 0 and 30) and, thus, variations in each parameter can noticeably influence the

final value of the index.

Other formulas may be derived from this basic expression if practical applications demonstrate it to be necessary. Purely as a preliminary exercise, substituting the denominator for $(\sqrt{n} + \sqrt{2a})^2$ may signify the assumption that porosity and swelling exhibit some kind of synergy, potentiating each other. This and other theoretically acceptable expressions were not pushed too far for lack of available real data for validation and calibration.

3 FIRST ATTEMPTS AT VALIDATION AND CALIBRATION

The literature is not prodigal of data that could be used for the proposed index. Most times, when one of the parameters is considered, one or two of the other are lacking. Strength could be dispensed with, and could be replaced by a normalised parameter for comparative purposes, but swelling strain and porosity are absolutely essential for this purpose. In this context, it is difficult to find reliable data outside our own to be used for independent validation and calibration. It is hoped that this job could be undertaken by our readers should this proposal deserve their interest.

During the study of some greywackes used as riprap material in Beliche dam (South Portugal), we had the opportunity to undertake two series of accelerated ageing tests, one under wetting and drying and the other under cycles of salt crystallization. The first method hardly introduced any perceptible modification in the tested specimens, whereas salts were able to discriminate the specimen quality. Fig. 2 shows the results. As this test can be considered a method for assessment of rock durability, we tried to superimpose on that figure the values of IRD computed for each of the tested samples. These values are also included in the figure.

One first conclusion is that the sequence of IRD between 2.9 and 22.4 is similar to that obtained in the ageing test, and clearly points towards the validity of the index. Calibration of these values cannot be definitive after this exercise, but they suggest that greywackes whose IRD is lower than 4 should be discarded for riprap purposes or, at least, used with care.

Application to the specimens that were used in a precedent work for defining a geotechnical classification of carbonate rocks (Delgado Rodrigues, 1988) allows us to formulate the following conclusions:

- One of the best Portuguese limestone ("Lioz" limestone) has an IRD of about 94.

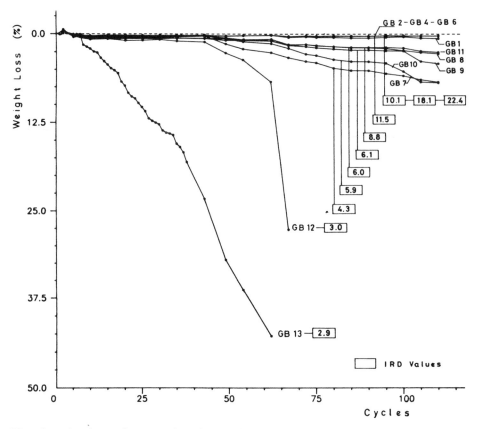

Fig. 2 Behaviour of greywackes in accelerated ageing tests and IRD values

and some porous clayey specimens with extremely bad behaviour have IRD lower than 1, going down to 0.2;

- "Ançã" limestone, a very porous though clay free rock, decays very rapidly, in particular when in contact with salts. This rock is very well known from some Portuguese masterpiece monuments, particulary for its softness and high decay rates. The values of IRD are of about 1.

-Common Portuguese micritic limestones, known to be good construction materials, have IRD above 5, and most of them above 15.

In the impossibility of a definitive appreciation, it was considered, in a first approach, that values of IRD lower than 2 signify low durability, while values greater than 10 correspond to materials having good to excellent durability. A larger number of results is necessary to clarify the transition gap between 2 and 10.

Although incomplete as far as our objective is concerned, a few other facts can be advanced for the validation of the proposed index. The strongest material so far reported for Portugal is a gabbro-diorite rock from Sines. With the published data (Oliveira et al. 1978) values above 1000 are obtained for the IRD, this meaning exceptional quality, as it was, therefore, considered in the paper quoted.

Felix (1987b) published a series of results obtained in some Swiss molassic rocks. Computed values of IRD fall in the range 1.2 to 3.7. These index values seem fairly well to express some differences in the outdoor behaviour of the stones concerned, but it was considered too early (and based on a too short number of results) for drawing any definite conclusion as regards calibration of the index for this type of rock.

4 FINAL COMMENTS

The above presented index only contains intrinsic parameters in its mathematical expression. This means that extrinsic factors, such as climate, topography, hydrological conditions and other environment factors are not taken into account in this procedure for assessing stone durability. Real practical application should, therefore, make a critical analysis of all these manifold possible situations before making use of this index (this is also applicable to the generality of similar indices).

It is also likely that the index will not have the same significance for all rock types. Future experimental and real work data will be essential for the development of adequately calibrated scales of rock durability, adapted to each rocktype (or to groups of rock types).

This index is intended to be applied to rock bodies whose representativeness is guaranteed to an adequate extent. This means that it should not be applied, acritically, to singular specimens or to anomalous values of any of the three constitutive parameters.

Preferentially, average values should be used and a good representativeness is a "must" condition for accurate assessment.

Compressive strength test is highly dispersive and is very dependent on the occurrence of small inhomogeneities of the material, on small deviations in the specimen's geometric form and on small uncorrections in the testing conditions. In addition to the necessity of testing an adequate number of specimens per sample (usually not less than 5), anomalous results should be discarded and average values should be used with the index.

When the ultimate compressive strength is obtained by correlation with the point load test, this should be stated in the report together with the indication of the correlation procedure that was used.

If absolute values are required, only the directly determined value of ultimate compressive strength should be used.

Preference should be given to correlation obtained with rock types identical to those under study. The work reports should state if the correlations were obtained from the research concerned or taken from the bibliography.

Determination of porosity is easy to carry out and regular faced specimens are recommended for this purpose.

Swelling should be determined, at least, in two specimens per sample. When large differences are obtained, some other specimens should be tested and the anomalous results discarded. However, it should be pointed out that anomalous results very frequently express real differences in the stone quality and this fact should be analyzed systematically before dismissing any of them. If this is the case, an average value for IRD may be computed, but, at the same time, it should be stressed that stone quality may present some lower values,

which can affect stone performance. A satisfactory explanation for these situations should always be looked for.

REFERENCES

Delgado Rodrigues, J. 1976. Estimation of the content of clay minerals and its significance in stone decay. Proc. 2nd Int. Symp. on the Deterioration of Building Stones. Athens, 1976.
- 1988. Proposed geotechnical classification of carbonate rocks based on Portuguese and Algerian examples. Engineering Geology, 25, (1), p.33-43.

Farran, J. & Thénoz, B. 1965. L'altérabilité des roches, ses facteurs, sa prévision. Annales de l'ITBTP, XVIIIéme année, Nov. 1965.

Felix, C. 1987a. Essais et critères de choix pour des grès (molasses) de substitution lors de travaux de restauration. Chantiers/Suisse, 18, (5), p.419-423.
- 1987b. Propriétés physiques de huit grès, dont celui dit "Grès de la Fontaine", provenant d'Alemagne. Chantiers Suisse, 18, (5), p.711-720.

Fookes, P. G., Gourley, C. S. & Oikere, C. 1988. Rock weathering in engineering time. Quat. Jour. of Engineering Geology, 21:33-57.

Franklin, J. A. & Chandra, R. 1972. The slake-durability test. Int. Jour. Rock Mech. Min. Sci. 9:325-341.

Gamble, J. C. 1971. Durability-plasticity classification of shales and other argillaceous rocks. PhD Thesis, University of Illinois.

Nascimento, U., Oliveira, R. & Graça, R. 1968. Rock swelling test. Proc. Int. Symp. on Rock Mech., Madrid, 1968.

Oliveira, R., Delgado Rodrigues, J. & Gomes Coelho, A. 1978. Engineering geological studies for the Sines harbour (Portugal). Proc. 3rd Int. Cong. of the IAEG, Madrid, Sept. Sec. III, vol.2, p.131-142.

Olivier, H. J. 1976. Importance of rock durability in the engineering classification of Karroo rock masses for tunnelling. Proc. Symp. on Exploration for Rock Engineering, Johannesburg. Edited by Z. T. Bieniawski, vol.1, p.137-144.

RILEM, 1980. Recommended tests to measure the deterioration of stone and to assess the effectiveness of treatment methods - - Tentative Recommendations. Matériaux et Constructions, Vol. 13, Nº5, 1980.

Scott, L. 1955. Secondary minerals in rock as a cause of pavement and base failure. Highway Res. Board, Proc., 1955, p.412-417.

Smith, T., MaCauley, M. L. & Mearns, R. W. 1970. Evaluation of rock slope protection material. Highway Res. Board (323). Nat. Res. Council, US Nat. Ac. of Science.

Stapel, E. E. & Verhoef, P. N. W., 1989. The use of the methylene blue adsorption test in assessing the quality of basaltic tuff rock aggregate. Engineering Geology, 26:233-246.

Struillou, R. 1969. Prévision de l'alterabilité des matériaux en fonction de leurs caractéristiques propres et de leurs utilisations. Colloque de Géotechnique, Toulouse, Mars, 1969.

Tran Ngoc Lan, 1980. L'essai au blue de méthylène un progrès dans la mesure et le controle de la propreté des granulats. Bull. Liasion Lab. Ponts et Chaussées, 107:130-135.

Weinert, H. 1964. Basic igneous rocks in road foundations. CSIR Research Report 218. Bulletin of the National Institute of Road Research, Pretoria, 5.
- 1968. Engineering petrology for roads in South Africa. Engineering Geology, 2, (6), p.363-395.

6th International IAEG Congres / 6ème Congrès International de AIGI, © 1990 Balkema, Rotterdam. ISBN 90 6191 130 3

The effects of lithologic characteristics on mudrock durability

Les effets des caractéristiques lithologiques sur la durabilité des roches argiliques

Jeffrey C. Dick & Abdul Shakoor
Department of Geology and Water Resources Research Institute, Kent State University, Kent, Ohio, USA

ABSTRACT: A large number of lithologically distinct mudrocks showing a wide range of field durability behavior were sampled and investigated for their lithologic characteristics, engineering properties, and durability. The mineralogic composition, fabric, cementation, and the degree of microfracturing were investigated using X-ray diffraction, scanning electron microscopy, and optical examination. The engineering properties that relate to the lithologic characteristics were also investigated. These included void ratio, dry density, absorption, adsorption, and plasticity characteristics. The relative durability of the mudrocks was determined using the slake durability test. The data were analyzed statistically to evaluate the relationships between lithology, engineering properties, and slake durability.

Preliminary results indicate that for shales the slake durability is closely related to the rock fabric as expressed by void ratio and absorption. These relationships are corroborated by SEM studies. The slake durability of mudstones is found to be closely related to the degree of microfracturing.

RÉSUMÉ: Un grande nombre des roches argiliques qui sont distinctes lithologiquement et qui montrent une grande variation de la vitesse de la désagrégation physique ont été échantillonés. Les caractéristiques qui ont été mesuré sont le lithologie, les propriétés géotechniques, et la durabilité La composition minéralogique, l'étoffe, la cimentation, et la degré des microfissures ont été étudiés par la technique de la diffraction aux rayons X, du microscope électronique à balayage, et de l'examen opticale. Les propriétés géotechniques mesurés sont la raison de vide (e), la densité sèche, l'absorption, l'adsorption, et les limites d'Atterberg. La durabilité relative a été déterminée par utiliser de l'essai de durabilité en immersion. Les data sont statistiquement analysés pour evaluer la relation entre de la lithologie, des propriétés géotechniques, et de l'essai de durabilité en immersion.

Les resultats préliminaires indiquent que pour les argiles, l'essai de durabilité en immersion est un fonction de l'étoffe du roche qui est mesuré de raison de vide et de l'absorption. Ces relations sont corroborées par les études au microscope électronique à balayage. L'essai de durabilité en immersion des roches argiliques est un fonction de la degré des microfissures.

INTRODUCTION

Mudrocks are sedimentary rocks of which greater than 50 percent of the grains have a diameter less than 0.075 mm. Mudrocks can be distinguished on the basis of percent clay (grains less than 0.004 mm) and structural laminations (Potter et al., 1979). Within the mudrock group, five different classes of rocks are recognized: shales, mudstones, claystones, siltstones, and argillites. Shales contain more than 33 percent clay and have laminations less than or equal to 1 cm thickness. Mudstones contain more than 33 percent clay and are either non-laminated or have laminations greater than 1 cm thickness. Claystones contain more than 66 percent clay. Siltstones contain less than 33 percent clay. Argillites are low grade metamorphosed mudrocks that have retained their sedimentary rock character. These individual types of mudrocks are sufficiently distinct from one another to warrant their treatment in this study as individual rock types.

Mudrocks are the most common sedimentary rock type. It has been estimated that mudrocks account for as much as 70 percent of all sedimentary rocks (Picard, 1971). Consequently, mudrocks are frequently encountered in all types of geotechnical engineering projects.

Where mudrocks are excavated and left exposed to weather, many apparently well indurated mudrocks rapidly slake to produce a soil-like material having significantly inferior engineering properties to those of the original rock. In these situations, the durability of mudrocks becomes their most important engineering property.

Much of the focus of mudrock research has been towards understanding the processes that control mudrock weathering and towards devising tests that can be used to predict mudrock durability. It is clear from this research that the lithologic characteristics are the dominant factors controlling mudrock durability behavior. However, no widely accepted system for predicting mudrock durability has emerged from these efforts. This lack of an acceptable durability prediction system can be attributed to one or more of three basic reasons: 1) the research was based on too few samples that were not representative of the mudrock group as a whole; 2) the durability tests utilized may not reflect the lithologic characteristics controlling durability; and 3) the validity of the research findings are questionable as the findings have not been corroborated with observed field durability behavior.

The objective of our research is to establish a mudrock durability classification system that is applicable to all mudrocks. We intend to achieve this by collecting a large number of lithologically distinct mudrocks that are representative of the mudrock group as a whole. The relationships between the lithologic characteristics, relevant engineering properties, and the documented field durability behavior of each mudrock will be statistically analyzed in order to establish durability classification criteria.

RESEARCH METHODS

Sample Collection

Mudrock samples were obtained from highway and railway cuts as well as from clay and coal surface mines. The field durability characteristics of each mudrock sampled were carefully observed and documented. To date, 54 samples, representing all classes of mudrocks, have been collected from six different North American physiographic provinces. These samples span geologic time from the Tertiary to the Cambrian.

Laboratory Investigations

Previous research has indicated that the durability of mudrocks is dependent on certain lithologic characteristics. These lithologic characteristic include the relative proportion of detrital quartz and clay mineral content, types of clay minerals, grain size, structural laminations, and degree of induration (Skempton, 1964; Spears and Taylor, 1972; Burnett and Fookes, 1974; Russell and Parker, 1979; Russell, 1982; Smart et al., 1982; Perry and Andrews, 1982; and Shakoor and Brock, 1987). Induration can be characterized by rock fabric (grain size, shape, and arrangement) and cementation. In addition, microfractures appear to be an important lithologic characteristic controlling durability. All of these lithologic characteristics have been investigated using X-ray diffraction, scanning electron microscopy, grain size analysis, and optical examination.

The relative proportions of clay and non-clay minerals and their respective grain sizes were determined according to ASTM method D422. Samples were disaggregated by alternate wetting and drying, ultra-sonic vibration, and agitation. The distribution of grain sizes greater than 0.045mm were determined by wet sieving. Grain sizes from 0.045 to 0.001 mm were determined using a hydrometer. The material retained on the sieves was optically examined for mineral composition. In all cases, quartz was found to be the dominant non-clay mineral with mica, pyrite, calcite, fossil fragments, and organic matter present in minor amounts.

The clay mineral component of the mudrocks was analyzed using a Rigaku Geigerflex X-ray diffractometer. Oriented ceramic tile mounts were prepared and the clay mineral composition determined according to diagnostic basal diffraction peaks. Where necessary, glycolation and heat treatment procedures were employed.

An International Scientific Instruments model SX40A scanning electron microscope (SEM) was used to examine the fabric of selected mudrock samples. Although qualitative in nature, the SEM studies have allowed us to examine the rock fabric and relate it to the void ratio and absorption determinations.

Through the course of field sampling and laboratory testing, it became evident to us that small scale discontinuities or micro-fractures were a common lithologic characteristic of many of the mudrocks, particularly the mudstones. The slaking characteristics of the mudrocks possessing the micro-fractures suggested that the micro-fractures were a dominant durability controlling lithologic characteristic. The frequency of the micro-fractures was determined by counting the fractures across the faces cut in the rock samples at right angles to each other and perpendicular to the bedding.

Several engineering properties were investigated. These include mudrock void ratio and dry density, which indicate fabric and degree of induration; absorption (ASTM C97), which is an alternate indicator of fabric; adsorption, an indicator of fabric and clay mineralogy; and Atterberg limits (ASTM D423 and D424), an indicator of clay mineralogy and clay/non-clay mineral proportions.

Mudrock void ratio and dry density are index properties determined from rock phase relationships. The rock phases were calculated using natural water content (ASTM D2216), bulk specific gravity of solids (ASTM D854), and bulk density. The bulk density was measured at natural water content using a jolly balance.

Mudrock adsorption, expressed as percent water content, was determined using a temperature and humidity controlled room. Oven dried samples were left exposed for 72 hours at 100 percent relative humidity and 20 degrees celsius. The samples were weighed at 24 hour intervals and the weight increase recorded. The increase in water content for each measurement was calculated.

The 2-cycle slake durability test (ISRM, 1979) was performed on all mudrock samples as a standardized method for ranking mudrock durability. Slake durability is an important test as it has been successfully utilized in mudrock durability classifications for compacted embankments (Deo, 1973; Strohm, 1980).

RESULTS

Although a total of 54 mudrock samples have been collected, only 34 have been studied in sufficient enough detail to include their results in this paper. These 34 samples include 11 shales, 17 mudstones, and 6 argillites. Unfortunately, there are no siltstones or claystones represented in this group. The clay mineral component of the samples is comprised of only non-swelling clay minerals belonging to the illite, kaolinite, and chlorite species.

The 2-cycle slake durability index for the mudrock samples tested ranged from 3 percent to 100 percent, indicating highly non-durable to very resistant mudrocks. Within the individual mudrock classes, the slake durability index ranged from 40 percent to 97 percent for the shales with a mean value of 79 percent. For the mudstones, the slake durability index ranged from 3 percent to 93 percent with a mean value of 50 percent. The argillites exhibited a very narrow range from 99 percent to 100 percent. These findings illustrate the variability of mudrock durability and they indicate that it is the mudstones that are particularly problematic.

The void ratios of the mudrocks exhibit a high degree of variation ranging from a low of 0.01 to a high of 0.32. The shales had void ratios ranging from 0.07 to 0.25 with a mean of 0.13. The void ratios for the mudstones ranged from 0.05 to 0.32 with a mean of 0.19. The argillites ranged from 0.01 to 0.09 with a mean of 0.035.

The void ratio is the ratio of the volume of the voids in the rock to the volume of the solids. Void ratio can therefore serve as an indicator of the mudrock fabric. The range of void ratio values supports what we had already anticipated from observations of the field durability behavior of the mudrocks and from the slake durability test results. The argillites are very durable and have low void ratios. This is logical since the argillites have been metamorphosed and their constituent minerals have been recrystallized, thus eliminating void spaces and interlocking the mineral grains to produce a very durable rock.

The shales show considerable variation in void ratio, with the mean skewed towards the low end. This variation may be attributed to varying degrees of cementation. SEM examination revealed a tendency towards an ordered or aligned mineral grain arrangement, possibly explaining the relatively low mean void ratio.

The mudstones have the largest range of void ratios reflecting the variable fabric of the mudstone group. Mudstones are transitional between shales and siltstones, having the ordered mineral grain arrangement and the hint of structural laminations of shales at one extreme and the distinct lack of structural laminations and dispersed arrangement of mineral grains at the siltstone extreme. The mean void ratio of mudstones is high in comparison to the other mudrocks, reflecting the dispersed fabric, and the lower limit of the void ratios is comparable to that of the shales, reflecting some degree of ordered mineral grain arrangement.

Absorption provides an alternative indicator of mudrock fabric. The results of the absorption test are in general agreement with the void ratio determinations. The absorption ranges from a low of 0.5 percent to a high of 34 percent. The absorption values for the shales range from 4 percent to 20 percent with a mean of 8 percent. The mudstones have absorption values ranging from 5 percent to 34 percent with a mean of 12 percent. The argillites range from 0.5 to 4 percent with a mean of 1 percent. The low absorption values of the argillites reflect the recrystallization of the mineral constituents and the consequent reduction of void space. The high degree of variation in the shales reflects the variation in the degree of cementation, with the mean skewed to the low end reflecting the ordered arrangement of mineral grains. The variation in the mudstones is explained in the same manner as that of the void ratios. Due to inherent slaking problems associated with the mudstones when immersed in water, the validity of the values exceeding 24 percent is questionable.

The micro-fracture frequency was measured for 12 of the 17 mudstones. The other five samples were omitted because of their small size and disintegration problems. The frequency of micro-fractures for the mudstones ranged from a minimum of 1.0 per cm to a maximum of 8.2 per cm.

Data Analysis

This research was initiated on the presumption that it is the lithologic characteristics of the mudrocks that control their durability. To evaluate this, the slake durability results were compared to the results of the other tests and analyses using bivariate statistical methods. To evaluate the effect of rock fabric on durability, we plotted void ratio and absorption against the second cycle slake durability index (Figures 1 and 2) for all the mudrock samples. Without distinguishing between shales, mudstones, and argillites, there is no clear relationship between fabric and slake durability. This finding is not surprising as each of the three mudrock lithotypes have considerably different fabric. When each lithotype group and their respective properties are evaluated individually, a much more meaningful assessment can be made. It is clear from Figures 1 and 2 that the slake durability of the argillites is independent of the void spaces present within the argillite fabric. This reiterates the fact that the mineral grains have been recrystallized and the resulting mineral intergrowths are the probable factors controlling durability.

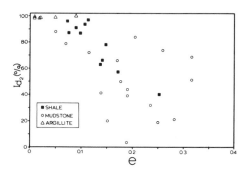

Figure 1. Relationship between void ratio and slake durability index for all mudrocks studied.

In Figures 3 and 4, respectively, the void ratio and percent absorption have been plotted against the slake durability index for the shales

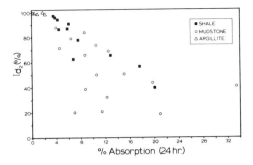

Figure 2. Relationship between absorption and slake durability index for all mudrocks studied.

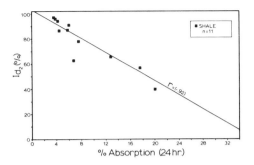

Figure 4. Relationship between absorption and slake durability index for shales only.

only. A very strong relationship between void ratio and slake durability index is indicated by the linear regression, having a correlation coefficient of (-.90). Similarly, a very strong relationship between percent absorption and slake durability index is indicated by the linear regression with a correlation coefficient of (-.92). Both absorption and void ratio show a strong correlation with slake durability. We believe these relationships can be attributed to the ordered arrangement of the mineral grains in the rock and the presence of structural laminations. The laminations are, in part, a reflection of this ordered rock fabric. The laminations are also a definite factor controlling the mode in which the shales slake. Both field observations and slake durability test results show that shales break down into tabular shaped particles.

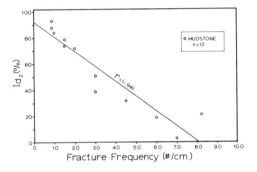

Figure 5. Relationship between micro-fracture frequency and slake durability index for mudstones only.

The poor relationship between fabric and slake durability for the mudstones (Figures 1 and 2) suggests that the fabric is not a dominant lithologic factor controlling durability. It was noted, however that micro-fractures were present in all the mudstones. The effect of micro-fracturing on durability was evaluated by comparing the micro-fracture frequency with the slake durability index as shown in Figure 5. The figure shows that indeed there is a very strong relationship between the two variables as indicated by the linear regression having a correlation coefficient of (-.94). This finding makes good sense because the micro-fractures are discontinuities along which mudstone slaking is initiated.

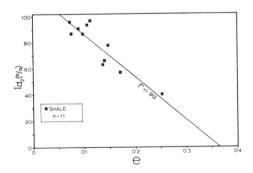

Figure 3. Relationship between void ratio and slake durability index for shales only.

CONCLUSION

The prediction of mudrock durability is complex because there are multiple lithologic factors controlling durability. This complexity is aggravated by the very fine-grained nature of mudrocks which makes the lithology difficult to study. As demonstrated here, the complexity of the problem can be simplified by separating the mudrocks into classes based on dominant lithologic characteristics. Shales are distinguished on the basis of their structural laminations. The durability of shales is related to the rock fabric as expressed by void ratio and absorption. Mudstones are distinguished on the basis of an absence of well developed structural laminations and the presence of micro-fractures. The durability of mudstones is related to the frequency of the micro-fractures. The durability of argillite is not a problem. The mineral grains in argillites have been recrystallized, resulting in a very durable rock.

The results presented here are not conclusive at this point. The samples studied so far represent only three of the five major mudrock classes. Claystones and siltstones remain to be studied. Moreover, the samples used contained only non-swelling clay minerals. Obviously, the presence of swelling clays will have an effect on durability. During the course of this work, twenty additional samples were collected that included many claystones and siltstones, additional shales and mudstones. Many of the samples contain swelling clay minerals. Work on these samples is in progress.

Many of the lithologic characteristics being studied can be quantified and directly compared with the engineering properties, slake durability index, and observed durability behavior. It is anticipated that when all the analyses are completed, a more comprehensive classification of mudrock durability, based on lithologic characteristics, will emerge.

REFERENCES

American Society for Testing and Materials (ASTM) (1987). Soil and Rock, Building Stones, Geotextiles: Annual Book of ASTM Standards, sect.4, vol. 4.08, 1189 pp.Philadelphia, Pennsylvania.

Burnett, A.D. & Fookes, P.G. (1974). A regional engineering geological study of the London Clay in the London and Hampshire basins. Quart. J. Eng. Geol. 7, 257-295.

Deo, P. (1972). Shales as embankment materials. Unpublished Ph.D. Thesis, Purdue University, West Lafayette, Indiana.

International Society for Rock Mechanics (ISRM) (1979). Suggested methods for determining water content, porosity, density, absorption and related properties and swelling and slake durability index properties. Int. J. Rock Mech. Min. Sci. and Geomech. Abstr. 16 (2), 143-156.

Perry, E.F. & Andrews, D.E. (1982). Slaking modes of geologic materials and their impact on embankment stabilization. Trans. Res. Record 837, 22-27. Transportation Research Board, Washington, D.C.

Picard, M. D. (1971). Classification of fine-grained sedimentary rocks. J. Sed.Pet. 41 (1), 179-195.

Potter, P.E., Maynard, J.B. & Pryor, W.A. (1979). Sedimentology of shale. 303 pp. Springer- Verlag, New York.

Russell, D.J. & Parker, A. (1979). Geotechnical, mineralogical and chemical inter-relationships in weathering profiles of an over-consolidated clay. Quart. J. Eng. Geol. 12, 107-116.

Russell, D.J. (1982). Controls on shale durability: the response of two Ordovician shales in the slake durability test. Can. Geotech. J. 19(1), 1- 13.

Shakoor, A. & Brock, D. (1987). Relationship between fissility, composition, and engineering properties of selected shales from northeast Ohio. Bull. Assoc. Eng. Geol. 24 (3), 363-379.

Skempton, A.W. (1964). Long-term stability of clay slopes. Geotechnique 14, 77-101.

Smart, G.D., Rowlands, N. & Isaac, A.K. (1982). Progress toward establishing relationships between the mineralogy and physical properties of coal measures rocks. Int. J. Rock Mech. Min. Sci. and Geomech. Abstr. 19(2), 81-89.

Spears, D.A. & Taylor, R.K. (1972). The influence of weathering on the composition and engineering properties of in situ coal measures rocks. Int. J. Rock Mech. Min. Sci. and Geomech. Abstr. 9, 729-756.

Strohm, W.E. (1980). Design and construction of shale embankments summary. U.S. Dept. Trans. Rept. FHWA-TS-80-219, Washington, D.C.

Caractérisation de la microdureté Vickers des roches par le microduromètre
Différence entre la dureté de la couche superficielle et des couches profondes des minéraux

Characterization of rocks Vickers microhardness with a microhardness tester
Hardness difference between superficial and deep layers of minerals

C.Galle, D.Leca, R.Struillou & C.Lapeyre
Ecole des Mines de Paris, France

RESUME: L'essai de microdureté Vickers est un essai statique qui consiste à appliquer sur le matériau à tester un pénétrateur en diamant de forme pyramidale à base carrée, sous une charge P, puis à mesurer la diagonale de l'empreinte laissée sur la surface. Les matériaux testés sont des échantillons de granite et de laitier de mangano-silicium polis au carbure de silicium à 12, 5, 2, 1, et 0.25 microns. Le polissage entraîne la formation d'une couche superficielle de dureté différente de celle des couches profondes. En fonction de la charge appliquée, on observe une très forte variation des duretés mesurées. Pour le feldspath et le quartz, le polissage conduit à une forte augmentation de la dureté en surface, surtout lorsqu'on utilise des abrasifs fins. Pour le laitier, au contraire la dureté superficielle est plus faible que la dureté de la couche profonde. Ces observations traduisent des comportements différents. L'utilisation de faibles charges ne permettrait d'accéder qu'à la dureté superficielle du matériau. La dureté de la masse serait obtenue avec des fortes charges. L'épaisseur moyenne de la couche superficielle a été évaluée à 2-4 microns. La dureté superficielle est de 850 kg/mm² pour le feldspath, 1500 pour le quartz et de 500 pour le laitier. Pour les couches profondes, les duretés sont respectivement de 600,1050 et 750 kg/mm².On discute de l'importance de ces modifications de dureté superficielle dans l'optique de l'abrasivité des matériaux, par exemple pour le concassage, et de la résistance à l'abrasion des roches polies, en particulier ornementales.

ABSTRACT: Vickers microhardness test is a static test consisting in puting on the material to be tested a square-based pyramidal diamond indenter under a P load and then in measuring the diagonal of the print left on the surface. The tested materials are granit and mangano-silicon foundry slag samples polished with silicon carbide at 12, 5, 2, 1 and 0.25 microns. With mechanical polishing, a superficial hardness layer appears which is different from that of deep layers. According to the load impressed, a considerable variation in the measured hardness can be observed. For felspar and quartz, the polishing involves a high increase of surface hardness above all when using fine abrasive. For foundry slag, on the contrary, the superficial hardness is lower than the deep layers hardness. These observations are significant of different behaviours. The use of small loads would only give access to the material superficial hardness. The bulk hardness would be obtained with heavy loads. The average thickness of the superficial layer was evaluated at 2-4 microns. The superficial hardness is 850 kg/mm² for feldspar, 1500 for quartz and 500 for foundry slag. For the deep layers, the respective hardnesses are 600, 1050, and 900 kg/mm².These modifications in superficial hardness are interesting with a view to materials abrasivity, for exammple for crushing and to the abrasion resistance of polished rocks, such as ornemental rocks.

1 INTRODUCTION

La mesure de la dureté d'un corps consiste à déterminer la résistance mécanique qu'il oppose à la pénétration d'un autre corps plus dur. La valeur de la dureté dépend des caractéristiques élastiques et mécaniques de la matière ainsi que de la forme, de la nature du pénétrateur et du mode opératoire. Parmi tous ces facteurs, l'état de surface du matériau testé, c'est à dire son degré de polissage, va influer de manière très importante sur les valeurs de dureté. Dans cet article, nous nous intéressons à l'essai de microdureté Vickers qui est un essai de dureté Vickers réalisé avec des charges petites (5 à 1000 g). l'essai de microdureté peut être considéré comme l'examen local de la dureté, intéressant le grain du matériau. L'empreinte laissée par le pénétrateur étant de l'ordre du micromètre, l'essai de microdureté Vickers, permet de déterminer les caractéristiques mécaniques à l'échelle du minéral.

Dans cette étude, nous avons travaillé avec deux types de matériaux : du granite, dans lequel nous avons sélectionné des plages de feldspath et de quartz et du laitier de mangano-silicium. Nous avons essayé de mieux comprendre l'influence du polissage sur la microdureté Vickers de ces matériaux. Nous montrons l'existence d'une couche superficielle dont nous déterminons l'épaisseur en fonction des paramètres d'essai : charge et polissage.

2 DESCRIPTION DE L'ESSAI ET PREPARATION DES ECHANTILLONS

Dans le cas de microdureté Vickers, le pénétrateur est un diamant en forme de pyramide droite à base carrée dont les faces forment un angle de 136°. L'essai consiste à faire pénétrer, à vitesse quasi nulle, cette pyramide dans l'échantillon à examiner, en appliquant une charge connue P et à mesurer la diagonale moyenne (d=(d1+d2)/2) de l'empreinte (Fig.1).

L'essai de microdureté Vickers est réalisé au moyen d'un microduromètre DURIMET fabriqué par E. LEITZ, Ltd., Wetzlar.

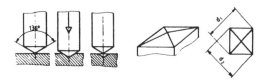

Fig.1 Schématisation de l'essai de microdureté Vickers et géométrie de l'empreinte

Le microduromètre est placé sur une table antivibrations ; la charge devant être appliquée sans choc ni vibration. L'essai est effectué à température ambiante ; Les temps de descente et de contact du diamant étant respectivement fixés à 17 et 13 secondes selon les spécifications du constructeur.

2.1 Définition de la microdureté Vickers

La valeur de la microdureté Vickers Hv est définie comme étant le rapport de la charge P sur la surface de l'empreinte. Connaissant la forme et les dimensions de celle-ci, on exprime facilement Hv en fonction de la diagonale moyenne d et de la charge P :

$$Hv = \frac{2*P*\sin 68°}{d^2} \qquad (1)$$

Soit

$$Hv = C * \frac{P}{d^2} \qquad (2)$$

Avec C = 1854.4

La profondeur d'empreinte h (µm) s'exprime en fonction de la diagonale d de l'empreinte de la manière suivante :

$$h = \frac{d*\sqrt{2}}{4*tg 68°} \qquad (3)$$

2.2 Définition de l'indice de Meyer

Si l'on considère que la valeur de la dureté Vickers d'un matériau a été obtenue à partir de mesures réalisées sur une surface homogène, alors cette dureté est théoriquement indépendante des charges utilisées. Dans ce cas, la relation liant la charge, la dureté et l'empreinte est la suivante :

$$\log P = \frac{Hv}{C} + 2*\log d \quad (4)$$

La pente de cette droite est rigoureusement égale à 2
Expérimentalement il s'avère que l'on obtient une droite d'équation :

$$\log P = \log a + n*\log d \quad (5)$$

Cette expression est équivalente à :

$$P = a*d^n \quad (6)$$

Cette équation correspond à la droite de Meyer ; la pente de la droite se nomme indice de Meyer. La valeur de ce dernier exprime les différences de dureté superficielle du matériau testé.

Si n<2 la dureté diminue avec l'augmentation de la charge.
Si n=2 la dureté est constante et indépendante de la charge.
Si n>2 la dureté augmente avec la charge.

2.3 Préparation et sélections des échantillons

Pour chaque type de matériau nous avons sélectionné des zones de même nature. Pour le feldspath nous avons travaillé sur 6 échantillons de granite polis respectivement à 12, 5, 2, 1 et 0.25 µm et un échantillon présentant uniquement la surface de sciage. Pour le quartz nous avons travaillé sur 4 échantillons de granite polis à 5, 2, 1 et 0.25 µm. Enfin pour le laitier, 4 échantillons également, polis à 5, 2 et 0.25 µm et un échantillon non poli (surface de sciage).
Les échantillons sont successivement soumis à des polissages de 50, 30, 18, 12, 5, 2, 1 et 0.25 µm. Chaque étape de polissage dure en moyenne entre 15 et 20 minutes. A titre d'exemple, un échantillon dont le polissage final est 5 µm

aura été poli à 50, 30, 18, 12 et 5 µm totalisant ainsi un temps de polissage moyen de 1ᴴ30. Un échantillon dont le polissage final est 0.25 µm aura été poli à 50, 30, 18, 12, 5, 2, 1, et 0.25 µm totalisant ainsi un temps de polissage moyen de 2ᴴ20. Le polissage est effectué avec du carbure de silicium.

2.4 Acquisition des données

Pour chaque finesse de polissage final et pour chaque échantillon de matériau testé, nous avons étudié un nombre de zones représentatives. Pour les feldspaths nous avons toujours sélectionné 5 zones, pour le quartz 2 zones et pour le laitier 2 zones. A partir des résultats relatifs à chaque zone nous avons déduit un comportement moyen du matériau et ceci pour chaque type de polissage.
Les tests de microdureté Vickers ont été réalisés avec des charges comprisent entre 15 g et 1000 g. Pour chaque charge nous avons effectué trois essais à partir desquels nous avons calculé une valeur moyenne de Hv. Nous avons ensuite calculé la moyenne des Hv pour chaque zone et pour chaque charge utilisée.

3 RESULTATS ET DISCUSSION

Les résultats (variation de la microdureté Vickers en fonction de la charge et de la profondeur) sont représentés par les figures 2 et 5 pour le feldspath, 3 et 6 pour le quartz et 4 et 7 pour le laitier.
Nous donnons dans les tableaux 1, 2, et 3, les résultats moyens respectivement pour le feldspath, le quartz et le laitier. Dans la grande majorité des publications traitant de dureté, la microdureté Vickers est exprimée en kg/mm². Par souci de clarté, nous adopterons cette unité de pression en rappelant les correspondances d'unité suivantes :

$1 \ kg/mm^2 \approx 1 \ bar \approx 10^{-5} \ Pa \approx 0.1 \ MPa$

Hvs : microdureté Vickers moyenne en Kg/mm² (charge ≤ 100 g)

Hv : microdureté Vickers moyenne en Kg/mm²

Hvp : microdureté Vikers moyenne

en Kg/mm² (charge ≥ 385 g)
h : profondeur en µm
n : indice de Meuer (sans unité)

NB : Les finesses de polissage sont reporté en µm dans la colonne de gauche (* pas de polissgae).

Tab.1 Microdureté Vickers, profondeur (valeurs moyennes) et indice de Meyer en fonction du polissage pour le feldspath.

	Hvs	Hv	Hvp	h	n
*	194	332	417	5.62	2.86
12	464	530	535	4.53	2.13
5	420	547	606	4.37	2.34
2	748	687	620	4.10	1.83
1	847	723	638	4.04	1.80
0.25	943	756	635	4.01	1.74

Tab.2 Microdureté Vickers, profondeur (valeurs moyennes) et indice de Meyer en fonction du polissage pour le quartz.

	Hvs	Hv	Hvp	h	n
5	380	569	680	4.23	2.58
2	1410	1187	1027	3.16	1.77
1	1575	1224	1043	3.17	1.71
0.25	1543	1245	1075	3.10	1.75

Tab.3 Microdureté Vickers, profondeur (valeurs moyennes) et indice de Meyer en fonction du polissage pour le laitier.

	Hvs	Hv	Hvp	h	n
*	118	170	289	5.22	3.17
5	431	524	727	3.03	2.33
2	429	555	807	2.91	2.71
0.25	750	726	676	2.74	1.91

L'ensemble des résultats obtenus nous amènent à constater plusieurs faits importants. Lorsque le polissage est fin, 1 et 0.25 µm et moyen 2 µm (sauf pour le laitier), la microdureté Vickers diminue lorsque la charge augmente. Au contraire, lorsque le polissage est grossier, 12, 5, voir 2 µm pour le laitier), la microdureté Vickers augmente avec la charge. D'autre part, à charge égale faible (inférieure à 100g), la microdureté Vickers augmente avec l'intensité du polis-

sage. Pour des charges plus forte (supérieures ou égales à 385g), la valeur de la microdureté a tendance à se stabiliser quelque soit le degré de polissage. C'est avec le feldspath et le quartz que les résultats sont les plus cohérents (cf. Fig.2, Fig.3 et Fig.4) Le laitier dans son ensemble se comporte de la même manière que le feldspath et le quartz excepté pour le polissage à 2 µm.

Des phénomènes semblables sont observés lorsque l'on étudie la variation de la microdureté Vickers en fonction de la profondeur de l'empreinte (cf. Fig.5, Fig.6 et Fig.7).

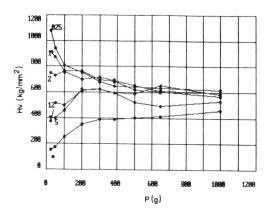

Fig.2 Variation de la microdureté Vickers Hv avec la charge P pour différentes finesses de polissage (12, 5, 2, 1, 0.25 µm et * sans polissage) pour le feldspath

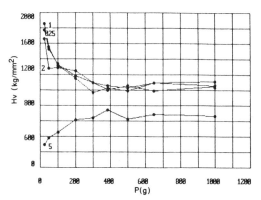

Fig.3 Variation de la microdureté Vickers Hv avec la charge P pour différentes finesses de polissage 5, 2, 1, 0.25 µm) pour le Quartz

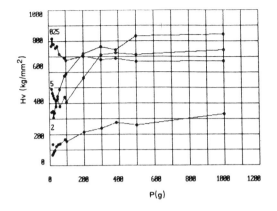

Fig.4 Variation de la microdureté Vickers Hv avec la charge P pour différentes finesses de polissage (5, 2, 0.25 µm et * sans polissage) pour le laitier

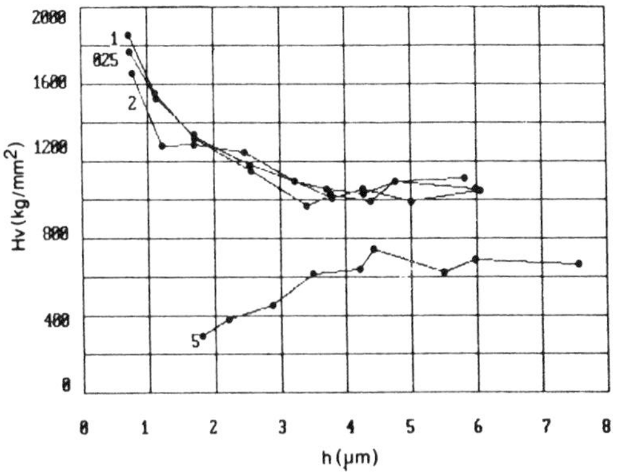

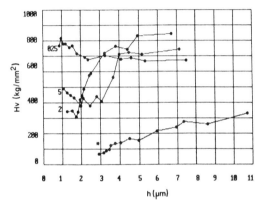

Fig.6 Variation de la microdureté Vickers Hv avec la profondeur h de l'empreinte pour différentes finesses de polissage (5, 2, 1, 0.25 µm) pour le quartz

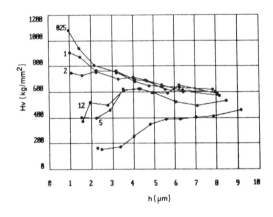

Fig.5 Variation de la microdureté Vickers Hv avec la profondeur h de l'empreinte pour différentes finesses de polissage (12, 5, 2, 1, 0.25 µm et * sans polissage) pour le feldspath

Fig.7 Variation de la microdureté Vickers Hv avec la profondeur h de l'empreinte pour différentes finesses de polissage (5, 2, 0.25 µm et * sans polissage) pour le laitier

Nous donnons dans les tableaux 4 et 5 les moyennes de Hvs (microdureté moyenne de la couche superficielle) et Hvp (microdureté Vickers moyenne de la masse) Hvs et Hvp en kg/mm² ainsi que les profondeurs moyennes atteintes en µm en

fonction du degré de polissage :

Finesse de polissage :
A, fin, 0.25 et (ou) 1 µm
B, moyen, 2 µm
C, grossier, 5 et (ou) 12 µm

Tab.4 Microdureté Vickers superficielle en fonction du polissage.

	Feldspath		Quartz		Laitier	
	Hvs	h	Hvs	h	Hvs	h
A	895	1.55	1559	1.18	750	3.87
B	748	1.64	1410	1.22	429	1.95
C	442	2.16	380	2.88	431	1.99

Tab.5 Microdureté Vickers de la
masse en fonction du polissage.

	Feldspath		Quartz		Laitier	
	Hvp	h	Hvp	h	Hvp	h
A	637	6.07	1059	4.69	676	5.81
B	620	6.18	1027	4.78	607	5.36
C	571	6.43	680	5.87	727	5.60

On constate que les valeurs de
microdureté de la couche superficielle du feldspath et du quartz
sont assez bien groupées pour des
polissages fin et moyen. Cette
dispersion devient très faible pour
la microdureté de la masse. Pour un
polissage grossier, on peut consi-
derer que les mesures sont réalisées sur une surface hétérogène ;
la dispersion est plus importante
dans ce cas. On peut donc conclure
que la microdureté de la masse
caractérise le matériau, à condition que la finesse de polissage
soit au moins de 2 µm.

En ce qui concerne le laitier,
nous n'observons pas de regrou-
pement pour les polissages fins et
moyens ; celui-ci est observé pour
les polissages moyens et grossiers.
Pour la microdureté de la masse la
dispersion des valeurs est faible.
La moindre cohérence des résultats
relatifs au laitier peut être
expliquée par le fait que c'est un
matériau artificiel amorphe finement stratifié. Cette stratification délimiterait des micro-
couches de compositions chimiques
et de caractéristiques mécaniques
différentes.

Le polissage entraîne la formation d'une couche superficielle
dont les caractéristiques mécaniques sont différentes de celles
du matériau en profondeur. Pour un
polissage fin la surface du matériau s'indure et la dureté de la
couche superficielle est supérieure

à celle de la masse du matériau.
Dans le cas d'un polissage grossier, la surface du matériau se
fragilise et l'on observe le
comportement inverse. Ces phénomènes ont également été observés
sur des métaux comme l'acier, le
cuivre et l'aluminium qui ont fait
l'objet d'études nombreuses.

A faible charge, et quelque soit
le polissage il n'est donc pas
possible d'accéder à la dureté
réelle du matériau car l'influence
de la couche superficielle est
prépondérante celle-ci n'ayant été
que partiellement pénétrée. Afin
d'atteindre le matériau en profondeur et de minimiser l'effet
perturbateur de cette couche superficielle sur la valeur de la microdureté, Il est souhaitable de travailler avec des charges plus
élevées (à partir de 400g).

Relation entre la microdureté
Vickers et la profondeur d'empreinte.

A partir des expressions (2), (3)
et (6), on peut exprimer la microdureté Vickers Hv en fonction de la
profondeur de l'empreinte h :

$$Hv = C*a*\left[\frac{4*t_g\ 68°}{\sqrt{2}}\right]^{n-2} * h^{n-2} \qquad (7)$$

Soit

$$Hv = C * a * 7^{n-2} * h^{n-2} \qquad (8)$$

Avec n, indice de Meyer et a
ordonnée à l'origine de la droite
de Meyer. Ainsi cette dernière
expression peut se mettre sous la
forme plus générale :

$$Hv = k*h^{n-2} \qquad (9)$$

Hv en kg/mm², h en µm,
n indice de Meyer

L'équation (9) correspond à loi
polynomiale de puissance n-2.

Nous avons déterminé cette équation pour les différents types de
matériau en fonction de la finesse
de polissage. Parallèlement nous
avons calculé l'équation de la
courbe qui ajuste au mieux les
points expérimentaux. La corrélation au niveau des expressions
est bonne (cf.ci-après). Nous véri-

fions ainsi la Loi de Meyer. Par contre le calage des points expérimentaux sur les courbes ne peut être considéré comme favorable que dans le cas des polissages fins (0.25 et 1 µm) en particulier pour le feldspath et le quartz. Pour les polissages moyens (2 µm) et grossiers (5 et 12 µm) le calage est moins satisfaisant. Nous donnons un exemple de ces calages pour un polissage de 0.25 µm avec les figures 8, 9, et 10

Résultats : L'expression A est tirée de la relation (9) ; l'expression B est tirée de la courbe d'ajustement
NB : Les finesses de polissage sont reportées en µm dans la colonne de gauche (* pas de polissage).

Corrélation des équations pour le Feldspath :

	A	B
*	$Hv=74*h^{0.857}$	$Hv=70*h^{0.901}$
12	$Hv=431*h^{0.134}$	$Hv=442*h^{0.124}$
5	$Hv=326*h^{0.343}$	$Hv=351*h^{0.315}$
2	$Hv=849*h^{-0.171}$	$Hv=808*h^{-0.136}$
1	$Hv=911*h^{-0.203}$	$Hv=923*h^{-0.209}$
0.25	$Hv=1008*h^{-0.262}$	$Hv=1033*h^{-0.275}$

Corrélation des équations pour le Quartz :

	A	B
5	$Hv=250*h^{0.580}$	$Hv=251*h^{0.579}$
2	$Hv=1463*h^{-0.231}$	$Hv=1455*h^{-0.223}$
1	$Hv=1577*h^{-0.291}$	$Hv=1582*h^{-0.292}$
0.25	$Hv=1550*h^{-0.252}$	$Hv=1553*h^{-0.251}$

Corrélation des équations pour le Laitier :

	A	B
*	$Hv=24*h^{1.165}$	$Hv=25*h^{1.150}$
5	$Hv=367*h^{0.327}$	$Hv=372*h^{0.322}$
2	$Hv=260*h^{0.714}$	$Hv=266*h^{0.708}$
0.25	$Hv=775*h^{-0.087}$	$Hv=777*h^{-0.089}$

Nous avons essayé d'évaluer l'épaisseur de la couche superficielle (µm) (cf Fig.8, Fig.9, Fig.10) en déterminant la profondeur correspondant à la rupture de pente des courbes que nous avons

calculées (intersection des tangentes). Les résultats sont donnés dans le tableau 5 qui suit.
NB : Les finesses de polissage sont reportées en µm dans la colonne de gauche.

Tab.5 Estimation de la Profondeur de la couche superficielle

	Feldspath	Laitier	Quartz
0.25	2.5±0.5	2.4±0.5	2.0±0.5
1	3.1±0.5	/	2.0±0.5
moyenne	2.8±0.5	2.4±0.5	2.0±0.5
Hv moyen (Kg/mm²)	637	676	1059

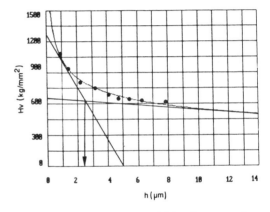

Fig.8 Evaluation de l'épaisseur de la couche superficielle (polissage 0.25 µm) pour le feldspath (2.5 µm)

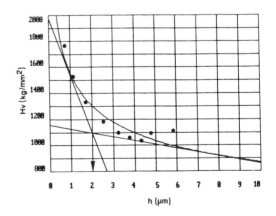

Fig.9 Evaluation de l'épaisseur de la couche superficielle (polissage 0.25 µm) pour le quartz (2 µm)

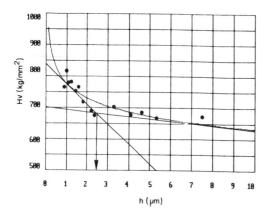

Fig.10 Evaluation de l'épaisseur de la couche superficielle (polissage 0.25 µm) pour le laitier (2.4 µm)

L'évaluation de l'épaisseur de la couche superficielle est délicate en raison du positionnement des tangentes. D'autre part, il semble que le passage de la couche superficielle à la masse du matériau soit progressif (pas de rupture de pente brutale au niveau des courbes) et que la limite entre ces deux domaines reste diffuse. Cela signifie que l'influence de la couche superficielle persiste sur 0.5 à 1 µm au-delà de son épaisseur moyenne. Les résultats montrent cependant une bonne corrélation entre la microdureté Vickers et la profondeur moyenne de la couche affectée par le polissage. Plus le matériau est dur et plus l'épaisseur de la couche superficielle est faible.

4 Conclusion

L'essai de microdureté Vickers est simple et rapide à mettre en oeuvre. Il permet de caractériser mécaniquement un matériau à l'échelle du grain. Néanmoins certaines précautions doivent être prises quant à la préparation des échantillons. La finesse de polissage et le choix des charges sont des facteurs primordiaux. Un polis-

sage fin entraînera une induration du matériau d'oú une augmentation très importante de la microdureté Vickers dans la couche superficielle. Un polissage grossier provoquera l'effet inverse en diminuant la microdureté dans cette même couche. Dans ces deux cas, la mauvaise évaluation de la dureté pourra avoir de graves conséquences quant à l'utilisation du matériau notamment dans les domaines de la résistance à l'abrasion et du concassage (exemple :le laitier).

Afin d'avoir une valeur significative de la microdureté Vickers, il est nécessaire de travailler avec des charges supérieures ou égales à 400 g afin de traverser totalement la couche superficielle et donc de minimiser son effet perturbateur sur le résultat final. La réalisation d'essais à faibles charges entraînera donc une surestimation de la microdureté du matériau. Nous avons évalué l'épaisseur moyenne de la couche superficielle à 3 µm pour le feldspath, 2 µm pour le quartz et de 2.5 µm pour le laitier. dans le cas des polissages fins (0.25 et 1 µm). Pour avoir un confort de mesure optimum (lecture des empreintes) le polissage devra être de bonne qualité et fin, 0.25 µm si possible ou 1 µm.

L'utilisation de charges proches de 1000 g est déconseillée, la lecture des empreintes devenant excessivement difficile. En effet, l'impact de telles charges provoque l'apparition d'une auréole fragilisée perturbatrice autour de l'empreinte.

Pour obtenir des valeurs de microdureté Vickers représentatives du feldspath et du quartz, il est nécessaire d'effectuer les essais avec des charges comprises entre 400 g et 600 g.

REFERENCES

BUCKLE, H. 1960. L'essai de microdureté et ses applications. Publications scientifiques et techniques du Ministère de l'air.

Mott, B.W. 1956. Micro-indentation Hardness Testing. London, Butterworths scientific publication.

6th International IAEG Congres / 6ème Congrès International de AIGI, © 1990 Balkema, Rotterdam. ISBN 90 6191 130 3

Settlement due to saturation

Tassement par saturation

K.-D. Hauss & M. H. Heibaum
Federal Waterway Engineering and Research Institute (BAW), Karlsruhe, FR Germany

ABSTRACT: Settlement due to saturation is one of the reasons for deformation of earth- or rockfill dams as well as of natural ground. In Germany the dam failures at two canals in 1976 and 1979 led to investigations on sandy fill materials. Research on rockfill material has been undertaken since the sixties and collapse of fine grained soil when inundated has been known even longer. In the BAW coal mining wastes and silty soils were tested on their behaviour during first or repeated saturation. The paper reviews earlier results concerning settlement due to saturation of distinct materials and presents our own laboratory tests. A procedure is proposed to estimate the tendency of a given soil towards settlement due to saturation.

RESUME: Le tassement par saturation est une raison de la déformation des barrages et des remblais en terre ou en enrochement ainsi que celle du sol naturel. En Allemagne, à la suite de la rupture des barrages de deux canaux en 1976 et 1979 on entreprit des recherches sur les remblais sableux. Dès les années soixante on a déjà poursuivis des recherches sur l'enrochement; quant à l'écroulement des sols cohérent après inondation, il est connue depuis plus longtemps. A l'institut BAW, on a examiné l'enrochement des roches minières et le sol d'un cône d'éboulis lors d'une saturation unique et lors des saturations répetées. Ce rapport donne un résumé des recherches connues, il presente les propres essais et il propose un procédé d'estimation de la tendance au tassement après la saturation d'un remblai.

1 INTRODUCTION

From experience of dam and dike construction and of placement of very high embankments as well as of large fills, e.g. restored opencast mining sites, it is known that soils and rock fragments may show a considerable amount of settlement (subsidence), when drowned or wetted the first time after being placed. Settlements may reach 35 % depending on the kind of soil or rock, on the compaction and on the pre-wetting.

This kind of volume decrease accompanying increases in water content at essentially unchanged total stresses may be termed '(wetting-induced) collapse', 'hydro-consolidation', 'hydro-compaction', 'settlement due to saturation'. The reason for this is a collapse of the grain skeleton, being metastable and suffering a certain total stress. Settlement increases with decreasing quality of compaction of the fill. Besides inundation, also dynamic loads or loads exceeding a critical col-

lapse load may create the sudden volume decrease, whereby all critical loads depend on the initial void ratio and the initial water content.

Settlement due to saturation may be hazardous with embankment dams. An embankment fill is placed with a water content far from saturation. Because of the leaking of a sealing or because of an abnormal rise of the ground water table, originally only humid parts of the fill may be saturated, leading to the collapse of the original structure of the grain skeleton. A significant volume decrease and thus a lowering of the dam crest may be the result, creating the danger of overtopping, this causing the erosion of the fill material and thus leading to an inundation of the downstream land.

The problem of settlement due to saturation is not a new one. Nevertheless, scientific research has concentrated always on one example or one specific kind of soil. In Germany there is a first hint in a soil mechanics manual (Schulze/ Muhs,

1950), where a settlement of 5 - 12% is mentioned as a design value. Already they have said that prevention measures may be taken by compacting the fill properly and thus reducing the settlement to 0.3 - 2 %.

2 REVIEW OF THE PHENOMENON

2.1 Collapse of cohesive soils

Research on loessian soils was published by Kezdi (1964, 1973). Kezdi reports 22 - 34 % of volume decrease because of a collapse of the mineral structure of a loessian deposit (fig.1). Fig.2 shows the severe volume decrease of such a collapse and emphasizes the need to avoid this reason of subsidence. Collapse may be initiated even by rising capillary water.

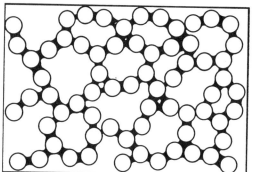

Fig.1: Collapsible soil structure

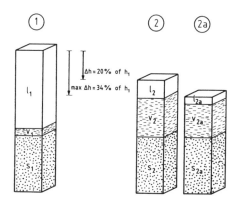

Fig.2: Volume decrease of loessian soils

Lawton et al. (1989) report on compacted clayey sand which exhibits a significant settlement due to saturation. The authors observed that at those stress levels where wetting-induced volume changes occurred, the amount of collapse varied inversely

with the compaction water content, when the degree of compaction was kept constant. Maximum collapse was found when the soil was allowed to take in water under an overburden stress equal to a so called compactive prestress. To evaluate the collapse potential for a given soil, the authors recommend a double-oedometer test, one test with an as-compacted soil sample, the other one with an inundated sample. This kind of testing was already proposed by Kezdi (1973).

Serious settlements because of a rise in the water table is reported of Thomson/ Schulz (1985). The lower parts (depth over 20 m) of a backfill of a restored opencast mining site were subsequently saturated by the rising groundwater table and thus leading to settlements up to 15 cm.

2.2 Settlement due to saturation in sandy soils

In 1981, Hellweg published research results on settlement due to saturation of different sands: The volume decrease of a soil is a combination of collapse and compression (fig.3).

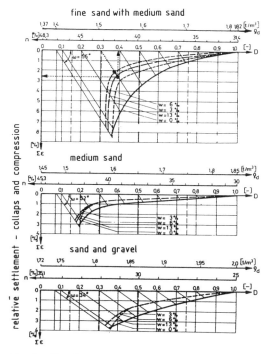

Fig.3: Total settlement $\Sigma \varepsilon$ versus compactness D (from: Hellweg, 1981)

Under low stress conditions, the major part is collapse while compression acts only to a smaller extent. The total settlement is independent of the overburden stress, when original void ratio and water content are kept constant. Maximum values were got with loose materials in dry state. Repeated drowning was only 1/10 as effective as the first one.

The German recommendations of water front structures (1985) give design values of settlement caused by inundation of 8 % as to a loose fill and 1 - 2 % after proper compaction. Rounded grains exhibit more settlement than angular ones, but such differences are obtained only when the material is placed loosely.

Similar results were gained by Prühs (1982) with clayey sands which were tested during the construction works for the Main-Donau-Canal. What there was found was a very strong relationship of the degree of settlement due to saturation and the compaction of the material (fig.4). There is also a significant strong curvature of the proctor compaction curve of that material.

als), and of the grain skeleton (grain size distribution, void ratio, effective stress) have to be investigated. A maximum was found with loosely placed, poorly graded rock fill, since fills which are well graded, are easier to compact, thus reducing the settlement due to saturation (fig.5).

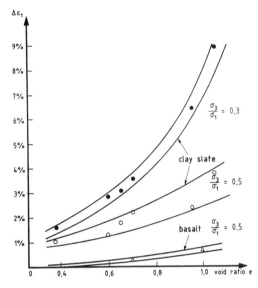

Fig.5: Settlement due to saturation $\Delta\varepsilon_1$ versus void ratio e (from: Schade, 1984)

The basic reason seems to be the stress release of latent stress and very high edge stresses, thus leading to fragmentation of the grains and to volume decrease. Rounded grains exhibit much less fragmenting and correspondingly less settlement. Pre-wetting of the rock fill reduces the settlement due to saturation considerably.

Angular rock fill material has to be placed in comparably thin layers and a high service weight of the compactor is necessary. Material of low grain strength is more easy to compact and shows a beneficial grain size distribution after compaction.

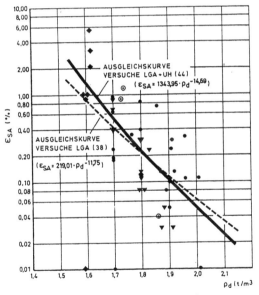

Fig.4: Settlement due to saturation ε_{SA} versus dry density ϱ_d (from: Prühs, 1982)

2.3 Rock materials

Schade (1984) pointed out some basic differences in settlement due to saturation with earth fill and rock fill. Considering rock fill, the characteristics of the single grain (size, strength, miner-

2.4 Conclusion of the different research reports

As has been found out by the above mentioned research reports (and many others), there seem to be three more or less different reasons for settlement due to saturation: (1) Cohesive soils show collapse effects because of their mineralogical structure, (2) sandy soils exhibit an apparent cohesion which vanishes, when the material is saturated, (3) rock fill suffers stress release of latent stress which is supported by access of water and of fragmentation because of extremely high edge stresses.

Since nearly every researcher had a special target, the evaluation of the results differs very much and even results of nearly the same materials aren't comparable. An attempt is made in section 4 to get comparable information on all of the tested materials to estimate the amount of settlement to be expected.

3 BAW LABORATORY TESTS

It is quite difficult to separate the three influences mentioned above, when using a fill material which contains rock fragments as well as fine grained soils. The Federal Waterway Engineering and Research Institute (BAW) was faced with this problem of the subsoil at the river Saar (talus) and of the problem of the mining waste materials which are used frequently as fill material of the embankment dams of the canals in the Federal Republic of Germany. Mining subsidence leads to the necessity of heightening dams. So the lower parts of the dam, originally dry, is then submerged into the ground water (fig.6). Since these submerged parts are the oldest ones, they are the layers with the poorest compaction. Settlement due to saturation therefore is almost inevitable.

A laboratory testing device was developed, comparable to an oedometer apparatus enlarged 20 times (fig.7). The material tested was placed on top of a granular filter and then loaded by an air

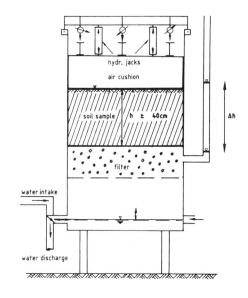

Fig.7: Testing device

cushion, which was able to follow all deformations of the surface of the soil specimen. The specimen could be submerged at any desired velocity, duration and frequency.

Depending on the degree of compaction, a certain amount of settlement due to saturation was observed. But it was impossible, to detect one distinct mechanism. Volume decrease was caused by fragmenting as well as loss of apparent cohesion and, to a lesser extent, of collapse of the cohesive parts.

Fig. 8 shows the settlement due to saturation versus the degree of compaction, i.e. the ratio of dry density and

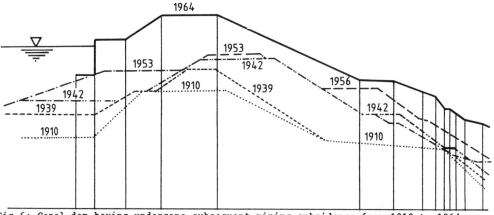

Fig.6: Canal dam having undergone subsequent mining subsidence from 1910 to 1964

proctor density, versus the relative
settlement due to saturation. Both soils
(talus and mining wastes) were tested with
50 kPa and without surface load. Both
soils seemed to be loaded lower than the
compactive prestress mentioned by Lawton
et al. (1989), since the surface load
leads to an increase in settlement due to
saturation.

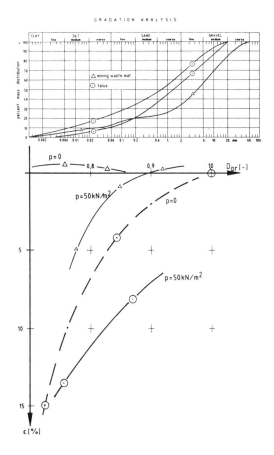

Fig.8: Settlement due to saturation of two
susceptible soils (BAW laboratory tests)

The waste material showed considerable
swelling, an effect which was observed by
other researchers with cohesive materials
only. Even though the waste material
exhibits a very wide grain size distribution, the cohesive fracture seems to be
dominant, since the swelling is of dominant influence without surface load. With
a surface load, swelling is monitored only
when the material is well compacted which
is the usual result with fills containing
clay minerals.

4 ESTIMATION OF SETTLEMENT

A measure was looked for, to judge a fill
material as to its tendency towards settlement due to saturation, independent of
the three possible effects mentioned in
section 2.4. And a scheme was needed to be
able to compare different research results. As has been shown in figures 3 to 5
every other uses his own parameters. To be
able to compare the results, the degree of
compaction was chosen as it has been done
presenting the BAW results in fig.8. Since
it was not possible to get further than
the published data, not all published
results could be transformed to that
relationship.

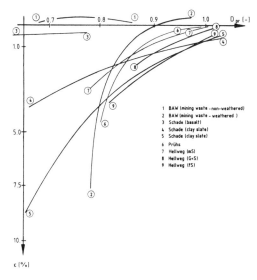

Fig.9: Settlement due to saturation of
selected soils

Fig.9 shows the degree of compaction
versus the observed settlement of most of
the authors cited above. Most of the
materials show the typical nonlinear
increase of settlement with decreasing
degree of compaction.
 Fig.10 gives the proctor compaction
curves of the same materials. It is evident, that a strongly bent proctor compaction curve in fig.10 is accompanied by a
definite curvature on the settlement
diagram (fig. 9). If the radius of curvature of the settlement curve is compared
to the maximum rising of the proctor
compaction curve at an interval of
$\Delta w = 3\%$, a strong relation is to be found,
when a logarithmic scale is used (fig.11).

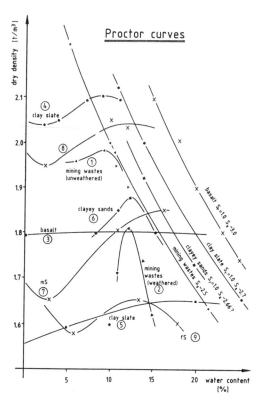

Fig.10: Proctor compaction curves of selected soils

Thus, at least a quantitative measure has been found to evaluate a tendency towards settlement due to saturation when comparing different fill materials. Usually numerous Proctor tests are carried out when testing a fill material, so there is little additional effort involved.

5 CONCLUSIONS

Settlement due to saturation may be created by different mechanisms which may occur separately depending on the soil or rock material of a fill of a subsoil, but which may be combined when materials are used that contain all particle sizes from clay to rock fragments (e.g. mining waste material). To estimate the possible settlement due to saturation of a certain material it was found that the curvature of the proctor compaction curve gives a quantitative measure, thus providing additional information without additional effort. The double oedometer test (Lawton et al., 1989) also seems to be a reliable method, but that's asking for extra tests. Till now it has not been tested on the soil materials discussed in section 3.

Since all the mentioned effects are predominant on the dry side of proctor density, one of the most important countermeasures is the placing of the fill at a water content on the wet side of proctor density. Rockfills should be pre-wetted to release latent stresses. The only objection is an increased swelling of materials

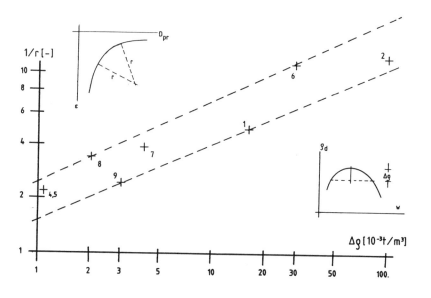

Fig.11: radius of curvature of the settlement curves versus the rising of proctor compaction curves

with clay minerals when increasing the
water content.

Generally, all fills should be placed
with proper compaction, which may be
difficult if poorly compacted old layers
underlie high fills as in restored opencast mining sites or in dams several
times heightened because of mining subsidence. Then laboratory tests are necessary
to estimate the possible settlement due to
saturation to be able to protect relevant
structures.

With the methods mentioned above it is
possible to consider countermeasures from
the very beginning and to reduce settlements due to saturation to a negligible
amount.

REFERENCES

German recommendations on waterfront
    structures (Empfehlungen des Arbeitsaus-
    schusses 'Ufereinfassungen') 1985.
    Berlin, Verlag Ernst und Sohn.
Hellweg, V. 1981. Ein Vorschlag zur Ab
    schätzung des Setzungs- und Sackungsver-
    haltens nichtbindiger Böden bei Durch-
    nässung. Mitt.Inst.f.Grundbau, Bodenme-
    chanik und Energiewasserbau der Univer-
    sität Hannover, Heft 17.
Kezdi, A. 1964. Handbuch der Bodenmecha-
    nik, Bd.I. Berlin, VEB Verlag für Bauwe-
    sen.
Kezdi, A. 1973. Handbuch der Bodenmecha
    nik, Bd.III. Berlin, VEB Verlag für
    Bauwesen.
Lawton, E.D.; Fragaszy, R.J.; Hardcastle,
    J.H. 1989. Collapse of compacted clayey
    sand. Journal of Geotechnical Enginee-
    ring, Vol.115, No.9, pp 1252-1265.
Prühs, H.; Stenzel, G.; Feile, W. 1982.
    Untersuchungen von Sackungen an Keuper-
    sanden. Vorträge der Baugrundtagung in
    Braunschweig. Essen, DGEG.
Schade, H. 1984. Die Sättigungssetzungen
    von Steinschüttungen. Dissertation
    Technische Hochschule Darmstadt (D 17).
Schulze, E. & Muhs, H. 1050. Bodenuntersu
    chungen für Ingenieurbauten. Berlin &
    Heidelberg, Springer.
Thomson, S. & Schulz, T.M. 1985. Settle
    ment studies of an open pit mine back-
    fill in Western Canada. In Geddes, J.D.
    (ed.), Ground movements and structures,
    p.423-442. London, Pentech Press.

Clay resources in the Netherlands – An inventory
Inventaire des ressources en argile aux Pays-Bas

R. Hillen
Geological Survey of the Netherlands, Haarlem, Netherlands

J.Th. van der Zwan
Public Works Department, Road and Hydraulic Engineering Division, Netherlands

ABSTRACT: In The Netherlands more than 6 million tons of clay are used annually in the ceramic industry and for dike construction. To evaluate the potential clay resources, an inventory was taken during 1988/89. Phase 1 of this inventory involved evaluating the material requirements and test methods used in the ceramic industry and the dike construction industry, and an evaluation of the data and relevant test methods available at the mapping institutes. Recommendations were formulated for a 1:250.000 reconnaissance map showing clay resources suitable for the ceramic industry and dike construction. Phase 2 involved production of this map by means of a Geographical Information System and calculation of the available clay reserves. The results of the inventory will be used for the national policy on surficial raw materials.

RESUME: Chaque année, plus de 6 millions de tonnes d'argile sont utilisées aux Pays-Bas par l'industrie céramique et pour la construction de digues. Un inventaire des ressources potentielles en argile a été réalisé en 1988/89. La première phase de cet inventaire a consisté en une évaluation des exigences matérielles et des méthodes de tests utilisées par l'industrie céramique, ainsi que par les compagnies de construction de digues d'une part, et une comparaison entre les données disponibles dans les instituts cartographiques et les méthodes de tests d'autre part. Des recommandations sont proposées en ce qui concerne une carte de reconnaissance au 1:250.000^{eme} des ressources en argile propre a l'industrie et à la construction. La seconde phase a consisté dans la réalisation de cette carte par l'intermédiaire d'un Système d'Information Géographique, ainsi que dans le calcul des réserve d'argile disponibles. Les résultats de cet inventaire sont utilisés pour une réglementation nationale sur l'utilisation des matières premières de surface.

1 INTRODUCTION

In The Netherlands excavation of construction materials is essentially restricted to unconsolidated deposits at or near the surface (De Mulder, 1984). The country is situated in the delta of the rivers Rhine, Meuse and Scheldt and is underlain by several hundreds of metres of sand, loam, clay and peat. Hard rock near the surface occurs only in the southeast and extreme east of the country.

The principal construction materials in The Netherlands are sand, clay, gravel and limestone. Annually, about 2,5 million tons of limestone from the southeast are used for cement and agricultural purposes.

Gravel is used mainly for concrete and asphalt (10 million tons/year); sand is used for filling (75 million tons/year), for concrete and cement (20 million tons/year) and for the glass industry (0,4 million tons/year). Clay is used mainly for ceramic purposes and for dike construction (6 million tons/year). These quantities do not include imported construction materials.

In 1989 the Ministry of Transport and Public Works presented a plan to Parliament outlining the national policy on surficial raw materials up to the year 2010. Some of the main conclusions were (Min. Transport & Public Works, 1989):

1. a good coordination between excavation companies and the town and country planning authorities is required;
2. with regard to the need for raw materials in The Netherlands, sufficient exploitation oppertunities for private companies should be guaranteed;
3. the application of offshore raw materials is encouraged;
4. the application of alternative aggregates is encouraged.

The above conclusions have certain implications for clay resources. Most of the clay for ceramic purposes used to be excavated in the forelands of the large rivers, i.e., in the areas in between the river dikes that are only flooded at very high water levels. However, clay resources in these areas are limited and because of ecological and environmental reasons permits for clay excavation in the river forelands are difficult to obtain. To ensure the long-term future of the brick factories in the river area, clay resources have to be found beyond the river forelands.

Table 1: Estimates of the quantity of clay required annually in the ceramic industry and for dike construction between 1989 - 2000.

year	quantity (10^6 m^3/year)	
	ceramic ind.	dike constr.
1989	5,1	2,1
1990	5,3	1,6
1991	5,1	1,1
1992	5,0	0,5
1993	5,1	0,5
1994-2000	4,8	0,3

The expected demand for clay until the year 2000 is shown in Table 1. In the period 2000-2010 the demand for clay for ceramic purposes is expected to decline slightly as a result of the increased use of alternative aggregates such as the "Euroclay" (ripened dredging mud from the Rotterdam harbour area). The demand for clay for dike construction will decrease significantly after 1991 when most of the riverdikes along the Rhine and Meuse will have been raised to the level stipulated in the preliminary Water Defence Bill.

At the request of the Public Works Department a working group was established in 1988 to evaluate potential clay resources for the ceramic industry and dike

Figure 1: Principal occurrences of clay at or near the surface. The area investigated in the present study is outlined.

construction in the river area and the eastern and southern provinces of The Netherlands (Figure 1). This working group consisted of representatives of the ceramic and dike construction industries as well as the mapping institutes (Geological Survey and Soil Survey). The working group conducted the inventory in two phases. Phase 1 involved evaluating the data and relevant test methods available at the mapping institutes on the one hand, and the material requirements and test methods of the users on the other hand. Phase 2 involved production of a 1:250.000 reconnaissance map of clay resources and calculation of the potential clay reserves (Public Works Department, 1989). In this paper, the results of the clay inventory are presented.

2 EVALUATION OF MATERIAL REQUIREMENTS AND TEST METHODS

Evaluation of the material requirements of the ceramic and dike construction industries is relevant only if it concerns data that are readily available at mapping institutes. To describe soil samples in the field a minimum set of data is needed (Table 2). For the description of soil samples the Geological Survey and the Soil

3084

Survey use the standardized "classification of unconsolidated soil samples". This classification scheme is the official standard in The Netherlands and is used by geotechnical and geological institutes (NSI, 1989).

Table 2: Soil data available at mapping institutes.

	Geol.S.	Soil S.
grainsize distribution		
fraction < 2 μm	x	x
fraction < 63 μm	x	x
other		
organic matter	x	x
CaCO$_3$	x	x
thickness layers	x	x
colour	x	x
shells/shell remains	x	
groundwater level		x
soil ripening		x
genesis deposits	x	x

Field descriptions are primarily estimates, often according to certain classes. For example, the organic matter content of a sample is estimated with about 5% accuracy. Field estimates are supported by laboratory tests.
It is very important to indicate the genesis of the deposits. In The Netherlands clays of fluviatile, marine and glacial origins occur. The physical and geotechnical characteristics of a deposit are determined to a large extent by its origin.

2.1 Ceramic industry

The material requirements in the ceramic industry are clearly related to production processes. Generally, a clay depot is formed by thoroughly mixing various clay layers. The composition of the homogenized clay determines the quality of the ceramic product. Therefore, the composition of individual clay layers may vary between certain limits.
    To the ceramic industry, grainsize distribution and mineralogical and chemical composition are important as well as several requirements relating to the manufacturing process (shaping, drying, firing). The soil data that are relevant to the ceramic industry and that are usually available in the databases of the mapping institutes, are listed in Table 3. The

requirements listed in Table 3 are not necessarily áll the requirements set by the ceramic industry. The wide range for the fraction < 10 μm is the result of common mixing practises. For the production of regular bricks, for example, a clay mixture with 40-42% particles < 10 μm is used; for rooftiles clay with 50-57% particles < 10 μm is required. The requirement regarding organic matter cannot be assessed in the field observations. In the field, the smallest quantity of organic matter that can be detected is 5%.

Table 3: Requirements set by the ceramic industry

grainsize distribution	
fraction < 10 μm	8 - 70 %
fraction 63 - 250 μm	< 40 %
fraction > 250 μm	< 20 %
other	
organic matter	< 3 %
CaCO$_3$	< 25 %
shell content	nil
thickness layers	> 0.5 m

An evaluation of the test methods used at the mapping institutes and the test laboratories of the ceramic industry showed that the results of the organic matter tests differ significantly. However, the requirement regarding CaCO$_3$ content is that vague (< 25%, Table 3), whereas the CaCO$_3$ content of clay samples is usually less than 10%, that deviations resulting from different test methods are not relevant.
    On the basis of the research results of the test laboratories of the ceramic industry over the past few years it is possible to correlate the fraction < 2 μm (standard at the mapping institutes) to the fraction < 10 μm (standard in the ceramic industry):

$$\% \text{ fraction} < 2 \mu m = \frac{\% \text{ fraction} < 10 \mu m}{1.5}$$

The above correlation (illustrated in Figure 2) has been applied succesfully to "translate" the requirements of the ceramic industry to the data of the mapping institutes (Geological Survey, 1984).

2.2 Dike construction

The criteria for clay to be used in dike construction have been formulated by the

Figure 2: Correlation (R = 0.97) of the fraction < 2 μm and the fraction < 10 μm (from test results of the laboratory for the ceramic industry).

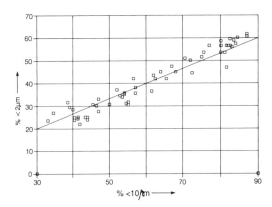

% <10 μm ⟶

Technical Advisory Committee for Water Defence (1985). These criteria are related to the erodibility of the clay, the impermeability, and the sensitivity to deformation (Table 4).

Table 4: Requirements for clay in the dike construction industry.
```
-------------------------------------
grainsize distribution
fraction < 2 μm          17.5 - 40 %
fraction > 63 μm              < 40 %

other
organic matter               <  3 %
CaCO₃                        < 25 %
plasticity index             > 18 %
Note: only ripened clay can be used
-------------------------------------
```

The plasticity index (liquid limit - plastic limit) cannot be derived from the field data, because at present geotechnical parameters are not yet determined on a regular basis during field surveys conducted by the Geological Survey or the Soil Survey. It should be noted, however, that plasticity indices of > 18% were calculated from the results of geotechnical tests on seven bulk samples from the river area which meets the requirement for dike construction (Table 5).

In practice, the requirement that the sand fraction should not exceed 40% is hardly a limiting factor. An evaluation of the results of more than 250 grainsize distributions in river clay samples (17.5-40% fraction < 2 μm) showed that only about

Table 5: Plasticity indices for 7 bulk samples from the river area (Delft Geotechnics, 1988).

sample	liq.lim.	pl.lim.	pl.ind.
1	31.6	12.0	19.6
2	33.9	14.3	19.6
3	41.2	18.9	22.3
4	32.7	13.6	19.1
5	43.2	20.0	23.2
6	40.9	17.7	23.2
7	47.0	21.2	25.8

4% of the samples contain over 40% sand particles.

A full comparison of the test methods of the dike construction industry and the mapping institutes is not possible as yet because of a very limited quantity of test data. The first results do not indicate the need for correction factors. Further research will be necessary in this field.

3 THE 1:250.000 RECONNAISSANCE MAP

On the basis of the results of phase 1, a legend was designed for a reconnaissance map of surficial clay deposits on a 1:250,000 scale. The map shows only the clay layers that consist of over 8% particles in the fraction < 2 μm, < 5% organic matter, virtually no shells or shell fragments, and a thickness of over one metre. Figure 3 shows part of this map in a simplified form.

The thickest clay layers in the river area that meet the demands of the ceramic and dike construction industries are the levee deposits along present and former river channels. Most of the overbank deposits consist of very heavy clay and/or contain too much organic matter.

In the eastern provinces, the clay is either of glacial origin (boulder clay) or of marine origin (Tertiary deposits). The boulder clay may attain considerable thicknesses but is usually inhomogeneous and generally too sandy. The Tertiary clays are pre-consolidated, rather heavy clays that were deformed by glacial activity during Saalien times.
   In the southern provinces, clay and loam of both fluviatile and local origin is encountered. The fluviatile clay deposits are of Pleistocene age and usually only a few metres thick. Part of the local clay/loam deposits are formed in an aquatic

Figure 3: Detail of the 1:250.000 reconnaissance map.

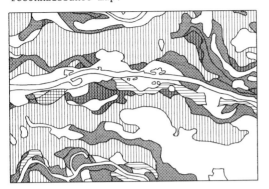

LEGEND

▤ clay/loam , 8-17.5% < 2μm

▦ clay/loam , 17.5-35% < 2μm

▥ clay , >35% < 2μm

The figure has been plotted through the GIS "Arc Info". Original maps are in colours and contain more detailled information.

environment, and part as aeolian deposits. These two types of deposits are not suitable for dike construction (< 17.5% particles < 2 μm) and can be used only as a mixing material in the ceramic industry.

The reconnaissance map has been produced by means of a Geographic Information System (Burrough, 1986). The application of a GIS is particularly relevant here because:

1. the clay reserves available for the ceramic industry and for dike construction had to be calculated;

2. the reconaissance map of clay resources will be used in combination with other maps, e.g. town and country planning maps;

3. the map will require updating every few years.

The minimum quantity of clay that is potentially suitable for ceramic products has been calculated at $1.8 * 10^7$ m^3. For dike construction around $0.5 * 10^9$ m^3 is potentially suitable. These quantities have been calculated from the map, i.e., only the requirements listed in Tables 3 and 4 are taken into account. In practice, not all this clay can be used in the ceramic industry or for dike construction, since these two branches have additional, very user-specific, requirements. It is important to realize that no limitations that might result from town and country

planning are taken into account, nor economic or environmental considerations.

The reconnaissance map is displayed at the poster session of the present IAEG-Conference.

4 FINAL REMARKS

The results of the clay inventory study will be used for a national policy on surficial raw materials. Each of the 12 Dutch provinces is expected to supply its own clay resources. In 1990, the provinces will develop plans regarding the excavation of raw material, which will be incorporated in the provincial schemes for town and country planning.

The present map is not suitable for the selection of locations for the excavation of clay. For this purpose far more detailed information will be required.

ACKNOWLEDGEMENTS

The authors present this paper on behalf of the working group on the national clay inventory. The various contributions of the members of this working group are gratefully acknowledged.

REFERENCES

Burrough, P.A. (1986). Principles of Geographical Information Systems for land resources assessment. Clarendon Press, Oxford.

Delft Geotechnics, (1988). Suitability of 10 clayey materials in road engineering, report for the working group on alternative fill materials (in Dutch). Report no. CO-281321.

De Mulder, E.F.J. (1984). A geological approach to traditional and alternative aggregates in The Netherlands. Bull. IAEG 29; pp. 49-57.

Ministry of Transport and Public Works, (1989). National policy on surficial raw materials (in Dutch). Parliament report no. 21100.

Netherlands Standardization Institute, (1989). Geotechnics, Classification of unconsolidated soil samples (in Dutch). NEN 5104.

Public Works Department, (1989). Report of the working group clay inventory (in Dutch). Report no. TW-R-89-4; Division of Roads and hydraulic engineering, Delft.

Geological Survey of The Netherlands, (1984). Research report on clay deposits for the ceramic industry (in Dutch). Report no. OP 5202; Haarlem.

6th International IAEG Congres / 6ème Congrès International de AIGI, © 1990 Balkema, Rotterdam. ISBN 90 6191 130 3

Rock aggregate suitability – A simple approach to creating classification schemes

Aptitude de roche pour granulat – Une approche simple à l'établissement d'une classification géotechnique

Paul Nathanail
Geoenvironmental Group, Wimpey Environmental Limited, Hayes, Middlesex, UK

ABSTRACT: A simple approach to creating a site specific classification scheme using the information procured from a site investigation to assess the suitability of rock for use as aggregate is presented. The application of the approach is illustrated with a scheme developed for a quarry extracting Carboniferous basalt.

RESUME: Une approche simple à l'établissement d'une classification géotechnique pour évaluer l'aptitude de roche pour granulat est presentée. L'application de l'approche est illustrée avec une classification qui était dévelopée pour une carrière en basalte carbonifère.

1. INTRODUCTION

Rock mass classification schemes have been developed for designing tunnels, slopes and foundations (Bieniawski, 1989). Rock is quarried worldwide for use as aggregate in concrete, roadstone, armour stone or fill.

The diverse uses of rock as an aggregate coupled with varying specifications of what aggregate quality is acceptable make a single all-encompassing classification scheme for assessing the suitability of rock for use as aggregate impractical. In this paper a simple approach to creating a site specific classification scheme using the information procured from a site investigation is presented. A case history from a quarry extracting Carboniferous basalt is described.

The scheme classifies rock according to its suitability for various applications enabling estimates of reserves to be made at an early stage and expensive but essential laboratory testing to be more effectively scheduled. It may also be used during the course of a site investigation to design the later stages of the investigation.

2. LIMITATIONS OF EXISTING ROCK MASS CLASSIFICATION SCHEMES

Rock mass classification schemes have been proposed for a multitude of engineering purposes in various ground conditions (Kircaldie, 1988; Bieniawski, 1989). Two have received widespread acceptance.

The Rock Mass Rating (RMR) System, or Geomechanics Classification, (Bieniawski, 1989) considers the uniaxial compressive strength of rock material, the rock quality designation (RQD), discontinuity spacing, condition and orientation and ground-water conditions. A number of points depending upon the value of each parameter is awarded. The final rating is obtained by summing the points from each parameter.

The Q-system of rock mass classification was proposed on the basis of 212 tunnelling case histories in Scandinavia (Barton et al., 1974; Bieniawski, 1989) and considers RQD, number of joint sets, roughness of the most unfavourable joint or discontinuity, degree of alteration or infilling along the weakest joint, water inflow and stress condition. Points are awarded for each parameter and the value of Q derived by combining the above parameters into three quotients and calculating the product of the quotients (Barton et al., 1974) .

Both the RMR and Q systems rely on a single overall rating to classify the rock mass. However for use as an aggregate the rock mass must have properties which are ALL above certain threshold values. If any one property does not exceed that threshold, then the rock mass is not suitable for the proposed use and must be rejected. In both the RMR and Q systems a low score for an individual parameter, rendering the rock unsuitable for use as an aggregate, may not result in a low overall rating. Therefore neither the RMR or Q system are applicable to assessing the suitability of rock for use as aggregate.

The Unified Rock Classification Scheme (Williamson, 1984) has been used to create a Rock Material Field Classification Procedure (Kircaldie et al., 1987). This includes a classification of rock masses according to construction quality and uses a threshold value approach. However only four parameters, degree of weathering, strength, unit weight and discontinuities (combining orientation, spacing and aperture in one parameter) are considered. The absence of lithology and the generalised way of taking into account discontinuity properties restrict the usefulness of the Procedure.

From the above it can be seen that it is unlikely that a universally applicable rock mass classification to assess the suitability of rock for use as aggregate can be set up. Instead a method for setting up a site specific classification scheme is proposed.

3.   AGGREGATE SUITABILITY CLASSIFICATION SCHEME

3.1   Factors to consider

In setting up a site specific rock mass classification scheme to assess the suitability of rock for use as aggregate the following factors should be considered:-
1. Anticipated use of aggregate
2. Method of working and processing
3. Site investigation methods
4. Availability of outcrops (natural and man-made)

The anticipated uses of the aggregate may include concrete, mortar, unbound pavement construction, bituminous pavements, railway track ballast, filter media or erosion protection (Collis and Fox, 1985).

The method by which the rock is won from the face and then processed to produce a final product determines how much the quality of the final product can be modified. If a simple series of screens is to be used to grade crushed rock then the final quality will largely be determined by the in situ quality of the source material. The use of washing plant would allow dust coating or clay discontinuity infill to be removed. For the supply of sand to the glass making industry both chemistry and grading must be very tightly controlled and complex grain size sorting processes are currently in use in some sandstone quarries.

The site investigation methods used will determine the type of information available. Rotary coring, while providing high quality information is expensive. Rotary open holes provide poorer information at lower cost. Geophysical logging of boreholes is a cheap and fast way of obtaining quantitative information while tomographic techniques can provide a three-dimensional visualisation of the rock mass.

Rock outcrops, both natural and man-made, provide by far the cheapest information. However the relevance of the information from any particular outcrop to the rock mass at depth must always be considered.

The properties of the rock mass which are to be used in the classification scheme must be measurable using the site investigation methods available and relevant to the anticipated use of the aggregate.

The properties should be drawn from the engineering geological description of outcrops or rock core and could include lithology, colour, grain size, strength, degree of weathering, fabric and discontinuity type, orientation, persistence, spacing, aperture, surface roughness and waviness and nature of infill. Parameters such as rock quality designation (RQD), point load index or Schmidt hammer rebound index could also be used.

3.2   Setting up the classification scheme

The properties which control the suitability of the rock for the planned use and which will be used in the classification scheme are chosen on the basis of experience from neighbouring sites or sites in similar geological settings.

Having determined the properties to be used in the classification scheme, minimum acceptable values, or thresholds, for each property are selected for each anticipated use of the material. Once again experience both of the geological setting of the quarry and of the requirements for the anticipated use of the aggregate are essential in setting these threshold values. The threshold values for the various properties are grouped to define aggregate suitability classes for the anticipated uses. The rock must satisfy all the criteria for a given use to be placed in the class for that use. The criteria for the highest quality material are applied to the engineering geological description. If the material does not meet one of the criteria, the criteria of the next are applied and so on until the material is placed into one of the classes.

4.   CASE HISTORY

4.1   Introduction

The setting up of a rock mass classification scheme for a quarry extracting Carboniferous basalt is described. The quarry has historically provided top grade aggregate for use as railway ballast. However the principal high quality markets are now for dry and coated roadstone and concrete aggregate.

A site investigation involving small diameter core drilling, open hole drilling, geophysical logging of boreholes and extensive field mapping was carried out to determine the feasibility of extending the quarry.

The classification scheme presented below assisted in providing estimates of the quantities of useful reserves in the proposed extension area.

The production process at the quarry includes a sequence of screens that sort crushed rock into various sizes by a combination of selection and rejection. As no washing or other processing plant are used, the quality of the in situ rock is the prime control on the quality of the final product.

4.2 Choosing the properties to use in the classification scheme

Extensive use was made of the faces being worked in present day operations in choosing which geotechnical properties to include in the classification scheme and in deciding what threshold values to set.

The highest quality rock was observed to be obtained from fresh to faintly decomposed, extremely to very strong, dark grey basalt with moderately widely to very widely spaced, tight or cemented joints. Therefore the properties included in the classification scheme were lithology, degree of decomposition (the result of the combined effects of weathering and hydrothermal alteration), unconfined compressive strength (estimated or based on point load test), nature of joint infill and joint spacing.

4.3 Describing the rock mass

The terminology used to describe the rock mass was based on that proposed by the Geological Society Engineering Group Working Party (EG WP) (1977) and BS5930 (Anon, 1981). The categories for lithology, Table 1, reflect the geology of the site. The term decomposition rather than weathering was used, Table 2, since changes to the rock mass had occured both due to near-surface weathering and deeper hydrothermal alteration. Unconfined compressive strength has been classified according to BS5930 in order to allow the inclusion of the category 'extremely strong', Table 3. The numbering of the categories of discontinuity infill, Table 4, differred slightly from those of the EG WP in

order to fit in with the ground conditions on site. The categories of discontinuity spacing, Table 5, was as proposed by the EG WP.

Table 1. Classification of Lithology.

Lithology	Class
Basalt	1
Brecciated Basalt	2
Other (Including tuff and volcanic breccia)	3

Table 2. Classification of Degree of Decomposition

Term	Class
Fresh	1
Faintly Decomposed	2
Slightly Decomposed	3
Moderately Decomposed	4
Highly Decomposed	5
Completely Decomposed	6
Residual Soil	7

Table 3. Classification of unconfined compressive strength

Description	Range (MN/m^2)	Class
Extremely strong	>200	1
Very strong	100 – 200	2
Strong	50 – 100	3
Moderately strong	12.5 – 50	4
Moderately weak	5 – 12.5	5
Weak	1.25 – 5	6
Very weak	<1.25	7

Table 4. Classification of discontinuity infill

Nature of infill	Class
None	1
Surface staining only	2
Cemented (quartz or calcite)	3
Non cohesive	4
Chlorite, talc	5
Inactive clay or clay matrix	6
Swelling clay or clay matrix	7
Other	8

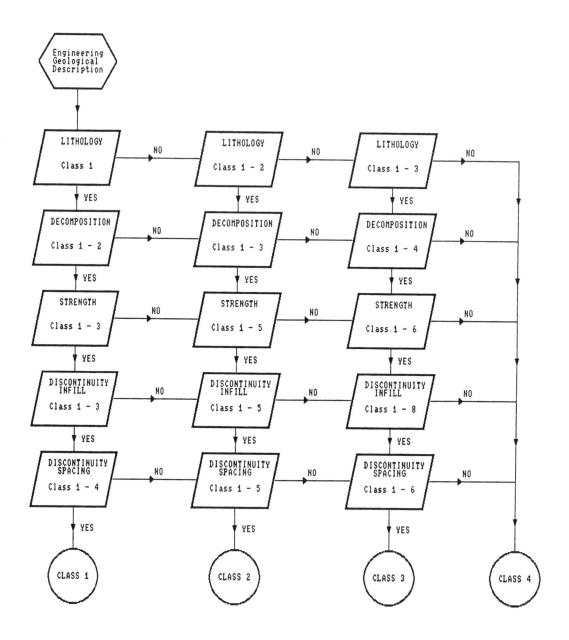

Figure 1   Flow diagram illustrating the use of a rock mass classification scheme to assess
suitability of rock for use as aggregate. See text and accompanying tables for
explanation of Classes

Table 5. Classification of discontinuity spacing

Term	Spacing (mm)	Class
Extremely wide	>2000	1
Very wide	600 - 2000	2
Wide	200 - 600	3
Moderately wide	60 - 200	4
Moderately narrow	20 - 60	5
Narrow	6 - 20	6
Very narrow	<6	7

4.4 Classifying the rock mass

Four classes of aggregate quality were defined on the basis of mapping of currently worked faces in the present workings. The threshold values for each of the classes are shown on Figure 1.

Class 1 material comprised fresh to faintly decomposed extremely to very strong dark grey basalt with moderately widely to very widely spaced tight or cemented joints and represented high grade rock potentially suitable for use a coated or uncoated roadstone and concrete aggregate.

Class 2 material included slightly decomposed, moderately weak basalt or brecciated basalt representing poorer quality material potentially suitable for use as general fill.

Class 3 comprised moderately decomposed, weak basalt, brecciated basalt, tuff and volcanic breccia which would have only limited uses such as landscaping.

Class 4 comprised material which would behave as a soil.

The use of the classification scheme is illustrated in a flow diagram, Figure 1. The geologist input the engineering geological desription, based on Tables 1 through to 5, and successively applied the criteria for Class 1 material. If the material did not meet one of the criteria, the criteria of Class 2 were applied and so on until the material was placed into one of the four classes.

On the basis of the classification scheme horizons of Class 1 material were identified. The quarry operator was able to plan his working method to avoid contaminating Class 1 horizons with poorer quality rock.

5.   CONCLUSIONS

Rock mass classification schemes such as the RMR and Q systems do not lend themselves to aggregate suitability classification. A simple site specific rock mass classification scheme based on threshold values for those properties of the rock mass which control the quality of the final product should be used to assess the suitability of rock for use as aggregate instead.

It must be borne in mind that no classification scheme is universally applicable. Any classification scheme based upon the approach described in this paper must be preceded by good engineering geological descriptions which allow modification and reclassification should factors such as working method or use of the aggregate change.

REFERENCES

Anon (1981). BS5930:1981. Code of practice for site investigations. British Standards Institution.

Barton N., Lien R., Lunde J. (1974). Engineering Classification of Rock Masses for Design of Tunnel Support. Rock Mech. Vol 6, pp 183 - 236.

Bieniawski Z.T. (1989). Engineering Rock Mass Classifications. John Wiley & Sons, New York.

Collis L. & Fox R.A. (1985), Aggregates: Sand, gravel and crushed rock for construction purposes. Geological Society Engineering Geology Special Publication No 1.

Geological Society Engineering Group Working Party (1977). The description of rock masses for engineering purposes. Q. J. Eng. Geol. Vol 10, No 4, pp 355 - 88.

Kircaldie L., Williamson, D. , Patterson, P. (1987). Rock Material Field Classification Procedure, Second Edition. US Dept. of Agriculture Soil Conservation Service Engg Division, Technical Release No 71.

Kircaldie L. (1988) Rock Classification Systems for Engineering Purposes. ASTM Special Technical Publication 984.

Williamson, D. A. (1984). Unified Rock Classification System. Bull. Assoc. Eng. Geol. Vol 21, No 3, pp 345 - 54.

The selection of construction materials for basic access streets in developing areas

La sélection des máteriaux de construction pour voies d'accès dans les région de développement

P. Paige-Green & L. R. Sampson
Low Volume Roads Programme, Division of Roads and Transport Technology, CSIR, Pretoria, South Africa

ABSTRACT: With the rapid urbanisation of many parts of the developing world, the need for the construction of increasing lengths of new access roads every year is becoming significantly greater. The development of most of these areas is usually hampered by the limited availability of funding. The use of standards pertaining to and derived for developed areas results in roads with a low risk of not performing satisfactorily, but which are usually disproportionately costly for the area. The use of local materials without the experience of precedent results in low cost roads but with an increased risk of premature failure. For this reason, an analysis of the potential use of a number of typical, relatively common construction materials as wearing courses for unpaved roads and base for paved roads has been carried out. The important aspects which need to be defined for each material type are discussed and possible limits are provided. Although the limits are based on experience in southern Africa, the local materials are often similar to materials in parts of South America and India. The theoretical basis, however, is applicable to the material type more than the prevailing milieu.

RÉSUMÉ: La demande pour une expansion des voies d'accès par les pays en voie de développement augmente chaque année dû à l'urbanisasion rapide dans plusieurs régions de ces pays. Or, le développement de la plus part de ces régions est habituellement limité par faute de resources financière suffisantes. L'usage des normes standardisés dérivés ou concus pour utilisation dans les pays en voie de développement, porte à croire que le risque de la structure n'ayant pas performé satisfaisante est relativement base. Toutefois, ceci, et l'expérience l'a démontré, est une solution coûteuse. Par contre, l'usage des matériaux locaux sans expérience présédente présente une solution peu coûteuse mais dons la performance s'est fait montrer questionnable et souvant insatisfaisante. Pour cette raison, une analyse a été conduite dans l'usage potentiel d'un nombre de matériaux de construction typique et relativement commun, utilisé dans les bases des routes pavées et dan les couches d'usure des routes non-pavées. Les aspects importants et typique aux différents materiaux ont été étudiés et définiés et des limites possibles ont été fournies. Bien que ces limites sont basées sur l'expérience Sud Africaine, les matériaux locaux sont souvent similaire à ceux trouvéx dans les régions de l'Amerique du Sud et l'Inde. La base theorique est toutefois plus applicable à la composition des matériaux qu'au milieu prévalant.

1 INTRODUCTION

With the rapid population growth in Third World and developing areas and the concomitant urbanisation necessary to accommodate and provide employment for this population, significant attention needs to be directed towards the provision of services in the urban areas. These include the provision of water, energy, sanitation and transportation routes. This paper deals solely with the latter which is considered by many to be one of the most important services. The reasoning behind this is that if the residents cannot be guaranteed regular, reliable transport to their places of employment, they may lose their jobs and the other services then cannot be afforded.

This is particularly true in parts of Africa and South America where road-based public transport is the primary mode of commuter travel. It is probably also relevant to southern and South East Asia.

With urbanisation comes a general aspiration for an increased quality of life in the residential areas. Prolonged periods of muddy roads during the wet season and excessive dustiness during the dry season are socially unacceptable. Extremely rough roads are even less desirable in these areas as vehicles are often fairly aged and many mechanical failures due to rough roads may render the vehicles temporarily or permanently unserviceable due to difficulties in obtaining replacement parts.

Compounding these problems is the general lack of funding for the development of services in these areas.

This paper is based on research carried out in southern Africa, and although the material and climatic conditions may be unique to this area, the general principles should be applicable to many other parts of the world.

2   STANDARDS AND TEST METHODS

Many of the specifications used in southern Africa and generally in Africa as a whole have been adapted from those developed in the northern hemisphere. Specifications developed for instance in the United Kingdom, France and the United States have often been applied to tropical and sub-tropical countries. Although these specifications are often suitable, they are usually very conservative with respect to the risk involved. However, they are seldom the most cost-effective solution in areas with generally strong subgrades and few readily accessible quarries producing high quality aggregate. In addition the prevalence of deeply weathered materials with thick soils and plentiful pedogenic materials (eg laterite, silcrete, calcrete) complicates the establishment of quarries.

The lack of standardisation of test methods is a problem which needs to be considered from the initial design stages. The British Standard method for example specifies a softer rubber base for the Casagrande liquid limit device than the ASTM method (Sampson and Netterberg, 1984; 1985). This results in a liquid limit and plasticity index determined with the BS device, on average 4 points lower than that determined with the ASTM device. Other aspects such as the temperature of drying prior to testing for the Atterberg limits or whether the material is soaked during preparation also affect the plasticity (Netterberg, 1979).

Other commonly used tests such as the determination of the California Bearing Ratio (CBR) use various compaction energies and different treatments of the oversize material. All these factors affect the transferrability of test methods and specifications from one region to another.

All of the test results used in this paper are based on standard South African test methods (NITRR, 1986) unless stated otherwise. The local satandards generally comply closely with AASHTO requirements.

3   BACKGROUND

3.1   Traffic

The majority of streets and ways in a developing urban area can be classified as very lightly trafficked. The material parameters discussed in this paper concentrate on the lowest street category carrying traffic, namely basic access streets.

Basic access streets are defined for the purpose of this paper as access streets in the early stages of development and are designed and constructed subject to constraints imposed by the administrative authority owing to limited budgets or other circumstances. Designs for such urban residential streets in developing communities with low traffic volumes (up to 75 vehicles per day of which only about five are heavy vehicles) should make maximum use of in situ materials and realistic material standards. Where possible these streets are constructed and maintained with labour intensive methods.

3.2   Climate and drainage

The performance of basic access streets depends to a major extent on the prevailing climate. In an arid area with minimal rainfall almost any material will provide a satisfactory road for light traffic. However, in wet areas many moisture sensitive materials which provide an acceptable performance during dry weather will rapidly fail should they become saturated.

Effective drainage is thus essential to good pavement performance and is therefore a prerequisite in the structural design of basic access streets.

The climate (Weinert, 1980) largely determines the processes of weathering of natural rock, the durability of weathered, natural street building materials and also, · depending on drainage conditions, the stability of untreated materials in the pavement. The climate may also influence the moisture content at which the material within a sealed road equilibrates. Southern Africa has been divided into three general climatic regions based on N-values (Weinert, 1980). These are significant from an engineering geological viewpoint as different modes of weathering occur depending on the N-value. Different geological materials can be grouped together in terms of their weathering products within each of these zones. However, different microclimates can occur in local areas. This is particularly important where such microclimates result in a high moisture content locally. This will have a direct influence on moisture susceptible materials in basic access streets and specific drainage considerations will be necessary.

It is recommended that only one, of either the drainage or the material standards are relaxed in any one street, and only then after careful consideration and engineering judgement. The presence of excess water in the pavement layers will usually result in failure irrespective of the quality of the materials and construction (Netterberg and Paige-Green, 1988a). The drainage standards should not be relaxed for unpaved roads.

If it is not possible to install adequate drainage in the very lightly trafficked access streets, consideration should be given to limiting the access of larger vehicles during periods of moisture build-up within the road.

3.3 Construction materials

The construction materials commonly used for low volume streets in southern Africa may generally be classified as crushed or natural, weathered rock, residual or transported gravels and soils derived from the following major material groups (Weinert, 1980):

Basic crystalline rocks - eg dolerite, andesite, basalt
Acid crystalline rocks - eg granite, gneiss
High silica rocks - eg quartzite, chert, hornfels
Arenaceous rocks - eg sandstone, conglomerate
Argillaceous rocks - eg mudstone, shale, slate
Carbonate rocks - eg limestone, dolomite
Diamictites - eg tillite, breccia
Metalliferous rocks - eg ironstone
Pedocretes - eg calcrete, laterite, ferricrete, silcrete

Unweathered materials need extensive processing (eg blasting, crushing, screening, and blending) and are thus generally not cost-effective for the lightly trafficked streets in developing areas, although they are necessary for the more heavily trafficked arterials and collector streets. Only weathered and partly weathered materials (with or without added transported soil) are thus considered in this paper. It is important to classify the potential construction materials correctly, as each group has a characteristic range of properties and problems.

Guidelines on the identification of these material groups (Weinert, 1980) exist and are not dealt with here. General specifications (some incorporating judicious relaxation of certain properties traditionally employed) follow, but these should be carefully considered with respect to the traffic, climate, drainage (surface and sub-surface) and the risk and implications of possible failure. Means of handling this situation of increased risk are discussed by Visser and Van Niekerk (1987). Other factors such as supervision during construction, maintenance and material variability (Netterberg and Paige-Green, 1988a) should be considered prior to the use of these appropriate specifications.

Specific recommendations for the use of basic igneous rocks (Weinert, 1980), argillaceous rocks (Venter, 1984), tillites (Paige-Green, 1984) and calcretes (Netterberg, 1971; 1982) in roads and streets have been made and summarised by Netterberg and Paige-Green (1988a).

Guidelines for the use of the different materials in various layers of basic access streets are described fully in the following sections but local experience and engineering judgement should always be used. If unweathered materials are utilised the cost of processing them will be such that compliance with TRH 14 (NITRR, 1985) standards should be achieved.

The use of the California Bearing Ratio as an indicator of the bearing capacity (or shear strength) of the foundation and pavement layers is standard practice for lightly trafficked streets. This test is, however, fairly time-consuming and requires large samples. Rapid estimates of the in situ CBR can be obtained from other material characteristics and devices such as the Dynamic Cone Penetrometer (Kleyn, 1975) and Clegg Hammer (Clegg, 1983).

3.4 Risk

The use of in situ and low quality materials brings with it a greater risk of premature distress. The instilling of a greater awareness in the public authorities and road user towards this possibility and acceptance of the consequences of using low cost designs is extremely important.

This is best brought about by increasing the public awareness and understanding of the alternatives of using higher design standards and obtaining a low percentage of functional roads or relaxing the standards and obtaining more kilometres of road but with a higher probability of failure. The risk should, however, be quantifiable and justifiable.

4 UNPAVED STREETS

Unpaved streets are the lowest quality basic access street acceptable. Although undesirable from the point of view of dust generation, the production of mud when wet and the necessity for constant maintenance, financial constraints usually preclude the use of higher quality streets for streets carrying the very low traffic typical of basic access streets. The optimum use of available materials is thus necessary to reduce the undesirable properties as far as possible.

The requirements of an ideal wearing course material for unpaved roads have been discussed by Paige-Green and Netterberg (1987) and are summarised as follows:

Resistance to deformation;
Provide adequate protection to the subgrade;
Provide an acceptably smooth and safe ride;
Shed water without excessive scouring;
Freedom from excessive dust;
Freedom from excessive slipperiness;
Resistance to abrasion by traffic;
Freedom from oversize stones;
Low cost and ease of maintenance.

Unpaved basic access streets in urban areas should be designed to satisfy these requirements as far as possible, especially with respect to providing adequate protection of the subgrade. Aspects such as slipperiness, dust and oversize material should be attended to wherever possible but the prevailing traffic and local situation should be taken into account and reduction in standards may be allowed if the cost of alternative materials is prohibitive.

If the in-situ material (subgrade) conforms with the specifications of normal wearing course materials (described below) no imported layer is necessary. This is the most economic solution in many cases but it is important to ensure that the material is capable of performing adequately. The in-situ material often needs only to be compacted to a density of

not less than 95% Mod AASHO. Problems with drainage and maintenance may, however, occur if the street is not built-up to some extent, and it is recommended that, as a minimum, the top-soil (containing vegetable matter) be removed over a width equal to the street and side-drains, the side-drains are excavated and the gravel from the side-drains is used to form the street. The side-drains should be below the level of the adjacent properties to avoid flooding of these properties during heavy storms.

Although numerous specifications for gravel wearing course materials exist in southern Africa (Netterberg and Paige-Green, 1988b), none appears to be applicable to all local materials under all the prevailing environmental conditions. The standard specification for wearing course gravels in TRH 14 (NITRR, 1985) is seen as rather harsh, rejecting many suitable materials and accepting some unsuitable materials (Paige-Green, 1989).

The following performance-related specifications have thus been developed (Paige-Green, 1989) for basic access streets carrying less than 75 vehicles per day:

Maximum size: 37.5mm
Oversize index (I_o): 0
Shrinkage product (S_p): 100 - 240
Grading coefficient (G_c): 16 - 34
CBR: > 15 at 95% Mod AASHO
density and OMC

where

I_o - mass of material larger than 37.5mm in a bulk sample as a percentage of the total mass (Paige-Green, 1988);
S_p - product of bar linear shrinkage and per cent passing 0.425mm (calculated as a percentage of the material finer than 37.5mm);
G_c - product of the percentage retained between the 26.5mm and 2.0mm sieves and the percentage passing the 4.75mm sieve/100.

It is often very difficult to obtain materials which will provide a dust-free surface. Dust palliatives are chemical or bituminous agents which are mixed into the upper part or sprayed on the surface of a gravel wearing course and bind the finer portions of the gravel reducing dust. A variety of dust palliatives is available but they are seldom economic unless the material is a waste product and is used close to its source (Visser et al, 1983). However, in the urban environment, other factors may take precedence over economics. Research currently in progress suggests that certain dust palliatives may become increasingly more viable.

5 PAVED STREETS

The implications and problems with relaxing material specifications for paved streets have been fully discussed by Netterberg and Paige-Green (1988a). Each layer in the street should be treated as a separate entity with respect to the materials used and the construction procedures.

As the base is the major component of basic access streets (often this is palced directly on the subgrade) only the material aspects regarding this layer are discussed. Aspects regarding the other layers ahave been fully described by Horak et al (1988).

The base is the most important structural layer of a lightly trafficked street and must have both adequate strength and durability to perform satisfactorily during the life of the street and provide a firm and strong upper layer which will allow a good bond between the base and surfacing. Structurally, a material with a CBR strength of 30 to 50 at field moisture and density is adequate for lightly trafficked streets (Kleyn and van Zyl, 1987). A compaction of 95% Mod AASHO is necessary to limit traffic-associated compaction to an acceptable amount. The durability of the material, however, must be ascertained for many aggregates especially basic igneous rocks.

It is considered necessary to provide different requirements for each material group. The bearing capacity is the major criterion in each case, and this should be related to the prevailing environmental conditions. The bearing capacity of bases for lightly trafficked streets is best specified in terms of the CBR value at 95% Mod AASHO compaction at the expected field moisture and density conditions. In poorly drained and wet areas it may be necessary to use the soaked values, while in arid areas a test at the OMC (Emery, 1984) may be used for the design.

The risk of using an unsoaked CBR for the design can be quantified and expressed as the probability of the street failing before the design traffic has been carried for varying moisture contents (Emery, 1987). The example discussed by Emery (1985) indicates that even with an EMC/OMC ratio of more than 1.5, the probability of the street not carrying 0.2 million E80s is less than about 6 per cent.

The specific requirements of each of the material groups according to the Weinert Classification are discussed separately below.

(a)   Basic crystalline rocks

The pyroxenes (and possibly some feldspars to a lesser extent) in basic crystalline rocks generally decompose to expansive smectite clays. The presence of smectites is not always adequately shown by plasticity or strength testing and a Durability Mill test (Sampson and Netterberg, 1989), is necessary to release the clays held in the aggregate. A maximum DMI of 125 should be permitted with not more than 35 % passing the 0.425mm sieve after any treatment. In addition a dry modified Aggregate Impact Value (AIV) (Sampson and Roux, 1982) should be carried out and a maximum of 39 is acceptable. In addition, two samples should be soaked, one in water and the other in ethylene glycol before AIV tests are carried out on them. The soaked/dry ratio should not exceed 1.14. If the increase in AIV from the water soaked to the glycol soaked samples is more than 4 percentage units the material should be rejected.

The fineness product (Sampson, 1990) is another useful indicator of performance. This is the product of the plasticity index and the percentage passing the 0.425 mm sieve and a maximum value of 150 is specified.

(b)   Acid crystalline rocks

These materials generally weather to a sandy material comprised of quartz, illite and kaolinite, all of these being relatively stable components. As long as the CBR strength at the expected moisture and density regimes is in excess of 50 and the plasticity index is less than 15 (Richards, 1978) the material will perform satisfactorily in basic access streets. It is, however, important not to include lumps of oversize material which will break down under traffic and a maximum AIV of 30 is recommended. A number of problems with acid crystalline rock bases have been attributed to the addition of excessive plastic fines to improve workability. The common practice of adding plastic fines to non-plastic materials is discouraged and should be used only under strict supervision and with careful judgement. A fineness product of up to 420 is permissible.

(c)   High silica rocks

High silica materials consist mainly of quartz, possibly with minor amounts of feldspar and clays. The materials are generally strong (provided the grading is

3099

not too uniform) and durable and if the CBR strength at the expected in situ density and moisture exceeds 50 the material should perform adequately. It is highly unlikely that durability problems will be encountered with these materials unless a plastic material is added to improve workability.

(d) Arenaceous rocks

Arenaceous rocks consist mainly of quartz, feldspars and clays, with the clays generally being inert kaolinites and illites. These materials, apart from having a CBR of 50 should have a 10% FACT value of not less than 110 kN (AIV not more than 30) and a soaked value of not less than 75% of the dry value where N < 5 and 55% where N > 10 in order to identify potentially degrading materials. A maximum Fineness product of 125 with a maximum of 35 per cent passing the 40 mesh sieve indicates an adequately durable material (Sampson, 1988). A maximum dry modified AIV of 31 is permitted and the soaked/dry ratio should not exceed 1.14.

(e) Argillaceous rocks

Argillaceous rocks should not be used in basic access streets if possible. The only argillaceous rocks which may be suitable are those unweathered, slightly baked mudrocks which require extensive treatment (including blasting and crushing) before use and are therefore not cost effective for lightly trafficked roads. All other mudrocks are likely to slake and disintegrate in service, usually leading to distress. If no other materials are available, mudrocks can be considered, but their durability according to the recommendations of Venter (1988) should be investigated. A maximum Fineness Product (Sampson, 1988) of 125 with not more than 35 per cent passing the 40 mesh sieve will result in adequately durable material.

(f) Carbonate rocks

Carbonate rocks are generally unsuitable for base course for basic access streets as the unweathered rocks are extremely hard and require blasting and crushing. The weathered materials, on the other hand, proceed directly into solution and do not form a residual gravel suitable for base construction.

(g) Diamictites

Diamictites (tillites, breccias, etc) have been investigated (Paige-Green, 1984) and have been shown to perform well, even where existing specifications are not met. The durability of some of these materials has sometimes been a problem and the

following limits for tillites are proposed (mostly after Paige-Green, 1984):

Maximum PI: 13%
Minimum CBR: 80%
Min 10% FACT: 180 kN
Min soaked/dry ratio (10% FACT):
   0.70 if N < 2;
   0.56 if N > 2:
Min Washington Degradation Value: 55

A maximum Fineness Product and percent passing the 40 mesh sieve after any treatment of 125 and 35 respectively identify materials unlikely to degrade to plastic fines in service (Sampson, 1988). The dry modified AIV should not exceed 22 and the soaked/dry ratio 1.15.

(h)   Metalliferous rocks

These materials are primarily available as the waste products of iron and magnesium mines. Their very high specific gravities result in uneconomically high haulage costs and they are probably thus not viable construction materials for basic access streets. If they are, however, close to the construction site, they can be used with no problems, as they are generally unweathered, already crushed and inert.

(i)   Pedocretes

Calcretes and ferricretes (with possibly small quantities of silcretes) are the main pedocretes used in construction in South Africa. Extensive work has been carried out on calcretes (Netterberg, 1971; 1982; Lionjanga et al, 1987a; 1987b) while local work on the other materials has so far been minimal.

A maximum fineness product of 480 with up to 55 per cent passing the 40 mesh sieve is allowable for pedogenic base materials (Sampson, 1988). However a minimum CBR of 50 should be obtained.

It is necessary to test the aggregate strength and a dry modified AIV value of 39 is the maximum allowable to ensure against particle breakdown (Sampson, 1990).

6 CONCLUSIONS

Basic access streets in developing areas generally carry low traffic with minimal heavy vehicles. In order to optimise their construction and conserve higher quality materials for use in future higher standard roads, maximum use should be made of local materials. A system of using local materials with specific requirements for each group has been described. This effectively relates the propensity of the material to release deleterious fines or aggregates to disintegrate to their constituent particles, in service, to the the material group and a number of simple tests.

ACKNOWLEDGEMENTS

The work reported in this paper was carried out at the Division of Roads and Transport Technology and is published with the permission of the Director.

REFERENCES

Clegg, B. 1983. Application of an impact test to field evaluation of marginal base course materials. Trans. Res. Rec. 898: 174-181.
Emery, S.J. 1984. Prediction of pavement moisture content in southern Africa. Proc. 8th Reg. Conf. Africa on Soil Mech. and Foundn Engn, Harare:239-250.
Emery, S.J. 1985. Prediction of moisture content for use in pavement design. PhD thesis, University of Witwatersrand.
Emery, S.J. 1987. Unsoaked CBR design to reduce the cost of roads. Proc. Annual Transportation Convention, Pretoria.
Horak, E., Paige-Green, P., Sampson, L.R. and Visser, A.T. 1988. Guidelines for the design and maintenance of low volume residential streets in developing communities. DRTT Report DPVT-C32.1, CSIR, Pretoria.
Kleyn, E.G. 1975. The use of the dynamic cone penetrometer (DCP). Transvaal Roads Department, Report L2/74, Pretoria.
Kleyn, E.G. and van Zyl, G.D. 1987. Application of the dynamic cone penetrometer (DCP) to light pavement design. Report L4/87, Transvaal Roads Dept, Pretoria.
Lionjanga, A.V, Toole, T.A. and Newill, D. 1987a. Development of specifications for calcretes in Botswana. Transportation esearch Record 1106: 1, 281-304
Lionjanga, A.V, Toole, T.A. and Greening, P.A.K. 1987b. The use of calcrete in paved roads in Botswana. Proc. 9th Reg. Conf. Africa Soil Mech. Foundn Engn., Lagos, 1: 489-502.
National Institute for Transport and Road Research. 1985. Guidelines for road construction materials. TRH 14, CSIR, Pretoria.
National Institute for Transport and Road Research. 1986. Standard methods of testing road construction materials. Technical Methods for Highways No 1, 2nd Edition, CSIR, Pretoria.
Netterberg, F. 1971. Calcrete in road construction. Bulletin 10, NITRR, CSIR, Pretoria.
Netterberg, F. 1979. Effect of drying temperature on the index properties of calcretes. Trans. Res. Rec., 675:24-32.
Netterberg, F. 1982. Behaviour of calcretes as flexible pavement materials in southern Africa. Proc. 11th Austral. Road Research Board Conf.,3: 60-69.
Netterberg, F. and Paige-Green, P. 1988a. Pavement materials for low volume roads in southern Africa: A review. Proc. 8th Quinquennial Convention of S. African Inst. Civil Engnrs and Annual Transportation Convention, Vol 2D, Paper 2D2, Pretoria.
Netterberg, F. and Paige-Green, P. 1988b. Wearing courses for unpaved roads in southern Africa: A review. Proc. 8th Quinquennial Convention of S. African Inst. Civil Engnrs and Annual Transportation Convention, Vol 2D, Paper 2D5, Pretoria.
Paige-Green, P. 1984. The use of tillite in flexible pavements in southern Africa. Proc. 8th Reg. Conf. Africa Soil Mech. Foundn Engng, Harare, 1:321-327
Paige-Green, P. 1988. A revised method for the sieve analysis of wearing course materials for unpaved roads. DRTT Report DPVT-C14.1, CSIR, Pretoria.
Paige-Green, P. 1989. The influence of geotechnical properties on the performance of gravel wearing course materials. PhD thesis, University of Pretoria, Pretoria.
Paige-Green, P. and Netterberg, F. 1987. Requirements and properties of wearing course materials for unpaved roads in relation to their performance. Trans. Res. Rec., 1106: 208-214.
Richards, R.G. 1978. Lightly trafficked roads in southern Africa. A review of practice and recommendations for design. NITRR Report RP/8/78, CSIR, Pretoria.
Sampson, L.R. 1988. Material dependent limits for the Durability Mill Test. DRTT Report DPVT-C57.1, CSIR, Pretoria.
Sampson, L.R. 1990. Recommended durability tests and specification limits for base course aggregates for road construction. DRTT Report in preparation, CSIR, Pretoria.
Sampson, L.R. and Netterberg, F. 1984. A cone penetration method for measuring the liquid limits of SouthAfrican soils. Proc. 8th Reg. Conf. Soil Mech. and Found. Engng, Harare, 1: 105-114.
Sampson, L.R. and Netterberg, F. 1985. The cone penetration index: A simple new soil index test to replace the plasticity index. Proc. 11th Int. Conf. Soil Mech. and Foundn Engng, San Fransisco, 1041-1048.
Sampson, L.R. and Netterberg, F. 1989. The durability mill: A performance-related durability test for basecourse aggregates. The Civil Engineer in South Africa, September: 287-294.

Sampson, L.R. and Roux, P.L. 1982. The aggregate impact value test as an alternative method for assessment of aggregate strength in a dry state. NITRR Technical Note TS/3/82, CSIR, Pretoria.

Venter, J.P. 1984. Mudrock (shale and mudstone) in road construction - A review of practice. Proc 8th Reg. Conf. Africa Soil Mech. Foundn Engng, Harare, 1:343-351.

Venter, J.P. 1988. Guidelines for the use of mudrock (shale and mudstone) in road construction in South Africa. NITRR Technical Note TS/7/87, CSIR, Pretoria.

Visser, A.T. and Van Niekerk T. 1987. The implementation of appropriate technology in the design of light pavement structures. Trans. Res. Rec., 1106: 222-231.

Visser, A.T., Maree, J.H. and Marais, G. P. 1983. Implications of light bituminous surface treatments on gravel roads. Trans. Res. Rec., 898: 336-343.

Weinert, H.H. 1980 The natural road construction materials of southern Africa. Academica, Cape Town.

De la stabilité des assises de chaussées réalisées avec des granulats alluvionnaires, plus ou moins anguleux, traités au bitume

About the stability of sub-bases and road-bases made with gravel aggregates which have different degrees of angularity and which are mixed with bitumen

A. Panis & C. Tourenq
Laboratoires des Ponts et Chaussées, France

RÉSUMÉ : La quantification de l'angularité de matériaux alluvionnaires plus plus ou moins concassés est faite à l'aide d'un essai d'écoulement qui rend parfaitement compte de la proportion d'arêtes vives, nécessaires à la stabilité interne des matériaux de chaussées, créées par le concassage. Les résultats sont comparés aux définitions utilisées en France de l'indice et du rapport de concassage.
Après traitement au bitume à différents dosages (grave bitume et béton bitumineux), l'influence du paramètre angularité est étudié vis-à-vis de la compacité, de la résistance mécanique statique et du fluage (orniérage) et de la sensibilité à l'eau. Des informations sont données sur le concassage minimal nécessaire pour chaque type d'usage et sur l'inutilité du concassage excessif qui est parfois préconisé.

ABSTRACT : The assessement of the angularity of gravel material crushed to varying degrees is carried out by means of a flow test which shows exactly the proportion of sharp edges created by crushing. The results are compared to the definitions of the index and crushing ratio used in France.
After treatement with different percentages of bitumen (gravel stabilized with bitumen, bituminous concrete) the effect of the angularity factor is studed regarding compactness, static mecanical strength, creeping (rutting) and sensitivity to water. Information is given about the minimum crushing needed for each use and the ineffectiveness of the excessive crushing which is sometimes recommended.

Cette étude a pour objectif de définir où se situe la limite raisonnable du concassage des alluvions pour assurer une bonne stabilité des enrobés et d'essayer d'améliorer l'imprécision de la notion d'angularité définie soit par l'indice de concassage (Ic) et le rapport de concassage (Rc), soit par une autre définition plus liée à la granularité des éléments supérieurs à D, soit par un essai simple de type écoulement.
- on appelle indice de concassage (Ic), la proportion en poids d'éléments supérieurs au D du granulat élaboré, contenue dans le matériau d'origine soumis au concassage;
- on appelle rapport de concassage (Rc) le rapport entre le d du matériau soumis au premier concassage et le D du granulat obtenu.
d et D étant la plus petite et la plus grande
dimension d'un granulat.
Deux matériaux ont été choisis : une alluvion entièrement silicatée de Moselle et un silico-calcaire du Rhône.

Les modes de préparation des matériaux ont été aussi voisins que possible pour avoir des Ic et Rc ayant la même signification d'un matériau à l'autre (tableau 1).

Tableau 1 Ic et Rc des matériaux fabriqués

		Ic	35	75	100	Rc2
Dimension à 50% du matériau à concasser (en mm)	Moselle	11	28	33	38
	Rhône	10,5	22	23	42
Module de finesse*	Moselle	4,9	7,7	8,7	9,6
	Rhône	4,7	7,4	8,5	10

* adapté à ces classes granulaires

Les granularités utilisées pour chaque type de traitement sont les mêmes quelque soit Ic ou Rc.

Les alluvions de Moselle utilisées comportent 20 à 25% de quartzite à grains fins et 75 à 80% de roches éruptives plus grenues.
Les alluvions du Rhône contiennent 45 à 75% de quartzite et 45 à 55% de calcaire lithographique.
Les temps d'écoulement Ec de ces matériaux sont très voisins, quel que soit le degré de concassage (tableau 2 et figure 1).

Tableau 2 Temps d'écoulement (en secondes) Ec des matériaux fabriqués

| | Gravillons | | | | Sables | |
| | Moselle | | Rhône | | Mos | Rhô |
Ic	% faces cassées	Ecg	% faces cassées	Ecg	Ecs	Ecs
0	0	78	0	70	14,9	15,1
35	46	89	39	90	18,8	25
75	65	103	63	102	19,2	
100	70	107	71	111	19,2	24
Rc2	77	112	78	116	20,7	23

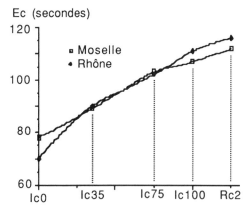

Ec (secondes)

Fig. 1 Temps d'écoulement (Ec) en fonction de Ic et Rc

1- Etude des bétons bitumineux

La plage d'angularité couverte va de Ic30 à Rc2.

Pour les deux matériaux étudiés, la granularité correspond à un 0/14 à 50% de sable 0/5 et 8% de fines. le dosage du bitume 60/70 est de 5,7%.

1. 1 - Matériau Rhône

Les résultats sont regroupés dans le tableau 3. On constate, au début du compactage, à la presse à cisaillement giratoire (PCG), une difficulté à compacter lorsque l'angularité croît (Fig. 2). Cette tendance disparaît pour un compactage plus poussé (200 girations).

Tableau 3. Béton bitumineux Rhône.

| | Ic.% | | | Rc | C% |
	30	60	100	2	
Compacité C % P.C.G 20 gir.	83	82,8	80,5	80	
200 gir.	98	99	98	98,5	
Orniérage (mm) à 10^3 cycles	5		2,4	2,3	94
à $3\ 10^4$ cycles	13		7	7,2	
Orniérage (mm) à 10^3 cycles	5,7		5,8	4,7	89
à $3\ 10^4$ cycles	14,2		13,4	11,2	
R (MPa)	7,6			8,2	
r (MPa)	6,7			7,2	
r/R	0,88			0,88	

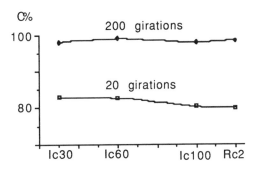

Figure 2 Compacité PCG en fonction de Ic et Rc

Les figures 3 et 4 montrent nettement le rôle de la compacité sur l'orniérage et de l'angularité, où à partir de Ic100 il y a une stabilisation du phénomène.

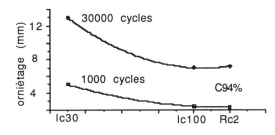

Figure 3. Orniérage en fonction de Ic et Rc pour une compacité de 94%

Ce rôle de la compacité et de l'angularité apparaît encore mieux sur la figure 5 (Tableau 4). Il y a toujours une compacité optimale pour limiter l'orniérage et ce pour chaque valeur d'angularité. Dans ce cas aussi, Ic100 et Rc2 donnent des résultats très voisins.

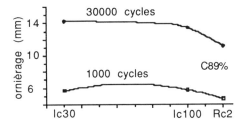

Figure 4. Orniérage en fonction de Ic et Rc pour une compacité de 89%

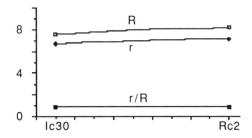

Figure 6 Résistances R et r (après immersion dans l'eau) et rapport $\frac{r}{R}$ en fonction de Ic et Rc

Tableau 4 Orniérage en fonction de la compacité C de l'enrobé. La limite spécifiée en France est de 10 mm

	C %						
	90	92	93	93,5	94	95	
Orniérage (mm) à 3.10⁴ cycles	10	7,8	7,2	7	7,5	12	RC 2
	11	8,7	7,4	7	7,2	8	IC100
	14,4	12,4	11,8	11,8	12,3	13	IC 30

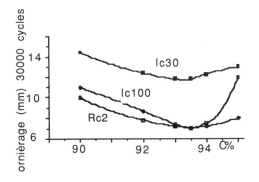

Figure 5 Orniérage en fonction de la compacité C de l'enrobé

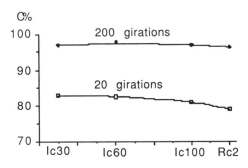

Fig. 7 Compacité PCG en fonction de Ic et Rc

Sur le plan de la résistance mécanique et de la sensibilité à l'eau (Tableau 3), les résultats sont bons, si on les compare aux valeurs spécifiées : R > 6 MPa et r/R > 0,75 pour Rc ≥ 2.MPa La différence entre les valeurs correspondantes pour Ic30 et Rc2 est faible (Fig.6).

1. 2 - Le matériau de Moselle

Les résultats obtenus figurent dans le tableau 5.
On constate une diminution de la compacité (PCG) quand Ic croît pour 20 girations. Ce phénomène s'estompe à 200 girations (Fig.7). Ce point est important, car la compacité a une forte influence sur l'orniérage (Fig.8). Il y a même un optimum de compacité pour chaque valeur de Ic : Fig.9.

Tableau 5.Résultats obtenus sur le matériau de Moselle

	IC%			RC
	30	60	100	2
PCG compacité à 20 gir.	83	82	81	80
à 200 gir.	97	97	97	96,5
Orniérage mm 10³ cycles C 94%	3,2	2,5	2,2	1,7
3.10⁴ cycles C 94%	12	9,2	6,3	6,2
Orniérage 10³ cycles C 90%	4,5	4,5	4,7	4,4
3.10⁴ cycles C 90%	10,5	10,3	10	9,8
C%	93,6			92,9
R (MPa)	8,7			8,6
r (MPa)	7,6			7,1
r/R	0,87	0,83		

Des essais complémentaires ont montré l'influence de la teneur en sable 0/2. Il y a une nette liaison avec la compacité,

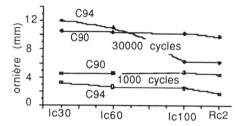

Fig. 8 Ornière en fonction de Ic et Rc

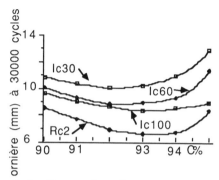

Fig. 9 Ornière en fonction de la
compacité de la plaque d'enrobé
La résistance et la sensibilité à l'eau
sont bonnes et indépendantes de Ic (Fig.
10).

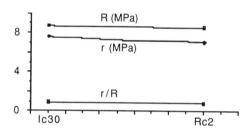

Fig. 10 Résistances R et r (après
immersion dans l'eau) et rapport   (r/R)
en fonction de Ic et Rc

avec Ici aussi un optimum (vers 35% de
sable). Le couple granularité-compacité
est important, mais il faut reconnaître
que le risque d'orniérage est moindre
pour Ic100 ou Rc2, lorsque la compacité
varie (Fig.9).

2 - Etude des graves-bitume

La granularité utilisée est constante, il
s'agit d'un 0/14 mm à 52% de sable 0/4 mm
et 7,5% de fines. La teneur en bitume
60/70 est de 4,4%.
Deux angularités ont été étudiées, Ic30
et Ic100, qui sont les valeurs extrêmes
figurant dans les Directives.

2. 1 - Matériau Rhône

La plus faible teneur en bitume que pour
le béton bitumineux, donne beaucoup plus
d'importance au squelette minéral. On
trouve certes une influence de la
compacité, mais l'orniérage reste
toujours faible, traduisant une très
bonne stabilité de ce matériau, quelle
que soit l'angularité (Tableau 6,
Fig.11).

Tableau 6. Grave-Bitume RHONE

	IC %		C %
	30	100	
R (MPa)	6,6	7,2	
r (MPa)	4,9	5,3	
r/R	0,74	0,74	
orniérage (mm) à : 10^3 cycles	1,5	1,3	89
$3 \ 10^4$ cycl.	5,2	2,3	
orniérage (mm) à : 10^3 cycles	3,2	2,7	85
$3 . 10^4$ cycl.	6,7	5,8	

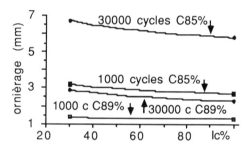

Figure 11.Ornière en fonction de Ic

La résistance en compression et la
sensibilité à l'eau (Tableau 6, Fig.12)
sont conformes aux spécifications les
plus sévères : pour Ic 100/T0 : R > 5
MPa, r/R > 0,65.

2. 2 - Matériau Moselle

Les résultats sont regroupés dans le
tableau 6. Les compacités obtenues sont
toujours bonnes et indépendantes de
l'angularité. La résistance à la
compression est élevée et la sensibilité
à l'eau, bien que limite, est acceptable
(Fig. 13).

L'orniérage reste faible (Fig.13), comme
pour le matériau Rhône, l'angularité ne
joue pas de rôle.

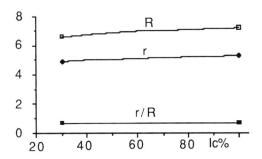

Figure 12 Résistances R et r (après immersion dans l'eau) et rapport $\frac{r}{R}$ en fonction de Ic et Rc

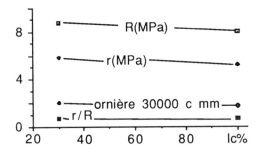

Fig.13 Evolution de R, r, $\frac{r}{R}$ et de la profondeur de l'ornière en fonction de Ic et Rc

3 - **La mesure de l'angularité**

Les définitions actuelles de Ic - Rc ont deux défauts :

- le contrôle a posteriori est impossible,

- on ne tient pas compte de la granularité, réellement concassée, d'où une pénalisation des gisements à gros éléments.

Dans le cadre de cette étude, nous nous sommes affranchis du second défaut. Il est donc normal de trouver de bonnes corrélations entre Ic et toutes les autres mesures que l'on a effectuées.

Un calcul simple, partant de sphères, montre que la surface lisse initiale restant une constante, quel que soit le niveau de réduction, la création de surfaces de rupture tend vers une limite qui se situe vers 90% pour les plus grands rapports de concassage exigés (Rc6). Les pourcentages de surfaces de ruptures calculés, à condition de

considérer au moins 4 classes granulaires consécutives, sont assez voisins des comptages effectués en laboratoire. Tableau 7 et Figure 14.

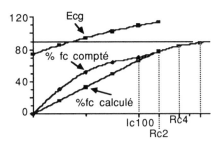

Fig. 14 Variation du % de faces cassées calculé, mesuré et de Ecg (secondes) en fonction de Ic et Rc

Ce calcul permet de compenser le % de faces de rupture d'un 0/10 à Ic = 100 obtenu par le concassage d'un 10/40 (63%) et d'un 10/80 (76%). Un tel écart correspond presque à 20% de Ic, c'est à dire qu'un matériau jugé à Ic80 peut-être à Ic100, si sa granularité est plus grossière. Il n'y a donc pas intérêt à conserver la définition actuelle de Ic et de Rc.

L'introduction du module de finesse (adapté, surface ABC pour faire un 0/10 à Ic = 100), constitue un progrès théorique, puisque l'on tient compte des éléments > D (Fig.15), mais on peut atteindre la même valeur de module avec un matériau contenant 30% de roulés et plus de gros éléments (surface A', B, C', D de la Fig. 15). D'autre part, il faudrait poursuivre l'analyse granulométrique au-delà de 80 mm.

Tableau 7 - Valeurs moyennes obtenues sur les 2 matériaux

Ic %	% faces cassées théorique	% faces cassées comptées	Module de finesse	Ecg	Ecs
0	0	0		74	15
25	16,5	30	4,3	85	20
50	33	52	5,8	94	21,1
75	49,5	64	7,5	102,5	20,7
100	66	70	8,6	109	21
Rc 2	77	77	9,1	114	23
4	84				
6	88				
8	89				

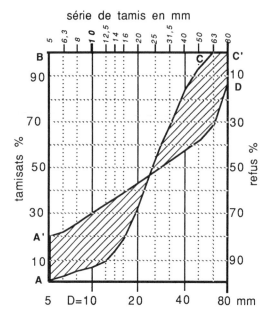

série de tamis en mm

Figure 15 Evaluation du module de finesse

L'essai d'écoulement, ou le fastidieux comptage des faces cassées, permettent de s'affranchir de toutes ces critiques. Les deux matériaux étudiés ayant montré des angularités très voisines (Tableau 2), nous regrouperons l'ensemble des données pour l'interprétation. Nous présenterons donc les valeurs moyennes obtenues sur les deux matériaux confondus.

On notera l'assez bonne corrélation entre Ecg et le % de faces cassées (Fig.16).

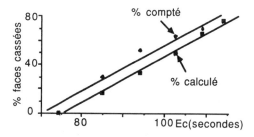

Fig. 16 % de faces cassées en fonction de Ecg

Par contre, pour le sable, les différences sont peu significatives entre Ic 25 et Rc2, cela est dû au fait que même pour Ic = 35, le sable produit est en réalité un Ic100 ou plus. Donc, mis à part Ic0, où il est roulé, dans toute l'expérience le sable est concassé et ne joue donc pas de rôle particulier.

A noter que le mélange sable roulé - sable concassé conduit à une loi d'écoulement simple, en fonction de la proportion roulé - concassé.

Les corrélations (bonnes) obtenues, entre l'écoulement (gravillons), la proportion de faces cassées et Ic, montrent que l'essai d'écoulement est au moins aussi bon que les autres. Il faut donc préférer cet essai rapide, qui différencie bien les matériaux en fonction de la granularité réelle soumise au concassage. L'exemple des alluvions de l'Allier (Tableau 8) montre un écart significatif pour Ic100 entre le concassage du 10/30 et du 10/60. Il est intéressant de noter que cet écart va diminuer à Rc2 et disparaître à Rc4, mais pour ces valeurs l'augmentation des faces de ruptures devient très faible et montre le peu d'intérêt d'utiliser de tels rapports de concassage.

Pour ce qui est du sable, il ne semble pas que l'essai d'écoulement mérite d'être maintenu.

Tableau 8 Alluvions de l'Allier

	gravillons concassés	Ecg 6/10	gravillons concassés	Ecg 10/14
Ic100	10/30	113	14/30	113
	10/60	119,4	14/60	119
Rc2	20/40	119	30/60	122
	20/100	121	30/100	121
Rc4	40/60	124		
	40/100	124		

4 - EssaiEssai d'écoulement et spécifications

Les graves bitume ne présentent pas de différences notables de comportement entre Ic30 et Ic100, les performances sont bonnes et les spécifications devraient se limiter à Ic30 pour les plus forts trafics (Ic100 actuellement). Ceci correspond à des Ec gravillons ≥ 90 secondes pour les plus forts trafics. Peut-être pourrait t-on spécifier Ecg ≥ 70 secondes pour les plus faibles trafics (Tableau 9).

Tableau 9 - Graves bitume, spécifications possibles de Ecg pour les assises de chaussées en fonction du trafic (T)

Ecg en secondes	< T1	≥ T1
couche de base	80	90
couche de fondation	70	80

Pour les sables, la limite est plus délicate à fixer, il serait raisonnable d'exiger un Ecs ≥ 18 secondes, ce qui correspond à un sable de Ic30 facile à obtenir en carrière.

Les bétons bitumineux commencent à être trop déformables en-deçà de Ic80, mais Ic100 donne de bons résultats dans tous les cas. Il est donc inutile de rechercher Rc2 pour des bétons bitumineux de liaison ou support de cloutage.
La figure 17 montre la tendance moyenne pour l'orniérage, qui seul pose un problème.

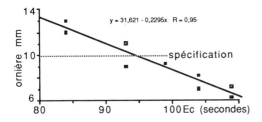

Fig. 17 Orniérage en fonction de Ecg (secondes)

Les Ecg à proposer pourraient être ceux du Tableau 10.

Tableau 10 - Bétons bitumineux Ecg en fonction des classes de trafic

Ecg en secondes	< T3	T2/T3	≥ T1
béton bitumineux liaison et support de cloutage	90	100	110

Pour le sable, la limite Ecs ≥ 18 secondes est à conserver.

On peut donc conclure qu'il est préférable d'utiliser Ec, qui ne pénalise pas les matériaux plus riches en gros éléments. et que l'on peut réduire sensiblement les spécifications sur les enrobés pour ce qui est de leur stabilité interne.

REFERENCES

AGARWAL, R.B. 1973. Influence of aggregate shape and size and binder modulus on flexible concrete. Public works, Ridgewood (USA), n°3.
BODNAR, G. 1978. Influence de la forme des grains sur la stabilité des bétons bitumineux. Symp. RILEM matériaux granulaires, Budapest.
ISHAI, J. 1984. Propriétés des granulats affectant le comportement des enrobés. Symp. Int. Granulats, AIGI, Nice, vol.2.
LIVNEH, M.and GREENSTEIN, J. 1972. Influence of aggregate shape on engineering properties of asphaltic paving mixtures. Highway Research Record, Washington, n°404.
PIKE, D.C. and JORDAN, JPR. 1973. Technical restrictions on the use of sand and gravel in dense bituminous mixtures for roads. Technical Paper n°3, London.
RAKHMATULLAEV, N. and YUMASHEV, V.M. 1979. Teneur admissible en grains roulés dans les graves bitume. Trudy Soyuzdornii, n°112, Moscou.
SCHMIDT, H. and SCHÜTTER, H. 1958. Rapport entre la forme des grains de sables et la résistance des mélanges bitume granlats. Straase und Autobahn, n°7.
STEPHENS, J.E. 1974. Effect of aggregate shape on bitiminous Report JHR 74-87, Storrs, Connecticut, USA.
SWANSON, L.H. 1974. Morainic gravels in bituminous surfacing, results from six scottish road experiments.TRRL, report L616, Berkshire (UK).

6th International IAEG Congress / 6ème Congrès International de AIGI, © 1990 Balkema, Rotterdam. ISBN 90 6191 130 3

Adhesion between bitumen and aggregate

Adhérence entre les bitumes et les granulats

K.P.K. Pylkkänen & P.H. Kuula-Väisänen
Tampere University of Technology, Finland

ABSTRACT: Adherence of two bitumens to the surface of ten aggregates was studied by three moisture susceptibility test methods. The aggregates were characterized by determining elemental and mineral composition, specific surface area and water adsorption of grain surfaces. It was found that bitumens adhered weakly to granite and granodiorite. The amount of adsorbed water vapour versus specific surface area correlated best with the water resistance test results.

RESUME: L'adhérence de deux bitumes à la surface de 10 qualités de granulats a été étudiée par trois procédés permettant de tester la résistance à l'eau. Les propriétés des granulats étaient caractérisées par la spécification de leur composition en matières premières et en minéraux, de leur surface spécifique, de l'adsorbtion de l'eau de la surface des grains. On a constaté que les liaisons entre les bitumes et le granite ou la granodiorite sont faibles. La meilleure corrélation a été trouvée entre les résultats des tests de résistance à l'eau et la quantité de la vapeur adsorbée par la surface spécifique.

1 INTRODUCTION

Damages of asphalt pavements caused by so called stripping -phenomena have been studied in a number of research projects all over the world. Stripping is a water induced damage, where the adhesion between binder and aggregate is weakened because of water penetrating between aggregate surface and covering binder film. Little by little this leads to the entire detachment of the aggregate grain from the binder. The result is often seen as raveling of the pavement surface.

In recent years also Nordic countries have shown special interest to stripping problems. It has been seen as one possible factor influencing on the wear resistance of asphalt pavements. For example in southern Finland icing of the main roads and motorways is inhibited by salting. This keeps the pavement surface wet for a quite long period of the year. So the circumstances are very favourable for stripping and especially for wear caused by studded tires.

In 1987 The Finnish Road Administration authorities introduced the most extensive pavement research program ever realized in Finland. One of the numerous themes within this so called ASTO -project is the adhesion between bitumen and aggregate. The main part of this research work is carried out at Tampere University of Technology in the Laboratory of Engineering Geology.

The aim of the study was to find out if there were differences in adhesion properties between two bitumens and ten aggregates. These differences were then tried to explain by determining some chemical and physical properties related to aggregate surface. The bitumens were characterized according to their acidity and basicity.

2 MATERIALS

2.1 Bitumens

Petroleum bitumen is obtained as a residual fraction in crude oil distillation and is mainly used in road construction. Bitumen is an amorphous, dark coloured and cementious substance composed principally of high molecular mass hydrocarbons. The characterization of bitumens can be done after their rheology and acid and base properties.

In this study we used bitumens distilled from Venezuelan and Arabian crude oils. Bitumens are called respectively Laguna (LAG) and Arabian Heavy (AH). In Finland bitumens are classified after their penetration grade at 25 °C. Bitumens used in this study were of penetration grade 120 1/10 mm, which is marked simply B-120. The viscosities (at 60 °C) of these two bitumens were 129 Pas for LAG and 115 Pas for AH.

The acidity and basicity of bitumens is determined after a ASTM -procedure of potentiometric titration. The titration curves of bitumens used in this study are presented in Figure 1.

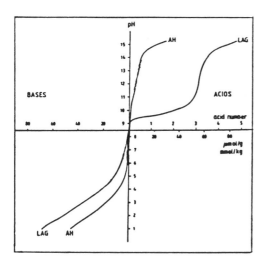

Fig.1 Titration curves of bitumens

These bitumens differ from each other mainly in their acid parts: bitumen LAG contains a large amount of carboxylic acids. The total acid and base numbers in mgKOH/g are calculated from the titration results. The total acid numbers were 3.3 mgKOH/g (LAG) and 0.48 mgKOH/g (AH). The total base numbers were respectively 0.98 mgKOH/g and 0.38 mgKOH/g.

2.2 Aggregates

The aggregates were chosen mainly on the basis of variation in their silicon dioxide content. Also the differences in grain surface structure and mineral composition was demanded. Following these

demands four minerals and six rock types were chosen: quartz, alkali feldspar, calcite, olivine, quartzite, granite, granodiorite, diabase, gabbro and peridotite.

The chemical composition of the aggregates was determined with x-rayfluorescence method. In addition to chemical compounds reported in Table 1 the aggregates contained smaller amounts of titanium and Na_2O.

Table 1. Chemical composition of aggregates.

Aggregate	SiO_2	K_2O	CaO	Al_2O_3	MgO	Fe
Quartz	100					
Alkali feldspar	67	11		18		
Calcite	2		54		2	
Olivine	30		2	5	17	27
Quartzite	99					
Granite	74	5	1	14	1	
Granodiorite	70	3	2	15	1	3
Diabase	48	2	8	18	5	10
Gabbro	46	5	6	18	9	7
Peridotite	46		8	8	19	9

The mineral composition of six aggregates was determined from thin sections. In addition to in Figure 2 presented minerals olivine contained 12 % opaque. Granodiorite contained also 7 % mica, which was mainly biotite.

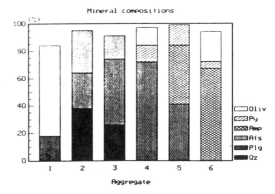

1 = Olivine   4 = Diabase
2 = Granite   5 = Gabbro
3 = Granodiorite   6 = Peridotite

Fig.2 Mineral composition of aggregates according to thin section determination

Specific surface area gives an estimation of the roughness of the grain surface: the greater the area is the more there are pores or ununiformities on the surface. The specific surface area in m^2/g was measured with a nitrogen adsorption method. To avoid the effect of the difference in grain size distribution a fraction of 74 - 125 μm was used. The surface areas varied up to 11.5 m^2/g. Most of the areas were smaller than 2 m^2/g with the exception of olivine (11.5 m^2/g).

Aggregate grains adsorb different amounts of moisture from the surrounding athmosphere. The amount of adsorbed water in g/kg was determined with a simple desiccator method developed in our laboratory. In this method samples are put in glass dishes and dried overnight in an oven at 105 °C. The dry weights of samples are determined after drying. Then the samples are laid in a desiccator , which has water on the bottom. When the desiccator lid is closed tightly a relative humidity of 100 % is formed inside the desiccator. After seven days of equilibration the samples are reweight and the amount of adsorbed water is calculated. Most of the ten aggregates adsorbed water less than 8 g/kg when fraction 74 - 125 μm was used. Only two samples adsorbed water more than that: olivine 17.6 g/kg and diabase 10.3 g/kg.

One way to characterize a surface is to determine whether it is hydrophilic or hydrophobic. A surface is hydrophilic when it "likes" water and hydrophobic when it "dislikes" water. However there is no accurate method or limit on the basis of which to answer that question. That is why we desided that it is better to use only the term hydrophilic and try to measure its intensity.

The amount of adsorbed water could be used as a measure of hydrophilicity unless it was influenced by specific surface area. To define hydrophilicity more accurately the amount of adsorbed water has to be related to specific surface area. The results mg/m^2 obtained by this procedure are presented in Table 2. The greater the reported value is the more hydrophilic the aggregate is. The most hydrophilic aggregates were quartz, alkali feldspar, granite and granodiorite. When these results are compared with the results in Table 1 a connection with few exceptions is found: the more silicon dioxide the aggregate contains the more hydrophilic it is.

Table 2. Hydrophilicity of aggregates.

Aggregate	Adsorbed water in mg/m^2
Quartz	8.66
Alkali feldspar	14.12
Calcite	9.09
Olivine	1.53
Quartzite	7.00
Granite	11.81
Granodiorite	11.86
Diabase	5.85
Gabbro	5.68
Peridotite	4.17

3. STRIPPING EXPERIMENTS

There is a quite large variety of test methods developed for predicting the water susceptibility of bituminous pavements. In most of them the bitumen-aggregate mixtures or compacted specimens are at first stressed by mixing, evacuating or storing them for a certain time in the presence of water. The water resistance is then determined for example by comparing the strength of dry and wet stored specimens.

In this study we used three different methods: so called Hallberg's test, rolling flask test and indirect tensile test. The test methods and results are described below.

3.1 Hallberg's test

The test is based on the idea presented in Figure 3. Due to capillary forces a binder column of height h_b has risen to a pore of aggregate. It is loaded by a pressure caused by a water column of increasing height h_w. The interface between these columns starts to decrease when the pressure exceeds the force the binder is adhered to pore surface. The adhesion force in mN/m is calculated by equation (1) from the height of water column at that moment.

$$\sigma_{wa} - \sigma_{ba} = 245.25 \, (h_w + 0.5) \quad (1)$$

In practice the pores are formed from voids between even graded aggregate grains. Crushed aggregate fractions of 0.074-0.125 mm and 0.5-1 mm are compacted (the coarse below the fine) in a glass tube. A solute of bitumen (1/3 part) and kerosene (2/3 part) is let to be absorbed from below to the compacted layers. After a certain ageing time water columns of different heights are poured above the top layer. After 24 h the highest water column remained stable is observed. The height of that column is used to calculate the result.

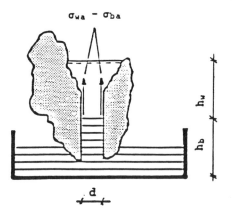

$\sigma_{wa} - \sigma_{ba}$ = adhesion force
d = diameter of capillary
h_b = height of binder column in capillary
h_w = height of water column in capillary

 binder

aggregate

water level

Fig.3 Theoretical model of Hallberg's test

3.2 The rolling flask test

This test has been developed in the Nordic countries to evaluate the moisture susceptibility of bitumen-aggregate mixes. The method is based on dynamic stressing of bitumen coated aggregate grains. This is accomplished by mixing them in a rolling flask filled with water. The proportion of the surface area still covered by bitumen is assessed visually after predetermined times. This estimation of the degree of coverage (%) is then regarded as the measure of moisture susceptibility.

In this study the sample consists 600 aggregate grains. A fraction of 6.3-8 mm was used and the grains were chosen on the basis of their form. Assuming them spherical the average diameter of the grains was calculated. With the aid of supposed surface area of each sample the amount of bitumen was quantified to form a 0.1 mm thick coating on the surface of each grain.

The sample was divided to three parallel samples containing 200 stones each. One day after mixing bitumen and aggregate grains they were inserted one by one in a flask containing deionised water at a temperature of +5 oC. A glass rod was put in, too and the flask was sealed. After 3 days of rolling at a speed of 40 rpm the degree of coverage was evaluated. It was performed by five observers from dry samples spread on a white cardboard.

3.3 Indirect tensile test

The indirect tensile test is developed for testing the moisture susceptibility of compacted asphalt specimens. Half the number of the manufactured test bodies are satured with water by evacuating the voids of the specimens. After conditioning the saturated specimens in water and the rest as dry they are subjected to loading as shown in Figure 4.

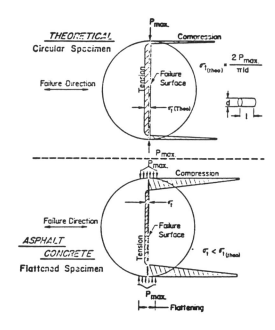

Fig.4 Loading of specimen in indirect tensile test

When loaded a compression stress is developed in the cross section near the pressing steel bars. This stress is changed into a tension near the centre of the specimen. As the tensile stress exceeds the capacity of the specimen it is broken along the failure surface. The moisture susceptibility is determined as the ratio (in %) between the indirect tensile strength of wet and dry stored specimens.

In this study fairly open-graded aggregate was mixed with 5.4 % bitumen from the weight of aggregate. Ten specimens were compacted with Marshall -apparatus (2*50 blows) at the temperature of 160 °C. After cooling overnight five specimens were saturated with water in vacuum at the residual pressure about 3 kPa for 3 h. Then they were conditioned for 70 h at 40 °C. Before loading the test specimens were cooled by storing them at 10 °C. The pressing was carried out at that temperature at the deformation rate of 50 mm/min, which is normally used in Marshall -test, too.

3.4 Results

Results of the stripping experiments for all the tested bitumen aggregate combinations are reported in Table 3.

Table 3. Results of the stripping experiments

| Aggregate | AH | | | LAG | | |
	HT mN/m	RFT %	ITT %	HT mN/m	RFT %	ITT %
Quartz	5.4	83	63	0	40	43
Alkali feldspar	0	64	54	0	43	38
Calcite	9.5	53	57	7.2	66	61
Olivine	11.3	51	81	10.6	73	96
Quartzite	11.3	85	65	0.9	43	52
Granite	0	10	39	1.4	11	45
Granodiorite	0	10	44	1.2	17	50
Diabase	3.8	63	71	5.8	72	64
Gabbro	5.9	63	74	4.2	65	75
Peridotite	10.1	48	78	7.7	62	76

HT = Hallberg's test
RFT = rolling flask test
ITT = indirect tensile test

According to the results above granite and granodiorite showed very weak adhesion to both bitumens in all test methods.

Asphalt mixes with olivine or peridotite as aggregate seem to be quite water resistant in HT and ITT -tests. Low degree of coverage in RFT -tests is mainly due to their low resistance against abrasive wear.

4 DISCUSSION

Adhesion between bitumen and aggregate is mainly a chemical process/mechanism where the chemical qualities of these materials affect. We have assumed that the most important qualities in this process are the chemical composition and the hydrophilicity of aggregates. Also it seems that the acidity of bitumen has a certain role in adherence.

The silicon dioxide content of aggregates has been widely accepted one of the most important quality in adhesion. When it is correlated with the results of the stripping experiments with Laguna -bitumen a fairly good agreement is obtained. The correlation coefficient is negative in all three cases. The poorest correlation is achieved with indirect tensile test (Figure 5). The correlation coefficients with the other test methods are −0.64 (rolling flask test) and −0.82 (Hallberg's test).

The correlation between silicon dioxide content and stripping results with AH -bitumen is not as good as above. An example is shown in Figure 5. However the correlation is much more better (−1.00) if the monomineral aggregates (quartz, alkali feldspar, calcite and quartzite) are not taken into account. In spite of the previous contradictions it seems that a high silicon dioxide content causes poor adhesion between bitumen and aggregate.

Other interesting chemical compounds on aggregate surface are magnesium oxide and ferrous compounds. The only problem in explaining the effect of these compounds on adhesion is their small amount in studied ten aggregates. Figure 6 presents the effect of MgO and Fe to the indirect tensile test results with AH -bitumen. The correlation is clearly positive, the correlation coefficients are 0.69 for Fe and 0.85 for MgO. Also with other test methods a same kind of correlation is obtained. So it seems that adhesion between bitumen and aggregate is improved when the amount of magnesium oxide and ferrous compounds in aggregate increases.

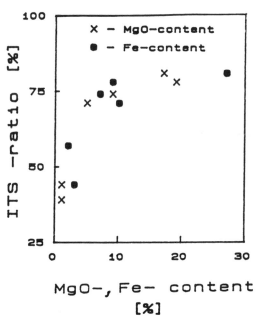

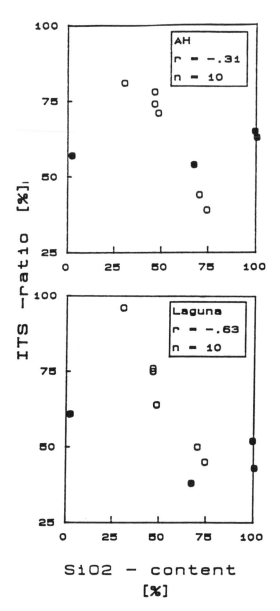

Fig.6 Connection between MgO and Fe
-content and indirect tensile strength
-ratio

SiO2 — content
[%]

Fig.5 Connection between silicon dioxide
content and indirect tensile strength
-ratio (• = monomineral aggregate)

In the litterature the consepts of
hydrophobicity or -philicity of the
aggregate has been quite often related to
water susceptibility of bituminous
pavements. In that context the aggregates
are usually classified hydrophilic or
-phobic according to their silicon dioxide
content. Generally it has been assumed
that hydrophilic aggregates are more
susceptible to stripping than hydrophobic
aggregates.

This assumption seems to agree also in
this study, when the hydrophilicity was
determined by water adsorption related to
specific surface area. For example in the
case of indirect tensile test the
correlation coefficients between
hydrophilicity and test results were
extremely good for both bitumens (Figure
7). Results of the other test methods did
not correlate so well; the correlation
coefficients varied from -0.32 to -0.80.
If the monomineral aggregates are
disqualified the results agree much
better. Anyway, it seems that when the
hydrophilicity of the aggregate is
increased its resistance to stripping is
decreased.

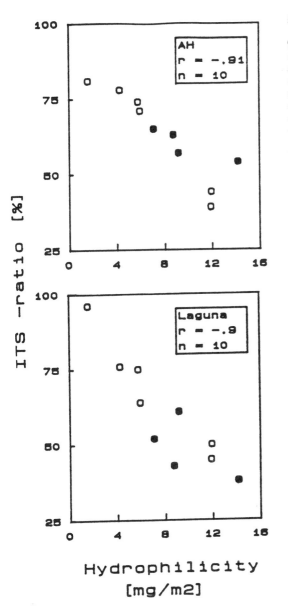

Fig.7 Connection between hydrophilicity
and indirect tensile strength -ratio (● =
monomineral aggregate)

5 CONCLUSIONS

This study shows that there are great
differences in the adherence of different
aggregates to bitumens. Especially two
granitic rock types showed extremely poor
adhesion to bitumens. Also such minerals
as quartz and alkali feldspar seemed to
adhere quite poorly. That is why the
adhesion of bitumen to aggregates
containing considerable amount of above
mentioned elements should be pretested to
take precautions to avoid stripping.

The water susceptibility of asphalt
mixes can in some respect be evaluated by
hydrophilicity determination of the
aggregate. This method is more
timeconsuming but not as troublesome as
the other test methods.

REFERENCES

Annual Book of ASTM Standars 1981.Part 23.
Petroleum Products and Lubricants (I):
D56-D1660. Standard Test Method for
Neutralization Number by Potentiometric
Titration. D664-81.

Fromm,H.J.(1974).The mechanisms of asphalt
stripping from aggregate surfaces.Proc.
AAPT,Vol 43,191-223.

Isacsson,U. and Jörgensen,T.(1987).Labora-
tory methods for determination of the
water susceptibility of bituminous
pavements. Swedish Road and Traffic
Research Institute. Report 324 A.
Linköping.Sweden.

Pylkkänen,K.P.K.(1989).Adhesion between
bitumen and aggregate.Proc. of ASTO
-conference 2nd November.Technical
Research Centre of Finland.Road and
Traffic Laboratory No.750.Espoo.

Scott,J.A.N.(1978).Adhesion and disponding
mechanisms of asphalt used in highway
construction and maintenance.Proc.
AAPT,Vol 47,19-48.

6th International IAEG Congress / 6ème Congrès International de AIGI, © 1990 Balkema, Rotterdam. ISBN 90 6191 130 3

Evaluation of strength of irregular rock lumps for characterization of rockfills

Détermination de la résistance des fragments rocheux irréguliers pour la caractérisation des enrochements

M. Quinta Ferreira
Geology Department, University of Coimbra, Portugal

J. Delgado Rodrigues, A. Veiga Pinto & F.T. Jeremias
Geotechnique Department, Laboratório Nacional de Engenharia Civil (LNEC), Lisbon, Portugal

ABSTRACT: During the preliminary study of rockfills index properties are usually used to allow a fast and inexpensive characterization of the rock material. Once one knows the strength of irregular rock lumps it is possible to make a prediction of the mechanical properties of the rockfill using some available correlations. In Portugal, it is a common practice to use the crushing test to evaluate strength. In this work the use of the point load test with the same objective is proposed. In spite of the small number of tests carried out, the analysis of the available data indicates that both tests are related. The point load test can be used extensively for this purpose because it is fast, easy to carry out, and the required equipment is light and can be carried anywhere in the field.

RESUME: Pendant l'étude préliminaire des enrochements les propriétés indices sont généralement utilisées pour obtenir une caractérisation rapide et pas couteuse des matériaux rocheux. Dés qu'on connaisse la résistance des éléments rocheux de forme irregulierè il est possible de prévoir les proprietés mécâniques des enrochements en utilisant quelques correlations préalablement définies. Au Portugal, il est fréquent d'utiliser des essais de écrasement pour évaluer la résistance. Dans ce travail il est proposé de réaliser des essais de charge pontuelle avec le meme objectif. Malgré le petit nombre d'essais exécutés, l'analise des résultats indique que les deux essais sont en rapport. On considère que l'essai de charge pontuelle peu être utilisé extensivement pour cette finalité parce qu'il est facile à exécuter, et l'équipement nécessaire est léger et peut être transporté facilement en site.

1 INTRODUCTION

The behaviour of rockfill materials depends significantly on the strength of individual rock elements. In order to determine the strength of individual rock elements and to assess the strength variation with size, the crushing strength test (Marsal and Resendiz, 1975) has been frequently used in Portugal.

When searching for rock materials for construction of rockfill dams it is most useful to be able to estimate, since the beginning, what could be the shear strength and the deformability modulus of the compacted rockfill. For this purpose some relations between shear strength, deformability modulus and some index properties, such as crushing strength and slake durability have been used in Portugal (Delgado Rodrigues et al., 1982; Veiga Pinto et al., 1986; Quinta Ferreira et al., 1987). At LNEC, the shear strength of compacted rockfill is obtained by means of large triax-

ial tests and the deformabillity modulus is obtained in large unidimensional compression tests (Veiga Pinto, 1983). Both tests are very expensive and time consuming, while index tests are intended to be inexpensive, easy and fast to execute.

Point load test is one of the index tests more widely executed on irregular lumps, and it was chosen for comparison with the crushing test. The point load test can be executed even in the field, because the equipment is light and easy to operate, overriding the need of bringing the samples to the laboratory as happens with the crushing strength test and most of other tests.

2 METHODS AND RESULTS

2.1 Tested materials

Tested materials are porphyritic coarse grained granite mainly constituted of quartz, potash feldspar, plagioclase (oligoclase), biotite and small amounts of muscovite and chlorite. The megacrystals are of potash feldspar. For both the crushing test and the point load test, irregular lumps of granite were used, without any special preparation.

Part of the research project also used graywackes sampled in the riprap of the upstream slope protection of Beliche rockfill dam. Graywackes are fine-grained, low-grade metamorphic rocks occurring interbedded with phyllites in the flysch sequence of the carboniferous series of South Portugal. Quartz is the prevailing mineral, feldspars (mainly albite) being relatively abundant. Lithic fragments, micas (mainly muscovite) are subordinate compounds.

Table 1 presents some relevant physical properties of the tested materials.

2.2 Tests and results

2.2.1 Crushing test

In the crushing test (Marsal, 1969, Marsal and Resendiz, 1975) a load is applied to three rock pieces approximately of the same dimension (dm) by means of a steel plate. The load (F) that produces the break of the first grain, and the number of contacts (Nc) of the grain with the plate are used to compute the crushing load (Pa) by the ratio F/Nc. The test is performed on three sets of lumps, with different nominal diameters. The dimensions for each lump are obtained by measuring the three principal axes of the lump. In order to reduce the dispersion of the results, the mean values of load and dimension should be computed using a minimum of ten lumps.

Marsal (1969) suggests that results can be expressed by the equation:

$$Pa = \eta \, d_m^{\lambda} \qquad [Eq. 1]$$

where,
Pa = average crushing load
d_m = actual average dimension of the test sample
η and λ = empirical correlation factors.

Parameter η usually varies between 20 and 150 while λ can take values between 1.2 and 2.2. Their actual values are obtained by correlation analysis of laboratory data.

If reported to the dimension (d_m) of 50mm the correspondent crushing load is called $Pa_{(50)}$.

In this work a different procedure was also used to report the results from the crushing test. For each lump the crushing strength (CS) was computed dividing the load (Pa) by the cross section area (A) of the lump aiming at giving this parameter the dimension of a stress. For each sample the value reported for CS is the average of the individual lumps.

$$CS = \frac{Pa}{A}$$

where,
CS = crushing strength,
Pa = crushing load;
A = area of the lump cross-section.

Considering lumps nearly spherical the area is $A = \pi \, r^2$. Substituting the radius (r) by the mean diameter (dm):

$$CS = \frac{Pa}{\frac{\pi}{4} dm^2} \approx 1,27 \frac{Pa}{dm^2} \qquad [Eq. 2]$$

Crushing strength (Pa) is usually reported to a conventional diameter (d=50mm) by Eq. 1. Then, if $Pa_{(50)}$ is divided by the corresponding area (A_{50}) a parameter (σ_{50}) with dimension of a stress is obtained. This parameter (σ_{50}) has, thus, the same meaning as CS but here, Pa is an experimental value obtained for each individual lump, while for σ_{50}, Pa is a computed value.

2.2.2 Point load test

Point load test was executed following the procedure for irregular lumps suggested by the International Society of Rock Mechanics (ISRM, 1985).

The load (P) is applied by the conical points and rupture occurs through one or more planes with the same direction of the applied load.

The result of the test is expressed as a strength value (Is) computed by the expression:

$$Is = \frac{P}{De^2} \qquad [Eq. 3]$$

where,
Is = strength index,
P = applied load,
De = equivalent diameter

For the equivalent diameter $De^2 = 4A/\pi$ is used, A being the minimum cross-section. The standard result for the test

TABLE 1. Physical properties of tested samples.

Samples		Porosity (%)	Dry Unit Weight (kN/m³)	Ultrasonic velocity (m/s)	Slake durability (% loss)
G	P1	1.0	25.7	3010	1.42
R	P3	1.2	25.6	3750	1.81
A	P4	3.7	25.0	2260	3.59
N	P6	5.0	24.3	3290	1.95
I	P8	1.7	25.5	2060	1.76
T	P9	1.6	25.5	3200	2.26
E	P12	0.8	25.7	3690	0.90
S	P13	1.7	25.3	4260	1.31
	L1	0.9	26.0	3950	1.11
	L2	1.6	25.8	2420	2.78
G	GB1	1.5	26.5	5040	0.45
R	GB6	1.1	26.5	4960	0.44
A	GB7	3.5	25.8	4860	0.37
Y	GB8	2.8	25.9	4210	0.58
W	GB9	3.5	25.8	4700	0.53
A	GB10	4.3	25.7	5070	0.58
C	GB11	3.5	25.9	4630	0.61
K	GB12	6.1	25.1	4330	0.20
E	GB13	7.7	24.8	3370	1.01
S					

(Is(50)) is obtained for a dimension of 50mm.

For comparative purposes the method suggested by Guifu and Hong (1986) was also used. In this case the strength value is:

$$PLS = \frac{P}{Af} \qquad [Eq. 4]$$

where,

PLS = point load strength.
P = applied load
Af = area of the rupture surface.

The advantages of using this last procedure is that the computed area corresponds to the rupture surface, which may be quite different from the minimum cross section assumed in ISRM (1985). It also avoids the need of any correction factor.

Fig. 1 represents some results obtained for both procedures. It is apparent that a good correlation exists, but PLS tends to give higher strength results. However, several tested rock lumps gave smaller PLS, normally when the specimens break in more than two pieces and when the rupture surface follows evident anisotropy planes. Because these planes are clearly less resistant, the values thus obtained seem to characterize them in a better way than by using a hypothetical minimum surface of rupture.

2.2.3 Results

The results obtained for the tested rocks are presented in tables 1 and 2. Table 2 presents the results of the mechanical tests.

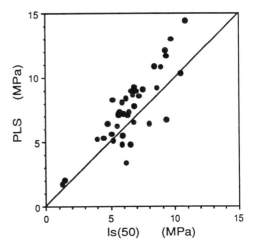

Fig. 1   Relation between the strength index Is(50) and PLS.

TABLE 2. Results of the mechanical tests

SAMPLES		CRUSHING TEST				POINT LOAD TEST	
		η	λ	Pa(50)	$CS = \dfrac{Pa}{A}$	Is(50)	$PLS = \dfrac{P}{Af}$
		(kg/cm^λ)		(kN)	(MPa)	(MPa)	(MPa)
G R A N I T E S	P1	47.70	1.75	7.82	4.78	4.16	5.13
	P3	33.97	1.90	7.10	3.96	3.57	3.98
	P4	17.86	1.68	2.62	1.68	1.41	1.71
	P6	25.28	1.85	4.87	2.77	1.97	2.37
	P8	23.63	1.97	5.51	2.84	3.29	3.78
	P9	39.59	1.59	5.02	3.52	3.01	3.42
	P12	75.42	1.63	10.19	6.89	5.17	6.70
	P13	66.40	1.70	10.04	6.59	3.53	5.00
	L1	66.57	1.80	11.84	6.75	4.63	6.86
	L2	31.10	2.14	9.55	4.55	3.09	4.24
G R A Y W A C K E S	GB1	-	-	-	14.14	9.86	11.20
	GB6	-	-	-	12.81	9.25	10.30
	GB7	-	-	-	7.54	6.20	7.40
	GB8	-	-	-	8.91	6.33	6.70
	GB9	-	-	-	9.57	7.12	8.80
	GB10	-	-	-	8.70	6.15	7.00
	GB11	-	-	-	7.64	5.91	7.30
	GB12	-	-	-	7.17	5.35	6.30
	GB13	-	-	-	3.96	3.60	4.60

3 DISCUSSION

In the point load test the load is applied by two conical points allowing to test samples with different shapes without changing significantly the area where the load is applied. According to Broch and Franklin (1972) this test is more tolerant to irregularities of the samples than the non pontual tests, because the critical stress area lies inside the sample where the influence of surface irregularities vanishes.

A different situation occurs in the crushing strength test because the number of points in contact with the plate and the contact area change with the shape of the specimen, with the way it is placed on the plate and with the increasing crushing of contact points. These test conditions are very difficult to quantify in the analysis of the results.

While conditions of loading are reasonably known in both tests, in rockfills the load distribution is far more complex because the contact between rock elements is made by multiple points that vary in number with the size of the elements and with the compacting state.

Parameters η and λ in Eq. 1 are somewhat difficult to work with. Parameter η is considered a characteristic of the rock material and is expressed in kgf/cm^λ. The exponent λ changes from sample to sample, creating some difficulty to relate η between different samples. On the other hand, both CS and PLS are physically well defined, having the dimensions of a stress. Despite the differences referred to there is some relation between CS or PLS and η, as shown in Fig. 2. The similar trends are justified by the close relation existing between PLS and CS presented in Fig. 3. The trend of the average values is close to the line 1:1 despite the different modes of measuring and testing. The use of two similar formulas to compute the results (Eq. 2 and Eq. 4) also justifies this correlation. In both tests the values obtained are approximations to the tensile strength of the rock.

The use of the point load test for the characterization of irregular lumps is widely used. The usefulness of the test increases computing the PLS, because it is easier to compute and it does not require any correction factor. Having PLS, for

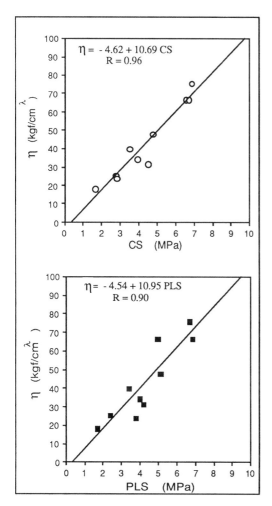

Fig. 2   Relation between CS or PLS and the parameter η for granites.

computing the load necessary to cause the rupture of any lump or rock block it is sufficient to multiply this value by the area of the minimum cross section of the rock piece, but taking into account that rupture always occurs through the section of minimum strength and this is not necessarily that having the lower area.

4   CONCLUSIONS

In the study of rockfills, there is a large tradition of using $Pa_{(50)}$ (crushing load of rock pieces having conventional diameters of 50mm). Experimental and theoretical reasons led the authors to the progressive

replacement of that parameter by other more currently used. Point load strength obtained by the method suggested by Guifu and Hong (1986) has the preference.

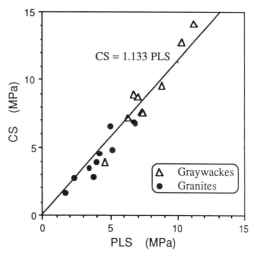

Fig. 3   Relation between PLS and CS.

As regards the results presented in this paper, the following conclusions may be pointed out:

1. Crushing strength and point load values are related, this meaning that tensile strength of rock lumps can be adequately estimated by both methods;

2. point load test can be advantageously used in replacement of the crushing test because it is better defined and easier to carry out, even in the field;

3. PLS is considered the best parameter to quantify the point load strength.

In the sequence of this work, the classification scheme of Marsal and Resendiz (1975) will be analysed aiming at introducing eventual improvements in the number and boundaries of the auxiliary parameters.

REFERENCES

BROCH, E. and FRANKLIN, J.A. (1972) "The point-load strength test" Int. J. Rock Mech. Min. Sci., Vol. 9, pp. 669-697.

DELGADO RODRIGUES, J., VEIGA PINTO, A. and MARANHA DAS NEVES, E. (1982) "Rock index properties for prediction of rockfill behaviour" Proc. 4th Int. Cong. of the IAEG, New

Delhi; LNEC Technical Paper Nº581.

GUIFU, X. and HONG, L. (1986) "On the statistical analysis of data and strength expression in the rock point load tests" Proc. 5th Int. Cong. of the IAEG, 1.5.7, pp. 383-394, Buenos Aires.

ISRM (1985) "Suggested method for determining point load strength" Int. J. Rock Mech. Vol. 22, Nº 2, pp. 51-60.

MARSAL, R.J. (1969) "Particle breakage in coarse granular soils" Proc. 7th ICSMFE, Specialty Session Nº 13, pp. 155-165.

MARSAL, R.J. and RESENDIZ, D. (1975) "Presas de tierra y enrocamiento" Limusa, S.A., México.

QUINTA FERREIRA, M., ANA QUINTELA and VEIGA PINTO, A. (1987) "Barragem de Paradela. Estudo dos enrocamentos e análise do comportamento estrutural", LNEC Report 230/87-NF/NP, October 1987, Lisboa, (In Portuguese).

VEIGA PINTO, A. (1983) "Previsão do comportamento estrutural de barragens de enrocamento", LNEC Research Officer Thesis, Lisboa, (In Portuguese).

VEIGA PINTO, A., DELGADO RODRIGUES, J. and MARANHA DAS NEVES, E. (1986) "Some improvements in the characterization of basalts and limestones for rockfill structures" Proc. 5th Int. Cong. of the IAEG, 4.5.6, pp. 1469-1475, Buenos Aires.

6th International IAEG Congress / 6ème Congrès International de AIGI, © 1990 Balkema, Rotterdam. ISBN 90 6191 130 3

Determination of mechanical properties of rockfill on model samples with reduced grain size distribution curve

Détermination des propriétés mécaniques d'enrochement sur des essais de modèles avec une courbe de distribution d'un granuleux réduit

A. Rozsypal
Stavebni geologie Praha Czechoslovakia

ABSTRACT: The natural particle size distribution in the rockfill dams is characteristic by the fact that the biggest grains may have very often the decimeter and or meter dimensions. In this case, even though the large-scale testing equipments are available, it is possible to perform the experiments on samples with a reduced gradation curve. At the sampling testing and evaluating the results, it is preferable to take into account the similarity theory requirements. Examples of such methods and results of model tests are presented.

RESUMÉ: En barrages des enrochement, la granulométrie du matériel est charactérisée par le fait que le diamétre de les plus grands grains atteint quelques décimètres. Essayer ce matériel est possible seulement par l'entremise des samples avec la granulometrie modifiée. En préparant tels samples, en réalisant des essais et en évaluant leurs résultats, il est indispensable de respecter les lois de similarité. On présente les examples de cette méthode et la évaluation de résultats corréspondants.

1. GENERAL

A necessary and sufficient condition of similarity of two processes is the equality of all dimensionless arguments. Under the dimensionless argument it is understood such a dimensionless group of physical quantities which significantly affect any process. For modelling in the rock mechanics the following dimensionless arguments K_F; K_L were derived by Kohoutek (1971)

$$\frac{m_m}{m_s} = \frac{P_m}{P_s} \cdot \frac{l^3_m}{l^3_s} = K_F$$

$$\frac{G_m}{G_s} = \frac{P_m}{P_s} \cdot \frac{l_m}{l_s} = K_G$$

Legend: m = mass (kg)
p = density (kg/ m^3)
l = length dimension (m)
G = mechanical quantity having a dimension force on surface

The symbol "m" specifies the values related to a model, while the symbol "s" indicates the natural quantities.

Besides these arguments it shall be valid:

$$\frac{Y_m}{Y_s} = K \quad \frac{l_m}{l_s} = K_L$$

y - dislacement
ε - unit strain

$$\frac{Y_m}{l_m} = \varepsilon_m = \frac{Y_s}{l_s} = \varepsilon_s$$

The unit strain - because of the dimensionless argument -, shall be always the same at the model, as well as in reality.

2. CORRECTION OF LOADING

For the preparation of the physical models, which are loaded with their proper "dead" load, the mechanical properties of the model mass shall be selected so to meet the above mentioned criteria. This means that when preparing the model there cannot be applied the natural rocks, but the artificially prepared materials with the required mechanicals properties; i.e. the so-called models from the equivalent materials. In case of rockill tests with a reduced gradation curve we work with a "model", the mechanical properties of which are given by the properties of a natural rock representing the grains of the "model". The possible non-keeping to the similarity criterion should therefore be compensated by such a change of model loading that its behaviour would be equivalent with respecting these criteria. According to the stated relations the criterion for loading the model sample of aggregates is as follows:

$$\mathsf{G_m} = \mathsf{G_s} \cdot \frac{Pm}{Ps} \cdot \frac{lm}{ls} \qquad (a)$$

Simultaneously it shall be respected the theoretical condition

$$\frac{Emt}{Est} = K_{Et} = K_G = K_L = \frac{lm}{ls} \qquad (b)$$

Where the symbol E_t is the required theoretical modulus of deformation of the rock forming the rockfill grains, or the quantity having the same dimension. This condition connot be, however, fulfilled. The relation of the model and the real natural rock moduli

$$\frac{Ems}{Ess} = K_{Es}$$

can be determined by experiment. This enables to correct the criterion for loading the aggregates model sample (a) as follows:

$$\mathsf{G_m} = \mathsf{G_s} \cdot \frac{Pm}{Ps} \cdot \frac{lm}{ls} \cdot \frac{K_{Es}}{K_{Et}} \qquad (c)$$

where K_{Es} is the relation of the natural rock moduli forming the aggregates grains in a sample and in reality.
K_{Et} is the relation of the theoretical moduli of the model rock forming the aggregates grains in the sample and in reality originating from the similarity criterion

$$K_{Et} = \frac{Emt}{Est} = \frac{lm}{ls} = K_L$$

After simplification

$$\mathsf{G_m} = \mathsf{G_s} \cdot \frac{pm}{ps} \cdot K_{Es} \qquad (d)$$

a) Dependence_of_rock_mechanical_properties_on_grain_size_
It is a known fact from the rock mechanics that the increasing dimensions of a rock sample results in decrease of its strength and its strain moduli are reducing. This effect is explained by a greater occurence

probability of various microstrural defects, discontinuity surfaces etc. in a larger sample than in a smaller one.

With a change of grain sizes in the aggregates sample arises simultaneously the unneglectable change of its mechanical properties. Therefore it is necessary to take them into consideration when performing the tests of aggregates samples with a reduced particle size distribution curve. As far as it is not possible to determine during the tests the rock properties of the "model", i.e. of smaller and natural grains, we have to take into consideration the following relation

$$K_{Es} = \frac{E_{ms}}{E_{ss}} = 1$$

then the expression (d) is simplified on:

$$\mathsf{G_m} = \mathsf{G_s} \cdot \frac{pm}{ps}$$

b) Dependence_of_the_density_of_aggregates_on_granularity_
The diameter change of the isometric grains in a mixture from the regular structures does not affect the volumetric mixture mass. This can be proved by a simple calculation. Lushnov (1975) proved by means of an experimental method on a greater quantity of samples of various aggregates types that this also concerns the samples prepared from the mixtures with the different grain sizes, if they have a mutually parallel displaced gradation curve. In addition to this fact the conditions for a model similariry of the natural and model aggregates structures shall be naturally fulfilled. The author verified the correctness of the Lushnov´s results on the model aggregates from Klecany - see Figure 1.

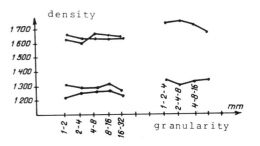

Fig.1. Granularity influence on the density of aggregates

Two samples with a displaced granulation curve but with an identity structure, fulfilling the conditions of the model similarity, will, therefore, have the desity approximately the same. The relation (e) will be then simplified on:

$$\mathcal{G}_m = \mathcal{G}_s$$

Under the simplifying presumptions that the mechanical properties of rock forming the aggregates grains in the sample and structure, as well as the values of aggregates density in the sample and structure, are the same, then the load of a model must be - on fulfilling the other similarity conditions - identical with the structure load.

3. SIMILARITY CRITERIA OF THE AGGREGATES STRUCTURES

It is recommended that the similarity conditions of the aggregates structures are divided on the conditions concerning the individual grains, conditions concerning the composition of grains and finally conditions concerning the spatial arrangement of grains.

a) Similarity conditions for the individual grains
It mus be valid:

$$\frac{\bar{d}_m}{\bar{d}_s} = K_L \qquad d = diametr$$

the relation of the average grain sizes "d" of the relevant fractions shall correspond to the geometrical similarity condition K_L

$$Z_{K_m} = Z_{K_S} :$$

the sphericity of grains of the particular fractions in the model and in the reality must be the same

$$Z_{om} = Z_{os} :$$

the roundness of the model and natural grains shall be the same for the grains of the relevant fractions (the grains shall correspond geometrically)

b) Conditions of the gradation similarity
Conformably to this condition the mutual relation of a relative representation of the individual fractions in the sample and in the natural aggregates must be the same. That means that the particle size distribution curve of the model and natu-

ral aggregates must be parallel,i.e. the equality of coefficients of uniformity and curvature and other dimensionless parameters analytically defining the shape of the granulation curve.

It must be also valid that the relation of the volumetric mass of the non-compact and max.compact aggregates must be the same both for the model and for the natural aggregates. (In this case it must be kept the dynamic similarity condition when selecting the compacting intensity, which means a lower compacting intensity of the fine-grain materials. At the equality of the compacting intensity the fine-grain materials with the parallel granulation curve are more compacted than the coarse-grain materials. It is not much practical to keep this condition and therefore the relation of the density   of the model and compact aggregates is not too suitable as a similarity criterion.)

c) Similarity conditions of the spatial grains arrangement
Among these conditions there are equality of density   , porosity and porosity numbers. If one sondition is valid then the other one must be valid too.It would also be possible to incorporate the relative compactness.To this type of conditions. The limitation is the same as in case of the relation of the noncompact and compact aggregates, because of the max.achievable compactness is a function of the compacting process effect, which must be bound by a dynamic similarity condition.

The other structural parameters, as the orientation of the contact normals   number of contacts, orientation of the longest grain axes etc., and or their relative localization inside the samples and the natural aggregates are important for the final aggregates properties. All these parameters can create the similarity conditions in a form of the dimensionless parameters. Practically   it is not possible to specify the natural aggregates in the dam body and to model them in a sample. For the engineering purposes it is sufficient to presume that all those parameters are in connection with the way of the structure origin,i.e. with the compacting method. If the same compacting method is used at

the preparation of the sample as in
the reality, the same surface grain
orientation of the model and natural
aggregates can be also assumed.

4. PREPARATION OF TESTS

a) Granularity_of_sample
In order to fulfill the similariry
condition for the grain composition
of the model aggregates, the granulation curve must be parallel with
the granulation curve in structure.
The limitation of the proposed metho-
dics represents a danger that by
displacing the granulation curve into the field of smaller grains, the
smallest fraction loses its decisive structural character of the coarse-grain bulk material.Then there
is no option than to remove from the
sample the grains not meeting the
condition on the max.graing size in
the sample. This approach is admissible when the great grains do not
form a skeleton, but "float" in a
more fine-grain material. As a rule,
the sample with the modified proper
ties, may be obtained, however,when
removing the position of great grai-
ns.The size of change shall be un-
conditionally found by means of the
model research on the more fine-grain samples. How the removal of the
coarse-grain admixture may affect
the results of the shear test,this
can be proved on the fine gravel
tests from Klecany performed by the
author himself (see Figure 2).

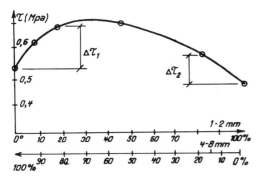

Fig.2. Mutual influence of the fine-
       grain and coarse-grain frac-
       tion on the shear test result

The shear tests were carried out on
a sample with a granularityof 4-8
mm. The portion of the fine-grain
fraction of 1-2 mm was gradually increased till it achieved 100%.

b) Sampling
When selecting and preparing the representative sample, it is necessary to take into consideration that
- the quality of the rock massif
  is sometimes considerably variable. This is projected both into
  the variability of the extracted
  aggregates granularity and into
  the quality of the individual grains;
- the aggregates structure in a construction is affected by the compacting process technology. During this procedure it occurs the
  crushing of grains particularly
  on surface of the layers. Therefore it is suitable to take samples for the laboratory tests -
  at which it is not possible to
  keep the same way of compacting
  as in the dam - from the testing
  compacting fields after compaction
  or from the dam ilself.
When compacting the sample, it is
necessary to reach the same
density   as it is at the aggregates in a structure.

5. SHEAR TEST RESULTS ON SAMPLES
WITH THE DISPLACED GRANULATION
CURVES

The author verified the test methods
with a displaced granulation curve
on the aggregates from Klecany and
Jirkov. In the Figure 3 there are
the granulation curves of the tested samples. The same dimensional
relations of the mean grains were
kept in the sample and of the shear
boxes dimensions during the tests.

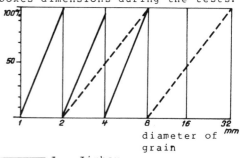

diameter of mm
grain

——————— J - Jirkov
— — — — K Klecany

Fig.3. Mutually parallel displaced
       granulation curves of the
       tested samples

Summary of test results with the displaced granulation curve is indicated in the following Table.

3128

Loca-lity	Granu-larity (mm)	Vol. density (kg/m³)	Shear Strength	
			φ (°)	γ_o (MPa)
J	4-8	1480	42.5	0.03
J	2-4	1530	41	0.023
J	1-2	1500	41	0.012
J	0.5-1	1550	42.5	0.0
J	4-8	1290	37	0.005
J	2-4	1230	36	0.004
J	1-2	1200	37.5	0.0
J	0.5-1	1300	35	0.0
K	8-32	1670	47-48	0.06
K	2-8	1600	47	0.039
K	2-8	1605	48	0.029

The graphic from of the results are illustrated in the Figures 3 and 4.

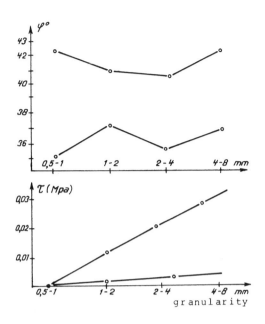

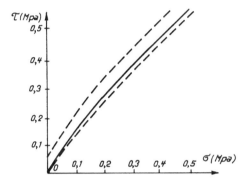

Fig.4. Envelopes of the Mohr´s circles of the Klecany aggregates and these with the mutually displaced granulation curves

The received results confirm that
the shear tests on the samples with
the reduced gradation curve by using
the method of the parallel displacement give the satisfactory results.
A better conformity is reached at
the angle of the internal friction.
The changes in its value does   exceed the method´s accuracy.
   The value of the inicial shear
strength γ_o is a little affected by
the granularity. With regard to the
relatively small density
it can be supposed that it is an influence of tension concentration
on the relatively large grains in
the proximity of the box faces placed in the gap between them. This
applies more to the coarse-grain
samples, if the gap width between
the boxes is the same.

Fig.5. Angle of internal friction (a)
and the initial shear strength
(b) of the Jirkov aggregates
with the mutually displaced
granulation curves

6. CORRECTION OF TEST RESULTS

a) Correction of compressibility line

When it is proved that a change of
the rock properties forming the grains of the model aggregates occurs
at the granularity reduction of the
aggregates sample, it is possible
to correct the affecting loading or
the compressibility line by means
of relation (d). (See Figure 5).

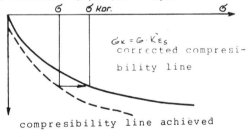

Fig.6. Correction draft of the compressibility line when not
keeping the model criterion
for the rock properties forming the model aggregates
grains

b) Correction of the shear strength
The shear strength has two components: internal friction and dilatancy component of the shear strength.
The internal friction expressed as
tg φ_m is a dimensionless argument
and cannot be affected by the granularity reduction. Otherwise a dilatancy component tg φ_d is expressing volumetrically the deformation
of a sample. The model criterion for
the properties of the rock forming
the aggregates grains shall ensure
that the relation of the contact
forces and contact strengths is the
same both in the sample and in the
natural aggregates. As far as the
model criterion is not kept accordingly to the increased strength of
smaller grains of a model sample,
then the relation of contact forces
an contact strength on grains is
modified. This results in another
character of crushing and grain deformation and therefore in a change
of dilatancy component of the shear
strength. Therefore, when not keeping the model criterion for the
rock properties of model grains, it
is important to correct the shear
strength conformably to the formula

$$\text{tg } \varphi_{kr} = \text{tg } \varphi_m + \text{tg } \varphi_d \cdot {}^{1/}K_{Es}$$

The range of the research works
has not yet made possible the sufi-
cient experimental verification of
the theoretically derived correction
relations. As a result it is recom-
mendede to consider temporarily the
proposed way of the correction of
the compressibility line and of
shear strength as a work hypothesis
and to verify it gradually through
the next tests. On the same principle it will be possible to correct
the test results with respect to
another relation of densities
     or aggregates loading in the
sample or structure. (For ex.in case, that the same density
of the aggregates in the structure
or such a high loading degree is
not reached in the test.

7. CONCLUSION

A. Consequently to the similarity theory the similarity conditions
are stated, which shall be respected
for the preparation, performance
and evaluation of tests of the aggregates with a reduced granulation
curve.

B. By means of the structural approach the work hypothesis was formulated, enabling to derive the
theoretical relation for the correction of the shear strength and compressibility line of aggregates.
The correction may be applied in
case when the similarity conditions
for the rock, forming the grains of
a sample with a reduced granulation
curve, could not be fulfilled.

C. The similarity conditions were
specified for the aggregates structure. The conditions result from
the criterion of the geometrical
similarity, equality requirement of
all dimensionless arguments defining the aggregates structure (sphericity, roundness and surface friction of the individual grains, spatial arrangement of grains as porosity, relative compactness). The
next conditions are equality of
densities   and parallelity
of granulation curves.

D. The shear tests on the Jirkov
and Klecany aggregates confirmed
that the parallel displacement of
the granulation curve did not affect
the test results. On the contrary
it was found that not only the elimination of the fine-grains but also
of coarse-grain admixtures in the
sample might affect its mechanical
behaviour.

REFERENCES

Dobr J., Rozsypal A.(1974):Rockfill
     testing in a large direct shear
     device. 2nd IC IAEG Sao Paolo
Kohoutek J. (1971): Modelování v me-
     chanice hornin (Modeling in rock
     mechanics) Acta Montana No.17
     (in Czech)
Lušnov N.P.(1975): Podbor plotnosti
     krupnooblomočnovo materiala v na-
     broske (Density of rockfill mate-
     rial in dam), Strojizdat Moskva,
     (in Russian)
Marsal R.J.(1975): Pressas de tiera
     y enrocamiento Limusa, Mexico,
Oda N. (1978):Significance of fabric
     in granular mechanics. Japan semi-
     nar on Continuum mechanical and
     statistical Approaches
Rozsypal A., (1983): Mechanical pro-
     perties of rockfill (in Czech)
     Phd thezis Academy of Science
     Prag

Petrographic and textural analyses of crystalline silicate aggregate for road pavement subjected to wear by studded tires

Analyse de la pétrographie et de la structure du granulat destiné pour l'enrobé bitumineux, exposé à l'usure des pneus cloutés

R.Uusinoka, P.Ihalainen & P.Peltonen
Institute of Engineering Geology, Tampere University of Technology, Tampere, Finland

ABSTRACT: Studded tires are regarded as the most important factor affecting the wear of bituminous pavements. Special attention is therefore paid to the quality of the aggregate concerning its resistance to blows and stratching of studs.

The aggregate material is crushed from the Precambrian bedrock of Finland. In addition to the wearing, brittleness, abrasion and adhesion tests, petrographic and structural analyses will be made from the samples of rocks planned for pavement aggregate. These analyses comprise the macro- and microscopic inspection as well as X-ray diffraction studies to find out:
- stage of alteration and presence of secondary minerals (carbonates etc) and weathering products (clay minerals)
- textural details like grain size, fabric and microfracturing
- mineralogy, especially the minerals prone to weathering like micas whose presence in large quantities and as bundle-like concentrations renders the rock unsuitable for road pavement aggregate

RÉSUMÉ: Les pneus cloutés sont considérés comme le facteur le plus important produisant l'usure à la couche bitumineuse des chaussées. Pour cette raison, une attention particulière est portée à la qualité du granulat, en tenant compte surtout de la résistance aux chocs et au grattement produits par les pneus cloutés.

Le granulat est extrait par concassage du gisement Précambrien finlandais. Outre les tests d'usure, de friabilité, d'abrasion et d'adhésion, on examine la pétrographie et la structure du granulat. Ces dernières analyses comprennent un examen macro- et microscopique comme aussi l'étude par diffraction à rayons X, dont l'objectif est de déterminer:
- taux de dégradation de la roche et la présence des minéraux secondaires (carbonates etc.) et des produits d'altération (minéraux d'argiles)
- détails de la texture comme la taille de grains, la fabrique et les microfractures
- minéralogie, surtout les minéraux susceptibles de l'altération comme le mica, dont la concentration ou la présence en quantités élevées rend le granulat impropre pour le revêtement bitumineux des chaussées.

1. INTRODUCTION

In Finland the use of studded tires is allowed as a rule from Noverber 1 to March 31. During this winter period most private cars wear studded tires, while trucks and buses have unstudded tires all year long. Because of the salt strewn on the road surface, the highways remain free of snow and ice, except during and a day or two after a heavy snowfall. The bare road is thus prone to strokes and scratching of studs. Studded tires are regarded as the most important factor affecting the wear of bituminous pavements in Finland. Other important factors affecting the condition of bituminous road pavements are (cf. Kauranne 1970, Niemi 1983, Lappalainen 1987, Peltonen 1985, Pylkkänen & Kuula-Väisänen 1990), the amount and kind of traffic, climate, especially frost action, the type of binder, properties of adhesion, the grading and the way of crushing of the aggregate, the shape and size of the individual grains, compaction of the lower courses and the subbase of the road, etc.

Special attention is paid to the quality of the pavement in countries allowing the use of studded tires. Wearing of road pavement by studded tires is due to stroke, compression and scratching through tire-studs as well as to abrasion caused by the combined effect of studs and tire. Pavement wear is directly proportional to the rotation velocity of studded tires.

According to Niemi (1983), the first stages of wearing of the aggregate grains in the pavement surface are increase of roundness in the edges and their splitting off from the grains. As wearing proceeds the grains break apart and finally get detached from the pavement. Studies by means of test surfacings in Sweden have been reported by Backman & Höbeda (1984). According to them, the grains are worn more flaky in the pavement surface and the highest initial macrotexture depth is given by cubical particles.

The pavement aggregate used in Finland is crushed from crystalline silicate rocks of the Precambrian. The few deposits of sedimentary rocks and non-silicate crystallites are ruled out because of their weak soundness.

The methods of testing the aggregate material include wearing, brittleness, abrasion and adhesion tests (Kauranne 1970, Uusinoka & Peltonen 1988, Pylkkänen & Kuula-Väisänen 1990) together with petrographic and textural analyses. The latter has greatly increased in importance along with the greater attention paid to the role of the mineralogical composition and texture of the rocks used in road pavement (Uusinoka & Peltonen 1988). These analyses comprise the study of rocks both macroscopically and in thin sections. To find out the clay minerals, X-ray diffraction analysis is used. Other rock properties affecting the quality of the aggregate, such as the specific surface area and pore properties (Nieminen & Uusinoka 1986) can be measured by N_2-adsorption and/or mercury porosimetry.

The quarries from which rock is removed are commonly open cuts in hills of outcropped bedrock. Waste material like country rock from ore mines is also used as rock aggregate, but it is mostly ruled out because of the largerly amounts of sulphides, soft alteration products etc it often contains.

2. PETROGRAPHIC AND TEXTURAL ANALYSES

2.1 Field investigation and macroscopic study of rocks

Prior to sampling and laboratory studies selection of rock aggregate material is often started in the field by using geological maps. These maps have proved their usefulness in areas showing a great variation in the composition of bedroc. Outcrops consisting of rocks known to be of lesser quality (limestones and marbles, mica, chlorite and talc schists, phyllites etc) will already be ruled out in this initial phase. Field investigation on the outcrops of potential aggregate reserves comprises determination of joint and fracture frequency, stage of weathering, occurrence of veins of secondary minerals, grain size of the rocks and variation in the composition of the outcrop.

As regards the different weathering-classifications in the characterization of rock presented by e.g. Niini (1968), Uusinoka (1975) and Dearman (1976), the rocks suitable for pavement aggregate must belong to the class I (fresh and hard). Some fine-grained and compact rocks, however, might be accepted as slightly weathered (class II), too. Any weathered sample must also be subjected to X-ray analysis because of the possible presence of certain clay minerals.

Outcrops containing fractures, weathered rock, coarse-grained material like pegmatite veins, components of strongly schistose and/or mica-rich rock, occurrence of sulphides, graphite, etc will be ruled out by macroscopic inspection. Sampling will thus be concentrated on outcrops consisting of more homogeneous and hard rock without any notable defects observed by naked eye.

2.2 Microscopic inspection

Thin sections are made of samples chosen for closer studies and tests. Following properties are observed:

2.2.1 Chemical decomposition

Signs of possible weathering and/or hydrothermal alteration of minerals like feldspar, amphibole and mica will be examined. Weathering of rock fragments with a texture already loosened by decomposition may proceed rapidly in road pavement under the influence of wetting and drying, freeze and thaw as well as salt action.

The so-called sericitisation of feldspar, unless too intense, is generally allowed in fine-grained and compact rocks. Practically no difference has been observed in the values of the mechanical tests between rocks with fresh feldspar and those having this mineral even moderately sericitised. The weathered rock must be analysed through X-ray diffraction. The presence of other clay minerals, especially smectite or kaolinite, is not allowed.

2.2.2 Microfractures

Cracks and fissures, either open or filled with secondary minerals or pigmentation of ferric iron, are often met with in brittle minerals like feldspars and in grain boundaries. This microfracturing is best observed by viewing the object with parallel nicols (Fig. 1).

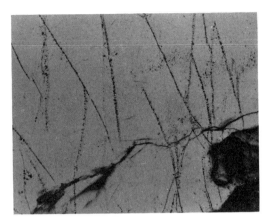

Figure 1. Microfracturing of feldspar in coarse-grained granite with smooth mineral grain boundaries; unsuitable for pavement aggregate. Parallel nicols, 40 x.

2.2.3 Textural features

2.2.3.1 Features promoting the soundness

2.2.3.1.1 Grain size of the individual minerals < 5 mm

Fine- and medium-grained rocks consist of of minerals with grain sizes much smaller than those of the pavement aggregate. The maximum grain size of the latter is generally 16 - 25 mm (Lappalainen 1987). With the exception of certain schistose and mica-bearing rocks, the fine- and medium-grained magmatic- and metamorphic rocks have showed their resistance against mechanical forces better than coarse-grained rocks. This fact has been proved by the tests made in the engineering-geological laboratory of Tampere University of Technology (Uusinoka & Peltonen 1988) as well as other field and laboratory studies (Niini 1968, Uusinoka 1968, Niini & Uusinoka 1978, Lappalainen 1987). The finer the grain size of the rock the smaller amount of cracks and fissures are found in the mineral grains and their boundaries. The micro-fractures in feldspar and the perfect cleavage along certain crystallographic surfaces of feldspar and mica, for example, have a less profound effect on the total fracturing of a fine grained rock than on that of a coarse-grained rock.

2.2.3.1.2 Moderate sorting in graind size distribution of the individual minerals

As regards the sorting of the mineral grains, best aggregate is found among the rocks consisting of evenly distributed grains of different minerals with grain sizes gently deviating from each other. Certain gabbros, diabases and porphyrites provide best examples (Peltonen 1985).

2.2.3.1.3 Isotropic and compact fabric

A homogeneous texture with un- or slightly oriented fabric having its mineral grains wedged and seamed between each other so that the strength of the rock is uniform or nearly so and the porosity minute (see Niini & Uusinoka 1978) seems to be a feature to be observed among most rocks accepted as pavement aggregate on the basis of the tests made in our laboratory.

2.2.3.2 Features reducing the soundness

2.2.3.2.1 Grain size of the individual minerals > 5 mm

As to the grain size of the pavement aggregate made of coarse-grained rock, the individual grains consist of one or at most three minerals only. Properties like the perfect cleavage along certain faces of feldspar and mica as well as the smooth boundary faces between the mineral grains (cf. also Kauranne 1970, Uusinoka 1968, Niini & Uusinoka 1978) abruptly reduce the soundness of the fragments of coarse-grained rocks like pegmatites and rapakivi granites.

2.2.3.2.2 Anisotropic and/or deformed fabric

Rocks with a moderate or strong schistosity are unsuitable for pavement aggregate because of:
- oriented mineral grains with smooth boundary faces
- microfracturing mostly along the orientation of the mineral grains
- soft or brittle minerals grouped in bands, zones, lenses etc

These facts strongly promote the cleavage of the rocks along the plane of schistosity. Further-more, wearing resistance is small in rocks having the so-called mortar texture (Uusinoka & Pelto-nen 1988). This texture is produced by dynamic metamorphism upon some granites and gneisses: small crushed grains of quartz and feldspar occupy the interstices between larger grains (Fig. 2).

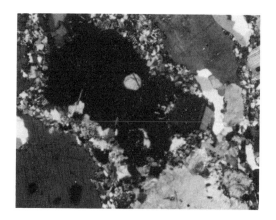

Figure 2. Mortar texture in a medium-grained gneissose granodiorite; unsuitable for pavement aggregate. Crossed nicols, 40 x.

Figure 3. Band of biotite in a medium-grained mica gneiss; unsuitable for pavement aggregate. Parallel nicols, 40 x.

2.2.4 Mineralogical composition

Soft rock-forming minerals, such as talc chlorite and mica are resistant against strokes by studs but not against scratching. As regards the hard minerals like quartz, feldspars, pyroxene and garnet the reverse is true, especially with the feldspars because of their perfect cleavage in two directions perpendicular to each other. Thus, small quantitites of soft components like mica should be scattered among hard and brittle minerals like quartz and feldspar. As to mica (biotite and/or muscovite), however, amounts exceeding 15 % mostly renders the coarse-grained rock unsuitable for pavement aggregate. If the amount of mica is 15 - 25 % in a fine-grained rock, the rock may be used as second-class aggregate. Micas tend to occur as bands (Fig. 3) or bundle-like concentrations when present in large amounts. Muscovite tends to reduce the soundness more than biotite. Weathering or breaking apart is often observed to start at these sites of inhomogeneity. Secondary minerals like epidote, carbonates, sulphides etc are not allowed in quantities exceeding a few per cents altogether.

2.3 X-ray analysis

The most convenient method to determine certain fine-grained secondary minerals, i.e. the clay minerals, is X-ray diffraction analysis. In Finnish Precambrian bedrock the most important products of weathering and/or hydrothermal alteration of the rock-forming minerals are illite (sericite), smectite, chlorite and kaolinite. Certain mixed-layer minerals like smectite-illite and smectite-chlorite are also met with (Uusinoka 1975).

Sericite (or illite) is a common weathering product of feldspar. Its presence up to 10 % can be allowed in fine- and medium-grained rocks and up to 5 % in coarse-grained rocks. The same is true with chlorite. When occurring together, the total amount of illite and chlorite must not exceed 10 and 5 %, respectively. Smectite, a decomposition product of feldspar and amphibole under stagnant groundwater conditions has the notorious ability to swell when rehydrated. The presence of smectite is not allowed in rock aggregate. Kaolinite, a weathering product of illitised or smectitised feldspar, mostly occurs in strongly weathered rocks. That is why neither kaolinite-bearing rocks are allowed. X-ray anaysis might be done instead of microscopic study if the textural properties are known, and the mineralogical composition only needs to be checked.

3. CONCLUSIONS

Petrographic and textural analyses generally start testing the applicability of rock aggregate crushed from Finnish Precambrian bedrock. The complicated and time-consuming abrasion, brittleness and wearing tests might be ruled out, especially, if the results of the microscopic and/ or X-ray diffraction analyses are obviously negative.

In this respect the most important defects are listed as follows:

By macroscopic inspection:
- fracture or joint frequency of the bedrock more than 3/m
- degree of weathering above the class II (slightly weathered)
- strong schistosity

- the average grain size of the individual minerals above 10 mm
- bands or large concentrations of micaceous minerals observable
- occurrence of sulphides, carbonates, graphite etc

By microscopic inspection of rocks with the average grain size of minerals 5 - 10 mm:
- smooth boundary faces between the mineral grains
- fine-grained sericite covering more than about 30 % of the feldspar grains
- fractures in most of the grains of brittle minerals and between the mineral grains
- clearly observable mortar texture
- mica content exceeding 15 - 25 %, depending on the grain-size of the minerals
- bands or large concentrations of mica
- occurrence of sulphides, carbonates etc more than 5 - 10 % altogether

By X-ray diffraction analysis:
- illite (or sericite) and chlorite more than 5 - 10 % together, depending on the grain-size of the minerals
- occurrence of smectite or kaolinite

Discovery of any defect listed above generally makes any further test unnecessary, especially, if rock aggregate of better quality might be available. As to fine-grained and compact rocks, unless too weathered, fractured or schistose as well as smectite-bearing, a few of these defects will be allowed, which finally depends on the results of the mechanical tests.

REFERENCES

Backman, C. & Höbeda, P. 1984. The performance of aggregates in a single surface dressing, subjected to wear by studded tyres. Bull. IAEG 30: 11 - 16.

Dearman, W.R. 1976. Weathering classification in the characterisation of rock: a revision. Bull. IAEG 13: 123 - 127.

Kauranne, L.K. 1970. Comparison of some methods used in the testing of road surfacing aggregates. Valtion teknillinen tutkimuslaitos, Tielaboratorio, Tiedonantoja 3.

Lappalainen, K. 1987. On aggregate factors influencing wear resistance of pavements (In Finnish). Tie ja liikenne 1 - 2: 16 - 29.

Niemi, A. 1983. Factors contributing to pavement wear. Technical Research Centre of Finland. Road and Traffic Laboratory.

Nieminen, P. & Uusinoka R. 1986. Influence of the quality of fine fractions on engineering-geological properties of crushed aggregate. Bull. IAEG 33: 97 - 101.

Niini, H. 1968. A study of rock fracturing in valleys of Precambrian bedrock. Fennia 97: 6,

Pap. Eng.-Geol. Soc. Finland 3: 26.

Niini, H. & Uusinoka R. 1978. Weathering of Precambrian rocks in Finland. Proc. III Int. Congr. IAEG, Sec. II, Vol. 1: 77 - 83.

Peltonen, P. 1985. Suitability of the bedrock material from North Tavastland to road pavement (In Finnish). University of Turku, Dept. of Quaternary Geology, Unpubl. MSc-thesis.

Pylkkänen, K. & Kuula-Väisänen, P. 1990. Adhesion between bitumen and aggregate. Proc. VI Int. Congr. IAEG (In Print).

Uusinoka, R. 1968. Mechanical weathering of Finnish Precambrian rocks with the influencing rock properties (In Finnish). University of Helsinki, Dept. of Geology, Unpubl. MSc-thesis.

Uusinoka, R. 1975. A study of the composition of rock-gouge in fractures of Finnish Precambrian bedrock. Comm. Phys.-Math., Soc. Sci. Fennica 45: 1.

Uusinoka, R. & Peltonen, P. 1988. On the testing methods and quality requirements of rock aggregates (In Finnish). Vuoriteollisuus 46: 102 - 105.

Standardization of soil description and classification
Sur les normes de description et de classification des sols

H. Wiegers
Traffic, Roads and Bridges Department Drenthe, Assen, Netherlands

ABSTRACT: The author, confronted with the standardization in description and classification of soil samples, made an inventory of the development of the usefulness of agricultural, geological and civil engineering types of soil classification all over the world. He opts for a lithological classification in the context of local geology, as part of a standard methodology for soil description in geotechnics. A universal geotechnical classification turns out to be merely an utopia.

RESUME: Confronté, à la normalisation de la description et la classification des échantillons de sol, l'auteur a fait un inventaire du développement de l'utilité des classifications de sols du point de vue agronomique, géologique et de terrassements. à travers le monde. Il en tire la conclusion d'utiliser une classification lithologique dans le cadre de la géologie locale, en tant que partie d'une méthodologie normée pour la description des soils en géotechnique. Une classification géotechnique intégrale et universelle ne se trouve d'être qu'une utopie.

HISTORICAL OUTLINE ON USCS

About 1870 a system of so-called soil classification was devised by Russian agricultural engineers. Around 1900 this system was adopted by the U.S. Department of Agriculture (USDA), which has been classifying and mapping soils in the United States ever since. Highway engineers thought that the valuable information on soils, resulting from this classification could be used in identifying soils for engineering purposes. Consequently, since 1931, the USDA soil classification system for agriculture has been used as a soil identification system by highway engineers.

When the Americans became involved in World War Two, the U.S. Army Corps of Engineers adopted a classification system developed by Dr. A. Casagrande that uses texture as description criterion. This classification has been expanded and in 1956 it was called the Unified Soil Classification System (USCS). In addition to these two systems, others have been designed. After the USCS had been expanded and revised in cooperation with the U.S. Bureau of Reclamation, it has been applied in embankments and foundations as well as in road and airfield construction (PCA 1956).

The USCS is based on the identification of soils according to (simplified H.W.) textural and plasticity qualities and groups are based on their expected performance as construction materials for engineering. The original identification method became obsolete and the mode of formation of soils out of the range of vision. The USCS is now a specialized system used in several countries all over the world, mostly with modifications, e.g. in B.S. 5930 (1981), B.S. 6031 (1981), I.S. 1498 (1970)and in the IAEG report on rock and soil description and classification for engineering geological mapping (IAEG 1981).

DISADVANTAGES OF THE CURRENT USCS

As a consequence of its development the USCS has a number of disadvantages:

1. The mode of formation of a soil is not taken into account.
2. Fine-grained soils are not characterized by their grain-size composition; the clay/silt character of the fines is not discriminated since the 2-micron limit is not used.

3. The subdivision of soils is artificial, since there is no feedback between the classification and the mode of origin of the soils.
4. The USCS is only useful in temperate climate zones.
5. Tropical and subtropical soils (e.g., laterites, gypsiferous/calcareous sands and calcrete sands) cannot be classified and identified by the USCS.
6. The USCS cannot be subdivided easily; when it is used for mapping purposes (IAEG 1981), it is too general to be applied at the project site the same time.

LITHOLOGICAL CLASSIFICATION AS A BASIS

Because soil and rock are both materials of the earth's crust it is feasible to use the same basis for their discrimination and description as far as possible. This implies that for all natural materials the first step is a lithological classification that takes two aspects into account: the mode of formation (= manufacturing) and the composition.

Such a geological classification is necessary for interpreting the geological structure of an area. It may also be important for construction purposes, for example when natural material is required for building stone, concrete aggregate, foundation material or mass.

A lithological classification does not include engineering properties. In practice, however, such properties are often related to lithological characteristics and engineers with local experience may interpret the lithological classification to some extent in terms of likely engineering characteristics.

A lithological classification for rocks is taken for granted (e.g. IAEG 1981, BS 5930) and it fits well in their geotechnical description and classification. A fine example of such a description of rocks is given in Geoguide 3 (GCO 1988) in the context of the local geology of Hong Kong.

DISTINCTION BETWEEN ROCKS AND SOILS

The classification of rocks that an engineer would call "engineering soils", inevitably brings up the distinction between rock and soil. In the IAEG report on mapping (IAEG 1981) we find this quote: "Soil is an aggregate of mineral grains that can be separated by such gentle means as agitation in water". In engineering practice the composition of the soil also includes other components such as moisture, organic matter and air.

Furthermore soil properties depend specially on e.g. density, moisture content and composition of the mixture of the soil components.

SIGNIFICANCE OF MODE OF FORMATION OF SOILS

The mode of formation offers a natural starting point for the classification of soils. It decides among other things grain-size distribution, mineral composition, grain shape and relative density of the soil. A principal division in this respect could be the nature of the material from which a soil is derived or in which it came into being (e.g. residual soils), taking into account sorting of the material by transport and deposition.

A world-wide inventory of all possible modes of formation is unrealistic. A useful table listing different modes of formation and the related soil types depends on lithology and can only be established in a local or national geological context i.e., for a single basin, delta or any other geological "province"). A fine example is the Swedish paper on soil classification and identification (Karlsson and Hansbo 1981). All modes of formation mentioned in their list are explained in the context of the geology of Sweden in separate paragraphs together with pictures and a few maps. However, grain-size distribution curves are not presented and neither are their places in a classification triangle, as is normal in sedimentology (Wiegers 1974b).

THE DUTCH LITHOLOGICAL SOIL CLASSIFICATION.

1. Sands.

In the Netherlands, the classification of enigeering soils started with the classification of sand samples, on the basis of the grading curve and the alternative use of a sand triangle (six types) (Wiegers 1970). Next, the mode of origin of sand was drawn in (seven types) and in this way a lithological sand classification was created (Wiegers 1971). In addition, the concept was introduced that sand (soil) is a mixture of components. The sand classification was expanded into a soil classification by incorporating the silt/clay triangle used in sedimentology and the gravel triangle of the Geological Survey of the Netherlands, including the boundaries of the soil fields in these triangles. Thus a system of soils with detailed subdivisons was established (Wiegers 1974a). Later, Wiegers (1974b) decided that soils that are coarse according to the USCS (gravel and sand) could be placed in a part of the gravel triangle, albeit by means of an artificial and unnatural subdivision. In the USCS the grain-size curve is not used for the classification and identification of fine soils, solely certain soil grain properties, i.e. the Atterberg limits (Liu 1967). This demonstrates the ambiguity of the USCS and its modifications.

In 1975 a complete lithological sand classification was published (Wiegers et al.). This system corresponds well with the lithostratigraphy of The Netherlands, that was developed by the Geological Survey of The Netherlands. This explains the compatability of the classification and the legends of the new geological and pedological maps.

The last stage of the geotechnical classification and description of sands, was completed in 1978 and published in the record on sands that can be used in highway engineering (Lubking, Van der Vring and Wiegers 1978). A main conclusion is that standardization is to be rejected for design purposes, since the properties of a sand type ought to be used to the full. In each individual design case standardized tests can only assist in part of the design and variations in relative density, moisture content, grain shape and clay content have to be taken into account.

In the publication of Lubking et al. (1978) a table is included for the rapid assessment of mechanical and physical properties of a sand, with the objective of achieving an economical road pavement. (See also Wiegers (1982). A geotechnical classification for using in the design of roads is thus not feasible, but geotechnical descriptions may be used for various purposes for the very details of several soil-aggregate properties (Liu 1967) for various purposes.

In the following years a standardized basic classification according to grain-size distribution and organic matter content was developed (NEN 5104, 1988). Expanding this system into a lithological classification is possible and thus it may be used in studying various other types of soils not just sand (Post 1989).

2. Silts and clays

The NEN 5104 standard was drafted by the Committee on Geotechnological Standards, on which as many disciplines as possible were represented, i.e., civil engineering, agriculture (pedology), hydraulic engineering, and geology. The task of this committee was greatly alleviated by the fact that during the previous twenty years the Geological Survey of The Netherlands and the Institute for Pedological Mapping had been using a similar classification for the grain-size composition of soils. A triangle, such as used in sedimentology is part of this classification. The division of this triangle is based on the differences between natural sediments, while an extension serves to classify sands more in detail.

The need for a classification of engineering soils was first put forward simultaneously, around 1970. The civil engineering standards on soils which dated from 1939, were out of date. Moreover, these standards were restricted to a single type of soil sample, sand, in which the proportion of particles smaller than 16 micron was 10% (m/m) or less. The Geological Survey at that time received a growing number of requests for studies to indicate sites for the winning of sand for several purposes. Reporting of these studies was hampered by the lack of a classification that is understood by civil engineers and geologists alike. This eventually led to the development of the NEN 5104 standard (1988), described in the preceding chapter. As a result of the interdisciplinary approach this standard is now employed for the description of soils in three earth sciences: geotechnics, geology and pedology. When used in combination with the mode of formation the classification is lithological.

The NEN 5104 standard was found to be a very good tool in the investigation programme of a working group of the CROW (Centre for Research and Contract Standardization in Civil and Traffic Engineering). Its task was to find out which natural materials can be used as alternatives to sands, for environmental reasons (Post 1988). CROW investaged six soil types, i.e. sandy clay, Brabant loam, loess, boulder clay (till), fluviatile clay and marine clay. Their fields in the classification triangle are presented in figure 1. Two series of laboratory tests were performed:

1. identification tests (on soil grain properties) concerning:
   - grain size distribution
   - organic matter content
   - calcium carbonate content
   - liquid limit
   - plastic limit
   - plasticity index

2. tests on properties that are significant in embankments and filling (soil-aggregate properties):
- compactibility
- friction properties
- Californian bearing ratio (CBR)
- natural water content
- swelling
- compressibility
- shrinkage

Various relations between these soil properties could be established and it is possible to fit these into an extension of the lithological classification. Others will remain part of the description of the soils. However, it is not possible to tie soil properties directly to functional properties fur use in engineering projects, let alone fit them into a geotechnical classification. The CROW working group came to the conclusion that functional properties are governed by the type and dimensions of a project, its topographical and hydrological situation and the geology of the site.

So, a universal geotechnical classification seems to be impossible. Other geotechnical investigators (Samejouand 1975, Buisson 1976) came to the same conclusion. They argue that a full geotechnical or engineering geological classification can only be drafted for a limited area or even

for a single engineering site. Such a classification should be based on formal regional lithological/pedological data, the physical state of the soil, and the physical and mechanical properties of the soil. Only in this way is it possible to obtain the correct connection between classification data and laboratory or field tests.

RECENT INTERNATIONAL DEVELOPMENTS IN (LITHOLOGICAL) SOIL CLASSIFICATION

There is definitively a tendency towards developing existing classifications and creating new ones where current classifications are not satisfactory or need to be expanded. The present intensity in explorations of the earth's crust requires this. The examples below serve to illustrate this tendency. At the end, some recommendations are made to promote a generally acceptable methodology. Such a methodology can provide better possibilities for communication and exchange of ideas and data in geotechnics.

<u>The Swedish classification</u>

In Sweden, four disciplines are involved in setting up an approach to soil classification and identification, i.e., forestry,

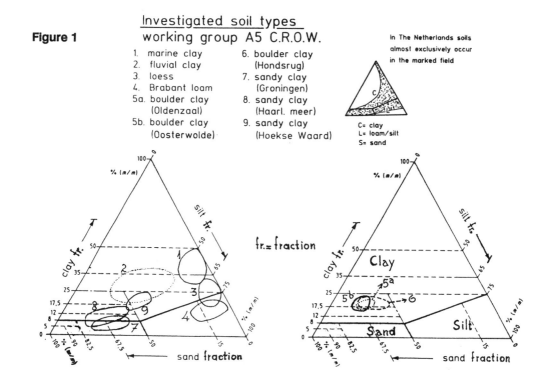

Figure 1

Investigated soil types
working group A5 C.R.O.W.

1. marine clay
2. fluvial clay
3. loess
4. Brabant loam
5a. boulder clay (Oldenzaal)
5b. boulder clay (Oosterwolde)
6. boulder clay (Hondsrug)
7. sandy clay (Groningen)
8. sandy clay (Haarl. meer)
9. sandy clay (Hoekse Waard)

In The Netherlands soils almost exclusively occur in the marked field

C= clay
L= loam/silt
S= sand

fr.= fraction

geotechnics, road research and geology (Karlsson and Hansbo, 1981). So it is no surprise that mode of formation, composition and the 2-micron limit are part of that soil classification. Although the various classifications have not been combined in one system, it is still possible to arrive at a lithological classification by integrating the parts concerned. This is demonstrated in figure 2, which shows the place of the Swedish tills in the composition triangle. This triangle is comparable with the Dutch first level classification triangle and can consequently be subdivided according to this classification down to a second level (Figure 3), and if desired even to a third one (classification of sands). Feedback between composition and mode of origin seems unavoidable in a good lithological classification. This is feasible in this classification system.

The classification of soil-grain properties according to geotechnical properties is restricted to plasticity and composition. The soil-aggregate properties are classified according to the six properties listed in Figure 3. A unified classification system in which each class would represent a soil unit with a narrow range of engineering properties is not presented. In this respect, the geotechnical part of the Swedish approach is essentially only descriptive. Important is the fact that this classification does function in the context of the geology of Sweden.

The British classification

The well-known British soil Classification System (BSCS) which is part of the BS5930: 1981 is used almost worldwide.

It was originally established for Great Britain as a modification of the USCS. Concerning earth works it refers to BS 6031: 1981: Code of practice for earthworks. The foreword of the BS 6031 clearly states: "the code has been prepared in relation to conditions existing in the British Isles. Some of the recommendations may not be appropriate for work in countries overseas where climate and other factors require different design and construction techniques." I suppose this passage has been overlooked all too often.

Typical for the classification is the paragraph 5.1.4., which reads: "Empirical relationships have been established between classification groups and factors such as pavement thickness, frost susceptibility, shrinkage and swelling, drainage characteristics and compaction methods. Some of these factors are set out in Tables 2 and

4".

Since disadvantages 1 to 3 that have been listed above for the USCS apply implicity to the BSCS one wonders if this extension of the classification can be sufficiently reliable when used outside Great Britain. In that case the relationships would concern appraisals of properties of unnatural lithological soil groups. Furthermore, some soil-aggregate properties can never be determined exactly in the framework of a classification. The relative density, for instance strongly influences the shear resistance and compressibility (Karlsson and Hansbo, 1981), as well as the moisture content (Lubking et al. 1978). Such uncertain factors make it impossible tu use a full classification in geotechnical engineering. Soil-aggregate properties can be determined more accurately in the laboratory but it remains to be seen to what extent they agree with field values. In any case, the values that are found are part of a detailed geotechnical description only, not of a geotechnical classification. (Also Vargas 1988).

Cooperation between pedology and civil engineering

Atlan and Feller (1983) published a report on pedological applications for geotechnics in road works in Senegal. This paper clearly demonstrates the possibility of borrowing information from pedological data that may be useful in geotechnics. The authors also used results of similar investigations in tropical Africa (Gidigasu 1971, 1972; Madu 1977). Wiegers (1985) studied the geotechnical application of a Dutch pedological report (Van der Sluijs 1971) and, although not concerned with tropical soils, he reached the same conclusion. Thus the use of a lithological classification is comparable with a pedological one.

This phenomenon hds attracted general attention, especially in the tropics. Therefore, the ISSMFE established The Technical Committee on Tropical and Residual Soils, during its congress in Brasila, Brazil, in 1985. This Committee organised the Second Conference on Tropical Soil (2 ICOTS) in Singapore in 1988. The main purpose of the conference was to review the progress that had been achieved since 1985; for example on the geotechnical properties and the classification of the named soils, their use as a construction material for roads, airfields and earthdams and as a foundation material (Broms 1988).

The well-known author on laterite

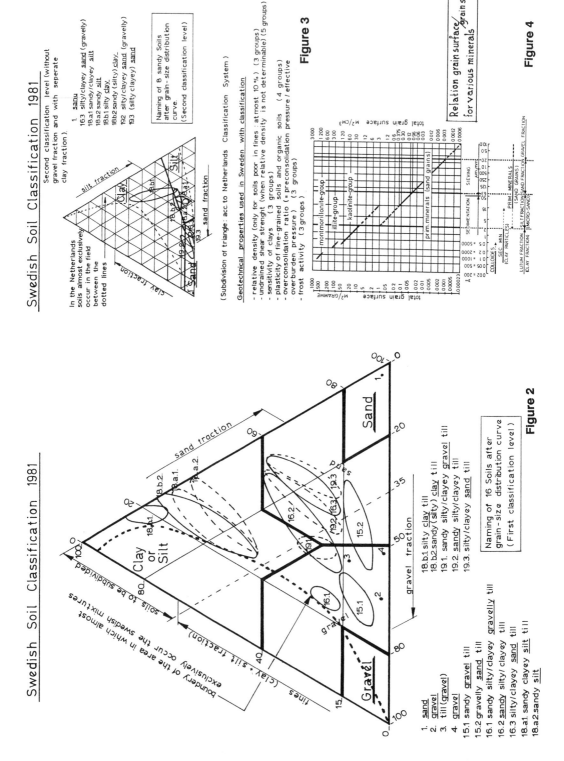

Figure 2

Figure 3

Figure 4

soils Gidigasu (1988) stated at 2 ICOTS that, in general European and North-American engineers are familiar with the geotechnics of temperate soils only (e.g., using the USCS) and that a new form of tailoring may be necessary for tropical and subtropical materials and soils (see also: Aitchison 1979 (or later) and Akpokodje 1985). The USCS cannot be used in tropical countries, according to Gidigasu. Vargas (1988) draws attention to the fact that it is necessary to subdivide the fine-grained soil groups, to be able to include the type of clay mineral and the mineralogical nature of the silt fraction. Fig. 4 illustrates this (Wiegers 1975).

Cook and Newill (1988) point out the difference in approach in the field description of tropical soils and of soils in temperate climate zones, in view of soil engineering. They propose a framework for the field description, examination and identification of tropical soils. Such a concept could also be used as a starting point for a revised description in geotechnics in temperate climate zones. Cook and Newill (1988) call also attention to the sensitivity of clays to sample treatment. This phenomenon is a general one, occurring as well in temperate climates. This is stated by Hough (1957).

It is a strong argument in favour of the use of the 2-micron limit in soil classifications.

CONCLUSIONS AND RECOMMENDATIONS

1. At present, the earth's crust is being explored very intensely by various earth sciences related to civil engineering, e.g., geology and pedology. It would be beneficial to use the relevant data obtained by these sciences in geotechnics. This is usually omitted as the current soil classifications have mostly not been designed for geotechnical use.
2. Developing a lithological classification of soil samples in the context of national or local geology will provide a sound basis for a geotechnical classification. This will enable a reliable soil description, especially if feedback of the test results is incorporated. A classification involving several levels offers the possibility to go into the very details that are useful for the investigation of the project site.
3. Because a world-wide classification of soils is not feasible it is recommended to aim at a standard methodology for soil description instead. Such a methodology should be universally applicable and offer the possibility of incorporating a regional soil classification that may be used in engineering projects.

REFERENCES

- Aitchison, G.D. (1979, or later).
  Soil engineering in Australia, Chapter 54 in: Soils, an Australian viewpoint, 857-871, CSIRO/Academic Press, Australia.
- Akpokodje, E.G. (1985).
  The Engineering Classification of some Australian Arid Zone Soils. Bull of the IAEG 31, 5-8.
- Atlan, Y. and Feller, C.(1983).
  Applications à la Géotechnique routière. Exemple du Sénegal. Documents du BRGM 56; 88 pp. Editions du BRGM, Orléans.
- British Standards Institution (1981).
  Code of practice for site investigations BS 5930, 147 pp. B.S.I., London.
- British Standards Institution (1981).
  Code of practice for earthworks BS 6031, 86 pp. B.S.I., London.
- Buisson, J.L. (1976). Le fichier des données géotechniques. Bull de Liaison LPC 84, 84,150.
- Broms, B.B. (1988).
  Proc. 2 ICOTS: Forword, Vol 1, V. Balkema, Rotterdam.
- Cook, J.R. and Newill, D. (1988)
  The field description and identification of tropical residual soils. Proc. 2 ICOTS: Geomechanics in Tropical Soils, Vol. 1, 3-10, Balkema, Rotterdam
- Geotechnical Control Office (GDO) (1988).
  Guide to Rock and Soil Descriptions, 189 pp. GCO, HongKong.
- Gidigasu, M.D. (1971).
  The importance of soil genesis in the engineering classification of Ghana soils. Engineering Geology 5, 117-161.
- Gidigasu, M.D. (1972).
  Mode of formation and geotechnical characteristics of laterite materials of Ghana in relation to soil forming factors. Engineering Geology 6, 79-150.
- Gidigasu, M.D. (1988).
  Potential application of engineering pedology in shallow foundation engineering on tropical residual soils. Proc. 2 ICOTS: Geomechanics in Tropical Soils, Vol. 1, 19. Balkema, Rotterdam.
- Hough, B.K. (1957).
  Basic Soils Engineering, 45,46. The Ronald Press Company, New York.
- IAEG (1981).
  Report by the IAEG commission on Engineering Geological Mapping on Rock and Soil Description and Classification.

Bull of the IAEG 24, 235-274,238.
- Indian Standards Institution (1970).
  Classification and identification of
  soils for general engineering purposes.
  First revision. IS 1498, 24 pp. I.S.I.,
  New Delhi.
- Karlsson, R. and Hansbo, S. (1981).
  Soil classification and identification,
  49 pp. Swedish Council for Building
  Research, Stockholm.
- Liu, T.K. (1967).
  Classification, Safety Factor, Terrain
  and Bearing. Highway Res. Rec. 156, 1-
  22.
- Lubking, P. Van der Vring, J.J.M., and
  Wiegers, H. (1978).
  Various Properties on Natural Sands for
  Netherlands Highway Engineering. 114
  pp. Stichting Studie Centrum Wegenbouw,
  Arnhem (now CROW, Ede), The Nether-
  lands.
- Madu, R.M. (1977).
  An investigation into the geotechnical
  and engineering properties of some la-
  terites of Eastern Nigeria, Engineering
  Geology 11, 101-125.
- Nederlands Normalisatie Instituut (1989).
  Geotechnics - Classification of soil
  samples NEN 5104, 23 pp. Delft.
- Portland Cement Association (1956).
  Soil Primer, 30-31, 43-46. Chicago.
- Post, H.J., (1989).
  Resten zijn géén afval (meer); Bijzon-
  dere ophoogmaterialen. Publikatie 16,
  92 pp. CROW, Ede.
- Sanejouand, R. (1975). Fichiers de
  données géotechniques sur ordinateur
  dans les LPC. Bull de Liaison LPC 76,
  109, 115.
- Van der Sluijs, P. (1971). Beschrijving
  en analyseresultaten van 15 standaard-
  profielen; intern rapport 985 Stiboka.
  No continuous paging. Wageningen, The
  Netherlands.
- Vargas, M. (1988).
  Characterization, identification and
  classification of tropical soils. Proc.
  2 ICOTS: Geomechanics in Tropical
  Soils, Vol. 1, 75. Balkema, Rotterdam.
- Wiegers, H. (1970).
  Dénomination et division uniforme des
  sables Néerlandais comme un point de
  départ pour leur classification.
  Proc. 1st. Int. Congress IAEG, 222-233.
  Paris.
- Wiegers, H. (1971).
  Classification, à base technique et
  géologique, des sables Néerlandais.
  Bull of the IAEG 4, 42-51.
- Wiegers, H. (1974 a).
  Considerations on the interdisciplinary
  basis of a Netherlands classification
  of unconsolidated sediments, Bull of

the IAEG 10, 23-26.
- Wiegers, H. (1974b).
  The interaction between classification
  and terminology in engineering geology
  and associated disciplines, particularly with relation to unconsolidated
  sediments. Proc. Sec. Int. Congress
  IAEG, Vol. 1, Contribution IV. 12. São
  Paulo.
- Wiegers, H. (1975).
  SCW-Classificatie van Zand, 88 pp.
  Stichting Studie Centrum Wegenbouw,
  Arnhem (now CROW, Ede), The Netherlands.
- Wiegers, H. (1982).
  The relation between the Netherlands
  geological classification of unconsolidated sediments and selection criteria
  for road and railway embankments. Proc.
  4th Congress of IAEG, Vol. 6, theme 3,
  1-9. New Delhi.
- Wiegers, H. (1985).
  Plaats en betekenis van de huidige bo-
  demkunde tussen de technische-aardkun-
  dige disciplines. Intern rapport werk-
  groep A5 "Bijzondere ophoogmaterialen"
  27 pp. plus figuren. CROW, Ede, The
  Netherlands.

7.4 Problem materials

Matériaux problématiques

Strength gain stages of soil-cement
L'accroissement de la résistance de sol-ciment

R. Angelová & D. Evstatiev
Geotechnical Laboratory of the Bulgarian Academy of Sciences, Sofia, Bulgaria

ABSTRACT: It has been established that the compressive strength of loess-cement undergoes four stages in a period of two years of hardening. Experimental data from x-ray diffraction, thermal, SEM, mercury porosimetry and complex chemical analyses prove that the stage character is due to changes in microstructure, composition and morphology of the new binding phases.

RESUME: On a établi que la résistance à la compression simple des mélanges "loess-ciment" passe 4 stades dans son périod de deux ans de maturation. Les données obtenues par les annalyses diffractométrique et thermodifférentieles ainsi que par la microskopie électronique, porosimétrie à mercure et des analyses chimiques prouvent les liens de la succession mentionée ci-dessus avec les changement dans la microstructure, la composition et la morphologie des phases cimentaires néoformées.

INTRODUCTION

In the recent twenty years the stabilization of soils by mixing with cement, lime and fly ash has found a wide application not only in road construction but also in hydrotechnical and civil engineering, melioration and underground construction (Minkov & Evstatiev 1975, Rzhanitsin 1986, etc.). The development of the jet-grouting during the last ten years provided the possibility for the construction of various three-dimensional soil-cement structures, situated at a considerable depth, and broadened the scope of application of soil-cement still further (Saitoh et al. 1985, Welsh, Ed. 1987, etc.).

However the available data concerning the kinetics of strength of soil-cement for longer periods of time are rather scarce in reference literature (Handy 1958, Voronkevich et al. 1981, Rzhanitsin 1986, Angelova & Evstatiev 1989, etc.). The same holds true for the mechanism of fabric formation.

Although a considerable progress has been made during the last twenty years in the research on cement paste hydration and hardening processes as it is reflected in the Proceedings of the last Congresses on the Chemistry of Cement, it is still considered

that "the hardened cement paste represents one of the most complex system among the manufactured materials" (Mehta 1986). The interraction between soil and cement is more complicated and is influenced by the following factors: soil and cement composition, degree of compaction, water content, cement quantity, temperature-moisture conditions, curing time, etc.

Investigations of the processes which occur in stabilized soils started with more simple mixes, including some pure soil minerals (quartz, kaolinite, montmorillonite, etc.) and monomineral binding agents (lime, alite, etc.). It was established that the strength gain of soil-cement undergoes some stages of increasing, retention and even decreasing in certain cases (Voronkevich et al. 1981, Rzhanitsin 1986). According to other authors the relationship between compressive strength R_c and curing time is linear in log - log scale (Handy 1958, Brandl 1981).

In Bulgaria loess soils are most often subjected to cement stabilization. They occupy about half of the plain territory of the country. Their collapsibility and permeability present serious constructional problems. The considerable volume of loess-cement foundation cushions (more than 500 000 m^3) as well as the prospects for application of the jet-grouting, require a comprehensive knowledge of the mechanism of the soil-cement strength formation.

The purpose of the present investigation is to study the variation with time of the microstructure and composition in the loess-cement mixtures and their influence on the strength formation.

MATERIALS AND INVESTIGATION METHODS

Three typical loess varieties which are most often subjected to cement stabilization were investigated: sandy, silty and clayey loess. They differ in granulometric, chemical and mineral composition (Table 1).

The semi-quantitative assessment of the mineral composition has been performed on the basis of the x-ray diffraction analysis. The fraction below 2 μm has been decarbonized before separation. Its composition is shown in brackets (Table 1). In sandy and silty loess carbonates are presented mainly as dolomite. Montmorillonite and mixed-layer clay minerals prevail in the fraction below 2 μm. The absence of clay fraction and a carbonate-sulphate type of salting are characteristic of sandy loess. Clayey loess contains the greatest quantity of carbonates, humus and amorphous SiO_2 and possesses a highest cation exchange capacity in alkaline medium and a highest pozzolanic activity. The values for the optimum moisture content w_0 and the maximum dry density ρ_{max} are obtained for specific compaction work of 2.25 J/cm^3.

Table 1. General properties of loess soils used

Main mineral components, %	Sandy loess	Silty loess	Clayey loess
Quartz	33	26 (6)	29 (4)
Feldspars	20	14 (4)	19
Calcite	5	7	15
Dolomite	16	14	1
Mica and hydromica	15	23 (22)	11 (11)
Chlorite	11	15 (6-12)	18 (18)
Montmorillonite	–	1 (48-53)	6 (58)
Kaolinite	–	– (7-9)	1 (9)
Amorphous SiO_2	0.47	0.56	0.96
Carbonates (after Scheibler)	8	6.4	15.1
Humus, %	0.23	0.18	0.36
pH	7.7	8.2	8.35
Cation exchange capacity in alkaline medium, mg equ/100 g	6.8	11.1	18.6
Quantity of the adsorbed Ca^{2+} for a 30 days period /hydraulic activity/, mg equ/100 g	22.4	27.7	42.1
Grain size distribution, %			
sand 2-0.05 mm	52	34	20
silt 0.05-0.002 mm	48	64	71
clay < 0.002 mm	0	2	9
Consistency limits, %			
liquid limit w_L	26.9	25.2	34.2
plastic limit w_p	25	21.1	18.6
plasticity index I_p	1.9	4.1	15.6
Optimum water content w_0, %	16	17.5	18
Maximum dry density ρ_{max}, g/cm^3	1.62	1.69	1.72
Density of solid particles ρ_s, g/cm^3	2.76	2.74	2.74

Sandy loess has the lowest degree of compaction in comparison with the other loess varieties. The sand and silt particles are usually in direct contact. However cohesive bonds formed by clay or carbonate particles can be observed in separate zones (Fig. 1a). Pores have irregular form with their size varing from 20 to 50 μm. In silty loess besides sand and silt particles clay aggregates can be observed. They create a greater number of contacts and hence a denser microstructure with dominating pore sizes 10-30 μm (Fig. 1b). Clayey loess is characterized by the most dense structure in which the main structural elements are clay particles adhered to silt grains or aggregated (Fig. 1c). The interaggregate pore size is 2-5 μm.

The loess soils were stabilized with Portland cement PC 35 with content of the basic clinker minerals: C_3S - 50-60%, C_2S - 17-18%, C_3A - 8-9%, C_4AF - 12%. Specific surface area is 3500-3800 cm^2/g.

Cylindrical samples of 5 cm in diameter and height were used in establishing the compressive strength R_c of soil-cement. The cement content was 5, 7, 10 and 15%. Sandy loess samples were compacted at w_0= 16% to ρ_d = 1.55; 1.58; 1.62 and 1.65 g/cm^3; silty loess samples - at w_0=17.5% to ρ_d= 1.55; 1.60; 1.65 and 1.70 g/cm^3; clayey loess samples - at w_0=18% to ρ_d=1.60; 1.64; 1.68 and 1.73 g/cm^3. After corresponding periods (2, 7, 30, 180, 270 and 720 days) of curing in air-moisture medium at t=20°C the samples were water-soaked for 24 h and then the compressive strength was determined. Representative quantities from the crushed samples were dried in vacuum oven at 40°C and used for the different kinds of analysis: x-ray diffraction, thermal, mercury porosimetry, SEM, pH determination, free CaO and carbonates content, etc. A more detailed description of the test procedure has been presented elsewhere (Angelova 1987).

TEST RESULTS AND DISCUSSION

During the two-year period of investigation loess-cement strength gain advances at different rate (Angelova 1987, Angelova & Evstatiev 1989). Strength gain passes through stages of intensive growth, retention, slower growth and decrease in some cases (Fig. 2). The statement that the relationship between compressive strength R_c and time is linear in log - log scale probably represents a special case. Even comparatively small differences in the loess composition greatly influence this relationship. In has been established that it is linear

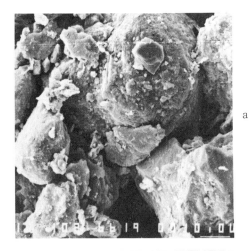

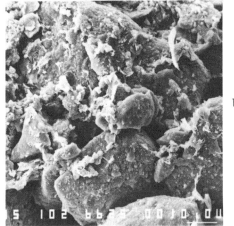

Fig. 1 SEM micrograph of compacted loess. Magn.: x 1000
a - sandy loess; b - silty loess;
c - clayey loess

3149

only for sandy loess mixtures (Fig. 3).
It was determined that loess-cement
strength gain is most intensive in the

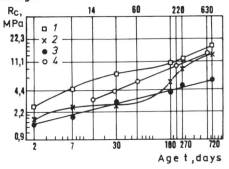

Unconfined
Compressive
Strength

Fig. 3 Relationship between time and compressive strength in log - log scale for soil-cement with 10% Portland cement
1 - clayey loess; 2 - silty loess;
3 - sandy loess; 4 - silty loess, according to Handy (1958)

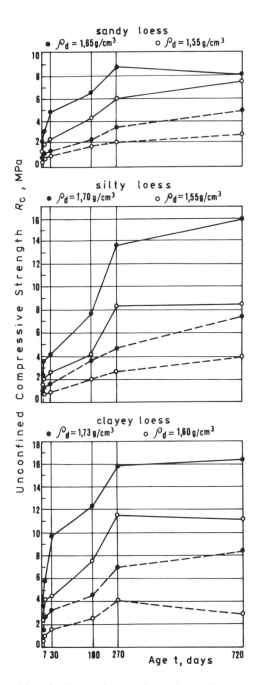

Fig. 2 Strength gain kinetics of loess-cement mixtures
- - - 5% Portland cement; ——— 15% Portland cement

first week of curing, especially for clayey loess containing 10-15% of cement (Fig. 2). Strength gain proceeds at a rather high rate till the 30th day, silty loess being an exception, where a certain retention is observed.

During the first month of curing needle-like phases (1.0-1.5 μm long & appr. 0.2 μm thick) occur in sandy loess-cement mixtures (Fig. 4). The sulphate salting of this loess presents a favourable condition for the formation of ettringite $C_3AH_{31}.3\ CaSO_4$. However the established diffraction peaks combined with the small size can possibly determine needle-like phases as a calcium silicate hydrate C-S-H I type (Diamond 1976). A needle-like phase of this type (2-6 μm long & 0.15-0.30 μm thick) is formed as early as the 1st day of curing in the pores and on the surface of the soil particles in the silty loess mixtures (Fig.5). Larger quantities are observed in the samples with lower ρ_d and higher cement content. The needle-like's I type C-S-H are the dominating structural element during the first days of curing in sandy and especially silty loess, while in clayey loess mixtures they can't be found. In a month's period they also appear in clayey loess and their length does not exceed 1.5 um. On one hand this fact could be explained with the more dense structure, hampering the growth of this phase. On the other hand the greater quantity of active clayey fraction alters hydration conditions due to the increased Ca^{2+} adsorption. The latter assumption is confirmed by the x-ray diffraction data and by the analysis

3150

of the free CaO content. The characteristic of Ca(OH)$_2$ peaks with d=0.260-0.263; 0.192; 0.180-0.184 nm are registered already on the 2nd day in the silty and sandy loess mixtures. In the sandy loess mixtures Ca(OH)$_2$ can be found till the 6th month while in the silty loess mixtures - till the third month. In clayey loess during the first days of hardening no Ca(OH)$_2$ diffraction peaks are observed. After a month slight peaks with d=0.264 and 0.178 nm appear. The peak with d=0.264 nm is registered after a year of hardening too. The mixtures of sandy loess with 10% Portland cement contain free CaO till the 6th month. In silty and clayey loess samples no free CaO is found after the 1st month of hardening.

In sandy and silty loess well shaped hexagonal crystals (2-4 μm in diameter) are observed up to the 30th day (Fig. 4). The x-ray diffraction charts of these mixtures for the same period exhibit an explicitly shaped peak with d=0.748-0.754 nm, which accounts for the calcium aluminate hydrate (Hilt & Davidson 1961) or for the calcium carboaluminate hydrate $C_3AH_{12}.CaCO_3$ (Gorshkov et al. 1981). It is possible that the hexagonal crystals may also belong to the separated in the cement hydration lime.

About the end of the 1st month of hardening in all the three loess varieties starts the formation of a network-like binding substance (Fig. 6), which is probably a II type C-S-H (Diamond 1976). In clayey loess mixtures another structural element is a fine-grained mass, which is observed for longer periods of curing (90 and 270 days).

Till the 30th day the mixtures'alkalinity is high (pH=11.5-13). In silty and sandy loess a slight increase of pH is registered between the 2nd and the 7th day, while in clayey loess pH decreases.

On the basis of thermal analysis it has been established that the degree of bonding of water molecules increases 2-3 times during the 1st month, which probably affects favourably the strength (Angelova 1987).

The second stage of strength gain is between the 30th and the 180th day. During this period R_c increases at a slower rate. This is accompanied by the initiation of phase transformations in the basic binding mass of calcium silicate hydrates C-S-H, which finds expression in decreasing the volume of needle-like phase (I type) and formation of network-like (II type) and gel-like (III type) compounds. In sandy loess mixtures during this period separate zones can be observed with soil particles covered by platy new formations of irregular shape and a size of 1-3 μm. A new peak with d=0.752 nm appears in clayey loess

mixtures, which as has been stated above corresponds to calcium carboaluminate hydrate. Series of slight peaks, inherent to C-S-H differing in composition are observed in the investigated loess-cement samples. This is an indication that the system is metastable and that the main binding substance - the calcium silicate hydrates, undergoes a number of phase transformations in the course of time. During this stage pH continues to decrease and reaches the values of 11-11.5.

The third stage (between the 180th and the 270th day) is characterized by a second period of relatively rapid increase of the strength R_c (Fig. 2). This is more pronounced for the silty loess containing more than 5% of Portland cement. After 9 months of hardening one can observe in the silty loess samples a gel-like mass, connecting the loess particles in a dense structure (Fig. 7). This mass is morphologically similar to the III type C-S-H (Diamond & Lachowski 1980). The clayey loess mixtures show a different kind of structure, depending on the degree of compaction - e.g. maximum dry density (ρ_{max}=1.73 g/cm^3) or lower density (ρ_d=1.6 g/cm^3). In denser samples the main binding substance are the gel-like compounds. In single regions hexagonal-shaped crystals (1 to 5 μm in diameter) can be found. In the lower density samples many regions with I type C-S-H 2-5 μm long are observed. They fill the pore space, form bridges between the soil particles and aggregates or grow as coatings on the surface of some quartz grains. Another element of the structure during this period are round grain particles 1-2 μm by size, densely packed to each other. It is probable that they have a carbonate composition. It was also observed II type C-S-H.

The processes presumably determining the more intensive strength gain in this stage are: activation of the fine quartz particles surface, which improves the adhesion between them and cement hydrates; additional pozzolanic interaction and increase of the binding mass volume. The latter has been proved by the pore size distribution data in mixtures of silty loess with 15% Portland cement (Angelova & Evstatiev 1985). The integral porosity during the period from the 2nd till the 270th day decreases insignificantly - 1.87%, but the amount of the pores greater than 0.1 μm decreases almost twice, while the amount of the pores < 0.1 μm increases. The explanation of this phenomenon is the growth of the new phases. They fill the larger micropores, but at the same time between the newly formed phases (with sizes of several μm or less) an additional ultracapillary porosity develops. Similar results, concerning the importance

Fig. 4 SEM micrograph of sandy loess with 15% PC 35, compacted at ρ_{max}, after 7 days. Magn.: x 7800
a - needle-like C-S-H; b - hexagonal crystal

Fig. 6 Network-like C-S-H II type. Magn.: x 10 000

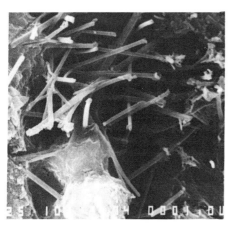

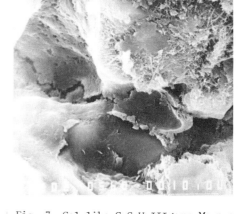

Fig. 5 Needle-like C-S-H I type in sample of silty loess with 15% PC 35, compacted at ρ_{max}, after 2 days. Magn.: x 10 000

Fig. 7 Gel-like C-S-H III type Magn.: x 2000

of pore size distribution for the strength were obtained for various materials, based on $Ca(OH)_2$, C_3S, C_3A and silicates (Jambor 1973).

In the fourth stage with duration from the 270th to the 720th day, an insignificant strength gain or strength retention are established (Fig. 2). Only the clayey loess mixtures containing less than 10% of cement and compacted to a lower than the maximum dry density exhibit a decrease in R_c from 0.5 to 1.5 MPa (Angelova & Evstatiev 1989). This fact could be explained with the formation of metastable C-S-H when active clay fraction is available and

cement content is low. These C-S-H are transformed during the hardening and reduce the quantity of the strong bonds. SEM analysis confirms this supposition. After two years of curing in the clayey loess samples containing 7% of cement and compacted to $\rho_d = 1.6$ g/cm³, the observed at the earlier stages needle-like formations are absent. On their places exist small needles shorter than 1 μm, resembling the remnants of the previous large needles (Fig. 8). Many sections with hexagonal (1-3 μm) and rhombic (1-2 μm) crystals of indefinite composition have been registered (Fig. 9). The common feature of the samples with decreasing strength is their relatively less compacted fabric.

The network- and gel-like C-S-H are the

Fig. 8 SEM micrograph of clayey loess with 7% PC 35, compacted at ρ_d=1.6 g/cm³ ($< \rho_{max}$) after 2 years. Magn.: x 5000

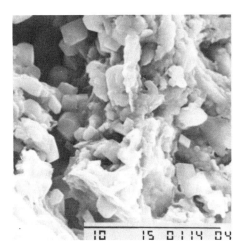

Fig. 9 Rhombic and hexagonal crystals in less compacted clayey loess with 7% PC 35, after 2 years. Magn.: x 5000

main binding elements during the fourth stage, but the needle-like C-S-H is encountered too. The x-ray diffraction charts of silty and clayey loess mixtures are with well-shaped peaks with d=0.965 and 1.037 nm, typical for C-S-H. The values of pH decrease from 12-13 during the first week to 9.4-9.7 two years later.

The influence of the initial dry density on R_c is much stronger in later periods of hardening (especially after two years). The influence of cement quantity (5-15%) is felt more considerably till the 270th day. This is due mainly to the intensive cement hydration in this period. In a year alite is almost completely hydrated and belite - to the extent of about 70% (Larionova et

al. 1977).

For practical uses it is important that the investigated loess varieties, stabilized with 5-15% PC 35 and compacted to ρ_{max} after two years increase their strength R_c 1.7-5.4 times in comparison with the standard age of testing (1 month). These results confirm the obtained data concerning strength R_c of loess-cement samples from water leveller screens after 6-13 years of exploitation. Their strength is 2.5-5.0 times as big as the standard test period strength. SEM observations of these samples showed that the main binding mass consists of gel- and network-like C-S-H. A new fabric element in this mass are the densely interwoven needle-like formations, organized in cellular structure (Fig. 10a). The fabric of the old lime-sand mixtures (Fig. 10b) used in the ancient capitals of Bulgaria - Pliska and Preslav (Evstatiev 1987) is similar.

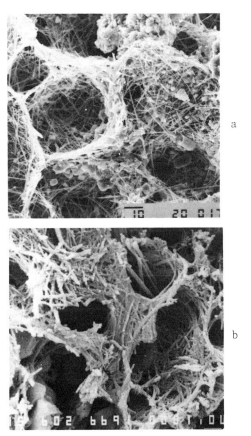

Fig. 10 Interwoven needle-like phases, organized in cellular structure
a - loess-cement from 8 years used water leveller screen. Magn.: x 750; b - ancient lime-sand mortar. Magn-: x 6000

CONCLUSIONS

• Strength development of soil-cement is a stage process. The intensity of strength gain or decrease in the single stages as well as their duration depend on the peculiarities of soil composition, type and quantity of binding agent, density and water content of the mixture and curing conditions.

• For the most widely applied soil-cement in Bulgaria, prepared by sandy, silty and clayey loess and 5-15% of Portland cement, four stages in strength formation have been determined: I (up to the 30th day) - a quick increase of R_c; II (30th to 180th day) - a slower increase of R_c; III (180th to 270th day) - a second increase of R_c at a higher rate; IV (270th to 720th day) - an insignificant gain or retention of R_c. During the fourth stage for the clayey loess mixtures with cement content below 10% and compacted at lower than the maximum dry density the strength R_c exhibited a decrease of about 0.5-1.5 MPa.

• In the two years period of investigation continuous changes in the type and volume of the new phases take place. C-S-H are the basic binding substance and they are most important for the strength formation. Most of them are present in semi-amorphous or amorphous state, which hampers the determination of the relative content of the various compounds. In the course of time C-S-H undergo complex phase modifications. The volume of the needle-like C-S-H (I type) considerably decreases, while the volume of the network-like (II type) and gel-like (III type) increases. These phase modifications influence the stage strength development in time. The products of pozzolanic reactions are morphologically similar to the cement hydration products, which makes them difficult to be distinguished. Mercury porosimetry data confirm that the volume of the new binding phases increases in the time.

• The established increase of loess-cement strength R_c (2-5 times in 2 years) is important for the practical application of this material as foundation cushions and screens of different constructions.

REFERENCES

Angelova, R. & Evstatiev, D. (1985). On the development of strength in cement soils. Eng. geology 1,64-74, (in Rus.).

Angelova, R. (1987). General relationships of kinetics of structure formation of cement stabilized loess from North Bulgaria. Ph.D. Thesis, 230 pp. University of Moscow, (in Rus.).

Angelova, R. & Evstatiev, D. (1989). Strength changes of cement-loess mixes in time. Eng. Geol. and Hydrogeol. 19, 3-14, (in Rus.).

Brandl, H. (1981). Alteration of soil parameters by stabilization with lime. Proc. 10th ICSMFE, vol. 3,587-594. Balkema, Rotterdam.

Diamond, S. (1976). Cement paste microstructure - an overview at several levels. Proc. Conf. on hydraulic cement pastes: their structure and properties, 2-30. University of Sheffield.

Diamond, S. & Lachowski, E. (1980). On the morphology of type III C-S-H gel. Cem. and Concr. Res. 10,703-705.

Evstatiev, D. (1987). Mechanism of stabilization of soils with hydraulic binders. Sc.D. Thesis, 302 pp. Bulg. Acad. of Sci., Sofia, (in Bulg.).

Gorshkov, V., Timashev, V. & Saveliev, V. (1981). Methods for physico-chemical analysis of binding substances, 335 pp. V.Shkola, Moscow, (in Rus.).

Handy, R. (1958). Cementation of soil minerals with Portland cement of alkalis. Highway Res.Board Bull. 198,55-64

Hilt, G. & Davidson, D. (1961). Isolation and investigation of a lime-montmorillonite crystalline reaction product. Highway Res. Board Bull. 304,51-64

Jambor, J. (1973). Influence of phase composition of hardened binder pastes on its pore structure and strength. Proc. Conf. on pore structure and properties of materials, vol. II, D75-D96. Prague.

Larionova, Z., Nikitina. L. & Garashin, V. (1977). Phase composition, microstructure and strength of cement stone and concrete, 264 pp. Stroyizdat, Moscow, (in Rus.).

Mehta, P. (1986). Hardened cement paste - microstructure and its relationship to properties. Proc. 8th Int. Congress on chem. of cement, vol. 1,113-121, Rio de Janeiro.

Minkov, M.& Evstatiev,D. (1975). Foundations, linings and screens from stabilized loess soils, 189 pp. Technika,Sofia, (in Bulg).

Rzhanitsin, B. (1986). Chemical stabilization of soils in construction, 264 pp. Stroyizdat, Moscow, (in Rus.).

Saitoh, S., Suzuki, Y. & Shirai, K. (1985). Hardening of soil improved by deep mixing method. Proc. 11th ICSMFE, vol. 3, 1745-1748. Balkema, Rotterdam.

Voronkevich, S., Evdokimova, L., Larionova, N. & Ogorodnikova, E.(1981). Structure formation processes in ash-cement-soil systems and low active ash addition. Eng. geology 6,95-103, (in Rus.).

Welsh, J., Ed.(1987). Soil improvement - a ten year update. Geotechn. Spec. Publ. 12, 331 pp. ASCE, New York.

Investigation of the suitability of some coal measures mudrocks for brickmaking: A case history

Enquête sur l'aptitude de certains pélites des filons houillers pour la fabrication de briques: un cas spécifique

F.G. Bell – *Department of Geology & Applied Geology, University of Natal, Durban, South Africa*

J.M. Coulthard – *Department of Civil Engineering, Teesside Polytechnic, Middlesbrough, UK*

M.G. Culshaw – *Engineering Geology Research Group, British Geological Survey, Keyworth, Nottingham, UK*

J.C. Cripps – *Department of Geology, University of Sheffield, Sheffield, UK*

ABSTRACT: Coal Measures mudrock is used extensively in the United Kingdom for brickmaking. The mudrock generally is interbedded with siltstone and sandstone and one of the factors which any site exploration for a new quarry has to determine is the quantity of waste which will result from quarrying operations and the quantity of material available for brickmaking. The quality of the mudrock should be ascertained by chemical and mineralogical analyses. The qualitative mineralogical composition can be determined indirectly by using Keeling's method or, directly, by differential thermal analysis. Neither of these methods give data relating to the relative proportions of the minerals present. However, X-ray diffraction methods provide a semi-quantitative analysis of the amounts of quartz and clay minerals in the mudrock. An assessment, using these methods, of the potential of a sequence of Coal Measures strata for development for brick clay is discussed.

RESUME: Le mudrock Coal Measures est largement utilisé au Royaume-Uni pour la fabrication des briques. Le mudrock alterne généralement avec du siltstone et du grès et l'un des facteurs que toute exploration du site d'une nouvelle carrière doit déterminer est la matériaux utilisables pour la fabriction des briques. La qualité du mudrock devrait être déterminée par des analyses chimiques et minéralogiques. La composition minéralogique qualitative peut être déterminée indirectement par la méthode de Keeling ou directement par des analyses thermiques différentielles. Aucune de ces méthodes ne donne d'information sur les proportions relatives des minéraux présents. Cependant, les méthodes de diffraction par rayons fournissent une analyse semi-quantitative de la présence de quartz et de minéraux argileux dans le mudrock. Une évaluation, utilisant ces méthodes, du potentiel d'une séquence de stratification Coal Measures pour le développement pour l'argile à briques est discutée.

1 INTRODUCTION

In the United Kingdom bricks are made from diverse mudrocks using a variety of production processes. The general properties which the raw material should possess are well known but because mudrocks vary appreciably in mineralogy, texture and degree of lithification, each deposit will require detailed investigation.

The nature of brick production has changed with time. Most of the small units which produced bricks for local needs have closed so that brickmaking is now concentrated in large units which supply bricks over much wider areas and which require large sources of material. The technology of brickmaking has changed in response to economic pressures which favour the concentration of brickmaking in large, highly mechanized units. These use the "extrusion" or "wire-cut" process in which the raw material is forced through a die, the continuous column of material being cut by wires to form "green" bricks. These are then dried before firing in a tunnel kiln. By varying the raw material used, common, facing or engineering bricks can all be produced by this process. The increasingly sophisticated market has demanded, over recent years, attractive, better quality bricks, hence the number of common bricks produced is declining and that of facing bricks increasing. Because of the changes in aesthetic requirements and the advances in technology, mudrocks previously considered unsuitable for brickmaking are now being exploited.

The Coal Measures are widely distributed in the United Kingdom and the mudrock from them has been worked in all the major coalfields for the manufacture of bricks. A quarter of the country's production currently comes from that source. The Coal Measures were deposited as generally muddy sediments in deltaic environments but can display appreciable local facies variations such as channel sands and silts. This variability increases the difficulty and cost of working these deposits. In particular, a clay pit will be less economically viable as the amount of waste material increases. Furthermore, the mineral composition of mudrocks from the Coal Measures is variable. For example, quartz and disordered kaolinite are abundant in Coal Measures shale, while the amounts of illite, chlorite, organic matter, iron oxides and sulphides vary. As a result, tests carried out on bricks made from mudrocks from various parts of the Coal Measures have shown significant variations in compressive strength, porosity and water absorption, and content of soluble salts.

2 EVALUATION OF BRICKMAKING MATERIALS

The suitability of a raw material for brick making is determined by its physical, chemical and mineralogical character and the changes which occur when its is fired. The unfired properties such as plasticity, workability (that is, the ability of the clay to be moulded into shape without fracturing and to maintain its shape when the moulding action ceases), dry strength, dry shrinkage and vitrification range are dependant upon the source material but the fired properties such as colour, strength, total shrinkage on firing, porosity, water absorption, bulk density and tendency to bloat are controlled by the nature of the firing process. The price that can be charged for a brick depends largely upon its attractiveness, that is, its colour and surface appearance. The ideal raw material should possess moderate plasticity, good workability, high dry strength, total shrinkage on firing of less than 10% and a long vitrification range. However, the suitability of a mudrock for brick manufacture can be determined only by running it through a production line or by pilot plant firing tests.

2.1 Mineral composition

The mineralogy of the raw material influences its behaviour during the brickmaking process and hence the properties of the finished product. Mudrocks consist of clay minerals and non-clay minerals, mainly quartz. The clay mineralogy varies from one deposit to another. Although bricks can be made from most mudrocks, the varying proportions of the different clay minerals have a profound effect on the processing and on the character of the fired brick. For example, mudrocks containing significant amounts of disordered kaolinite tend to have moderate to high plasticity and, therefore, are easily workable. They produce lean clays which undergo little shrinkage during brick manufacture. They also possess a long vitrification range and produce a fairly refractory product. However, mudrocks containing appreciable quantities of well-ordered kaolinite are poorly plastic and less workable. Illitic mudrocks are more plastic and less refractory than those in which disordered kaolinite is dominant, and fire at somewhat lower temperatures. Smectites are the most plastic and least refractory of the clay minerals. They show high shrinkage on drying since they require high proportions of added water to make them workable.

The presence of quartz, in significant amounts, gives strength and durability to a brick. This is because during the vitrification period quartz combines with the basic oxides of the fluxes released from the clay minerals on firing to form glass, which improves the strength. However, as the proportion increases, the plasticity of the raw material decreases.

The accessory minerals in mudrocks play a significant role in brickmaking. The presence of carbonates is particularly important and can influence the character of the bricks produced. When heated above $900°C$ carbonates break down yielding carbon dioxide and leaving behind reactive basic oxides, particularly those of calcium and magnesium. The escape of carbon dioxide can cause "lime popping" or "bursting if large pieces of carbonate, for example shell fragments, are present, thereby pitting the surface of the brick. To avoid "lime popping" the material must be finely ground to pass a 20 mesh sieve. The residual lime and magnesia form fluxes which give rise to low viscosity silicate melts. The reaction lowers the temperature of the brick and hence, unless additional heat is supplied, lowers the firing temperature and shortens the range over which vitrification occurs. The reduction in temperature can result in inadequately fired bricks. If excess oxides remain in the brick it will hydrate on exposure to moisture, thereby destroying the brick. The expulsion of significant quantities of carbon dioxide can increase the porosity of bricks, reducing their strength. Engineering bricks must be made from a raw material which has a low carbonate content.

Sulphate minerals in mudrocks are detrimental to brickmaking. For instance, calcium sulphate does not decompose within the range of firing temperature of bricks. It is soluble and, if present in trace amounts in the fired brick, causes efflorescence when the brick is exposed to the atmosphere. Soluble sulphates dissolve in the water used to mix the clay. During drying and firing they often form a white scum on the surface of a brick. Barium carbonate may be added to render such salts insoluble and so prevent scumming.

Iron sulphides, such as pyrite and marcasite, frequently occur in mudrocks. When heated in oxidizing conditions, the sulphides decompose to produce ferric oxide and sulphur dioxide. In the presence of organic matter oxidation is incomplete yielding ferrous compounds which combine with silica and basic oxides, if present, to form black glassy spots. This may lead to a black vitreous core being present in some bricks which can reduce strength significantly. If the vitrified material forms an envelope around the ferrous compounds and heating continues until this decomposes, then the gases liberated cannot escape causing bricks to bloat and distort. Under such circumstances the rate of firing should be controlled in order to allow gases to be liberated prior to the onset of vitrification. Too high a percentage of pyrite or other iron bearing minerals gives rise to rapid melting which can lead to difficulties on firing.

Pyrite, and other iron-bearing minerals such as hematite and limonite, provide the iron which primarily is responsible for the colour of bricks. The presence of other constituents, notably calcium, magnesium or aluminium oxides, tends to reduce the colouring effect of iron oxide, whereas the presence of titanium oxide enhances it. High original carbonate content tends to produce yellow bricks. The colour also is influenced by the firing conditions. For example, it is darkened by increasing the firing temperature; oxidizing conditions give rise to red bricks, reducing conditions to blue bricks.

Organic matter commonly occurs in mudrock. It may be concentrated in lenses or seams, or be finely disseminated throughout the mudrock. Incomplete oxidation of the carbon upon firing may result in black coring or bloating. In the latter case, even minute amounts of carbonaceous material can give black coring in dense bricks if it is not burned out.

2.2 Mineral and chemical analysis and physical testing

Mineralogical and chemical information is essential for determining the brickmaking characteristics of a mudrock. Differential thermal analysis and thermogravimetric analysis can identify clay minerals in mudrocks but provide only very general data on relative abundance. X-ray diffraction methods are used to determine the relative proportions of clay and other minerals present. The composition of the clay minerals present can be determined also by plotting ignition loss against moisture absorption (that is, by using Keeling's Method [Keeling 1961]). The moisture absorption characterizes the type of clay mineral present while ignition loss provides some indication of the quantity present.

The presence of most of the accessory minerals can be determined by relatively simple analysis. Wet chemical methods can be used to determine total calcium, magnesium, iron and sulphur. The organic matter can be estimated by oxidation or calorimetric methods. Carbon and sulphur are important impurities, which are critical to the brickmaking process, and the quantities present are always determined.

Physical tests can provide some indication of how the raw material will behave during brick manufacture. The particle size distribution affects the plasticity of the raw material and, hence its workability; plasticity is related to shrinkage and ignition loss during firing. Plasticity can be assessed in terms of the Atterberg limits or, less frequently, by means of the Pfefferkorn test. Shrinkage, loss on ignition and the temperature range for firing are determined by the production of briquettes in a laboratory furnace. Other tests which are carried out in relation to brickmaking such as water absorption, bulk density, strength, liability to efflorescence are described in B.S. 3921 (Anon 1985).

2.3 Resource assessment

Sufficient quantities of suitable raw material must be available at a site before a brickfield can be developed. The volume

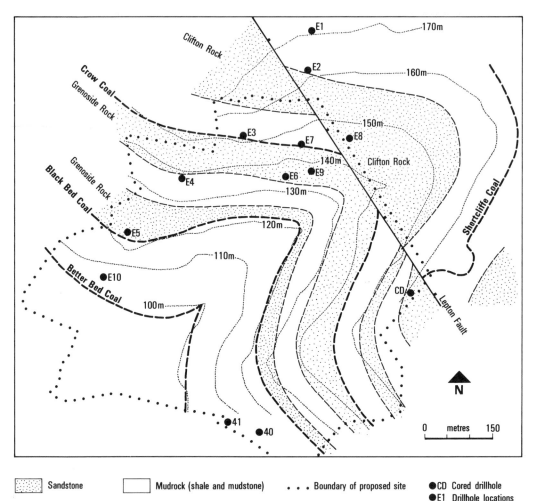

Fig. 1a Generalised geology of the proposed site with drillhole locations.

Sandstone   Mudrock (shale and mudstone)   . . . Boundary of proposed site   ●CD Cored drillhole
●E1 Drillhole locations

of suitable mudrock must be determined and the amount of waste, that is, the overburden and unsuitable material within the sequence that is to be extracted. The first stage of the investigation are topographical and geological surveys, followed by a drillhole programme. This leads to a lithostratigraphic and structural evaluation of the site. It also should provide data on the position of the water table and the stability of the slopes which will be produced during excavation of the brick pit.

3 CASE HISTORY

The site for the proposed extraction of mudrock for brickmaking is in West Yorkshire. The precise location cannot be

given for reasons of confidentiality. It extends over an area of approximately 30 Ha and rises from a height of some 90 to 95 m AOD towards the north and east where it is between 165 and 170 m AOD (Fig. 1a).

3.1 Geology of the site and surrounding area

The strata within the area of the proposed site are of Lower Coal Measures age, falling mainly within the Carbonicola communis Zone, and range upwards from the Elland Flags (Anthraconaia lenisulcata Zone), which crop out to the south of the site, to the Clifton Rock to the north (Figs 1a and b). Mudrock constitutes most of the succession between these two sandstone horizons but a silty sandstone, the Grenoside Rock, together with

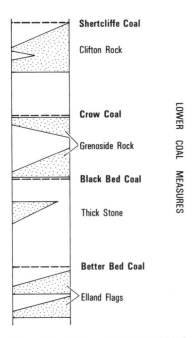

Shertcliffe Coal

Clifton Rock

Crow Coal

Grenoside Rock

Black Bed Coal

Thick Stone

Better Bed Coal

Elland Flags

LOWER COAL MEASURES

Fig. 1b Litho-stratigraphical succession for the rocks in the vicinity of the proposed site (note: the Thick Stone [sandstone] is absent at the site).

three principal coal seams also occur within the site. The latter from lowest upward are the Better Bed Coal, the Black Bed Coal and the Crow Coal.

The Elland Flags usually consist of two sandstone divisions separated by some 15 to 20 m of mudrock and constitute one of the most prominent horizons in the Lower Coal Measures giving rise to striking topographic features.

The Better Bed Coal is usually between 0.3 and 1 m thick and rests on a fireclay which varies from 1 to over 1.5 m in thickness. The latter rests on a varying thickness of mudrock (mainly shale) but locally rests directly upon the Elland Flags.

Some 20 to 35 m of mudrock (shale and mudstone), with occasional thin silty sandstone and siltstone, separate the Better Bed and Black Bed Coals.

The Black Bed Coal is normally between 0.3 m and 1 m thick. It is overlain by mudstone and the Grenoside Rock which occurs as a silty sandstone, split into two horizons, each about 5 m thick, which are separated by mudstone.

The Crow Coal occurs above the Grenoside Rock and is between 0.1 and 0.3 m thick. It is succeeded by 20 to 37 m of mudrock which contain bands of ironstone nodules, thin sandstones and siltstones, and thin seams of

coal. To the north east of the site the succeeding Clifton Rock forms one bed of thick sandstone which splits into two to the immediate north of the site.

The major geological structure of the region is the asymmetric Pennine anticline which dips steeply on its western flank and more gently to the east. In the area, the Coal Measures strata have a very low and generally uniform inclination towards the east, and across the site dip at shallow angles, between 2° and 5°, in a north easterly direction.

Two principal fault systems occur in this area, one trending in a general north west to south east direction, and the other running in a direction between east to west and north east to south west (and lying outside the site, hence it is not shown on Fig. 1a). The Lepton Fault is the major fault, belonging to the former system and trending in a north west to south east direction. It lies on the north east boundary of the site (Fig. 1a), where it has an estimated throw to the north east of approximately 60 m.

Two major joint systems are present. These joints incline at between 75 and 90° whilst minor joints, which are also present, have curvilinear forms. Both major systems comprise two joint orientations which are at right angles to each other. These trend at N70°E and N340°E and (the others) at approximately N20°E and N120°E respectively. The major joints have a minimum separation of 0.5 m. Their spacing and orientation will influence slope stability and groundwater flow and, hence, affect the design of the excavation. The rock masses tend to break away along these discontinuities to form relatively clean sharp faces (Fig. 2).

Fig. 2 Discontinuities in pit face.

Table 1. Proportion of sandstone to mudstone in drillholes

Drillhole No.	Height (AOD)	Depth sunk to (AOD)	Depth (m)	Amount Sandstone (m)	Amount Sandstone (%)	Amount Mudstone (%)
E1	171.5	139.5	32	6.4	20	80
E2	162.0	122	40	8.7	22	78
E3	152.0	112	40	2.6	6.5	93.5
E4	137.5	100	37.5	0	0	100
E5	124.5	84.5	40	0.1	0.25	99.75
E6	135.5	101.5	34	5.35	15.7	84.3
E7	146.0	123.3	22.7	10.5	46.5	53.5
E8	146.7	106.7	40	10.35	25.9	74.1
E9	138.0	128	10	3.3	33	67
E10	93.5	71.5	22	4.3	19.5	80.5

3.2 Site investigation and estimation of resources

Ten uncored drillholes were sunk during the initial investigation (Fig. 1a). The total thickness of strata proved was about 100 m, the youngest beds being proved in drillhole E1 which was located at 171.5 m AOD and started above the Clifton Rock, and the oldest, which were proved to a depth of 71.5 m AOD, in drillhole E10. Precise stratigraphic correlation was not possible. Subsequently, a cored hole (CD) was drilled in the eastern part of the site to a depth of 90 m from 158.8 m AOD to prove the sequence to be evaluated.

Table 2. Semi-quantitative representation of XRD data

Sample No.	Depth below top of drillhole (m)	Description	Mineral						
			Q	S	F	C	K	M	Mxd-layer illite-smectite
1	5	Grey-black hard, laminated mudstone	****	tr	tr	**	**	*	tr
2	28	Dark grey-black hard laminated mudstone	****	*	tr	**	**	*	tr
3	45	Dark grey-black hard laminated mudstone	****	**	tr	**	**	**	tr
4	60	Dark grey-black hard laminated, micaceous mudstone	****	**	nd	*	***	**	tr
5	75	Grey hard laminated mudstone	****	tr	tr	*	**	**	nd
6	87	Dark grey-black hard laminated mudstone	****	tr	tr	*	**	**	nd

Q =	Quartz	****	= Dominant (possibly over 30%)
S =	Siderite	***	= Major (possibly between 15-20%)
F =	Feldspar	**	= Minor (possibly 5% or less)
C =	Chlorite	*	= Present (possibly 1 or 2%)
K =	Kaolinite	tr	= Trace
M =	Mica	nd	= Not detected

3.3 Estimation of resource quantity

The sandstone present is regarded as wastage, the remaining mudrock representing the potential proportion of material available for brickmaking purposes. The percentages of sandstone in the ten drillholes are given in Table 1. These figures indicate that in the western part of the site about 90% of material should be available for brickmaking purposes. In the cored drillhole, some 15.5% is sandstone and roughly 2.5% coal. Therefore, approximately 82% of material is available for brickmaking on the eastern half of the site.

Taking these various estimates together, it could reasonably be assumed that about 86% of the material which would be quarried could be used for brickmaking. The total volume of material which could be excavated was estimated as in excess of 16.25 M tonnes of which around 14.1 M tonnes would be suitable for brickmaking. It was estimated that the productive life of the pit would be over 40 years.

3.4 Estimation of resource quality

Six samples of mudstone core from different depths in drillhole CD were prepared for analysis. One third of the sample was used for whole-rock X-ray diffraction analysis, another third for detailed clay mineral analysis and the remaining material for surface area determination. Table 2 gives a summary of the X-ray diffraction (XRD) analysis data. The XRD peak intensity data show that quartz was the dominant mineral comprising approximately 30% of these mudrocks. This amount of quartz should provide strength and durability to bricks made from this material. Kaolinite is the principal clay mineral and chlorite and mica sometimes occur in significant amounts. Fig. 3 shows that glycolation has no effect on the trace, indicating a lack of swelling minerals.

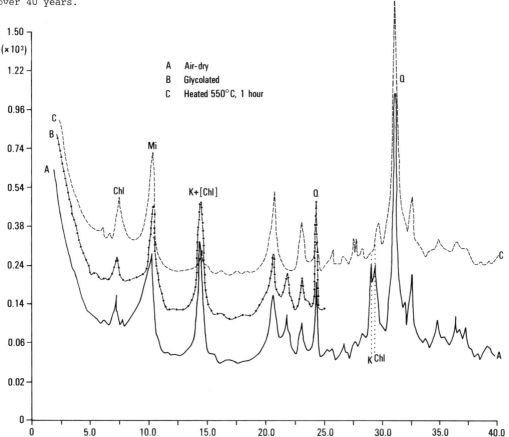

Fig. 3 X-ray diffraction traces for typical sample from the site. (Chl = chlorite, Mi = mica, K = kaolinite, Q = quartz).

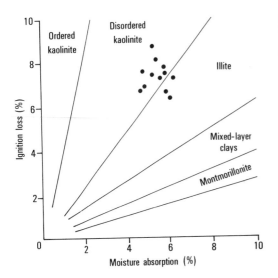

Fig. 4 Clay mineral determination, Keeling's Method.

Examination of results from the use of Keeling's Method shows that the majority of the values fell within the field of disordered kaolinite, confirming the results of the XRD analysis (Fig. 4).

The smectite group of minerals has a much higher surface area than other clay minerals, hence its measurement affords a means of determining the amount of smectite (or an interlayed clay containing smectite). Surface areas were determined using a procedure based on the absorption of the organic reagent 2-ethoxyethanol onto the clay mineral surface under vacuum (Carter et al. 1965). The absence of smectites was confirmed by the results of surface area measurements, which were as follows:

Sample No.	Surface area (m²/g)
1	55
2	64
3	47
4	36
5	60
6	55

Standard smectite ("Paddock Blue") = 765 m²/g

The clay-type minerals present are those with non-expansive lattices, hence the mudrocks should produce a lean raw material with low shrinkage factors. On firing, the clay minerals and micas present will liberate fluxes such as potash, soda and iron which will facilitate the firing

process, helping to produce dense bricks of low porosity.

Deleterious accessory minerals, such as calcite and pyrite, were not detected. Gypsum was absent also. Iron was present, mainly in the form of siderite.

Analyses were made of the carbon and the sulphur content of samples taken at intervals of approximately 5 to 7.5 m throughout the cored drillhole (Figs 5a and 5b respectively). Fig. 5a shows that, in most cases, the carbon content is below 1.5%. Those locations where the carbon content is higher, notably around 60 m depth, are from thin layers of shales. Some of these shales are associated with thin coals, as at 60 m, or with an organic layer, as at 56 m, and are not representative of the mudrock as a whole. Fig. 5b shows that the sulphur content only once exceeds 0.2%. The chemical analyses of the material from the site are similar to the analyses of other Coal Measures mudrocks which are used elsewhere for brickmaking (Table 3).

A number of small green bricks were made from material obtained from pits dug at the site and from the drillhole core in order to assess their fired properties. The fired briquettes have an attractive reddish pink colour. The average drying and firing

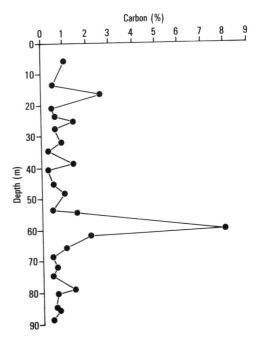

Fig. 5a Carbon content of mudrock with depth below top of drillhole CD, as determined from core material.

3162

Table 3. Chemical analysis of principle oxides in mudrocks from site
in West Yorkshire compared with those from other Coal Measures mudrocks used for
brickmaking

	1	2	3	4	5
SiO_2	61.2	62.5	62.4	65.3	62.5
Al_2O_3	17.6	17.8	17.6	15.9	21.5
Fe_2O_3	7.65	6.47	6.70	6.36	7.40
TiO_2	0.96	1.00	0.87	–	1.15
CaO	0.64	0.42	0.40	1.66	1.00
MgO	1.62	1.84	1.50	1.46	2.03
K_2O	2.87	2.88	3.11	2.28	2.90
Na_2O	1.11	1.15	0.45	1.74	0.77

1, 2 and 3 from site
4 - Ripley, Derbyshire (from Ridgway, 1982)
5 - Wardley, Durham (from Ridgway, 1982)

shrinkages obtained were 2.6% and 4.3%
respectively. Their compressive strength
ranged from 94.2 to 114.1 MPa with a mean
value of 104.8 MPa and moisture absorption
had an average value of 6.2% with a range
from 4.5 to 8.9%. The bulk density of these
briquettes varied from 2.29 to 2.56 Mg/m³,
the average being 2.48 Mg/m³, and the
average porosity was 8.3% (range, 6.5 and
15.8%). Tests on the briquettes were
carried out in accordance with B.S. 3921
(Anon. 1985) and showed that their liability
to effluoresce was nil to slight, that their
suction rate was 0.42 kg/mm²/min and that
they were durable. These results meet the

requirements for bricks of engineering grade
quality (specifications for Engineering
Grade A are - over 69 MPa compressive
strength, 4.5% or less moisture absorption;
Engineering Grade B - between 48 and 69 MPa
compressive strength, 4.5 to 7.0% moisture
absorption [Anon. 1985]).

4 CONCLUSIONS

The proposed site for the extraction of
brickmaking material comprised Coal Measures
rocks of which over 80% were mudrocks. The
investigation indicated that over 14 M
tonnes of mudrock was available at the
proposed site for the manufacture of bricks.
If the site was put into production it was
estimated that the proposed brickworks would
have a life of over 40 years.
The mudrock material was found to consist
primarily of quartz and clay minerals,
mainly disordered kaolinite. Only traces of
mixed-layer clays were present. Hence, they
are a lean clay with low shrinkage factors.
When fired the raw material produces a dense
engineering grade quality brick with an
attractive reddish pink colour and low
moisture absorption and high strength.
These Coal Measures mudrocks constitute a
good quality brickmaking material.

ACKNOWLEDGEMENTS: The authors gratefully
acknowledge the helpful comments of Albert
Horton. This paper is published with the
permission of the Director of the British
Geological Survey (NERC).

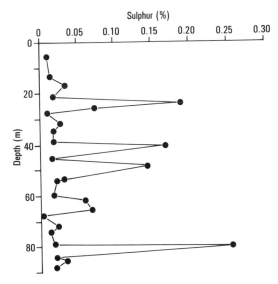

Fig. 5b Sulphur content of mudrock with
depth below top of drillhole CD, as
determined from core material.

REFERENCES

Anon. 1985. Specification for clay bricks.
B.S. 3921. British Standards Institution,

London.

Carter, D. L., Heilman, M. D, and Gonzalez, F. L. 1965. Ethylene glycol monoethyl ether for determining surface area of silicate minerals. Soil Science, 100, 356-360.

Keeling, P. S. 1961. The examination of clays by IL/MA. Transactions of the British Ceramic Society, 60, 217-242.

Ridgway, J. M. 1982. Common Clay and Shale. Mineral Dossier No. 22. Institute of Geological Sciences. HMSO, London.

6th International IAEG Congress / 6ème Congrès International de AIGI, © 1990 Balkema, Rotterdam. ISBN 90 6191 130 3

Characteristics of the Korean briquette ash and its potential as a usable material

Caractéristiques de la cendre des briquettes de charbon coréennes et son potentiel comme un matériau utilisable

Daesuk Han
Korea Institute of Energy and Resources, Daejeon, ROK

ABSTRACT: The Korean coal briquette used as a fuel mainly for heating is 15 cm in diameter and 14 cm in height, weighing about 3.6 kg and having 22 holes. The ash residue resulting from the combustion of the briquette weighs about 1.5 kg and is one of the three solid by-products of the coal industry in the Republic of Korea. Little attention has been paid to such waste products as a potential replacement of natural material.

The ash was tested for its specific gravity, gradation, compaction characteristics, CBR, permeability, thermal conductivity, chemistry, and mineralogy. The test procedures and ash characterization were discussed.

The ash waste was classified as SM according to the Unified Soil Classification System. Most of the ash particles were porous, giving rise to their lightness and relatively low thermal conductivity.

The main inorganic constituents of a representative sample were Si, Al, and Fe: the oxides of these three elements comprised 80% of the ash composition. The minerals identified were mullite, quartz, magnetite, ilmenite, glass, biotite, and amphibole.

Based on the test results, the ash waste can be used as a cover material for sanitary landfill as well as a construction material for road sub-base.

RÉSUMÉ: Les briquettes de charbon coréennes utilisées comme un fuel principalement pour le chauffage domestique sont de 15 cm de diamètre et de 14 cm de hauteur pesant près de 3,6 kg avec 22 trous. La cendre résiduelle venant de la combustion d'une briquette pèse près de 1,5 kg et qui est un des trois sous-produits solides de l'industrie charbonneuse dans la République de Corée. Peu d'attention a été attirée sur ce genre de déchets comme un remplacement potentiel du matériau naturel.

La cendre a été analysée pour son poids spécifique, granulométrie, caractère de compactage, CBR, perméabilité, conductivité thermique, chimie, et minéralogie. La procédure de l'essai et la caractérisation de la cendre ont été discutées.

Le déchet en cendres a été classifié comme SM d'après le Système de Classification des Sols Unifiés. La plupart des grains de cendre sont poreux, donnant lieu à la légèreté et une conductivité thermique relativement basse.

Les principaux constituants inorganiques d'un échantillon représentatif étaient Si, Al, ete Fe; les oxydes de ces trois éléments se composent 80% de la composition de la cendre. Les minéraux identifiés étaient la mullite, le quartz, la magnétite, l'ilménite, le verre, la biotite et les amphiboles.

Basé sur les résultats de l'essai, la cendre peut être utilisée comme un matériau de couverture pour le remblai sanitaire aussi bien que comme un matériau de construction pour la sous-couche de base de la chaussée.

1 INTRODUCTION

The Korean briquette is made of the mixture of 85 to 90 percent domestic anthracite and 10 to 15 percent imported bituminous coal by weight, being used as a fuel mainly for heating. It is 15 cm in diameter and 14 cm in height, weighing about 3.6 kg and having 22 holes as shown in Figure 1. The ash residue resulting from the combustion of the briquette is also shown in Figure 1. This residue is pale pink in color and one of the three coal-related wastes in the Republic of Korea, two of which are coal mine refuse and fly ash. The briquettes used in the year of 1988 amounted to approximately

23 million metric tons according to the Korea Mining Promotion Corporation, becoming approximately 9.6 million metric tons of ashes. The ashes have been frequently disposed of anywhere without any environmental consideration as shown in Figure 2.

A laboratory investigation was undertaken during the period from January to October, 1989 for the following purposes: (1) to find out the physical, engineering, thermal, chemical, and mineralogical characteristics of the crushed ash material, (2) to evaluate its utilization potential as replacements of conventional materials, and (3) to recommend further research on it.

Figure 1. Korean coal briquette (right) and ash (left).

Figure 2. Pile of briquette ashes disposed of in a residential area, Seoul, 1989.

2 SAMPLE PROCUREMENT AND PREPARATION

One hundred and fifty coal briquette ashes were collected from the city of Seoul.

In order to have all kinds of laboratory tests be performed on the samples with same grain-size distribution, all the briquette ashes collected were crushed by the following manner. They were first divided into 30 groups. Each group comprising 5 briquette ashes was then placed in a Los Angels abrasion machine together with three steel balls averaging approximately 4.7 cm in diameter and each weighing approximately

0.4 kg and crushed by being subjected to
50 revolutions. Each of the crushed group
samples was subdivided into 16 portions by
using a riffle box. By mixing each portion
of a group sample with that taken from each
of the rest, 30 samples weighing approximately 14 kg each were finally prepared.
A riffle box was used whenever necessary
during the course of the subsequent laboratory testing.

3 LABORATORY INVESTIGATION

Throughout this and subsequent chapters,
unless stated otherwise, the term "ash material" refers to the crushed briquette
ashes prepared as described in the preceding chapter.

3.1 Physical characteristics

The grain-size distribution for the ash
material was determined in accordance with

ASTM D 422. A sample of the ash material
was first wet sieved using 3/4" (19.0-mm),
3/8" (9.5-mm), No. 4 (4.75-mm), and No. 10
(2.0-mm) sieves. The portion passing No. 10
sieve was then tested following a hydrometer method. As the grain-size distribution curve (Figure 3) illustrates, the
ash material is composed of 12 percent
gravel-size particles, 65 percent sand-size
particles, and 23 percent fines by weight.

The specific gravity determination for
the ash material was conducted following
the Gas Jar Method of BS 1377: 1975, Test
6(A), which can be used for any material
containing particles up to 37.5 mm size.
The specific gravity was determined at
2.36.

The moisture content of 50 briquette ashes ranged from 0.22 to 1.32 percent with
an average value of 0.58 percent.

The determination of the liquid and plastic limits for the ash material was conducted in accordance with ASTM D 423 and
D 424, respectively, producing nonplastic
value.

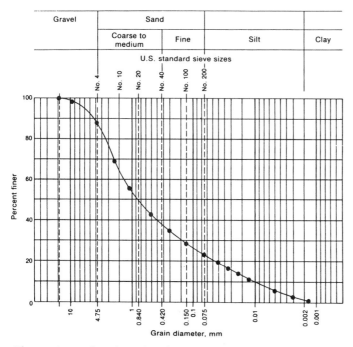

Figure 3. Grain-size distribution curve for the ash material.

3.2 Engineering characteristics

The moisture-density relationship for the ash material was determined following the procedure of ASTM D 698, "Standard Test Methods for Moisture-Density Relations of Soils Using 2.5-kg Rammer and 304.8-mm Drop." A 152-mm diameter mold was used to evaluate the moisture-density curve. As illustrated in Figure 4, the optimum moisture content and maximum dry density were 19.4 percent and 1.337 g/cm³, respectively.

Five specimens were compacted over a range of moisture contents with standard compactive effort and tested in accordance with ASTM D 1883 to obtain their CBR values. They were not soaked prior to determining penetration resistance. The test results are presented in Figure 5. A specimen compacted to its optimum moisture-density condition was subjected to soaking prior to determining its CBR value. The values of swell and CBR were determined at 0.03 percent and 44.7 percent, respectively.

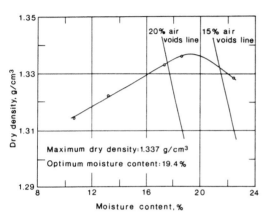

Figure 4. Moisture-dry density relationship for the ash material.

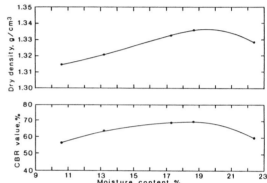

Figure 5. CBR value and dry density versus moisture content for the ash material.

Two specimens were compacted dry and wet of their optimum moisture content in accordance with ASTM D 698, Method C and used to evaluate the permeability of the ash material following a generally accepted procedure of falling head permeability test. The coefficient of permeability at 20°C was determined at the values appearing in Table 1.

Table 1. Coefficient of permeability for the ash material.

Specimen number	Dry density, g/cm³	Moisture content, %	k_{20}, m/s
1*	1.335	18.0	5.7×10^{-7}
2**	1.336	19.9	5.2×10^{-7}

* Compacted dry of OMC.
**Compacted wet of OMC.

3.3 Thermal conductivity

Thirty one specimens were prepared using a 2.5-kg rammer and CBR mold at the five different levels of compaction, which were obtained by applying 2 to 6 layers and 56 blows per layer. For each level of compaction, 3 to 10 specimens were prepared so as to cover a range of moisture contents, 8.0 to 23.0 percent; the dry density of the specimens varied from 1.314 to 1.369 g/cm³. Each specimen was tested to obtain its thermal conductivity using Quick Thermal Conductivity Meter, Showa Denko. The thermal conductivity ranged from 1.68×10^{-3} to 2.84×10^{-3} cal/cm s °C depending upon both moisture content and dry density. All the test data were utilized to construct a diagram of thermal conductivity as a function of moisture content and dry density as shown in Figure 6. Using this diagram, it is possible to estimate the thermal conductivity of the ash material.

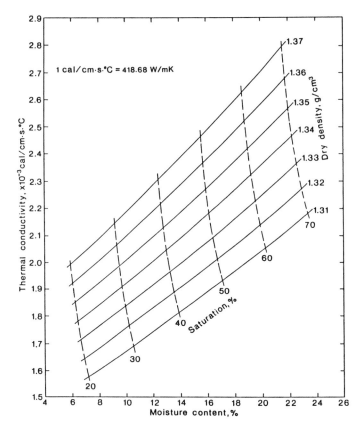

Figure 6. Thermal conductivity for the ash material as a function of moisture content and dry density.

3.4 Chemical characteristics

The determination of pH was conducted using an electric pH meter and following the Electrometric Method, BS 1377: 1975, Test 11(A). A 30 g sample was taken from the ash material crushed to pass through a 3.35-mm sieve and tested for its pH. The value was measured at 6.0.

The water-soluble sulfate content for the ash material, which was crushed to pass through a 475-μm sieve, was determined at 1.44 g/liter when it was tested in accordance with the Ion-exchange Method of BS 1377: 1967, Test 10.

Seven chemical constituents were determined from the ash material crushed to pass through a 475-μm sieve. The determinations of SiO_2, Al_2O_3, and Fe_2O_3 were made by colorimetric methods, TiO_2, MgO, and CaO by a spectrochemical method, and SO_3 by a gravimetric method. The unburnt fuel was determined by loss on ignition. The contents of the chemical constituents are presented in Table 2.

Table 2. Chemical composition of the ash material.

Chemical constituent	Content, %
SiO_2	46.90
Al_2O_3	29.57
Fe_2O_3	3.19
TiO_2	1.21
MgO	0.88
CaO	0.33
SO_3	0.34
Loss on ignition	10.50
Others (undetermined)	7.08

3.5 Mineralogy

The minerals identified were quartz, mullite, magnetite, ilmenite, biotite, amphibole, and glass. The characterization of the X-ray powder diffraction patterns (Figure 7) revealed the presence of quartz, mullite, biotite, and amphibole. Magnetite and glass were identified by the examination of thin sections under a polarizing microscope, whilst ilmenite by electron microprobe analysis. Figure 8 shows a lot of opaque magnetite crystals formed in some of the ash particles which were separated by a magnet.

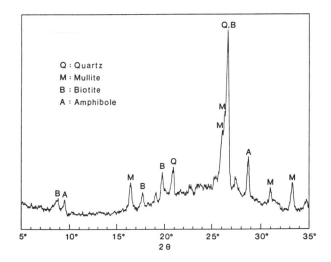

Figure 7. X-ray powder diffraction patterns for the ash material.

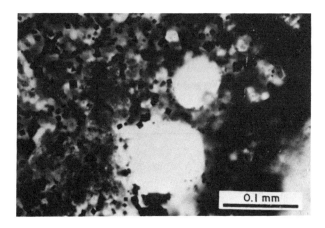

Figure 8. Tiny magnetite crystals (black) scattered throughout the groundmass of aluminum silicate glass. Note the bubble-shaped cavities (white).

4 DISCUSSION

Based on the results of the determinations of gradation and Atterberg limits, the ash material is classified as SM according to the Unified Soil Classification System.

Most of the ash particles are porous as shown in Figure 9. The ash material containing the cavities is characterized by 17.4 percent air voids by volume when compacted to its optimum moisture-maximum density condition, giving rise to that thermal conductivity is lower for the ash material than for other earth materials with similar grain-size distribution and moisture content.

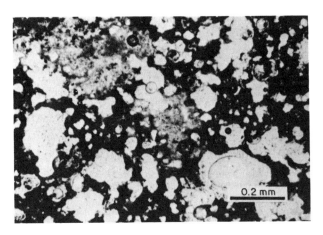

Figure 9. Tiny round and irregular cavities (white) developed in the ash particles.

Grain-size analyses were conducted both before and after the ash material was subjected to standard compaction. The results were plotted to see the degree of particle breakdown (Figure 10). According to the figure, the material did not suffer extensive degradation.

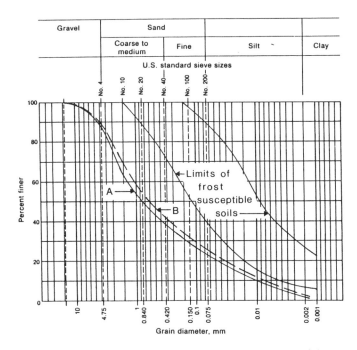

Figure 10. Gradation curves before (A) and after (B) compaction and limits of frost-susceptible soils according to Croney (1949).

3171

The ash material is non-frost-susceptible when its grain-size distribution is compared with the gradation limits shown in Figure 10 and those of the Canadian Department of Transport (Armstrong and Csathy 1963).

The CBR value for the unsoaked specimen compacted to its optimum moisture-density condition is 70.5 percent; that for the soaked one 44.7 percent. The strength loss due to saturation is about 37 percent.

The ash material is characterized by very low degree of permeability according to the permeability classification by Terzaghi and Peck (1970).

The ash material is slightly acidic according to the pH value measured. This degree of acidity does not appear so high as to cause corrosion of metals.

Based on the content of either water-soluble or total sulfates, ordinary cement can be used for any burried concrete in the ground filled with the ash material.

5 CONCLUSIONS

In 1988, the municipal solid wastes in the large cities of the Republic of Korea amounted to 2.2 kg per capita per day, whilst those in the rural communities to 1.5 kg per capita per day. About 50 percent of the wastes are coal briquette ashes. In general, the wastes generated in both communities are presently disposed of in some type of landfill, being deposited with little regard for pollution control or aesthetics. The coal briquette ashes collected separately from other wastes can be used as a cover material for sanitary landfill provided that they are crushed to be classified as SM according to the Unified Soil Classification System and compacted to have the coefficient of permeability in the range of 1×10^{-6} to 1×10^{-8} m/s. The ash material may be economically advantageous over the conventional cover materials which cannot be obtained from the areas adjacent to disposal sites.

The ash material is non-frost-susceptible on the basis of its grain-size characteristics and characterized by relatively low thermal conductivity. Based on these properties along with the CBR values, the ash material can also be used as a material for sub-base, which is generally situated above the level of frost penetration in the Republic of Korea where approximately 98 percent of the land area is affected by seasonal freezing.

It is recommended that the ash material compacted to its optimum moisture-density condition be subjected to several cycles of wetting and drying as well as freezing and thawing prior to determining its CBR value in order to evaluate its durability. It is also recommended that the quality of leachate from the ash material be determined to see whether it is environmentally acceptable or not.

6 ACKNOWLEDGEMENT

The writer wishes to express his sincere appreciation to the Ministry of Science and Technology for its financial support.

REFERENCES

American Society for Testing and Materials 1980. Annual book of ASTM standards, Part 19.
Armstrong, D.M. & T.I. Csathy 1963. Frost design practice in Canada. Highway Research Record, no.393, p.12-18.
Bowles, J.E. 1978. Engineering properties of soil and their measurement, 2nd ed. New York: McGraw-Hill.
British Standard Institution 1975. Methods of test for soils for civil engineering purposes, BS 1377: 1975.
Chamberlain, E.J. 1981. Frost susceptibility of soil - Review of index tests. CRREL Monograph 81-2, p.16-20.
Croney, D. 1949. Some cases of frost damage to roads. Department of Science and Independent Research, Road Research Lab., Berkshire, England, Road Note no.8.
Han, D. 1987. Road damage due to ground freezing in the Republic of Korea and a method to predict frost depth for pavement design. Proceedings of the symposium on environmental geotechnics and problematic soils and rocks, p.115-122. Rotterdam: Balkema.
Head, K.H. 1980. Manual of soil laboratory testing vol.1. London: Pentech.
Head, K.H. 1982. Manual of soil laboratory testing vol.2. London: Pentech.
Kim, J.D., et al 1983. A study on the utilization of the coal ash. Korea Institute of Energy and Resources Research Report KE-83-8, p.56-60.
Terzaghi, K. & R.B. Peck 1967. Soil mechanics in engineering practice. New York: Wiley.
The Asphalt Institute 1962. The asphalt handbook, Chapt.5.

Stability of rhyolites used in a breakwater core

Stabilité de la rhyolite dans un noyau de digues de protection

E. Dapena
Laboratorio de Geotecnia, CEDEX, Madrid, Spain

M. Romana
Universidad Politécnica, Valencia, Spain

E. Hernández
Dragados & Construcciones, S.A., Madrid, Spain

ABSTRACT: This work studies the stability of a considerable volume of rhyolite in the core of the breakwaters in a port in the South of Iran subjected to the external agents affecting the core in this type of works. For this purpose rock samples were analysed and subjected to laboratory tests consisting of seawater alteration cycles. Once this study was made and taking into consideration the functions of the different elements composing a breakwater, the works were rated in terms of the materials to be utilized in the core and in terms of their possible future behaviour.

RESUME: On étudit ici la stabilité de la rhyolite utilisée dans une proportion très importante dans le noyau des digues d'un port situé au sud de l'Iran face aux agents extérieurs agissant sur ces dernières dans ce type de travaux. A cette fin on analyse des échantillons de roches que l'on soumet en laboratoire à des cycles d'altération par eau de mer. Une fois effectuée cette étude, et en tenant compte des fonctions des différents éléments qui constituent une digue, on les califie comme matériaux à utiliser dans le noyau et on définit leur comportement futur éventuel.

1 INTRODUCTION

In the port for the Bandar Abbas Shipyard the interior protection layers are 2500 m long and consist of a quarry run and two outside layers of rockfill jetty.

The external protection layers and breakwaters consist of a quarry run core and different thicknesses of rockfill layers as a function of the design wave characteristics (Fig. 1) and are approximately 1360 and 2430 m long respectively.

The quarry run was composed of two types of material, rhyolites and limestone.

This research studies the stability of the rhyolites under the environmental conditions existing in the core of the breakwaters and protection layers.

2 MINERALOGICAL COMPOSITION

The samples taken were rhyolites with a porphydic texture and a colour ranging

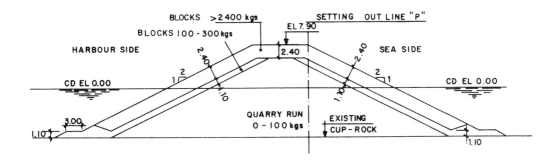

Fig. 1 Section 4-4 of the East Breakwater

from greenish-white to dark grey in which scattered white nodules could be seen and in some cases considerable cavities.

The polarized light optical microscope analysis indicated that the essential components were quartz and plagioclase, with potassium feldspar being seen in some cases. Next in the quantity came phyllo-silicates (micas and opaque matter) which either appeared throughout the rock or were related to the weathering of the ferromagnesium particles. Carbonates were also found (calcite and dolomite) as products of alteration. Zircon was the most common accessory mineral found.

3 PHYSICAL PROPERTIES OF THE ROCK

3.1 Density

3.1.1. Apparent specific weight

The distribution of the apparent specific weight of the rock fragments tested, according to ASTM C-97-83, is shown in Figure 2. It ranges from 2 g/cm3 to 2.35 g/cm3 with a minimum value of 1.72 g/cm3. The representative average value of apparent specific weight taken for these rocks is γ_{ap} = 2.15 g/cm3.

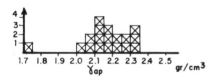

Fig. 2 Distribution of the apparent density values γ_{ap}

3.1.2 Specific weight of the solid

The specific weight of the solid part of the rock could be determined using ASTM-C-97-33 if all the voids were accessible. One of the results obtained from this test, known as the real specific weight test, refers to the specific weight of the material with non accessible cavities.

The values obtained for this real specific weight range between 2.44 g/cm3 and 2.59 g/cm3 (Fig. 3) with the majority of them being above 2.54 g/cm3.

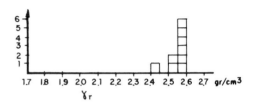

Fig. 3 Distribution of the real specific weight values γ_r

The mineralogical analysis carried out proved that the minerals present in greatest proportion were quartz, plagioclases and potassium feldspar. The quartz had a specific weight of 2.55, potassium feldspar one of 2.56 g/cm3 and the plagioclases, depending on their sodium and calcium content, ranged between 2.60 and 2.76 g/cm3.

The presence of micas, carbonates, etc. with higher specific weights than the preceding ones increased the specific weight of the solid part of the rock.

Quantifying the mineralogical composition in an approximate way using the results of chemical analysis, Table 1 takes the average specific weight of the rock solid as γ_r and this was the value taken to determine the total volume of voids.

Table 1 Chemical composition of the rhyolites

Ref.	SiO₂	Al₂O₃	Fe₂O₃*	TiO₂	CaO	MgO	Na₂O	K₂O	H₂O com.
GAFd3	70.1	13.5	4.1	0.28	0.52	1.20	1.7	6.8	1.5
GAFd6	74.8	9.6	4.2	0.27	-	0.72	3.9	4.4	1.0
GAFd7	67.9	12.8	6.1	0.26	0.65	0.72	4.0	6.9	1.4
GAFw2	72.6	11.4	2.5	0.30	0.23	0.42	2.8	6.5	0.33
GAFw5	73.7	11.1	4.3	0.34	0.19	0.73	5.3	4.9	0.23
GAFw8	71.2	13.1	3.9	0.30	0.39	0.84	4.7	5.5	0.54
GAFd9	70.8	12.6	3.7	0.28	1.2	1.2	4.7	5.2	0.98

* corresponds to the total iron in the sample

3174

3.2 Porosity

3.2.1 Absorption

Figure 4 gives the distribution of the values obtained for the quantity of water absorbed through immersion of the rocks for a period of 24 hours.

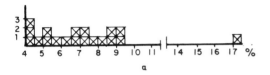

Fig. 4 Distribution of the absorption a values

The results vary between 4 and 9.5% with the top figure being 17.3% which corresponded to the lowest density sample, GAFw4.

Figure 5 shows the ratio between the apparent specific weight and the absorption of each sample. These pairs of values adapt to the curve

$$\frac{a}{100} = \frac{1}{\gamma_{ap}} - \frac{1}{2.55} \qquad (1)$$

According to this law, an absorption value of a=7.3% corresponds to the average apparent specific weight representative of these rocks, γ_{ap} = 1.25 g/cm3, and this absorption value was taken as the average absorption rate representative of the rhyolites under study.

3.2.2 Moisture in the saturated sample

Starting from the value taken as specific weight of the solid γ_r = 2.63 g/cm3, the moisture content in the saturated sample, w, as a function of the apparent density of each sample, will be:

$$\frac{w}{100} = \frac{1}{\gamma_{ap}} - \frac{1}{2.63} \qquad (2)$$

The water content corresponding to the inaccessible voids, w_i, would have an average of:

$$\frac{w_i - a}{100} = \frac{1}{2.55} - \frac{1}{2.63} \qquad (3)$$

$$w_i = 1.2\%$$

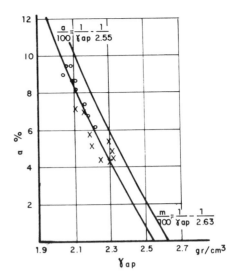

Fig. 5 Apparent density γ_{ap}:a absorption ratio

4 STABILITY OF THE RHYOLITE IN THE ENVIRONMENTAL CONDITIONS AFFECTING THE CORE

4.1 Environmental conditions affecting the breakwater core

Figure 1 shows a standard section of the breakwaters and protection layers. The breakwater in this particular section is built on the bottom coquina and is 16 m high. The core, however, is only 13.6 m high with the top 2.4 m corresponding to the rockfill layer.

Of these 13.6 m of core, 8.6 m in this section stand below the equinoxial low tide elevation and the top 5 m are above this, so that it can be assumed that the former 8.6 m are submerged during the life of the works.

Bearing in mind that the maximum tidal amplitude of the Persian Gulf is 4 m, although there are local variations which drop to as far as 1.5 m in Abu-Dhabi, it means that only the top core crest section thicker than 1 m remains out of the tidal area.

Furthermore these breakwaters and protection layers have been deliberately built so that the protection layers stop

3175

wave action on the core, therefore the rocks of which it is made will be mainly affected by the climate in the 1 m crest, the tides in the 4 m section inbetween and seawater immersion in the area permanently submerged.

4.1.1 Climate action

The port stands in one of the hottest areas in Iran, with high relative humidity. The annual temperature variation goes from 5 to 48°C and the average annual rainfall is 150 mm which usually occurs during winter and spring. Winds can reach 80 km per hour.

In a climate such as this, according to Peltier (1950), chemical weathering is very low and mechanical weathering as a result of ice action nil. However, the mechanical action on the breakwater crest owing to salt crystallisation can be fairly important and it is induced by the evaporation which can take place, even though the protection layers will stop the direct action of the sun, stabilizing the environment in the area above the core, so that the possible crystallisation effect of the salts will be mitigated.

4.1.2 Sea action

The action of seawater on the core material of the breakwaters can be specified in three aspects, in addition to the crystal-lisation of salts already mentioned. For one, the wetting-drying cycles to which the intertidal area is subjected, secondly the constant hydration which the material of the submerged area experiences and thirdly the chemical transformations which the minerals can undergo.

The crystallisation of the salts and the chemical transformations which can affect the minerals are highly conditioned by the composition of the seawater.

The waters of the Persian Gulf, accord-ing to Purser & Seibold (1973), have a surface salinity varying from a rate of 36.5 per thousand in the contact area with the Straits of Hormuz up to rates of over 100 per thousand in the South coast, Qatar Peninsula. These surface variations also extend deep down into the water. The Bandar-Abbas area lies in a zone of lower salinity and, as will be seen later, is similar in its concentration of salts to some areas of the Mediterranean.

4.2 Stability of the rhyolites under the processes of NaCl crystal formation

The mechanism whereby crystals form and grow in a supersaturated solution is described in detail in the works of Wellman & Wilson (1965) and Sperling & Cook (1980).

The crystallisation pressure, M, of a crystal with a supersaturated concentra-tion, C, above the saturation concentra-tion, C_S is given by the equation:

$$M = \frac{R.T.}{V} \cdot Ln \left(\frac{C}{C_S}\right) \tag{4}$$

where R is the gas constant, V the molar volume of the crystal and C the saturation concentration at the absolute temperature T.

When the solutions are subjected to pressures lower than M, the crystals can grow and develop, however a crystal can dissolve when placed in a solution subject-ed to a pressure greater than M.

In one of the lab tests we carried out to reproduce this process the formation of NaCl crystals was provoked by saturating the rock with seawater and precipitating out the crystals by stove drying.

The seawater used in the tests came from the Mediterranean and was taken from the Port of Valencia since it was proved that its composition was similar to that of the Persian Gulf waters in the Bandar-Abbas Port, as shown by the results of analysing both waters which are given in Table 2.

Table 2   Composition of the seawater

Results	In Bandar-Abbas Port mg/l	In Valencia Port mg/l
(Cl$^-$)	22,000	24,100
(SO$_4$=)	3,000	3,000
(CO$_3$=)	10.60	(*)
(HCO$_3^-$)	139.00	171.20
(Br$^-$)	(*)	(*)
(NO$_3$)	(*)	(*)
H$_4$SiO$_4$	20.80	6.40
Al(OH)$_4$	(*)	(*)
(Na^{++})	11,000	17,000
(K$^+$)	510.00	540.00
(Ca^{--})	457.00	450.00
(Sr^{++})	(*)	(*)
(Mg^{++})	3,700	1,500
(Fe^{++})	(*)	(*)
(Fe^{+++})	0.05	0.40
Dens.(17°) g/cm3	1.031	1.031

(*) No traces

Two test methods were used on the rock samples. One consisted of immersing the fragments for 12 hours in seawater and provoking the crystallisation of the dissolved salts by evaporating off the water in a stove at 100 ± 5°C for 16 ± 2 hours. The process was repeated for 20 cycles to provoke the growth of salt crystals and even to detect the possible fatigue effect as a result of the cycles repeated.

This method subjected the rock samples to the simultaneous action of the processes, salt crystallisation and wetting-drying cycles so that the results were one level higher for each of them.

The possible effect of this crystal formation and growth can translate into an increase in fissuration and in the crumbling away of material which should be reflected in loss of weight.

The variation in the state of fissuration was determined by obtaining the absorption of the samples during the test process and the weight loss was determined by monitoring this at fixed cycle intervals in each of the samples.

The test results, measuring the variation in absorption and weight loss after the tenth and twentieth cycle, are shown in Table 3.

Only two of the nine samples tested, 22% of the total, were affected by the cycles to any significant extent, involving weight losses after 20 cycles of 19 and 11.7%. In another two samples weight loss was below 2.5% and in the five remaining samples it stayed below 1%. The average weight loss we took as representative of the rhyolites in the core was around 3% after ten cycles and 4% after 20,

which would be the maximum average value to be obtained as a result of salt crystallisation.

The absorption increases were less significant. The samples to experience the greatest weight loss also displayed the greatest variation in absorption, increasing by 2.3 and 1.5%. The absorption variation in the remaining samples stayed below 0.4% and in the most resistant samples the absorption even decreased, which we put down to the filling of the voids caused by the salt crystallisation.

The other test method consisted of keeping the base of the rock fragments in contact with a saturated solution of sodium chloride for 21 days, the rest of the rock in each sample being kept in an environment with relative humidity below 60%.

Two rock fragments weighing 311.2 and 152.8 g each were tested. The crystals gradually formed throughout the surface over the 21 day period, increasing the dry weight at the end of the test to 69.2 and 57 g respectively.

After washing the fragments, a slight weight increase was observed in them due to the fact it was impossible to wash out the sodium chloride which had filled the scarcely accessible voids and that the fissuration, measured as an increase in absorption, was on the level of the lowest increases recorded by the samples studied by cycles (Table 4).

Table 3   Results of the test to determine the effect of sodium chloride crystallisation

Ref.	Weight g	γ_{ap}	γ_r	a	a_{10}	x_{10}	Δa_{10}	a_{20}	x_{20}	Δa_{20}
				%	%	%	%	%	%	%
GAFd1	1000	2.16	2.57	7.4	7.3	1.4	-0.1	7.8	2.4	0.4
GAFd1	350	2.08	2.59	9.5	10.5	17.7	+1	11	19	1.5
GAFd4	300	2.04	2.50	9	8.6	-0.2	-0.4	9	10.4	0.1
GAFd5	650	2.22	2.57	6.1	5.8	-0.3	-0.3	6	0.3	-0.1
GAFw1	950	2.11	2.58	8.6	8.0	-0.6	-0.6	8.7	0.5	0.1
GAFw3	1300	2.18	2.54	6.8	6.7	0	-0.1	7	0.7	0.2
GAFw4	800	1.72	2.44	17.3	17.9	6.6	0.6	19.6	1.7	2.3
GAFw6	1750	2.11	2.56	8.2	7.7	-0.1	0.6	8.3	0.7	0.1
GAFw8	1450	2.09	2.55	8.6	8.2	0.8	-0.4	8.9	1.7	0.3

a_{10} = absorption after ten cycles
x_{10} = weight loss after ten cycles
Δa_{10}= absorption increase after ten cycles
a_{20} = absorption after 20 cycles
x_{20} = weight loss after 20 cycles
Δa_{20}= absorption increase after 20 cycles

Table 4 Result of sodium chloride crystal formation test, keeping the base 21 days in a saturated solution

	M-A		M-B	
	Before	After	Before	After
x g	311.2	313.4	152.3	152.8
γ_{ap} (g/cm3)	2.392	2.420	2.130	2.155
γ_r (g/cm3)	2.559	2.594	2.573	1.630
a%	2.7	2.8	8.1	8.4
Δa%	-	0.1	-	0.3

x = dry weight of rock sample

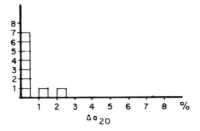

Fig. 7 Absorption increase distribution after 20 cycles Δa_{20}

4.3 Stability of the rhyolites against hydration processes and wetting-drying cycles

The behaviour study carried out was based on the results of the tests on the rock samples to determine how sensitive they were to salt crystallisation since, in addition to this phenomenon, the type of test involved meant it also included a wetting and drying process in each cycle, therefore the results obtained are a combination of all the processes and the values obtained should be considered the highest level for each one of them.

Weight loss is the parameter best capable of differentiating the behaviour of the rock (Fig. 6) and provided three different answers, whereas the variation in absorption recorded was very small and gave only two types of answer (Fig. 7). In terms of weight loss a group of two samples were found with losses of over 10% and these can be considered to be affected by the phenomenon, another two samples recorded losses of between 1.5 and 2.5% which were therefore little affected, and a third group of five samples occurred recording losses below 1% which can therefore be considered stable behaviour.

The average weight loss after 20 cycles was 4% and this can be considered the maximum level in the set of rhyolites in the core.

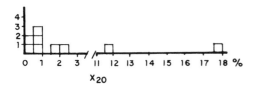

Fig. 6 Weight loss distribution after 20 cycles x_{20}

4.4 Chemical stability of the rhyolites

Studying the mineralogical composition of these rocks proved that they fundamentally consist of quartz and plagioclase with feldspar occasionally also present. Micas, opaque matter and carbonates also appeared, already linked to the degree of alteration of the rock in the quarry.

Of the minerals mentioned the plagioclases displayed the least chemical stability against acid or slightly acid water such as rainwater, and this was particularly the case if they were calcitic, since these can turn into clays and, depending on the environmental conditions prevailing, can give rise to cabonates.

The sea, with its basic medium, has a different level of chemical aggression from the slightly acid atmospheric water so that the minerals composing the rocks studied will react differently from how they would react to weathering. Quartz, whose solubility increases under high pH conditions, continues to have a low solubility level for values of up to pH = 8.5 and seawater never goes beyond this level. Plagioclases are more stable in a marine environment than in an atmospheric one.

These phenomena translate into a weakening of the rock and increased disintegration. It is difficult to quantify this disintegration by, for instance, the increase in the fines content but, as a result of the analysis we carried out, it is thought that the disintegration of these rocks should not be above 2%.

5 QUALIFICATION OF THE RHYOLITES USED IN THE CORE OF THE BREAKWATERS AND PROTECTION LAYERS

The grading of the material that has been

used in the core of the breakwaters is shown in Figure 8, obtained by analysing the materials extracted from the samples taken in the actual breakwaters.

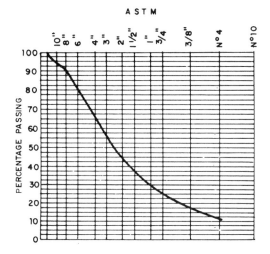

Fig. 8 Current grading of the materials forming the core of the breakwaters erected

The fines existing in the core are below 12% and the continuous grading can be considered suitable for the functions the core must fulfil.

There is no risk of this material dissolving to any appreciable extent by seawater.

In accordance with the laboratory tests made, the majority of the fragments making up the core show good resistance to the cycles to which they could be subjected during the life of the structure and only approximately 20% experienced any appreciable weight loss, with the fines content capable of increasing to 4%.

The least chemically stable minerals are the plagioclases which however stand up better to the basic marine medium than they do to the acid one on land. The result is that any chemical transformations the rocks undergo will only be small, if difficult to actually quantify. The increased fines content this can give rise to can be assumed to be under 2%.

Consequently, in line with these results, the fines content during the service life of the structure should be kept below 20% in which case it is acceptable to consider these rhyolites as apt material for use in the core of the port breakwaters.

6 SUMMARY AND CONCLUSIONS

The material studied consists of porphydic volcanic rocks of the rhyolite family. The fundamental components are quartz and plagioclase and in some cases potassium feldpar. Next in importance come micas and opaque matter which are the outcome of the alteration of the ferromagnesium minerals.

The apparent density of these rocks can be considered to be around $\gamma_{ap} = 2.15 \pm 0.2$ g/cm3 and the absorption rate to be a = 7.3 ± 3%.

The way these materials behave under external environmental action, sodium chloride salt crystallisation wetting-drying cycles and mineral hydration can be measured by an increase in the fines content up to 4%.

As regards the chemical stability of the minerals forming the rhyolites, solution of the quartz is not expected to occur. The potassium feldspars and especially the plagioclases may alter with time, despite the fact that they show more stable reactions to basic media like seawater. It is precisely as an outcome of this alteration that we consider disintegration could increase slightly, taking a maximum level of 2% over the 40 year service life of the structure.

The material to be used in the core of the breakwaters, as part of good port construction practice, is the quarry run from the rockfill operation with no qualities or grading specified. In terms of grading it is recommended keeping the fines content below 20% or so and, since the fines of the core at the time of construction did not exceed 12%, this value, together with any fines generated by alteration if the maximum forecasts are fulfilled, would not go beyond the recommended 20% fines.

For all the above it can be concluded that these rhyolites used in the construction of the core of the breakwaters and protection layers made out of rockfill are an acceptable material for the technique used in port construction.

ACKNOWLEDGEMENTS   Thanks are due to Francisco Enriquez and Victoriano Fernandez Dupuy, Drs of Civil Engineering, and to Salvador Ordoñez, Dr of Geology, for their support throughout the work, especially in the port engineering and geological and mineralogical aspects, and to the firm Dragados & Construcciones, SA for permission to print this work.

BIBLIOGRAPHY

Peltier, L. (1950) The geographic cycle
    in periglacial regions as it is related
    to climatic geomorphology, Ann. Assoc.
    Amer. Geog. 40, 214-236.
Purser, G.H. & Seibold, E. (1973) The
    principal environmental factors
    influencing holocene sedimentation and
    diagenesis in the Persian Gulf. In B.H.
    Purser (ed) The Persian Gulf, 1-9,
    Springer Verlag, New York.
Sperling, C.H.B. & Cooke, R.V. (1980) Salt
    weathering in arid environments,
    Geography 8, 4-46, Bedford College,
    London.
Wellman, H.W. & Wilson, A.T. (1965) Salt
    weathering, a neglected erosive agent in
    coastal and arid environments, Nature,
    205, 1097-1098, London.

Rhenish tuff – A widespread, weathering-susceptible natural stone
Le tuf de la Rhénanie – Une pierre naturelle très répandue et sensible à l'altération

B. Fitzner & L. Lehners
Institute of Geology, Aachen Technical University, FR Germany

ABSTRACT: Rhenish tuff is a widespread, weathering-susceptible natural stone which has been used for constructing important historical buildings in western Germany and neighbouring countries since Roman times. The poor condition of these tuff buildings requires immediate protection measures. The effectiveness of such measures depends upon an accurate knowledge of the causes and state of the damage.
A method is presented which enables the exact recording of the state of damage and gives information on the exogenetic factors producing the deterioration of this natural stone. Selected results from building investigations and laboratory tests are explained.

RESUME: Le tuf de la Rhénanie est une pierre naturelle très répandue et très sensible à l'altération. Elle a déjà été utilisée dès la période romaine pour la construction des bâtiments historiques importants. Le mauvais état de ces bâtiments en tuf exige à prendre des mesures de protection et d'assainissement immédiates. L'efficacité de telles mesures dépend de la connaissance parfaite de l'état et de la cause des dégâts.
Une méthode sera présentée qui permet un enregistrement de l'état des dégâts des tufs et qui donne des renseignements sur les facteurs exogènes qui provoquent principalement l'altération des pierres. Quelques exemples choisis portant sur les études des bâtiments et sur des essais de laboratoire seront expliqués.

1 INTRODUCTION

Tuffaceous rocks are consolidated pyroclastics which represent a link between igneous and sedimentary rocks. There are various criteria to distinguish them. Particle size is used to differentiate the fine- and coarse-grained varieties, and the fine-texture of the pyroclastic constituents is used to differentiate crystal tuff, vitric tuff and lithic tuff (PFEIFER et al.,1981; FISCHER & SCHMINKE,1984; FUECHTBAUER & MUELLER,1970).

Tuffs have been used as a natural building stone world-wide. The Quaternary Rhenish tuffs from the western German area range from about 11,000 to 600,000 years. They occur in the Laach volcanic region, west of Koblenz on the Rhine.

Rhenish tuffs are easy to quarry, transport and work with. For these reasons, they have been used as building stones in western Germany and the neighbouring countries since Roman times (KIESOW et al.,1987).

Many of the historical buildings constructed out of this weathering-susceptible natural stone are showing signs of damage, which makes preservation measures a matter of great urgency. Such measures, however, definitely require detailed knowledge of the condition and causes of the damage to the tuffs.

Using a classification scheme developed for the weathering forms of tuffaceous rocks, the condition of these rocks can be phenomenologically recorded and mapped at the building. A map, made in this way, of an important Middle Ages building from the Rhine area is presented below.

The state of weathering can be quantified using petrographical investigations of mineral composition and textural characteristics in the weathering profile. A combination of various investigation methods for determining the characteristic porosity values proves to be particularly meaningful when assessing material alteration that is dependent on weathering. The influence of various factors on the weathering behaviour of tuffaceous rocks is investigated by means of weathering simulation tests.

By comparing laboratory results and observations made at the building, the weathering processes which make a major contribution to the destruction of the Rhenish tuffs can be identified and quantified.

2 METHODOLOGY

In order to be able to reproducibly characterise the state of damage of historical buildings made from Rhenish tuffs according to phenomenological criteria, investigations are carried out at the building. By means of these investigations a facade-mapping of the natural stone varieties used in building and the damage suffered by them becomes possible.

As first step, all natural stones in the building are noted and documented photographically. They are differentiated according to usual petrographical classification schemes.

Next, the distribution of the noted natural stone varieties in the masonry is mapped. Lithological mapping such as this provides an overall view of the inventory of the natural stones used in the building.

The second stage of the work involves registering and describing the weathering forms. The photographic documentation of these weathering forms serves to visualise and document the damage of the Rhenish tuffs. This work is necessary in order to develop a classification

scheme of the weathering forms. A classification such as this constitutes the most important aid for the mapping of weathering forms on a tuff building according to objective and reproducible criteria.

In accessible areas the mapping proceeds on a "stone by stone" basis, taking into account all weathering forms and their intensities. When surveying a very large facade area an overall picture proves to be sufficient in many cases. In such a survey only the predominant weathering form per ashlar or facade area is mapped.

The results of mapping lithology and weathering forms are presented as a colour-coded layout and/or by plotting symbols on a plan of the building.

The most suitable areas for measurements and sampling can be located using these mapping results.

The aim of the petrographical investigations is to record the mineral composition and porosity characteristics of the diverse tuff varieties. By comparing the results of investigations on weathered and unweathered zones of a natural stone it is possible to determine the material alterations caused by weathering and to quantify the damaged state of the tuff.

Using macroscopic observations and microscopic analysis methods the mineral composition as well as the structural and textural characteristics are recorded.

X-ray diffraction investigations provide additional information on the mineral composition of the tuffs.

The pore space of a rock has a considerable influence on the type and intensity of the weathering processes, and is at the same time altered by these processes itself. The change in pore geometry and pore space is, therefore, a quantitative measure of the state of weathering.

Using the microscope in combination with image analysis, and mercury porosimetry, the pore space of unweathered and weathered tuff samples is investigated.

The first method is used to directly record pores with a radius of 3um to 700um.

Mercury porosimetry facilitates the calculation of pore throat radii between $0.0019\mu m$ and $250\mu m$ and additional porosity values.

A laboratory-based experimental programme on the water absorption and evaporation behaviour of tuffs provides, on one hand, additional information about pore geometry, and, on the other hand, first information about the quantitative influence of water as a main weathering factor.

The influence of combined weathering factors on the texture of the Rhenish tuffs is investigated using frost-thaw cycle experiments, following DIN 52104.

SO_2-weathering experiments and salt crystallization tests provide information about the ability of the tuff to form gypsum, and on the type and amount of mechanical destruction of the texture by salt-wedging forces.

3 RESULTS AND DISCUSSION

3.1 Building investigations

A classification scheme of weathering forms is essential when mapping a building. Such a scheme has been developed on the basis of the results of work carried out on a great number of tuff buildings in the Rhine area (table 1). The following requirements were considered:

1. The most important weathering forms are taken into account in the classification scheme.

2. Levels of intensity were taken into account.

3. Damage categories can be assigned to the weathering forms.

The classification scheme contains five main weathering forms, which correlate with the development of a weathering profile and material loss. Five categories of damage are assigned to the main weathering forms. Each category is further differentiated according to degree of intensity. The weathering forms can , therefore, be evaluated using these categories.

It is possible to produce a modified and more exact record of the condition of the building using six additional separate weathering forms. They involve mineral and

biogenic layering which weakens the rock surface, causes disaggregation and can result in loss of material. In addition, there are separate weathering forms which are attributed to mechanical stress. The separate weathering forms, some of which are further differentiated according to level of intensity, can be assigned to the five categories of damage.

Table 1: Classification scheme of the weathering forms of Rhenish tuffs:

Main weathering forms	Category of damage
UNWEATHERED	No damage
Stone surface completely intact	1
Some detachment of small components	1-
DETACHMENT OF SMALL GRANULAR ELEMENTS	Slight damage
Detachment to 5mm over < 50% of stone surface	2+
Detachment to 5mm over > 50% of stone surface	2
Detachment to > 5mm, start of surface receding	2-
FLAKES,SPALLING	Moderate damage
Loosening to 10mm over < 50% of stone surface	3+
Loosening to 10mm over > 50% of stone surface	3
Loosening to > 10mm, start of surface receding	3-
SCALES	Severe damage
Scales > than 20mm over < 50% of stone surface	4+
Scales > than 20mm over > 50% of stone surface	4
Scales > than 20mm, start of surface receding	4-
SURFACE RECEDING	Very severe damage
Surface receding to 10mm over > 50% of stone surface	5+
Surface receding > than 10mm over > 50% of stone surface	5
Separate weathering forms	
SOOT AND DUST BUILD UP	1-
BIOGENIC LAYERING	2
moss, lichens, algae, higher plants	
EFFLORESCENCES	2-3
slight efflorescences	2
severe efflorescences, salt crusts	3
CRUSTS	2-3
attached crusts	2
detached crusts	3
FISSURING	3-5
slight fissuring	3
moderate fissuring	4
severe fissuring	5
OUTBURSTS	4-5
small outbursts	4
large outbursts	5

As an example, a map of the west facade of the St. Quirinus Minster in Neuss is described (figs. 1-3).

This late romanic vaulted basilica dating from the thirteenth century has been subjected to numerous material and structural chan-

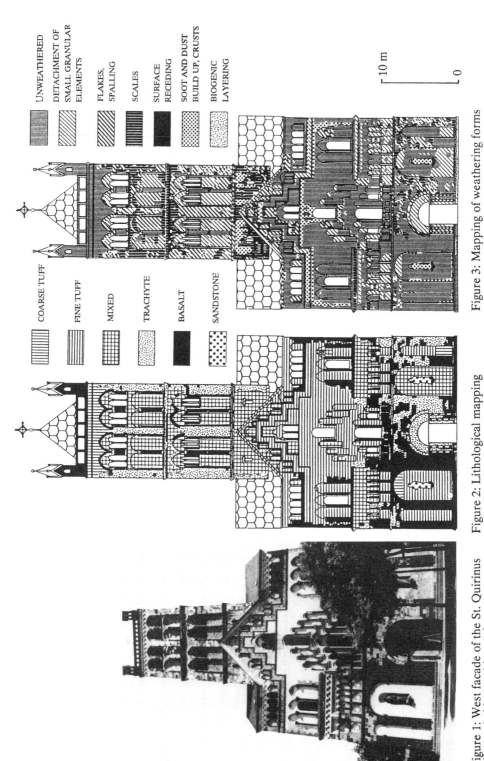

UNWEATHERED

DETACHMENT OF
SMALL GRANULAR
ELEMENTS

FLAKES,
SPALLING

SCALES

SURFACE
RECEDING

SOOT AND DUST
BUILD UP, CRUSTS

BIOGENIC
LAYERING

10 m

0

Figure 3: Mapping of weathering forms

COARSE TUFF

FINE TUFF

MIXED

TRACHYTE

BASALT

SANDSTONE

Figure 2: Lithological mapping

Figure 1: West facade of the St. Quirinus
Minster in Neuss, thirteenth century

ges. A total of twelve clearly distinguishable natural stone varieties were used when building the outer facade, whose surface area is approximately 5000m². They are, for the most part, volcanic rocks such as tuffaceous rocks, trachytes and basaltes. Sandstones and carbonate rocks only occur occasionally.

The distribution of the different natural stones in the west facade of the Minster can be seen in figure 2. According to this, more than 95% of the facade is of volcanic rock, about 65% being constructed out of the Roman and Weibern tuffs, and some 30% being out of trachytes and basaltes. The main surfaces of the facade and some cornices are built out of tuffaceous rocks. Corner pillars, facings, columns, the socle and a part of the cornice are built out of trachyte and basalt. Only the sculptures and individual portal columns of the west facade are of sandstone.

The mapping of the weathering forms on the west facade of the St. Quirinus Minster in Neuss was carried out using the classification scheme shown in table 1. Although this scheme was developed specifically for tuffs, the categories of weathering forms can also be applied to other volcanic rocks used in building.

The distribution of weathering forms observed on the west facade of the Minster is shown in figure 3. Approximately 30% of the stone in the tower is severely or very severely damaged, some 30% of the ashlar shows moderate damage and only about 40% shows slight or no damage. About 80% of the stone in the transept is unweathered or slightly damaged and only about 20% shows moderate or severe damage. The damage to the tower is more severe and more extensive than on the transept.

Moderate and severe damage appear both on facade areas which are constructed out of a mixture of the Roman and Weibern tuffs, and on sections built out of trachytes.

The mapping of lithology and weathering forms makes it possible to produce a precise description of the condition of the building. It therefore is a fundamental requirement for the formulation of effective preservation measures for tuff buildings, as well as for a programme of continuous long-term monitoring of a building.

3.2 Petrographical investigations

The different tuffs taken from buildings and quarries can be petrographically categorised into three tuff types. One important tuff variety of each of these three groups is presented below (table 2):
Tuff types:
-coarse-grained Selbergittuff (tuff variety = Ettringer tuff)
-fine-grained Selbergittuff (tuff variety = Weiberner tuff)
-trachytic tuff (tuff variety = Roman tuff)

Table 2: Petrographical composition and porosity values of the tuff varieties

Tuff-varieties	Ettringer	Weiberner	Roman
	components [vol.-%]		
Matrix	42.4	59.0	50.2
Rock fragments	38.0	17.2	30.5
Pumice	13.6	11.2	13.1
Identifiable mineral composition	6.0	12.6	6.2
Leucite	1.5	2.5	-
Biotite	0.3	1.3	-
Sanidine	0.2	4.5	4.4
Titanaugite	1.8	0.3	0.1
Amphibole	0.9	0.8	1.1
Olivine	0.2	0.1	0.4
Ore	0.8	1.4	0.1
Carbonate	0.3	1.6	-
Others	-	0.1	0.1
Predominant Zeolite Type	Phillipsite	Analcite	Analcite/Chabasite
Bulk density $[g \cdot cm^{-3}]$*	1.66	1.32	1.35
Density $[g \cdot cm^{-3}]$*	2.40	2.43	2.39
Total porosity [vol.-%]*	30.74	45.62	43.44
Mean pore radius [µm]*	0.3847	0.9627	1.3512
Pore surface area $[m^2 \cdot g^{-1}]$**	19.95	8.94	5.32

* mercury porosimetry   ** nitrogen adsorption method

All tuff varieties display a cryptocrystalline to vitreous and porous matrix. This consists for the most part of zeolite, as well as volcanic glass, and to a lesser degree of clay minerals and chlorite. The matrix makes up between 40% and 60% by volume of the tuffs. Microscopically, only 6%-13% by volume of the mineral content can be clearly identified.

The tuff varieties are characterised according to differences in the amounts of xenocrysts they contain, such as mineral fragments, rock fragments and pumice.

The rock fragments mainly originate from clastic sediments, such as graywackes and slates, as well as basalt. The proportion of rock fragments in the tuffs varies between 17% and 38% by volume. The very porous pumice xenocrysts make up between 11% and 14% by volume. In other tuff varieties the pumice xenocrysts have been established as making up 27% by volume. The xenocrysts are embedded in the predominantly glassy matrix. The distribution of the xenocrysts is irregular, the space infilling incomplete. Only in the Ettringer tuff microfissures were detected as a distinguishing feature.

All of these tuffs are notable for the very high total porosity , between 30% and 50% by volume, and a wide spectrum of pore radii, between 0.002um and 1000um. Pores having a radius less than 10um are in the greatest abundance.

The trachytic tuffs have the highest proportion of large pores. The coarse-grained Selbergittuff has the highest proportion of small pores.

3.3 Petrographical investigations within the weathering profile

One of the main weathering forms which commonly occurs with Rhenish tuffs is "scale". This can be subdivided into three zones within the weathering profile using the results from the investigations (figures 4-6):

outer zone - scale
middle zone - zone of detachment
inner zone - unweathered area

Scales are layers of stone, several mm to cm thick, which peel from an underlying zone of detachment. As a rule, the scales differ comparatively little in their texture from unweathered zone of the rock. In contrast, the mm to several cm thick zone of detachment displays severe textural destruction. Textural alterations within the microregion of the weathering profile occur predominantly as variously-sized fissures in the matrix of the tuffs, which run parallel to the outer surface. There is a direct connection between the process of scale development and salt accumulation within the weathering profile. As has been shown by the laboratory investigations on weathered tuffs from the building area, sulphate accumulation, in particular new formation of gypsum, can be detected in the majority of the weathering profiles.

As an example, the alterations in the complex porosity ratios of a scale profile from the Weiberner tuff is explained (figures 5,6):

In the zone of detachment, a clear increase of larger pores emerges. This displacement of the pore radius spectrum in favour of large pores is related to the formation of fractured pores in the zone of detachment, these running parallel and perpendicular to the rock surface (figure 6).

The scale shows less large pores than the unweathered area. This compaction effect results from the near-surface accumulation of gypsum and mirabilite aggregates.

3.4 Laboratory investigations of weathering behaviour

The results of various sets of laboratory experiments on the water absorption and evaporation behaviour of Rhenish tuffs correlate with those of porosity analyses (ch. 3.2).

Experiments on capillary water uptake show that, because of their high proportion of very small pores, coarse-grained Selbergittuffs take considerably longer to achieve water saturation than do the fine-grained Selbergittuffs and trachytic tuffs.

The high proportion of very small pores in the coarse-grained Selbergittuffs enables up to 25% of the pore space to be water-filled through capillary condensation. In comparison, under these conditions only some 15% of the pore space of the other tuffs is water-filled.

Evaporation experiments reveal that the fine-grained Selbergittuffs and trachytic tuffs also show a comparably lower residual water content than the coarse-grained Selbergittuffs. This is due to

WEATHERING PROFILE - WEIBERNER TUFF

SCALE

ZONE OF DETACHMENT

UNWEATHERED

Figure 4: Weathering profile - Weiberner Tuff

SCALE

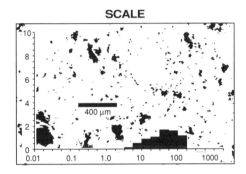

ZONE OF DETACHMENT

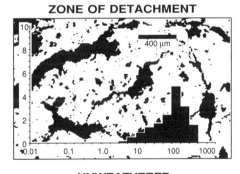

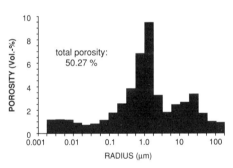

UNWEATHERED

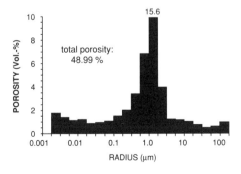

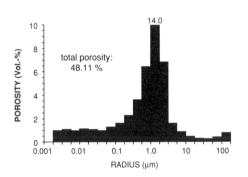

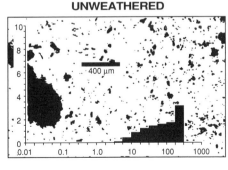

Figure 5: Pore radius distribution
- mercury porosimetry

Figure 6: Computer enhanced images of macropores
and pore radius distribution - image analysis

their low proportion of very small pores.

Provided that there is an adequate supply of water available, tuffs naturally absorb such a large quantity of water in a relatively short time that 85% of the available pore space is filled.

The weathering forms of the tuffs which are developed from a frost-thaw experiment are similar to those observed at the building. In principle, all tuff varieties must be said to be susceptible to frost-thaw stress.

The stability of the tuff types in relation to this stress can be shown as a sequence: trachytic tuffs > fine-grained Selbergittuffs > coarse-grained Selbergittuffs. This sequence is the result of the various pore radii distributions of the three tuff types and the related water absorption and evaporation behaviour.

In a further test, different tuff varieties were exposed to a water-saturated atmosphere containing SO_2. Even after only a short experimental time newly-formed gypsum could be detected on all the tuff varieties. Gypsum formation such as this, which is also to be observed on the buildings, constitutes a chemical alteration within the rock, which contributes to the mechanical destruction of the rock texture.

In salt recrystallization experiments, the tuff varieties do show a very slight variation in their resistance to salt wedging forces, but they are all severely damaged in a relatively short time.

4 CONCLUSIONS

Information about the Rhenish tuffs, their state of damage and the causes of damage can be obtained by combining investigations at the building and in the laboratory. This information can be used as a basis for decisions on preservation measures.

Systematic building investigations permit a mapping of the type, extent and distribution of the weathering forms which occur. The combination of mapping lithology and weathering forms, following defined classification schemes, ensures the reproducibility and high degree of accuracy of the survey.

The petrographically heterogeneous composition of Rhenish tuffs, compared to other natural stones, causes their variable reaction and explains their susceptibility to climatic and anthropological weathering influences.

Petrographical investigations of the tuff within the weathering profile make it possible to register material alterations caused by weathering and to quantify the state of damage. The state of damage of a tuff can be particularly accurately defined by the changes in certain porosity characteristics. A combination of petrographical measuring techniques is used to record the pore space and pore geometry, and thus provides exact material values.

Simulation tests show that weathering factors like frost-thaw stress and salt wedging forces are involved in the destruction of Rhenish tuffs to a very high extent.

REFERENCES

ANONYM (1982): DIN 52104, Teil 1; Prüfung von Naturstein, Frost-Tau-Wechsel Versuch; Beuth Verlag GmbH, Berlin 30

FISCHER, R.V. & SCHMINCKE, H.U. (1984): Pyroclastic Rocks.- S. 89-128, Springer Verlag, Berlin-Heidelberg-New York-Tokyo

FITZNER, B. (1988): Porosity analysis - a method for characterization of building stones in different weathering states.- Proceedings of an International Symposium organized by the Greek, "The Engineering Geology of Ancient Works, Monuments and Historical Sites, Preservation and Protection", vol.4, editors: P.G. Marionos & G.C. Koukis, Balkema-Rotterdam-Brookfield 1988, in press

FITZNER, B. & KOWNATZKI, R.(1989): Object studies on natural stone monuments - methodology and examples.- European Symp. "Science, Technology and European Cultural Heritage, Bologna 1989, in press

FITZNER, B. & KOWNATZKI, R. (1990): Bauwerkskartierung - Schadensaufnahme an Natursteinen. - Der Freiberufliche Restaurator, Heft 4, S.25-40, Hrsg. Deutscher Verband Freiberuflicher Restauratoren, Kiel, in press

FUECHTBAUER, H. & MÜLLER, G. (1970): Sediment-Petrologie, Teil II: Sedimente und Sedimentgesteine.- 726 S., E. Schweizerbart`sche Verlagsbuchhandlung, Stuttgart

KIESOW, G. et al. (1987): Naturwerkstein in der Denkmalpflege; Handbuch für den Steinmetzen und Steinbildhauer, Architekten und Denkmalpfleger.- Hrsg. Berufsbildungswerk des Steinmetz- und Bildhauerhandwerkes e.V., Wiesbaden; Ebner Verlag Ulm

PFEIFFER, L.; KURZE, M. & MATHE, G. (1981): Einführung in die Petrologie. - 632 S., Akademie Verlag, Berlin

Désordres dans les chaussées, dûs aux sulfures des granulats

Damages caused to pavements by the sulphides of the aggregates

Alain Jigorel & Jacques Estéoule
Minéralogie et Géotechnique, INSA, Rennes, France

RESUME: Des études récentes effectuées sur des chaussées à structures légères de la région de Rennes, montrent que les désordres observés en surface : intumescences, fissuration, faïençage, sont imputables à la présence de sulfures altérables dans les granulats non traités de la couche de fondation. Les granulats incriminés sont issus de cornéennes qui renferment des minéralisations en pyrite et marcasite. Ces sulfures donnent par oxydation des sulfates hydratés complexes de la famille de l'halotrichite-pickeringite. Les déformations des enrobés sont dues aux pressions de cristallisation des sulfates. Les conséquences économiques de tels désordres sont importantes car les coûts d'entretien des chaussées sont élevés et leur réfection totale parfois nécessaire pour des raisons de sécurité. A cela, il convient d'ajouter les dégâts causés par les eaux de drainage acides et chargées de sels qui provoquent une oxydation accélérée des réseaux de distribution souterrains.

ABSTRACT: Recent studies about different pavements with light structures in the area of Rennes (FRANCE) show that the disorders of the road-surfaces, such as intumescences and cracks, often result from the presence of weathering sulphides in the unbound aggregates of the base course. These aggregates come from hornfels which contain some pyrite and marcasite. These sulphides give, through oxydation, hydrated sulphates belonging to the halotrichite-pickeringite family. The disorders in the bituminous macadam wearing courses are caused by the cristallisation-pressure of the sulphates. The economic consequences of such disorders are always important because of the high costs of pavement-maintenance and because they sometimes involve a complete refection of the whole road for security-reasons. Besides, one must take into account the damages resulting from such acid and salt waters which always bring about a faster oxydation of the underground supply-networks.

1 INTRODUCTION

Le rôle nocif des sulfures de fer dans les granulats, utilisés pour la confection de bétons, a souvent été mis en évidence (Bérard, Roux & Durand 1975), (Oberholster, Du Toit & Pretorius 1984). Par altération, les sulfures de fer se transforment en sulfates ; ceux-ci provoquent une expansion et une fissuration du béton qui peut, à l'extrême, entraîner la destruction totale des ouvrages (Vasquez & Toral 1984).

Des études ont également montré que la présence de pyrite dans les matériaux pouvait faire chuter d'une manière importante les performances mécaniques des bétons de terre stabilisés à la chaux (Toubeau 1987). Des observations récentes, effec-

tuées en Bretagne, dans la région de Rennes, montrent que la présence de sulfures dans les granulats issus de cornéennes et utilisés sans traitement dans les chaussées à structures légères, est à l'origine des désordres observés dans la couche de surface. A notre connaissance de tels phénomènes, qui peuvent avoir des conséquences économiques importantes, n'ont à ce jour, jamais été signalés.

2 ASPECT DES DESORDRES DANS LES CHAUSSEES

Les chaussées étudiées, qui présentent des désordres en relation avec la présence de sulfures dans les granulats, sont situées dans la région de Rennes. Dans tous les cas, ces chaussées ont une structure lé-

Fig. 1 : Intumescence due à l'oxydation des sulfures des granulats de la chaussée

est constituée d'un bicouche de gravillons, la dégradation de la chaussée se traduit par l'apparition de nombreux nids de poule. En période de beau temps, ils sont tapissés, par une poudre blanchâtre formée des produits d'altération des sulfures contenus dans les granulats.

Les désordres ont donc un aspect banal et en l'absence d'étude approfondie, leur origine a pu être attribuée dans le passé à divers facteurs : présence de racines dans le corps de chaussée, caractéristiques insuffisantes des granulats ou du sol support, mise en oeuvre défectueuse, trafic trop intense... mais jamais la nature minéralogique des granulats n'avait été évoquée.

gère et leur couche de fondation est constituée de granulats non traités issus de cornéennes. Les ouvrages concernés sont essentiellement des cours d'école, des plateaux sportifs, des trottoirs et des routes de desserte dans les lotissements.

L'aspect des désordres varie en fonction de la structure de la chaussée et de la nature du trafic. En l'absence de trafic de véhicules, il se développe dans l'enrobé, des intumescences semblables à de petits volcans, associées à une fissuration plus ou moins dense (fig. 1). Sous un trafic régulier de véhicules, la croissance d'intumescences bien individualisées, est entravée: la couche de roulement subit une déformation sous forme d'ondulation légère, associée à un véritable faïençage (fig. 2). Lorsque la couche de roulement

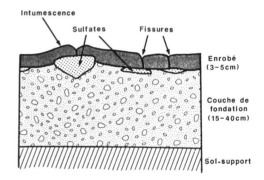

Fig. 3 : Coupe schématique d'une chaussée

L'étude des sondages réalisés dans les chaussées, a permis de montrer que les désordres observés en surface étaient toujours dûs à l'altération des sulfures présents dans les granulats de la couche de fondation. La coupe schématique présentée sur la figure 3 permet de bien visualiser ces phénomènes.

Toutes les chaussées étudiées montrent une structure voisine. La couche de surface est généralement constituée d'un enrobé dense de 3 à 5 cm d'épaisseur ou parfois d'un bicouche de gravillons. Elle repose sur une couche de fondation formée selon les cas, d'une grave 0/31,5 mm ou d'un macadam. L'épaisseur de la couche de fondation varie de 15 à 40 cm selon la

Fig. 2 : Faïençage d'un enrobé de surface supportant un faible trafic

destination de la chaussée. Le sol-support a une texture limono-argileuse, qui correspond à l'horizon d'altération des schistes briovériens du bassin de Rennes.

Dans tous les cas, des concrétions minérales blanchâtres ou jaunâtres observées sous l'enrobé de surface , forment un revêtement quasi continu, mais d'épaisseur variable. Elles sont plus volumineuses au niveau des intumescences et des fissures où elles occupent totalement l'espace ménagé par la déformation de l'enrobé. Cette disposition montre nettement que les concrétions sont directement responsables des déformations observées.

Fig. 4 : Plaque d'enrobé retournée montrant les concrétions de sulfates →

Tous les cailloux de la couche de fondation présentent des revêtements d'altération de couleur jaune-orangé à rouille. Lorsqu'ils sont bien cristallisés, les sulfates qui enrobent les cailloux ont la forme de fines aiguilles jaunâtres très longues, bien visibles à l'oeil nu et qui hérissent la surface des granulats. Dans les chaussées particulièrement bien drainées, dont la couche de fondation est formée d'un macadam, nous avons pu remarquer que les concrétions minérales blanchâtres les plus remarquables par leur épaisseur étaient situées au sommet de cailloux très altérés, devenus totalement pulvérulents. Dans ce cas, les concrétions blanchâtres formées de sulfates sont directement issues de l'altération des cailloux sous-jacents.

3 NATURE PETROGRAPHIQUE DES GRANULATS

Tous les granulats qui constituent le corps de chaussée des différents sites étudiés, sont du point de vue pétrographique, des cornéennes au sens large. Il s'agit de roches issues d'un métamorphisme de contact dans la séquence pélitique; ce sont, selon le degré de métamorphisme, soit des cornéennes véritables, soit des schistes tachetés, soit enfin des schistes à peine métamorphisés. Les matériaux incriminés ont été fournis par plusieurs carrières de la région de Rennes qui exploitent des roches situées dans l'auréole de métamorphisme de granites cadomiens. La mise en place de ces massifs granitiques intrusifs a provoqué une métasomatose des schistes briovériens encaissants, qui s'est traduite par un enrichissement de la roche en sulfures.

L'examen au microscope pétrographique montre que les sulfures sont concentrés dans des microfractures subparallèles au plan de schistosité de la roche. Ces microfractures minéralisées ont selon les échantillons, un espacement tantôt millimétrique, tantôt centimétrique. Il faut de plus, ajouter à ces sulfures d'aspect lamellaire, des petits amas cubiques et sphériques disséminés dans la masse de la roche.

L'examen aux rayons X a montré que les amas sulfurés présents dans les fissures minéralisées sont essentiellement constitués de pyrite martiale associée à une faible quantité de marcasite (raie n°2).

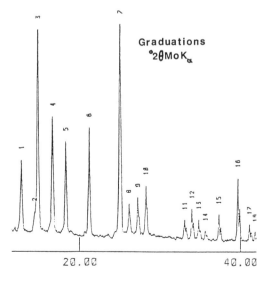

Graduations °2θMoK$_\alpha$

20.00   40.00

Fig. 5 : Diagramme X des pyrites

4 CARACTERISATION DES CONCRETIONS MINE-RALES NACREES PRESENTES SOUS LA COUCHE DE SURFACE DES CHAUSSEES

Les concrétions minérales nacrées les plus importantes, ont à l'oeil nu et au micro-scope, un aspect fibreux nettement domi-nant. L'examen au Microscope Electronique à Balayage montre que les concrétions sont constituées de fines aiguilles enchevê-trées qui sont souvent assemblées entre-elles pour former des faisceaux fibreux (fig. 6). La fibre élémentaire semble toujours avoir un diamètre voisin de 3 microns.

Fig. 6 : Sulfates d'aspect fibreux

De plus, juxtaposés ou intimement mêlés à ces fibres, on observe soit des cristaux bien formés en tablettes parallélépipédi-ques allongées, ou en lattes d'apparence monoclinique, soit des masses mal cristal-lisées de forme trapue ou d'aspect fondu.

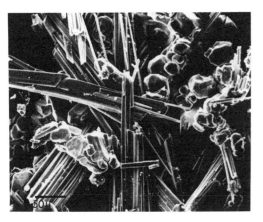

Fig. 7 : Sulfates à l'aspect de sucre

Les concrétions qui sont constituées de tels cristaux, présentent un aspect de "sucre" (fig. 7).

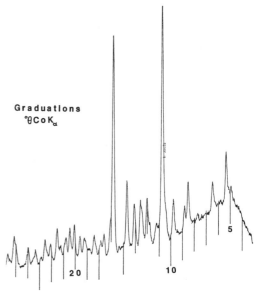

Fig. 8 : Diagramme X des sulfates

L'examen de ces produits d'altération par diffraction des rayons X indique qu'ils sont bien cristallisés et appar-tiennent aux sulfates doubles de la famil-le pickeringite-halotrichite comme le mon-tre le diffractogramme de la figure 8.

5 ORIGINE DES DESORDRES DE LA COUCHE DE SURFACE DES CHAUSSEES

Ce sont les réactions d'oxydation de la pyrite contenue dans les granulats de la couche de fondation qui sont à l'origine des désordres de la couche de surface des chaussées. La chaîne de réaction d'oxyda-tion des sulfures de fer est bien connue. Elle fournit globalement des sulfates, des hydroxydes ferriques et de l'acide sulfu-rique. Si la pyrite est réputée relative-ment stable, elle peut toutefois dans certaines conditions s'altérer en présence d'air humide. Ainsi nous avons pu observer que des gravillons de cornéennes renfermant des amas sulfurés et conservés au laboratoire dans une atmosphère humide, se recouvrent en quelques jours d'efflorescences blanchâtres qui correspondent à des sulfates hydratés : l'air humide présent dans les chaussées suffit donc à déclencher le processus d'oxydation des sulfures. Il est vraisemblable que la marcasite qui a la réputation d'être moins stable que la pyrite s'altère la première. Les produits de son altération et en particulier l'acide sulfurique libéré favorisent à leur tour l'oxydation de la pyrite. Lorsque le processus est engagé, il s'accélère rapidement et fournit en fin de réaction des hydroxydes ferriques et de l'acide sulfurique. Les valeurs de pH, voisines de 3, mesurées dans les granulats témoignent de cette acidité sulfurique.

Les sulfates ferreux et l'acidité sulfurique formés par oxydation des sulfures peuvent être partiellement évacués avec les eaux de drainage de la chaussée. Les sulfates présents en solution dans les eaux interstitielles du corps de chaussée, vont pendant les périodes climatiques favorables, se concentrer puis cristalliser à la surface des granulats. L'interface enrobé-couche de fondation constitue la zone préférentielle de concentration des solutions de sulfates. A la belle saison, l'enrobé ou le bicouche de surface subit des élévations importantes de température, entraînant ainsi par évaporation, une concentration des solutions. La pression de cristallisation est suffisante pour soulever et fissurer un enrobé de faible épaisseur. Les fissures ainsi formées contribuent à drainer vers le haut, par évaporation, les solutions sous-jacentes. Les premières concrétions formées et situées au niveau des fissures, sont ensuite nourries d'une manière préférentielle, par les eaux intersticielles du corps de chaussée. Leur croissance progressive entraîne la formation d'intumescences bien individualisées. Les chaussées qui supportent un trafic de véhicules subissent une déformation accompagnée d'un faïençage de la couche de roulement. La vitesse de dégradation des chaussées est étroitement liée à l'importance du trafic.

6 CONCLUSION

Les désordres qui apparaissent au niveau de la couche de surface des chaussées à structures légères sont dûs à l'altération des sulfures présents dans les granulats de la couche de fondation. Ce sont les contraintes de cristallisation des sulfates produits par oxydation des sulfures qui engendrent la déformation des enrobés et la dégradation des bicouches.

En raison de leur aspect banal, de tels désordres sont souvent attribués à de multiples causes, sans rapport avec la présence de sulfures dans les granulats. Bien que peu spectaculaires, ils entraînent des coûts élevés de maintenance des voiries et des réseaux et obligent souvent pour des raisons de sécurité à la réfection totale des ouvrages. A la différence de ce qui s'observe dans les bétons, l'acidité sulfurique produite par l'oxydation des sulfures n'est pas neutralisée, dans les granulats naturels, par la basicité du ciment. De ce fait, le processus d'altération et les désordres qui en découlent apparaissent rapidement avec une grande intensité, même lorsque la teneur en sulfures des granulats est relativement peu élevée.

En conclusion, il apparaît que les restrictions relatives aux teneurs en sulfates et sulfures des granulats naturels, fixées par les normes françaises pour les bétons hydrauliques, doivent être adaptées aux granulats non traités, utilisés pour la construction des chaussées à structures légères et des ouvrages comportant des structures métalliques au contact de ces granulats.

REFERENCES

Bérard J., Roux R. & Durand M. (1975) : "Performance of concrete containing a variety of black shale". Can. J. Civ. Eng. 2,58.
Toubeau P. (1987) : Optimisation des caractéristiques géotechniques et minéralogiques dans le traitement des sols. Application aux bétons de terre. Thèse de Doctorat de l'Université de Paris 6.
Oberholster R.E., Du Toit P. & Pretorius J.L. (1984) : Deterioration of concrete containing a carbonaceous sulphide-bearing aggregate. National Building Research Institute, C.S.I.R., Pretoria, South Africa.
Vasquez E., Toral T. (1984) : Influence des sulfures de fer des granulats du Maresme (Barcelone) sur les bétons. C. R. du Symp. Int. sur les granulats, Nice. Bull. de l'A.I.G.I., n° 30, p 297-300.

The non-destructive investigation of a pavement residual strength in a tropical savannah environment of Nigeria

L'investigation non-destructive d'une force résiduelle d'un pavement dans l'environnement d'une savane tropicale du Nigeria

N.S. Nwogute
Department of Geological Sciences, Anambra State University of Technology, Anambra State, Nigeria

ABSTRACT: The non-destructive in-situ investigation of the residual strength of a 126-6 km non-flexible pavement in a developing tropical Savannah environment is discussed. The problems associated with this technique of investigation in this environment are highlighted. It is found that (i) the geology, and (ii) pavement course material type, affect the measured pavement residual strength and (iii) a properly articulated mobilisation process is essential for a successful testing programme in this environment.

RESUME: L'investigation en place pas destructif d'une pavement inflexible de 126.6 km dans l'environment d'une savana tropicale en voie du developpement est discuté. Les problèmes relatifs `a cette technique d'investigation dans cet environment sont indiqués. On a trouvé que (i) la géologie et (ii) le type de material du pavé touche la force residuelle mesurée et (iii) un procédé d'une mobilisation bien articule est épreuve du succès dans cet environment.

KEY WORDS: Pavement, deflexion beam, mobilisation, soil cement, base, residual strength, black-cotton soil, crushed rock

INTRODUCTION

Gombe and Biu are two towns joined by a 126.6 km road and located within a fertile Sudan Savannah region of Nigeria. Many food processing factories are located along this road and in these towns. When the Federal government of Nigeria decided to boost the agricultural produc-- tion of this area, it anticipated increased traffic of trucks and other commercial vehicles plying the road, hence it was decided to strengthen this existing non-flexible road pavement.

While modern new pavement design techniques are based on standard axle loadings derived from counts and measure- ments of the numbers of vehicles that will use the road over the chosen life, existing pavements have a residual stre- ngth which must be taken into account for the design of strengthening overlays. There, therefore, arose the need for a rapid non-destructive and cost-effective method of assessing the current pavement residual strength. This resi- dual strength can be measured by subje- cting the pavements to a standard axle loading and measuring the deflection of the surface. The Beckenham Deflexion Beam was used for this measurement while the pavement surface condition was assessed both visually and the straight edge and wedge method. This paper describes the methods used in data acquisition, the associated problems and findings.

2 GEOLOGY AND SOILS AND PAVEMENT PERFOR-MANCE

The geology of the area between Gombe and Biu is complex. The investigated road crosses Basement metamorphic crystalline rocks, continental and marine sediments, alluvium, and finally at Biu, it crosses a low plateau built up by a series of thin lava flows (fig. 1). The superficial deposits of these formations comprise mainly leached sandy soils formed in-situ from sandstone or weathered granite.

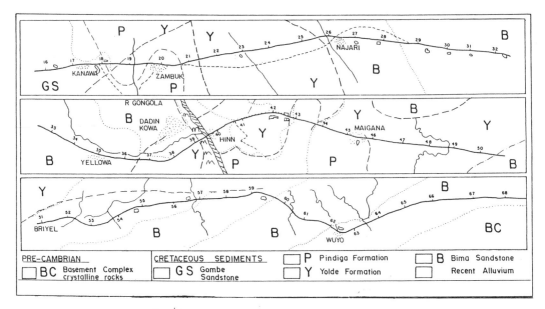

PRE-CAMBRIAN
☐ BC Basement Complex crystalline rocks

CRETACEOUS SEDIMENTS
☐ GS Gombe Sandstone

☐ P Pindiga Formation
☐ Y Yolde Formation

☐ B Bima Sandstone
☐ Recent Alluvium

Fig. I GEOLOGY OF A SECTION OF THE GOMBE-BIU ROAD

However, the Pindiga and Yolde shale formations weather to form thick strata of expansive highly montmorillonitic clay-soil called 'black cotton' soil.

The Gombe and Bima sandstones and the rocks of the Basement complex have not given rise to soils that appear to have affected the pavement performance. However the high swelling and shrinkage properties of the 'black cotton' soil and the accompanying suction forces lead to horizontal and vertical movement of the subgrade and the embankment and pavement in turn. The pavement has performed differently throughout its entire length and this reflects the different construction methods adopted and available geological materials.

3 MOBILISATION AND PROBLEMS

The Savannah region of Nigeria has extensive land areas. Farmlands stretch over several tens of kilometres to separate urban areas linked by a network of roads. The Gombe-Biu non-flexible road pavement is 126.6 km long while Gombe, the starting point of the investigations is about 500 kilometres away from the work base of the investigation crew.

Nigeria is a member of OPEC but the inefficient distributive system of petroleum products causes their scarcity frequently. These facts, together with the demands and nature of this investigation itself, were considered in articulating a mobilisation programme for the in-situ investigations. Aspects considered were Personnel Accommodation, Transport and Fuel, Finance and Health.

3.1 Personnel

Major site investigation work is a team activity (Robb, 1982). The team comprised the leader who recorded the results, ensured accuracy of tests, carried out the surface condition rating, analysed results and generally managed the team;

two technicians that operated the beams;
two drivers for the test Bedford lorry
and a landcruiser used for location of
test points and staff movement; a mechanic who took care of the vehicle,fuel
supply and spare parts; and two casual
workers who helped with tape measurements,
painting of test locations, and traffic
warning system. It is important to evolve
a systematic working approach.

3.2 Accommodation

The long distance and the rural environment of the test stretch together with
the character of tests influenced the
choice of camp sites at the few available
service centres and towns (Gombe, Dandin
Kowa, Kwaya Tera and Biu) that are about
40 km apart and having water supply,func-
tional dispensaries to supplement the
teams First Aid Kit and a market. To
ensure safety of staff and equipments, the
village heads were kindly requested to
provide the accommodation.

3.3 Finance

To avoid disruption of work enough money
to cover all costs had to be taken to
site or a financing relay system evolved
in view of the great distance to the base
office. Prudent financial management was
very necessary.

3.4 Problems

Since the deflection beam was devised in
1953 and reported by Lee and Croney
(1963); and Salt (1968) a great deal of
deflection tests have been performed in
the United Kingdom, the United States, and
over very short distances (generally less
than 10km) in Malysia and Zambia, over
bituminous premix pavement surfaces. The
Gombe-Biu pavement test is the first to
be performed in Africa, South of the
Sahara and over the longest stretch of
non-bituminous surface in the world.
The problems encountered, therefore,
included lack of previous experience,
by the indigenous testing staff, and
pertinent literature about testing over
great distances of this type of pavement
and in the Savannah environment. Most of
these problems were solved by the properly
articulated mobilisation process for the
investigations and adaptations from
experiences in the temperate regions.

4. BEAM SURVEY PROCEDURE

Deflection measurement for pavement
evaluation is based on the principle that
when a loaded wheel passes over a
pavement, a small transient vertical
depression or deflection or deflexion of
the pavement surface occurs, the magnitude
of which is a function of the wheel load,
the area of contact between the tyre and
the road, the speed of the wheel, and the
stress-strain characteristics and thickness of the various pavement layers and
the subgrade (Smith 1973). IF, therefore,
a standard wheel load, tyre size and
measurement method are adopted, the
resulting surface deflection enables
comparison between the stiffness of
different pavements and their changes
over time.
The techniques adopted for the deflection survey of the Gombe-Biu road are
adapted from the British Transport and
Road Research Laboratory (TRRL)'s LR 525
of Smith (1973)
A 5-ton Bedford lorry with two twin rear
tyres of sizes 7.50 x 20, each pair having
a spacing of 45 mm and inflated to 585
KN/m², is loaded to 3175 kg on each pair
of twin rear wheels. The lorry is
equipped with pointers, 1.3m in front of
each rear wheel and are bent to be within
the gap between the rear wheels and lorry
chasis.

4.1 The in-situ tests

The tests were carried out at 100mm intervals as recommended by LR 525. First
the test locations are measured out and
small crosses are painted in the wheel
paths of both lanes. The lorry is then
driven at creep speed (50 cm/Sec.) so
that pointers are over the small painted
crosses. The technicians place the beams
between the rear lorry wheels clear of
tyres with beam ends directly under the
pointers, zero the dial gauges on
each beam and while tapping it with
fingers and vibrating it continuously
(Fig 2) the lorry is driven slowly
forward until the rear wheels are about
1.2m past the end of the beams.

Fig. 2 Deflexion beam testing

Fig. 3 Road condition test

The maximum and final dial gauge readings
are recorded by the leader. If this is
properly done, the highest dial gauge
readings of the two wheel tracks is taken
as the reading for that lane at that test
location. False readings may be caused by
the lorry wheels touching the beam,
friction in the pivot or dial gauge storm.
Ten deflexions are summed up in each
wheel track and the mean taken. If the
maximum and mean values differ by more
than the mean divided by three, then
additional tests are carried out 50m each
side of the maximum or minimum reading
point.

As the lorry is driven to the next marked
test location the pavement surface condition test for that measurement point is
taken. This is done by placing an allumi-
nium straight edge across the lane and the
wedge is used to measure the deformation,
rotting, and the degree of cracking (fig3).
It was observed, for this investigation,
that the worst surface condition rating was
measured in the near side wheel track.

5 PAVEMENT CONSTRUCTION MATERIALS AND
   MEASURED DEFLEXIONS

The pavement was constructed with three
types of base materials, namely naturally
occuring materials, crushed stone and soil
cement, with variations in thickness
(fig. 4)

5.1 Naturally occuring material base

This construction expedient was adopted
in the stretches indicated on fig. 4 which
also lie mostly over the weak Gombe and
Bima sandstones. The pavement base course
material in these areas is a mixture of
locally won plastic and non-plastic soils
laid 150 mm thick over 150 mm subbase.
The exception to this is chainage 19904
to 26341 where the subbase is 300 mm
presumably because of the occurence here
of weak`black-cotton´soil subgrade. A
single coat surface dressing was applied
on all naturally occuring base.

The in-situ tests in these base course
type areas show a highly flexured
pavement condition, lowest and more
consistent deflexions and therefore,
higher residual strength than the adjoining differently constructed pavements.-
Further, the extent of deflexions increased immediately this base type changed
from an average of about 40 x 10² mm,
to the much variability-of the crushed
stone base. This result is unexpected
and shows that these sections were much
stronger than expected.

5.2 Cement -stabilised base

The pertinent areas as shown in fig. 4 lie
mainly over the Pindiga and Yolde shale
formations. The pavement base is 125mm
thick cement stabilised naturally
occuring material and a 150mm unstabilised
subbase with a single coat surfacing.
The investigations showed that the pavement

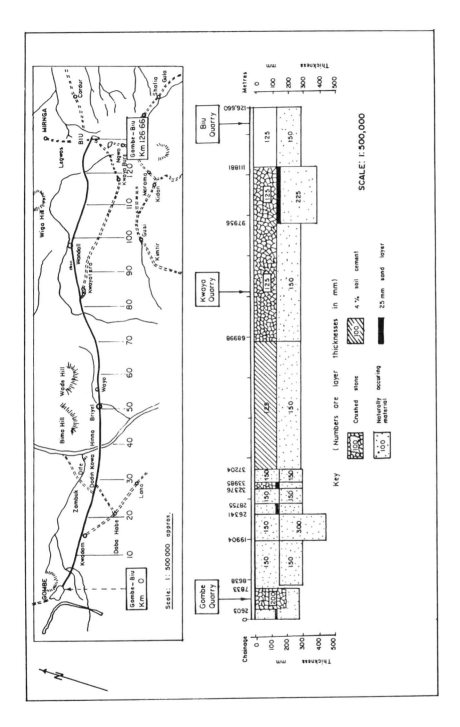

Fig. 4   GOMBE–BIU ROAD PAVEMENT LOCATION AND DETAILS

3199

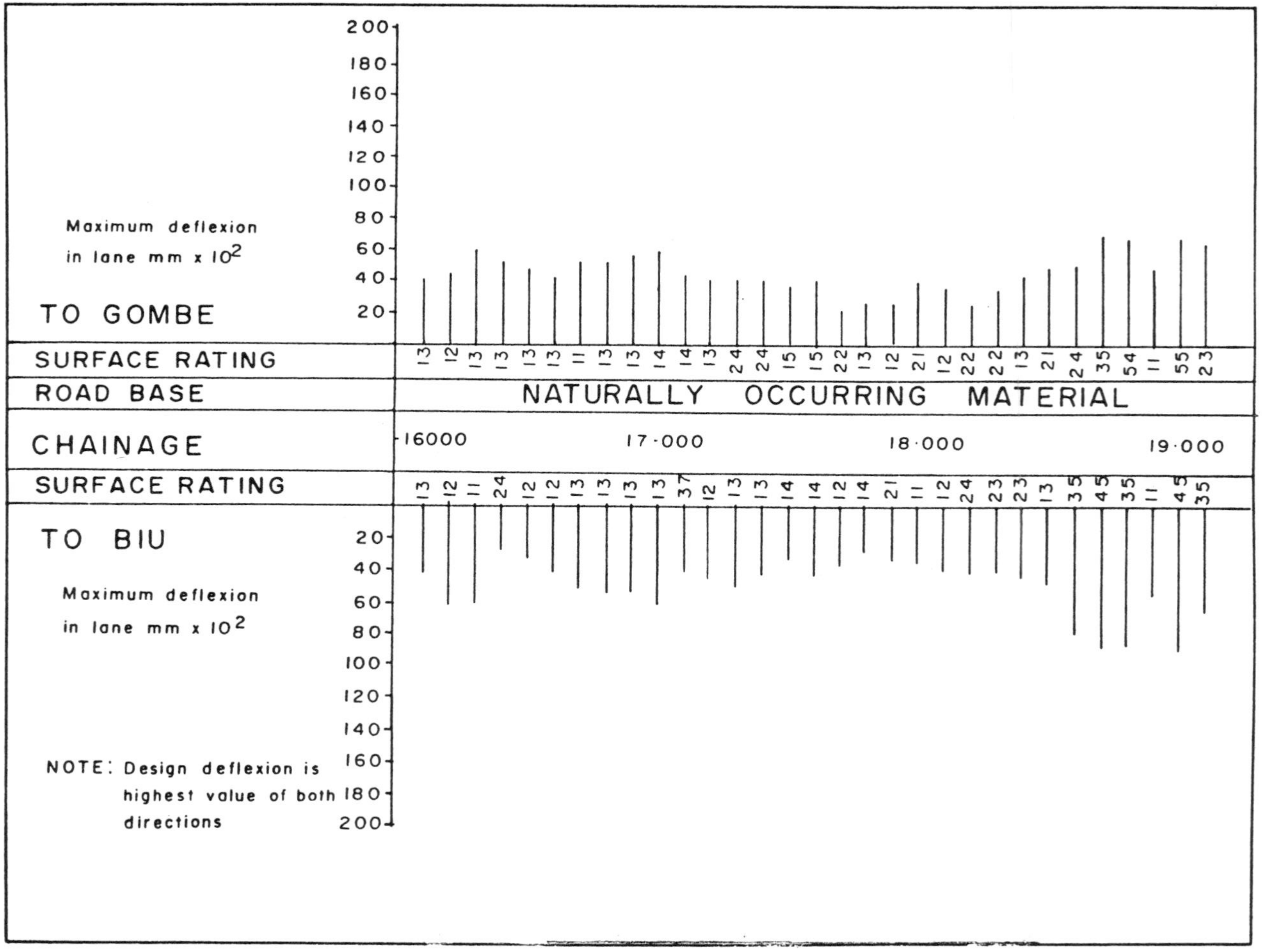

Fig. 5A   DEFLECTION READINGS AND SURFACE
RATINGS ON GOMBE – BIU ROAD

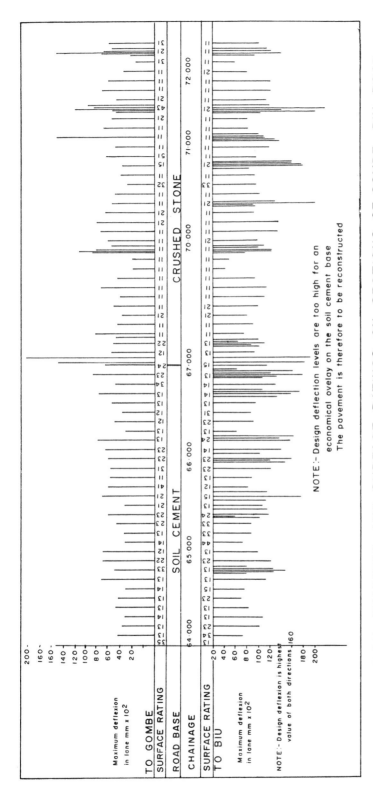

Fig. 5B DEFLECTION READINGS AND SURFACE RATINGS ON SECTIONS OF GOMBE – BIU ROAD

was in very poor surface condition exhibiting extensive parallel longitudinal and lesser transverse cracking thus splitting the base into numerous roughly rectangular 'blocks'. The deflexions showed greater variability than over the naturally occuring base stretches, and no value can be taken as typical (fig. 5). Deflexions are very high for a base of this nature though the subgrade soil is highly plastic. The main cause of this low residual strength here could be attributed to the thin base course.

5.3 Crushed Stone Base

The stretches lie on the Gombe and Bima sandstones areas. This 125 mm thick base was laid by the dry process over a 25 mm sand blanket with a 150mm subbase. A two-coat surface dressing was applied on all crushed stone base. Prevalent potholes extending through the full thickness of the base indicated a lack of stability of the stones due perhaps to removal of fines and thus increased porosity and a possible inadequate protection by the surfacing. Deflexions and surface here show much greater variability than even the soil cement base.

It is, therefore, the base material type and construction method, rather than the ground geology that have direct effect on the measured residual strength of a pavement. It was further observed that (1) a high density surface dressing aggregate produces low value deflexions and thus high residual strength and vice versa (2) a low density surface dressing aggregate may give high residual strength if the area is well-drained; in a low embankment; and underlain by well-graded base course materials.

6 CONCLUSION

The residual strength of a surface-dressed road pavement in a tropical savannah environment has been investigated using the Beckenham Deflexion Beam with the aim of acquiring information for designing strengthening overlays. The investigation has shown that:
(1) That soil cement base developes low residual strength over time, a weakness that renders it little stronger than subbase.
(2) The road pavement section constructed with naturally occuring base materials, even over weekest subgrades, have higher residual strength than expected.

(3) Crushed stone base areas have the lowest residual strength and exhibit the worst surface condition.
(4) The ground geology of the route corridor has no direct effect on the residual strength developed by a pavement.
(5) The base course material type and construction method directly affect the residual strength developed by a pavement.

The investigation has achieved its aim in that having obtained the residual strength of pavement sections, strengthening overlays or reconstruction designs can be undertaken more confidently, and it is therefore, recommended for road condition monitoring in developing countries where there is scarce development and redevelopment funds. The result of these investigations can be a useful guide for road pavement design and construction in similar environments

ACKNOWLEDGEMENT: The author is very grateful to Messers Ove Arup & Partners (Nig.) and the Federal Government of Nigeria for funding this investigation, and particularly to Ove Arup (Nig.) for kindly granting the permission to publish this material.

REFERENCES

Highway Research BOard (1953). The WASHO Test, Part 1. Design, Construction and testing procedures. Highway Research Board Special Report No. 18, Washington.
Lee, A.R. & D. Croney (1963). British full-scale pavement design experiments. Proc. Int. Conf. Structural Design of Asphalt Pavements at Univ. Michigan, Ann Arbor, USA 1962, An Arbor (University of Michigan)
Robb, A.D. (1982) ICE Works construction guides: Site investigation. Thomas Telford Ltd., London.
Salt, G.F. (1968). Recent full-scale Flexible pavement design experiments in Britain. Proc. 2nd Int. Conf. Struc. Design of Asphalt Pavements at Univ. Michigan, Ann Arbor, USA 1967, Ann Arbor (University of Michigan)
Smith, H.R. (1973). A deflection survey technique for pavement evaluation in developing countries, Department of the Environment TRRL Report 525, Crowthone, Berkshire.

Strength formation of stabilized lagooned ashes

Formation de la résistance de résidus stabilisés

Kristio Todorov & Dimcho Evstatiev
Geotechnical Laboratory of the Bulgarian Academy of Sciences, Sofia, Bulgaria

ABSTRACT: Investigations on the peculiarities in the strength formation over a long period of time of lagooned ash stabilized with 3-15% Portland cement are described. An attempt has been made to elucidate the role of the cement hydration processes and the pozzolanic reactions for the strength development. Possibilities have been found for improving the strength of stabilized lagooned ashes without increasing the binding substance content.

RESUME: Sont décrites les études sur les particularités dans la formation de la résistance de résidus des bassins des centrales thermoélectriques, stabilisés par 3-15% de ciment. Les études sont effectuées dans un période de temps continué. Le rôle de la hydratation du ciment et de la réaction pouzzolanique est éclairci. Sont montrées les possibilités pour l'accroissement de la résistance de résidus stabilisés sans augmentation du liant hydraulique.

INTRODUCTION

Lagooned ashes from thermo-electric power plants are transported hydraulically to precipitation basins, which occupy large areas of utilizable land. This disturbs the balance in Nature and exerts unfavourable influence on the environment (Todorov 1989). The construction of precipitation dumps, the removal and the safe storage of waste products consume resources, labour and energy. A successful decision of the ecological and economical problems is the application of such technogenous soils as road construction materials. The coarser lagooned ashes after compaction possess satisfactory strength and deformation properties and are suitable for embankments and bases of roads of little traffic, streets, parkings, grain storage grounds, side-walks, etc. (Todorov et al. 1986). After stabilization with 5-10% of Portland cement the lagooned ashes have inconfined compressive strength R_c above 4,0 MPa and modulus of elasticity E above 1000 MPa and are suitable for all kinds of road bases (Evstatiev & Todorov 1990). Unfortunately they are in comparatively smaller quantities and very often access to them in the dumps is obstructed. The predominant finegrained lagooned ashes after compaction find restricted application in road construction because of their sharply decreased strength and bearing capacity under conditions of high water level and because of the danger of liquefaction under dynamic loading. Therefore they are more often stabilized with cement than coarser lagooned ashes. The compacted mixtures of fine-grained lagooned ashes with 5-15% of Portland cement possess R_c=1,0-2,0 MPa and E=250-500 MPa (Evstatiev & Todorov 1990). They were used in the construction of street bases. In order to obtain stabilized lagooned ashes with strength and deformation properties according to the specific requirements, it is necessary to know the structure formation processes. The study of these processes is a hard task because of the complex composition of the interacting components and of the continuous changes in the newly formed mineral phases with the elapse of time. They are not fully elucidated even for the comparatively well studied fly ash-cement, i.e. for the mixtures with dominating cement content (Kokubu 1968, Diamond 1986). Still more scarce is the information concerning the processes, taking place in low cement content mixtures (Montgomery et al. 1981). In the present paper have been described the peculiarities in strength formation of lagooned ashes stabilized with up to 15% of Portland cement within a long period of time. The main objective is to stimulate the wider application of these technogenous soils for constructional purposes.

INVESTIGATION METHODS

Compacted mixtures of lagooned ashes from different kinds of coal and Portland cement were cured under moisture and room temperature conditions. After different periods of time the samples were subjected to the following analyses: x-ray diffraction, thermal, electron microscopy and chemical investigations. X-ray diffraction tests were accomplished with monochromatic Co k-α radiation. Differential thermal analysis was performed using a standard of Al_2O_3, the heating rate being $10^{\circ}C/min$ in a temperature range of 0° to $1000^{\circ}C$ and sensitivity 200. Electron microscope investigations were carried out on scaning electron microscope (SEM) attached to a microsound "Jeol superprobe - 733". Freshly broken surfaces, covered with a 20 nm thick gold film were observed. Chemical investigations cover defining of free calcium oxide by extraction with a mixture of acetacetic ester and isobutyl alcohol and pH value of transparent filtrate, separated from a 1:5 (solid:liquid) suspension after 15 minutes of shaking. The above mentioned investigations were performed on preliminary desiccated samples in an vacuum oven below $30^{\circ}C$.

For each of the corresponding curing ages of the investigated mixtures the unconfined compressive strength R_c was determined on cylindrical samples with diameter and height of 0,05 m, the speed of deformation being 0,5 mm/min. The samples were prepared by compaction at optimum moisture content w_o to the laboratory established maximum dry density ρ_{max}. The samples were water soaked for 24 h before testing.

TEST RESULTS

In the mixtures of lagooned ash, Portland cement and water intensive hydration of the cement clinker starts initially. On the x-ray diffractograms in the earlier stages the lines of hydrated cement can be traced together with the lines of lagooned eash and non-hydrated cement. In this case hydration proceeds more vigorously than in a pure cement paste. The intensity of the characteristic for the initial clinker minerals x-ray lines decreases more tangibly in the course of time and some of them completely disappear already on the 90th day. However, the presence of x-ray lines typical for non-hydrated cement minerals after 3 months of curing confirms the fact, that cement hydration is not completed during this period and that even after long curing alongside the newly formed mineral phases can be observed also unchanged cement grains. In the first weeks as a result of cement hydration ettringite and numerous needle-like formations resembling the I type C-S-H according to the classification of Diamond (1976) are registered (Fig. 1). The lagooned ash particles are germ nuclei and a considerable part of them are covered by a layer of these hydration products (Fig. 2). In the investigated low cement mixtures the lagooned ash particles are in close proximity to each other. Since cement hydration proceeds predominantly around them the outgrowing needle-like formations are interwoven and this intertwining contributes to the strength growth of the formed structure in the earlier stages (Fig. 3). The SEM observations proved also that the fracture surface never passes through the intertwined needles. They are strong and neither of them has been encountered to be broken. Most often the fracture surface passes between the lagooned ash particles and the needle-like phases attached to them, which proves that the strength of this bond is comparatively small in the early stages of hardening (Fig. 4).

Secondary pozzolanic interaction starts during the initial stages too and the amount of free calcium oxide rapidly decreases. The diffractograms for the age of seven days show the most typical lime x-ray line at 0.262 nm. On the 28th day this line is less intensive while on the 90th day it is almost unnoticeable. The decrease of lime content is observed in DTA investigations too. The endothermal effect at $500^{\circ}C$ due to $Ca(OH)_2$ decomposition is reduced with ageing and the 90th day it is practically absent. The pH value of pure cement paste exceeds the pH value of a concentrated lime solution and does not change in time. In the investigated mixtures as a result of the interaction between lime and lagooned ash particles pH value are lower already on the first day and continuously decrease during the investigated period of 3 months. Lime diminution in the course of time is probably due not only to its inclusion in early pozzolanic reactions but also to carbonization. The simultaneous determination of the free calcium oxide content in the investigated mixtures and in cured and tested under the same conditions cement paste provides evidence that lime reduction can not be attributed to carbonization alone. It was established the cement-lagooned ash mixture contains but a part of

Figure 1. Needle-like silicate hydrates in stabilized lagooned ash after 28 days
Magn.: x 2000

Figure 3. Interwoven needle-like formations in stabilized lagooned ash after 28 days
Magn.: x 10 000

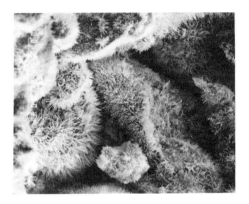

Figure 2. Products of cement hydration on lagooned ash grains
Magn.: x 4000

Figure 4. Surface of breaking in stabilized lagooned ash after six months
Magn.: x 10 000

that quantity of free calcium oxide which has been expected during the normal setting of Portland cement and at absence of physico-chemical and chemical interaction between cement and lagooned ash. Pozzolanic interaction is convincingly substantiated by electron microscopy analysis. Globular particles with strongly eroded surfaces are observed on the micrographs at later curing periods (Fig. 5). The errosion observed, ussually affects the smaller spherical grains and is obviously caused by pozzolanic reactions. SEM investigations confirmed the opinion of Halse et al. (1984), that as a consequence of the pozzolanic interaction the surfaces of many lagooned ash spheres are eroded and the less active lower layers are partially or fully exposed as aerly as the first months

of curing. A thin coating around these spheres is formed, whose internal envelope consists of a dense layer of $Ca(OH)_2$, while the external one consists of perpendicularly oriented poorly developed needle-like phase (Fig. 6). Elongated crystals with tapering ends have been found in the surface of some spherical grains. The x-ray spectrum analysis has shown that they contain Al and Si in approximately equal quantities. This result was published but these crystals were not indentified (Evstatiev & Todorov 1983). Halse et al. (1984) found out that this is mulitte and that with the passage of time, the cement hydrates interact with it, with quartz or other minerals deposited on the inner nonactive surfaces completely destroying them. The early start-

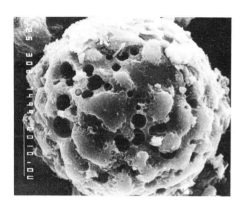

Figure 5. Sphere eroded as a result of
pozzolanic rections in stabilized lagooned
ash after 12 years
Magn.: x 3000

ing pozzolanic reactions proceed for a long
time and most probably follow the mechanism established in fly ash-cements. The pozzolanic interaction products are amorphous
and are difficultly registered by x-ray
analysis. They resemble the morphological
products of cement hydration.

After 90 days appear network-like formations, covering and linking the particles
(Fig. 7). After a year their quantity is
visibly increased. Morphologically they resemble the II type C-S-H. In later periods
(after a year or more and especially after
10-12 years) a dense gel-like mass of sma-
shed irregular-shaped particles of size
below 1 μm are observed. They resemble the
III type C-S-H (Fig. 8).

Strength tests showed that the rapid
strength growth during the 1st month is
followed by a long period of delayed
strength gain. Sometimes either strength
retention between the 28th and the 90th
day or strength decrease between the 180th
and 360th day is recorded. Irrespective of
the complex character of the strength curves in the later periods of curing the
strength is always higher than the 28th day
one (Todorov 1985).

TEST RESULTS DISCUSSION

The cement hydration and the pozzolanic
interaction products contribute mostly to
the strength formation in the investigated
lagooned ash-cement mixtures. The unitial
compounds are ettringite and more or less
developed needle-like phases similar to the
I type C-S-H. During the first months the
structure consists of single particles of

a

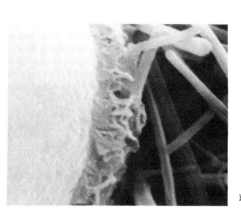

b

Figure 6. "Halo" of pozzolanic products
around a spheric grain in stabilized
lagooned ashes after 28 days
Magn.: a - x 10 000; b - x 30 000

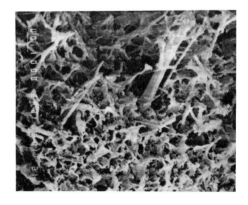

Figure 7. Network-like silicate hydrates
in stabilized lagooned ash after 360 days
Magn.: x 10 000

3206

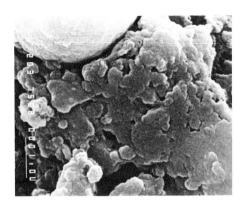

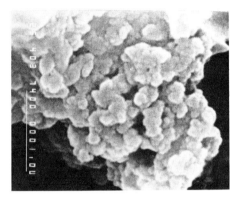

Figure 8. Gel-like mass in stabilized lagooned ash after 12 years
Magn.: a - x 20 000; b - x 40 000

changed surface, connected with each other to a different extent and on different areas by interwoven needle-like products. Pozzolanic reactions perform a substantial role in strength formation. They produce additional quantities of cementing material. It has a binding function and contributes to the establishment of more cohesive contacts. The released by the cement hydration lime in the investigated mixtures is insufficient and even its full participation in pozzolanic reactions does not produce binding material enough for promotion of a considerable strength growth. We suppose that the pozzolanic reaction has another important role. The newly formed hydration products serve as epitaxonomic basis for the similar cement hydrates. In the course of time a strong bond is created between the particles affected by pozzolanic reactions and the products of cement hydration precisely via the "halo"

of the pozzolanic products.

Because of the intensive cement hydration and pozzolanic interaction in the early stages, no great quantities of new mineral phases are formed in later periods. The further slow strength gain is due most probably to the observed phase transition of the needle-like formations into network-like and gel-like ones (Evstatiev & Todorov 1983). The transformation of C-S-H from type I to types II and III resemble the processes established by Dent Glasser et al. (1978) in cement pastes. It entails the formation of a binding substance of greater density and volume and more even distribution. The structure becomes denser and more homogenous. The lagooned ash particles are bound in a uniformly distributed gel-like mass, the smallest of them disappearing as a result of the pozzolanic reactions. The described bond strengthening contributes to the strength increase in later periods.

The presented mechanism of strength formation is similar to that of soil-cements and soil-limes (Evstatiev 1984), but there are a number of differences too. Cement hydration and the subsequent pozzolanic reactions are very intensive during the early stages in the case of cement-lagooned ash mixtures. Physico-chemical processes account for a considerable share in strength formation of stabilized soils. Unlike the soils lagooned ashes do not contain active clayey fractions and humus substance, and in their cement mixtures these processes are not so clearly expressed. Here the separated lime is mostly involved in pozzolanic reactions. They are more distinctly manifested and their products are better differentiated than in the case of soil-cement. The pozzolanic reaction is not only cement hydration accelerator and a binding substance source, but also contributes to the strengthening of lagooned ash particles and cement hydrates bonds. In later periods only the finest grains disappear, while most of the particles are preserved as building elements of the structure.

Most lagooned ashes have a composition conductive to normal cement hydration. However, they differ in hydraulic activity, which depends on coal particles content, dispersity and on the quantity, type and state of the glass-like mass, as well as on the alkalinity of the medium. Strength tests confirmed that stabilization is more effective for lagooned ashes possessing a higher hydraulic activity. Therefore the corrections in composition (elimination of unburnt coal particles, additional grinding, introduction of calcium admixtures) and creation of favourable conditions (compaction, introduction of alkaline admixtures), can produce a high strength structure without increasing the cement content. Such an approach is appropriate for the stabilization of lagooned ashes of lower quality.

In conclusion it should be emphasized that lagooned ashes can be successfully stabilized with small quantities of Portland cement. The formation of high strength affords the opportunity for more broad application of stabilized lagooned ashes in construction and hence for solving a number of ecological and industrial problems, connected with these waste products.

REFERENCES

Dent Glasser, Z.S., Lachowski, E.E., Mohan, K. & Tayler, H.F.W.(1978). A multi-method study of C_3S hydration. Cement and Concret Res. 8, 773-740.

Diamond, S.(1976). Cement paste microstructure-an overview at several levels. Proc. Conf. on hydraulic cement pastes: their structure and properties, 2-30. University of Sheffild.

Diamond, S.(1986). The microstructure of cement paste in concrete. Proc. 8th Inter. Congress on the chemistry of cement, Special Reports, Vol. 1, 122-147. Rio de Janeiro.

Evstatiev, D. & Todorov, K.(1983). Microstructural investigations of cement-lagooned ash mixtures. Proc. 8th ECSMFE, 763-766. Balkema, Rotterdam.

Evstatiev, D.(1984). Strength formation of soil-cement, 94 pp. Bulg. Acad. Sci., Sofia (in Bulgarian).

Evstatiev, D. & Todorov, K. Eds.(1990). New applications of thermo-electric power plant wastes in construction in Bulgaria. Bulg. Acad. Sci., Sofia (in print).

Halse, J., Pratt, P.L., Dalzyel, J.A., Gutteriage, W.(1984). Development of microstructure and other properties in fly ash OPC systems. Cement and Concrete Res. 14, 491-498.

Kokubu, M.(1968). Fly ash and fly ash-cement. Proc. 5th Inter. Congress on the chemistry of cement, Part IV, 105 pp. Tokyo.

Montgomery, D.G., Hughes, D.C. & Williams, I.T.(1981). Fly ash in concrete-a microstructure study. Cement and Concrete Res. 4, 591-603.

Todorov, K.(1985). Engineering geological characteristics of lagooned ashes in Bulgaria. Doctoral thesis, 175 pp. Bulg. Acad. Sci., Sofia (in Bulgarian).

Todorov, K., Evstatiev D. & Slavov, P.(1986). Investigating lagooned ashes utilization in road construction. Proc. 5th ICIAEG, 1693-1696. Balkema. Rotterdam.

Todorov, K.(1989). Ecological problems related to technogenic soils. Proc. 14th Congress CBGA, 1130-1133. University of Sofia.

Mineral aggregate from lead mine tailings

Granulats préparés des déchets de mine de plomb

Branka Zatler-Zupančič
Institute for Testing and Research in Materials and Structures, Ljubljana, Yugoslavia

ABSTRACT: The suitability of lead mine tailings for use as all - purpose mineral aggregate was proved. The general behaviour of mortars and concretes with sandstone - dolostone aggregate (containing approximately 3% of zinc and lead compounds with total sulphur content sometimes greater than 1%) was studied. Special attention was paid to the influence of non - carbonate admixtures in the aggregate on the setting time of cement concretes.

RÉSUMÉ: La convenance des déchets de mine de plomb pour 1' application comme des granulats généralement utilisables a été prouvée. Le comportement général des mortiers et des bétons avec des granulats de grès et de dolomite - contenant environ de 3% des combinaisons de zinc et de plomb, et de plus qu' 1% de souffre total a été etudié. L'attention specialle a été adressé a l'influence des ingrédients non-carboniques dans ces granulats au temps de prise des bétons de ciment.

1 INTRODUCTION

The lead and zinc mine Mežica is older than 300 years. It is situated in the river Črna valley - tributary to the river Drava - not far from Yugoslav - Austrian border.

Galena and sphalerite lodes are in sandstone and dolostone. The mine is of hydrothermal origin as published by A.Zorc and I.Štrucl. The mine tailings contain also: cerrusite, anglesite, smithsonite and hydrozincite.

In Žerjav there is a deposit of some millions m³ of mine tailings. They are in the form of grains from 4 to 45 mm size.

The usability of wastes from deposit and from everyday mine production was studied.

2 EXPERIMENTAL

The basic properties of solid wastes as a resource for aggregate production was analysed (Table 1). The noncarbonate admixtures were deter- mined with chemical and minerological analysis (Table 2 and 4).

The gradings 4/8, 8/16 and 16/32mm were prepared by sieving of tailings. Sand (0/4 mm) was prepared by crushing and sieving. The main properties of cement mortars and concretes were studied (Table 3).

As known from the work of W.Koenne, W.Lieber and H.G.Midgley, watersoluble compounds of lead and zinc retard the setting time of portland cement mortars or concretes but have no influence on their final strength. The setting time of concrete was determined according to the Yugoslav standard U.M1.019. The influence of water/cement ratio, type of aggregate and quantity of retarding admixtures in aggregate on the setting time is presented in figures 1 and 2 and table 4.

3 DISCUSION

The quantity of total sulphur in Mežica tailings varies during production from 0,7 to 1,3%. This consti-

Table 1: The main properties of tailings

```
Radioactivity                         : lower than criteria in Yugoslav
                                        specification
Admixtures                            : (per cent by weight)
   lead  .........................................  0,11 - 0,20
   zinc  .........................................  0,66 - 1,03
   chlorides  ....................................  0,00
   total sulphur as SO_3 .........................  0,73 - 1,26
           sulphides ...............  0,20 - 0,50
           water soluble sulphate .....  0,00 - 0,01

Physical properties                   :
   particle size  ................................  4/45 mm
   particles smaller than 0,09 mm ................  0,1%
   particle shape - Faury coefficient ............  0,23 - 0,26
   abrasion resistance Los Angeles
           grading B  .............................  32,7
                   C  .............................  28,5
                   D  .............................  26,6
density  .........................................  2760 - 2800 kg/m^3
water absorption  ................................  0,7 - 1,0%
soundness
           5 cycles in Na_2SO_4 ....................  0,4 - 1,2%
          25 cycles in Na_2SO_4 ....................  1,6 - 5,9%
friable particles  ...............................  2,0 - 12,4%
```

Table 2: The results of mineralogical-petrographic analysis of tailings

Particle size (mm)	:	0/4	4/8	8/16	16/32
Constituents	:	(per cent by weight)			
sandstone		36,0	22,1	46,0	86,0
dolostone		63,0	75,0	51,1	12,9
noncarbonate admixtures	:	1,0	2,9	2,9	1,0
quartz	:	0,55	0,84	2,04	0,06
sphalerite	:	0,33	1,93	0,44	0,97
hematite	:	0,03	0,01	0,09	0,08
galena	:	0,12	0,17	0,35	0,01

tuent in aggregate will cause no troubles because the content of water soluble sulphates is lower than 0,1% and the main quantity of sulphur is in the form of sulphides which can be used as heavyweight aggregate for concrete as published by H.G.Midgley. There was no cracking and staining on white concrete thin specimens prepared from galena aggregate which were exposed to different humidity.

Concrete aggregates produced from Mežica tailings can be "problem materials" because of their variable contents of lead and zinc compounds.

4 CONCLUSION

Experiments have shown that it is possible to produce two groups of mineral aggregates from Mežica tailings:
- mineral aggregate for asphalt mixtures and
- mineral aggregate for cement concretes and mortars when the appropriate quality control is undertaken.

The determination of: lead and zinc

Tabel 3: The composition and properties of concrete

Constituents: mineral aggregate 0/32 mm - from mine tailings
             cement PC 30 dz 45 S Trbovlje

w/c ratio		0,40	0,55	0,70	0,80
Fresh concrete					
Constituents:	(kg/m^3)				
cement		450	370	300	260
aggregate		1848	1859	1904	1917
water		180	203	210	208
Consistency - slump	(cm)	11	12	12	13
Setting time	(hour/min)				
beginning		8/15	9/25	11/25	20/35
end		10/15	11/35	13/10	24/15
Hardened concrete					
Compressive strength	$(KPa)x10^3$				
age (day)	1	6,9	7,9	1,4	0
	3	22,1	19,7	11,5	13,2
	7	31,1	29,0	19,6	22,0
	28	47,5	42,7	35,3	36,5
	90	54,1	49,6	38,4	40,5
Water permeability	(cm)				
at the age of 90 days		3,3	5,2	6,9	11,5
Frost resistance - per cent of the modulus of elasticity					
$(-20 + 15^oC)$					
number of cyles	25	95,4			
	50	95,5			
	75	90,9			
	100	84,4			
	125	77,1			
	150	71,8			

Table 4: The influence of aggregate type and w/c ratio on the setting time
of concrete

Type of aggregate	sand and gravel		tailings sample 1		tailings sample 2	
Noncarbonate constituents of aggregate (%)						
lead	0,00		0,04		0,17	
zinc	0,00		0,45		0,92	
total sulphur as SO_3	0,03		0,54		0,69	
w/c ratio	0,40	0,80	0,40	0,80	0,40	0,80
Cement per m^3 of concrete	450	260	450	260	450	260
Setting time (hour/min)						
beginning	6/00	6/30	6/00	7/45	8/15	10/15
end	7/45	9/00	7/45	10/00	10/15	14/15

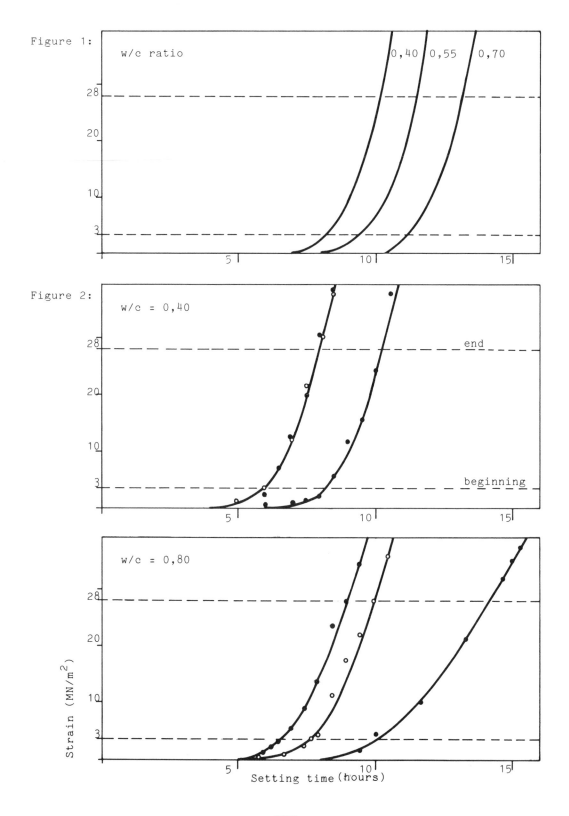

Figure 1:

w/c ratio

0,40 0,55 0,70

Figure 2:

w/c = 0,40

end

beginning

w/c = 0,80

Strain (MN/m^2)

Setting time (hours)

3212

Figure 1: The influence of w/c ratio on the setting time of concrete
          prepared by mine talings-sample 1

Figure 2: The influence of aggregate type and w/c ratio on setting time
          of concrete
          Zinc and lead compounds in aggregate:
          ⊗ 0,00%   sand and gravel
          ○ 0,49%   tailings-sample 1
          ● 1,19%   tailings-sample 2

compounds, sulphate and sulphide is necessary and the control of setting time of concrete must be carried out especially when the sum of lead and zinc in aggregate is higher than 0,5%.

REFERENCES

Koenne,W. (1961). Erstärungsstörungen durch Metaloxide, Zement-Kalk-Gips 4, 158-160.

Lieber,W. (1967). Einfluss von Zinkoxyd auf das Erstarren und Erhärten von Portlandzmenten, Zement-Kalk-Gips 3, 91-95.

Lieber,W. (1968). The influence of lead and zinc compounds on the hydration of portland cement, Proceedings of the Fifth International Symposium on the Chemistry of Cement, Tokio, Part II. Volume II., 444-454.

Midgley,H.G. (1970). The effects of lead compounds in aggregate upon the setting of Portland cement, Magazine of Concrete Research 22, 70, 42-44.

Mohan,R., R.Srivastava (1979). Utilization of waste zink tailings in making cellular concrete,Res. Ind. 24, 226-228.

Rai,M., G.S.Mehrotra, D.Chandra (1983). Use of zinc, iron and copper tailings as a fine aggregate in concrete,Durability of Building Materials 1, 377-388.

Štrucl,I. (1960). Geologische Verhältnisse der Lagerstätte Mežica mit besonderem Blick auf die Klassifizierung der Erzvorräte, Geologija 6, 251-278.

Štrucl,I. (1984). Geological and geochemical characteristics of ore and host rock of lead-zinc ores of the Mežica ore deposit, Geologija 27, 215-288.

YUS U.M1.019 (1981). Concrete. Determination of setting time of concrete mixtures by penentration.

Zorc,A. (1955). Mining geological features of the Mežica ore deposit, Geologija 3, 24-80.

AKNOWLEDGEMENTS

The financial support of the Research Community of Slovenia, Mežica Mine and building firm Kograd is gratefully acknowledged.

3213